D1031659

Large deviations and metastability

The book provides a general introduction to the theory of large deviations and a wide overview of the metastable behaviour of stochastic dynamics. With only minimal prerequisites, the book covers all the main results and brings the reader to the most recent developments. Particular emphasis is given to the fundamental Freidlin–Wentzell results on small random perturbations of dynamical systems. Metastability is first described on physical grounds, following which more rigorous approaches are developed. Many relevant examples are considered from the point of view of the so-called pathwise approach. The first part of the book develops the relevant tools, including the theory of large deviations, which are then used to provide a physically relevant dynamical description of metastability. Written to be accessible to graduate students, this book provides an excellent route into contemporary research.

ENCYCLOPEDIA OF MATHEMATICS AND ITS APPLICATIONS

40 N. White (ed.) *Matroid Applications*
41 S. Sakai *Operator Algebras in Dynamical Systems*
42 W. Hodges *Basic Model Theory*
43 H. Stahl and V. Totik *General Orthogonal Polynomials*
45 G. Da Prato and J. Zabczyk *Stochastic Equations in Infinite Dimensions*
46 A. Björner et al. *Oriented Matroids*
47 G. Edgar and L. Sucheston *Stopping Times and Directed Processes*
48 C. Sims *Computation with Finitely Presented Groups*
49 T. Palmer *Banach Algebras and the General Theory of *-Algebras I*
50 F. Borceux *Handbook of Categorical Algebra I*
51 F. Borceux *Handbook of Categorical Algebra II*
52 F. Borceux *Handbook of Categorical Algebra III*
53 V. F. Kolchin *Random Graphs*
54 A. Katok and B. Hasselblatt *Introduction to the Modern Theory of Dynamical Systems*
55 V. N. Sachkov *Combinatorial Methods in Descrete Mathematics*
56 V. N. Sachkov *Probabilistic Methods in Descrete Mathematics*
57 P. M. Cohn *Skew Fields*
58 R. Gardner *Geometric Tomography*
59 G. A. Baker, Jr., and P. Graves-Morris *Pade Approximants,* 2ed
60 J. Krajicek *Bounded Arithmetic, Propositional Logic and Complexity Theory*
61 H. Groemer *Geometric Applications of Fourier Series and Spherical Harmonics*
62 H. O. Fattorini *Infinite Dimensional Optimization and Control Theory*
63 A. C. Thompson *Minkowski Geometry*
64 R. B. Bapat and T. E. S. Raghavan *Nonnegative Matrices with Applications*
65 K. Engel *Sperner Theory*
66 D. Cvetkovic, P. Rowlinson, and S. Simic *Eigenspaces of Graphs*
67 F. Bergeron, G. Labelle, and P. Leroux *Combinatorial Species and Tree-Like Structures*
68 R. Goodman and N. Wallach *Representations and Invariants of the Classical Groups*
69 T. Beth, D. Jungnickel, and H. Lenz *Design Thoery I,* 2ed
70 A. Pietsch and J. Wenzel *Orthonormal Systems for Banach Space Geometry*
71 G. E. Andrews, R. Askey, and R. Roy *Special Functions*
72 R. Ticciati *Quantum Field Theory for Mathematicians*
73 M. Stern *Semimodular Lattices*
74 I. Lasiecka and R. Triggiani *Control Theory for Partial Differential Equations I*
75 I. Lasiecka and R. Triggiani *Control Theory for Partial Differential Equations II*
76 A. A. Ivanov *Geometry of Sporadic Goups 1*
77 A. Schinzel *Polynomials with Special Regard to Reducibility*
78 H. Lenz, T. Beth, and D. Jungnickel *Design Theory II,* 2ed
79 T. Palmer *Banach Algebras and the General Theory of *-Algebras II*
80 O. Stormark *Lie's Structural Approach to PDE Systems*
81 C. F. Dunkl and Y. Xu *Orthogonal Polynomials of Several Variables*
82 J. P. Mayberry *The Foundations of Mathematics in the Theory of Sets*
83 C. Foias et al. *Navier–Stokes Equations and Turbulence*
84 B. Polster and G. Steinke *Geometries on Surfaces*
85 R. B. Paris and D. Kaminski *Asymptotics and Mellin–Barnes Integrals*
86 R. McEliece *The Theory of Information and Coding,* 2ed
87 B. Magurn *Algebraic Introduction to K-Theory*
88 T. Mora *Systems of Polynomial Equations I*
89 K. Bichteler *Stochastic Integration with Jumps*
90 M. Lothaire *Algebraic Combinatorics on Words*
91 A. A. Ivanov and S. V. Shpectorov *Geometry of Sporadic Groups II*
92 P. McMullen and E. Schulte *Abstract Regular Polytopes*
93 G. Gierz et al. *Continuous Lattices and Domains*
94 S. Finch *Mathematical Constants*
95 Y. Jabri *The Mountain Pass Theorem*

Large deviations and metastability

ENZO OLIVIERI
Università di Roma Tor Vergata

MARIA EULÁLIA VARES
Centro Brasileiro de Pesquisas Físicas, Rio de Janeiro

PUBLISHED BY THE PRESS SYNDICATE OF THE UNIVERSITY OF CAMBRIDGE
The Pitt Building, Trumpington Street, Cambridge, United Kingdom

CAMBRIDGE UNIVERSITY PRESS
The Edinburgh Building, Cambridge CB2 2RU, UK
40 West 20th Street, New York, NY 10011-4211, USA
477 Williamstown Road, Port Melbourne, VIC 3207, Australia
Ruiz de Alarcón 13, 28014 Madrid, Spain
Dock House, The Waterfront, Cape Town 8001, South Africa

http://www.cambridge.org

First published 2005

Printed in the United Kingdom at the University Press, Cambridge

Typeface Times 10/13 pt. *System* LaTeX 2ε [TB]

A catalogue record for this book is available from the British Library

Library of Congress Cataloguing in Publication data

ISBN 0 521 59163 5 hardback

To Anna and Daniela, Juliana, Lucas and Vladas

CONTENTS

PREFACE

This book has germinated from the lecture notes of a course 'Large deviations and metastability' given by one of us at the 'CIMPA First School on Dynamical and Disordered Systems', at Universidad de la Frontera, Temuco, during the summer of 1992 [293].

Since then a large amount of new material on metastability has been accumulated, and our goal was to combine a basic introduction to the theory of large deviations with a wide overview of the metastable behaviour of stochastic dynamics.

Typical examples of metastable states are supersaturated vapours and magnetic systems with magnetization opposite to the external field. Metastable behaviour is characterized by a long period of apparent equilibrium of a pure thermodynamic phase followed by an unexpected fast decay towards the stable equilibrium of a different pure phase or of a mixture, e.g. homogeneous nucleation of the liquid phase inside a highly supersaturated vapour, due to spontaneous density fluctuations. The point of view of metastability as a genuinely dynamical phenomenon is now widely accepted. Approaches which aim to describe static aspects of metastability (such as determination of the metastable branch of the equation of state of a fluid) in the Gibbs equilibrium set-up are, in their 'naïve form', applicable only in a mean field context. In this case, the physically unacceptable assumption that the range of the interaction equals the linear dimension of the container gives rise to pathological behaviour of non-convex free energy that implies negative compressibility, namely, thermodynamic instability. It is this feature that gives rise to the idea of associating metastability with local minima of the free energy. Moreover, dynamical aspects such as the lifetime of the metastable state require an investigation that a static approach is programmatically unable to provide. Thus, metastability for short range systems is included in the field of non-equilibrium statistical mechanics. Since a general theory of non-equilibrium thermodynamic phenomena is still lacking, a particularly relevant role is played by the study of specific mathematical models, for instance the stochastic Ising model.

The first attempt to formulate a rigorous dynamical theory of metastability goes back to Lebowitz and Penrose (see [240, 241]). In their approach the decay from metastability to stability is essentially characterized by a slow irreversible evolution of the expected values of the observables during the process. In [48] another method was proposed, based on a pathwise analysis of the process. The single trajectories of the process are characterized by a long period of random oscillations in apparent equilibrium (with a relatively fast loss of memory of the initial condition) followed by a sudden decay towards another, different regime, corresponding to stable equilibrium. In this approach metastability becomes strictly related to the first exit problem from special domains. Characterization of the most probable exit mechanism involves comparison between different rare events – a typical problem in large deviation theory.

After describing metastability on physical grounds, we present the existing rigorous approaches, with a particular emphasis on the pathwise approach, the main object of our analysis. Large deviation theory is applied, in combination with specific tools, to provide a dynamical description of metastability.

The construction of a mathematical theory of metastability not only provides interesting and physically relevant applications of the already established large deviation theory, but also poses new problems.

The first part of the book provides a reasonably self-contained account of basic results about large deviation theory. In Chapter 1 we discuss the classical basic results in the frame of large deviations for sums of independent random variables. In Chapter 2 we concentrate on the results of Freidlin and Wentzell in the context of small random perturbations of deterministic flows. Chapter 3 is mainly dedicated to the treatment of large deviations for interacting systems, and to its role in equilibrium statistical mechanics. The first two sections contain a short summary of large deviations for Markov chains and the Gärtner–Ellis theorem. The third section provides a brief introduction to equilibrium statistical mechanics, and the last section discusses large deviations for Gibbs measures and its relation to thermodynamical formalism.

In Chapter 4 we start the description of the metastability phenomenon and the various rigorous approaches to its treatment. The pathwise approach, which is one of the main topics of the book, is introduced in Section 4.2. The next two sections contain two examples: first we consider the extremely simple mean field model of the Curie–Weiss chain. Though unphysical, this mean field model can be considered as an initial 'laboratory', due to the explicitness of computations. The second example is the one-dimensional Harris contact process, which presents a nontrivial spatial structure. In the final section, we briefly outline results on metastability for other mean field type dynamics as well as the multidimensional Harris contact process. In Chapter 5 we are concerned with the verification of metastability for Itô processes in the context of the Freidlin–Wentzell theory. This is done in Section 5.4, based on results of Freidlin and Wentzell combined with coupling techniques. The important example of a double well potential is discussed in detail in Section 5.2. Finally, extensions to infinite dimensional situations such as reaction–diffusion models are briefly discussed at the end of the chapter.

In Chapter 6 we study the long time behaviour of general reversible Freidlin–Wentzell Markov chains; these are characterized by a finite state space and transition probabilities exponentially small in an external parameter that in many applications is the inverse temperature. In particular we analyse the first exit problem from particular sets of states, called *cycles*, whose characteristic property is that all their points are typically visited before the exit. Various aspects that are relevant for the description of metastable behaviour are studied: the asymptotic exponentiality of properly renormalized first exit times, the conditional equilibrium (Gibbs) measure, the 'tube' of typical trajectories during the exit.

In Chapter 7 we study metastability and nucleation for various short range lattice spin models that can be seen as generalizations of the standard stochastic Ising model. We consider the asymptotic regime with fixed volume and coupling constants in the limit of very low temperature. From a physical point of view this corresponds to the study of local aspects of nucleation; from a mathematical point of view it corresponds to the study of some large deviation phenomena for a class of Freidlin–Wentzell Markov chains. To study these models we apply the general results of Chapter 6 and have to solve some specific model dependent variational problems.

A particular emphasis is given to the case of reversible stochastic evolutions. Under the reversibility condition, many different dynamics such as quite general mean field models, Itô stochastic differential equations of gradient type, and stochastic Ising models can be treated by the same methods.

Parts of this text have been used in graduate courses at IMPA, Rio de Janeiro, and at Università di Roma 'Tor Vergata'. We would like to thank D. Tranah for the invitation to write this book, for his patience, attention and professionalism which made the process run smoothly during all these years.

The authors wish to thank all the colleagues who contributed in various ways to the realization of this work. Special thanks to A. Bovier, R. Cerf, F. den Hollander, S. Friedli, V. Gayrard, A. Gaudilliere, Ch. Gruber, H. Kesten, R. Kotecky, T. Mountford, O. Penrose and G. Sewell, for discussions on specific points or on general aspects of metastability. We are deeply grateful to E. Andjel, S. Brassesco, E. Cirillo, S. Carmona, A. Hinojosa, F. Manzo and F. Nardi, for reading parts of the text, making corrections, comments and suggestions which helped us to improve the presentation and correct several defects. We are particularly indebted to M. Cassandro, C.-E. Pfister, P. Picco, E. Presutti, R. Schonmann, E. Scoppola and V. Sidoravicius for their criticism, for many stimulating conversations, and for the clarifications they offered us through many discussions. Finally we would like to express gratitude to our families: Anna and Daniela,

Juliana, Lucas and Vladas, for their patience and constant support, specially in the hardest moments of the long preparation of this book.

E. O. and M. E. V. acknowledge the warm hospitality of IMPA, Rio de Janeiro, and the Department of Mathematics at the Università di Roma 'Tor Vergata', respectively.

M. E. V. acknowledges financial support of CNPq and Faperj. This work has also received support from the CNPq-CNR scientific cooperation agreement.

Corrections that may appear in the future will be posted at
http://www.cbpf.br/~eulalia/

Illud in his rebus non est mirabile, quare,
Omnia cum rerum primordia sint in motu,
Summa tamen summa videatur stare quiete,
Praeterquam siquid proprio dat corpore motus.

Lucretius, *De rerum natura*

1

Large deviations: basic results

Introduction

In the analysis of a system with a large number of interacting components (at a microscopic level) it is of clear importance to find out about its collective, or macroscopic, behaviour. This is quite an old problem, going back to the origins of statistical mechanics, in the search for a mathematical characterization of 'equilibrium states' in thermodynamical systems. Though the problem is old, and the foundations of equilibrium statistical mechanics have been settled, the general question remains of interest, especially in the set-up of non-equilibrium systems. We could then take as the object of study a (non-stationary) time evolution with a large number (n) of components, where the initial condition and/or the dynamics present some randomness. One example of such a collective description is the so-called hydrodynamic limit. Passing by a space-time scale change (micro $\rightarrow$ macro) it allows, through a limiting procedure, the derivation of a reduced description in terms of macroscopic variables, such as density and temperature. Other limits, besides the hydrodynamic, may also appear in different situations, giving rise to macroscopic equations.

In all such cases the macroscopic equation indicates the *typical* behaviour in a *limiting situation* ($n \rightarrow +\infty$, and proper rescaling). Thus, it is essential to know something about:

(i) rates of convergence, i.e. how are the fluctuations of the macroscopic random fields (for example, the empirical density) around the prescribed value given by the macroscopic equation?

(ii) how to estimate the chance of observing something quite different than what is prescribed by the macroscopic equation. According to the prescription of the macroscopic equation these are 'rare events' and their probabilities will tend to zero, but *at which speed*?

In the above description we identify the three most basic limit theorems in classical probability: the macroscopic description corresponds to a 'law of large numbers'; the behaviour of the fluctuations, or 'moderate deviations', fits into the frame of a 'central limit theorem'; and the estimates of the probability of rare events constitute what are usually called 'large deviation principles'. The program for investigating the collective behaviour for evolutions given by Markov processes on $[0, 1]^{\mathbb{Z}^d}$ or $\mathbb{N}^{\mathbb{Z}^d}$, has grown since the 1980s (see [79]), and has taken definite forms for a class of them, cf. [78, 178, 283]. The situation is much less developed in the context of mechanical systems (see [283]).

The content of this book is closely related to questions such as (ii) above, and in particular to their connection with metastability, which will be discussed from Chapters 4 to 7. Perhaps we should say a few words on possible motivations for such estimates, bearing in mind the collective description of large systems. For example, if one wants to investigate the behaviour of the system at time scales longer than those for which the macroscopic equation is valid, then it is necessary to pay attention to such 'large fluctuations', since they will eventually occur. The ability to compare their probabilities becomes a crucial point in order to predict the long-term behaviour of the system. The classic example is a tunnelling event between two stable points of the macroscopic equation. Somehow, this comparison can be seen as a first step: one would believe that the large fluctuation should occur in the *least improbable* way. Nevertheless, carrying out this long time analysis may present (technical or serious) difficulties. One instance where this has been done quite completely is that in which the dynamics is, in some sense, already macroscopic; more precisely, it is obtained by the addition of a small external noise to a non-chaotic dynamical system. This is the object of Freidlin and Wentzell's theory [122], which will be studied in Chapters 2 and 5 of this book, also in connection with the phenomenon of metastability.

One should stress how closely related are the three mentioned problems: derivation of macroscopic equations/law of large numbers, fluctuations, and large deviations. Large deviation estimates yield stronger statements on the convergence of macroscopic density fields. On the other hand, a standard method for the derivation of large deviation estimates involves the validity of a large class of deterministic macroscopic limits (law of large numbers). A very important example of such a connection comes from equilibrium theory, through the possibility of applying large deviations to obtain the equivalence of ensembles, as pointed out in the fundamental articles of Ruelle [257], and Lanford [189], which have stimulated intense research. This goes far beyond the scope of this book, as for instance, the questions related to phase separation and surface large deviations. A brief discussion will appear in Chapter 3, with indications to recent research articles.

As a usual set-up for large deviations we could take a sequence of probability measures $(\mu_n)_{n\geq 1}$ on some metric space M, weakly converging to a Dirac point measure at some $m \in M$, in the sense that $\lim_{n\to+\infty} \int f d\mu_n = f(m)$ for all $f: M \to \mathbb{R}$ continuous and bounded. (In our previous discussion, μ_n should represent the law of an observable such as the empirical density, m representing a macrostate such as an equilibrium density.) The goal is to find the speed at which $\mu_n(A)$ tends to zero, when A is a fixed measurable set staying at positive distance from m. In particular, one wishes to detect whether a fast, exponential decay happens, in the sense that there exists $I(A) \in (0, +\infty]$ such that

$$\mu_n(A) \approx e^{-nI(A)}. \tag{1.1}$$

Throughout the text, $\approx$ denotes logarithmic equivalence, i.e. (1.1) means $n^{-1} \log \mu_n(A) \to -I(A)$ as $n \to \infty$. (*Notation*. $\log = \log_e$ everywhere in this text.)

Let us assume that (1.1) holds for a certain class of sets A; let A and B be two disjoint sets for which it holds. Since $\mu_n(A \cup B) = \mu_n(A) + \mu_n(B)$ it follows at once that $\mu_n(A \cup B) \approx e^{-n \min\{I(A), I(B)\}}$. This might suggest $I(A)$ of the form

$$I(A) = \inf_{x \in A} I(x), \tag{1.2}$$

for some point function I, which would then be called a 'rate function'. If so, we cannot expect (1.1) to hold for all measurable sets A; to see this, consider for example continuous measures, so that $\mu_n\{x\} = 0$ for all points $x \in M$. If (1.1) were true for such sets, this would force I to be identically $+\infty$, incompatible with (1.1) and (1.2) for $A = M$. This means that some restriction on the sets for which (1.1) holds is needed. This will be discussed in the next two sections, where a possible set-up will be presented.

It is natural to ask why one chooses the logarithmic equivalence $\approx$ instead of a sharper estimate like the usual equivalence ($a_n \sim b_n$ iff $a_n/b_n \to 1$). Significant advantages of the previous choice (allowing polynomial errors in (1.1)) include simplicity and a wide range of applicability. On the other hand, 'exact' results are essential in many applications though in this text we shall not pursue them.

Moreover, situations are expected to occur where $I(A)$ could vanish, meaning that the decay is less than exponential, and that (1.1) does not provide enough information. In such cases, one definitely needs a more precise asymptotics.

For a comparison with 'moderate deviations' (central limit theorems), let us take $M = \mathbb{R}^d$. According to the previous notation, these refer to the asymptotics of $\mu_n(A_n)$ where $A_n = m + \alpha_n A$, $\alpha_n \to 0$ suitably, and A is fixed.

1.1 Cramér–Chernoff theorem on $\mathbb{R}$

Let us start with the simplest situation: the microstates correspond to the results of n independent tosses of a fair coin, and μ_n represents the law of the proportion of 'heads' (the macroscopic observable). The microstates are thus uniformly distributed on $\mathcal{X}_n = \{(\omega_1, \ldots, \omega_n) \colon \omega_i = 1 \text{ or } \omega_i = 0, \forall\, i\} = \{0, 1\}^n$, i.e. all 2^n points have equal probabilities, and if $A \subseteq [0, 1]$ is a Borel set, then

$$\mu_n(A) = 2^{-n} \sum_{k: k/n \in A} \binom{n}{k}, \tag{1.3}$$

where $\binom{n}{k} = \frac{n!}{k!(n-k)!}$ if $k \in \{0, 1, \ldots, n\}$, $\binom{n}{k} = 0$, otherwise.

We know that the weak law of large numbers holds in this situation, i.e. for any $\varepsilon > 0$ we have $\lim_{n \to +\infty} \mu_n(1/2 - \varepsilon, 1/2 + \varepsilon) = 1$. Let us in fact check this, getting a stronger estimate. Let $0 < a < b < 1$ and consider the probability that the resulting proportion of 'heads' belongs to the interval $[a, b]$. From (1.3) we immediately have:

$$2^{-n} \max_{k: k/n \in [a,b]} \binom{n}{k} \leq \mu_n[a, b] \leq (n+1) 2^{-n} \max_{k: k/n \in [a,b]} \binom{n}{k}.$$

In particular, as $n \to +\infty$

$$\frac{1}{n} \log \mu_n[a, b] \sim -\log 2 + \frac{1}{n} \max_{k: k/n \in [a,b]} \log \binom{n}{k}. \tag{1.4}$$

Recall now the classical Stirling formula (see e.g. [115], p. 54):

$$\sqrt{2\pi} n^{n+\frac{1}{2}} e^{-n} e^{1/(12n+1)} \leq n! \leq \sqrt{2\pi} n^{n+\frac{1}{2}} e^{-n} e^{1/12n}, \tag{1.5}$$

or its weaker version (see e.g. [115], p. 52)

$$n! \sim \sqrt{2\pi} n^{n+\frac{1}{2}} e^{-n}. \tag{1.6}$$

Considering logarithmic equivalence let us now use the even weaker relation:

$$\log n! = n \log n - n + O(\log n), \tag{1.7}$$

where $a_n = O(b_n)$ means that a_n / b_n remains bounded in n, which follows at once from (1.6). Applying (1.7) to equation (1.4) and performing simple calculations we get:

$$\lim_{n \to \infty} \frac{1}{n} \log \mu_n[a, b] = -\inf_{x \in [a,b]} I(x), \tag{1.8}$$

where, for each $x \in (0, 1)$,

$$I(x) = \log 2 + x \log x + (1 - x) \log (1 - x). \tag{1.9}$$

It is easily seen that (1.8) extends for any $a < b$ with the convention that $0 \log 0 = 0$ and $I(x) = +\infty$ if $x \notin [0, 1]$. Thus, we have a large deviation estimate for μ_n in the previously announced frame, and (1.1) holds for all intervals with non-empty

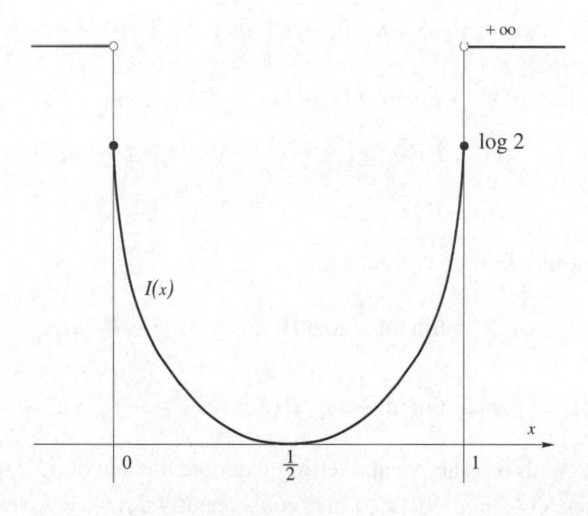

Figure 1.1

interior. The function I is continuous in the open interval $(0, 1)$, $I(x) = I(1 - x)$ for $0 \leq x \leq 1$, and it is strictly increasing in $(1/2, 1)$ (see Figure 1.1). In particular, $I(x) = 0$ iff $x = 1/2$, and for any $\varepsilon > 0$

$$\mu_n\big([0, 1] \setminus (1/2 - \varepsilon, 1/2 + \varepsilon)\big) \approx e^{-nI(1/2+\varepsilon)}.$$

The previous example, a very simple application of the Stirling formula, can be thought of as a particular case of the following: let $X_1, X_2, \ldots$ be independent and identically distributed (i.i.d.) real random variables on some probability space $(\Omega, \mathcal{A}, P)$. Let μ_n be the law of the sample average $\bar{X}_n := n^{-1} \sum_{i=1}^n X_i$. If X_1 is integrable and $m = EX_1$, the classical weak law of large numbers tells us that $\mu_n(\mathbb{R} \setminus (m - \varepsilon, m + \varepsilon)) = P(|\bar{X}_n - m| \geq \varepsilon)$ tends to zero as $n \to +\infty$, for any $\varepsilon > 0$. With proper conditions on the tails of the distribution of X_1, we again expect exponential decay of $\mu_n\big(\mathbb{R} \setminus (m - \varepsilon, m + \varepsilon)\big)$, for any given $\varepsilon > 0$, and we may hope to get something like (1.1) for a large class of sets A; the goal is to compute the rate, in the sense of logarithmic equivalence. This classical situation was treated by Cramér in 1937 (cf. [69]) for distributions with an absolutely continuous component, providing 'exact' asymptotics, and extended to the general case by Chernoff in 1952 [61] in the sense of logarithm equivalence. The result, as stated and proven by Chernoff, is the following.

Theorem 1.1 (Cramér–Chernoff) *Let $X_1, X_2, \ldots$ be i.i.d. real random variables with common law μ, and consider the sample average $\bar{X}_n = n^{-1} \sum_{i=1}^n X_i$. Let $\hat{\mu}$ denote the moment generating function of μ, i.e. $\hat{\mu}(\zeta) = E\, e^{\zeta X_1}$ for $\zeta \in \mathbb{R}$, and define:*

$$I_\mu(x) = \sup_{\zeta \in \mathbb{R}} (\zeta x - \log \hat{\mu}(\zeta)), \tag{1.10}$$

for $x \in \mathbb{R}$. *(*$\hat{\mu}$ *takes values on* $(0, +\infty]$ *and* I_μ *is* $[0, +\infty]$ *valued; we set* $\log(+\infty) = +\infty$.)

(a) Upper bound. If X_1 *is integrable and* $m = EX_1$*, then:*

$$\begin{aligned} (i) \quad & P(\bar{X}_n \geq x) \leq e^{-nI_\mu(x)}, \quad \text{if } x \geq m; \\ (ii) \quad & P(\bar{X}_n \leq x) \leq e^{-nI_\mu(x)}, \quad \text{if } x \leq m. \end{aligned} \tag{1.11}$$

(b) Lower bound. For any $x \in \mathbb{R}$:

$$(i) \quad \liminf_{n\to+\infty} \frac{1}{n} \log P(\bar{X}_n \geq x) \geq -I_\mu(x);$$

$$(ii) \quad \liminf_{n\to+\infty} \frac{1}{n} \log P(\bar{X}_n \leq x) \geq -I_\mu(x).$$

Remark 1.2 With μ as above and letting ν denote the law of $aX_1 + b$, for $a, b \in \mathbb{R}$, $a \neq 0$, then $\hat{\nu}(\zeta) = e^{b\zeta}\hat{\mu}(a\zeta)$ and consequently $I_\nu(x) = I_\mu((x-b)/a)$, for all $x \in \mathbb{R}$.

Proof of the upper bound We shall now see that optimization over a class of exponential Markov inequalities gives us the upper estimate.

Notice first that Remark 1.2 reduces the proof to the case $x = 0 \leq m$, all other cases being reduced to this by change of sign and considering $X_i' = X_i - x$.

If $\zeta \leq 0$ we apply a Markov exponential inequality to write:

$$P(\bar{X}_n \leq 0) = P(\zeta S_n \geq 0) \leq P(e^{\zeta S_n} \geq 1) \leq E\, e^{\zeta S_n} = (\hat{\mu}(\zeta))^n, \tag{1.12}$$

where $S_n = \sum_{i=1}^n X_i = n\bar{X}_n$. Thus we have:

$$P(\bar{X}_n \leq 0) \leq \big(\inf_{\zeta \leq 0} \hat{\mu}(\zeta)\big)^n. \tag{1.13}$$

On the other hand, if $\zeta \geq 0$, the Jensen inequality yields

$$\hat{\mu}(\zeta) = E\, e^{\zeta X_1} \geq e^{\zeta m} \geq 1 = \hat{\mu}(0), \tag{1.14}$$

that is

$$m \geq 0 \Rightarrow \inf_{\zeta \leq 0} \hat{\mu}(\zeta) = \inf_{\zeta \in \mathbb{R}} \hat{\mu}(\zeta) = e^{-I_\mu(0)}, \tag{1.15}$$

and the proof follows at once from (1.13). □

Remark 1.3 The previous argument with the Jensen inequality tells us that if $m = 0$ then $I_\mu(0) = -\log \hat{\mu}(0) = 0$ (if $m = 0$, (1.14) is applicable to any ζ), and by Remark 1.2 we conclude that $I_\mu(m) = 0$ whenever $m \in \mathbb{R}$.

The proof of part (b) of Theorem 1.1, as presented below, is exactly that given by Chernoff [61]. It does not fit so well in the general 'frame' to be used later; in the present discussion we wanted to stress the role of the Stirling formula and to give an idea of the historical development. We shall discuss this later.

Proof of the lower bound Again, by Remark 1.2, it suffices to prove

$$\liminf_{n\to+\infty} \frac{1}{n} \log P(\bar{X}_n \leq 0) \geq -I_\mu(0). \tag{1.16}$$

We first observe that the statement becomes trivial if $P(X_1 \leq 0) = 1$, in which case $n^{-1} \log P(\bar{X}_n \leq 0) = 0 \geq -I_\mu(0)$. If $P(X_1 \geq 0) = 1$, we have $n^{-1} \log P(\bar{X}_n \leq 0) = n^{-1} \log P(X_1 = 0, \dots, X_n = 0) = \log P(X_1 = 0)$. On the other hand, the monotone convergence theorem implies that in this case $\hat{\mu}(\zeta)$ decreases to $P(X_1 = 0)$ as ζ decreases to $-\infty$, so that $\log P(X_1 = 0) = \inf_{\zeta\in\mathbb{R}} \log \hat{\mu}(\zeta) = -I_\mu(0)$, and (1.16) becomes trivial too.

We now assume that $P(X_1 < 0) > 0$ and $P(X_1 > 0) > 0$. The main ingredient for the proof is the following lemma, which gives a stronger form of the lower bound in the case of random variables taking finitely many values, usually called simple. As in the coin-tossing example, this proof is based on the Stirling formula. □

Lemma 1.4 *Assume there exist real numbers $x_1, \dots, x_r$ such that*

$$\min_{1\leq i\leq r} x_i < 0 < \max_{1\leq i\leq r} x_i\,, \tag{1.17}$$

and that $p_i = P(X_1 = x_i) > 0 \quad (i = 1, \dots, r)$. Let us set $\psi(\zeta) = \sum_{i=1}^r p_i\, e^{\zeta x_i}$ and $\chi = \inf_{\zeta\in\mathbb{R}} \psi(\zeta)$. Then, there exist $c > 0$ and $N \in \mathbb{N}$ such that for all $n \geq N$ we may find positive integer numbers $n_1, \dots, n_r$ such that

$$\sum_{i=1}^r n_i = n,$$

$$\sum_{i=1}^r n_i x_i \leq 0, \tag{1.18}$$

$$p(n_1, \dots, n_r) := n! \prod_{i=1}^r \frac{p_i^{n_i}}{n_i!} \geq c\, n^{-(r-1)/2}\, \chi^n.$$

Before proving Lemma 1.4 let us examine its content. If $n = \sum_{i=1}^r n_i$ and $p_i = P(X_1 = x_i)$, then $p(n_1, \dots, n_r)$, defined in (1.18), gives the probability that each x_i appears exactly n_i times among the observations $X_1, \dots, X_n$. In particular, if $\sum_{i=1}^r n_i x_i \leq 0$, then

$$p(n_1, \dots, n_r) \leq P(S_n \leq 0) = P(\bar{X}_n \leq 0).$$

Now, if X_1 takes values on the finite set $\{x_1, \dots, x_r\}$, so that $p_1 + \cdots + p_r = 1$, we have $\psi(\zeta) = \hat{\mu}(\zeta)$ and $\log \chi = -I_\mu(0)$, and the lemma tells us: $P(\bar{X}_n \leq 0) \geq c\, n^{-(r-1)/2}\, e^{-nI_\mu(0)}$, clearly stronger than the lower bound stated at Theorem 1.1 for this particular case. (The constant c appearing in the previous formula will depend on $r, p_1, \dots, p_r, x_1, \dots, x_r$.)

Proof of Lemma 1.4 If $n = \sum_{i=1}^r n_i$ and $n_1, \ldots, n_r$ are positive integers, we can write, using the Stirling formula (1.5):

$$p(n_1, \ldots, n_r) \geq \sqrt{2\pi}\, n^{n+\frac{1}{2}} e^{-n} \prod_{i=1}^{r} \frac{(p_i\, e)^{n_i}}{\sqrt{2\pi}\, n_i^{n_i+\frac{1}{2}}\, e^{\frac{1}{12 n_i}}} \geq C(r)\, n^{-\frac{r-1}{2}}\, q(n_1, \ldots, n_r), \tag{1.19}$$

where $C(r) := e^{-\frac{r}{12}} (2\pi)^{-\frac{r-1}{2}}$ and, for $z_1, \ldots, z_r$ positive real numbers,

$$q(z_1, \ldots, z_r) := \prod_{i=1}^{r} \left(\frac{n\, p_i}{z_i} \right)^{z_i}. \tag{1.20}$$

Inequalities (1.17) imply that $\psi(\zeta) \to +\infty$ as $|\zeta| \to +\infty$. Thus we may take $\zeta_0 \in \mathbb{R}$ such that $\psi(\zeta_0) = \chi$, and since the function ψ is smooth, its derivative at the point ζ_0 vanishes. Taking $z_i = n\, p_i\, e^{\zeta_0 x_i}/\chi$ we have

$$\sum_{i=1}^{r} z_i = n, \qquad \sum_{i=1}^{r} z_i\, x_i = 0 \quad \text{and} \quad q(z_1, \ldots, z_r) = \chi^n. \tag{1.21}$$

The numbers z_i might fail to be integers, but they are large, for large n, since $p_i > 0$ for each i. Assuming, without loss of generality, $x_1 \leq x_j$ for $j = 2, \ldots, r$, let:

$$n_j = [z_j], \qquad j = 2, \ldots, r,$$

$$n_1 = n - \sum_{i=2}^{r} n_j,$$

where $[x] := \max\{k \in \mathbb{N} : k \leq x\}$ denotes the integer part of x, if $x \geq 0$. Thus, $n_1, \ldots, n_r$ are now positive integers verifying the first two relations in (1.18). Moreover, as we shall now see, there exist $\tilde{c} = \tilde{c}\,(r, p_1, \ldots, p_r, x_1, \ldots, x_r) > 0$ and $N \in \mathbb{N}$ so that if $n \geq N$

$$q(n_1, \ldots, n_r) \geq \tilde{c}\, q(z_1, \ldots, z_r),$$

which will conclude the proof of the lemma. One way to see this consists (following Azencott [7], Chapter 1) in observing that if J is a bounded interval and $a > 0$ is fixed, we can find $c_1 = c(a, J) > 0$ so that

$$\left(\frac{u}{au + b} \right)^{au+b} \geq c_1 \left(\frac{u}{au} \right)^{au}, \tag{1.22}$$

for each $u \geq 1$, with $au + b \geq 1$, and $b \in J$. (Indeed, for u, b as above, we have $(\frac{u}{au+b})^{au+b} (\frac{au}{u})^{au} = (1 + \frac{b}{au})^{-au} (\frac{u}{au+b})^b \geq c(a, J)$ as one easily checks.)

For each fixed $i = 1, \ldots, r$ we apply this for $u_i = np_i$, $a_i u_i = z_i$, $a_i u_i + b = n_i$, so that $a_i = e^{\zeta_0 x_i}/\chi$, for each i, $J_i = [-1, 0]$ if $i = 2, \ldots, r$, $J_1 = [0, r]$, and take $\tilde{c}$ as the rth power of the smallest of such $c(a_i, J_i)$. $\square$

Remark 1.5 As already noticed, the previous lemma gives a sharper estimate than (1.16) in the case of finitely valued random variables. Considering (1.19), to get (1.16) it would suffice to prove, for all large n, the existence of $n_1, \ldots, n_r$ verifying the first two equations in (1.18) and such that $\liminf_{n\to+\infty} n^{-1} \log q(n_1, \ldots, n_r) \geq \log \chi$. We shall come back to this in Section 1.3.

Proof of the lower bound (Continued) Assume $\mu(0, +\infty) > 0$ and $\mu(-\infty, 0) > 0$. If X_1 is a simple random variable, i.e. with values on a finite set, the result is contained in Lemma 1.4. Let us proceed to the general case.

Case 1. X_1 is a discrete random variable. In this case, let $x_1, x_2, \ldots$ be such that $p_i = P(X_1 = x_i) > 0$ for each i and $\sum_{i=1}^{\infty} P(X_1 = x_i) = 1$. By assumption we may take $r \geq 2$ such that

$$\min_{1\leq i\leq r} x_i < 0 < \max_{1\leq i\leq r} x_i \,. \tag{1.23}$$

Letting $\psi_k(\zeta) = \sum_{i=1}^{k} p_i \, e^{\zeta x_i}$ and using Lemma 1.4, for any $k \geq r$ we may take $c_k > 0$, $N_k \in \mathbb{N}$ such that if $n \geq N_k$, $n_1, \ldots, n_k$ are given by Lemma 1.4 then

$$P(\bar{X}_n \leq 0) \geq p(n_1, \ldots, n_k) \geq c_k \, n^{-\frac{k-1}{2}} \big(\inf_{\zeta\in\mathbb{R}} \psi_k(\zeta) \big)^n,$$

which implies:

$$\liminf_{n\to+\infty} \frac{1}{n} \log P(\bar{X}_n \leq 0) \geq \inf_{\zeta\in\mathbb{R}} \log \psi_k(\zeta),$$

for each $k \geq r$. It then suffices to check that

$$\lim_{k\to+\infty} \inf_{\zeta\in\mathbb{R}} \log \psi_k(\zeta) = \inf_{\zeta\in\mathbb{R}} \log \hat{\mu}(\zeta). \tag{1.24}$$

Since the sequence $0 < \psi_k(\zeta)$ increases to $\hat{\mu}(\zeta)$ for any given ζ, we have at once that

$$a := \lim_{k\to+\infty} \inf_{\zeta\in\mathbb{R}} \log \psi_k(\zeta) \leq \inf_{\zeta\in\mathbb{R}} \log \hat{\mu}(\zeta) \in (-\infty, 0].$$

It remains to check the reversed inequality and for this let

$$A_k := \{\zeta \in \mathbb{R} : \log \psi_k(\zeta) \leq a\}.$$

Due to the continuity of $\log \psi_k$ and to (1.23), A_k is a non-empty compact set, for each $k \geq r$. From the monotonicity of the sequence $\psi_k(\zeta)$ we have $A_r \supseteq A_{r+1} \supseteq \ldots$ and so their intersection is non-empty. But, if $\bar{\zeta} \in \bigcap_{k\geq r} A_k$, we must have $\log \hat{\mu}(\bar{\zeta}) = \lim_{k\to+\infty} \log \psi_k(\bar{\zeta}) \leq a$ proving (1.24) and thus concluding the proof of the lower bound for discrete random variables. (This last step corresponds to the Dini theorem applied to the functions $\log \psi_k(\cdot)$.)

General case. For each $k \in \mathbb{N}$ let

$$X_j^{(k)} = \sum_{i \in \mathbb{Z}} \frac{i}{k} \mathbf{1}_{\{X_j \in (\frac{i-1}{k}, \frac{i}{k}]\}},$$

where $\mathbf{1}_A$ denotes the indicator function of the set A ($\mathbf{1}_A(\omega) = 1$ if $\omega \in A$; $\mathbf{1}_A(\omega) = 0$ otherwise), and $\bar{X}_n^{(k)} = n^{-1} \sum_{j=1}^n X_j^{(k)}$. Since

$$X_j \le X_j^{(k)} \le X_j + 1/k,$$

we see that if $\mu_{(k)}$ is the law of $X_1^{(k)}$ and $\widehat{\mu_{(k)}}$ its moment generating function, we have:

$$\begin{aligned} \hat{\mu}(\zeta) \le \widehat{\mu_{(k)}}(\zeta) \le e^{\zeta/k}\, \hat{\mu}(\zeta), \qquad & \text{if } \zeta \ge 0, \\ e^{\zeta/k}\, \hat{\mu}(\zeta) \le \widehat{\mu_{(k)}}(\zeta) \le \hat{\mu}(\zeta), \qquad & \text{if } \zeta \le 0, \end{aligned}$$

so that $\widehat{\mu_{(k)}}(\zeta) \ge e^{-|\zeta|/k}\, \hat{\mu}(\zeta)$ for each ζ. Since both $\mu(0, +\infty)$ and $\mu(-\infty, 0)$ are assumed to be positive, there exist a and b positive constants for which

$$\hat{\mu}(\zeta) \ge a\, e^{b|\zeta|},$$

and with the same argument (Dini theorem) leading to (1.24) applied now to the functions $f_k(\zeta) = e^{-|\zeta|/k}\, \hat{\mu}(\zeta)$, we conclude that given $\delta > 0$, there exists $k \ge 1$ such that

$$\inf_{\zeta \in \mathbb{R}} \log \widehat{\mu_{(k)}}(\zeta) \ge \inf_{\zeta \in \mathbb{R}} \log \hat{\mu}(\zeta) - \delta. \tag{1.25}$$

From Case 1 and (1.25) we have

$$\liminf_{n \to +\infty} \frac{1}{n} \log\, P(\bar{X}_n \le 0) \ge \liminf_{n \to +\infty} \frac{1}{n} \log\, P(\bar{X}_n^{(k)} \le 0) \ge -I_\mu(0) - \delta,$$

and the proof follows, since $\delta > 0$ can be made arbitrarily small. □

Theorem 1.1 says that if $A = (-\infty, x]$ with $x \le m$, or $A = [x, +\infty)$ with $x \ge m$, the sequence $n^{-1} \log \mu_n(A)$ converges to $-I_\mu(x)$, where μ_n denotes the law of $\bar{X}_n$. (If instead, $A = (-\infty, x]$ with $x > m$, or $A = [x, +\infty)$ with $x < m$, then $\log \mu_n(A)$ converges to 0, by the weak law of large numbers.) As we shall see, the non-negative function I_μ is convex and takes the mininum at m ($I_\mu(m) = 0$) so that the previous statement might be rephrased by saying that (1.1) and (1.2) hold for all such intervals, with $I = I_\mu$.

Except for the integrability in part (a), no further assumptions on the tails of the distribution of X_1 were imposed to derive Theorem 1.1. In particular, we could have the situation when $\hat{\mu}(\zeta) = +\infty$ for all $\zeta \ne 0$, i.e. none of the tails of the distribution decays exponentially. From the expression (1.10) it follows at once that in such cases the function I_μ will vanish identically (the expression on which we take the supremum will be zero for $\zeta = 0$ and $-\infty$ otherwise). One then asks about the information contained in Theorem 1.1 in such a case. The upper bound becomes trivial, as it ought to, if all 'exponential moments' are infinite. On the other hand, the lower bound says that the decay of those probabilities is slower than any exponential $e^{-\delta n}$ with $\delta > 0$. A more refined analysis is required in such cases. (See the bibliographical remark at the end of this section.)

After the previous result we are naturally led to the study of:

(a) the basic properties of the function I_μ;
(b) the asymptotic behaviour of $n^{-1} \log \mu_n(A)$ as $n \to +\infty$, for a larger class of sets A.

In this first section we take up (b), based as much as possible on Theorem 1.1, and exploiting the basic properties of I_μ.

Later on we shall take up another (quite general) method, which is more natural for treating higher dimensional situations and is in fact applicable to a great variety of them. Also, this should lead to a better understanding of why the upper bound obtained by optimizing the use of Markov exponential inequalities is the correct one, when seeking for logarithmic equivalence.

Definition 1.6 A function $f\colon \mathbb{R} \to [-\infty, +\infty]$ is said to be lower (upper) semi-continuous if for any $x \in \mathbb{R}$ and any sequence $x_n \to x$ we have $f(x) \le \liminf_{n\to+\infty} f(x_n)$ ($f(x) \ge \limsup_{n\to+\infty} f(x_n)$, respectively).

It follows at once from the definition that any lower (upper) semi-continuous function must attain a minimum (maximum) value over any compact set.

Definition 1.7 (a) A function $f\colon \mathbb{R} \to (-\infty, +\infty]$ is said to be convex, if for any $x, y \in \mathbb{R}$ and any $0 < a < 1$ we have $f(ax + (1-a)y) \le af(x) + (1-a)f(y)$. If we have strict inequality unless $x = y$, we say that f is strictly convex. A function $f\colon \mathbb{R} \to [-\infty, +\infty)$ is said to be concave (strictly concave) if and only if $-f$ is convex (strictly convex, respectively).

(b) If M is a real vector space, a set $C \subseteq M$ is said to be convex if $ax + (1-a)y \in C$, for any $x, y \in C$, $a \in [0, 1]$. Part (a) extends in the obvious way if f is defined on a convex subset of M.

The function in Figure 1.1 is lower semi-continuous and convex; it is just a single example, but it illustrates the typical situation (except by the symmetry).

Proposition 1.8 *Let μ be the distribution of a real random variable X_1, $\hat\mu$ its moment generating function, and let $I_\mu(\cdot)$ be defined by* (1.10). *Set $\mathcal{D}_{\hat\mu} = \{\zeta : \hat\mu(\zeta) < +\infty\}$ and $\mathcal{D}_{I_\mu} = \{x : I_\mu(x) < +\infty\}$. Then*

(i) $\log \hat\mu\colon \mathbb{R} \to (-\infty, +\infty]$ and $I_\mu\colon \mathbb{R} \to [0, +\infty]$ are both convex and lower semi-continuous.

(ii) When $\mathcal{D}_{\hat\mu}$ has a non-empty interior, which we denote by $\mathcal{D}^\circ_{\hat\mu}$, then $\hat\mu$ is differentiable in $\mathcal{D}^\circ_{\hat\mu}$, with $(\hat\mu)'(\zeta) = E(X_1 e^{\zeta X_1})$ for any $\zeta \in \mathcal{D}^\circ_{\hat\mu}$. In

particular

$$(\log \hat{\mu})'(\zeta) = \frac{1}{\hat{\mu}(\zeta)} E(X_1 e^{\zeta X_1}) \tag{1.26}$$

for $\zeta \in \mathcal{D}^\circ_{\hat{\mu}}$, *and if* $x = (\log \hat{\mu})'(\zeta)$ *for one such* ζ *then* $I_\mu(x) = \zeta x - \log \hat{\mu}(\zeta)$.

(*iii*) *If there exists* $\delta > 0$ *so that* $\hat{\mu}(\zeta) < +\infty$ *for each* $\zeta \in (-\delta, \delta)$, *then* $m := EX_1 \in \mathbb{R}$ *and* $I_\mu(x) = 0$ *if and only if* $x = m$; I_μ *is strictly increasing in* $\mathcal{D}_{I_\mu} \cap (m, +\infty)$ *and strictly decreasing in* $\mathcal{D}_{I_\mu} \cap (-\infty, m)$.

Proof (i) Let $0 < a < 1$. Applying the Holder inequality we have that

$$\hat{\mu}(a\zeta + (1-a)\zeta') \le (\hat{\mu}(\zeta))^a (\hat{\mu}(\zeta'))^{1-a}$$

for any $\zeta, \zeta' \in \mathcal{D}_{\hat{\mu}}$. From this, and taking logarithms, we conclude the convexity of $\log \hat{\mu}$ on $\mathbb{R}$. Since $\log \hat{\mu}(0) = 0$ we have $I_\mu \ge 0$, and for its convexity just notice that for each ζ the expression $\zeta\, x - \log \hat{\mu}(\zeta)$ gives an affine function of x, either real or constantly equal to $-\infty$. Being the pointwise supremum, I_μ is convex. In more detail:

$$\begin{aligned} I_\mu(ax + (1-a)y) &= \sup_{\zeta \in \mathbb{R}} \{\zeta\, (ax + (1-a)y) - \log \hat{\mu}(\zeta)\} \\ &\le a \sup_{\zeta \in \mathbb{R}} \{\zeta\, x - \log \hat{\mu}(\zeta)\} + (1-a) \sup_{\zeta \in \mathbb{R}} \{\zeta\, y - \log \hat{\mu}(\zeta)\} \\ &= a I_\mu(x) + (1-a) I_\mu(y), \end{aligned}$$

for $0 < a < 1$. The lower semi-continuity of $\log \hat{\mu}$ follows from the Fatou lemma. Since I_μ is the supremum of a family of continuous maps, it must be lower semi-continuous.

(ii) The first statement is a consequence of the dominated convergence theorem: if ζ is in $\mathcal{D}^\circ_{\hat{\mu}}$, taking ε small, we see that $\varepsilon^{-1}(\hat{\mu}(\zeta + \varepsilon) - \hat{\mu}(\zeta)) = E\{e^{\zeta X_1}(e^{\varepsilon X_1} - 1)/\varepsilon\}$; as $\varepsilon \to 0$ the variables under the expectation sign converge to $X_1 e^{\zeta X_1}$, and for $|\varepsilon| \le \bar{\varepsilon}$ they are dominated by $e^{\zeta X_1}(e^{\bar{\varepsilon}|X_1|} - 1)/\bar{\varepsilon}$, which is integrable if $[\zeta - \bar{\varepsilon}, \zeta + \bar{\varepsilon}] \subseteq \mathcal{D}_{\hat{\mu}}$. Equation (1.26) follows at once. (The same argument applies to subsequent derivatives of $\hat{\mu}$ in $\mathcal{D}^\circ_{\hat{\mu}}$, this being the reason for calling it the 'moment generating function': $(\hat{\mu})^{(k)}(\zeta) = E(X_1^k e^{\zeta X_1})$.) As for the last statement, the function $g_x(\zeta) := \zeta x - \log \hat{\mu}(\zeta)$ is concave, upper semi-continuous and it is finite exactly on the set $\mathcal{D}_{\hat{\mu}}$. When $x = (\log \hat{\mu})'(\zeta)$ for some ζ in $\mathcal{D}^\circ_{\hat{\mu}}$ we have $(g_x)'(\zeta) = 0$ so that the maximun is achieved.

(iii) By the assumption, both tails of μ decay exponentially so that m is finite; as already noticed in Remark 1.3 we have $I_\mu(m) = 0$. Recalling Remark 1.2 and the convexity of I_μ, it now suffices to consider the case $m > 0$ and to prove that $I_\mu(0) > 0$. But, from (1.15) we have seen that $-I_\mu(0) = \inf_{\zeta \le 0} \log \hat{\mu}(\zeta)$ in this case. Since $\log \hat{\mu}(0) = 0$ and $(\log \hat{\mu})'(0) > 0$ this infimum is strictly negative, concluding the assertion. □

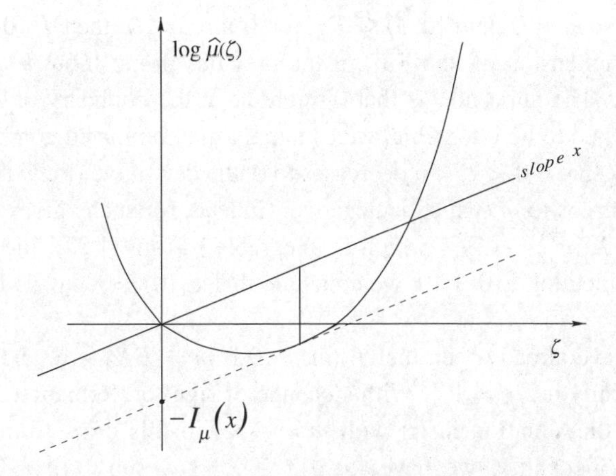

Figure 1.2

Figure 1.2 gives the geometric interpretation of $I_\mu(x)$.

Let us now complement Proposition 1.8 by examining a little more the situation when possibly only one of the tails of the distribution decays exponentially. By a change of sign, cf. Remark 1.2, we reduce to the positive tail as in part (i) below.

Proposition 1.9 *With the same notation as in Proposition* 1.8 *we have*:

(*i*) *if* $\hat{\mu}(\zeta) < +\infty$ *for some* $\zeta > 0$, *then* $m = EX_1$ *is well defined and* $m < +\infty$; *in this case, for* $x \geq m$

$$I_\mu(x) = \sup_{\zeta \geq 0}(\zeta x - \log \hat{\mu}(\zeta)), \tag{1.27}$$

I_μ *is non-decreasing on* $(m, +\infty)$, *and* $\lim_{x\to+\infty} I_\mu(x) = +\infty$; *when* $m \in \mathbb{R}$, I_μ *is strictly increasing in* $\mathcal{D}_{I_\mu} \cap (m, +\infty)$;

(*ii*) $\inf_{x\in\mathbb{R}} I_\mu(x) = 0$;

(*iii*) $\hat{\mu}(\zeta) < +\infty$ *for some* $\zeta > 0$ *iff* $I_\mu(x) \to +\infty$ *as* $x \to +\infty$;

(*iv*) *if* $\hat{\mu}(\zeta) < +\infty$ *for all* $\zeta > 0$, *then* $I_\mu(x)/|x| \to +\infty$ *as* $x \to +\infty$;

(*v*) *if* $\mu(x, +\infty) > 0$ *and* $\mu(-\infty, x) > 0$, *then* $x \in \mathcal{D}^\circ_{I_\mu}$;

(*vi*) *if* $\mu(x, +\infty) = 0$ *or* $\mu(-\infty, x) = 0$, *then* $I_\mu(x) = -\log \mu\{x\}$.

Proof (i) From the assumption we have $EX_1^+ < +\infty$, so that $m < +\infty$ is well defined. If $m \in \mathbb{R}$ and $x \geq m$ the expression (1.27) follows at once from the Jensen inequality (rewriting (1.15) with a change in sign, since now $x \geq m$). If $m = -\infty$, then $-\log \hat{\mu}(\zeta) = -\infty$ for all $\zeta < 0$ so that (1.27) also holds. If there exists $\zeta \in \mathcal{D}_{\hat{\mu}} \cap (0, +\infty)$, we have $\liminf_{x\to+\infty} I_\mu(x)/x \geq \liminf_{x\to+\infty}(\zeta - \log \hat{\mu}(\zeta)/x) = \zeta$. In particular we see that $I_\mu(x)$ tends to $+\infty$ as $x \to +\infty$, and also the statement (iv). To verify the last statement of (i), subtracting a constant (cf. Remark 1.2), and recalling the convexity of I_μ, we again reduce to checking that if $-\infty < m < 0$, and $[0, \delta) \subseteq \mathcal{D}_{\hat{\mu}}$ for some $\delta > 0$, then $I_\mu(0) > 0$, giving a one-sided extension of item (iii) of the previous proposition, and the proof is quite similar. The point now is that 0 might be at the boundary of $\mathcal{D}_{\hat{\mu}}$, but since $|X_1|$ is assumed to be integrable, we can apply the dominated convergence theorem to check that if $0 < \zeta < \delta$ decreases to 0, the derivative $(\log \hat{\mu})'(\zeta)$, given by (1.26), converges to m, which is negative. (Indeed, for such values of ζ we have $|X_1 e^{\zeta X_1}| \le X_1^+ e^{\delta X_1^+} + X_1^-$, which is integrable.) From (1.27), the convexity of $\log \hat{\mu}$, and since $\log \hat{\mu}(0) = 0$, we conclude that $I_\mu(0) = -\inf_{\zeta \ge 0} \log \hat{\mu}(\zeta) > 0$ in this case.

(ii) We have already seen that $I_\mu(m) = 0$ if $m = EX_1 \in \mathbb{R}$ and that I_μ vanishes identically if $\mathcal{D}_{\hat{\mu}} = \{0\}$. With a change of sign (cf. Remark 1.2), it remains to consider the situation in (i) with $m = -\infty$; in this case, from the Markov exponential inequality, we have $\log \mu[x, +\infty) \le -\sup_{\zeta \ge 0}(\zeta x - \log \hat{\mu}(\zeta)) = -I_\mu(x) \le 0$. Recalling (1.27) and that $\lim_{x \to -\infty} \log \mu[x, +\infty) = 0$, we conclude that $\lim_{x \to -\infty} I_\mu(x) = 0$, proving (ii).

The remaining statement in (iii) is easily obtained from previous considerations.

(v) By Remark 1.2 it suffices to consider $x = 0$. We may take $\delta > 0$ so that $\mu(\delta, +\infty) > 0$ as well $\mu(-\infty, -\delta) > 0$. For $\zeta > 0$ we then have $\log \hat{\mu}(\zeta) \ge \zeta\delta + \log \mu(\delta, +\infty)$ and, similarly, $\log \hat{\mu}(\zeta) \ge -\zeta\delta + \log \mu(-\infty, -\delta)$ for $\zeta < 0$. From (1.10) we then see that $I_\mu(y) < +\infty$ for $|y| < \delta$.

(vi) This was already seen at the begining of the proof of part (b) of Theorem 1.1. □

The function I_μ defined by (1.10) is usually called the Fenchel–Legendre transform of the convex and lower semi-continuous function $\log \hat{\mu}$. There is indeed a duality between both functions, i.e. a reciprocal relation holds: $\log \hat{\mu}(\zeta) = \sup_{x \in \mathbb{R}}(\zeta x - I_\mu(x))$ is the Fenchel–Legendre transform of the convex and lower semi-continuous function I_μ (see Theorem 12.2 in [255]). The geometric interpretation is clear from that of convex functions, and thus its connection to separating hyperplanes (lines).[1] A few examples of I_μ are given below. Figure 1.3 illustrates two typical situations.

Examples

(a) Exponential random variable with rate $\lambda > 0$: $\mu(dx) = \lambda e^{-\lambda x} \mathbf{1}_{[0,+\infty)}(x) dx$; $\log \hat{\mu}(\zeta) = \log(\lambda/(\lambda - \zeta))$ for $\zeta < \lambda$, $\log \hat{\mu}(\zeta) = +\infty$ otherwise; $I_\mu(x) = \lambda x - 1 - \log(\lambda x)$ for $x > 0$, $I_\mu(x) = +\infty$ otherwise.

(b) Gaussian with average m and variance σ^2: $\mu(dx) = (\sigma\sqrt{2\pi})^{-1} e^{-(x-m)^2/2\sigma^2} dx$; $\log \hat{\mu}(\zeta) = \zeta m + \zeta^2\sigma^2/2$ for all ζ; $I_\mu(x) = (x - m)^2/2\sigma^2$ for all x.

(c) Bernoulli with parameter $p \in (0, 1)$: $p = \mu\{1\} = 1 - \mu\{0\}$; $\log \hat{\mu}(\zeta) = \log(1 + p(e^\zeta - 1))$ for all ζ; $I_\mu(x) = x \log(x/p) + (1 - x) \log((1 - x)/(1 - p))$ for $x \in [0, 1]$, $I_\mu(x) = +\infty$ otherwise.

[1] Duality will play a crucial role in Chapter 3, also in infinite dimensional spaces.

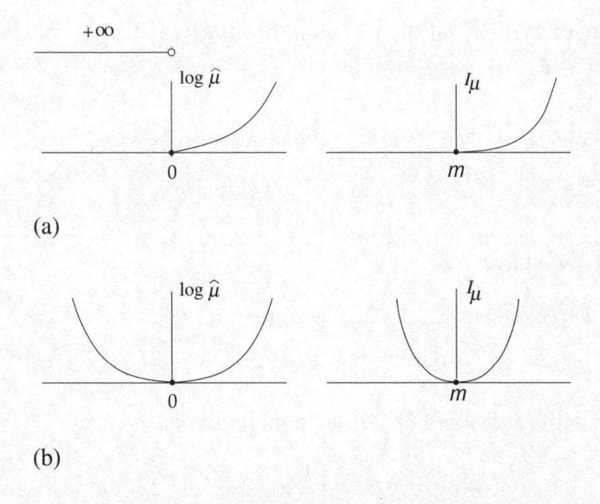

Figure 1.3

(d) Poisson distribution with rate λ: $\mu\{k\} = e^{-\lambda}\lambda^k/k!$ for all $k \in \mathbb{N}$; $\log\hat{\mu}(\zeta) = \lambda(e^\zeta - 1)$ for all ζ; $I_\mu(x) = \lambda - x + x\log(x/\lambda)$ if $x \geq 0$, $I_\mu(x) = +\infty$ otherwise.

Corollary 1.10 *Let μ be a probability on $\mathbb{R}$ and let μ_n denote the law of $\bar{X}_n = n^{-1}\sum_{i=1}^n X_i$, where $X_1, X_2, \ldots$ are i.i.d. according to μ. Then:*

(i) if $C \subseteq \mathbb{R}$ is closed, then

$$\limsup_{n\to+\infty} \frac{1}{n} \log \mu_n(C) \leq -\inf_{x\in C} I_\mu(x);$$

(ii) if $G \subseteq \mathbb{R}$ is open, then

$$\liminf_{n\to+\infty} \frac{1}{n} \log \mu_n(G) \geq -\inf_{x\in G} I_\mu(x),$$

where $I_\mu(\cdot)$ is given by (1.10), *with the convention that the infimum over the empty set is $+\infty$.*[2]

Proof (i) It suffices to consider non-empty closed sets C such that $\inf_{x\in C} I_\mu(x)$ is positive, otherwise the inequality becomes trivial. In particular, there exists $\zeta \neq 0$ for which $\hat{\mu}(\zeta) < +\infty$ and by Proposition 1.9, $m = EX_1$ is well defined. We first assume $m \in \mathbb{R}$, in which case $m \notin C$, cf. Remark 1.3. If $C \cap (m, +\infty) \neq \emptyset$ we take $x_+ := \min\{x \in C : x > m\}$ and Theorem 1.1 gives us:

$$\begin{aligned} \frac{1}{n} \log \mu_n(C \cap (m, +\infty)) &\leq \frac{1}{n} \log \mu_n[x_+, +\infty) \leq -I_\mu(x_+) \\ &= -\inf_{x\in C\cap(m,+\infty)} I_\mu(x), \end{aligned} \tag{1.28}$$

[2] This convention holds throughout the text.

using the convexity of I_μ at the last step. Similarly, if $C \cap (-\infty, m) \neq \emptyset$, we set $x_- := \max\{x \in C : x < m\}$, so that:

$$\begin{aligned}\frac{1}{n}\log \mu_n(C \cap (-\infty, m)) &\leq \frac{1}{n}\log \mu_n(-\infty, x_-] \leq -I_\mu(x_-)\\ &= -\inf_{x \in C \cap (-\infty, m)} I_\mu(x).\end{aligned} \tag{1.29}$$

Consequently, we have

$$\begin{aligned}\mu_n(C) &\leq \mu_n[x_+, +\infty) + \mu_n(-\infty, x_-]\\ &\leq e^{-nI_\mu(x_+)} + e^{-nI_\mu(x_-)} \leq 2e^{-n(I_\mu(x_+)\wedge I_\mu(x_-))},\end{aligned} \tag{1.30}$$

and together with (1.28) and (1.29) we conclude that

$$\frac{1}{n}\log \mu_n(C) \leq \frac{1}{n}\log 2 - \inf_{x \in C} I_\mu(x),$$

proving (i) in this case. (*Notation.* $a \wedge b = \min\{a, b\}$, $a \vee b = \max\{a, b\}$.)

If $m = -\infty$, due to our previous consideration and the proof of part (ii) of Proposition 1.9 we conclude that C must be bounded from below. Letting x_+ be the minimum of C we get (1.28); analogously when $m = +\infty$.

We now prove (ii). A simple observation tells us that it suffices to check that

$$\liminf_{n \to +\infty} \frac{1}{n}\log \mu_n(x - \delta, x + \delta) \geq -I_\mu(x) \tag{1.31}$$

for all $x \in \mathbb{R}$ and $\delta > 0$. Taking $X_i' = X_i - x$ and recalling Remark 1.2, we reduce to checking (1.31) with $x = 0$. In fact we shall prove the stronger result:

$$\liminf_{n \to +\infty} \frac{1}{n}\log \mu_n(-\delta, 0] \geq -I_\mu(0). \tag{1.32}$$

To prove this we first notice that (1.32) becomes trivial if $\mu(-\infty, 0) = 0$ or $\mu(0, +\infty) = 0$ since in such cases we have $-I_\mu(0) = \log \mu\{0\}$, cf. (vi) of Proposition 1.9 and, on the other hand,

$$\frac{1}{n}\log \mu_n(-\delta, 0] \geq \frac{1}{n}\log \mu_n\{0\} \geq \log \mu\{0\}.$$

Now suppose $\mu(0, +\infty) > 0$ and $\mu(-\infty, 0) > 0$; by part (v) of Proposition 1.9, and decreasing δ if needed, we may assume that $(-\delta, \delta) \subseteq \mathcal{D}_{I_\mu}$.

We first impose the extra condition that $\hat{\mu}(\zeta) < +\infty$ for all ζ in some open interval containing 0 so that $m \in \mathbb{R}$ and, from part (iii) of Proposition 1.8, we know that $I_\mu(\cdot)$ is strictly increasing (decreasing) on $\mathcal{D}_{I_\mu} \cap (m, +\infty)$ ($\mathcal{D}_{I_\mu} \cap (-\infty, m)$, respectively). If $m \geq 0$, we write

$$\mu_n(-\delta, 0] = \mu_n(-\infty, 0] - \mu_n(-\infty, -\delta]$$

and apply Theorem 1.1: part (a) is applied to $\mu_n(-\infty, -\delta]$, and part (b) to $\mu_n(-\infty, 0]$. Equation (1.32) follows since $I_\mu(-\delta) > I_\mu(0)$. If $m < 0$, we take

$0 < \tilde{\delta} < \delta \wedge -m$ and write

$$\liminf_{n\to+\infty} \frac{1}{n} \log \mu_n(-\delta, 0] \geq \liminf_{n\to+\infty} \frac{1}{n} \log \mu_n[-\tilde{\delta}, 0) \geq -I_\mu(-\tilde{\delta}),$$

by the previous case (inverting signs and shifting, cf. Remark 1.2). On the other hand, again by part (iii) of Proposition 1.8, $I_\mu(-\tilde{\delta}) \leq I_\mu(0)$ in the present case. This proves (1.32) under the extra condition of the finiteness of $\hat{\mu}$ in some open interval containing 0.

It remains to relax this extra assumption. We may use a truncation argument, setting ν as the distribution of X_1 conditioned to $|X_1| \leq k$, i.e.

$$\nu(dx) = \frac{\mathbf{1}_{|x|\leq k}\, \mu(dx)}{\mu[-k, k]},$$

where $k \in \mathbb{N}$ is taken so that $\mu[-k, 0) > 0$ and $\mu(0, k] > 0$. Observe that

$$\log \hat{\nu}(\zeta) + \log \mu[-k, k] = \log \int_{-k}^{k} e^{\zeta x}\, \mu(dx)$$

and that

$$\nu_n(-\delta, 0](\mu[-k, k])^n \leq \mu_n(-\delta, 0]$$

to write:

$$\frac{1}{n} \log \mu_n(-\delta, 0] \geq \frac{1}{n} \log \nu_n(-\delta, 0] + \log \mu[-k, k]. \tag{1.33}$$

By (1.33) and the previous analysis applied to ν_n we get

$$\liminf_{n\to+\infty} \frac{1}{n} \log \mu_n(-\delta, 0] \geq \inf_{\zeta\in\mathbb{R}} \log \int_{-k}^{k} e^{\zeta x}\, \mu(dx). \tag{1.34}$$

It remains to prove that as $k \to +\infty$ the infimum on the right-hand side of (1.34) tends to $\inf_{\zeta\in\mathbb{R}} \log \hat{\mu}(\zeta)$. This is just the same argument as in (1.24) and it concludes the proof. □

Remarks (a) The estimate coming from Theorem 1.1 is indeed stronger than what is stated as part (ii) of the previous corollary; according to (1.32) we can take intervals of the form $(x - \delta, x]$ or $[x, x + \delta)$ and still get the same lower bound.

(b) According to part (i) of Proposition 1.9, if $[0, \zeta) \subseteq \mathcal{D}_{\hat{\mu}}$ for some positive ζ and $m \in \mathbb{R}$, then the function $I_\mu(x)$ is strictly increasing on $(m, +\infty) \cap \mathcal{D}_{I_\mu}$, so that the one-sided estimates provided by the upper bound in Theorem 1.1 are indeed informative; analogously if there exists $\zeta < 0$ such that $\hat{\mu}(\zeta) < +\infty$.

In Section 1.4, while examining the higher dimensional case, it will be convenient to consider a different proof for lower estimates; we now sketch this for dimension one.

Another proof for (1.31) (Change of measure) We suppose that $\mathcal{D}_{\hat{\mu}} = \mathbb{R}$ (see comment below). The basic ingredient is part (ii) of Proposition 1.8. For this proof to work we need $\mu(x, +\infty) > 0$ as well as $\mu(-\infty, x) > 0$. Again, by shifting the variables (Remark 1.2) we make $x = 0$.

Due to our assumptions, $\log \hat{\mu}(\zeta)$ tends to $+\infty$ as $|\zeta| \to +\infty$, implying the existence of ζ_0 such that $\log \hat{\mu}(\zeta_0) = \inf_{\zeta \in \mathbb{R}} \log \hat{\mu}(\zeta) = -I_\mu(0)$, and (1.26) tells that $E(X_1 e^{\zeta_0 X_1})/\hat{\mu}(\zeta_0) = (\log \hat{\mu})'(\zeta_0) = 0$. That is, looking at transformed probability measures $\nu^{(\zeta)}(dz) = e^{\zeta z}\mu(dz)/\hat{\mu}(\zeta)$ for $\zeta \in \mathbb{R}$, we see that $\nu^{(\zeta_0)}$ (which we write simply ν) has expectation equal to 0. In other words, this transformation turns 0 into the 'central' value for the sample average. If ν_n denotes the law of the sample average $\bar{Y}_n$, for $Y_1, Y_2, \ldots$ i.i.d., distributed according to ν, the (weak) law of large numbers says that $\nu_n(-\delta, \delta)$ tends to one. Computing $\mu_n(-\delta, \delta)$ in terms of the transformed measures, and if $0 < \delta' < \delta$ we have:

$$\begin{aligned} \mu_n(-\delta, \delta) \geq \mu_n(-\delta', \delta') &= \int_{\{|\frac{1}{n}\sum_{i=1}^n x_i| < \delta'\}} \prod_{i=1}^n \mu(dx_i) \\ &= (\hat{\mu}(\zeta_0))^n \int_{\{|\frac{1}{n}\sum_{i=1}^n x_i| < \delta'\}} e^{-\zeta_0 \sum_{i=1}^n x_i} \prod_{i=1}^n \nu(dx_i). \end{aligned} \tag{1.35}$$

Taking logarithms and setting $n \to +\infty$ we get

$$\liminf_{n \to +\infty} \frac{1}{n} \log \mu_n(-\delta, \delta) \geq \log \hat{\mu}(\zeta_0) - \delta' |\zeta_0|.$$

Letting δ' tend to zero we recover (1.31) for $x = 0$. □

Comments (i) The previous argument applies to any $x \in \mathbb{R}$, if μ is not supported by any half-line, and $\mathcal{D}_{\hat{\mu}} = \mathbb{R}$. The remaining cases could then be solved as in the proof of Corollary 1.10.

(ii) *Uniqueness of the optimal* ζ_0. From (1.26) we check that $(\log \hat{\mu})''(\zeta)$ coincides with the variance corresponding to $\nu^{(\zeta)}$; thus, $\log \hat{\mu}$ is strictly convex in $\mathcal{D}^\circ_{\hat{\mu}}$ for non-constant X_1, implying the uniqueness of ζ_0. In such a case, I_μ is strictly convex in the interval $\{(\log \hat{\mu})'(\zeta) \colon \zeta \in \mathcal{D}^\circ_{\hat{\mu}}\}$. The previous argument yields (1.31) directly for x in this set.

(iii) With the method based on the change of measure and the law of large numbers, we have not recovered the lower estimates for $\mu_n(x - \delta, x]$ as in (1.32). Following the previous discussion, we need a suitable lower bound for $\nu_n(-\delta', 0]$ in (1.35). This may be achieved through the central limit theorem, which gives $\lim_{n \to +\infty} \nu_n(-\delta', 0] = 1/2$ (see e.g. Section 2.2.1 of [85]).

(iv) The above discussion somehow explains 'why' optimization over the exponential Markov inequality leading to the upper bound in Theorem 1.1 was efficient, i.e. gave the 'correct' estimate. (Under the previous assumptions, if ζ is optimal for $x \in \mathbb{R}$, then $I_\mu(x) = \int (d\nu^{(\zeta)}/d\mu) \log(d\nu^{(\zeta)}/d\mu) d\mu = \mathcal{H}(\nu^{(\zeta)} \mid \mu)$, cf. Sections 1.3 and 1.5.)

(v) *Steepness.* Let us assume that $0 \in \mathcal{D}^\circ_{\hat{\mu}}$, and that $|(\log \hat{\mu})'(\zeta)|$ tends to $+\infty$ as $\zeta \in \mathcal{D}^\circ_{\hat{\mu}}$ tends to $\partial\mathcal{D}_{\hat{\mu}}$, the boundary of $\mathcal{D}_{\hat{\mu}}$. This guarantees that whenever $I_\mu(x) < +\infty$ the supremum in (1.10) is attained at some $\zeta \in \mathcal{D}^\circ_{\hat{\mu}}$, allowing us to use the 'change of measure' argument at any x. The requirement corresponds to the 'steepness' of $\log \hat{\mu}$, to appear again in Section 1.4. It is worth trying out examples.

Remark 1.11 One of the main motivations for the development of estimates like those in Theorem 1.1 appeared in connection with problems in classical mathematical statistics, for instance determination of the asymptotic efficiency of some hypothesis tests. A very simple but illustrative example is the following: consider two hypotheses H_0 and H_1; under H_i, X is a Bernoulli random variable with p_i $(i = 0, 1)$ being the probability of success, and $p_1 \neq p_0$. That is,

$$P(X = 1 \mid H_i) = p_i = 1 - P(X = 0 \mid H_i), \qquad i = 0, 1.$$

Without loss of generality assume $0 < p_0 < p_1 < 1$. Based on n independent observations $X_1, \ldots, X_n$ the maximum likelihood tests have the form:

$$\text{reject} \qquad H_0 \quad \text{iff} \quad S_n := \sum_{i=1}^{n} X_i > k,$$

for some $k = \gamma n$, with $p_0 < \gamma < p_1$. To compare different tests, or in order to determine how large the sample must be, it is important to estimate the probabilities of wrong conclusion, i.e. rejecting H_0 when it is true, or not rejecting H_0 when H_1 is true. These error probabilities are:

$$e_0(n) = P(S_n > k \mid H_0) \quad \text{and} \quad e_1(n) = P(S_n \leq k \mid H_1).$$

Notice that for large values of n the difference between the averages, $n(p_1 - p_0)$, is much larger than that between the standard deviations, $\sqrt{n}(\sqrt{p_1(1-p_1)} - \sqrt{p_0(1-p_0)})$. It is then not reasonable in such cases to use a central limit approximation, and indeed $e_0(n)$ and $e_1(n)$ involve tail probabilities as in Theorem 1.1, since $p_0 < \gamma < p_1$.

More generally, consider hypothesis tests based on a sample of i.i.d. observations $Y_1, \ldots, Y_n$ of a given variable, which under H_i has a distribution $\mu^{(i)}$, $i = 0, 1$, with $\mu^{(0)} \neq \mu^{(1)}$ assumed to be mutually absolutely continuous. Taking the log-likelihood ratio $X_j = \log(d\mu^{(1)}/d\mu^{(0)})(Y_j)$, then $E_0(X_j) < 0 < E_1(X_j)$. Maximum likelihood tests take the form as above, and evaluation of the error probabilities involves large deviation estimates for $\sum_{j=1}^n X_j$. Thus, the Cramér–Chernoff theorem gives an answer to the important problem of obtaining good estimates for $e_0(n)$ and $e_1(n)$.

We have used an extremely simplified situation, just to illustrate the need for large deviation estimates; classical hypothesis testing usually involves 'compound' hypotheses, as for instance if some parametric value of the model is below or above a given threshold. A very common situation is with H_0 being a 'simple' hypothesis, say 'null effect' $H_0\colon \theta = \theta_0$, against an alternative $H_1 : \theta > \theta_0$ on a suitable parametric model (P_θ); the parameter θ could indicate efficiency of a treatment, etc. As for optimality criteria, the classical proposal of Neyman and Pearson started by attributing an asymmetric weight to the two errors: under the constraint that $e_0(n) \le \alpha$ (the level) one tries to minimize $e_1(n)$ (i.e. increase the power). According to this criterion, and for simple hypotheses, i.e. $H_0 : \theta = \theta_0$ versus $H_1 : \theta = \theta_1$ as in the previous example, the classical Neyman–Pearson lemma gives the optimality of likelihood tests, the determination $\lim n^{-1} \log e_1(n)$ being a classical large deviation problem. More details on this connection may be found in [71]. For a discussion related to statistical tests in the frame of asymptotically Gaussian statistical 'experiments', see Section 11.9 of [195].

Bibliographical remarks Cramér considered a special class of distributions (with an absolute continuous component), giving 'exact' asymptotics; in this context the method of exponential change of measure goes back to his work. In the classical situation of i.i.d. real variables, an extension by Feller [114] followed his article; lattice variables were treated by Blackwell and Hodges [21], and more general cases by Bahadur and Rao [10]. Another classical reference is [9]. For indications and basic references on more recent 'exact' results see [296]. The advantages of the weaker formulation, as set by Chernoff, are simplicity and broad applicability. We hope this will be clear throughout the text.

There is a vast literature dealing with estimates for $P(S_n \ge x)$, where $x \to +\infty$ as $n \to +\infty$, beyond the central limit theorem. Situations where $\hat{\mu}(t) = +\infty$ if $t \ne 0$ are also considered; see e.g. [27, 156], the review article [223], and references therein.

1.2 Abstract formulation

Estimates as in Corollary 1.10 constitute the basic content of what has usually been called a 'large deviation principle' (l.d.p.), according to the following definition introduced by Varadhan. An equivalent formulation was introduced by Borovkov (see Remark 1.16).

Definition 1.12 A sequence (μ_n) of probability measures on the Borel σ-field $\mathcal{B}(M)$ of a metric space M is said to satisfy a large deviation principle (l.d.p.), with scaling $a_n \nearrow +\infty$ and rate function I if:

(a) $I\colon M \to [0, +\infty]$ is lower semi-continuous (cf. extension of Definition 1.6);

(b) lower bound: $\liminf_{n\to+\infty} \frac{1}{a_n} \log \mu_n(G) \ge -\inf_{x\in G} I(x)$ for any open set $G \subseteq M$;

(c) upper bound: $\limsup_{n\to+\infty} \frac{1}{a_n} \log \mu_n(C) \le -\inf_{x\in C} I(x)$ for any closed set $C \subseteq M$.

For brevity, we usually say that (μ_n) satisfies a l.d.p. If condition (c) is weakened, so as to hold for all compact sets $C \subseteq M$, we say that the sequence (μ_n) satisfies a weak large deviation principle.

Remark The applications and results analysed in this text, except for a few comments, refer to cases when M is a Polish space, i.e. a separable metric space admiting a metric ϱ for which it is Cauchy complete. The restriction to Polish spaces avoids extra difficulties (see discussion at the end of Section 1.6).

As in Corollary 1.10, if we have in mind μ_n which converge weakly to δ_m, the Dirac point-mass at some $m \in M$, the situation when $I(x) > 0$ for any $x \ne m$ becomes of particular interest; if applicable, Definition 1.12 provides bounds on the rate at which $\mu_n(A)$ tends to zero if m stays at a positive distance from A. Indeed, for any sequence (μ_n) satisfying Definition 1.12 we have

$$-\inf_{x\in A^\circ} I(x) \le \liminf_{n\to+\infty} \frac{1}{a_n} \log \mu_n(A) \le \limsup_{n\to+\infty} \frac{1}{a_n} \log \mu_n(A) \le -\inf_{x\in \overline{A}} I(x) \tag{1.36}$$

for each Borel set $A \subseteq M$, where A° and $\overline{A}$ denote respectively the interior and the closure of the set A. In particular,

$$\lim_{n\to+\infty} \frac{1}{a_n} \log \mu_n(A) = -\inf_{x\in A} I(x), \tag{1.37}$$

whenever

$$\inf_{x\in A^\circ} I(x) = \inf_{x\in \overline{A}} I(x). \tag{1.38}$$

In the context of Theorem 1.1 we may have (1.37) without (1.38). In general, conditions (b) and (c) of Definition 1.12 should *not* be required for all Borel sets. (For instance, if each μ_n is non-atomic, (b) cannot hold for all singletons, since it would force $I(x) = +\infty$ for all x, but taking $C = M$ in (c) we see that $\inf_{x\in M} I(x) = 0$.) Stronger statements may hold in particular situations. Notice that (1.36) gives a way to modify Definition 1.12 when the measures μ_n are defined on a σ-field $\mathcal{B}$ not necessarily containing $\mathcal{B}(M)$.

Motivated by the considerations in Section 1.1, and since Definition 1.12 does not exclude the less informative case of I identically null, we are led to the following.

Definition 1.13 In the context of Definition 1.12, we say that I is a *good rate function* if for any $b < +\infty$ the level set $\mathbb{F}_I(b) := \{x : I(x) \le b\}$ is compact.

Notation $B_\delta(x) := \{y : \varrho(y, x) < \delta\}$, and $\bar{B}_\delta(x) := \{y : \varrho(y, x) \le \delta\}$, respectively the open and closed balls of radius δ, centred at x.

Remark 1.14 Let ϱ denote the metric in M. We can easily check the equivalence of the following properties:

(i) I is lower semi-continuous (i.e. if $x_n \to x$ then $\liminf_{n\to+\infty} I(x_n) \geq I(x)$);
(ii) $\lim_{\delta\to 0} \inf_{y\in B_\delta(x)} I(y) = I(x)$ for each x;
(iii) the level sets $\mathbb{F}_I(b)$ are closed, for each $b < +\infty$.

Thus, if I is a good rate function it must achieve the infimum over each closed set; in particular, since $\inf_{x\in M} I(x) = 0$, there exists $x \in M$ so that $I(x) = 0$.

Example In the set-up of the Cramér theorem on $\mathbb{R}$ (Corollary 1.10), if $0 \in \mathcal{D}^\circ_{\hat{\mu}}$ then $I_\mu(\cdot)$ is a good rate function, cf. Proposition 1.8.

It might be convenient to rephrase the validity of a l.d.p. with a good rate function in terms of a *local* lower estimate and a *global* upper estimate, as done by Freidlin and Wentzell:

Proposition 1.15 *In the frame of Definition* 1.12, *let* ϱ *be the metric in* M, *and set* $\varrho(x, A) := \inf\{\varrho(x, y)\colon y \in A\}$, *for* $x \in M$, $A \subseteq M$,
Consider the conditions:

(b)′ $\liminf_{n\to+\infty} \frac{1}{a_n} \log \mu_n(B_\delta(x)) \geq -I(x)$ *for any* $x \in M$ *and any* $\delta > 0$;
(c)′ $\limsup_{n\to+\infty} \frac{1}{a_n} \log \mu_n\{y\colon \varrho(y, \mathbb{F}_I(b)) \geq \delta\} \leq -b$, *for any* $b < +\infty$ *and any* $\delta > 0$.

Related to the conditions in Definition 1.12 *we have*: (b)′ ⇔ (b), *and* (c) ⇒ (c)′.
If I is a good rate function, then also (c)′ ⇒ (c).

Proof The equivalence (b)′ ⇔ (b) is completely simple. The implication (c)⇒(c)′ is trivial since $\{y\colon \varrho(y, \mathbb{F}_I(b)) \geq \delta\}$ is a closed set. As for the last statement, assume that I is a good rate function, and that condition (c)′ holds. We verify condition (c): take $C \subseteq M$ a closed set; if $\inf_{x\in C} I(x) = 0$ there is nothing to show. Otherwise, let $0 < b < \inf_{x\in C} I(x)$, so that the set $\mathbb{F}_I(b)$ is (non-empty) compact and disjoint from C. Therefore, there exists $\delta > 0$ so that $C \subseteq \{y\colon \varrho(y, \mathbb{F}_I(b)) \geq \delta\}$, and (c)′ implies that

$$\limsup_{n\to+\infty} \frac{1}{a_n} \log \mu_n(C) \leq -b.$$

Since this holds for any such b we get condition (c). □

Remark 1.16 (i) In the previous proposition, if I is a rate function (i.e. lower semi-continuous, not necessarily good), and conditions (b)′ and (c)′ hold, then (μ_n) satisfies a weak l.d.p. Indeed, if the set C in the previous proof is compact and disjoint from the closed set $\mathbb{F}_I(b)$, then $C \subseteq \{y\colon \varrho(y, \mathbb{F}_I(b)) \geq \delta\}$ for some $\delta > 0$, and we conclude as before.

(ii) *Uniqueness of the rate function.* Under the conditions of Definition 1.12, and having fixed a_n, the function I is uniquely defined. Indeed, let I and $\tilde{I}$ both satisfy Definition 1.12 with the same scaling (a_n), and let $x \in M$. Using condition

(c) for I and condition (b) for $\tilde{I}$ we get:

$$-\tilde{I}(x) \le -\inf_{y \in B_\delta(x)} \tilde{I}(y) \le \liminf_{n\to+\infty} \frac{1}{a_n} \log \mu_n(B_\delta(x))$$

$$\le \limsup_{n\to+\infty} \frac{1}{a_n} \log \mu_n(\bar{B}_\delta(x)) \le -\inf_{y\in\bar{B}_\delta(x)} I(y).$$

The lower semi-continuity of I implies that the rightmost term tends to $-I(x)$ as $\delta \downarrow 0$, cf. Remark 1.14. Thus $I(x) \le \tilde{I}(x)$. Reverting the roles of I and $\tilde{I}$, we get $I(x) = \tilde{I}(x)$, and

$$-I(x) = \lim_{\delta\downarrow 0}\liminf_{n\to+\infty} \frac{1}{a_n} \log \mu_n(B_\delta(x)) = \lim_{\delta\downarrow 0}\limsup_{n\to+\infty} \frac{1}{a_n} \log \mu_n(B_\delta(x)).$$

Thus, if M is locally compact, the validity of a weak l.d.p. with given scaling already determines I. The argument shows that if I satisfies conditions (a) and (c) in Definition 1.12 for a given scaling (a_n), then $I \le \tilde{I}$ for any function $\tilde{I}\colon M \to [0, +\infty]$ verifying condition (b) in Definition 1.12 for the same (a_n).

(iii) *Borovkov's formulation.* As we saw before, under Definition 1.12 we have:

(d) $\lim_{n\to+\infty} \frac{1}{a_n} \log \mu_n(A) = -\inf_{x\in A} I(x)$ if $\inf_{x\in A^\circ} I(x) = \inf_{x\in\bar{A}} I(x)$, and $A \in \mathcal{B}(M)$.

Conversely, if I is a good rate function and the above condition (d) holds, conditions (b) and (c) in Definition 1.12 are also satisfied. This provides an equivalent way to settle the l.d.p. with a good rate function, as introduced by Borovkov [26]. We omit the proof (see e.g. Theorem 3.4, Chapter 3 of [122]).

(iv) Definition 1.12 has been stated for a sequence (μ_n); we may as well consider a continuous parameter family $(\mu_\varepsilon)_{0<\varepsilon\le 1}$, with $\varepsilon \to 0$ replacing $n \to +\infty$.

The following property is useful in connection with large deviations, as exemplified below.

Definition 1.17 In the notation of Definition 1.12, the sequence (μ_n) is said to be exponentially tight (for the scaling a_n) if for each $b < +\infty$ there exists a compact set K_b so that

$$\limsup_{n\to+\infty} \frac{1}{a_n} \log \mu_n(K_b^c) < -b. \tag{1.39}$$

(A^c denotes the complement of the set A.)

Remarks (i) From the preceding arguments we easily deduce that if (μ_n) is exponentially tight and satisfies a weak l.d.p. with rate function I, then I is a good rate function and we have the full l.d.p. Indeed, to check condition (c) in Definition 1.12, it suffices to consider $C \subseteq M$ closed such that $\inf_{x\in C} I(x) > 0$. Let $0 < b < \inf_{x\in C} I(x)$; using (1.39) we have for all n large enough:

$$\mu_n(C) \le \mu_n(C \cap K_b) + \mu_n(K_b^c) \le \mu_n(C \cap K_b) + e^{-a_n b}. \tag{1.40}$$

Since $C \cap K_b$ is compact and disjoint of $\mathbb{F}_I(b)$, the weak l.d.p. tells us that

$$\limsup_{n\to+\infty} \frac{1}{a_n} \log \mu_n(C \cap K_b) \le -b,$$

which, together with (1.40), gives condition (c) in Definition 1.12.

On the other hand, from (1.39) and the lower bound applied to the open set $G = K_b^c$ we see that $\mathbb{F}_I(b) \subseteq K_b$, for any b, implying the goodness of the rate function.

(ii) After the previous remark (to be applied in Section 1.4) one is tempted to ask about the converse relationship, i.e. whether exponential tightness should follow from the validity of a l.d.p. with a good rate function. Though false in general (see e.g. Exercise 6.2.24 in [85], cf. also Section 1.6), it is true if M is a Polish space, cf. Lemma 2.6 in [204]. The basic sketch of the proof goes as follows: consider a countable dense set $\{x_1, x_2, \ldots\}$ and use the upper bound of the l.d.p. together with the compactness of the level sets $\mathbb{F}_I(b)$, $b < +\infty$ to prove that for each $b < +\infty$ and $k \ge 1$ there exists a finite number m_k such that $\mu_n\left((\cup_{i=1}^{m_k} B_{1/k}(x_i))^c\right) \le e^{-bk\,a_n}$ for all n. With this in hand, it is easy to get a totally bounded set K_b such that (1.39) holds. Since M is a Polish space, $\overline{K_b}$ is compact and we conclude the verification (see also Theorem (P) in [251]). For M locally compact the above implication is a simple exercise.

(iii) The last remark extends at once to continuous parameter families.

Large deviations and evaluation of integrals

Basic large deviation estimates are closely related to extensions of the Laplace method for determining the asymptotic behaviour of integrals such as $\int_a^b e^{nf(x)}\, g(x)\, dx$, where $a < b$ are real numbers, f and g are continuous, with g positive. For example, if f has a unique maximun at $x_0 \in (a, b)$, and $\delta > 0$ (small), let us split the integral into two parts: one corresponding to $[x_0 - \delta, x_0 + \delta]$ and the remaining $[a, b] \setminus [x_0 - \delta, x_0 + \delta]$. Fix $0 < \eta < \delta$. The first of these integrals can be bounded from below by $e^{n(\min_{|y-x_0|\le\eta} f(y))} 2\eta c$, and the remaining one can be bounded from above by $e^{n(\max_{|y-x_0|\ge\delta} f(y))}(b - a)C$, where $0 < c \le g(x) \le C < +\infty$ for all $x \in [a, b]$. Taking η small we see that as $n \to +\infty$, the main contribution comes from the first term, in the sense that the remaining one will be exponentially smaller. That is,

$$\liminf_{n\to+\infty} \frac{1}{n} \log \int_a^b e^{nf(x)}\, g(x)\, dx \ge f(x_0).$$

The validity of the reversed inequality for the upper limit is completely trivial, and we conclude the logarithmic equivalence $\int_a^b e^{nf(x)}\, g(x)\, dx \approx e^{nf(x_0)}$.

Of course, we are talking about very rough estimates, in the sense of logarithm equivalence; with differentiability assumptions and performing Taylor expansion one obtains better results.

As another example, let us now recall that

$$n! = \int_0^{+\infty} x^n e^{-x} \, dx = (n+1)^{n+1} \int_0^{+\infty} e^{n(\log y - y)} e^{-y} dy.$$

Some 'stretching' of the previous argument gives us $n^{-1}(\log n! - (n+1) \log(n+1)) \to -1$ or equivalently $\log n! \sim n \log n - n$, as $n \to +\infty$, which can be seen as a very weak version of the Stirling formula (compare with (1.7)). Recall, on the other hand, that the Stirling formula was the basis for our first example of a l.d.p. in Section 1.1.

Performing a Taylor expansion of $\log y$ around $y = 1$ one improves the estimate, and gets (1.6).

The following general result, due to Varadhan (cf. [291, 292]), can be seen as an extension of the Laplace method.

Theorem 1.18 (Varadhan lemma) *Let M and (μ_n) be as in Definition* 1.12, *with a good rate function I. If $F : M \to \mathbb{R}$ is continuous and bounded from above, then*

$$\lim_{n\to+\infty} \frac{1}{a_n} \log \int_M e^{a_n F} \, d\mu_n = \sup_{x \in M} (F(x) - I(x)) . \tag{1.41}$$

Moreover, (1.41) *continues to hold if the boundedness condition is relaxed to a 'super exponential estimate', such as*

$$\limsup_{L\to+\infty} \limsup_{n\to+\infty} \frac{1}{a_n} \log \int_{\{x:\, F(x) \geq L\}} e^{a_n F(x)} \, \mu_n(dx) = -\infty. \tag{1.42}$$

Proof We first prove that if F is lower semi-continuous and (μ_n) satisfies a l.d.p. (here we do not need the goodness of the rate function), then for each $x \in M$

$$\liminf_{n\to+\infty} \frac{1}{a_n} \log \int_M e^{a_n F} \, d\mu_n \geq F(x) - I(x). \tag{1.43}$$

For this it suffices to consider $x \in M$ such that $I(x) < +\infty$. Let $\delta > 0$ be given and by the lower semi-continuity of F we may take G, an open neighbourhood of x, such that $\inf_{y \in G} F(y) \geq F(x) - \delta$. Thus,

$$\int_M e^{a_n F} \, d\mu_n \geq \int_G e^{a_n F} \, d\mu_n \geq e^{a_n (F(x) - \delta)} \, \mu_n(G),$$

and consequently

$$\begin{aligned} \liminf_{n\to+\infty} \frac{1}{a_n} \log \int_M e^{a_n F} \, d\mu_n &\geq F(x) - \delta + \liminf_{n\to+\infty} \frac{1}{a_n} \log \mu_n(G) \\ &\geq F(x) - I(x) - \delta. \end{aligned}$$

Since δ may be taken arbitrarily small we get (1.43).

Now, assuming F to be bounded from above, upper semi-continuous, and I being a good rate function, we shall prove that

$$\limsup_{n\to+\infty} \frac{1}{a_n} \log \int_M e^{a_n F}\, d\mu_n \le \sup_x (F(x) - I(x)). \tag{1.44}$$

Let $\alpha = \sup_{x\in M} F(x) < +\infty$. Given $b < +\infty$ and $\delta > 0$, due to the upper semi-continuity of F and the lower semi-continuity of I, for each $x \in M$ we may take an open ball centred at x, called G_x, such that $\sup_{y\in\overline{G_x}} F(y) \le F(x) + \delta$ and $\inf_{y\in\overline{G_x}} I(y) \ge I(x) - \delta$. Since the level set $\mathbb{F}_I(b) = \{x\colon I(x) \le b\}$ is compact, it can be covered by finitely many such balls $G_{x_1}, \dots, G_{x_r}$ and we have

$$\int_M e^{a_n F}\, d\mu_n \le \sum_{i=1}^r e^{a_n(F(x_i)+\delta)}\, \mu_n(\overline{G_{x_i}}) + e^{a_n\alpha}\, \mu_n\Big((\bigcup_{i=1}^r G_{x_i})^c \Big). \tag{1.45}$$

We now observe that n_0 can be taken in a way that if $n \ge n_0$:

$$\frac{1}{a_n} \log \mu_n(\overline{G_{x_i}}) \le -I(x_i) + 2\delta,$$

$$\frac{1}{a_n} \log \mu_n\Big((\bigcup_{i=1}^r G_{x_i})^c \Big) \le -b + \delta,$$

where we have used condition (c) of Definition 1.12 in both inequalities (recall that $\inf_{y\in\overline{G_{x_i}}} I(y) \ge I(x_i) - \delta$ and that $(\cup_{i=1}^r G_{x_i})^c$ is a closed set, disjoint from $\mathbb{F}_I(b)$). Replacing in (1.45), if $n \ge n_0$:

$$\frac{1}{a_n} \log \int_M e^{a_n F}\, d\mu_n \le \frac{\log(r+1)}{a_n} + \max_{i\le r} (F(x_i) - I(x_i) + 3\delta) \vee (\alpha - b + \delta).$$

Since b can be made arbitrarily large and δ arbitrarily small we get (1.44), for F bounded from above. To complete the proof, let us assume the validity of (1.42) and consider the truncated functions $F_L(\cdot) = F(\cdot) \wedge L$. Writing

$$\int_M e^{a_n F}\, d\mu_n \le \int_M e^{a_n F_L}\, d\mu_n + \int_{\{x\colon F(x)\ge L\}} e^{a_n F}\, d\mu_n,$$

and using the preceding argument we have

$$\limsup_{n\to+\infty} \frac{1}{a_n} \log \int_M e^{a_n F}\, d\mu_n \le \sup_x (F_L(x) - I(x)) \vee \limsup_{n\to+\infty} \frac{1}{a_n} \log \int_{\{x\colon F(x)\ge L\}} e^{a_n F}\, d\mu_n.$$

Taking $L \to +\infty$ and using (1.42), we get (1.44). □

Remarks (i) Going back to the examples in Section 1.1, let μ_n be the law of the sample average of n i.i.d. random variables $X_1, \dots, X_n$, and take $F_\zeta(y) := \zeta y$. Now $a_n = n$ and $n^{-1} \log \int_M e^{nF_\zeta}\, d\mu_n = \log \hat{\mu}(\zeta)$. The above conditions hold if, e.g., X_1 is bounded (see also (iii) below); (1.41) exemplifies the Fenchel–Legendre duality between the convex functions I_μ and $\log \hat{\mu}$, mentioned in Section 1.1:

$$\log \hat{\mu}(\zeta) = \sup_{x\in\mathbb{R}}(\zeta\, x - I_\mu(x)).$$

(ii) *Tilting*. The Varadhan lemma provides a way to generate a new l.d.p. through the so-called tilting operation: the new measures are $\nu_n(dx) = Z_n^{-1} e^{a_n F(x)} \mu_n(dx)$, with $Z_n = \int_M e^{a_n F}\, d\mu_n$, and the new rate function is $\tilde{I}(x) = (I(x) - F(x)) - \inf_{y\in M}(I(y) - F(y))$, keeping the scaling (a_n). This will be used later (see [291, 292], Chapter II of [108]).

(iii) Using the Hölder inequality we see that (1.42) is satisfied if, for some $u \in (1, +\infty)$:

$$\sup_{n\geq 1}\left(\int_M e^{u\, a_n F(x)} \mu_n(dx)\right)^{1/a_n} < +\infty. \tag{1.46}$$

Comment A converse of Theorem 1.18 has been obtained by Bryc in [39]; it basically says that if (μ_n) is an exponentially tight sequence of probability measures on the metric space M (scaling a_n), and the limit $\mathcal{L}(F) = \lim_{n\to+\infty} a_n^{-1} \log \int e^{a_n F}\, d\mu_n$ exists (in $\mathbb{R}$) for each bounded continuous function $F\colon M \to \mathbb{R}$, then (μ_n) satisfies a l.d.p. with scaling a_n and rate function I given by

$$I(x) = \sup_{F\in C_b(M)} \big(F(x) - \mathcal{L}(F)\big),$$

where $C_b(M)$ denotes the space of bounded continuous functions from M to $\mathbb{R}$. As a consequence of this and the Varadhan lemma we would have $\mathcal{L}(F) = \sup_{x\in M}(F(x) - I(x))$, for $F \in C_b(M)$. (This is in analogy with the Fenchel–Legendre duality, without convexity.)

Generally, it suffices to verify the existence of the limit $\mathcal{L}(F)$ for a sufficiently rich class of functions F. This has great interest both for applications and from a theoretical point of view, since it points to interesting connections between large deviations and weak convergence. Such results might be seen as a very general version of the Gärtner–Ellis theorem, to be described in Section 3.1. We shall not develop or use this, referring to Section 4.4 of [85] and to [97] for the connection between the l.d.p. and weak convergence.

Contraction principle

Transformations of measures under the action of a (continuous) map appear quite often in probabilistic/statistical analysis purely reflecting the need to change observables. In this direction, and with respect to large deviations, the following theorem is very useful, and it will be applied at several occasions in this text.

Theorem 1.19 (Contraction principle) *Let (μ_n) be a sequence of probability measures on a metric space (M, ϱ), satisfying a l.d.p. with a good rate function*

I. Let also $T: M \to \tilde{M}$ be a continuous map into another metric space $(\tilde{M}, \tilde{\varrho})$, and consider the measures induced under T, $\tilde{\mu}_n := \mu_n \circ T^{-1}$ defined through $\tilde{\mu}_n(A) = \mu_n(T^{-1}(A))$ for any Borel set $A \subseteq \tilde{M}$. The sequence $(\tilde{\mu}_n)$ satisfies a l.d.p. with the same scaling and a good rate function $\tilde{I}$, given by

$$\tilde{I}(y) = \inf_{\{x:\, T(x)=y\}} I(x). \tag{1.47}$$

Proof Clearly, $\tilde{I}(y) \in [0, +\infty]$ and we now observe that conditions (b) and (c) in Definition 1.12 hold for $\tilde{\mu}_n$: since T is continuous, $T^{-1}(A)$ is closed (open) in M, if A is closed (open) in $\tilde{M}$ and from (1.47) it follows that

$$\inf_{y \in A} \tilde{I}(y) = \inf_{x \in T^{-1}(A)} I(x).$$

It remains to check that $\tilde{I}$ is a good rate function, and it is exactly here that we need the goodness of I: in this case, if y belongs to the image of T then the infimum in (1.47) is really attained and it follows that for any $b < +\infty$ the level set $\mathbb{F}_{\tilde{I}}(b) := \{y : \tilde{I}(y) \le b\}$ coincides with $T(\mathbb{F}_I(b))$. Since I is a good rate function and T is continuous, this is a compact subset of $\tilde{M}$. □

Remark The validity of conditions (b) and (c) in Definition 1.12 for the measures $\tilde{\mu}_n$ does not depend on the goodness of I. Such a property was used to ensure the lower semi-continuity of $\tilde{I}$, which will in fact have compact level sets. It is easy to give extreme counter-examples otherwise, for which the lower semi-continuity is lost.

This basic contraction principle has several useful extensions, as for instance when $\tilde{\mu}_n = \mu_n \circ T_n^{-1}$, and the maps T_n converge uniformly on compacts; if T is the limit, the above result extends to this situation with $\tilde{I}$ being defined by equation (1.47). For a proof of this see [292]. Further discussions, including the related question of whether the large deviations for $(\mu_n \circ T_n^{-1})$ and $(\mu_n \circ T^{-1})$ are described by the same rate function, can be found in Section 4.2.2 of [85]; this is important in several applications and is related to a sort of 'asymptotic exponential equivalence' of these measures. Related considerations appear naturally in Section 2.5 for the treatment of Freidlin and Wentzell systems with non-constant diffusion coefficient, as well as in Section 3.4.

1.3 Sanov theorem for finite 'alphabets'

Let us assume, similarly to Lemma 1.4, that $X_1, X_2, \ldots$ are i.i.d. random variables, taking values on a finite set $S = \{x_1, \ldots, x_r\}$. To fix the notation we call μ their common law, and denote by $(\Omega, \mathcal{A}, P)$ the underlying probability space. For notational convenience we let $S = \{1, \ldots, r\}$, and we assume that $p_x := \mu\{x\} > 0$ for each $x \in S$. If $M_n := \{(n_1, \ldots, n_r) \in \mathbb{Z}_+^r : \sum_{x=1}^r n_x = n\}$ and $(n_1, \ldots, n_r) \in M_n$, let $p(n_1, \ldots, n_r)$ denote the probability of observing each digit x exactly n_x times, among $X_1, \dots, X_n$. Thus $p(n_1, \dots, n_r) = n! \prod_{x:\, n_x>0} p_x^{n_x}/n_x!$ (a multinomial distribution on M_n) and we have:

$$e^{-\frac{r}{12}}(2\pi)^{-\frac{r-1}{2}} n^{-\frac{r-1}{2}} \prod_{x:\, n_x>0} \left(\frac{np_x}{n_x}\right)^{n_x} \le p(n_1, \dots, n_r) \le \prod_{x:\, n_x>0} \left(\frac{np_x}{n_x}\right)^{n_x}, \tag{1.48}$$

where the first inequality is exactly (1.19), obtained from the Stirling formula in the proof of Lemma 1.4, and the second one follows from the next simple observations: use that $p(n_1, \dots, n_r) \le 1$, with p_x replaced by $\alpha_x := n_x/n$, to see that $n!/n_1! \dots n_r! \le \prod_{x:\, n_x>0}(n/n_x)^{n_x}$. Inserting into the expression for $p(n_1, \dots, n_r)$ we have the second inequality in (1.48).

To interpret (1.48) in the frame of Section 1.2 we consider the following random measure on the set S, also called the empirical measure associated with the sample $X_1, \dots, X_n$:

$$L_n := \frac{1}{n} \sum_{j=1}^{n} \delta_{X_j},$$

with δ_x the Dirac point-mass at x.

Since S has r elements, the space $\mathcal{M}_1(S)$, of all probability measures on S, is identified with the simplex $\Theta_r := \{(\alpha_1, \dots, \alpha_r) \in [0,1]^r \colon \sum_{x=1}^{r} \alpha_x = 1\}$ in $\mathbb{R}^r$, through $\alpha_x := \alpha\{x\}$, $\alpha \in \mathcal{M}_1(S)$. (*Sometimes we write $\alpha(x)$, if needed to avoid notational confusion with sequence indexing.*) With this identification, $\mu \equiv (p_1, \dots, p_r)$, and $L_n(\omega)$ corresponds to the vector of observed frequencies of digits $(1, \dots, r)$ among $X_1(\omega), \dots, X_n(\omega)$, so that

$$p(n_1, \dots, n_r) = P\left(L_n = (n_1/n, \dots, n_r/n)\right), \qquad \forall (n_1, \dots, n_r) \in M_n.$$

On $\mathcal{M}_1(S)$ we consider the metric induced by the restriction of the ℓ_1-norm in $\mathbb{R}^r$: $|\alpha - \tilde{\alpha}|_1 := \sum_{x=1}^{r} |\alpha_x - \tilde{\alpha}_x|$; this corresponds to the total variation distance, here defined as $2\sup_{A \subseteq S} |\alpha(A) - \tilde{\alpha}(A)|$. The law of large numbers tells that $L_n \to \mu$ almost surely (a.s.). The relative entropy $\mathcal{H}(\cdot \mid \mu)$ is defined on $\mathcal{M}_1(S)$ as

$$\mathcal{H}(\alpha \mid \mu) := \sum_{x=1}^{r} \alpha_x \log(\alpha_x/p_x), \tag{1.49}$$

where $0 \log 0 = 0$. Since $p_x > 0$ for all x, $\mathcal{H}(\alpha \mid \mu) = \sum_x \chi(\alpha_x/p_x) p_x$, with $\chi(y) := y \log y$ for $y \ge 0$. The strict convexity of χ, the Jensen inequality, and simple observations, give us:

- $\mathcal{H}(\alpha \mid \mu) \ge 0$, $\mathcal{H}(\alpha \mid \mu) = 0 \Leftrightarrow \alpha = \mu$;
- $\mathcal{H}(\alpha \mid \mu)$ is continuous, finite and strictly convex (since $p_x > 0$, $\forall x$).

The range of L_n is $M_n/n = \{(n_1/n, \dots, n_r/n) \colon (n_1, \dots, n_r) \in M_n\}$, and if $\alpha \in M_n/n$, we rewrite (1.48) as:

$$e^{-\frac{r}{12}}(2\pi)^{-\frac{r-1}{2}} n^{-\frac{r-1}{2}} e^{-n\mathcal{H}(\alpha|\mu)} \le P(L_n = \alpha) \le e^{-n\mathcal{H}(\alpha|\mu)}. \tag{1.50}$$

The cardinality of M_n is bounded from above by $(n+1)^{r-1}$. Thus, setting $Q_n^{(1)}(A) := P(L_n \in A) = \sum_{\alpha \in \frac{M_n}{n} \cap A} P(L_n = \alpha)$, for $A \subseteq \mathcal{M}_1(S)$:

$$\max_{\tilde{\alpha} \in \frac{M_n}{n} \cap A} P(L_n = \tilde{\alpha}) \le Q_n^{(1)}(A) \le (n+1)^{r-1} \max_{\tilde{\alpha} \in \frac{M_n}{n} \cap A} P(L_n = \tilde{\alpha}) \qquad (1.51)$$

so that, from (1.50), and for any $A \subseteq \mathcal{M}_1(S)$:

$$\frac{1}{n} \log Q_n^{(1)}(A) \le - \inf_{\alpha \in A} \mathcal{H}(\alpha \mid \mu) + O\left(\frac{\log n}{n}\right). \qquad (1.52)$$

For the lower bound, if $\alpha \in \mathcal{M}_1(S)$ and $B_\delta(\alpha) := \{\tilde{\alpha} \in \mathcal{M}_1(S), |\tilde{\alpha} - \alpha|_1 < \delta\}$, (1.50) and (1.51) imply that

$$\liminf_{n \to +\infty} \frac{1}{n} \log P(L_n \in B_\delta(\alpha)) \ge - \limsup_{n \to +\infty} \inf_{\tilde{\alpha} \in \frac{M_n}{n} \cap B_\delta(\alpha)} \mathcal{H}(\tilde{\alpha} \mid \mu) \ge -\mathcal{H}(\alpha \mid \mu), \qquad (1.53)$$

where we have used the continuity of $\mathcal{H}(\cdot \mid \mu)$ and the density of $\cup_{n \ge 1} M_n/n$ in $\mathcal{M}_1(S)$. From (1.52) and (1.53) we conclude, for any $A \subseteq \mathcal{M}_1(S)$:

$$\begin{aligned} - \inf_{\alpha \in A^\circ} \mathcal{H}(\alpha \mid \mu) &\le \liminf_{n \to +\infty} \frac{1}{n} \log Q_n^{(1)}(A) \le \limsup_{n \to +\infty} \frac{1}{n} \log Q_n^{(1)}(A) \\ &\le - \inf_{\alpha \in A} \mathcal{H}(\alpha \mid \mu). \end{aligned} \qquad (1.54)$$

Thus:

(a) $(Q_n^{(1)})$ satisfies a l.d.p. with rate function $I^{(1)}(\cdot) = \mathcal{H}(\cdot \mid \mu)$, and scaling n.

(b) $\lim_{n \to +\infty} n^{-1} \log P(L_n \in A) = - \inf_{\alpha \in A} \mathcal{H}(\alpha \mid \mu)$ if A is open. Since $p_x > 0$, $\forall x$, the rate function is continuous, and this convergence holds if $A \subseteq \overline{A^\circ}$. In particular:

$$-I^{(1)}(\alpha) = -\mathcal{H}(\alpha \mid \mu) = \lim_{\varepsilon \downarrow 0} \lim_{n \to +\infty} \frac{1}{n} \log P(|L_n - \alpha|_1 < \varepsilon).$$

Remarks (i) We have seen a particular case of the Sanov theorem, to be discussed more generally in Section 1.6. Due to (1.50), and the polynomial growth of $\sharp M_n$ the estimates are stronger here. In particular, condition (c) in Definition 1.12 holds for all Borel sets.

(ii) Using Lagrange multipliers and the contraction principle, the previous discussion yields Corollary 1.10, for finite-valued X_1. Compare with the proof of Theorem 1.1.

(iii) If $p_x = 0$ is allowed, one extends $\mathcal{H}(\cdot \mid \mu)$ by setting $0 \log(0/0) = 0$ and $a \log(a/0) = +\infty$ for $a > 0$; $\mathcal{H}(\cdot \mid \mu)\colon \mathcal{M}_1(S) \to [0, +\infty]$ is lower semicontinuous and convex. The l.d.p. still holds. $\mathcal{H}(\alpha \mid \mu) = +\infty$ if $\alpha_x > 0$ for some x such that $p_x = 0$; also $P(L_n = \alpha) = 0$ for such α. (This is unnecessary in the case of a single measure μ on a finite set; instead, we just reduce S, stating the l.d.p. on a closed subset of $\mathcal{M}_1(S)$.)

(iv) Simple calculations give directly the following variational expression:

$$\mathcal{H}(\alpha \mid \mu) = \sup_{f \in \mathbb{R}^r} \left(\sum_{x=1}^{r} f_x \alpha_x - \log \sum_{x=1}^{r} e^{f_x} p_x \right). \tag{1.55}$$

When $\alpha_x > 0$, $p_x > 0$, for all x, the supremum is attained at the points $f = (f_1, \ldots, f_r)$ such that $f_x = \log(\alpha_x / p_x) + c$, $x = 1, \ldots, r$, where c is any constant.

(v) The above l.d.p. may be seen as an example of the Cramér theorem on $\mathbb{R}^r$, discussed in the next section. Identified with the simplex, $\mathcal{M}_1(S)$ is a closed subset of $\mathbb{R}^r$, outside which the rate function becomes $+\infty$. We have an average of i.i.d. $\mathcal{M}_1(S)$-valued variables, δ_{X_i}, $i = 1, \ldots, n$, and (1.55) is just (1.69) in this case.

(vi) Direct verification, using Lagrange multipliers, shows that

$$\log \sum_{x=1}^{r} e^{f_x} p_x = \sup_{\alpha \in \mathcal{M}_1(S)} \left(\sum_{x=1}^{r} f_x \alpha_x - \mathcal{H}(\alpha \mid \mu) \right), \tag{1.56}$$

and that $\alpha_x = e^{f_x} p_x / \sum_z e^{f_z} p_z$, $x = 1, \ldots, r$ is the unique point of maximum if $p_x > 0$ for all x. (In both (1.55) and (1.56), the functions are concave, so that the maximization is a simple calculation.) This is an example of Fenchel–Legendre duality, with $I^{(1)}(\cdot)$ extended to all $\mathbb{R}^r$ as $+\infty$ outside $\mathcal{M}_1(S)$.

Beyond the empirical measure. Finite words

The previous analysis can be extended as follows: besides estimating the probabilities on the frequencies with which each digit x is observed, we ask ourselves the same question for any given k-word $(x_1, \ldots, x_k)$ from the finite alphabet S. The asymptotic frequency is easily seen to be a.s. $p_{x_1} \cdots p_{x_k}$, due to the law of large numbers. More precisely, setting $Y_i = (X_i, X_{i+1}, \ldots, X_{i+k-1})$ we consider the empirical measure:

$$\tilde{L}_n^{(k)} = \frac{1}{n-k+1} \sum_{i=1}^{n-k+1} \delta_{Y_i}.$$

$\tilde{L}_n^{(k)}$ takes values on the space of probability measures on S^k, as before identified with the simplex: $\mathcal{M}_1(S^k) = \{\alpha \in [0,1]^{S^k} : \sum_{x \in S^k} \alpha(x) = 1\}$, endowed with the $|\cdot|_1$ distance. Applying the law of large numbers to the sequences $(Y_{i+jk};\, j \geq 0)$, with $i \in \{1, \ldots, k\}$ we easily see that

$$\tilde{L}_n^{(k)} \to \mu^k \text{ a.s.} \qquad (\mu^k := \mu \times \cdots \times \mu, \quad k \text{ times}).$$

Already at this point it might be useful to take the periodic extension of

$X_1, \ldots, X_n$. Let $L_n^{(k)} := L_n^{(k)}[X_1, \ldots, X_n]$, where

$$L_n^{(k)}[x_1, \ldots, x_n] := \frac{1}{n} \sum_{i=1}^{n} \delta_{(x_i, x_{i+1}, \ldots, x_{i+k-1})}, \quad \text{with}$$

$$x_{n+j} = x_j \text{ for each } 1 \leq j \leq k-1.$$

We check that

$$|L_n^{(k)} - \tilde{L}_n^{(k)}|_1 \leq \frac{2(k-1)}{n}, \qquad \text{for each } X_1, \ldots, X_n,$$

so that the l.d.p. for $\tilde{L}_n^{(k)}$ would follow from that for $L_n^{(k)}$, and vice versa.

Due to the periodic extension, for each realization of $X_1, \ldots, X_n$, the measure $L_n^{(k)}$ has its marginal on the first $k-1$ coordinates identical with that on the last $k-1$ coordinates (and given by $L_n^{(k-1)}$, where $L_n^{(1)} = L_n$). That is, $L_n^{(k)}$ takes values on

$$\hat{\mathcal{M}}_1(S^k) := \Bigg\{ \alpha \in \mathcal{M}_1(S^k) \colon \sum_{x \in S} \alpha(x_1, \ldots, x_{k-1}, x) \\ = \sum_{x \in S} \alpha(x, x_1, \ldots, x_{k-1}), \; \forall x_1, \ldots, x_{k-1} \Bigg\}. \tag{1.57}$$

Being a closed subset of $\mathcal{M}_1(S^k)$, $\hat{\mathcal{M}}_1(S^k)$ is a compact metric space, for the induced metric. In the study of a l.d.p. for the laws of $L_n^{(k)}$, which we denote by $(Q_n^{(k)})$, we may take $\hat{\mathcal{M}}_1(S^k)$ as the metric space M of Definition 1.12. As previously discussed, $Q_n^{(k)} \to \delta_{\mu^k}$ weakly.

Let $M_n^{(k)} = \{(n_x) \in {\mathbb{Z}_+}^{S^k} \colon (n_x/n) \in \hat{\mathcal{M}}_1(S^k)\}$. The range of $L_n^{(k)}$ is $n^{-1} M_n^{(k)}$; as for $k = 1$, this set gets quite dense on $\hat{\mathcal{M}}_1(S^k)$ as n grows, and its cardinality grows polynomially in n (bounded from above by $(n+1)^{r^k - 1}$). Therefore, the basic step consists in estimating

$$\sup_{\tilde{\alpha} \in \frac{M_n^{(k)}}{n} \cap B_\delta(\alpha)} P(L_n^{(k)} = \tilde{\alpha}), \tag{1.58}$$

for any $\alpha \in \hat{\mathcal{M}}_1(S^k)$, $\delta > 0$. The combinatorics is now more involved. We just sketch a proof, taking for simplicity $k = 2$. Since

$$M_n^{(2)} := \left\{ (n_z) \in {\mathbb{Z}_+}^{S^2} \colon \bar{n}_x := \sum_{y=1}^{r} n_{(x,y)} = \sum_{y=1}^{r} n_{(y,x)}, \; \forall x \in S, \; (\bar{n}_x) \in M_n \right\},$$

the main point, analogous to (1.50), is verification of the following (as usual $0! = 1$):

$$\frac{1}{n} \log P\left(L_n^{(2)}(z) = \frac{n_z}{n}, \; \forall z \in S^2 \right) = \frac{1}{n} \log \left(\frac{\prod_{x \in S} \bar{n}_x!}{\prod_{z \in S^2} n_z!} \prod_{x \in S} p_x^{\bar{n}_x} \right) + \frac{O(\log n)}{n}. \tag{1.59}$$

Indeed, together with considerations as in the case $k = 1$, (1.59) entails a l.d.p. for $Q_n^{(2)}$, and, as in (1.54), no closure is needed for the upper bound. The rate function is

$$I^{(2)}(\alpha) := \sum_{(x,y)\in S^2} \alpha(x,y) \log\left(\frac{\alpha(x,y)}{\bar{\alpha}_x\, p_y}\right) = \mathcal{H}(\alpha \mid \bar{\alpha} \times \mu),$$

for $\alpha \in \hat{\mathcal{M}}_1(S^2)$, $\bar{\alpha}$ denoting its marginal. ($I^{(2)}$ is finite and continuous on $\hat{\mathcal{M}}_1(S^2)$, since $p_x > 0$ for each x.)

Sketch of proof of (1.59) Any realization $(x_1, \ldots, x_n)$ of $(X_1, \ldots, X_n)$, compatible with the event on the left-hand side of (1.59) exibits each digit $x \in S$ exactly $\bar{n}_x$ times, i.e. $L_n[x_1, \ldots, x_n] = (\bar{n}_n^{-1}, \ldots, \bar{n}_r/n)$. Thus, $P(X_1 = x_1, \ldots, X_n = x_n) = \prod_{x\in S} p_x^{\bar{n}_x}$. It remains to estimate the number of such realizations:

$$N((n_z)) := \sharp\{(x_1, \ldots, x_n) \in S^n : L_n^{(2)}[x_1, \ldots, x_n](z) = n_z/n,\ \forall z \in S^2\}.$$

For this one can prove, given $(n_z) \in M_n^{(2)}$,

$$\frac{\prod_{x:\,\bar{n}_x>0}(\bar{n}_x - 1)!}{\prod_{z\in S^2} n_z!} \le N((n_z)) \le n\, \frac{\prod_{x\in S} \bar{n}_x!}{\prod_{z\in S^2} n_z!}$$

which readly implies (1.59). We omit this proof, referring to Chapter I.6 of [108], or to Lemma II.10 in [157], where a natural and convenient interpretation in terms of graphs is used: if $x, y \in S$, an oriented arrow (x, y) is added whenever the pair appears as (x_i, x_{i+1}) for $1 \le i \le n$ (recall that $x_{n+1} = x_1$); thus we have exactly n arrows, and for each vertex the number of ingoing arrows is the same as the number of outgoing arrows. The cardinality of interest is estimated in terms of the number of Euler circuits of such graphs, i.e. loops which use each arrow only once. □

Notation With a little abuse of notation we use the same symbol π_k to denote the projection $(x_1, \ldots, x_{k'}) \mapsto (x_1, \ldots, x_k)$ from $S^{k'}$ to S^k, for any $k' > k$. If $\alpha \in \mathcal{M}_1(S^{k'})$, let $\pi_k\alpha$ be the measure induced by α under π_k, i.e. the marginal distribution of the first k coordinates under α; we easily check that $\pi_k\alpha \in \hat{\mathcal{M}}_1(S^k)$ if $\alpha \in \hat{\mathcal{M}}_1(S^{k'})$.

The combinatorial analysis may be extended to any $k \ge 2$, and the l.d.p. which emerges, analogously to (1.54), is governed by the rate function $I^{(k)}$ on $\hat{\mathcal{M}}_1(S^k)$, given by:

$$\begin{aligned} I^{(k)}(\alpha) :&= \sum_{(x_1,\ldots,x_k)\in S^k} \alpha(x_1, \ldots, x_k) \log\left(\frac{\alpha(x_1, \ldots, x_k)}{\pi_{k-1}\alpha(x_1, \ldots, x_{k-1})\, p_{x_k}}\right) \\ &= \mathcal{H}(\alpha \mid \pi_{k-1}\alpha \times \mu). \end{aligned} \tag{1.60}$$

In this particular situation, as in (1.54), no closure is needed on the upper estimate. For each $\alpha \in \hat{\mathcal{M}}_1(S^k)$, we have:

$$-I^{(k)}(\alpha) = \lim_{\varepsilon \downarrow 0} \lim_{n \to +\infty} \frac{1}{n} \log P(|L_n^{(k)} - \alpha|_1 < \varepsilon).$$

(A more general treatment appears in Section 3.4, so we omit details now; see Corollary 3.41.)

Remark 1.20 (Simple observations. Basic properties of $I^{(k)}$ on $\hat{\mathcal{M}}_1(S^k)$)

(i) Let $\tilde{S} = S_1 \times S_2$, with S_1, S_2 finite. If $\alpha, \nu \in \mathcal{M}_1(\tilde{S})$, α_i, ν_i are the corresponding marginals in S_i, for $i = 1, 2$, and $\nu = \nu_1 \times \nu_2$, then

$$\mathcal{H}(\alpha \mid \nu) = \mathcal{H}(\alpha|\alpha_1 \times \alpha_2) + \mathcal{H}(\alpha_1 \mid \nu_1) + \mathcal{H}(\alpha_2 \mid \nu_2). \tag{1.61}$$

(The above relation is verified at once from the definition (1.49).)

(ii) $I^{(2)}(\alpha) = I^{(1)}(\pi_1\alpha) + \mathcal{H}(\alpha \mid \pi_1\alpha \times \pi_1\alpha)$, for any $\alpha \in \hat{\mathcal{M}}_1(S^2)$. $I^{(2)}(\alpha) \geq I^{(1)}(\pi_1\alpha)$; equality happens if and only if α is a product. (Apply (1.61) with $S_1 = S_2 = S$ and $\nu = \pi_1\alpha \times \mu$, and $\alpha \in \hat{\mathcal{M}}_1(S^2)$.)

(iii) (1.60) may be written as:

$$I^{(k)}(\alpha) = \sum_{x_1,\ldots,x_{k-1}} \pi_{k-1}\alpha(x_1, \ldots, x_{k-1}) I^{(1)}\left(\alpha(\cdot \mid x_1, \ldots, x_{k-1})\right), \tag{1.62}$$

where $\alpha(x_k \mid x_1, \ldots, x_{k-1})$ is the conditional probability:

$$\alpha(x_k \mid x_1, \ldots, x_{k-1}) = \alpha(x_1, \ldots, x_k)/\pi_{k-1}\alpha(x_1, \ldots, x_{k-1}),$$
$$\text{if } \pi_{k-1}\alpha(x_1, \ldots, x_{k-1}) > 0\,,$$

and otherwise, $\alpha(x_k \mid x_1, \ldots, x_{k-1}) = \gamma(x_k)$ for a chosen $\gamma \in \mathcal{M}_1(S)$. We see that:

- $I^{(k)}$ is finite, continuous and convex on $\hat{\mathcal{M}}_1(S^k)$; $I^{(k)}(\alpha) = 0$ if and only if $\alpha = \mu^k$;
- if $k \geq 2$, $I^{(k)}(\cdot)$ loses the strict convexity of $I^{(1)}(\cdot)$: if $\alpha, \tilde{\alpha} \in \hat{\mathcal{M}}_1(S^k)$, with $\alpha(\cdot \mid x_1, \ldots, x_{k-1}) = \tilde{\alpha}(\cdot \mid x_1, \ldots, x_{k-1})$ for each x, then $I^{(k)}$ is affine on the segment between α and $\tilde{\alpha}$, i.e., $I^{(k)}(a\alpha + (1-a)\tilde{\alpha}) = aI^{(k)}(\alpha) + (1-a)I^{(k)}(\tilde{\alpha})$, $\forall a \in [0, 1]$.

(iv) From (1.60), if $\alpha \in \hat{\mathcal{M}}_1(S^{k+1})$, $k \geq 2$, we have:

$$\begin{aligned} I^{(k+1)}(\alpha) = I^{(k)}(\pi_k\alpha) + \sum_{x_1,\ldots,x_k} \pi_k\alpha(x_1, \ldots, x_k) \\ \times \mathcal{H}(\alpha(\cdot \mid x_1, \ldots, x_k) \mid \alpha(\cdot \mid x_2, \ldots, x_k)). \end{aligned} \tag{1.63}$$

Let $Y_1, \ldots, Y_{k+1}$ denote the coordinates in S^{k+1}. Equation (1.63) shows that $I^{(k+1)}(\alpha) \geq I^{(k)}(\pi_k\alpha)$, with equality if and only if, under α, the conditional distribution of Y_{k+1} given $Y_1, \ldots, Y_k$ is the same as given $Y_2, \ldots, Y_k$, i.e. if Y_{k+1} and Y_1 are conditionally independent given $Y_2, \ldots, Y_k$ under α.

(v) $I^{(j)}(\alpha) = \inf\{I^{(k)}(\zeta)\colon \zeta \in \hat{\mathcal{M}}_1(S^k),\ \pi_j\zeta = \alpha\}$, if $1 \le j < k$, and $\alpha \in \hat{\mathcal{M}}_1(S^j)$, as we see from the contraction principle, or directly from (1.63).

The empirical process

To consider words of arbitrary length it is convenient to take the full periodic extension, i.e. the infinite sequence $X^{(n)} \in S^{\mathbb{N}}$ obtained by repeating $X_1, \ldots, X_n$ periodically over and over. We consider $\mathcal{X} := S^{\mathbb{N}}$ with the product topology (the coarsest one, on $\mathcal{X}$, making $x \mapsto x_i$ continuous, $\forall i$), where on S we take the discrete topology (all sets are open). This topology is metrizable, and

$$d(x, y) := \sum_{k \ge 1} 2^{-k} |x_k - y_k| \tag{1.64}$$

is a possible metric for it, where $x = (x_1, x_2, \ldots)$, $y = (y_1, y_2, \ldots) \in \mathcal{X}$, and $|\cdot|$ is a fixed norm on $\mathbb{R}$ (taking $S \subseteq \mathbb{R}$). In this way $\mathcal{X}$ becomes a compact metric space. The family of cylinder sets, $C^{x_1,\ldots,x_k} := \{z\colon \pi_k(z) = (x_1, \ldots, x_k)\}$, with $k \ge 1$, $x_1, \ldots, x_k \in S$, constitutes a basis of open sets. The Borel σ-field of $\mathcal{X}$, denoted by $\mathcal{B}(\mathcal{X})$, coincides with the σ-field generated by the family of cylinder sets $\{C^{x_1,\ldots,x_k}\colon k \ge 1, x_1, \ldots, x_k \in S\}$.

Let $\mathcal{M}_1(\mathcal{X})$ denote the space of all probability measures on $\mathcal{B}(\mathcal{X})$, endowed with the metric

$$\varrho(\alpha, \alpha') := \sum_{k \ge 1} 2^{-k} |\pi_k\alpha - \pi_k\alpha'|_1,$$

where, if $\alpha \in \mathcal{M}_1(\mathcal{X})$, $\pi_k\alpha$ denotes the distribution of the first k-coordinates under α, i.e. the measure induced by α under π_k. We see at once that $\varrho(\alpha^{(n)}, \alpha) \to 0$ if and only if for each $k \ge 1$, $|\pi_k\alpha^{(n)} - \pi_k\alpha|_1 \to 0$. For finite S this is just the convergence in law in $\mathcal{X}$.

$\mathcal{M}_1(\mathcal{X})$ is a Polish space; since $\mathcal{X}$ is compact we can in fact see that it is a compact metric space. (See [20] or [235] for a proof. Here we give a brief sketch for finite S: given a sequence $(\alpha^{(n)})$, by diagonalization one extracts a subsequence $(\alpha^{(n')})$, so that for each k, $\pi_k\, \alpha^{(n')}$ is convergent on $\mathcal{M}_1(S^k)$. The limits, call them α_k, are consistent, i.e. $\pi_k\alpha_{k+1} = \alpha_k$, for each k. The Kolmogorov extension theorem yields $\alpha \in \mathcal{M}_1(\mathcal{X})$ such that $\pi_k\alpha = \alpha_k$ for each k; thus $\varrho(\alpha^{(n')}, \alpha) \to 0$.)

On $\mathcal{X}$ we consider the shift to the left: $(\theta y)_i = y_{i+1}$ for each $i \in \mathbb{N}$ and $y \in \mathcal{X}$, and we define the empirical process

$$R_n := \frac{1}{n} \sum_{j=1}^{n} \delta_{\theta_j(X^{(n)})}, \tag{1.65}$$

with θ_j denoting the jth iteration of θ, i.e. the shift by j to the left. We have $\pi_k R_n = L_n^{(k)}$ for all $n \ge k$. In particular, for each k, $\pi_k R_n \to \mu^k$ a.s., as

$n \to +\infty$. Taking a countable union of null sets, we see right away that $\varrho(R_n, \mu^{\mathbb{N}}) \to 0$ a.s., $\mu^{\mathbb{N}}$ being the infinite product measure.

The previous observations make us suspect that all the previous l.p.d. for the laws of the $L_n^{(k)}$ might be lifted to a single l.d.p. on $\mathcal{M}_1(\mathcal{X})$, through consideration of the empirical field R_n.

Since $X^{(n)}$ is the periodic extension of $X_1, \ldots, X_n$, the empirical field R_n takes values on the space of probability measures which are invariant under θ,

$$\mathcal{M}_1^{\theta}(\mathcal{X}) = \{\alpha \in \mathcal{M}_1(\mathcal{X})\colon\ \theta\alpha = \alpha\},$$

where $\theta\alpha$ is the measure induced by α under θ: $\theta\alpha(A) := \alpha(\theta^{-1}(A))$, for each $A \in \mathcal{B}(\mathcal{X})$. $\mathcal{M}_1^{\theta}(\mathcal{X})$ is a closed convex subset of $\mathcal{M}_1(\mathcal{X})$.

Remark Let $\alpha \in \mathcal{M}_1(\mathcal{X})$. Then $\alpha \in \mathcal{M}_1^{\theta}(\mathcal{X}) \Leftrightarrow \pi_k\alpha \in \hat{\mathcal{M}}_1(S^k)$ for each $k \geq 1$.

We then consider the validity of a l.d.p. on $M = \mathcal{M}_1^{\theta}(\mathcal{X})$, equipped with the restriction of the metric ϱ.

Theorem 1.21 *Let S be finite and $X_1, X_2, \ldots$ be i.i.d. S-valued random variables with common law μ such that $\mu\{x\} > 0$ for each $x \in S$. The laws induced by R_n satisfy a l.d.p. on $\mathcal{M}_1^{\theta}(\mathcal{X})$, with scaling n and rate function given by*

$$I^{(\mathbb{N})}(\alpha) := \sup_{k\geq 1} I^{(k)}(\pi_k\alpha) = \sup_{k\geq 1} \mathcal{H}(\pi_k\alpha \mid \pi_{k-1}\alpha \times \mu) \tag{1.66}$$

according to (1.60).

Remarks (i) We see that $I^{(\mathbb{N})}(\alpha) = 0$ if and only if $\alpha = \mu^{\mathbb{N}}$.

(ii) Since $\varrho(\alpha^{(n)}, \alpha) \to 0$ if and only if $\pi_k\alpha^{(n)} \to \pi_k\alpha$ for each k, this lifting of the finite dimensional l.d.p. (for $L_n^{(k)}$) to an infinite dimensional principle may be thought of as a very particular example of the so-called 'projective limit' approach, which was set by Dawson and Gärtner in a more general situation (cf. Theorem 3.3 in [72]).

Proof For each k the function $\alpha \mapsto I^{(k)}(\pi_k\alpha)$ is lower semi-continuous, since π_k is continuous and $I^{(k)}$ is lower semi-continuous. The lower semi-continuity of $I^{(\mathbb{N})}$ follows at once from its definition. Being a closed subset of a compact metric space we have that $\mathcal{M}_1^{\theta}(\mathcal{X})$ is also compact. Consequently $I^{(\mathbb{N})}$ is a good rate function.

Lower bound. It suffices to show that for each $\alpha \in \mathcal{M}_1^{\theta}(\mathcal{X})$, and each $\delta > 0$:

$$\liminf_{n\to+\infty} \frac{1}{n} \log P(\varrho(R_n, \alpha) < \delta) \geq -I^{\mathbb{N}}(\alpha).$$

Recall that if $\tilde{\alpha} \in \mathcal{M}_1^{\theta}(\mathcal{X})$, then $\pi_k\tilde{\alpha} \in \hat{\mathcal{M}}_1(S^k)$, for any $k \geq 2$, and, if $|\pi_k\tilde{\alpha} - \pi_k\alpha|_1 < \delta/2$, then obviously $|\pi_j\tilde{\alpha} - \pi_j\alpha|_1 < \delta/2$, for each $j \leq k$, and so $\varrho(\tilde{\alpha}, \alpha) < \delta/2 + 2\sum_{j\geq k+1} 2^{-j} = \delta/2 + 2^{-(k-1)}$. Thus, taking $k = k_\delta$, large enough, we have:

$$\begin{aligned}
&\liminf_{n\to+\infty} \frac{1}{n} \log P(\varrho(R_n, \alpha) < \delta) \\
&\geq \liminf_{n\to+\infty} \frac{1}{n} \log P(|L_n^{(k)} - \pi_k\alpha|_1 < \delta/2) \\
&\geq -I^{(k)}(\pi_k\alpha) \geq -I^{\mathbb{N}}(\alpha).
\end{aligned}$$

Upper bound. We first prove that if $\alpha \in \mathcal{M}_1^\theta(\mathcal{X})$ then

$$\lim_{\delta\downarrow 0} \limsup_{n\to+\infty} \frac{1}{n} \log P(\varrho(R_n, \alpha) < \delta) \leq -I^{(\mathbb{N})}(\alpha). \tag{1.67}$$

From its definition, $\varrho(\tilde{\alpha}, \alpha) < \delta$ implies that $|\pi_k\tilde{\alpha} - \pi_k\alpha|_1 < 2^k\delta$, if $\delta > 0$, $k \geq 1$, for any α, $\tilde{\alpha} \in \mathcal{M}_1(\mathcal{X})$. Thus:

$$\begin{aligned}
\lim_{\delta\downarrow 0} \limsup_{n\to+\infty} \frac{1}{n} \log P(\varrho(R_n, \alpha) < \delta) &\leq \lim_{\delta\downarrow 0} \limsup_{n\to+\infty} \frac{1}{n} \log P(|L_n^{(k)} - \pi_k\alpha|_1 < 2^k\delta) \\
&\leq -I^{(k)}(\pi_k\alpha),
\end{aligned}$$

the last inequality following at once from the l.d.p. for the laws of $L_n^{(k)}$, and the lower semi-continuity of $I^{(k)}$.[3] Taking the infimum on the right, for $k \geq 1$, gives (1.67) (see also point (ii) of Remark 1.16).

Since $\mathcal{M}_1(\mathcal{X})$ is compact, the upper bound for closed sets is easily deduced from (1.67). Indeed, it suffices to prove that

$$\limsup_{n\to+\infty} \frac{1}{n} \log P(R_n \in F) \leq -\inf_{\alpha\in F} I^{(\mathbb{N})}(\alpha)$$

when F is closed, non-empty, and $\inf_{\alpha\in F} I^{(\mathbb{N})}(\alpha) > 0$, since otherwise the inequality becomes trivial. In this case, let $b < \inf_{\alpha\in F} I^{(\mathbb{N})}(\alpha)$. From (1.67), if $\alpha \in F$, we can find $\delta_\alpha > 0$ so that

$$\limsup_{n\to+\infty} \frac{1}{n} \log P(\varrho(R_n, \alpha) < \delta_\alpha) \leq -b.$$

Since F is compact there exists a finite set $\{\alpha_1, \ldots, \alpha_m\} \subseteq F$, so that $F \subseteq \cup_{i=1}^m B_{\delta_{\alpha_i}}(\alpha_i)$, and consequently

$$\frac{1}{n} \log P(R_n \in F) \leq \frac{1}{n} \log m + \max_{1\leq i\leq m} \frac{1}{n} \log P(\varrho(R_n, \alpha_i) < \delta_{\alpha_i})$$

from where we get

$$\limsup_{n\to+\infty} \frac{1}{n} \log P(R_n \in F) \leq -b,$$

[3] In the present case $\lim_{n\to+\infty} \frac{1}{n} \log P(|L_n^{(k)} - \pi_k\alpha|_1 < 2^k\delta)$ exists, but we did not use it.

which holds for any $b < \inf_{\alpha \in F} I^{(\mathbb{N})}(\alpha)$, leading to the conclusion of the proof. □

Comments (a) Let $\alpha \in \mathcal{M}_1^\theta(\mathcal{X})$, with $\mathcal{X} = S^{\mathbb{N}}$ as in this section. From (1.60) (or proper choices in (1.61)) we can check that for any $k \geq 2$:

$$\mathcal{H}(\pi_k \alpha \mid \mu^k) = I^{(k)}(\pi_k \alpha) + \mathcal{H}(\pi_{k-1} \alpha \mid \mu^{k-1}).$$

Iterating this relation, and recalling that $I^{(1)}(\pi_1 \alpha) = \mathcal{H}(\pi_1 \alpha \mid \mu)$, we have:

$$\mathcal{H}(\pi_k \alpha \mid \mu^k) = \sum_{j=1}^{k} I^{(j)}(\pi_j \alpha), \quad \text{if } k \geq 1.$$

But $j \mapsto I^{(j)}(\pi_j \alpha)$ is increasing in j for $\alpha \in \mathcal{M}_1^\theta(\mathcal{X})$, cf. (1.63), and we get the alternative representation

$$I^{(\mathbb{N})}(\alpha) = \lim_{k \to +\infty} \frac{1}{k} \mathcal{H}(\pi_k \alpha \mid \mu^k) =: h(\alpha | \mu^{\mathbb{N}}),$$

also called the specific relative entropy of $\alpha \in \mathcal{M}^\theta(\mathcal{X})$ with respect to the product measure $\mu^{\mathbb{N}}$, to be addressed in more general situations in Section 3.4. One also gets the Shannon–McMillan–Breiman formula on $\mathcal{M}_1^\theta(\mathcal{X})$:

$$I^{(\mathbb{N})}(\alpha) = -\sum_{x \in S} (\pi_1 \alpha)_x \log p_x - h_{Sh}(\alpha), \tag{1.68}$$

where $h_{Sh}(\alpha) = \lim_{k \to +\infty} k^{-1} \mathcal{H}(\pi_k \alpha)$ and, if ν is a probability measure on a finite set $\tilde{S}$,

$$\mathcal{H}(\nu) := -\sum_{x \in \tilde{S}} \nu_x \log \nu_x .$$

Notation. In statistical mechanics literature the standard notation is $\mathcal{S}(\alpha) = k\mathcal{H}(\alpha)$, $s(\alpha) = k h_{Sh}(\alpha)$ where k denotes the Boltzmann constant; we shall use this in Chapter 3.

(b) We have seen that $I^{(2)}$ is affine in certain segments. The analysis could be more carefully continued for all $I^{(k)}$ (see e.g. [108] or [157]). Using that $I^{(\mathbb{N})}(\alpha) = \lim_{k \to +\infty} k^{-1} \mathcal{H}(\pi_k \alpha \mid \mu^k)$ it is not hard to see that $I^{(\mathbb{N})}$ is an affine function on $\mathcal{M}_1^\theta(S^{\mathbb{N}})$. We shall discuss this again in Section 3.4 for more general measures.

(c) A further and important representation of $I^{(\mathbb{N})}$ might be naturally deduced from (1.60): given $\alpha \in \mathcal{M}_1^\theta(S^{\mathbb{N}})$, it has a unique extension to a θ-invariant probability measure on $S^{\mathbb{Z}}$, which we still denote by α. By the θ-invariance, and letting Y_z, $z \in \mathbb{Z}$, represent the coordinates on $S^{\mathbb{Z}}$, (1.62) (which is just the same as (1.60)) may be written as

$$I^{(k+1)}(\pi_{k+1} \alpha) = \int_{S^{\mathbb{Z}}} \mathcal{H}(\alpha(Y_0 \mid Y_{-k}, \ldots, Y_{-1})(\omega) \mid \mu) d\alpha(\omega).$$

Since $\alpha(Y_0 = x \mid Y_{-k}, \ldots, Y_{-1})(\omega)$ converges α a.s. to $\tilde{\alpha}_x(\omega) := \alpha(Y_0 = x \mid Y_{-1}, Y_{-2}, \ldots)(\omega)$ as $k \to +\infty$, it suffices to verify that the dominated convergence theorem applies (see e.g. Lemma IX.2.2, Chapter IX of [108], or p. 165 in [19]), to conclude that

$$I^{(\mathbb{N})}(\alpha) = \int_{S^{\mathbb{Z}}} \mathcal{H}(\tilde{\alpha}(\omega) \mid \mu) d\alpha(\omega) < +\infty.$$

1.4 Cramér theorem on $\mathbb{R}^d$

Our goal in this section is to discuss the higher dimensional version of Corollary 1.10, in the frame of Section 1.2. For this, let $X_1, X_2, \ldots$ be i.i.d. $\mathbb{R}^d$-valued random variables, with μ denoting the common law. As before, μ_n denotes the law of the sample average $\bar{X}_n = n^{-1} \sum_{i=1}^n X_i$ and we study the validity of a l.d.p. for the sequence (μ_n). The assumptions on μ will imply that $m := EX_1 \in \mathbb{R}^d$, so that according to the (weak) law of large numbers, the measures μ_n converge weakly to the Dirac point-mass at m, as in Section 1.1.

The original approach, based on ideas and methods developed by Ruelle and Lanford, and discussed systematically in [189], made evident the role of convexity. A little of this will be discussed in Chapter 3 in its 'natural scenario' of interacting systems from statistical mechanics. In fact, the i.i.d. case was presented in [189] as a 'digression'. The importance of the entropy function became evident there, providing basic ideas for several extensions including to infinite dimensional spaces. The application of this procedure to the i.i.d. case is postponed to the next section.

We first pursue, for $d \geq 2$, the discussion at the end of Section 1.1, based on a change of measure,[4] under the simplifying hypothesis that the moment generating function $\hat{\mu}(\cdot)$, defined as

$$\hat{\mu}(\zeta) := E\, e^{\langle \zeta, X_1 \rangle},$$

is finite everywhere. Here $\langle \zeta, x \rangle = \sum_{i=1}^d \zeta_i x_i$ denotes the Euclidean inner product on $\mathbb{R}^d$, with $\zeta = (\zeta_1, \ldots, \zeta_d)$, $x = (x_1, \ldots, x_d)$. As in the case $d = 1$ we set:

$$I_\mu(x) := \sup_\zeta \left(\langle \zeta, x \rangle - \log \hat{\mu}(\zeta) \right). \tag{1.69}$$

Some basic properties, analogous to those in Proposition 1.8, are summarized below, and similarly established ($|\cdot|$ denotes the Euclidean norm on $\mathbb{R}^d$).

Proposition 1.22 *Extending the notation of Proposition* 1.8, *one has the following.*

[4] Developed by Varadhan, cf. [291, 292].

(i) *Set* $\mathcal{D}_{\hat{\mu}} := \{\zeta \in \mathbb{R}^d : \hat{\mu}(\zeta) < +\infty\}$. *The function* $\log \hat{\mu} : \mathbb{R}^d \to (-\infty, +\infty]$ *is convex; it is differentiable in* $\mathcal{D}^\circ_{\hat{\mu}}$, *where* $\nabla(\log \hat{\mu})(\zeta) = (\hat{\mu}(\zeta))^{-1} \int x\, e^{\langle \zeta, x\rangle} \mu(dx)$ (∇ *denotes the usual gradient*).

(ii) I_μ *is convex; it is a good rate function if* $0 \in \mathcal{D}^\circ_{\hat{\mu}}$.

(iii) *If* $y = \nabla(\log \hat{\mu})(\zeta)$ *for some* $\zeta \in \mathcal{D}^\circ_{\hat{\mu}}$, *then* $I_\mu(y) = \langle \zeta, y\rangle - \log \hat{\mu}(\zeta)$.

Proof (i) This follows as in Proposition 1.8.

(ii) The convexity, lower semi-continuity, and that $I_\mu \geq 0$ follow as in Proposition 1.8. Observe that for any $0 < r < +\infty$ taking $\zeta = ry/|y|$ in (1.69) we get

$$I_\mu(y) \geq r|y| - \sup_{|\zeta|=r} \log \hat{\mu}(\zeta), \tag{1.70}$$

which implies the compactness of the level sets when $\hat{\mu}$ is finite on some neighbourhood of the origin.

(iii) Assume that $y = \nabla \log \hat{\mu}(\zeta)$ for some $\zeta \in \mathcal{D}^\circ_{\hat{\mu}}$, and let us see that for any $\eta \in \mathbb{R}^d$:

$$\langle \eta, y\rangle - \log \hat{\mu}(\eta) \leq \langle \zeta, y\rangle - \log \hat{\mu}(\zeta). \tag{1.71}$$

Since (1.71) is trivial if $\eta \notin \mathcal{D}_{\hat{\mu}}$, we take $\eta \in \mathcal{D}_{\hat{\mu}}$ and consider $f : [0,1] \to \mathbb{R}$ defined as

$$f(s) = \langle s\eta + (1-s)\zeta, y\rangle - \log \hat{\mu}(s\eta + (1-s)\zeta).$$

f is concave, and differentiable at $s = 0$, by part (i), since $\zeta \in \mathcal{D}^\circ_{\hat{\mu}}$. Thus, for any $s \in [0,1]$, $f(s) - f(0) \leq f'(0) = \lim_{u \downarrow 0}(f(u) - f(0))/u = \langle \eta - \zeta, y\rangle - \langle \eta - \zeta, \nabla \log \hat{\mu}(\zeta)\rangle = 0$. Consequently, $\langle \eta, y\rangle - \log \hat{\mu}(\eta) = f(1) \leq f(0) = \langle \zeta, y\rangle - \log \hat{\mu}(\zeta)$. □

The basic result of this section is the following theorem.

Theorem 1.23 *In the set-up of this section, and assuming that* $\mathcal{D}_{\hat{\mu}} = \mathbb{R}^d$, *the sequence* (μ_n) *satisfies a l.d.p. with scaling* n *and a good rate function* I_μ, *given by equation* (1.69).

Proof By Proposition 1.22 we already know that I_μ is a good rate function. We now check the upper and lower bounds in Definition 1.12.

Lower bound. According to Proposition 1.15 it suffices to prove that

$$\liminf_{n\to+\infty} \frac{1}{n} \log \mu_n(B_\delta(y)) \geq -I_\mu(y), \tag{1.72}$$

for any $y \in \mathbb{R}^d$ and $\delta > 0$, where $B_\delta(y)$ denotes the open Euclidean ball in $\mathbb{R}^d$, centred at y and with radius δ. By shifting the variables, as in Remark 1.2, it suffices to take $y = 0$. Moreover, we assume that $I_\mu(0) < +\infty$, since the inequality is trivial otherwise.

We first make an extra assumption: the existence of $\zeta_0 \in \mathbb{R}^d$ so that $\nabla(\log \hat{\mu})(\zeta_0) = 0$. In this case $I_\mu(0) = -\log \hat{\mu}(\zeta_0)$ by part (iii) of the previous proposition, while by part (i), letting $\nu(dx) = e^{\langle \zeta_0, x\rangle}\mu(dx)/\hat{\mu}(\zeta_0)$ we have $\int_{\mathbb{R}^d} x\, \nu(dx) = 0$. Using the argument already explained at the end of Section 1.1, and if $0 < \delta' < \delta$ we see that

$$\mu_n(B_\delta(0)) \geq \mu_n(B_{\delta'}(0)) = \int_{\{\frac{1}{n}\sum_{i=1}^n x_i \in B_{\delta'}(0)\}} \prod_{i=1}^n \mu(dx_i)$$

$$= (\hat{\mu}(\zeta_0))^n \int_{\{\frac{1}{n}\sum_{i=1}^n x_i \in B_{\delta'}(0)\}} e^{-\sum_{i=1}^n \langle \zeta_0, x_i\rangle} \prod_{i=1}^n \nu(dx_i).$$

Taking logarithms, setting $n \to +\infty$, and applying the (weak) law of large numbers we get

$$\liminf_{n\to+\infty} \frac{1}{n} \log \mu_n(B_\delta(0)) \geq \log \hat{\mu}(\zeta_0) - |\zeta_0|\delta' = -I_\mu(0) - |\zeta_0|\delta'\,,$$

and letting δ' tend to zero we recover (1.72). (Recall the discussion preceeding (1.35).)

To remove the extra assumption we follow [292], and consider $\mu^{(\varepsilon)} = \mu * \eta^{(\varepsilon)}$, the convolution of μ and a Gaussian measure with average zero and covariance matrix $\varepsilon\mathbb{I}$, where $\mathbb{I}$ is the $d \times d$ identity matrix and $\varepsilon > 0$ (i.e. we consider the variables $Y_n^{(\varepsilon)} = X_n + \sqrt{\varepsilon} Z_n$ where Z_n, $n \geq 1$, are i.i.d., standard Gaussian in $\mathbb{R}^d$, and X_n, Z_n , $n \geq 1$ are all independent). Observe that

$$\widehat{\mu^{(\varepsilon)}}(\zeta) = \hat{\mu}(\zeta)\, e^{\frac{\varepsilon}{2}|\zeta|^2} \geq \hat{\mu}(\zeta), \tag{1.73}$$

and so

$$I_{\mu^{(\varepsilon)}}(y) \leq I_\mu(y), \tag{1.74}$$

for any $y \in \mathbb{R}^d$. Since $0 \in \mathcal{D}_{\hat{\mu}}^\circ$, using the Markov inequality we see that $E|X_1| = \int_{\mathbb{R}^d} |x|\mu(dx) < +\infty$ and then the Jensen inequality implies that for any ζ

$$\log \hat{\mu}(\zeta) \geq \int_{\mathbb{R}^d} \langle \zeta, x\rangle \mu(dx) \geq -|\zeta| \int_{\mathbb{R}^d} |x|\mu(dx). \tag{1.75}$$

From (1.73) and (1.75) we have that for any $\varepsilon > 0$

$$\lim_{a\to+\infty} \inf_{|\zeta|>a} \log \widehat{\mu^{(\varepsilon)}}(\zeta) \geq \lim_{a\to+\infty} \inf_{|\zeta|>a} (-|\zeta|\, E|X_1| + \frac{\varepsilon}{2}|\zeta|^2) = +\infty, \tag{1.76}$$

which implies at once that the function $\log \widehat{\mu^{(\varepsilon)}}(\zeta)$ attains its mininum on $\mathbb{R}^d$ and, being differentiable everywhere, there exists $\zeta_0 \in \mathbb{R}^d$ for which $\nabla(\log \widehat{\mu^{(\varepsilon)}})(\zeta_0) = 0$. (We use that $\mathcal{D}_{\hat{\mu}} = \mathbb{R}^d$ to exclude the possibility of the mininum being attained at its boundary.) From the previously considered case and (1.74) we get

$$\liminf_{n\to+\infty} \frac{1}{n} \log \mu_n^{(\varepsilon)}(B_{\delta/2}(0)) \geq -I_{\mu^{(\varepsilon)}}(0) \geq -I_\mu(0), \tag{1.77}$$

where $\mu_n^{(\varepsilon)}$ denotes the law of $\bar{Y}_n^{(\varepsilon)} = \bar{X}_n + \sqrt{\varepsilon}\,\bar{Z}_n$, i.e. $\mu_n^{(\varepsilon)} = \mu_n * \eta_n^{(\varepsilon)}$, the convolution of μ_n with a Gaussian measure with average zero and covariance matrix $\varepsilon n^{-1}\mathbb{I}$. But

$$\mu_n(B_\delta(0)) \ge \mu_n^{(\varepsilon)}(B_{\delta/2}(0)) - \eta_n^{(\varepsilon)}\{z : |z| \ge \delta/2\}, \tag{1.78}$$

and

$$\eta_n^{(\varepsilon)}\{z : |z| \ge \delta/2\} \le 2d \int_{\frac{\delta}{2}\sqrt{n/\varepsilon d}}^{+\infty} \frac{e^{-u^2/2}}{\sqrt{2\pi}} du \le \frac{4d}{\delta\sqrt{2\pi}} \sqrt{\frac{\varepsilon d}{n}}\, e^{-\frac{\delta^2 n}{8d\varepsilon}},$$

according to the following easily checked relation, true for any $x > 0$ (see e.g. Lemma 2, Chapter VII in [115]):

$$\int_x^{+\infty} \frac{e^{-u^2/2}}{\sqrt{2\pi}} du \le \frac{1}{x} \frac{e^{-x^2/2}}{\sqrt{2\pi}}. \tag{1.79}$$

Since $I_\mu(0) < +\infty$, given $\delta > 0$ we may take $\varepsilon > 0$ so that $\delta^2/8d\varepsilon > I_\mu(0)$ and the last term on the right-hand side in (1.78) becomes negligible with respect to the first, and (1.72) follows at once from (1.77) and (1.78).

Upper bound. We prove the upper bound for compact sets C, and the exponential tightness, cf. remark (i) after Definition 1.17.

Upper bound for compact sets. Let $C \subseteq \mathbb{R}^d$ be a Borel set. The random variables $e^{\langle \zeta, X_i \rangle}$ are independent and non-negative, so that $\widehat{\mu_n}(\zeta) = E(e^{\langle \zeta, \bar{X}_n \rangle}) = E(\prod_{i=1}^n e^{\langle \frac{\zeta}{n}, X_i \rangle}) = (\hat{\mu}(\zeta/n))^n$ for any $\zeta \in \mathbb{R}^d$.

By the Markov exponential inequality, for any $\zeta \in \mathbb{R}^d$ we then have

$$\mu_n(C) \le e^{-\inf_{y \in C} \langle \zeta, y \rangle} \int_C e^{\langle \zeta, x \rangle} \mu_n(dx) \le e^{-\inf_{y \in C} \langle \zeta, y \rangle} (\hat{\mu}(\zeta/n))^n,$$

which we write as

$$\frac{1}{n} \log \mu_n(C) \le - \inf_{y \in C} \langle \frac{\zeta}{n}, y \rangle + \log \hat{\mu}(\zeta/n).$$

(Of course, this is useless if $\zeta/n \notin \mathcal{D}_{\hat{\mu}}$.) Setting $f(\zeta, y) := \langle \zeta, y \rangle - \log \hat{\mu}(\zeta)$, with values in $[-\infty, +\infty)$, and optimizing in ζ we have:

$$\frac{1}{n} \log \mu_n(C) \le - \sup_{\zeta \in \mathbb{R}^d} \inf_{y \in C} f(\zeta, y). \tag{1.80}$$

Equation (1.80) is valid for any Borel set C, but the term on its right-hand side is usually larger than $-\inf_{x \in C} I_\mu(x)$ (the order in which sup and inf are taken is interchanged!).

We now consider C compact, and derive the upper bound in Definition 1.12 for this particular case. The statement is trivial when $\inf_{y \in C} I_\mu(y) = 0$. Otherwise, let $0 < a < \inf_{y \in C} I_\mu(y)$, and for each $y \in C$, according to (1.69) we may take $\zeta_y \in \mathbb{R}^d$ so that $f(\zeta_y, y) > a$. This forces $\hat{\mu}(\zeta_y) < +\infty$; if $\delta_y |\zeta_y| < f(\zeta_y, y) - a$

and $|x - y| < \delta_y$, we must have $f(\zeta_y, x) > a$. Applying (1.80) we get, for each $n \geq 1$:

$$\frac{1}{n} \log \mu_n(B_{\delta_y}(y)) \leq -a. \tag{1.81}$$

Since C is compact we may take a finite set $\{y_1, \ldots, y_r\}$ (which depends on a) so that $C \subseteq \cup_{i=1}^r B_{\delta_{y_i}}(y_i)$. Thus,

$$\mu_n(C) \leq \sum_{i=1}^{r} \mu_n(B_{\delta_{y_i}}(y_i)) \leq r \max_{1 \leq i \leq r} \mu_n(B_{\delta_{y_i}}(y_i)),$$

so that from (1.81) we have

$$\frac{1}{n} \log \mu_n(C) \leq \frac{1}{n} \log r + \max_{1 \leq i \leq r} \frac{1}{n} \log \mu_n(B_{\delta_{y_i}}(y_i)) \leq \frac{1}{n} \log r - a,$$

and we obtain

$$\limsup_{n \to +\infty} \frac{1}{n} \log \mu_n(C) \leq -a.$$

But a may be taken arbitrarily close to $\inf_{y \in C} I_\mu(y)$, and we have checked the upper bound in Definition 1.12, restricted to compact sets. (*Remark.* No assumption on $\mathcal{D}_{\hat{\mu}}$ is required for this part.)

Exponential tightness. We now verify that given $a > 0$ there exists a compact set K_a such that for all n sufficiently large

$$\mu_n(\mathbb{R}^d \setminus K_a) \leq e^{-na}, \tag{1.82}$$

and for this it suffices that $0 \in \mathcal{D}^\circ_{\hat{\mu}}$, as we now check. This follows at once by considering the exponential Markov inequality, as in Theorem 1.1: indeed, if $k > E|X_1| \geq \max_{1 \leq j \leq d} |EX_1^{(j)}|$, where $X_1 = (X_1^{(1)}, \ldots, X_1^{(d)})$, we have

$$\mu_n(\mathbb{R}^d \setminus [-k, k]^d) \leq \sum_{j=1}^{d} \left\{ e^{-n I_\mu^j(k)} + e^{-n I_\mu^j(-k)} \right\},$$

where for $s \in \mathbb{R}$, $I_\mu^j(s)$ is the Fenchel–Legendre transform associated with the jth coordinate $X_1^{(j)}$. As seen in part (iii) of Proposition 1.9,

$$I_\mu^j(s) \to +\infty \quad \text{as} \quad |s| \to +\infty$$

and we immediately get (1.82), concluding the proof. □

Remarks (i) The proof of the upper bound in Theorem 1.23 uses only $0 \in \mathcal{D}^\circ_{\hat{\mu}}$.

(ii) It is worth observing that when C is compact and convex the upper bound is valid not only asymptotically, but for each $n \geq 1$. To see this, we recall (1.80). The function $(\zeta, y) \in \mathbb{R}^d \times C \mapsto f(\zeta, y) \in [-\infty, +\infty)$ is concave and upper semi-continuous in ζ, convex and lower semi-continuous in y. If, besides compact, the set C is convex, the classical minimax theorem says that

$$\sup_{\zeta\in\mathbb{R}^d}\inf_{y\in C} f(\zeta, y) = \inf_{y\in C}\sup_{\zeta\in\mathbb{R}^d} f(\zeta, y), \tag{1.83}$$

i.e. we can interchange the order of sup and inf on the right-hand side of (1.80). (For a proof of the minimax theorem see [180] or [281].) From (1.80) and (1.83) it follows that

$$\frac{1}{n}\log\mu_n(C) \le -\inf_{y\in C} I_\mu(y), \tag{1.84}$$

provided C is compact and convex.

(iii) Concerning the lower bound in Theorem 1.23, if $\mathcal{D}_{\hat\mu} = \mathbb{R}^d$ is not assumed, in principle we may have $I_\mu(y) < +\infty$, with the supremum in (1.69) not being achieved in $\mathcal{D}^\circ_{\hat\mu}$, and we cannot apply the given argument directly. To prevent this one may add the condition of $\log\hat\mu$ being 'steep' at $\partial\mathcal{D}_{\hat\mu}$, as mentioned briefly at the end of Section 1.1. The assumption $\mathcal{D}_{\hat\mu} = \mathbb{R}^d$ avoids this difficulty, allowing a simpler proof for Theorem 1.23, as we have discussed following [292]: with the addition of an arbitrarily small Gaussian term one reduces to a situation in which the supremum is attained.

(iv) With arguments based on subadditivity and convexity, as initially proven in [189], one arrives at the conclusion of Theorem 1.23 assuming only $0 \in \mathcal{D}^\circ_{\hat\mu}$. This will be seen in the next section in a more general set-up. (The lower bound will require no assumption.)

(v) While taking care of the second term on the right-hand side of (1.78) we used $|x| \le \sqrt{d}\max_i |x_i|$ for $x = (x_1, \dots, x_d) \in \mathbb{R}^d$, reducing to the one-dimensional case and applying (1.79). From the upper bound and the expression for the Fenchel–Legendre transform of the multivariate Gaussian we instead could see that

$$\limsup_{n\to+\infty}\frac{1}{n}\log\eta_n^{(\varepsilon)}\{z : |z| \ge \delta/2\} \le -\frac{\delta^2}{8\varepsilon}. \tag{1.85}$$

(vi) In contrast to the situation in (1.54), which is a particular case of the last theorem, here the lower bound in (b) of Definition 1.12 might be strict, for $d \ge 2$ (see [282]). Related construction gives an example for $d = 3$ and $0 \notin \mathcal{D}^\circ_{\hat\mu}$, cf. [89], where the upper bound for closed sets may fail.

1.5 Cramér theorem in infinite dimensional spaces[5]

At least two basic directions demand extensions of Theorem 1.23. Firstly, the introduction of dependence among the variables is essential for the analysis of a

[5] This section may be omitted at a first reading.

large class of processes, including Markov chains, and models arising in statistical mechanics. This will be discussed in Chapter 3. On the other hand, having in mind a further analysis of large deviations, it is important to extend Theorem 1.23, allowing $X_1, X_2, \ldots$ to take values in (suitably regular) infinite dimensional vector spaces. Examples of immediate relevance include the space of Borel measures on $\mathbb{R}^d$, which comes in naturally if we consider the empirical measures $L_n = n^{-1} \sum_{i=1}^n \delta_{Y_i}$ with $\mathbb{R}^d$-valued observables $Y_1, \ldots, Y_n$. The Sanov theorem (1957) is such an example. Other examples include the space of $\mathbb{R}^d$-valued continuous functions defined on some interval $[0, T]$ which appears in the analysis of sample path large deviations, to be developed in the next chapter in the context of Freidlin and Wentzell theory.

We now study an extension of Theorem 1.23 to such spaces, following the proposal of Ruelle and Lanford, based on convexity and subadditivity. For this let us start with a few basic definitions.

Definition 1.24 A topological space M with the property that for any two points $x, y \in M$ there exist disjoint open sets U_x, U_y such that $x \in U_x$, $y \in U_y$, and such that the singletons $\{x\}$ are closed sets, is said to be Hausdorff. A real vector space M with a topology under which $(x, y) \mapsto x + y$ and $(\alpha, x) \mapsto \alpha x$ are both continuous on $M \times M$, and $\mathbb{R} \times M$, respectively, is called a topological (real) vector space. The topological vector space is said to be locally convex if for each open set G and each $x \in G$ there exists an open convex set C_x with $x \in C_x \subseteq G$.

The extension of the Ruelle–Lanford approach to infinite dimensional spaces was initiated by Bahadur and Zabell in [11]. Further basic references where the material we now discuss is studied in much more detail are [87] and [85], from where conditions (a3) below have been taken. In [95], the method of the previous section is extended.

Let μ be a probability measure on the Borel σ-field $\mathcal{B}(M)$ of a locally convex Hausdorff topological (real) vector space M and let $X_1, X_2, \ldots$ be i.i.d. random vectors on some probability space $(\Omega, \mathcal{A}, P)$, with common law μ. (For instance, $\Omega = M^{\mathbb{N}}$, $\mathcal{A} = \mathcal{B}(M)^{\mathbb{N}}$, $P = \mu^{\mathbb{N}}$ the canonical product space.) In order to investigate the behaviour of

$$\bar{X}_n = \frac{1}{n} \sum_{i=1}^n X_i \, ,$$

we would like $\bar{X}_n$ to be a random vector (measurable) on $(\Omega, \mathcal{A}, P)$. Since M is a topological vector space, the map

$$(x_1, \ldots, x_n) \in M^n \quad \mapsto \quad \frac{1}{n} \sum_{i=1}^n x_i \in M$$

is continuous (product topology), and so measurable from $(M^n, \mathcal{B}(M^n))$ to

$(M, \mathcal{B}(M))$. But the inclusion $(\mathcal{B}(M))^n \subseteq \mathcal{B}(M^n)$ may be strict, in which case we cannot conclude the measurabilility of $\bar{X}_n$ from $(\Omega, \mathcal{A})$ to $(M, \mathcal{B}(M))$. To eliminate this trouble we recall that $\mathcal{B}(M^n) = (\mathcal{B}(M))^n$ if M is separable, i.e. if its topology can be generated by a countable family of open sets. As in the previously mentioned examples, and for many applications, μ will be supported by a convex set M_0 that can be turned into a Polish space with the induced topology, avoiding the just described problem. Based on this we set the following.

Assumptions 1.25

(a1) M is a locally convex Hausdorff topological (real) vector space; μ is a probability measure on $(M, \mathcal{B}(M))$. There exists a closed convex set $M_0 \subseteq M$ such that $\mu(M_0) = 1$, and which is a Polish space for the topology induced from M. (cf. remark following Definition 1.12.)

(a2) $X_1, X_2, \ldots$ are M_0-valued i.i.d. random variables, distributed according to μ, and defined on some probability space $(\Omega, \mathcal{A}, P)$.

Since $\mathcal{B}(M_0^n) = (\mathcal{B}(M_0))^n$ for each n, the maps $\omega \mapsto n^{-1} \sum_{j=m+1}^{m+n} X_j(\omega)$ are $\mathcal{A}$-measurable, for any m, n. It might be convenient to consider the canonical realization, i.e. $(\Omega, \mathcal{A}, P)$ given by $\Omega := M_0^{\mathbb{N}}$, $\mathcal{A} := \mathcal{B}(\Omega) = \mathcal{B}(M_0)^{\mathbb{N}}$, the product σ-field, and $P = \mu^{\mathbb{N}}$, the product measure, with the coordinate variables $X_1, X_2, \ldots$, i.e. $X_i(\omega) = \omega_i$ for each $\omega = (\omega_1, \omega_2, \ldots) \in \Omega$.

While (a1) and (a2) will be assumed throughout this section, for some results we also need the following.

(a3) The closed convex hull of each compact set K in M_0 is also compact.

Here, the convex hull of a set A, denoted by env(A), is defined as follows:

$$\operatorname{env}(A) = \left\{ \sum_{i=1}^{n} a_i x_i : n \geq 1, a_1, \ldots, a_n \geq 0, \sum_{i=1}^{n} a_i = 1 \text{ and } x_1, \ldots, x_n \in A \right\};$$

its closure is the closed convex hull of A, denoted by $\overline{\operatorname{env}(A)}$. Thus, $\overline{\operatorname{env}(A)}$ is the smallest closed convex set, containing A.

Under (a1), (a2), we let μ_n denote the law of $\bar{X}_n := \frac{1}{n} \sum_{i=1}^{n} X_i$ on M_0.

Remarks (i) An altenative way to have a realization as in (a2) is to take any complete probability space $(\Omega, \mathcal{A}, P)$ on which we have $X_1, X_2, \ldots$ i.i.d. M-valued, and distributed according to μ. The completeness here is in the usual measure-theoretical sense (i.e. if $A \in \mathcal{A}$ and $P(A) = 0$, then $\mathcal{A}$ contains all subsets of A). This guarantees the $\mathcal{A}$-measurability of the sample averages $n^{-1} \sum_{j=m+1}^{m+n} X_j$, due to the separability of M_0 and the fact that $\mu(M_0) = 1$. Obviously it does not represent a restriction.

(ii) Two important examples satisfying Assumptions 1.25 are: (1) $M_0 = M$, a separable Banach space (or more generally a Fréchet space), in which case condition (a3) is a classical result (cf. Theorem 3.25, Chapter 3 of [256]); (2) $M_0 = \mathcal{M}_1(S)$ the space of probability measures on the Borel σ-field of a Polish space S, with $M = \mathcal{M}(S)$ the space of (signed) finite measures on the Borel σ-field of S, considered with the weak topology induced by the space of bounded continuous functions from S to $\mathbb{R}$ (cf. Section 1.6). The validity of condition (a3) in both cases may be seen through the following lemma, taken from [85], and which extends the classical Mazur lemma (cf. Theorem B13 in [85]).

Lemma 1.26 *Under* (*a*1), *the following metric condition implies* (*a*3):

(*a*3)′ *There exists a Cauchy-complete metric* ϱ *on* M_0 *generating the topology induced from* M, *and satisfying*:

$$\varrho\,(ax_1 + (1-a)x_2, ay_1 + (1-a)y_2) \le \varrho(x_1, y_1) \vee \varrho(x_2, y_2),$$

for all $x_1, x_2, y_1, y_2 \in M_0$, *and all* $a \in [0, 1]$.

Proof Since M_0 is a Polish space, it suffices to check that $\mathrm{env}(K)$ is totally bounded, whenever K is a compact set in M_0. To see this, first observe that under (a3)′ the ϱ-balls are convex. Moreover if K is compact in M_0 and $\delta > 0$ we may take $n \ge 1$ and $x_1, \dots, x_n \in K$ so that $K \subseteq \bigcup_{i=1}^n B_\delta(x_i)$, where $B_r(x)$ denotes the open ϱ-ball with centre x and radius r. Thus

$$\mathrm{env}(K) \subseteq \mathrm{env}\left(\bigcup_{i=1}^n B_\delta(x_i)\right)$$

and since each $B_\delta(x_i)$ is convex we immediately see that any $y \in \mathrm{env}(K)$ may be written as $y = \sum_{i=1}^n a_i y_i$ where $y_i \in B_\delta(x_i)$, $a_i \ge 0$ and $\sum_{i=1}^n a_i = 1$. From (a3)′, $\varrho\big(y, \sum_{i=1}^n a_i x_i\big) < \delta$, so that

$$\mathrm{env}(K) \subseteq (\mathrm{env}\{x_1, \dots, x_n\})_{(\delta)} \tag{1.86}$$

with the notation $G_{(\delta)} = \{z \colon \varrho(z, G) < \delta\}$.

But $\mathrm{env}\{x_1, \dots, x_n\}$ is clearly compact, and so totally bounded. Since δ is arbitrary, it follows that $\mathrm{env}(K)$ is also totally bounded, concluding the proof. □

We need to introduce a few definitions.

If μ, M are as above and M^* denotes the dual of M, i.e. the space of continuous linear functionals on M, we define the moment generating function $\hat\mu \colon M^* \to (0, +\infty]$, via:

$$\hat\mu(\zeta) = \int_M e^{\langle \zeta, x\rangle}\, \mu(dx)$$

for each $\zeta \in M^*$, where $\langle \zeta, x\rangle := \zeta(x)$, if $\zeta \in M^*$, $x \in M$.

On M^* we shall use the weak* topology, denoted by w^*: it is the weakest (coarsest) topology which renders the maps $\zeta \mapsto \langle \zeta, x\rangle$ continuous from M^* to $\mathbb{R}$, for any $x \in M$.

Just as in the finite dimensional case, $\log \hat\mu(\cdot)$ is seen to be convex and lower semi-continuous for the w^*-topology on M^*. We may define its

Fenchel–Legendre transform $I_\mu\colon M \to [0, +\infty]$:

$$I_\mu(x) = \sup_{\zeta \in M^*} (\langle \zeta, x\rangle - \log \hat{\mu}(\zeta)), \tag{1.87}$$

which, as in the finite dimensional case, is also seen to be convex and lower semi-continuous on M.

(We may state the following result, due to Bahadur and Zabell [11].)

Theorem 1.27 *Under Assumptions* 1.25 *we have*:

(*a*) I_μ *is a convex rate function on* M_0;

(*b*) $\liminf_{n\to+\infty} n^{-1} \log \mu_n(G) \geq -\inf_{x\in G} I_\mu(x)$, *for each open set* $G \subseteq M_0$;

(*c*) $\limsup_{n\to+\infty} n^{-1} \log \mu_n(K) \leq -\inf_{x\in K} I_\mu(x)$, *for each compact set* $K \subseteq M_0$;

(*d*) $\lim_{n\to\infty} n^{-1} \log \mu_n(C) = -\inf_{x\in C} I_\mu(x)$, *if* $C \subseteq M_0$ *is open and convex.*

Items (a)–(c) of Theorem 1.27 characterize a weak l.d.p. on M_0, with rate function I_μ as defined in Section 1.2. (Recall that M_0 is considered with the induced topology from M; all the statements involving subsets of M_0, as in the previous theorem, refer to this topology, i.e. G' is open in M_0 if and only if $G' = G \cap M_0$, for some G open in M.)

Remark 1.28 The following facts will be used later on.

(i) Since M is locally convex and M_0 is convex, it follows at once that if $G \subseteq M_0$ is open and $x \in G$, there exists W_x convex, open in M_0, such that $x \in W_x \subseteq G$.

(ii) Since M_0 is a metric space, it is regular, i.e. if G is open in M_0 and $x \in G$, there exists V_x a neighbourhood of x such that $\overline{V_x} \subseteq G$ (it suffices to take a ball with centre x and small radius). Consequently, the set W_x in (i) may be taken so that $\overline{W_x} \subseteq G$. (In our case, M is also regular, cf. Theorem 1.11, Chapter 1 of [256].)

The closed set M_0 is convex, so that $\mu_n(M_0) = 1$ for all n. Theorem 1.27 might be read as a result on M, but obviously no aditional information would be gained ($I_\mu(x) = +\infty$ if $x \notin M_0$).

The proof of Theorem 1.27 will be achieved after several lemmas.

Lemma 1.29 *Assuming* (a1) *and* (a2), *for any open, convex set* $C \subseteq M_0$ *we have*:

$$\lim_{n\to+\infty} \frac{1}{n} \log \mu_n(C) = \sup_n \frac{1}{n} \log \mu_n(C) =: \ell(C) \in [-\infty, 0].$$

The main ingredient for the proof of Lemma 1.29 is subadditivity:

Definition 1.30 We say that a function $f\colon \mathbb{N} \to [0, +\infty]$ is subadditive if $f(m+n) \leq f(m) + f(n)$ for each $m, n \in \mathbb{N}$.

Lemma 1.31 *Let $f\colon \mathbb{N} \to [0, +\infty]$ be a subadditive function such that there exists $n_0 < +\infty$ with $f(n) < +\infty$ for all $n \geq n_0$. Then*

$$\lim_{n\to+\infty} \frac{f(n)}{n} = \inf_{n\geq n_0} \frac{f(n)}{n} < +\infty.$$

Proof of Lemma 1.31 Let $n \geq n_0$ be fixed, and consider $m \geq 2n$. We can write $m = kn + r$ where $r \in \{n, \ldots, 2n-1\}$ and $k \in \mathbb{N}$. Since $f(m) \leq f(kn) + f(r)$, we may write

$$\frac{f(m)}{m} \leq \frac{k\,f(n)}{m} + \frac{f(r)}{m} \leq \frac{f(n)}{n} + \frac{b_n}{m},$$

where $b_n = \max\{f(j) : n \leq j < 2n\} < +\infty$, since $n \geq n_0$. Thus

$$\limsup_{m\to\infty} \frac{f(m)}{m} \leq \frac{f(n)}{n},$$

for each $n \geq n_0$, proving the lemma. □

Proof of Lemma 1.29 Let $C \subseteq M_0$ be open and convex, and let $f(n) = -\log \mu_n(C)$, so that $f\colon \mathbb{N} \to [0, +\infty]$. With $X_1, X_2, \ldots$ i.i.d. and C convex, for each $n, m \geq 1$ we have:

$$\begin{aligned}
\mu_n(C)\mu_m(C) &= P\left(\frac{1}{n}\sum_{i=1}^{n} X_j \in C\right) P\left(\frac{1}{m}\sum_{j=n+1}^{n+m} X_j \in C\right) \\
&\leq P\left(\frac{n}{n+m}\bar{X}_n + \frac{m}{n+m}\cdot\frac{1}{m}\sum_{j=n+1}^{n+m} X_j \in C\right) \\
&= P(\bar{X}_{n+m} \in C) = \mu_{n+m}(C).
\end{aligned}$$

It follows that f is subadditive and to apply Lemma 1.31 we need to check that if $\mu_m(C) > 0$ for some m then there exists $n_0 \geq 1$ such that $\mu_n(C) > 0$ for all $n \geq n_0$. (The case $\mu_n(C) = 0$ for all $n \geq 1$ clearly yields $\ell(C) = -\infty$.) To see this we shall use condition (a1) and that C is open.

We first observe that having fixed such an m, there must exist a point $x_0 \in C$ so that for any neighbourhood U for x_0 we have $\mu_m(U) > 0$. In effect, if this were not true, using the separability of M_0, we would be able to cover C with countably many sets of measure zero for μ_m, contradicting $\mu_m(C) > 0$. Let us then fix such a point $x_0 \in C$.

On the other hand, since M_0 is a Polish space, a classical result (due to Ulam) tells us that each μ_i is tight (cf. Theorem 1.4 in [20], Remark 1.32 below). We see that a compact set $K \subseteq M_0$ may be taken so that $\mu_r(K) > 0$, for $r = 1, \ldots, m$.

Since C is open, a simple continuity argument gives that for each $y \in M_0$ we may take V_y a neighbourhood of y, U_y a neighbourhood of x_0, and $\varepsilon_y > 0$ so that

$$(1-\varepsilon)U_y + \varepsilon\, V_y \subseteq C,$$

for each $0 \le \varepsilon \le \varepsilon_y$. (At this point we are using that the addition and multiplication by scalars are continuous.) Moreover, according to Remark 1.28(i) we may assume U_y to be convex. The compactness of K allows us to take $y_1, \dots, y_s$ so that

$$K \subseteq \bigcup_{i=1}^{s} V_{y_i}.$$

In this case we set $U = \bigcap_{i=1}^{s} U_{y_i}$, and $\varepsilon' = \min_{i=1,\dots,s} \varepsilon_{y_i}$. It is easy to see that if $\varepsilon < \varepsilon'$

$$(1-\varepsilon)U + \varepsilon K \subseteq C.$$

Thus, if $n \ge n_0 := \lceil m/\varepsilon' \rceil + 1$, so that $m/n < \varepsilon'$, writing $n = km + r$ with $1 \le r \le m$ and $k \ge 1$, we see that

$$\begin{aligned} 0 &< \mu_m(U)^k \, \mu_r(K) \le \mu_{km}(U)\mu_r(K) \\ &= P\left(\bar{X}_{km} \in U, \frac{X_{km+1} + \cdots + X_n}{r} \in K \right) \\ &\le P\left(\frac{km}{n} \bar{X}_{km} + \frac{r}{n} \frac{\sum_{j=km+1}^{n} X_j}{r} \in \left(1 - \frac{r}{n}\right) U + \frac{r}{n} K \right) \\ &= \mu_n\left(\left(1 - \frac{r}{n}\right) U + \frac{r}{n} K \right) \le \mu_n(C), \end{aligned}$$

where the convexity of U is also used. The proof is complete. □

Remark 1.32 A probability measure μ on the Borel σ-field of a metric space is said to be tight if for each $\delta > 0$ there exists a compact set K_δ with $\mu(K_\delta) > 1 - \delta$. (It clearly implies regularity, i.e. $\mu(B) = \sup\{\mu(K) \colon K \subseteq B, K \text{ compact}\}$, for each Borel set B.)

Lemma 1.29 gives a clear indication of the basic procedure; since at any point x we can take a local basis of open convex neighbourhoods, we are led to consider the following natural candidate as rate function:

$$J(x) := \sup\{-\ell(C) : C \text{ open, convex}, x \in C \subseteq M_0\}, \tag{1.88}$$

for any $x \in M_0$. In the case of a metric with convex balls, as under (a3)$'$, this becomes $J(x) = -\lim_{\delta \to 0} \ell(B_\delta(x))$.

We now check that in the present set-up J is indeed a convex rate function.

Lemma 1.33 *Assuming* (a1) *and* (a2), *the function J defined by* (1.88) *is convex and lower semi-continuous from M_0 to* $[0, +\infty]$.

Proof The lower semi-continuity of J follows at once from (1.88). Indeed, if $x \in M_0$ and $a < J(x)$ we may take C, an open, convex neighbourhood of x such that $a < -\ell(C)$. Still from (1.88), $-\ell(C) \le J(y)$ for any $y \in C$ so that $a < \inf_{y \in C} J(y)$ verifying the lower semi-continuity, cf. Remark 1.14.

As for the convexity of J, due to the lower semi-continuity, it now suffices to check that for each $x, y \in M_0$

$$J\left(\frac{x+y}{2}\right) \leq \frac{1}{2}J(x) + \frac{1}{2}J(y). \tag{1.89}$$

For this, and by (1.88), it suffices to show that if C is an open convex neighbourhood of $(x+y)/2$ then there exists C_x, C_y open convex sets such that $x \in C_x$, $y \in C_y$ and such that

$$\ell(C) \geq \frac{1}{2}(\ell(C_x) + \ell(C_y)).$$

But, from the continuity we can take C_x, C_y open convex sets such that $x \in C_x$, $y \in C_y$ and

$$\frac{1}{2}C_x + \frac{1}{2}C_y := \left\{\frac{1}{2}u + \frac{1}{2}v\colon\ u \in C_x, v \in C_y\right\} \subseteq C.$$

Thus,

$$\left\{\bar{X}_n \in C_x\,,\ \frac{1}{n}\sum_{i=n+1}^{2n} X_i \in C_y\right\} \subseteq \left\{\bar{X}_{2n} \in C\right\},$$

which implies $\mu_{2n}(C) \geq \mu_n(C_x)\mu_n(C_y)$ and so

$$\frac{1}{2n}\log \mu_{2n}(C) \geq \frac{1}{2}\left(\frac{1}{n}\log \mu_n(C_x) + \frac{1}{n}\log \mu_n(C_y)\right),$$

from which we get (1.89). □

The next proposition tells that indeed the sequence (μ_n) satisfies a weak large deviation principle, with rate function J and scaling n.

Proposition 1.34 *Under* $(a1)$ *and* $(a2)$ *of Assumptions* 1.25, *and if J is given by* (1.88), *we have*:

- *(i)* $\liminf_{n\to+\infty} n^{-1}\log \mu_n(G) \geq -\inf_{x\in G} J(x)$ *for any open set* $G \subseteq M_0$;
- *(ii)* $\limsup_{n\to+\infty} n^{-1}\log \mu_n(K) \leq -\inf_{x\in K} J(x)$ *for any compact set* $K \subseteq M_0$.

Proof (i) Let $x \in G$. Due to the local convexity (cf. Remark 1.28(i)) we may take an open, convex set C such that $x \in C \subseteq G$. Thus

$$\liminf_{n\to+\infty} \frac{1}{n}\log \mu_n(G) \geq \ell(C) \geq -J(x),$$

proving (i).

(ii) Let $a < \inf_{x\in K} J(x)$. Then, for each $x \in K$ there exists C_x an open, convex set, with $x \in C_x$ and $\ell(C_x) < -a$. From the compactness of K, there exist $m \geq 1$ and $x_1, \ldots, x_m$ such that $K \subseteq \cup_{i=1}^m C_{x_i}$, and so

$$\mu_n(K) \leq m \max_{i=1,\ldots,m} \mu_n(C_{x_i}),$$

from which we have

$$\limsup_{n\to+\infty} \frac{1}{n} \log \mu_n(K) \le \max_{i=1,\dots,m} \lim_{n\to+\infty} \frac{1}{n} \log \mu_n(C_{x_i}) < -a,$$

and (ii) follows. □

Remark The previous proof shows that the validity of Lemma 1.29 for a sequence (μ_n) yields a weak l.d.p. in quite general situations. In particular, the method should extend beyond independence. We shall see a little of this in Chapter 3.

Before completing the proof of Theorem 1.27 we need the following lemma, where for the first time we shall use condition (a3) of Assumptions 1.25, whose main point is to guarantee that the probability of an open convex set C is well approximated by those of compact convex sets contained in C, i.e. a 'convex-regularity', or 'convex-tightness' property.

Lemma 1.35 *Under Assumptions* 1.25 *let the function J be defined by equation* (1.88). *Then, for any open and convex set $C \subseteq M_0$ we have*

$$\ell(C) = -\inf_{x\in C} J(x).$$

Proof That $\ell(C) \ge -\inf_{x\in C} J(x)$ is automatic from the definition of J. For the reverse inequality, it is enough to look at the case $-\infty < \ell(C)$. Thus, given $\varepsilon > 0$ we may take n_0 so that, if $n \ge n_0$

$$\ell(C) - \varepsilon \le \frac{1}{n} \log \mu_n(C).$$

At this point we use (a3): by Lemma 1.36, below, there exists a compact, convex set $K \subseteq C$ so that

$$\ell(C) - 2\varepsilon \le \frac{1}{n_0} \log \mu_{n_0}(K).$$

The convexity of K implies that $f(n) := \log \mu_n(K)$ is superadditive (i.e. $f(m+n) \ge f(m) + f(n)$) and so

$$\ell(C) - 2\varepsilon \le \frac{1}{n_0} \log \mu_{n_0}(K) \le \lim_{k\to\infty} \frac{f(kn_0)}{kn_0} \le \limsup_{n\to\infty} \frac{1}{n} \log \mu_n(K).$$

Using Proposition 1.34(ii) we have

$$\ell(C) - 2\varepsilon \le -\inf_{x\in K} J(x) \le -\inf_{x\in C} J(x)$$

and since ε can be taken arbitrarily small the proof follows. □

To complete the previous proof it remains to verify the property of approximation by compact convex sets, which we now state and prove. Without the convexity, it would be just the regularity or tightness of probability measures, true in Polish spaces (cf. proof of Lemma 1.29, Remark 1.32).

Lemma 1.36 *Under Assumptions* 1.25, *let* $C \subseteq M_0$ *be open and convex. Then, for any* $\varepsilon > 0$, *there exists a set* $K \subseteq C$ *compact and convex so that* $\mu(C \setminus K) < \varepsilon$.

Proof Since M_0 is a Polish space, μ must be regular, i.e. a compact set $F \subseteq C$ may be taken so that $\mu(C \setminus F) < \varepsilon$. (The point is that F may fail to be convex!) From condition (a3), $\overline{\text{env}(F)}$ is compact contained in M_0. This could in principle fail to be contained in C, so we need to perform some adjustment. (Notice that this adjustment is not necessary under (a3)$'$, in which case, $\overline{\text{env}(F)} \subseteq C$ automatically! see (1.86).)

For the needed adjustment, take for each $x \in F$, an open, convex neighbourhood W_x so that $\overline{W_x} \subseteq C$ (here we use Remark 1.28(ii)). Covering the compact set F with finitely many $\overline{W_{x_1}}, \ldots, \overline{W_{x_n}}$ let

$$F_0 = \overline{\text{env}(F)} \cap \bigcup_{i=1}^{n} \overline{W_{x_i}} = \bigcup_{i=1}^{n} (\overline{W_{x_i}} \cap \overline{\text{env}(F)}).$$

Since $\overline{W_{x_i}} \cap \overline{\text{env}(F)}$ is convex and compact we see that

$$\text{env}(F_0) = \left\{ \sum_{i=1}^{n} a_i y_i : \ \sum_{i=1}^{n} a_i = 1, a_i \geq 0, \ y_i \in \overline{W_{x_i}} \cap \overline{\text{env}(F)}, \forall i \right\}$$

is already compact, verifying our conditions, since $F \subseteq \text{env}(F_0) \subseteq C$. □

Remark As seen from Proposition 1.34, we do not need condition (a3) for the validity of a weak l.d.p. for (μ_n), with a convex rate function. This assumption came into play in Lemma 1.35, which is important, as we see next, for identification of the rate function with the Fenchel–Legendre transform of $\log \hat{\mu}$, according to (1.87).

The identification of J with I_μ is obtained through the following reduction to one-dimensional problems, cf. [11].

Proposition 1.37 *Under Assumptions* 1.25, *and for any* $x \in M_0$:

$$J(x) = \sup\{-\ell(H \cap M_0) : H \text{ is open semi-space and } x \in H\}, \tag{1.90}$$

where H *is said to be an open semi-space iff* $H = H(\zeta, r) = \{y \in M : \langle \zeta, y \rangle > r\}$ *for some* $\zeta \in M^*, r \in \mathbb{R}$.

Proof For notational convenience we write $H_0(\zeta, r) = \{y \in M_0 : \langle \zeta, y \rangle > r\}$ with $\zeta \in M^*$, $r \in \mathbb{R}$. If $x \in H_0(\zeta, r)$, the inequality $J(x) \geq \sup\{-\ell(H_0(\zeta, r)) : \zeta \in M^*, r \in \mathbb{R}\}$ is immediate from the definition, cf. (1.88). Now let $a < J(x)$, so that the set $C := \{y : J(y) \leq a\}$ is closed, convex (since J is lower semi-continuous and convex), and $x \notin C$. Under such conditions, the Hahn–Banach theorem (see e.g. Theorem 3.4, Chapter 3 of [256]), implies the existence of $\zeta \in M^*, r \in \mathbb{R}$ so that $x \in H(\zeta, r)$ and $H(\zeta, r) \cap C = \emptyset$. Thus, by Lemma 1.35, $-\ell(H_0(\zeta, r)) \geq a$, proving the proposition. □

Proof of Theorem 1.27 It remains to verify that J, defined by (1.88), coincides with I_μ, the Fenchel–Legendre transform of $\log \hat{\mu}$. This is a consequence of the Cramér theorem on $\mathbb{R}$ and the previous proposition, as we see next.

If $M_0 = M = \mathbb{R}$ we have $J = I_\mu$ as follows from Corollary 1.10, Proposition 1.34 and Remark 1.16(ii) (since $\mathbb{R}$ is locally compact).

Turning to the general case, for each $\zeta \in M^*$, consider the real variables $Y_n(\omega) = \langle \zeta, X_n(\omega) \rangle$, $n \geq 1$ and let $\mu^{(\zeta)}$ be their common law. If $\mu_n^{(\zeta)}$ is the law of $\overline{Y}_n := n^{-1} \sum_{i=1}^n Y_i$, let

$$\ell_\zeta(C) := \lim_{n \to \infty} \frac{1}{n} \log \mu_n^{(\zeta)}(C),$$

for C an open interval in $\mathbb{R}$, cf. Lemma 1.29, and $J_{\mu^{(\zeta)}}$ given by (1.88), in this case. Obviously, $\ell(H_0(\zeta, r)) = \ell_\zeta(r, +\infty) = \ell_{-\zeta}(-\infty, -r)$, for each $\zeta \in M^*, r \in \mathbb{R}$, i.e. for each $u \in \mathbb{R}$:

$$\begin{aligned} J_{\mu^{(\zeta)}}(u) &= \max\Big\{ \sup_{r<u} -\ell_\zeta(r, +\infty),\ \sup_{r>u} -\ell_\zeta(-\infty, r) \Big\} \\ &= \max\Big\{ \sup_{r<u} -\ell(H_0(\zeta, r)),\ \sup_{r>u} -\ell(H_0(-\zeta, -r)) \Big\}. \end{aligned} \tag{1.91}$$

By the previous observation $J_{\mu^{(\zeta)}} = I_{\mu^{(\zeta)}}$. On the other hand, we may rewrite (1.90) as:

$$\begin{aligned} J(x) &= \sup_{\zeta \in M^*} \sup_{r < \langle \zeta, x \rangle} -\ell(H_0(\zeta, r)) \\ &= \sup_{\zeta \in M^*} \max\Big\{ \sup_{r < \langle \zeta, x \rangle} -\ell(H_0(\zeta, r)),\ \sup_{r > \langle \zeta, x \rangle} -\ell(H_0(-\zeta, -r)) \Big\}, \end{aligned}$$

and the previous considerations give us

$$\begin{aligned} J(x) &= \sup_{\zeta \in M^*} J_{\mu^{(\zeta)}}(\langle \zeta, x \rangle) \\ &= \sup_{\zeta \in M^*} \sup_{u \in \mathbb{R}} (u \langle \zeta, x \rangle - \log \widehat{\mu^{(\zeta)}}(u)) \\ &= \sup_{\zeta \in M^*} \sup_{u \in \mathbb{R}} (\langle u\zeta, x \rangle - \log \hat{\mu}(u\zeta)) \\ &= \sup_{\zeta \in M^*} (\langle \zeta, x \rangle - \log \hat{\mu}(\zeta)) = I_\mu(x), \end{aligned}$$

concluding the proof. □

Remarks (i) Theorem 1.27 provides a weak l.d.p. To get a l.d.p., one could search for conditions ensuring exponential tightness for the sequence (μ_n), cf. Definition

1.17. When $M = \mathbb{R}^d$, a sufficient condition for that is the finiteness of $\hat{\mu}$ in a neighbourhood of the origin, as seen in Section 1.4.

In infinite dimensional spaces, there is no general way to tackle this; if M is the dual of a separable Banach space, the study of exponential tightness becomes easier due to properties of the w^*-topology, and it can be done under the assumption that the moment generating function of the norm $||X_1||$ is finite in some neighbourhood of the origin (see e.g. Section 6 of [7]). For the discussion of such an extension within the frame of separable Banach spaces, classical references include [7, 11, 95]. Further results have been developed by several authors, for which we refer to [85].

(ii) The proof is simpler if $M = \mathbb{R}^d$, in the sense that 'convex tightness' (Lemma 1.36) is immediate, since the closed balls are convex and compact, and the separation through hyperplanes, used in Proposition 1.37, is much simpler too. Thus, the Ruelle–Lanford method provides an alternative proof of Theorem 1.23, under the assumption that $0 \in \mathcal{D}^\circ_{\hat{\mu}}$. No assumption is needed for the weak l.d.p.

(iii) Under conditions (a1) and (a2) of Assumptions 1.25, the upper bound given in item (c) of Theorem 1.27, for K compact and convex, may be obtained using the minimax theorem and optimizing over the exponential Markov inequality, as in Section 1.4.

(iv) Part (d) in Theorem 1.27 clearly applies to any finite union of open convex sets.

(v) Theorem 1.18 requires the validity of a full l.d.p. with a good rate function. If applicable here, the identity $J = I_\mu$ would follow by convex duality, since J (set as $+\infty$ on $M \setminus M_0$) is convex and lower semi-continuous on M. More precisely, if we could apply the Varadhan lemma to the functions $F_\zeta(\cdot) = \langle \zeta, \cdot \rangle$, with $\zeta \in M^*$, we would get $\log \hat{\mu}(\zeta) = \sup_{x \in M}(\langle \zeta, x \rangle - J(x))$, for each ζ. The duality theorem would then assert that $J(x) = \sup_{\zeta \in M^*}(\langle \zeta, x \rangle - \log \hat{\mu}(\zeta))$, for each x. A priori, we do not have the conditions of Theorem 1.18. For the situation in Section 3.4 this will be automatic, due to compactness. It also applies in the next section. For a very general discussion, see Section 2.4.5 of [85].

Duality theorem *Let M be as in* (a1) *above. Let $f : M \to (-\infty, +\infty]$ be lower semi-continuous and convex. If $g : M^* \to (-\infty, +\infty]$ is defined as*

$$g(\zeta) := \sup_{x \in M}(\langle \zeta, x \rangle - f(x))$$

then $f(x) = \sup_{\zeta \in M^*}(\langle \zeta, x \rangle - g(\zeta))$, for each $x \in M$.

We omit the proof of the duality theorem, cf. e.g. Theorem 2.2.15 in [87] or Lemma 4.5.8 in [85]. As Proposition 1.37, used for the identification of J, it is also based on the separation through hyperplanes, which in the infinite dimensional case is obtained using the Hahn–Banach theorem (see also Theorem 3.13 in [38]).

1.6 Empirical measures. Sanov theorem

We now apply the discussion of the last section to the study of empirical measures. The problem was initially considered by Sanov in [263], for i.i.d. [0, 1]-valued random variables, and in Section 1.3 we analysed the simplest case of finite alphabets. If $Y_1, Y_2, \dots$ are i.i.d. random variables, with values on a Polish space $(S, \mathcal{B}(S))$, and defined on a probability space $(\Omega, \mathcal{A}, P)$, we consider the empirical measure:

$$L_n(\omega, \cdot) := \frac{1}{n} \sum_{i=1}^{n} \delta_{Y_i(\omega)}(\cdot) \tag{1.92}$$

for each $\omega \in \Omega$ where, as before, δ_y denotes the Dirac point-mass at y. Thus, if $A \in \mathcal{B}(S)$, $L_n(\omega, A) = n^{-1} \sum_{i=1}^n \mathbf{1}_A(Y_i(\omega))$, the proportion of observations which fall in A.

To fit into the framework of the last section, we consider $M = \mathcal{M}(S)$, the space of finite signed measures on $(S, \mathcal{B}(S))$ (i.e. $\eta \in \mathcal{M}(S)$ if and only if there exist η', η'' finite (σ-additive) measures on $(S, \mathcal{B}(S))$, such that $\eta = \eta' - \eta''$), and $M_0 = \mathcal{M}_1(S) := \{\eta \in \mathcal{M}(S) : \eta \text{ is a probability}\}$.

On $\mathcal{M}(S)$ we consider the topology determined by the seminorms $\eta \mapsto \left|\int f \, d\eta\right|$, with f varying in $C_b(S)$, the space of bounded and continuous functions from S to $\mathbb{R}$. Namely, if $\eta \in \mathcal{M}(S)$ the family $\{W(\eta, f_1, \dots, f_k, \delta) : \delta > 0, f_1, \dots, f_k \in C_b(S), k \geq 1\}$ constitutes a system of open neighbourhoods of η, where

$$W(\eta, f_1, \dots, f_k, \delta) := \{\nu \in \mathcal{M}(S) : \max_{1 \leq i \leq k} |\langle f_i, \nu \rangle - \langle f_i, \eta \rangle| < \delta\}, \tag{1.93}$$

adopting the usual notation for the duality bracket:

$$\langle f, \nu \rangle := \int f \, d\nu, \qquad f \in C_b(S), \nu \in \mathcal{M}(S). \tag{1.94}$$

The weak law of large numbers gives us $\lim_{n \to +\infty} P(L_n \notin W(\mu, f_1, \dots, f_k, \delta)) = 0$, for each $f_1, \dots, f_k \in C_b(S)$ and each $\delta > 0$, so that $L_n \to \mu$, in probability, as $n \to +\infty$. To investigate the rate of such decay, and the validity of the l.d.p., we use the result of the last section, with $M = \mathcal{M}(S)$, $M_0 = \mathcal{M}_1(S)$, and $X_i = \delta_{Y_i}$ for $i \geq 1$, which are i.i.d. $\mathcal{M}_1(S)$-valued random variables, and $L_n = n^{-1} \sum_{i=1}^n X_i$.

When restricted to $\mathcal{M}_1(S)$ the above topology corresponds to the usual weak convergence of probabilities on S; it is indeed the weak* topology inherited by $\mathcal{M}_1(S)$ considered a subset of the dual of the Banach space $C_b(S)$ (with $||f||_\infty := \sup_x |f(x)|$), and so we call it simply the w^*-topology. An important point is that this restriction to $\mathcal{M}_1(S)$ is metrizable, it may be obtained by the following

Lévy–Prohorov metric:

$$\varrho(\eta, \nu) = \inf\{\delta > 0 : \eta(F) \le \nu(F_{(\delta)}) + \delta, \text{ and } \nu(F) \le \eta(F_{(\delta)}) + \delta \text{ for each closed } F \subseteq S\}, \tag{1.95}$$

where $F_{(\delta)} := \{x : d(x, F) < \delta\}$ and d is a complete metric in S.

Remark This type of metric was first defined by Lévy when $S = \mathbb{R}$; writing $f_\eta(x) = \eta(-\infty, x]$, $x \in \mathbb{R}$, the cumulative distribution function of η, the usual Lévy metric on $\mathcal{M}_1(\mathbb{R})$ becomes:

$$\varrho(\eta, \nu) = \inf\{\delta > 0\colon f_\eta(x - \delta) - \delta \le f_\nu(x) \le f_\eta(x + \delta) + \delta, \ \forall x\}. \tag{1.96}$$

The extension to the present set-up is due to Prohorov.

To check the validity of Assumptions 1.25 we need to recall a few classical results.

Lemma 1.38 *Let S be a Polish space.*

(*a*) *With the w^*-topology, $\mathcal{M}(S)$ is a locally convex, Hausdorff topological vector space. Its topological dual may be identified with $C_b(S)$ via the functionals p_f defined as*

$$\nu \mapsto \langle f, \nu \rangle = \int_S f \, d\nu \,, \qquad \nu \in \mathcal{M}(S), \tag{1.97}$$

with $f \in C_b(S)$.

(*b*) *$\mathcal{M}_1(S)$ is a closed convex subset of $\mathcal{M}(S)$; it is a Polish space for the induced topology; the Lévy–Prohorov metric* (*defined by* (1.95)) *is compatible with this topology and Cauchy-complete.*

(*c*) *The convexity condition* (*a*3)$'$ *given in Lemma* 1.26 *is satisfied by the Lévy–Prohorov metric.*

Proof (Sketch) (a) The first statement follows at once since the class of linear functionals given by $\eta \mapsto \langle f, \eta \rangle$ with $f \in C_b(S)$ is separating, that is, if $\eta \neq \eta'$ then $\langle f, \eta \rangle \neq \langle f, \eta' \rangle$, for some $f \in C_b(S)$. As for the representation of the topological dual of $\mathcal{M}(S)$, one inclusion is trivial: each $f \in C_b(S)$ determines an element p_f of such a dual, via the relation $p_f(\eta) = \langle f, \eta \rangle$. It remains to prove that if ψ is a linear functional on $\mathcal{M}(S)$, continuous for the w^*-topology, then $\psi = p_f$, for certain $f \in C_b(S)$. For this, set $f(x) := \psi(\delta_x)$, for each $x \in S$. One easily checks that f is continuous. Since ψ is linear and continuous, there exists a neighbourhood of $\eta = 0$, cf. (1.93), where $|\psi(\cdot)| < 1$. By the linearity of ψ this provides a finite set $R \subseteq C_b(S)$ so that $|\psi(\nu)| \le \sum_{g \in R} |\langle g, \nu \rangle|$ for each ν, from which the boundedness of f follows. From the construction $\psi(\cdot) = p_f(\cdot)$ on the set of finite linear combinations of Dirac point measures; this set is dense in $\mathcal{M}(S)$ for the w^*-topology, as one easily checks (Theorem 4 Appendix III of [20]), and we get $\psi = p_f$.

(b) This is a classical result, due to Prohorov and Varadarajan, appearing in many graduate courses in probability. We omit most of it, referring to Appendix III of [20] (cf. also Lemma 3.2.2 in [87]), and just recall the most delicate point.[6] To check the Cauchy-completeness one may prove that each ϱ-Cauchy sequence (η_n) in $\mathcal{M}_1(S)$ must be tight, i.e. given $\varepsilon > 0$ we can find a compact set $K \subseteq S$ so that $\eta_n(K) > 1 - \varepsilon$ for each n. According to the Prohorov theorem (Remark 1.39 below) this would imply the existence of a convergent subsequence, and from the Cauchy property, the convergence of the full sequence. To construct such a compact set, we first use the Cauchy property to take for each $j \geq 1$ an index n_j so that $\varrho(\eta_n, \eta_m) < \varepsilon/2^{j+1}$ for all $n, m \geq n_j$. The space S being Polish, we know that each finite set in $\mathcal{M}_1(S)$ must be tight (Theorem 1.4 in [20], Remark 1.32), i.e. there exists a compact set $K_j \subseteq S$ so that $\eta_i(K_j) \geq 1 - \varepsilon/2^{j+1}$ for each $1 \leq i \leq n_j$. From these two properties we have $\eta_n((K_j)_{(\varepsilon/2^{j+1})}) \geq 1 - \varepsilon/2^j$ for each n. The set $K := \overline{\cap_{j=1}^{+\infty}(K_j)_{(\varepsilon/2^{j+1})}}$ is compact (closed and totally bounded for the complete metric d in S) and has the desired property.

(c) This is very simple to check. □

Remark 1.39 $\mathcal{F} \subseteq \mathcal{M}_1(S)$ is said to be tight iff for each $\delta > 0$ there exists a compact set K_δ so that $\mu(K_\delta^c) < \delta$ for each $\mu \in \mathcal{F}$. The previous proof uses the *direct part* of the Prohorov theorem (Theorem 6.1 in [20]), valid for any metric space S: if $\mathcal{F}$ is tight then it is (sequentially) relatively compact in $\mathcal{M}_1(\mathcal{X})$, i.e. each sequence in $\mathcal{F}$ must have a w^*-convergent subsequence (in $\mathcal{M}_1(S)$). The *converse part* requires S to be Polish, i.e. (sequential) relative compactness implies tightness, if S is Polish (Theorem 6.2 in [20]).

As a consequence of Lemma 1.38, Theorem 1.27 applies, yielding the following.

Theorem 1.40 (Sanov theorem) *Let S be a Polish space, and let μ be a probability on $(S, \mathcal{B}(S))$. If $Y_1, Y_2, \ldots$ are i.i.d. S-valued random variables with law μ, and $Q_n \in \mathcal{M}_1(\mathcal{M}_1(S))$ denotes the law of the empirical measure $L_n = n^{-1}\sum_{i=1}^n \delta_{Y_i}$, then:*

(a) *the sequence (Q_n) satisfies a weak l.d.p. with a convex rate function $I_{Q_1}(\nu)$ defined according to* (1.87), *i.e.*

$$I_{Q_1}(\nu) = \sup_{f \in C_b(S)} \left(\langle f, \nu\rangle - \log\langle e^f, \mu\rangle\right), \tag{1.98}$$

for each $\nu \in \mathcal{M}_1(S)$;

(b) *if C is a finite union of open convex sets in $\mathcal{M}_1(S)$, then*

$$\frac{1}{n}\log Q_n(C) \underset{n\to+\infty}{\longrightarrow} -\inf_{\nu\in C} I_{Q_1}(\nu);$$

[6] That the Prohorov metric generates the w^*-topology in $\mathcal{M}_1(S)$ depends only on the separability of S; the choice of d a complete metric gives ϱ also complete.

(c) *extending* (1.49), *the relative entropy of* $\nu \in \mathcal{M}_1(S)$ *with respect to* μ *can be defined as*

$$\mathcal{H}(\nu \mid \mu) := \begin{cases} \int h \log h \, d\mu & \text{if } \nu \ll \mu \text{ and } h = \frac{d\nu}{d\mu} \\ +\infty & \text{otherwise,} \end{cases} \tag{1.99}$$

and $I_{Q_1}(\cdot) = \mathcal{H}(\cdot \mid \mu)$ *on* $\mathcal{M}_1(S)$. *(*$\mathcal{H}(\nu \mid \mu)$ *is also called the Kullback–Leibler information.)*

Proof (a) and (b) follow at once from Theorem 1.27 and Lemma 1.38, after observing that if $f \in C_b(S)$ and $\langle f, \cdot \rangle$ is the element of $\mathcal{M}(S)^*$ given by (1.97), then

$$\int_{\mathcal{M}_1(S)} e^{\langle f, \eta \rangle} \, Q_1(d\eta) = E \, e^{f(Y_1)} = \int_S e^{f(x)} \, \mu(dx).$$

It remains to check item (c), cf. Lemma 2.1 in [93]. First observe that $\mathcal{H}(\nu \mid \mu)$ is well defined, taking values on $[0, +\infty]$, as follows at once from the convexity of the function $x \mapsto x \log x$ on $(0, +\infty)$ and the Jensen inequality. Now let $\nu \in \mathcal{M}_1(S)$. To check that $I_{Q_1}(\nu) \le \mathcal{H}(\nu \mid \mu)$ we may assume that $\mathcal{H}(\nu \mid \mu) < +\infty$. In particular, there exists $h \ge 0$ Borel measurable such that $\nu(A) = \int_A h \, d\mu$ for each $A \in \mathcal{B}(S)$. We write (1.99) as $\mathcal{H}(\nu \mid \mu) = \int \log h \, d\nu$, and if $f \in C_b(S)$ we apply the Jensen inequality to the identity $\int f \, d\nu - \mathcal{H}(\nu \mid \mu) = \int (f - \log h) \, d\nu$, to get

$$\exp\left(\int f \, d\nu - \mathcal{H}(\nu \mid \mu) \right) \le \int e^f \, d\mu.$$

It follows that

$$\int f \, d\nu - \log \int e^f \, d\mu \le \mathcal{H}(\nu \mid \mu),$$

for each $f \in C_b(S)$, proving the inequality.

For the converse, we may fix $\nu \in \mathcal{M}_1(S)$ with $I_{Q_1}(\nu) < +\infty$. We first claim that the supremum in (1.98) does not change if we replace $C_b(S)$ by $B_b(S)$, the space of bounded, Borel measurable functions from S to $\mathbb{R}$.

Indeed, having fixed $\nu \in \mathcal{M}_1(S)$ and $f \in B_b(S)$, we may take a sequence (f_n) in $C_b(S)$ so that $(\nu + \mu)(|f - f_n| > \delta) \to 0$ and moreover, f being bounded, we may also assume $\sup_n \|f_n\|_\infty < +\infty$. Thus $\int f_n \, d\nu \to \int f \, d\nu$, and $\int e^{f_n} \, d\mu \to \int e^f \, d\mu$. Consequently,

$$\sup_{f \in B_b(S)} \left[\int f \, d\nu - \log \int e^f \, d\mu \right] \le I_{Q_1}(\nu). \tag{1.100}$$

The reverse inequality being trivial, we have checked this claim.

We need to prove that $\nu \ll \mu$ and that $\mathcal{H}(\nu \mid \mu) \le I_{Q_1}(\nu)$. For this, let $A \in \mathcal{B}(S)$ with $\mu(A) = 0$. Taking $f = n \mathbf{1}_A$ in (1.100) and varying n, we immediately get $\nu(A) = 0$, i.e., $\nu \ll \mu$. Let then $h = d\nu/d\mu$, so that $h \ge 0$ μ a.s., $\int_S h$

$d\mu = 1$. If there exist constants $0 < c < C < +\infty$ so that $c \le h(\cdot) \le C$ μ a.s. we may assume $\log h \in B_b(S)$ and (1.100) implies

$$\mathcal{H}(\nu \mid \mu) = \int \log h \, d\nu \le I_{Q_1}(\nu).$$

To cover the general case we use approximation. For example, first assume h to be bounded away from zero, and let $h_n = h \wedge n$, so that $\log h_n \in B_b(S)$ and from (1.100) together with the monotone convergence theorem we get

$$\begin{aligned} \mathcal{H}(\nu \mid \mu) &= \int \log h \, d\nu = \lim_{n\to\infty} \int \log h_n \, d\nu \\ &\le I_{Q_1}(\nu) + \lim_{n\to\infty} \log \int h_n \, d\mu = I_{Q_1}(\nu). \end{aligned} \tag{1.101}$$

The restriction of h being bounded away from zero remains. But, if $\nu_\varepsilon = \varepsilon\mu + (1-\varepsilon)\nu$, then $h_\varepsilon = d\nu_\varepsilon/d\mu = \varepsilon + (1-\varepsilon)h$ μ a.s. by the uniqueness of the Radon–Nikodym derivative, so that by the previous case, together with the convexity of I_{Q_1}, we have

$$\mathcal{H}(\nu_\varepsilon \mid \mu) \le I_{Q_1}(\nu_\varepsilon) \le \varepsilon\, I_{Q_1}(\mu) + (1-\varepsilon) I_{Q_1}(\nu) = (1-\varepsilon) I_{Q_1}(\nu).$$

We now check that $\mathcal{H}(\nu_\varepsilon \mid \mu) \to \mathcal{H}(\nu \mid \mu)$, as $\varepsilon \to 0$.

Since $\nu \mapsto \mathcal{H}(\nu \mid \mu)$ is also convex (easy to see from the convexity of $x \mapsto x \log x$, as in Section 1.3), and $\mathcal{H}(\mu \mid \mu) = 0$, we see that $\mathcal{H}(\nu \mid \mu) \ge \limsup_{\varepsilon \to 0} \mathcal{H}(\nu_\varepsilon \mid \mu)$. We are really interested in the converse and for this observe that $x \mapsto \log x$ is concave and increasing in $(0, +\infty)$, so that:

$$\log h_\varepsilon \ge (1-\varepsilon) \log h \vee \log \varepsilon \quad \mu \text{ a.s.}$$

From this:

$$\begin{aligned} \mathcal{H}(\nu_\varepsilon \mid \mu) &= \int h_\varepsilon \log h_\varepsilon \, d\mu \\ &= \varepsilon \int \log h_\varepsilon \, d\mu + (1-\varepsilon) \int h \log h_\varepsilon \, d\mu \\ &\ge \varepsilon \log \varepsilon + (1-\varepsilon)^2 \mathcal{H}(\nu \mid \mu), \end{aligned}$$

which allows us to conclude the proof. □

To get a full l.d.p. (with a good rate function) it now suffices to prove the exponential tightness of (Q_n), cf. Definition 1.17 in Section 1.2. This is originally due to Donsker and Varadhan, whose proof (cf. [93–95]) we essentially follow.

Proposition 1.41 *Under the conditions of Theorem* 1.40, *the sequence* (Q_n) *is exponentially tight.*

Proof Let (b_j) be a sequence of positive numbers, to be fixed later. Since S is a Polish space, the measure μ is tight. Thus we take, for each $j \ge 1$, a compact set

$A_j \subseteq S$ such that $\mu(S \setminus A_j) \le b_j$. Now let, for each $j \ge 1$:

$$F_j = \{\nu \in \mathcal{M}_1(S) : \nu(S \setminus A_j) \le 1/j\}.$$

We easily see that each F_j is closed; indeed, if $(\nu_m) \subseteq F_j$ and $\nu_m \to \nu$, from the portmanteau theorem (Theorem 2.1 in [20]) we have, for each j:

$$\nu(S \setminus A_j) \le \liminf_{m \to +\infty} \nu_m(S \setminus A_j) \le \frac{1}{j}.$$

As a consequence $K_J := \cap_{j \ge J} F_j$ is also closed, if $J \ge 1$. Moreover, by the Prohorov theorem (see Remark 1.39) and since A_j is compact each K_J is relatively compact.[7] Thus for each $J \ge 1$, K_J is a compact subset of $\mathcal{M}_1(S)$. We now estimate $Q_n(K_J^c)$:

$$Q_n(F_j^c) = P(L_n \notin F_j) = P(L_n(S \setminus A_j) > 1/j) \le e^{-a/j}\, E\, e^{aL_n(S \setminus A_j)}$$

for each $a > 0$. The choice $a = 2nj^2$ yields:

$$\begin{aligned} Q_n(F_j^c) &\le e^{-2nj}\, E\, e^{2j^2 \sum_{i=1}^n \mathbf{1}_{S \setminus A_j}(Y_i)} \\ &= e^{-2nj} \big(E\, e^{2j^2 \mathbf{1}_{S \setminus A_j}(Y_1)}\big)^n \\ &= e^{-2nj} (e^{2j^2} \mu(S \setminus A_j) + \mu(A_j))^n \\ &\le e^{-2nj} (e^{2j^2} b_j + 1)^n, \end{aligned}$$

for each $n, j \ge 1$. Thus, a proper choice would be $b_j = e^{-2j^2}$, in which case we would end up with

$$Q_n(F_j^c) \le e^{-2nj}\, 2^n \le e^{-nj},$$

so that

$$Q_n(K_J^c) \le \sum_{j=J}^{+\infty} Q_n(F_j^c) \le \sum_{j=J}^{+\infty} e^{-nj} \le 2 e^{-nJ},$$

concluding the proof. □

As remarked immediately after Definition 1.17, the last two results entail the following.

Corollary 1.42 *Under the conditions of Theorem* 1.40, (Q_n) *satisfies a l.d.p. on* $\mathcal{M}_1(S)$, *with the good rate function* $I_{Q_1}(\cdot) = \mathcal{H}(\cdot \mid \mu)$ *given by* (1.99) *on* $\mathcal{M}_1(S)$.

Comment The approximation at the end of the proof of Theorem 1.40 is taken from Section 3.2 in [87]. For a more in depth study see Chapter 3 of [87] or Section 6.2 of [85].

[7] $\mathcal{M}_1(S)$ is metrizable; thus sequential relative compactness and relative compactness are the same.

Other topologies The original version of the Sanov theorem (cf. [263]) uses a stronger topology. He proved large deviation estimates for the empirical distribution of i.i.d. real random variables in terms of the Kolmogorov distance on $\mathcal{M}_1(\mathbb{R})$, defined by

$$|\nu - \eta|_K = \sup_{x\in\mathbb{R}} |\nu(-\infty, x] - \eta(-\infty, x]|.$$

The topology generated by this metric is strictly stronger (finer) than the w^*-topology and it is not separable. Therefore, Sanov's result, extended to $S = \mathbb{R}^d$ by Borovkov [26], Hoeffding [156], Hoadley [155] and Stone [284], is not contained in Theorem 1.40.

The consideration of stronger topologies arises naturally: if L_n denotes the empirical measure, as in Theorem 1.40, the weak law of large numbers says that for each $f \in B_b(S)$, the space of bounded Borel measurable functions from S to $\mathbb{R}$, we have

$$L_n(f) = \frac{1}{n}\sum_{i=1}^{n} f(Y_i) \quad \xrightarrow{P} \quad \int f\,d\mu.$$

This immediately suggests the consideration of the so-called τ-topology, with a system of open neighbourhoods of $\eta \in \mathcal{M}_1(S)$ given by

$$\{U(\eta, f_1, \ldots, f_k, \delta) \colon \delta > 0, f_1, \ldots, f_k \in B_b(S), k \geq 1\},$$

where, analogously to (1.93) ($\langle f, \nu\rangle = \int f d\nu$):

$$U(\eta, f_1, \ldots, f_k, \delta) := \{\nu \in \mathcal{M}_1(S) : \max_{1\leq i\leq k} |\langle f_i, \nu\rangle - \langle f_i, \eta\rangle| < \delta\}.$$

Groeneboom, Oosterhoff and Ruymgaart [147] seem to have been the first to study large deviations using the τ-topology, which was also used by Csiszár [70], and later by many authors. Clearly, a l.d.p. for a stronger topology refines the estimates in Definition 1.12. Besides this potential improvement, the main motivation comes from the range of applicability, as the need to treat tansformations which are not continuous for the weak topology. Examples in [147] include 'trimmed' means and other statistics important for their robustness.

This analysis does not fit into the framework of Theorem 1.40, since generally $\mathcal{M}_1(S)$ will be neither separable nor metrizable, with the τ-topology. Measurability problems could also arise if we consider the Borel σ-field for the τ-topology on $\mathcal{M}_1(S)$: simply, the map $x \in S \mapsto \delta_x$ might not be measurable. A natural σ-field (smaller but containing the $U(\eta, f_1, \ldots, f_k, \delta)$ above) is that generated by the maps $p_f(\nu) := \langle f, \nu\rangle$, with $f \in B_b(S)$, i.e. the smallest σ-field making measurable (from $\mathcal{M}_1(S)$ to $(\mathbb{R}, \mathcal{B}(\mathbb{R}))$ each of the maps p_f. If S is a Polish space, this indeed coincides with the Borel σ-field of $\mathcal{M}_1(S)$ for the w^*-topology, as we easily see from Part (b) of Lemma 1.38.

If S is a Polish space, it is not hard to extend the conclusion of Theorem 1.40 to the τ-topology (see e.g. Theorem 3.2.21 in [87] for a proof). If $c < +\infty$, the set $\{\nu \in \mathcal{M}_1(S) \colon \mathcal{H}(\nu|\mu) \le c\}$ is τ-compact (and τ-sequentially compact).

The proof can also be achieved through the projective limit approach formalized by Dawson and Gärtner in [72], and in some sense initiated in [147]. More general spaces can be treated, provided we endow $\mathcal{M}_1(S)$ with the σ-field generated by the maps p_f, $f \in B_b(S)$. One may get rid of topological considerations on S and take $(S, \mathcal{B}(S))$ a standard Borel space. Let $B_b(S)^*$ be the algebraic dual of $B_b(S)$ (a Banach space with $\|\cdot\|_\infty$) with the weak topology induced by $B_b(S)$, i.e. the coarsest one under which the evaluation maps $\nu \in B_b(S)^* \mapsto \nu(f)$ are continuous. Identifying $\nu \in \mathcal{M}(S)$ with the map $f \mapsto \int f d\nu$, $\mathcal{M}_1(S)$ is embedded in the unitary ball of $B_b(S)^*$. The previous considerations based on subadditivity provide a weak l.d.p. for L_n. Since the unitary ball is compact for the topology under consideration we get indeed a full l.d.p. on this larger space. In order to get a l.d.p. on $\mathcal{M}_1(S)$ with the restricted topology, it suffices to know that the rate function is $+\infty$ outside $\mathcal{M}_1(S)$. This is usually the delicate point (see Section 3.5 of [72], Section 4.6 of [85]).

Concerning empirical processes, we study only the case of finite S. Another proof of the results in Section 1.3 will be seen in Chapter 3, with extension to interacting systems on $\mathbb{Z}^d$.

2

Small random perturbations of dynamical systems. Basic estimates of Freidlin and Wentzell

This chapter deals with an example of large deviation estimates in infinite dimensional spaces. It refers to the trajectories of a class of continuous stochastic processes obtained by imposing a small 'white noise' on certain simple deterministic systems. The basic set-up is that of a stochastic process $X^{x,\varepsilon} = (X_t^{x,\varepsilon} : t \geq 0)$ formally described as

$$\begin{aligned} dX_t^{x,\varepsilon} &= b(X_t^{x,\varepsilon})dt + \varepsilon\alpha_t, \qquad t \geq 0, \\ X_0^{x,\varepsilon} &= x, \end{aligned} \tag{*}$$

where $b(x)$ is a vector field on $\mathbb{R}^d$ with proper regularity and growth condition and α is a 'white noise' in $\mathbb{R}^d$. Formally, α_t would be the 'derivative' of a Brownian motion and as such it cannot be realized as an $\mathbb{R}^d$-valued function (Theorem 2.16); we should consider equation (*) in the following integral form:

$$X_t^{x,\varepsilon} = x + \int_0^t b(X_s^{x,\varepsilon})\, ds + \varepsilon W_t, \tag{**}$$

where (W_t) is an $\mathbb{R}^d$-valued Brownian motion. Independently of precise definitions, as $\varepsilon \to 0$ we expect $X^{x,\varepsilon}$ to converge to the deterministic system $X^{x,0}$, the unique solution of

$$\begin{aligned} dX_t^{x,0} &= b(X_t^{x,0})dt, \qquad t \geq 0, \\ X_0^{x,0} &= x. \end{aligned} \tag{2.1}$$

The problem of estimating the probability of 'rare events' for $X^{x,\varepsilon}$, for $\varepsilon > 0$ small, fits into the frame of Section 1.2. These estimates may be obtained as an application of the Cramér theorem in the separable Banach space $C([0, T], \mathbb{R}^d)$ (Theorem 1.27). This is discussed briefly at the end of Section 2.4. Instead, we now follow a more direct proof, as developed by Freidlin and Wentzell in [295], and which has been systematically discussed and extended in their book [122]. Earlier related results were obtained by Borovkov [26] and Schilder [264].

In order to provide some minimal self-sufficiency to this text, we start by recalling one of the many classical constructions and a few basic properties of the Wiener process, Itô integrals, and simple Itô processes. The reader already familiar with such topics should go directly to Section 2.4. On the other hand, for more detail on the material to be reviewed briefly in Section 2.2 and Section 2.3, we refer to [173] or [138]. Together with these books, some of the many available references for the material in Section 2.1 are [121, 153, 181], which we follow at some points.

2.1 Brownian motion

In this section we quickly review a construction and some basic properties of Brownian motion (also called the Wiener process). They will be used for our discussion of large deviation principles within the theory of Freidlin and Wentzell (Section 2.4), and its application to metastability (Chapter 5).

Definition 2.1 A collection of random variables $(W_t\colon 0 \le t \le T)$ defined on some probability space $(\Omega, \mathcal{A}, P)$ is called a standard Brownian motion on $\mathbb{R}$, starting at 0, if:

(a) $P(W_0 = 0) = 1$;

(b) for each $m \ge 1$, each $0 = t_0 < t_1 < \cdots < t_m \le T$, $W_{t_{i+1}} - W_{t_i}$, $0 \le i \le m-1$ are independent random variables and $W_{t_{i+1}} - W_{t_i}$ has Gaussian distribution with zero average and variance $t_{i+1} - t_i$, with density given by

$$\frac{1}{\sqrt{2\pi(t_{i+1}-t_i)}} \exp\left\{-\frac{1}{2(t_{i+1}-t_i)} y^2\right\}, \qquad y \in \mathbb{R}; \tag{2.2}$$

(c) for each $\omega \in \Omega$, the trajectory (or path) $t \mapsto W_t(\omega)$ is continuous on $[0, T]$.

For the reader not yet familiar with the subject, it is natural to ask: does such a process exist? is it 'unique' in some sense?

Observe first that conditions (a) and (b) of Definition 2.1 determine what the finite dimensional distributions should be, and can be replaced by the following:

(b)′ For each $m \ge 1$, each $0 < t_1 < \cdots < t_m \le T$, the distribution of $(W_{t_1}, \ldots, W_{t_m})$, denoted by $p_{t_1,\ldots,t_m}$, is a Gaussian measure on $\mathbb{R}^m$ with zero average and covariance matrix $\Sigma := ((t_i \wedge t_j)_{i,j})$, i.e.

$$p_{t_1,\ldots,t_m}(dx) = \frac{1}{(2\pi \det \Sigma)^{m/2}} \exp\left\{-\frac{1}{2}\langle x, \Sigma^{-1} x\rangle\right\} dx. \tag{2.3}$$

Notation. $t \wedge s = \min\{s, t\}$, $t \vee s = \max\{s, t\}$.

It is quite simple to check the consistency of the family $\{p_{t_1,\dots,t_m}\}$, i.e.

$$p_{t_1,\dots,t_{k-1},t_{k+1},\dots,t_{m+1}}(A)$$
$$= p_{t_1,\dots,t_{m+1}}\{(x_1,\dots,x_{m+1})\colon (x_1,\dots,x_{k-1},x_{k+1},\dots,x_{m+1}) \in A\},$$

for each $m \geq 1$, each $t_1 < \cdots < t_{m+1}$, each $A \in \mathcal{B}(\mathbb{R}^m)$. Thus, the extension theorem of Kolmogorov (see e.g. Theorem 2.2, Chapter 2 of [173]) tells us that there exists a probability measure $\tilde{P}$ on the functional space $\tilde{\Omega} = \mathbb{R}^{[0,T]}$ whose finite dimensional projections are the $p_{t_1,\dots,t_m}$. More precisely, let $\tilde{\mathcal{A}}$ denote the Borel σ-field for the product topology on $\tilde{\Omega}$ or, equivalently, the σ-field generated by the field of the finite dimensional cylinder sets of the form

$$\{\tilde{\omega} \in \tilde{\Omega} : (\tilde{\omega}(t_1),\dots,\tilde{\omega}(t_m)) \in A\},$$

where $m \geq 1$, $0 \leq t_1 < \cdots < t_m \leq T$, $A \in \mathcal{B}(\mathbb{R}^m)$. There exists a unique probability $\tilde{P}$ on $\tilde{\mathcal{A}}$ such that

$$\tilde{P}\{\tilde{\omega} \in \tilde{\Omega} : (\tilde{\omega}(t_1),\dots,\tilde{\omega}(t_m)) \in A\} = p_{t_1,\dots,t_m}(A),$$

for each cylinder set as above. This construction is nevertheless very poor. It does not even allow the verification of condition (c) in Definition 2.1, since

$$C([0,T],\mathbb{R}^d) := \{f \in \tilde{\Omega} : f \text{ is continuous}\}$$

is *not* an element of $\tilde{\mathcal{A}}$.

There are many ways to go beyond the above construction. For an overview we refer to some of the books on this subject, for example [121, 153, 173]. Let us now see a particular construction, first provided by Lévy (1948) and later simplified by Ciesielski (1961). It is based on the following simple but important observation.

Lemma 2.2 (Lévy) *Assume (a) and (b) of Definition 2.1. If $t_1 < t < t_2$, the conditional distribution of W_t given W_{t_1} and W_{t_2} is Gaussian with average*

$$\mu(t) = \frac{t_2 - t}{t_2 - t_1} W_{t_1} + \frac{t - t_1}{t_2 - t_1} W_{t_2} \tag{2.4}$$

and variance

$$\sigma^2(t) = \frac{(t - t_1)(t_2 - t)}{t_2 - t_1}. \tag{2.5}$$

Consequently, for any such t, W_t may be represented as

$$W_t = \mu(t) + \sigma(t)\xi, \tag{2.6}$$

where ξ is a standard Gaussian random variable (i.e. average zero and unit variance) independent of $(W_s : s \in [0,t_1] \cup [t_2,T])$.

Proof The last statement follows at once from the first, since due to the property of independence of increments, the processes $(W_u : 0 \le u < t_1)$, $(W_u : t_1 < u < t_2)$ and $(W_u : t_2 < u \le T)$ are conditionally independent given W_{t_1} and W_{t_2}.

The computation leading to the first statement is quite simple. We first consider the case $t_1 = 0$ (and so $W_{t_1} = 0$ a.s.). In this situation (W_t, W_{t_2}) has a density given by

$$f(x, y) = \frac{1}{\sqrt{2\pi t}} e^{-\frac{1}{2}\frac{x^2}{t}} \frac{1}{\sqrt{2\pi(t_2 - t)}} e^{-\frac{1}{2}\frac{(y-x)^2}{(t_2-t)}},$$

and the conditional density of W_t given $W_{t_2} = y$ can easily be checked to be

$$\frac{f(x, y)}{\int_{\mathbb{R}} f(u, y)\, du} = \frac{1}{\sqrt{2\pi t \frac{(t_2-t)}{t_2}}} \exp\left\{ \frac{-1}{2t(t_2 - t)/t_2}(x - \frac{t}{t_2} y)^2 \right\},$$

which agrees with (2.4), (2.5) when $t_1 = 0$, $W_{t_1} = 0$.

For the general case, notice that under our assumptions W_{t_1} is independent of $(W_t - W_{t_1}, W_{t_2} - W_{t_1})$, which has the same law as $(W_{t-t_1}, W_{t_2-t_1})$. We are easily reduced to the previous situation. □

The main idea behind Lévy's construction is to build a process with continuous trajectories satisfying conditions (a) and (b) of Definition 2.1 as a limit of an interpolation procedure, where we repeatedly use the expression (2.6).

Let us consider a probability space $(\Omega, \mathcal{A}, P)$ on which we have defined an infinite sequence of i.i.d. standard Gaussian random variables (ξ_n) (this can be done for example on the interval $[0, 1]$ with the usual Borel σ-field and uniform distribution).

For simplicity let us fix $T = 1$. For n a non-negative integer we set

$$D_n := \{k/2^n : k = 0, 1, \ldots, 2^n\}, \quad D_\infty := \bigcup_{n \ge 0} D_n,$$

and define inductively a sequence of processes $X^{(n)}$, $n \ge 1$.

If $n = 0$ we set $X_0^{(0)} = 0$, $X_1^{(0)} = \xi_0$, and interpolate linearly in between:

$$X_t^{(0)} = t\xi_0, \qquad 0 \le t \le 1. \tag{2.7}$$

Having constructed $X^{(n)}$, we define $X^{(n+1)}$ in such a way that: on D_n it coincides with $X^{(n)}$; on $D_{n+1} \setminus D_n$ we use the construction suggested by Lemma 2.2;

finally on $[0,1] \setminus D_{n+1}$ we do linear interpolation. That is

$$X_t^{(n+1)} = \begin{cases} X_t^{(n)} & \text{if } t \in D_n, \\ X_t^{(n)} + 2^{-(n/2+1)}\xi_{i(t)} & \text{if } t \in D_{n+1} \setminus D_n, \\ & i(t) = 2^n + (2^{n+1}t - 1)/2, \\ (2^{n+1}t - k)X_{(k+1)/2^{n+1}}^{(n+1)} & \text{if } k/2^{n+1} < t < (k+1)/2^{n+1}, \\ \quad + (k+1-t\,2^{n+1})X_{k/2^{n+1}}^{(n+1)} & k = 0, 1, \ldots, 2^{n+1}-1. \end{cases}$$

From Lemma 2.2 and proceeding by induction, we see that the distribution of $(X_t^{(n)} : t \in D_n)$ coincides with that prescribed by conditions (a) and (b) of Definition 2.1, for each n.

The main point is then the following.

Lemma 2.3 (Lévy) *There exists $\Omega_0 \in \mathcal{A}$ with $P(\Omega_0) = 1$ and such that for $\omega \in \Omega_0$, $X_t^{(n)}(\omega)$ converges uniformly on $t \in [0,1]$, as $n \to +\infty$.*

Before proving this lemma, notice that it gives a construction of Brownian motion, according to Definition 2.1. Indeed, if Ω_0 is the full set in Lemma 2.3 we may set

$$W_t(\omega) = \begin{cases} \lim_{n\to+\infty} X_t^{(n)}(\omega) & \text{if } \omega \in \Omega_0, \\ 0 & \text{otherwise.} \end{cases}$$

Due to the uniform convergence, $t \mapsto W_t(\omega)$ is continuous; condition (a) of Definition 2.1 is immediate and, according to previous observations, it satisfies the restriction of condition (b) to $t_1, \ldots, t_m \in D_\infty$. But, since D_∞ is dense, the continuity implies condition (b) of Definition 2.1.

Proof of Lemma 2.3 From the construction we have

$$\max_{0\le t\le 1} \left| X_t^{(n+1)}(\omega) - X_t^{(n)}(\omega) \right| = 2^{-(n/2+1)} \max_{2^n \le j < 2^{n+1}} |\xi_j|$$

so that

$$\begin{aligned} P\left(\max_{0\le t\le 1} \left| X_t^{(n+1)} - X_t^{(n)} \right| \ge \varepsilon_n \right) &\le 2^n P(|\xi_1| > 2^{n/2+1}\varepsilon_n) \\ &\le 2^{n+1} \frac{1}{\sqrt{2\pi}} \frac{1}{2^{n/2+1}\varepsilon_n} e^{-2^{n+1}\varepsilon_n^2} \qquad (2.8) \\ &= \frac{1}{\sqrt{2\pi}} \frac{2^{n/2}}{\varepsilon_n} e^{-2^{n+1}\varepsilon_n^2}, \end{aligned}$$

where we have used the simple inequality (1.79). Taking $\varepsilon_n = n^{-r}$ with $r > 1$ we already see that

$$\sum_{n=1}^{+\infty} P\left(\max_{0\le t\le 1}\left|X_t^{(n+1)} - X_t^{(n)}\right| \ge \varepsilon_n\right) < +\infty.$$

Since $\sum_{n=1}^{+\infty} \varepsilon_n < +\infty$, by the Borel–Cantelli lemma we get

$$\sum_{n=1}^{+\infty} \max_{0\le t\le 1}\left|X_t^{(n+1)} - X_t^{(n)}\right| < +\infty \quad \text{a.s.},$$

which proves the lemma. □

A better estimate is achieved if ε_n is replaced by $\varepsilon_n' = (c(n+1)\log 2)^{1/2}/2^{n/2+1}$, with $c > 2$. (*Notation.* $\log = \log_e$ everywhere in the text.) For this choice the right-hand side of (2.8) is bounded from above by

$$\frac{2^{(2-c)n/2}2^{-c/2}}{\sqrt{c(n+1)\log 2}},$$

which is summable for $c > 2$, in which case

$$P\left(\max_{0\le t\le 1}\left|W_t^{(n+1)} - W_t^{(n)}\right| \ge \varepsilon_n' \text{ i.o.}\right) = 0.$$

Notation $[A_n \text{ i.o.}] = \bigcap_{n\ge 1}\bigcup_{k\ge n} A_k = \limsup_{n\to+\infty} A_n$.

Remark Of course, the consideration of $T = 1$ is just a simplification. The same works on any compact interval $[0, T]$.

A simple extension of Definition 2.1 consists in the following.

Definition 2.4 A stochastic process $W = (W_t : 0 \le t \le T)$ defined on a given probability space $(\Omega, \mathcal{A}, P)$ is said to be a Brownian motion with constant drift $b \in \mathbb{R}$, diffusion coefficient $\sigma^2 > 0$, and initial point $x \in \mathbb{R}$ iff:

(a) $P(W_0 = x) = 1$;
(b) if $n \ge 1$ and $0 = t_0 < t_1 < \cdots < t_n \le T$, $W_{t_{i+1}} - W_{t_i}$, $0 \le i \le n-1$, are independent Gaussian random variables with averages $(t_{i+1} - t_i)b$ and variances $(t_{i+1} - t_i)\sigma^2$;
(c) for each ω, the function $t \mapsto W_t(\omega)$ is continuous.

($b = 0$, $\sigma^2 = 1$ corresponds to a standard Brownian motion starting at x.)

Remark 2.5 If W^0 is a standard Brownian motion starting at 0, then $W_t = x + bt + \sigma W_t^0$ describes a Brownian motion with constant drift b, diffusion coefficient σ^2 and initial point x, and conversely.

Remark 2.6 It is natural to extend Definitions 2.1 and 2.4 to the time interval $[0, +\infty)$. Given Lemma 2.3, the construction of such a process presents no problem. For example, taking two independent sequences (ξ_n'), (ξ_n'') as in the previous construction, and applying the previous prescription to each of them, we get two independent standard Brownian motions $(W'_t : 0 \le t \le 1)$ and $(W''_t : 0 \le t \le 1)$. We may now define $W_t = W'_t$ for $0 \le t \le 1$ and $W_t = W'_1 + tW''_{1/t} - W''_1$ for $t > 1$. This process $(W_t : 0 \le t < +\infty)$ satisfies the requirements, but we omit the details (see e.g. Section 2.3 in [153]).

Remark 2.7 (Wiener measure. The canonical process) As already mentioned, there are several different constructions of Brownian motion which, besides revealing different aspects of the process, might indeed give different objects. Nevertheless, under the conditions of Definition 2.1 we may always consider the map $R : \Omega \to C([0, T], \mathbb{R})$ defined as

$$R(\omega)(t) = W_t(\omega). \tag{2.9}$$

We may endow the space $C_{0,T} := C([0, T], \mathbb{R})$ with its Borel σ-field $\mathcal{B}(C_{0,T})$ for the usual uniform topology. It is simple to see that $\mathcal{B}(C_{0,T})$ coincides with the σ-field generated by the projections $(\pi_t : 0 \le t \le T)$, where

$$\pi_t(\psi) = \psi(t), \tag{2.10}$$

if $t \in [0, T]$, $\psi \in C_{0,T}$. That is, $\mathcal{B}(C_{0,T})$ is the σ-field generated by the field of the finite dimensional cylinder sets of the form

$$D^A_{(t_1,\ldots,t_n)} = \{\psi \in C([0, T], \mathbb{R}) : (\psi(t_1), \ldots, \psi(t_n)) \in A\}, \tag{2.11}$$

with $n \ge 1$, $0 \le t_1 < \cdots < t_n \le T$, and $A \in \mathcal{B}(\mathbb{R}^n)$. In particular, the map R defined by (2.9) is measurable. If $\mathbb{P}_0$ denotes the measure it induces on $C([0, T], \mathbb{R})$, then

$$\mathbb{P}_0(D^A_{t_1,\ldots,t_n}) = p_{t_1,\ldots,t_n}(A), \tag{2.12}$$

cf. (2.3), for $t_1 > 0$. Therefore the measure $\mathbb{P}_0$ is completely determined by Definition 2.1. It is usually called the (standard) Wiener measure.

If $b \in \mathbb{R}$ and $\sigma \ne 0$, the affine map

$$\psi \mapsto x + bt + \sigma\psi(t), \tag{2.13}$$

transforms $\mathbb{P}_0$ into the law of a Brownian motion starting at x, with constant drift b and diffusion coefficient σ^2, that we denote by $\mathbb{P}_{x,b,\sigma^2}$. When $b = 0$, $\sigma^2 = 1$ we write simply $\mathbb{P}_x$.

On the space $(C_{0,T}, \mathcal{B}(C_{0,T}))$ we may then consider the canonical process $(\pi_t)_{0 \le t \le T}$, and the family of measures $(\mathbb{P}_x)_{x \in \mathbb{R}}$ as the 'canonical' Brownian motion family. Similarly, for the time interval $[0, +\infty)$ and $b \ne 0$, $\sigma^2 > 0$ we take the space $C([0, +\infty), \mathbb{R})$.

We need to extend this to higher dimensions.

Definition 2.8 Let $d \ge 1$ be an integer.

(i) A d-dimensional standard Brownian motion starting at $x = (x_1, \dots, x_d) \in \mathbb{R}^d$ is an $\mathbb{R}^d$-valued stochastic process $W_t = (W_t^{(1)}, \dots, W_t^{(d)})$, $0 \le t \le T$ for which the coordinates $(W_t^{(i)} : 0 \le t \le T)$ are mutually independent standard real Brownian motions starting at x_i, for $i = 1, \dots, d$ (similarly for the time interval $[0, +\infty)$).

(ii) If $b \in \mathbb{R}^d$ and σ is a $d \times d$ matrix with real entries, we say that (W_t) is a Brownian motion with constant drift b and diffusion matrix $a := \sigma \cdot \sigma^{\mathbb{T}}$ when it can be written as $W_t = \sigma \tilde{W}_t + t\, b$, with $(\tilde{W}_t)$ a standard Brownian motion on $\mathbb{R}^d$. Here $\sigma^{\mathbb{T}}$ denotes the transpose of σ, so that $a_{i,j} = \sum_{k=1}^d \sigma_{i,k}\sigma_{j,k}$, and $(\sigma \tilde{W})^{(i)} := \sum_{k=1}^d \sigma_{i,k} \tilde{W}^{(k)}$. To have a genuinely d-dimensional process we assume σ invertible and let $\mathcal{S}^+_{d\times d} := \{a : a \text{ is a } d \times d \text{ symmetric positive definite matrix (real entries)}\}$.

Remark 2.9 (W_t) is a d-dimensional Brownian motion starting at $x \in \mathbb{R}^d$, with drift $b \in \mathbb{R}^d$ and diffusion matrix $a \in \mathcal{S}^+_{d\times d}$, if and only if:

(a) $P(W_0 = x) = 1$;

(b) for each $s < t$, $W_t - W_s$ is independent of $\mathcal{A}^W_s := \sigma(W_u : u \le s)$ and has a Gaussian distribution with average $(t-s)b$ and covariance matrix $(t-s)a$, i.e. for each $A \in \mathcal{B}(\mathbb{R}^d)$,

$$P(W_t - W_s - (t-s)b \in A \mid \mathcal{A}^W_s) = \frac{1}{\sqrt{(2\pi(t-s))^d \det a}} \times \int_A e^{-\frac{1}{2(t-s)}\langle y, a^{-1} y\rangle}\, dy \quad \text{a.s.},$$

where $\langle \cdot, \cdot \rangle$ denotes the usual Euclidean inner product in $\mathbb{R}^d$ and $P(\cdot|\cdot)$ stands for the conditional probability;

(c) W has continuous paths.

Some basic properties of Brownian motion

Martingale properties

Definition 2.10 Given a probability space $(\Omega, \mathcal{A}, P)$, a non-decreasing family of sub σ-fields of $\mathcal{A}$, $(\mathcal{A}_t)$, here indexed by $t \ge 0$, is called a filtration in $\mathcal{A}$. A stochastic process (X_t), understood as a family of random variables in this space, is said to be $(\mathcal{A}_t)$-adapted if, for each s, X_s is $\mathcal{A}_s$-measurable. The process (X_t) is said to be a martingale with respect to $(\mathcal{A}_t)$ if and only if:

(i) (X_t) is $(\mathcal{A}_t)$-adapted and X_t is integrable, for each t;

(ii) $E(X_t|\mathcal{A}_s) = X_s$ a.s., for all $s < t$.

Examples (1) Let $(\mathcal{A}^W_t)$ be the filtration generated by a standard one-dimensional Brownian motion, as defined in Remark 2.9. A simple computation shows that (W_t) and $((W_t)^2 - t)$ are both martingales with respect to $(\mathcal{A}^W_t)$.

(2) From Remark 2.9, and since the Gaussian distribution on $\mathbb{R}^d$ with average $\tilde{b}$ and covariance matrix $\tilde{a} \in \mathcal{S}^+_{d\times d}$ is characterized by its moment generating

function $\exp\{\langle \zeta, \tilde{b} \rangle + \frac{1}{2}\langle \zeta, \tilde{a}\zeta \rangle\}$, we see at once that (W_t) is a d-dimensional Brownian motion starting at $x \in \mathbb{R}^d$, with drift $b \in \mathbb{R}^d$ and diffusion matrix $a \in \mathcal{S}^+_{d\times d}$, if and only if it has continuous paths, $P(W_0 = x) = 1$ and for any $\zeta \in \mathbb{R}^d$, $s < t$:

$$E\left(\exp\{\langle \zeta, W_t - W_s \rangle\} \mid \mathcal{A}^W_s\right) = \exp\left((t-s)\{\langle \zeta, b\rangle + \frac{1}{2}\langle \zeta, a\zeta\rangle\}\right) \quad \text{a.s.} \tag{2.14}$$

The validity of (2.14) for all $s < t$ corresponds to the martingale property of

$$Z^\zeta_t(W) := \exp\left\{\langle \zeta, W_t\rangle - t\langle \zeta, b\rangle - \frac{t}{2}\langle \zeta, a\zeta\rangle\right\}, \quad t \geq 0, \tag{2.15}$$

with respect to $(\mathcal{A}^W_t)$.

From the previous example we conclude the following.

Proposition 2.11 *The law of a d-dimensional Brownian motion with constant drift b, diffusion matrix $a \in \mathcal{S}^+_{d\times d}$ and initial point x is the unique probability $\mathbb{P}_{x,b,a}$ on the Borel σ-field $\mathcal{B}^0_\infty := \sigma(\pi_s : s \geq 0)$ of $C([0,+\infty), \mathbb{R}^d)$ such that:*

(i) $\mathbb{P}_{x,b,a}(\pi_0 = x) = 1$;

(ii) under $\mathbb{P}_{x,b,a}$, for each $\zeta \in \mathbb{R}^d$, the process

$$Z^\zeta_t(\pi) = \exp\left\{\langle \zeta, \pi_t\rangle - t\langle \zeta, b\rangle - \frac{t}{2}\langle \zeta, a\zeta\rangle\right\}, \quad t \geq 0, \tag{2.16}$$

is a martingale with respect to the canonical filtration $(\mathcal{B}^0_t)$, with $\mathcal{B}^0_t := \sigma(\pi_s : s \leq t)$.

Markov property

Remark 2.9 implies the Markov property of Brownian motion. Usually understood as the conditional independence of the 'future' and the 'past', given the 'present', it is also described by the property that

$$P(W_{t+\cdot} \in B \mid \mathcal{A}^W_t) = P(W_{t+\cdot} \in B \mid W_t) \quad \text{a.s.} \quad \text{for} \quad B \in \mathcal{B}^0_\infty. \tag{2.17}$$

Due to the independence of the increments one has indeed more than (2.17): the process $(W_{t+s} - W_t : s \geq 0)$ is independent of $\mathcal{A}^W_t$, and due to the stationarity of the increments, its distribution is the same as that of $(W_s : s \geq 0)$ starting at 0. In other words, the measures $\mathbb{P}_{x,b,a}$ on the space $C([0,+\infty), \mathbb{R}^d)$ are such that for each t, $\mathbb{P}_{\pi_t,b,a}$ is a regular conditional distribution for $\pi_{t+\cdot}$ given $\mathcal{B}^0_t$ (see e.g. [121] or [173]).

Markov processes with suitable regularity usually share a strong form of the Markov property, which consists in the extension of the Markov property to a certain class of random times; knowing the position of the process at such time we can forget the past, the process starts afresh from this position. Before introducing the necessary definitions let us observe that this cannot hold for all random times: as a prototype of a random time for which it should not hold consider a one-dimensional Brownian motion and the time of the *last* visit to the origin before time 1. It is clear that the process *does not start afresh* at this time.

Definition 2.12 Let $(\mathcal{A}_t)$ be a filtration on some probability space $(\Omega, \mathcal{A}, P)$.
(a) A random variable $\tau\colon \Omega \to [0, +\infty]$ is said to be a stopping time if $[\tau \leq t] \in \mathcal{A}_t$ for each $t > 0$. In such a case we define $\mathcal{A}_\tau = \{A \in \mathcal{A}\colon A \cap [\tau \leq t] \in \mathcal{A}_t,$ for each $t > 0\}$, which represents the information 'up to τ'.
(b) If instead the event $[\tau < t]$ belongs to $\mathcal{A}_t$ for each $t > 0$ we say that τ is an optional time. In such a case we define $\mathcal{A}_{\tau+} = \{A \in \mathcal{A}\colon A \cap [\tau < t] \in \mathcal{A}_t,$ for each $t > 0\}$, which represents the information 'up to right after τ'.

Remarks (i) It easily follows from the definition that τ is an optional time if and only if it is a stopping time for the filtration $(\mathcal{A}_{t+})$, where $\mathcal{A}_{t+} := \cap_{\varepsilon>0}\mathcal{A}_{t+\varepsilon}$. In particular, each stopping time is an optional time and the concepts coincide when the filtration is right-continuous, i.e. when $\mathcal{A}_{t+} = \mathcal{A}_t$ for each $t \geq 0$. We shall work mostly with such filtrations, so that the concept of optional time will not be important.

(ii) If τ is a stopping time, $\mathcal{A}_\tau \subseteq \mathcal{A}_{\tau+}$ are σ-fields and τ is $\mathcal{A}_\tau$-measurable. The same measurability applies to X_τ, provided (X_t) is an $\mathbb{R}^d$-valued stochastic process adapted to the given filtration, and with right-continuous trajectories. (If τ is optional, $\mathcal{A}_{\tau+}$ is still a σ-field on which τ is measurable.) For more details, see Section 1.2 of [173].

Examples Let (X_t) be an $\mathbb{R}^d$-valued stochastic process with continuous trajectories, adapted to a given filtration. If $A \subseteq \mathbb{R}^d$ is a closed set, it is easy to verify that the hitting time to A, defined as $\tau(A) = \inf\{t \geq 0\colon X_t \in A\}$, is a stopping time for the given filtration. ($\tau(A)$ is optional if A is open, provided the trajectories are right-continuous.) This example will be mostly used later. When $A = \{x\}$ we write simply $\tau(x)$.

Proposition 2.13 *Let $(W_t : t \geq 0)$ be a d-dimensional Brownian motion with drift b and diffusion matrix a, defined on some probability space $(\Omega, \mathcal{A}, P)$ and let $(\mathcal{A}_t)$ be a filtration such that (i) $\mathcal{A}_t \supseteq \mathcal{A}_t^W$ for each $t \geq 0$; (ii) $(W_{t+s} - W_t : s \geq 0)$ is independent of $\mathcal{A}_t$, for each $t \geq 0$. Then for any finite optional time τ $(W_{\tau+s} - W_\tau : s \geq 0)$ is a d-dimensional Brownian motion independent of $\mathcal{A}_{\tau+}$, with drift b and diffusion matrix a, starting at $x = 0$.*

We omit the proof of this proposition, but just point out the basic idea: approximate τ by $\tau_n(\omega) := \min\{j/2^n : j/2^n > \tau(\omega)\}$ and use the usual Markov property together with the (right) continuity of trajectories. This is essentially the standard argument used for the proof of the Doob optional sampling theorem for (right) continuous martingales, from which the proposition can indeed be deduced, once we use the exponential martingales $Z_t^\zeta(W)$ given by (2.15). (For a proof see e.g. Section 2.6, Chapter 1 in [173], or Section 1.3 in [121].)

An application of the strong Markov property to one-dimensional Brownian motion yields the following classical result due to Désiré André and which goes under the name of 'reflection principle' due to its geometric interpretation.

Theorem 2.14 (Reflection principle) *Let* (W_t) *be a standard one-dimensional Brownian motion, starting at* 0. *Then, if* $t \geq 0$, $y > 0$:

$$P(\max_{s \leq t} W_s \geq y) = 2P(W_t \geq y).$$

Proof Let

$$\tau(y) = \inf\{t \geq 0 \colon W_t = y\}.$$

We can verify that $\tau(y)$ is an a.s. finite stopping time. Since $P(W_t = y) = 0$ we write

$$P(\max_{s \leq t} W_s \geq y) = P(\tau(y) < t) = P(\tau(y) < t, W_t > y) + P(\tau(y) < t, W_t < y).$$

Clearly $P(\tau(y) < t, W_t > y) = P(W_t > y)$. As for the second term, we recall that the process starts afresh at $\tau(y)$ and that (by symmetry) the probability of a negative increment in the remaining time is the same as that of a positive one. This gives

$$P(\tau(y) < t, W_t < y) = P(W_t > y).$$

(To be fully rigorous, we condition on $\tau(y)$, which is $\mathcal{A}^W_{\tau(y)}$-measurable, and use Proposition 2.13.) □

Remark The same argument gives $P(\max_{s \leq t} W_s > y, W_t < y - x) = P(W_t > y + x)$, for $x > 0$, $y > 0$.

A few sample path properties

Theorem 2.15 below summarizes a classical result of Lévy (1940) which describes the behaviour of the quadratic variation of the trajectories, along a sequence of increasing partitions. This is important for the consideration of Itô's stochastic integral.

If $q = \{0 = t_0 < \cdots < t_m = T\}$ represents a finite partition of the interval $[0, T]$ into subintervals with endpoints t_i $(i = 0, \ldots, m)$, we call $\delta q = \max_{0 \leq i \leq m-1}(t_{i+1} - t_i)$, and for any function $\varphi \colon [0, T] \to \mathbb{R}$, and $r > 0$ we set

$$\mathrm{var}_q^{(r)}(\varphi) = \sum_{i=0}^{m-1} |\varphi(t_{i+1}) - \varphi(t_i)|^r, \tag{2.18}$$

the rth variation of φ, relative to q. When $r = 2$ we call it the squared variation.

Having fixed this notation we may state and prove the following.

Theorem 2.15 (Lévy) *Let W be a one-dimensional Brownian motion, with diffusion coefficient $\sigma^2 > 0$. If (q_n) is a sequence of finite partitions of $[0, T]$ such that $q_1 \subseteq q_2 \subseteq \ldots$ and $\delta q_n \to 0$ as $n \to +\infty$, then*

$$\operatorname{var}^{(2)}_{q_n}(W) \underset{n\to+\infty}{\longrightarrow} \sigma^2 T \qquad a.s.$$

Remark A related natural question refers to the true quadratic variation of Brownian paths on $[0, T]$, i.e. $\lim_{\varepsilon\to 0} \sup_{\delta q<\varepsilon} \operatorname{var}^{(2)}_q(W)$. The classical law of the iterated logarithm is a way to see that this should be $+\infty$ a.s. (see also Section 1.4 of [121], or Section 2.3 of [181]). When considered in the strong sense, the rth variation of the trajectories of W equals a.s. 0 if $r > 2$ and $+\infty$ if $r \le 2$. To get a finite and non-zero limit the power must be replaced by a different function (see e.g. Chapter 5 in [181]). We will not use such properties in this text.

Proof Using (2.13) and the path continuity we reduce to the case of a standard Brownian motion starting at 0. Let $q_n = \{0 = t^n_0 < t^n_1 < \cdots < t^n_{k_n} = T\}$ and $v_n = \operatorname{var}^{(2)}_{q_n}(W)$. Let us start by observing that the limit, if it exists, must be as in the statement. Indeed, by the independence of the increments and since $E(Z^4) = 3\sigma^4$, for Z a Gaussian variable with zero average and variance σ^2, we have

$$E(v_n - T)^2 = E\left(\left|\sum_{i=0}^{k_n-1}\left\{\left|W_{t^n_{i+1}} - W_{t^n_i}\right|^2 - (t^n_{i+1} - t^n_i)\right\}\right|^2\right)$$

$$= 2\sum_{i=0}^{k_n-1}(t^n_{i+1} - t^n_i)^2 \le 2T\delta q_n,$$

which shows that v_n tends to T in L^2. The above inequality shows that v_n tends to T a.s. if $\sum_{n=1}^{+\infty}\delta q_n < +\infty$, by the Borel–Cantelli lemma. Nevertheless the theorem requires only δq_n tending to 0. This can be seen as an application of the Doob theorem on convergence of martingales. The main point, a consequence of the symmetry and the independence of increments of W_t, is contained in the following.

Claim *If $q_1 \subseteq q_2 \subseteq \ldots$ are finite partitions of $[0, T]$ then $E(v_n \mid v_{n+1}, v_{n+2}, \ldots, v_{n+k}) = v_{n+1}$ a.s. for each $n \ge 1$, $k \ge 1$.*

To check the above claim one constructs, for each n, a process $(Y_n(s, t)\colon (s, t) \in [0, 1] \times [0, T])$ via

$$Y_n(s, t) = \sum_{i=0}^{k-1} \eta_{i+1}(s)(W_{t^{n+1}_{i+1}} - W_{t^{n+1}_i}) + \eta_{k+1}(s)(W_t - W_{t^{n+1}_k}),$$

if $t \in [t^{n+1}_k, t^{n+1}_{k+1})$, for some $k = 0, \ldots, k_{n+1} - 2$, or $t \in [t^{n+1}_k, t^{n+1}_{k+1}]$ and $k = k_{n+1} - 1$, and where $\eta_1, \eta_2, \ldots$ form a Rademacher sequence in $[0, 1]$, namely

$$\eta_i(u) = 2\zeta_i(u) - 1,$$

with $\zeta_i(u)$ the ith digit in the binary expansion of u. Notice that η_i are i.i.d. random variables on $([0,1], \mathcal{B}([0,1]), \lambda_1)$ (λ_1 denoting the Lebesgue measure); each takes the values 1 and -1 with probability $1/2$ each. In other words, $(\eta_i : i \geq 1)$ forms an orthonormal system in $L^2([0,1], \mathcal{B}([0,1]), \lambda_1)$. We also take the sequence (η_i) as independent of W (e.g. replacing the original probability space Ω by $\Omega \times [0,1]$ with the product measure $P \times \lambda_1$).

We observe the following.

(i) Since we have assumed no drift, the distribution of $W_{t''} - W_{t'}$ is symmetric around 0. This implies that for each s fixed $Y_n(s, \cdot)$ is a standard Brownian motion.

(ii) If $t', t'' \in [t_i^{n+1}, t_{i+1}^{n+1}]$ $(i = 0, \ldots, k_{n+1} - 1)$, then

$$\left|Y_n(s,t'') - Y_n(s,t')\right|^2 = |W_{t''} - W_{t'}|^2,$$

for each $s \in [0,1]$. In particular:

$$\operatorname{var}^{(2)}_{q_{n+i}}(Y_n(s,\cdot)) = v_{n+i}, \qquad \text{if } i \geq 1.$$

(iii) $\int_0^1 \operatorname{var}^{(2)}_{q_n}(Y_n(s,\cdot))ds = v_{n+1}$ a.s.

To check (iii) we first assume that q_{n+1} is obtained from q_n by the addition of a single $t^* \in (t_i^n, t_{i+1}^n)$. In this case

$$\operatorname{var}^{(2)}_{q_n}(Y_n(s,\cdot)) - v_{n+1} = \left|\eta_{i+1}(s)(W_{t^*} - W_{t_i^n}) + \eta_{i+2}(s)(W_{t_{i+1}^n} - W_{t^*})\right|^2$$
$$- \left|W_{t^*} - W_{t_i^n}\right|^2 - \left|W_{t_{i+1}^n} - W_{t^*}\right|^2.$$

Integrating over $s \in [0,1]$, using the orthonormality of (η_j) as well as the independence of the variables (η_j) and W, we get (iii). A similar argument applies to the general case.

If $g : \mathbb{R}^k \to \mathbb{R}$ is a bounded Borel function we may write:

$$\begin{aligned} &E(v_n\, g(v_{n+1}, \ldots, v_{n+k})) \\ &\quad = E\left(\operatorname{var}^{(2)}_{q_n}(Y_n(s,\cdot))\, g(\operatorname{var}^{(2)}_{q_{n+1}}(Y_n(s,\cdot)), \ldots, \operatorname{var}^{(2)}_{q_{n+k}}(Y_n(s,\cdot))\right) \\ &\quad = E\left(\operatorname{var}^{(2)}_{q_n}(Y_n(s,\cdot))\, g(v_{n+1}, \ldots, v_{n+k})\right), \end{aligned}$$

for each $s \in [0,1]$, where we have used (i) in the first equality and (ii) in the second. Integrating in s and using (iii) we have:

$$\begin{aligned} &E(v_n\, g(v_{n+1}, \ldots, v_{n+k})) \\ &\quad = E\left(\int_0^1 \operatorname{var}^{(2)}_{q_n}(Y_n(s,\cdot))\, ds\;\, g(v_{n+1}, \ldots, v_{n+k})\right) \\ &\quad = E(v_{n+1}\, g(v_{n+1}, \ldots, v_{n+k})), \end{aligned}$$

proving the claim.

The claim above tells that (v_n) is a reversed martingale (i.e. taking the index set $\mathbb{N}$ with the reversed order and filtration $\tilde{\mathcal{A}}_n := \sigma(v_j : j \geq n)$, for each $n \geq 1$). As a consequence of the classical Doob convergence theorem for martingales (see e.g. [173]) it must converge a.s. as $n \to +\infty$. Due to the previous observation on the L^2 convergence, this concludes the proof. □

Concerning their irregular behaviour, another interesting property of Brownian paths is described by the following classical result whose proof we omit (see e.g. Chapter 1 of [121], or Chapter 2 of [173]).

Theorem 2.16 (Paley, Wiener and Zygmund (1933)) *For almost all ω the path $t \mapsto W_t(\omega)$ is nowhere differentiable.*

2.2 Itô's integral

Brownian motion plays a fundamental role in the construction of an important class of Markov processes in $\mathbb{R}^d$, also called diffusion processes. Loosely speaking, these correspond to those Markov processes with continuous paths which can be described in terms of local characteristics. Stochastic integration $\int_0^t \varphi_s dW_s$ is a basic tool in their description. Nevertheless, as a consequence of Theorem 2.15, almost all Brownian paths have unbounded variation on the interval $[0, t]$ and interpretation of the integral cannot be that of a pathwise Stieltjes integral. In this section we summarize the most basic facts of stochastic integration in the sense of Itô, as they will be needed later on. This is a very brief summary; for more details and discussion we refer to any of the many classical books on the subject. To make it easier for those who want to work out the details, we try to give precise references in terms of [173].

Let us start by fixing a probability space $(\Omega, \mathcal{A}, P)$ on which there is defined a standard one-dimensional Brownian motion (W_t). On this space we consider a filtration $(\mathcal{A}_t)$ so that:

Assumptions 2.17

(i) $\mathcal{A}_0$ contains all P-null sets for $\mathcal{A}_\infty$, where $\mathcal{A}_\infty := \vee_{t \geq 0} \mathcal{A}_t := \sigma(\cup_{t \geq 0} \mathcal{A}_t)$, the smallest σ-field containing.

(ii) $\mathcal{A}_t^W \subseteq \mathcal{A}_t$ for each t.

(iii) $E(W_t - W_s | \mathcal{A}_s) = 0$ a.s., $E((W_t - W_s)^2 | \mathcal{A}_s) = t - s$ a.s. for $0 \leq s \leq t$.

One example of such a filtration is the 'augmentation' of $\mathcal{A}_t^W$, also called the enlarged Brownian filtration, defined as

$$\tilde{\mathcal{A}}_t^W := \mathcal{A}_t^W \vee \mathcal{N} := \sigma(\mathcal{A}_t^W \cup \mathcal{N}), \tag{2.19}$$

where $\mathcal{N}$ is the class of all P-null sets for $\mathcal{A}_\infty^W$ (i.e. $N \in \mathcal{N}$ iff there exists $M \in \mathcal{A}_\infty^W$ with $P(M) = 0$ and $N \subseteq M$).

Another commonly used example is when in (2.19) $\mathcal{A}_t^W$ is replaced by $\mathcal{A}_t^W \vee \mathcal{A}_*$, where $\mathcal{A}_*$ is a sub σ-field of $\mathcal{A}$, independent of $\mathcal{A}_\infty^W$, and accordingly $\mathcal{N}$ is replaced by the class of P-null sets for $\mathcal{A}_\infty^W \vee \mathcal{A}_*$. This is particularly suitable for the introduction of a random starting point ξ_0 with some distribution μ. If the initial process (W_t) starts at 0, and is independent of ξ_0, the new process given by $\xi_0 + W_t$ has initial distribution μ. Its law is given by $\mathbb{P}_\mu(A) := \int_{\mathbb{R}} \mathbb{P}_x(A)\mu(dx)$.

Remark Example (2.19) has the property of right-continuity, i.e. $\tilde{\mathcal{A}}_t^W = \tilde{\mathcal{A}}_{t+}^W = \bigcap_{s>t} \tilde{\mathcal{A}}_s^W$. (Here we need the 'augmentation' or enlarged filtration.) A right-continuous filtration, which also satisfies condition (i) of Assumptions 2.17 is said to satisfy the 'usual conditions'. These are useful in martingale theory and for the development of a more general stochastic calculus, since they allow us to find 'separable' versions of martingales and to obtain the Doob–Meyer decomposition. Since we stay in the context of Brownian motion, we omit the general discussion here (see Section 1.2 of [173]).

We now define the class of 'integrands'. Having fixed a filtration verifying Assumptions 2.17, we simply say that a process is adapted if it is $(\mathcal{A}_t)$-adapted. Let $T > 0$ be fixed, and define

$$\begin{aligned}\mathcal{H}_{0,T} = \{\varphi\colon [0,T]\times\Omega \to \mathbb{R}\,,\ &(\mathcal{B}[0,T]\times\mathcal{A}_\infty)\text{-measurable, adapted,}\\ &\text{and } |||\varphi|||_{0,T}^2 := E\left(\int_0^T |\varphi_s|^2\, ds\right) < +\infty\}.\end{aligned} \tag{2.20}$$

Notation $\varphi_t(\cdot) = \varphi(t,\ \cdot)$. When no confusion arises we use the shorter notation φ_t.

Identifying any two processes φ and ψ if $|||\varphi - \psi|||_{0,T}^2 = 0$, $\mathcal{H}_{0,T}$ can be viewed as a subspace of $\mathrm{L}^2([0,T]\times\Omega, \mathcal{B}[0,T]\times\mathcal{A}_\infty)$ and $|||\cdot|||_{0,T}$ is the restriction of the L^2-norm for the product measure (Lebesgue $\times P$). (In this case 'adaptedness' corresponds to the existence of an adapted process in the equivalence class.)

On $\mathcal{H}_{0,T}$ let us consider the subspace $\mathcal{S}$ of the simple (predictable) processes, of the type

$$\varphi(t,\ \cdot) = \xi_0(\cdot)\mathbf{1}_{[0,t_1]}(t) + \sum_{i=1}^{m-1} \xi_i(\cdot)\mathbf{1}_{(t_i,t_{i+1}]}(t), \tag{2.21}$$

where $0 = t_0 < \cdots < t_m = T$ is a finite (deterministic) partition of $[0,T]$, ξ_i is $\mathcal{A}_{t_i}$-measurable and bounded, $i = 1,\ldots,m$, and $\mathbf{1}_B$ denotes the indicator function of the set B, as before.

For $\varphi \in \mathcal{S}$, we define the process $(\varphi \circ W)_t$, $0 \le t \le T$, by

$$(\varphi \circ W)_t = \sum_{i=0}^{i^*-1} \xi_i(W_{t_{i+1}} - W_{t_i}) + \xi_{i^*}(W_t - W_{t_{i^*}}),$$

where $i^* = i^*(t)$ is given by $t_{i^*} \le t < t_{i^*+1}$. Equivalently,

$$(\varphi \circ W)_t = \sum_{i=0}^{m-1} \xi_i (W_{t \wedge t_{i+1}} - W_{t \wedge t_i}), \tag{2.22}$$

for $0 \le t \le T$. The following properties are checked at once for $\varphi, \tilde{\varphi} \in \mathcal{S}$:

(a) the paths $t \mapsto (\varphi \circ W)_t$ are continuous;

(b) $((c\varphi + \tilde{c}\tilde{\varphi}) \circ W)_t = c(\varphi \circ W)_t + \tilde{c}(\tilde{\varphi} \circ W)_t$ P a.s., for $c, \tilde{c} \in \mathbb{R}$;

(c) $(\varphi \circ W)_t$ is an $(\mathcal{A}_t)$-martingale;

(d) $(\varphi \circ W)_t(\tilde{\varphi} \circ W)_t - \int_0^t \varphi_u \tilde{\varphi}_u \, du$ is an $(\mathcal{A}_t)$-martingale.

In particular

$$E(\varphi \circ W)_T^2 = |||\varphi|||_{0,T}^2 . \tag{2.23}$$

Since we have a continuous martingale, the classical Kolmogorov inequality (easily extended from the discrete time case) yields, if $\varphi, \tilde{\varphi} \in \mathcal{S}$ and $y > 0$,

$$P\left[\sup_{0 \le t \le T} |(\varphi \circ W)_t - (\tilde{\varphi} \circ W)_t| \ge y \right] \le \frac{1}{y^2} |||\varphi - \tilde{\varphi}|||_{0,T}^2 . \tag{2.24}$$

(See e.g. Theorem 3.8, Section 1.3 of [173] applied to $((\varphi - \tilde{\varphi}) \circ W)_t^2$.)

We need the following.

Lemma 2.18 *$\mathcal{S}$ is dense in $\mathcal{H}_{0,T}$ for $|||\cdot|||_{0,T}^2$.*

Postponing the proof of Lemma 2.18, we first see how to use it in order to extend the definition of the stochastic integral to all $\mathcal{H}_{0,T}$ and to check the desired properties. By Lemma 2.18, given $\varphi \in \mathcal{H}_{0,T}$ and $n \ge 1$ we may take $\varphi^{(n)} \in \mathcal{S}$ such that $|||\varphi^{(n)} - \varphi|||_{0,T} \le 1/n^3$ and by (2.24) we have that

$$P\left(\sup_{0 \le t \le T} \left| (\varphi^{(n)} \circ W)_t - (\varphi^{(n+1)} \circ W)_t \right| \ge \frac{1}{n^2} \right) \le n^4 \left(\frac{2}{n^3} \right)^2 = \frac{4}{n^2}. \tag{2.25}$$

Applying the Borel–Cantelli lemma we conclude that if A_n is the event on the left-hand side of (2.25) and $N = [A_n \text{ i.o.}]$, then $P(N) = 0$ so that $N \in \mathcal{A}_0$. On $\Omega \setminus N$ the sequence $(\varphi^{(n)} \circ W)_t(\omega)$ converges uniformly in t. We thus define

$$(\varphi \circ W)_t(\omega) = \begin{cases} \lim_{n \to \infty} (\varphi^{(n)} \circ W)_t(\omega) & \text{if } \omega \notin N, \\ 0 & \text{if } \omega \in N. \end{cases} \tag{2.26}$$

The definition is seen to be well posed. Due to the uniform convergence it follows that $t \mapsto (\varphi \circ W)_t(\omega)$ is continuous, for each ω. Moreover, $(\varphi \circ W)_t$ is $\mathcal{A}_t$-measurable (here we use that the filtration is enlarged) and for each t

$$\lim_{n \to +\infty} (\varphi^{(n)} \circ W)_t = (\varphi \circ W)_t \quad \text{in} \quad \mathrm{L}^2\text{-norm},$$

as follows at once from (2.23). It is then simple to check that the above properties (b), (c) and (d) extend to all $\varphi, \tilde{\varphi} \in \mathcal{H}_{0,T}$. In particular, for each $\varphi \in \mathcal{H}_{0,T}$ and $y > 0$

$$P\left(\sup_{0\le t\le T} |(\varphi \circ W)_t| \ge y\right) \le \frac{1}{y^2} |||\varphi|||^2_{0,T}. \tag{2.27}$$

Proof of Lemma 2.18 (Sketch) The appearance of ds in the definition of $|||\cdot|||^2_{0,T}$ is due to the fact that $(W_t)^2 - t$ is a martingale. This particularity simplifies the proof compared to the case of general continuous martingales. (In the general case, the class of integrands is slightly more restricted, cf. Proposition 2.8, Section 3.2 of [173].) The proof in the present case is adapted from Lemma 4.4, Section 4.2 of [203], omitting some details (see also Problem 2.5, Section 3.2 of [173]).

Let $\varphi \in \mathcal{H}_{0,T}$. We want to prove the existence, for $k \ge 1$, of $\varphi^{(k)} \in \mathcal{S}$ so that

$$\lim_{k\to+\infty} \int_0^T E\left|\varphi_s - \varphi_s^{(k)}\right|^2 ds = 0. \tag{2.28}$$

We first assume that φ is bounded and for convenience we set $\varphi_t = 0$ for any $t \le 0$.

(i) Under the present conditions we first check that

$$\lim_{u\downarrow 0} \int_0^T E\,|\varphi_s - \varphi_{s-u}|^2\, ds = 0. \tag{2.29}$$

Observe that if φ is jointly measurable (not necessarily adapted) with P a.s. continuous trajectories, then (2.29) follows at once from the bounded convergence theorem. On the other hand, given $\varphi \in \mathcal{H}_{0,T}$ bounded, we may take a sequence of jointly measurable processes $\tilde{\varphi}_t^{(n)}$ with a.s. continuous trajectories, such that

$$\lim_{n\to+\infty} \int_0^T E\left|\varphi_s - \tilde{\varphi}_s^{(n)}\right|^2 ds = 0. \tag{2.30}$$

For this we may take $\tilde{\varphi}_t^{(n)} := n \int_{t-1/n}^t \varphi_s ds$ (recall that $\varphi_t = 0$ if $t \le 0$). Given $\delta > 0$, we fix n so that $|||\tilde{\varphi}_t^{(n)} - \varphi_t|||_{0,T} \le \delta$ and use the triangle inequality for $|||\cdot|||_{0,T}$ (Minkowski inequality) to conclude that $\limsup_{u\downarrow 0} \int_0^T E\,|\varphi_s - \varphi_{s-u}|^2$ $ds \le 2\delta$.

(If we could guarantee the above $\tilde{\varphi}^{(n)}$ also to be adapted, the approximation of φ by simple processes would follow at once. This requires more on the process φ and it is part of the general procedure, beyond the Brownian case, see Section 3.2 of [173].)

(ii) We now consider the processes $\varphi_s^{(n,r)} := \varphi_{\alpha_n(s-r)+r}$, where

$$\alpha_n(s) = \sum_{j\in\mathbb{Z}} \frac{j}{2^n} \mathbf{1}_{(\frac{j}{2^n}, \frac{j+1}{2^n}]}(s),\ s \in \mathbb{R}.$$

For each $r \geq 0$ and $n \geq 1$, $\varphi^{(n,r)} \in \mathcal{S}$ (for this observe that $s - 2^{-n} \leq \alpha_n(s - r) + r < s$), and one can verify that

$$\lim_{n\to+\infty} \int_0^T \left(\int_0^1 E \left| \varphi_s - \varphi_s^{(n,r)} \right|^2 dr \right) ds = 0. \tag{2.31}$$

Inverting the order of integration in dr and ds in (2.31), we conclude the existence of $r \geq 0$ and of a sequence of integers $n_k \uparrow +\infty$ so that (2.28) holds for $\varphi^{(k)} = \varphi^{(n_k,r)}$.

To lift the restriction of boundedness, let $\varphi_t^{(n)}(\omega) := \varphi_t(\omega)\mathbf{1}_{\{|\varphi_t(\omega)|\leq n\}}$. These are bounded, belong to $\mathcal{H}_{0,T}$, and applying the dominated convergence theorem we get

$$\lim_{n\to+\infty} \left|\left|\left| \varphi_t^{(n)} - \varphi_t \right|\right|\right|_{0,T}^2 = \lim_{n\to+\infty} E\left(\int_0^T (\varphi_t)^2 \mathbf{1}_{\{|\varphi_t(\omega)|>n\}} \right) = 0,$$

which completes the proof. □

Definition 2.19 Given $T > 0$ and $\varphi \in \mathcal{H}_{0,T}$, the process $(\varphi \circ W_t : 0 \leq t \leq T)$ is called Ito's stochastic integral of φ. The notation $\int_0^t \varphi_s dW_s$ is also used.

Remark 2.20 (i) *Extension to the time interval* $[0, +\infty)$. In the previous set-up, if (φ_t) is jointly measurable, adapted, with $|||\varphi|||_{0,T}^2 < +\infty$ for each $T > 0$, and we set $\varphi_t^{(n)} = \varphi_t \mathbf{1}_{\{t\leq n\}}$, we see that $P\{(\varphi^{(n)} \circ W)_t = (\varphi^{(n+1)} \circ W)_t$ for all $0 \leq t \leq n\} = 1$, for each n. This consistency property allows us to define the stochastic integral $(\varphi \circ W)_t$ for all $t > 0$ in the obvious way.

(ii) *Localization*. If $\varphi \in \mathcal{H}_{0,T}$ and $\tau \leq \tau' \leq T$ are stopping times for the filtration $(\mathcal{A}_t)$,

$$E((\varphi \circ W)_{t\wedge\tau'} | \mathcal{A}_\tau) = (\varphi \circ W)_{t\wedge\tau} \quad P \text{ a.s.}$$

We have:

$$(\varphi \circ W)_{t\wedge\tau} = (\varphi^{(\tau)} \circ W)_t \quad P \text{ a.s.}$$

where $\varphi_t^{(\tau)} := \varphi_t \mathbf{1}_{\{t\leq\tau\}}$ (for a proof see Proposition 2.10, Section 3.2 in [173]).

The extension to unbounded stopping times can be done according to the previous observation (i), provided $\int_0^T E \left| \varphi_s \mathbf{1}_{\{s\leq\tau\}} \right|^2 ds < +\infty$ for each $T < +\infty$.

This procedure, called localization, is essential to eliminate assumptions on square integrability, i.e. to go beyond the $\mathcal{H}_{0,T}$ space. A more natural assumption should be $\int_0^t (\varphi_s)^2 ds < +\infty$ a.s. for each $t > 0$; the extension can be done and it provides a 'local-martingale' (see Part D, Section 3.2 in [173]).

Notation $\int_u^v \varphi_s dW_s = (\varphi \circ W)_v - (\varphi \circ W)_u$, for $0 \leq u < v < +\infty$.

The extension of stochastic integration to higher dimensions is straightforward: let $W_t = (W_t^{(1)}, \ldots, W_t^{(\ell)})$, $t \geq 0$ be an ℓ-dimensional standard Brownian motion, and let $(\mathcal{A}_t)$ be a filtration satisfying Assumptions 2.17, with (iii) modified to (iii)′ $E(W_t^{(i)} - W_s^{(i)} | \mathcal{A}_s) = 0$ a.s., $E(W_t^{(i)} W_t^{(j)} - W_s^{(i)} W_s^{(j)}) | \mathcal{A}_s) = (t-s)\,\delta_{i,j}$ a.s., for $0 \le s \le t$, and each i, j, where $\delta_{i,j}$ is Kronecker's delta.

(As before, the enlarged Brownian filtration $(\tilde{\mathcal{A}}_t^W)$, cf. (2.19), satisfies the required conditions.)

We consider matrix valued processes

$$\varphi(s, \omega) = (\varphi_{i,j}(s, \omega))_{\substack{1 \le i \le d \\ 1 \le j \le \ell}}$$

where $\ell \ge 1$, and each $\varphi_{i,j} \in \mathcal{H}_{0,T}$. Then

$$(\varphi \circ W)_t = \int_0^t \varphi(s) dW_s$$

is the d-dimensional process, whose ith coordinate is

$$(\varphi \circ W)_t^{(i)} = \sum_{j=1}^{\ell} \int_0^t \varphi_{i,j}(s) dW_s^{(j)}.$$

For ψ, φ as above, and $i, k \in \{1, \ldots, d\}$ we have:

(c)′ $((\varphi \circ W)_t^{(i)} : 0 \le t \le T)$ is a continuous martingale;

(d)′ $(\varphi \circ W)_t^{(i)} (\psi \circ W)_t^{(k)} - \int_0^t \sum_{j=1}^{\ell} \varphi_{i,j}(s) \psi_{k,j}(s)\, ds,\ 0 \le t \le T$ is a martingale.

Notation If $d = 1$ and $\varphi(s) = (\varphi^1(s), \ldots, \varphi^\ell(s))$ we write

$$\int_0^t \langle \varphi(s), dW_s \rangle = \sum_{i=1}^{\ell} \int_0^t \varphi^i(s) dW_s^{(i)}. \tag{2.32}$$

It would be more proper to write W as a columm vector, but we refrain from that.

Itô's formula

Since the total variation of almost all Brownian paths is infinite, we get a different rule for change of variables for Itô stochastic integrals. This is summarized below.

An $\mathbb{R}^d$-valued stochastic process $(X_t : 0 \le t \le T)$ is called an Itô process if it is given by

$$X_t = X_0 + \int_0^t \varphi(s) dW_s + \int_0^t \psi(s)\, ds, \tag{2.33}$$

which we usually write as

$$dX_t = \varphi(t) dW_t + \psi(t) dt, \tag{2.34}$$

where X_0 is $\mathbb{R}^d$-valued and $\mathcal{A}_0$-measurable, (W_t) is an ℓ-dimensional standard Brownian motion, φ is a matrix-valued stochastic process and the stochastic integral has been defined above. Moreover, $\psi = (\psi^1, \ldots, \psi^d)$, where each coordinate ψ^i is a measurable and adapted process, whose trajectories are integrable on $[0, T]$.

Let $f\colon [0, T] \times \mathbb{R}^d \to \mathbb{R}$ be a bounded function for which the derivatives $(\partial f/\partial t)(t, x)$, $(\partial^2 f/\partial x_i \partial x_j)(t, x)$ $(1 \le i, j \le d)$ exist, are continuous and bounded. Then

$$\begin{aligned} df(t, X_t) = &\sum_{j=1}^{\ell}\sum_{i=1}^{d} \frac{\partial f}{\partial x_i}(t, X_t)\varphi_{i,j}(t) dW_t^{(j)} \\ &+ \left\{ \frac{\partial f}{\partial t}(t, X_t) + \sum_{i=1}^{d} \psi^i(t)\frac{\partial f}{\partial x_i}(t, X_t) \right. \\ &\left. + \frac{1}{2}\sum_{j=1}^{\ell}\sum_{i,k=1}^{d} \varphi_{i,j}(t)\varphi_{k,j}(t)\frac{\partial^2 f}{\partial x_i \partial x_k}(t, X_t) \right\} dt. \end{aligned} \tag{2.35}$$

We omit the proof of (2.35), but in order to see why we expect this formula to hold, just take $\ell = d = 1$ and $X_t = W_t$ a standard one-dimensional Brownian motion. We write down $f(W_t) - f(W_0) = \sum_{k=0}^{2^n-1}(f(W_{(k+1)t/2^n}) - f(W_{kt/2^n}))$, expand the telescopic sum using Taylor expansion, and recall the behaviour of the quadratic variation, described by Theorem 2.15. Taking proper account of the limits we get (2.35). For a proof see e.g. Theorems 3.3 and 3.6, Section 3.3 of [173].

Remark The assumption on the boundedness of f and of its derivatives may be relaxed provided one has extended the stochastic integration beyond $\mathcal{H}_{0,T}$, as mentioned in Remark 2.20(ii).

Examples Applying Itô's formula to $X_t = W_t$, the standard one-dimensional Brownian motion starting at 0 and $f(x) = x^2$ we have (compare with Example (1) after Definition 2.10)

$$(W_t)^2 = 2\int_0^t W_s dW_s + t.$$

Comment The development of stochastic integration beyond Brownian motion was proven to be very convenient, as soon as the theory was established in the 1960s. In the context of square integrable martingales this was settled with the article of Kunita and Watanabe [186], where a proper Hilbertian structure in the space of such martingales provided an orthogonal decomposition, allowing a square integrable martingale to be written as a sum of two (orthogonal) terms: one which could be treated with Stieltjes integral and a square integrable continuous martingale.

For M a square integrable continuous martingale, the delicate point in the extension of the stochastic integral to $\int_0^t \varphi_s dM_s$ is the behaviour of its quadratic variation; all follows as in the Brownian case if we take for granted the existence of a continuous increasing process $\langle M\rangle_t$, with $E(\langle M\rangle_T) < +\infty$ and such that $(M_t)^2 - \langle M\rangle_t$ is a martingale. The process $\langle M\rangle_t$ is associated with the limiting quadratic variation in a sense that can be made precise and this result constitutes the classical Doob–Meyer decomposition of $(M_t)^2$ (see e.g. Sections 1.4 and 1.5 in [173]). In particular, when $t \mapsto \langle M \rangle_t$ is absolutely continuous, if we define the space $\mathcal{H}_{0,T}$ as in (2.20), replacing the ds there by $d\langle M \rangle_s$ with the proper modification in $|||\cdot|||_{0,T}$, then the density of the subspace $\mathcal{S}$ is checked in the same way, and the same procedure for defining the integral can be followed. The extension to multidimensional processes is straightforward: if $M = (M^{(1)}, \ldots, M^{(\ell)})$ one sets $\langle M^{(j)}, M^{(j')} \rangle_t = 4^{-1}[\langle M^{(j)} + M^{(j')} \rangle_t - \langle M^{(j)} - M^{(j')} \rangle_t]$, so that $M_t^{(j)} M_t^{(j')} - \langle M^{(j)}, M^{(j')} \rangle_t$ is a martingale. The change of variables formula, applied as before to a process given by an expression like (2.33) where $W = (W^{(1)}, \ldots, W^{(\ell)})$ is replaced by $M = (M^{(1)}, \ldots, M^{(\ell)})$, is obtained as in (2.35) by replacing $dW^{(j)}$ by $dM^{(j)}$ and the term with second derivatives

$$\frac{1}{2} \sum_{j=1}^{\ell} \sum_{i,k=1}^{d} \varphi_{i,j}(t) \varphi_{k,j}(t) \frac{\partial^2 f}{\partial x_i \partial x_k}(t, X_t)\, dt$$

by

$$\frac{1}{2} \sum_{j,j'=1}^{\ell} \sum_{i,k=1}^{d} \varphi_{i,j}(t) \varphi_{k,j'}(t) \frac{\partial^2 f}{\partial x_i \partial x_k}(t, X_t) d\langle M^{(j)}, M^{(j')} \rangle_t.$$

This extension can be readily applied to give a proof of the following characterization of Brownian motion, initially proven by Lévy (see [197]), and which we shall need later.

Using the notation of Proposition 2.11 one has the following.

Theorem 2.21 (Lévy) *The law of a standard d-dimensional Brownian motion starting at x is the unique probability measure* $\mathbb{P}_x$ *on* $C([0, +\infty), \mathbb{R}^d)$ *(with its Borel* σ*-field* $\mathcal{B}_\infty$*) under which:*

(i) $\pi_t^{(i)}$ *and* $\pi_t^{(i)} \pi_t^{(j)} - \delta_{i,j} t$ *are martingales with respect to* $(\mathcal{B}_t^0)$ *for each* $1 \le i, j \le d$, *where* $\delta_{i,j}$ *is the Kronecker delta;*

(ii) $\mathbb{P}_x(\pi_0 = x) = 1$.

Proof (Sketch) The validity of conditions (i) and (ii) for the standard d-dimensional Brownian motion follows at once from Example (1) following Definition 2.10, and the independence of coordinates. The main point is the converse. According to Remark 2.9 it suffices to prove that if $\mathbb{P}_x$ verifies conditions (i) and (ii), then for $i = \sqrt{-1}$, any $s < t$, and any $\zeta \in \mathbb{R}^d$:

$$\mathbb{E}_x(\exp \langle i\zeta, \pi_t - \pi_s \rangle \mid \mathcal{B}_s^0) = \exp(-|\zeta|^2 (t-s)/2) \quad \mathbb{P}_x \text{ a.s.},$$

where $\mathbb{E}_x$ denotes the expectation with respect to the probability measure $\mathbb{P}_x$. Considering the martingales $(\pi_t^{(j)})$ and the stochastic integration as previously discussed, we apply the change of variables formula, which due to (i) and (ii) is just as in (2.35). We apply it to the real and imaginary parts of the function $f(u) = \exp i\langle\zeta, u\rangle$. Recalling the martingale property of the stochastic integrals, and performing some simplifications, for $A \in \mathcal{B}_s^0$, $t > s$, $\zeta \in \mathbb{R}^d$, we get

$$\mathbb{E}_x(e^{i\langle\zeta,\pi_t-\pi_s\rangle}\mathbf{1}_A) = \mathbb{P}_x(A) - \frac{1}{2}|\zeta|^2 \int_s^t \mathbb{E}_x(e^{i\langle\zeta,\pi_v-\pi_s\rangle}\mathbf{1}_A)\,dv. \tag{2.36}$$

For s and A fixed the unique solution to (2.36) on $t \geq s$ is $\mathbb{P}_x(A)e^{-\frac{1}{2}|\zeta|^2(t-s)}$, which proves the statement (for full details see Theorem 3.16, Section 3.3 in [173].) □

Remark 2.22 The previous proof shows that if (M_t) is a continuous martingale and $(M_t)^2 - t$ is a martingale as well, then $M_t - M_0$ is a standard Brownian motion starting at 0. The assumption of continuity cannot be eliminated, as we see if $M_t = N_t - t$ where (N_t) is a unit rate Poisson process.

It follows that if W and $(\mathcal{A}_t)$ satisfy Assumptions 2.17 and W is continuous, then $(W_{t+s} - W_t : s \geq 0)$ is a Brownian motion independent of $\mathcal{A}_t$, for each t (analogously for the higher dimensional situation).

Girsanov theorem

In Chapter 1 we noticed that a useful tool in establishing large deviation principles involves suitable changes between equivalent measures and control of the Radon–Nikodym derivatives. For the present situation this is closely related to the following analysis.

Let (W_t) be a standard d-dimensional Brownian motion, with a filtration $(\mathcal{A}_t)$ verifying Assumptions 2.17. Let $T > 0$ be given and let $\varphi = (\varphi^{(1)}, \ldots, \varphi^{(d)})$ be a process with each coordinate $\varphi^{(i)} \in \mathcal{H}_{0,T}$. Thus we may define, for $0 \leq t \leq T$:

$$Z_t^\varphi = \exp\left\{\int_0^t \langle\varphi_s, dW_s\rangle - \frac{1}{2}\int_0^t |\varphi_s|^2\ ds\right\}. \tag{2.37}$$

If we apply (2.35) formally with $f(t,x) = e^x$ we get that Z_t^φ verifies the equation

$$Z_t^\varphi = 1 + \int_0^t \langle Z_s^\varphi \varphi_s, dW_s\rangle. \tag{2.38}$$

(We said 'formally' because of the boundedness restrictions under which we stated (2.35). As we have pointed out before, the use of stopping times allows us to go beyond $\mathcal{H}_{0,T}$ relaxing such boundedness assumptions; in this case the stochastic integrals yield only 'local martingales', see e.g. [173].)

We shall be particularly interested in situations when (Z_t^φ) is really a martingale (e.g. if φ is bounded). Since $Z_t^\varphi \geq 0$ and $Z_0^\varphi = 1$, we then have $EZ_T^\varphi = 1$ for each T, and the equation

$$\tilde{P}_T(A) = \int_A Z_T^\varphi \, dP \tag{2.39}$$

for $A \in \mathcal{A}_T$, defines a new probability on $(\Omega, \mathcal{A}_T)$. (If $T \le T'$, $\tilde{P}_T$ coincides with the restriction of $\tilde{P}_{T'}$ to $\mathcal{A}_T$, i.e. we have a consistent family.)

At this point we need to recall the following basic and important result.

Theorem 2.23 (Cameron-Martin (1944), Girsanov (1960)) *Let us assume that* $(Z_t^\varphi : 0 \le t \le T)$, *defined through equation* (2.37), *is a martingale. Define* $\tilde{P}_T$ *by* (2.39), *and set*

$$\tilde{W}_t^{(i)} = W_t^{(i)} - \int_0^t \varphi_s^{(i)} \, ds \tag{2.40}$$

$i = 1, \dots, d$. *Under the new measure* $\tilde{P}_T$, *the process* $(\tilde{W}_t : 0 \le t \le T)$ *is a standard d-dimensional Brownian motion.*

The Girsanov theorem is very important in a large range of problems. The proof, due to Girsanov, is based on the martingale characterization of Brownian motion (Proposition 2.11), but it will not be discussed here. We refer to Section 3.5 in [173].

We shall only prove the very particular and simpler case of deterministic φ, which is a classical result due to Cameron-Martin. This particular situation is exactly what we need in the next section. It tells us that if $\int_0^T |\varphi_s|^2 \, ds < +\infty$ then $\tilde{\mathbb{P}}_0$, the law of $(W_t + \int_0^t \varphi_s ds)$, is absolutely continuous with respect to $\mathbb{P}_0$, the law of (W_t) on $C([0,T], \mathbb{R}^d)$ and $d\tilde{\mathbb{P}}_0/d\mathbb{P}_0 = \exp\{\int_0^T \langle \varphi_s, d\pi_s \rangle - 1/2 \int_0^T |\varphi_s|^2 \, ds\}$ on $\mathcal{B}_T^0$. To understand 'why' we should get the above expression, consider two finite dimensional Gaussian distributions with identity covariance matrix and compute the Radon–Nikodym derivative of one with respect to the other.

Cameron-Martin formula If $\varphi \colon [0,T] \to \mathbb{R}^d$ is measurable and $\int_0^T |\varphi(u)|^2 \, du < +\infty$, then Z_t^φ is a martingale and the statement in the Girsanov theorem is true in this case.

Proof of the Cameron-Martin formula To simplify let us take $\mathcal{A}_t = \tilde{\mathcal{A}}_t^W$, the enlarged Brownian filtration cf. (2.19). The same proof applies under Assumptions 2.17 (see Remark 2.22).

Since φ is deterministic and $(W_{t+u} - W_t : u \ge 0)$ is independent of $\mathcal{A}_t$, we easily check that if $0 \le s < t \le T$ then $(\varphi \circ W)_t - (\varphi \circ W)_s$ is independent of $\mathcal{A}_s$ and its distribution is Gaussian with zero average and variance $\int_s^t |\varphi(u)|^2 \, du$ (first take φ simple, then pass to the limit). Thus:

$$E\left(e^{\{(\varphi \circ W)_t - (\varphi \circ W)_s\}} \mid \mathcal{A}_s\right) = e^{\frac{1}{2}\int_s^t |\varphi(u)|^2 du},$$

from which it follows at once that (Z_t^φ) is a martingale with respect to $(\mathcal{A}_t)$. We may then define $\tilde{P}_T$ on $\mathcal{A}_T$ through (2.39). Using Remark 2.9 or Proposition 2.11

it suffices to show that for each $\zeta \in \mathbb{R}^d$, $0 \le s < t \le T$, and $A \in \mathcal{A}_s$:

$$\int_A \exp\left(\langle \zeta, \tilde{W}_t - \tilde{W}_s \rangle\right) d\tilde{P} = \exp\left(\frac{|\zeta|^2}{2}(t-s)\right) \tilde{P}(A). \tag{2.41}$$

Using (2.39) and (2.40) we write the left-hand side of (2.41) as

$$\exp\left(-\int_s^t \langle \zeta, \varphi(u) \rangle du - \frac{1}{2}\int_s^t |\varphi(u)|^2 du\right) \int_A \exp(M_t^\zeta - M_s^\zeta) Z_s^\varphi \, dP,$$

where $M_t^\zeta = \int_0^t \langle \zeta + \varphi(u), dW_u \rangle$. As just observed, under the measure P, $M_t^\zeta - M_s^\zeta$ is independent of $\mathcal{A}_s$ (and so of $Z_s^\varphi \mathbf{1}_A$), and has Gaussian distribution with zero average and variance $\int_s^t |\zeta + \varphi(u)|^2 \, du$. From this, standard computations give (2.41), proving the Cameron-Martin formula. □

Remark If $\varphi \in \mathcal{H}_{0,T}$ the process Z^φ is a local martingale and a supermartingale. It is a martingale if and only if $EZ_t^\varphi = 1$ for each t positive. An important sufficient condition for this to happen is due to Novikov: under the previous usual conditions on the filtration $(\mathcal{A}_t)$, if the process φ is measurable, $(\mathcal{A}_t)$-adapted and

$$E(e^{\frac{1}{2}\int_0^T |\varphi(u)|^2 \, du}) < +\infty, \tag{2.42}$$

then $(Z_t^\varphi : 0 \le t \le T)$ is a martingale with respect to $(\mathcal{A}_t)$ (see e.g. Proposition 5.12, Chapter 3 of [173]).

2.3 Itô's equations

For the discussion of Freidlin and Wentzell theory in the next sections and its application to metastability (Chapter 5) it is convenient to recall a few basic facts on stochastic differential equations of the type

$$\begin{aligned} dX_t^x &= b(X_t^x)dt + \sigma(X_t^x)dW_t, \\ X_0^x &= x, \end{aligned} \tag{2.43}$$

where W is an ℓ-dimensional standard Brownian motion on a given probability space and the filtration $(\mathcal{A}_t)$ is fixed as in Section 2.2 (Assumptions 2.17).

In (2.43), $b(x) = (b^i(x))_{1 \le i \le d}$ is a vector field on $\mathbb{R}^d$, $\sigma(\cdot)$ takes values on the space of $d \times \ell$ matrices with real entries $\sigma(x) = (\sigma_{i,j}(x))$ and the precise meaning of equation (2.43) is, with probability one

$$X_t^x = x + \int_0^t b(X_s^x)\, ds + \int_0^t \sigma(X_s^x)\, dW_s, \tag{2.44}$$

for each $t \ge 0$, with X^x being $(\mathcal{A}_t)$-adapted and the integrals meaningful, with the stochastic integral taken in Itô's sense. This is usually called a 'strong solution'.

The coefficients $b^i(\cdot)$ and $\sigma_{i,j}(\cdot)$ are assumed to satisfy a global Lipschitz condition: there exists $\kappa < +\infty$ such that

$$|b(x) - b(y)| + |\sigma(x) - \sigma(y)| \le \kappa\, |x - y|\,, \tag{2.45}$$

for all $x, y \in \mathbb{R}^d$. (For any $k \ge 1$ and $x \in \mathbb{R}^k$ we use $|x|$ to denote the Euclidean norm and $|\sigma| = (\sum_{i,j} \sigma_{i,j}^2)^{1/2}$, if $\sigma = (\sigma_{i,j})$ is a matrix.)

Under condition (2.45) it is not hard to prove existence and pathwise uniqueness of the strong solution of (2.43). More precisely, the following is well known.

Theorem 2.24 (Itô) *Let $b(\cdot)$ and $\sigma(\cdot)$ satisfy condition* (2.45), *and let W be a standard ℓ-dimensional Brownian motion on a given probability space $(\Omega, \mathcal{A}, P)$, with $(\tilde{\mathcal{A}}_t^W)$ its natural enlarged filtration* (2.19). *Then, for each $x \in \mathbb{R}^d$, there exists an $\mathbb{R}^d$-valued process (X_t^x), with continuous trajectories, $(\tilde{\mathcal{A}}_t^W)$-adapted and such that with probability one* (2.44) *holds. If X^x, $\tilde{X}^x$ are two such (strong) solutions, then*

$$P\{\omega : X_t^x(\omega) = \tilde{X}_t^x(\omega),\ \forall t\} = 1.$$

Moreover, for each $t > 0$ there exists a constant $C < +\infty$ such that

$$E|X_s^x|^2 \le C(1 + |x|^2)e^{Cs}, \qquad \text{for each } 0 \le s \le t. \tag{2.46}$$

If ξ_0 is an $\mathbb{R}^d$-valued random variable, independent of $\mathcal{A}_\infty^W$, and $E|\xi_0|^2 < +\infty$, the initial condition x in (2.44) *may be replaced by ξ_0, provided the filtration $(\tilde{\mathcal{A}}_t^W)$ is replaced by the larger $(\mathcal{A}_t)$ verifying Assumptions* 2.17. *The previous statements continue to hold, with $|x|^2$ replaced by $E|\xi_0|^2$ in* (2.46).

Sketch of the proof (Uniqueness) Consider two solutions defined in terms of the same Brownian motion (W_t). Applying the classical Gronwall inequality to their difference the uniqueness follows very easily. (This argument works under local Lipschitz conditions, i.e. if for each n there exists a constant κ_n so that (2.45) holds for $\kappa = \kappa_n$ if $|x| \le n$, $|y| \le n$.)

(Existence) Under condition (2.45) this can be achieved following the Picard iteration method. Consider successive approximations given by $X_{1,t}^x = x$, and for $k \ge 1$:

$$X_{k+1,t}^x = x + \int_0^t b(X_{k,s}^x)ds + \int_0^t \sigma(X_{k,s}^x)dW_s.$$

Each of these processes is clearly continuous and $(\tilde{\mathcal{A}}_t^W)$-adapted. According to (2.45) we can take $\tilde{\kappa} < +\infty$ such that

$$|\sigma(x)|^2 \vee |b(x)|^2 \le \tilde{\kappa}(1 + |x|^2)$$

for all x. Using this linear growth condition one checks that (2.46) holds for all the processes X_k^x, with the same constant C.

Since $X_{k+1,t}^x - X_{k,t}^x = \int_0^t (b(X_{k+1,s}^x) - b(X_{k,s}^x))ds + \int_0^t (\sigma(X_{k+1,s}^x) - \sigma(X_{k,s}^x))dW_s$, we see that (2.45) and the validity of (2.46) for the approximating processes imply that the stochastic integral is indeed a square-integrable martingale. Using martingale inequalities, (2.45), and the Borel–Cantelli lemma, one can prove the uniform convergence of the sequence $X^x_{k,t}$ in $[0, T]$, with probability one. The validity of (2.46) follows from Fatou's lemma (see Theorem 2.5 and Theorem 2.9, Chapter 5 of [173] for full details). □

Remark (i) From the independence of $(W_{t+s} - W_t : s \geq 0)$ and $(\mathcal{A}_t)$ it is easy to see that the processes X^x are Markovian with respect to the fixed filtration. The strong Markov property is also verified and these processes constitute the basic examples of 'diffusion processes'. (Itô's integral was developed exactly for the construction of diffusions; see the given references for the proof as well as for historical comments.) The strong Markov property will be essential for the analysis in Section 2.6, as well as in Chapter 5.

(ii) Markov processes may be constructed in a more analytical way, based on semigroup theory and the Hille–Yosida theorem, starting from the infinitesimal generator. Such constructions indeed preceded that based on Itô's equations.

The infinitesimal generator L associated with the semigroup corresponding to the solutions of (2.43), when acting on smooth functions f, is given by:

$$Lf(y) = \frac{1}{2} \sum_{i,j} a_{i,j}(y) \frac{\partial^2}{\partial y_i \partial y_j} f(y) + \sum_i b^i(y) \frac{\partial}{\partial y_i} f(y), \qquad (2.47)$$

where $a(x) = \sigma(x) \cdot \sigma^{\mathbb{T}}(x)$, with $\sigma^{\mathbb{T}}$ denoting the transpose matrix, i.e.

$$a_{i,j}(x) = \sum_{k=1}^{\ell} \sigma_{i,k}(x) \sigma_{j,k}(x), \quad i, j = 1, \ldots, d.$$

If $(S_t : t \geq 0)$ is a Markovian semigroup associated with a generator L, i.e. $S_0 f = f$, $S_{t+s} f = S_t S_s f$, for f bounded and continuous, and it satisfies the semigroup equation $(\partial/\partial t) S_t f(x) = S_t L f(x)$ (f in the domain of L), the correspondence with a (homogeneous) Markov process is through the relation $Ef(X^x_{s+t} \mid X^x_u : u \leq s) = S_t f(x)$ for bounded continuous functions f. This would determine the finite dimensional distributions. The construction of a process with continuous paths may be obtained by appealing to the Kolmogorov extension theorem together with some criteria for the path continuity. This method (forward) usually requires stronger conditions than (2.45); another classical analytical alternative to obtain a semigroup is the so-called backwards equation, where one considers the adjoint operator L^* acting on measures, but one still needs stronger regularity than (2.45). The main point of Itô's equation is to provide directly a probabilistic construction of a continuous process with the given local characteristics, in the sense that

$$Lf(X^x_t) = \lim_{h \downarrow 0} E\left(\frac{f(X^x_{t+h}) - f(X^x_t)}{h} \mid X^x_u : u \leq t \right) \quad \text{a.s.}, \qquad (2.48)$$

for a large class of smooth functions f. Assuming that we can take $f_i(x) = x_i$, $f_{i,j} = x_i x_j$ (though unbounded), for $x = (x_1, \dots, x_d)$, we conclude the interpretation of the coefficients $b^i(x)$ and $a_{i,j}(x)$. Due to this, the functions $a_{i,j}(\cdot)$ are usually called diffusion coefficients and $a(\cdot)$ the diffusion matrix; the function $b(\cdot)$ is called the drift coefficient, extending the constant case. Lévy's proposal, involving (2.48) as probabilistic differentials obtained a precise formulation through Itô's method. For strong solutions this is done on the same probability space of a basic standard Brownian motion and adapted to $\tilde{\mathcal{A}}_t^W$.

(iii) Another concept of existence and uniqueness of the solution for equation (2.44) is through the law it induces on $C([0, T], \mathbb{R}^d)$. This is called the weak solution and it is particularly important because it allows us to go much beyond condition (2.45). The main difference is that one does not fix the probability space or the Brownian process (W_t); they are part of the solution and one looks for uniqueness in law. This was developed by Watanabe and Yamada (see [164]). We shall not discuss this aspect, since (most of) our examples will fit into the previous case (see Section 5.3 of [173]).

Closely related to the concept of weak solution is the construction of the law of a diffusion process as the unique solution of a martingale problem in $C([0, +\infty), \mathbb{R}^d)$. This was formulated precisely by Stroock and Varadhan (see [285]) and generalizes Lévy's characterization of the law of Brownian motion via martingale relations, as discussed in Section 2.1; in some sense, this is a probabilistic way to see the semigroup equation since it determines that $f(\pi_t) - \int_0^t Lf(\pi_s)ds$ is martingale under the given law, for a sufficiently rich class of test functions f and where (π_t) is the canonical process on $C([0, +\infty), \mathbb{R}^d)$. The method allows the construction of diffusion processes under more relaxed regularity conditions on the coefficients, provided $a(\cdot)$ is uniformly elliptic, i.e. if there exists $\alpha > 0$ so that

$$\langle a(x)u, u \rangle \geq \alpha |u|^2 \text{ for all } x, u \in \mathbb{R}^d \tag{2.49}$$

(see Chapter 5 of [173]). The relaxation of the ellipticity condition on $a(\cdot)$ is a very delicate point. (See [252] for recent research.)

2.4 Basic Freidlin and Wentzell estimates

The simplest situation corresponds to diffusions on $\mathbb{R}^d$, defined by the equation:

$$X_t^{x,\varepsilon} = x + \int_0^t b(X_s^{x,\varepsilon})\, ds + \varepsilon W_t, \quad t > 0, \tag{2.50}$$

where $x \in \mathbb{R}^d$ denotes the initial position, (W_t) is a standard d-dimensional Brownian motion starting at the origin, and $b\colon \mathbb{R}^d \to \mathbb{R}^d$ satisfies a global Lipschitz condition:

$$|b(x) - b(y)| \leq \kappa|x - y|, \quad \text{for all } x, y \in \mathbb{R}^d, \tag{2.51}$$

for some constant $\kappa \in [0, +\infty)$. Under such conditions (2.50) has a unique strong solution, as stated in Theorem 2.24: this solution has continuous trajectories and is adapted to the enlarged Brownian filtration $(\tilde{\mathcal{A}}_t^W)$, defined by (2.19); with such properties the solution is a.s. unique.

By $(X_t^{x,0})$ we denote the unique solution of the deterministic equation

$$\begin{aligned} \dot{\varphi}(t) &= b(\varphi(t)), \quad t \geq 0, \\ \varphi(0) &= x, \end{aligned} \tag{2.52}$$

and for $\varepsilon > 0$ small we may think of $X^{x,\varepsilon}$ as a small perturbation of $X^{x,0}$ in the sense that if $T \in (0, +\infty)$ is fixed and $\varrho_T(\varphi, \psi) := \sup_{0\leq t\leq T} |\varphi(t) - \psi(t)|$ is the uniform metric on $C([0, T], \mathbb{R}^d)$, then for any $\delta > 0$:

$$\lim_{\varepsilon\downarrow 0} \sup_x P(\varrho_T(X^{x,\varepsilon}, X^{x,0}) > \delta) = 0.$$

Indeed, from equations (2.50), (2.51) and (2.52) we get

$$|X_t^{x,\varepsilon} - X_t^{x,0}| \leq \kappa \int_0^t |X_s^{x,\varepsilon} - X_s^{x,0}|\, ds + \varepsilon W_t$$

and applying Gronwall's inequality we have:

$$\varrho_T(X^{x,\varepsilon}, X^{x,0}) \leq \varepsilon e^{\kappa T} \sup_{0\leq t\leq T} |W_t|.$$

Since $|W_t| \leq \sqrt{d} \max_{1\leq i\leq d} |W_t^{(i)}|$ we may use Theorem 2.14 to write

$$P(\varrho_T(X^{x,\varepsilon}, X^{x,0}) > \delta) \leq 4d \int_\eta^{+\infty} \frac{e^{-u^2/2}}{\sqrt{2\pi}}\, du,$$

where $\eta = e^{-\kappa T}\delta/\varepsilon\sqrt{Td}$. Recalling the standard inequality (1.79) we have the existence of $\tilde{C}_1, \tilde{C}_2 \in (0, +\infty)$, depending on d, T and κ, so that

$$\sup_x P(\varrho_T(X^{x,\varepsilon}, X^{x,0}) > \delta) \leq C_1 e^{-C_2/\varepsilon^2}, \tag{2.53}$$

for all $0 < \varepsilon \leq 1$, where $C_1 = \tilde{C}_1/\delta$ and $C_2 = \tilde{C}_2\delta^2$. (See also (1.85).)

This brings us to the situation described in Section 1.2: denoting by P_x^ε the law of $X^{x,\varepsilon}$, then for each $T > 0$, P_x^ε converge weakly on $C([0, T], \mathbb{R}^d)$ to the Dirac point-mass on the deterministic trajectory $X^{x,0}$, as $\varepsilon \to 0$; the validity of a l.d.p. on $C([0, T], \mathbb{R}^d)$ with scaling ε^{-2} and a rate function which vanishes only at $X^{x,0}$ would be consistent with (2.53). This is the content of the next theorem, for which we follow [122].

Theorem 2.25 (Freidlin and Wentzell) *Let $X^{x,\varepsilon}$ be the solution of equation* (2.50), *and P_x^ε its law, where b is assumed to satisfy equation* (2.51). *Then, for each x, T the family (P_x^ε) satisfies a l.d.p. on $C([0,T],\mathbb{R}^d)$ with scaling ε^{-2} and (good) rate function I_T^x, given by:*

$$I_T^x(\varphi) = \begin{cases} \frac{1}{2}\int_0^T |\dot\varphi(s) - b(\varphi(s))|^2 ds & \text{if } \varphi \text{ is absolutely continuous and } \varphi(0) = x, \\ +\infty & \text{otherwise.} \end{cases} \tag{2.54}$$

Proof Consider first the case $b \equiv 0$, $x = 0$, which is Schilder's theorem [264]. In this case P_0^ε is the law of (εW_t), where (W_t) is a standard d-dimensional Brownian motion with $W_0 = 0$, defined on some space $(\Omega, \mathcal{A}, P)$. We check the validity of a l.d.p. with (good) rate function:

$$I_T^0(\varphi) = \begin{cases} \frac{1}{2}\int_0^T |\dot\varphi(s)|^2 ds & \text{if } \varphi \text{ is absolutely continuous, with } \varphi(0) = 0, \\ +\infty & \text{otherwise.} \end{cases}$$

We start by checking that I_T^0 is lower semi-continuous. For this, let $(\varphi^{(n)})$ be a convergent sequence in $C([0,T],\mathbb{R}^d)$, i.e. $\varrho_T(\varphi^{(n)}, \varphi) \to 0$, for some $\varphi \in C([0,T],\mathbb{R}^d)$. Without loss of generality we may assume that the sequence $I_T^0(\varphi^{(n)})$ has a finite limit. On the other hand, letting

$$H_{0,T}^{(1)} = \{\varphi \in C([0,T],\mathbb{R}^d): \\ \varphi \text{ is absolutely continuous and } \textstyle\int_0^T |\dot\varphi(s)|^2 ds < +\infty\},$$

a classical result of Riesz tells us that $\varphi \in H_{0,T}^{(1)}$ iff

$$\sup_{\substack{0\le t_1<\cdots<t_N\le T\\ N\ge 1}} \sum_{i=1}^{N-1} \frac{|\varphi(t_{i+1}) - \varphi(t_i)|^2}{|t_{i+1} - t_i|} < +\infty.$$

Also, the supremum coincides with $\int_0^T |\dot\varphi(s)|^2 ds$ (see e.g. the lemma on p. 75 of [254]). Thus,

$$\begin{aligned} I_T^0(\varphi) &= \frac{1}{2} \sup_{\substack{0\le t_1<\cdots<t_N\le T\\ N\ge 1}} \sum_{i=1}^{N-1} \frac{|\varphi(t_{i+1}) - \varphi(t_i)|^2}{|t_{i+1} - t_i|} \\ &= \frac{1}{2} \sup_{\substack{0\le t_1<\cdots<t_N\le T\\ N\ge 1}} \lim_{n\to+\infty} \sum_{i=1}^{N-1} \frac{|\varphi^{(n)}(t_{i+1}) - \varphi^{(n)}(t_i)|^2}{|t_{i+1} - t_i|} \\ &\le \frac{1}{2} \lim_{n\to+\infty} \sup_{\substack{0\le t_1<\cdots<t_N\le T\\ N\ge 1}} \sum_{i=1}^{N-1} \frac{|\varphi^{(n)}(t_{i+1}) - \varphi^{(n)}(t_i)|^2}{|t_{i+1} - t_i|} \\ &= \lim_{n\to+\infty} I_T^0(\varphi^{(n)}). \end{aligned}$$

This implies that the function I_T^0 is lower semi-continuous, i.e. the level sets $\mathbb{F}_T^0(c) := \{\varphi : I_T^0(\varphi) \le c\}$ are closed. (The notation is modified slightly with respect to Chapter 1 in order to avoid too many subscripts.) To check that I_T^0 is a good rate function observe that given $c < +\infty$ the set $\mathbb{F}_T^0(c)$ is uniformly bounded and uniformly equi-continuous, as follows from the fact that if $\varphi \in \mathbb{F}_T^0(c)$, $\varphi(0) = 0$ and for $0 \le s < t \le T$

$$|\varphi(t) - \varphi(s)| = \left| \int_s^t \dot{\varphi}(u) du \right| \le \left((t-s) \int_s^t |\dot{\varphi}(u)|^2 \, du \right)^{1/2} \le \sqrt{2(t-s)c}.$$

Applying the classical Arzelà–Ascoli theorem we see that I_T^0 is a good rate function in this particular case.

We now check conditions (b)′ and (c)′ in Proposition 1.15.

(b)′ *Lower bound.* As in the proof of Theorem 1.23, we look for an absolutely continuous modification of the measure for which the event under consideration becomes typical. The Cameron-Martin formula (Theorem 2.23) furnishes a good guess for the proof of the lower bound. (Notice that $\varphi(t)$ here corresponds to $\int_0^t \varphi(s) ds$ in the Cameron-Martin formula.)

With this in mind let φ be such that $I_T^0(\varphi) < +\infty$, i.e. $\varphi \in H_{0,T}^{(1)}$ and $\varphi(0) = 0$. Since $\varphi(t) = \int_0^t \dot{\varphi}(s) ds$ and $\int_0^T |\dot{\varphi}(s)|^2 ds < +\infty$, by the Cameron-Martin formula we may define, with T fixed, a probability measure $\tilde{P}^\varepsilon$ on $(\Omega, \mathcal{A}_T^W)$ as $\tilde{P}^\varepsilon(A) = \int_A L_T^\varepsilon(\omega) dP(\omega)$, where

$$L_T^\varepsilon(\omega) = \exp\left\{ -\varepsilon^{-1} \int_0^T \langle \dot{\varphi}(s), dW_s \rangle - \frac{\varepsilon^{-2}}{2} \int_0^T |\dot{\varphi}(s)|^2 \, ds \right\}, \tag{2.55}$$

and under $\tilde{P}^\varepsilon$, the process $\tilde{W}_t := W_t + \varepsilon^{-1}\varphi(t)$, $0 \le t \le T$, is a standard Brownian motion, starting at 0. Thus

$$\begin{aligned} P\left(\varrho_T(\varepsilon W, \varphi) < \delta\right) &= \tilde{P}^\varepsilon\left(\varrho_T(\varepsilon \tilde{W}, \varphi) < \delta\right) = \tilde{P}^\varepsilon\left(\sup_{0\le t\le T} |\varepsilon W_t| < \delta\right) \\ &= \int_{[\sup_{0\le t\le T} \varepsilon |W_t| < \delta]} L_T^\varepsilon(\omega) dP(\omega) = \exp\{-\varepsilon^{-2} I_T^0(\varphi)\} \\ &\quad \times \int_{[\sup_{0\le t\le T} \varepsilon |W_t| < \delta]} \exp\left(-\varepsilon^{-1} \int_0^T \langle \dot{\varphi}(s), dW_s \rangle \right) dP. \end{aligned} \tag{2.56}$$

Given $\delta > 0$, $\lim_{\varepsilon \downarrow 0} P(\sup_{0\le t\le T} \varepsilon\, |W_t| < \delta) = 1$ by (2.53); we fix $\varepsilon_0 > 0$ so that for each $0 < \varepsilon \le \varepsilon_0$:

$$P\left(\sup_{0\le t\le T} \varepsilon\, |W_t| < \delta \right) \ge \frac{2}{3}.$$

Unlike the situation in Theorem 1.23, we do not have uniform control of the variable in the exponent inside the integral on the right-hand side of (2.56), but we apply the Markov inequality to get

$$P\left(\int_0^T \langle \dot\varphi(s), dW_s\rangle \geq z\sqrt{I_t^0(\varphi)}\right) \leq P\left(\left(\int_0^T \langle \dot\varphi(s), dW_s\rangle\right)^2 \geq z^2 I_T^0(\varphi)\right) \leq \frac{2}{z^2},$$

for any $z > 0$. Choosing $z = \sqrt{6}$ we can write

$$P\left(\sup_{t\leq T} \varepsilon\,|W_t| < \delta,\ \varepsilon^{-1}\int_0^T \langle \dot\varphi(s), dW_s\rangle < \varepsilon^{-1}\sqrt{6 I_T^0(\varphi)}\right) \geq \frac{1}{3}, \tag{2.57}$$

for each $\varepsilon \leq \varepsilon_0$. From (2.56) and (2.57), given $\eta > 0$ we may take ε_0' so that if $\varepsilon \leq \varepsilon_0'$:

$$\begin{aligned} P_0^\varepsilon\{\psi\colon \varrho_T(\psi,\varphi) < \delta\} &\geq \frac{1}{3}\exp\left\{-\frac{1}{\varepsilon^2} I_T^0(\varphi) - \frac{1}{\varepsilon}\sqrt{6 I_T^0(\varphi)}\right\} \\ &\geq \exp\left\{-\frac{1}{\varepsilon^2}(I_T^0(\varphi) + \eta)\right\}, \end{aligned} \tag{2.58}$$

verifying condition (b)′ of Proposition 1.15. (It suffices to consider φ such that $I_T^0(\varphi) < +\infty$ while checking (b)′.) Given T and c finite, condition (b)′ holds uniformly on $\mathbb{F}_T^0(c) = \{\varphi\colon I_T^0(\varphi) \leq c\}$, i.e. ε_0' may be chosen independently of φ, provided $\varphi \in \mathbb{F}_T^0(c)$. It is also uniform on $T \leq T_0$, for any fixed T_0.

(c)′ *Upper bound.* One useful tool is the consideration of a time discretization, with linear interpolation of the original process; as we shall see, compared with the exponential errors in the l.d.p. such approximations can be as good as we want; in other words, with respect to the l.d.p. estimates, the interpolation becomes ‘indistinguishable’ from the original process (see (2.61) below).

Let $m \geq 1$ and consider, for each $\varepsilon > 0$, the stochastic process $Y^{\varepsilon,m}$, whose trajectories are polygonal lines connecting the points $(kT/m, \varepsilon W_{kT/m})$, $k = 0, 1, \ldots, m$. For each $0 < c < +\infty$, $\delta > 0$ and $m \geq 1$:

$$P_0^\varepsilon\{\psi\colon \varrho_T(\psi, \mathbb{F}_T^0(c)) \geq \delta\} \leq P(I_T^0(Y^{\varepsilon,m}) > c) + P(\varrho_T(\varepsilon W, Y^{\varepsilon,m}) \geq \delta). \tag{2.59}$$

We now estimate each term on the right-hand side of (2.59). Due to the polygonal shape of $Y^{\varepsilon,m}$:

$$\begin{aligned} P(\varrho_T(\varepsilon W, Y^{\varepsilon,m}) \geq \delta) &\leq P\left(\max_{1\leq i\leq m}\ \max_{\frac{i-1}{m}T \leq s \leq \frac{i}{m}T} \left|\varepsilon W_s - Y^{\varepsilon,m}(s)\right| \geq \delta\right) \\ &\leq mP\left(\max_{0\leq s\leq T/m}\left|\varepsilon W_s - \varepsilon s\frac{m}{T}W_{T/m}\right| \geq \delta\right) \\ &\leq mP\left(\max_{0\leq s\leq T/m} |\varepsilon W_s| \geq \frac{\delta}{2\varepsilon}\right), \end{aligned}$$

where the first inequality is trivial, the second is a consequence of the property that $(W_{kT/m+u} - W_{kT/m} : u \geq 0)$ has the same law as $(W_u : u \geq 0)$ and the shape of $Y^{\varepsilon,m}$, and the third is a simple observation. Now, as in the argument leading to (2.53), we have

$$P\left(\max_{s \leq T/m} |W_s| \geq \frac{\delta}{2\varepsilon} \right) \leq 4d \int_{\eta}^{+\infty} \frac{e^{-u^2/2}}{\sqrt{2\pi}} \, du,$$

where $\eta = \frac{\delta}{2\varepsilon}\sqrt{\frac{m}{Td}}$, and using (1.79) we get

$$P(\varrho_T(\varepsilon W, Y^{\varepsilon,m}) \geq \delta) \leq \varepsilon \frac{8d}{\delta} \sqrt{\frac{mTd}{2\pi}} e^{-\frac{m\delta^2}{8\varepsilon^2 Td}}.$$

Given $c_0 \in (0, +\infty)$ let us fix $m \geq 8dTc_0/\delta^2$. Having fixed such m, we may take $\varepsilon_0 > 0$ so that

$$P(\varrho_T(\varepsilon W, Y^{\varepsilon,m}) \geq \delta) \leq \frac{1}{2} e^{-\frac{c_0}{\varepsilon^2}}, \tag{2.60}$$

for all $\varepsilon \leq \varepsilon_0$. Thus, the second term in (2.59) behaves as desired.

Remark We have seen that for each $\delta > 0$,

$$\lim_{m \to +\infty} \limsup_{\varepsilon \to 0} \varepsilon^2 \log P(\varrho_T(\varepsilon W, Y^{\varepsilon,m}) \geq \delta) = -\infty. \tag{2.61}$$

As for the first term on the right-hand side of (2.59) we observe that

$$I_T^0(Y^{\varepsilon,m}) = \frac{\varepsilon^2}{2} \sum_{k=1}^{m} \frac{m}{T} \left| W_{kT/m} - W_{(k-1)T/m} \right|^2.$$

Since $\sqrt{m/T}(W_{kT/m} - W_{(k-1)T/m})$, $1 \leq k \leq m$ are i.i.d. random variables, distributed as d-dimensional Gaussian with zero average and identity covariance matrix, we see that $I_T^0(Y^{\varepsilon,m})$ has the same law as $\frac{\varepsilon^2}{2} \sum_{i=1}^{dm} |Z_i|^2$ where $Z_1, \ldots, Z_{dm}$ are i.i.d. real standard Gaussian random variables (average 0 and variance 1). Thus, if $\zeta < 1$

$$E e^{\zeta Z_1^2/2} = \frac{1}{\sqrt{2\pi}} \int_{\mathbb{R}} e^{\frac{\zeta - 1}{2} x^2} dx := C_\zeta < +\infty$$

and by the exponential Markov inequality:

$$\begin{aligned} P(I_T^0(Y^{\varepsilon,m}) > c) &\leq e^{-\frac{\zeta c}{\varepsilon^2}} (C_\zeta)^{dm} \\ &= e^{-\frac{c}{\varepsilon^2}} e^{\frac{1-\zeta}{\varepsilon^2} c} (C_\zeta)^{dm}. \end{aligned}$$

Having fixed m as before, and $\gamma > 0$, we take ζ so that $0 < 1 - \zeta < \gamma/2c_0$ and we get, for any $c \in (0, c_0]$:

$$P(I_T^0(Y^{\varepsilon,m}) > c) \leq e^{-c/\varepsilon^2} e^{\frac{\gamma}{2\varepsilon^2}} (C_\zeta)^{dm}.$$

Taking ε_0' small we have $(C_\zeta)^{dm} \le \frac{1}{2} e^{\gamma/2\varepsilon^2}$ for all $\varepsilon \in (0, \varepsilon_0']$. Thus, if $c \in (0, c_0]$, $\varepsilon \in (0, \varepsilon_0']$:

$$P(I_T^0(Y^{\varepsilon,m}) > c) \le \frac{1}{2} e^{-\frac{c-\gamma}{\varepsilon^2}}. \tag{2.62}$$

Putting this together with (2.60) into (2.59) we get condition (c)′ of Proposition 1.15, completing the proof in this particular case $b = 0, x = 0$.

Under the global Lipschitz condition on $b(\cdot)$ the rest of the proof is simple. Under such a condition we have a continuous map $\mathbb{T}_x$ on $C([0, T], \mathbb{R}^d)$, given by $\mathbb{T}_x \varphi = \psi$, where ψ is the unique solution of

$$\psi(t) = x + \int_0^t b(\psi(s))\, ds + \varphi(t), \qquad 0 \le t \le T. \tag{2.63}$$

Indeed, using Gronwall's inequality we have

$$\varrho_T(\mathbb{T}_x \varphi, \mathbb{T}_x \tilde{\varphi}) \le e^{\kappa T} \varrho_T(\varphi, \tilde{\varphi}), \tag{2.64}$$

for any $\varphi, \tilde{\varphi} \in C([0, T], \mathbb{R}^d)$. Therefore we may apply the contraction principle given by Theorem 1.19. Moreover, the map $\mathbb{T}_x$ is injective and if ψ belongs to its range, $\mathbb{T}_x^{-1}\psi$ is given by

$$\mathbb{T}_x^{-1}\psi(t) = \psi(t) - x - \int_0^t b(\psi(s))\, ds,$$

from where we see that ψ is absolutely continuous if and only if $\mathbb{T}_x^{-1}\psi$ is, and $\int_0^T \left|\frac{d}{dt}(\mathbb{T}_x^{-1}\psi)(t)\right|^2 dt = \int_0^T \left|\dot{\psi}(t) - b(\psi(t))\right|^2 dt$. The conclusion follows from Theorem 1.19. □

Remark In situations like the present one, where the family of measures under consideration depends on some extra parameter, it is natural to check the dependence of the estimates on this parameter (the initial value x, here). It is then convenient to change the notation slightly: let $I_T(\varphi)$ (without the superscript x) denote the function on $C([0, T], \mathbb{R}^d)$, defined by (2.54) but eliminating the condition on $\varphi(0)$, i.e. $I_T(\varphi) = I_T^{\varphi(0)}(\varphi)$, and if $0 \le c < +\infty$:

$$\mathbb{F}_T^x(c) := \{\varphi \colon \varphi(0) = x,\ I_T(\varphi) \le c\} = \{\varphi \colon I_T^x(\varphi) \le c\}.$$

Corollary 2.26 (Freidlin and Wentzell) *The estimates in Theorem* 2.25 *hold uniformly on x for fixed T, in the following sense.*

$(a)_u$ *$I_T(\cdot)$ is lower semi-continuous and the set $\cup_{x \in A} \mathbb{F}_T^x(c) = \{\varphi \colon \varphi(0) \in A,\ I_T(\varphi) \le c\}$ is compact, for each $c < +\infty$ and $A \subseteq \mathbb{R}^d$ compact.*

$(b)_u'$ *For each $c_0 > 0$, $\delta > 0$, $\gamma > 0$ there exists $\varepsilon_0 > 0$ so that*

$$P_x^\varepsilon\{\psi : \varrho_T(\psi, \varphi) < \delta\} \ge \exp\{-\varepsilon^{-2}(I_T(\varphi) + \gamma)\},$$

for all $\varepsilon \in (0, \varepsilon_0]$, all x and all $\varphi \in \mathbb{F}_T^x(c_0)$.

(c)$'_u$ *For each $c_0 > 0$, $\delta > 0$, $\gamma > 0$ there exists $\varepsilon_0 > 0$ so that*

$$P_x^\varepsilon\{\psi : \varrho_T(\psi, \mathbb{F}_T^x(c)) \geq \delta\} \leq \exp(-\varepsilon^{-2}(c - \gamma)),$$

for all $\varepsilon \in (0, \varepsilon_0]$, all $c \leq c_0$ and all $x \in \mathbb{R}^d$.

Proof If restricted to $\mathbb{R}^d \times \{\varphi \in C([0, T], \mathbb{R}^d) : \varphi(0) = 0\}$, the map $(x, \varphi) \mapsto \mathbb{T}_x\varphi$, where $\mathbb{T}_x$ is the unique solution of (2.63), is in fact a homeomorphism to its image and (a)$_u$ follows at once from this. The validity of (b)$'_u$, (c)$'_u$ is an immediate consequence of the uniform continuity of the maps $\mathbb{T}_x$ as x varies (see (2.64)), the uniformity in condition (b)$'$ for $\varphi \in \mathbb{F}_T^0(c)$ and the uniformity in (c)$'$ in $c \leq c_0$ both in the case of $b \equiv 0$, $x = 0$. Details are left as an exercise. □

The next corollary is an immediate consequence of the proof of Proposition 1.15 and the previous corollary. (In fact, we only need (a)$_u$ and the uniformity in (b)$'_u$ and (c)$'_u$ on compact sets in x.)

Corollary 2.27 *Under the conditions of Theorem 2.25, for each $T < +\infty$ and each compact set $A \subseteq \mathbb{R}^d$ we have:*

(b)$_u$ *For each $G \subseteq C([0, T], \mathbb{R}^d)$ open we have*

$$\liminf_{\varepsilon \to 0} \varepsilon^2 \log \inf_{x \in A} P_x^\varepsilon(G) \geq - \sup_{x \in A} \inf_{\varphi \in G} I_T^x(\varphi);$$

(c)$_u$ *for each $F \subseteq C([0, T], \mathbb{R}^d)$ closed we have*

$$\limsup_{\varepsilon \to 0} \varepsilon^2 \log \sup_{x \in A} P_x^\varepsilon(F) \leq - \inf_{x \in A} \inf_{\varphi \in F} I_T^x(\varphi).$$

Comments (i) Our exposition in this section has essentially followed that of Freidlin and Wentzell in [122], with a construction given in [294, 295], in a more general set-up. As already mentioned, the particular case of $b = 0$, $x = 0$ is due to Schilder, cf. [264]. It may also be obtained as an application of the Cramér theorem in the separable Banach space $M = C([0, T], \mathbb{R}^d)$: if μ is the Wiener measure, corresponding to the standard Brownian motion starting at the origin, then the measure μ_n of Theorem 1.27 is the law of $n^{-1} \sum_{i=1}^n W_{(i)}$ where $W_{(1)}, \ldots, W_{(n)}$ are independent standard Brownian motions. We check at once that μ_n coincides with the law of $n^{-1/2}W$. It is easy to compute $\hat{\mu}$, and to determine the rate function in this case, according to (1.87). Since Theorem 1.27 gives us only a weak large deviation principle, one needs to prove exponential tightness. For proofs along these lines see [7] or [87] (Lemma 1.3.8), where a result of Fernique is used to verify exponential tightness.

(ii) The 'superexponential estimate' (2.61) is closely related to the concept of two exponentially equivalent measures. Based on this, a proof of Theorem 2.25 can be obtained as an example of the so-called projective limit approach of Dawson and Gärtner, cf. [72]: one first proves the validity of the l.d.p. for the family of finite dimensional distributions; together with the exponential tightness on $C([0,T],\mathbb{R}^d)$ one gets a l.d.p. for the uniform convergence (see e.g. Chapter 5 of [85]).

2.5 Freidlin and Wentzell basic estimates. Variable diffusion coefficients[1]

For technical simplicity, our discussion of metastability in the Freidlin and Wentzell set-up, to be done in Chapter 5, will be restricted to the case of a constant diffusion matrix, though it is interesting to analyse more general situations and we shall comment briefly on that. Already in their basic article [295], Freidlin and Wentzell considered a wide class of diffusion processes on compact manifolds and there exists a vast literature on this subject. To get an idea of the tools, let us consider $\sigma(\cdot)$ and $b(\cdot)$ satisfying the simplest regularity conditions for existence and uniqueness of the strong solution of Itô's equation and let us examine the validity of the l.d.p. for processes given by:

$$\begin{aligned} dX_t^{x,\varepsilon} &= b(X_t^{x,\varepsilon})dt + \varepsilon\sigma(X_t^{x,\varepsilon})dW_t, \\ X_0^{x,\varepsilon} &= x. \end{aligned} \tag{2.65}$$

Throughout this section we assume that $b\colon \mathbb{R}^d \to \mathbb{R}^d$ is bounded and uniformly Lipschitz, that $\sigma(\cdot)$ takes values on the space of $d \times d$ matrices with real entries $\sigma(x) = (\sigma_{i,j}(x))$, each $\sigma_{i,j}$ being bounded and uniformly Lipschitz. Moreover, the diffusion matrix $a(x) = \sigma(x)\cdot\sigma^{\mathbb{T}}(x)$, given by the product of $\sigma(x)$ and its transpose matrix $\sigma^{\mathbb{T}}(x)$ is assumed to be uniformly elliptic, i.e. we assume the validity of (2.49) for some $\alpha > 0$. Let $a^{-1}(x) = (a(x))^{-1}$.

Looking back at the argument in Section 2.4 the difficulty is the lack of continuity of the map which takes εW into $X^{x,\varepsilon}$, preventing a mere application of the contraction principle.

Remark One way to understand the lack of continuity is through the so called 'Wong and Zakai correction'; in [297] the authors first establish the relation between the limit of solutions of equations like (2.65), with (W_t) being replaced by a family of bounded variation trajectories (W_t^n) such that $\varrho_T(W_t^n, W_t)$ tends to zero. For $d = 1$, they obtain the precise correction term that appears in the limit and which corresponds to the difference between the solutions to (2.65) in Itô and Stratonovich senses (see Section 5.2 in [173] and references therein). For a class of approximations of W_t, including the linear interpolations of the previous section, the same convergence holds in higher dimensions. Since we have a factor ε multiplying the noise term W this correction is $O(\varepsilon^2)$ and we may expect it not to disturb the large deviation estimates, but a more careful analysis is

[1] This section may be omitted at a first reading.

needed. Various procedures to deal with this can be used, as in Theorem 2.4 in [7], Section 1 of [295], or Section 6 of [292].

One possible strategy could be to use a family of approximating processes $X^{x,\varepsilon,m}$ $(m \to +\infty)$ which are themselves continuous functions of εW, so that the l.d.p. for each of them can be worked out as in Section 2.4, and such that $\varrho_T(X^{x,\varepsilon}, X^{x,\varepsilon,m})$ becomes 'negligible' in terms of the large deviation estimates, analogously to (2.61).

Following [292] in this discussion, we take the approximating processes as solutions to equations analogous to (2.65), but with coefficients kept constant on small time intervals. More precisely, for $m \geq 1$ let $r_m(t) = [mt]/m$, $[\cdot]$ denoting the integer part, and consider the process $X^{x,\varepsilon,m}$ defined by:

$$\begin{aligned} dX_t^{x,\varepsilon,m} &= b(X_{r_m(t)}^{x,\varepsilon,m})dt + \varepsilon\sigma(X_{r_m(t)}^{x,\varepsilon,m})dW_t, \\ X_0^{x,\varepsilon,m} &= x. \end{aligned} \tag{2.66}$$

Thus, if $0 \leq t \leq 1/m$

$$X_t^{x,\varepsilon,m} = x + b(x)t + \varepsilon\sigma(x)W_t,$$

and inductively, if $k/m \leq t \leq (k+1)/m$:

$$X_t^{x,\varepsilon,m} = X_{k/m}^{x,\varepsilon,m} + b(X_{k/m}^{x,\varepsilon,m})(t - k/m) + \varepsilon\sigma(X_{k/m}^{x,\varepsilon,m})(W_t - W_{k/m}).$$

From the assumptions on $b(\cdot)$ and $\sigma(\cdot)$ we can see that $X^{x,\varepsilon,m} = \mathbb{T}_{m,x}(\varepsilon W)$, where the map $\mathbb{T}_{m,x}$ is continuous for the sup norm. In fact $(x, \varphi) \mapsto \mathbb{T}_{m,x}(\varphi)$ is jointly continuous on $\mathbb{R}^d \times C([0,T], \mathbb{R}^d)$. Applying the contraction principle, as in Section 2.4, we get a l.d.p. for the laws of $X^{x,\varepsilon,m}$, $P_x^{\varepsilon,m} := \bar{P}_0^{\varepsilon}\mathbb{T}_{m,x}^{-1}$ where $\bar{P}_0^{\varepsilon}$ is the law of εW starting at 0. The scaling is ε^{-2} and the rate function is given by:

$$I_{T,m}^x(\varphi) = \begin{cases} \frac{1}{2}\int_0^T \big\langle \dot\varphi(t) - b(\varphi(r_m(t))), a^{-1}(\varphi(r_m(t)))\,(\dot\varphi(t) - b(\varphi(r_m(t))))\big\rangle\, dt \\ \qquad\qquad \text{if } \varphi \in H_{0,T}^{(1)}, \quad \varphi(0) = x, \\ +\infty \qquad \text{otherwise.} \end{cases} \tag{2.67}$$

Moreover, for each fixed $m \geq 1$, the estimates are uniform on x in the sense of Corollary 2.26, with $I_{T,m}(\varphi) = I_{T,m}^{\varphi(0)}(\varphi)$.

To get a l.d.p. for $X^{x,\varepsilon}$ we want to see that $\varrho_T(X^{x,\varepsilon,m}, X^{x,\varepsilon})$ can be made smaller than the errors which can be felt by the l.d.p. A way to make this precise is through the following lemma. It also shows that the estimates are uniform on the initial point.

Lemma 2.28 *For any $T < +\infty$ and $\delta > 0$,*

$$\lim_{m\to+\infty} \limsup_{\varepsilon\to 0} \varepsilon^2 \sup_x \log P[\varrho_T(X^{x,\varepsilon,m}, X^{x,\varepsilon}) > \delta] = -\infty. \tag{2.68}$$

Before proving this lemma, usually called the 'superexponential estimate', we see how it allows us to extend Theorem 2.25 to the present situation.

Theorem 2.29 *Under the above conditions on $b(\cdot)$ and $\sigma(\cdot)$, let P_x^ε be the law of $X^{x,\varepsilon}$, the solution of the Itô equation (2.65), on the space $C([0,T],\mathbb{R}^d)$. The family $(P_x^\varepsilon)_{\varepsilon>0}$ satisfies a l.d.p. (as $\varepsilon \to 0$) with scaling ε^{-2} and good rate function I_T^x, given by:*

$$I_T^x(\varphi) = \begin{cases} \frac{1}{2}\int_0^T \langle \dot\varphi(t) - b(\varphi(t)), \quad a^{-1}(\varphi(t))(\dot\varphi(t) - b(\varphi(t)))\rangle dt, & \\ \qquad\qquad\qquad\qquad\qquad\qquad \text{if } \varphi \in H_{0,T}^{(1)}, \quad \varphi(0) = x, & \\ +\infty \qquad\qquad\qquad\qquad\qquad\quad \text{otherwise.} \end{cases} \tag{2.69}$$

Moreover, the estimates are uniform on the initial condition x, in the sense described in Corollary 2.26, with $I_T(\varphi) = I_T^{\varphi(0)}(\varphi)$.

Proof of Theorem 2.29 We first check that I_T satisfies condition $(a)_u$ in Corollary 2.26. Since $a(\cdot)$ is uniformly elliptic and uniformly bounded we immediately see that $a^{-1}(\cdot)$ is uniformly elliptic and we can take $\tilde\alpha > 0$ so that for any $\varphi \in H_{0,T}^{(1)}$

$$I_T(\varphi) \geq \tilde\alpha \int_0^T |\dot\varphi(t) - b(\varphi(t))|^2 \, dt. \tag{2.70}$$

On the other hand using the Cauchy–Schwarz inequality we get

$$\int_0^T |\dot\varphi(t) - b(\varphi(t))|^2 \, dt \geq \left(\left(\int_0^T |\dot\varphi(t)|^2 \, dt \right)^{1/2} - \left(\int_0^T |b(\varphi(t))|^2 \, dt \right)^{1/2} \right)^2 .$$

From (2.70), the boundedness of $b(\cdot)$ and the inequality $(u+v)^2 \leq 2(u^2+v^2)$ we get:

$$\int_0^T |\dot\varphi(t)|^2 \, dt \leq \frac{2}{\tilde\alpha} I_T(\varphi) + 2T \sup_y |b(y)|^2 .$$

Thus, for given T and $c < +\infty$, if $\bar\alpha = 2c/\tilde\alpha + 2T \sup_y |b(y)|^2$ we see that

$$\bigcup_{x \in A} \mathbb{F}_T^x(c) \subseteq \left\{ \varphi \in H_{0,T}^{(1)} : \varphi(0) \in A, \int_0^T |\dot\varphi(t)|^2 \, dt \leq \bar\alpha \right\} := \tilde C_{\bar\alpha, A}, \tag{2.71}$$

implying the relative compactness of $\bigcup_{x\in A} \mathbb{F}_T^x(c)$, if $c < +\infty$ and A is compact. Also

$$\lim_{m \to +\infty} \sup_{\varphi \in \tilde C_{\bar\alpha, A}} \left| I_{T,m}(\varphi) - I_T(\varphi) \right| = 0. \tag{2.72}$$

Since $I_{T,m}$ is lower semi-continuous, we get the compactness of $\bigcup_{x\in A} \mathbb{F}_T^x(c)$.

We now check condition (b)$'_u$ in Corollary 2.26. For this, we take $c_0 > 0$, $\eta > 0$, $\gamma > 0$ and show the existence of $\varepsilon_0 > 0$ so that

$$P_x^\varepsilon\{\psi\colon \varrho_T(\psi,\varphi) < \eta\} \geq \exp\{-\varepsilon^{-2}(I_T(\varphi)+\gamma)\}, \tag{2.73}$$

for all $\varepsilon \leq \varepsilon_0$, all x, all $\varphi \in \mathbb{F}_T^x(c_0)$. As in the proof of Theorem 2.25, we write

$$P_x^\varepsilon\{\psi\colon \varrho_T(\psi,\varphi) < \eta\} \geq P(\varrho_T(X^{x,\varepsilon,m},\varphi) < \eta/2) - P(\varrho_T(X^{x,\varepsilon},X^{x,\varepsilon,m}) \geq \eta/2), \tag{2.74}$$

and by Lemma 2.28 we take $m_0 \geq 1$, $\varepsilon_0 > 0$ in such a way that if $m \geq m_0$, $0 < \varepsilon \leq \varepsilon_0$:

$$\sup_x P(\varrho_T(X^{x,\varepsilon},X^{x,\varepsilon,m}) \geq \eta/2) \leq e^{-2c_0/\varepsilon^2}. \tag{2.75}$$

Using now the l.d.p. for $X^{x,\varepsilon,m}$ ($\varepsilon \downarrow 0$) with fixed $m \geq 1$ and the fact that for $\varphi \in H_{0,T}^{(1)}$,

$$I_{T,m}^x(\varphi) \underset{m\to+\infty}{\to} I_T^x(\varphi)$$

uniformly on x, we can conclude the validity of (2.73). (Recall (2.72), where A may be taken as $\mathbb{R}^d$.)

To check condition (c)$'_u$ in Corollary 2.26, let $c_0 > 0$, $\eta > 0$, and for $0 \leq c \leq c_0$ we write:

$$\begin{aligned} P_x^\varepsilon\{\psi\colon \varrho_T(\psi,\mathbb{F}_T^x(c)) \geq \eta\} &\leq P(\varrho_T(X^{x,\varepsilon},X^{x,\varepsilon,m}) \geq \eta/2) \\ &\quad + P(\varrho_T(X^{x,\varepsilon,m},\mathbb{F}_T^x(c)) \geq \eta/2). \end{aligned}$$

Let us first take ε_0, m_0 so that (2.75) holds for all $\varepsilon \in (0,\varepsilon_0]$, all $m \geq m_0$, thus controlling the first term. As for the second, since $C_x := \{\varphi\colon \varphi(0) = x, \varrho_T(\varphi,\mathbb{F}_T^x(c)) \geq \eta/2\}$ is a closed set, by the argument leading to (2.72) we may take $\tilde m_0 > 0$, $\eta' > 0$ in such a way that for all $m \geq \tilde m_0$, $x \in \mathbb{R}^d$

$$\inf_{\varphi\in C_x} I_{T,m}^x(\varphi) \geq c + \eta',$$

implying that $\eta'' > 0$ and $\tilde m_0 > 0$ may be taken so that for each $m \geq \tilde m_0$ and each x,

$$C_x \subseteq \{\varphi\colon \varphi(0) = x, \varrho_T(\varphi,\mathbb{F}_{T,m}^x(c)) \geq \eta''\}.$$

With this and the validity of condition (c)$'_u$ for the process $X^{x,\varepsilon,m}$ we conclude the proof of Theorem 2.29. □

It remains to prove Lemma 2.28. This is a technical point and we sketch it.

Proof of Lemma 2.28 Let $Y_t^{x,\varepsilon,m} := X_t^{x,\varepsilon} - X_t^{x,\varepsilon,m}$. Considering equations (2.65) and (2.66), we may write

$$Y_t^{x,\varepsilon,m} = \int_0^t \tilde{b}^{\varepsilon,m}(s)ds + \varepsilon \int_0^t \tilde{\sigma}^{\varepsilon,m}(s)dW_s, \tag{2.76}$$

where

$$\begin{aligned}\tilde{b}^{\varepsilon,m}(s) &= b(X_s^{x,\varepsilon}) - b(X_{r_m(s)}^{x,\varepsilon,m}),\\ \tilde{\sigma}^{\varepsilon,m}(s) &= \sigma(X_s^{x,\varepsilon}) - \sigma(X_{r_m(s)}^{x,\varepsilon,m}).\end{aligned} \tag{2.77}$$

Given η and ρ positive, let us consider the following stopping times:

$$\begin{aligned}\tau_x'(\eta) &= \inf\{s > 0\colon \left|Y_s^{x,\varepsilon,m}\right| \geq \eta\} \wedge T,\\ \tilde{\tau}_x(\rho) &= \inf\{s > 0\colon \left|X_s^{x,\varepsilon,m} - X_{r_m(s)}^{x,\varepsilon,m}\right| \geq \rho\} \wedge T.\end{aligned}$$

The statement to be proven is that for any $\eta > 0$:

$$\lim_{m\to+\infty} \limsup_{\varepsilon\to 0} \sup_x \varepsilon^2 \log P(\tau_x'(\eta) < T) = -\infty. \tag{2.78}$$

But, given any $\eta, \rho > 0$, dropping their dependence from the notation we write:

$$P(\tau_x' < T) \leq P(\tilde{\tau}_x < T) + P(\tau_x' < T, \tilde{\tau}_x = T);$$

the first term is estimated quite simply, since

$$P(\tilde{\tau}_x < T) = P\left(\max_{i<mT} \sup_{i/m\leq s<T\wedge(i+1)/m} \left|X_s^{x,\varepsilon,m} - X_{i/m}^{x,\varepsilon,m}\right| \geq \rho\right) \tag{2.79}$$

and since b and σ were assumed to be bounded, if $C := \sup_x\{\max_{i,j} d\left|\sigma_{i,j}(x)\right| \vee |b(x)|\}$, we have

$$\sup_{i/m\leq s<(i+1)/m} \left|X_s^{x,\varepsilon,m} - X_{i/m}^{x,\varepsilon,m}\right| \leq C/m + \varepsilon C \sup_{i/m\leq s<(i+1)/m} \left|W_s - W_{i/m}\right|, \tag{2.80}$$

and consequently, for any $\rho > 0$, and $m > C/\rho$

$$\begin{aligned}P(\tilde{\tau}_x < T) &\leq m\,T\,P\left(\sup_{0\leq s\leq \frac{1}{m}} |W_s| \geq \frac{\rho - C/m}{\varepsilon C}\right)\\ &\leq C_1\, m\, T\, e^{-mC_2(\rho-C/m)^2/\varepsilon^2 C^2},\end{aligned}$$

where C_1, C_2 just depend on the dimension d (cf. proof of (2.53)). We conclude that for any $\rho > 0$,

$$\lim_{m\to+\infty} \limsup_{\varepsilon\to 0} \sup_x \varepsilon^2 \log P(\tilde{\tau}_x < T) = -\infty.$$

It now suffices to check that

$$\lim_{\rho \downarrow 0} \lim_{m \to +\infty} \limsup_{\varepsilon \to 0} \sup_x \varepsilon^2 \log P(\tau_x'(\eta) < T, \tilde{\tau}_x(\rho) = T) = -\infty. \tag{2.81}$$

But, if $s \le \bar{\tau}_x := \tau_x' \wedge \tilde{\tau}_x$ we use (2.77) and the uniform Lipschitz condition on $b(\cdot)$ and $\sigma(\cdot)$ to see that

$$\left|\tilde{b}^{\varepsilon,m}(s)\right| \vee \max_{i,j} \left|\tilde{\sigma}_{i,j}^{\varepsilon,s}(s)\right| \le \kappa\{\left|Y_s^{x,\varepsilon,m}\right| + \rho\}, \tag{2.82}$$

for κ a suitable fixed constant.

Taking $g(y) = (\rho^2 + |y|^2)^{1/\varepsilon^2}$ and applying Itô's formula (2.35) we write

$$g(Y_t^{x,\varepsilon,m}) = g(0) + \int_0^t \alpha_x(s) ds + \sum_{i,j=1}^d \int_0^t \frac{\partial g}{\partial y_i}(Y_s^{x,\varepsilon,m}) \tilde{\sigma}_{i,j}^{\varepsilon,m}(s) dW_s^{(j)}, \tag{2.83}$$

where

$$\alpha_x(s) = \langle \nabla g(Y_s^{x,\varepsilon,m}), \tilde{b}^{\varepsilon,m}(s) \rangle + \frac{\varepsilon^2}{2} \mathrm{Trace}(\tilde{a}^{\varepsilon,m}(s) D^2(g)(Y_s^{x,\varepsilon,m}))$$

with $\tilde{a}^{\varepsilon,m}(s) = \sigma^{\varepsilon,m}(s)(\sigma^{\varepsilon,m}(s))^{\mathbb{T}}$, $D^2(g)(y)$ denotes the matrix of second order derivatives of g at y (also called the Hessian of g at y) and we have the usual product and trace of matrices. With standard (but a little lengthy) calculations one gets for $s \le \bar{\tau}_x$,

$$\alpha_x(s) \le \frac{\tilde{C}\, g(Y_s^{x,\varepsilon,m})}{\varepsilon^2},$$

for some suitable constant $\tilde{C}$ that depends on κ and on the dimension d. Recalling (2.83) (or applying Itô's formula for $f(t, y) = e^{-\tilde{C}t/\varepsilon^2} g(y)$) and the classical Doob optional sampling theorem we see that

$$e^{-\tilde{C}(\bar{\tau}_x \wedge s)/\varepsilon^2} g(Y_{\bar{\tau}_x \wedge s}^{x,\varepsilon,m}), \quad 0 \le s \le T,$$

is a supermartingale. Since its value at $s = 0$ is ρ^{2/ε^2}, we get

$$E\left(e^{-\tilde{C}\bar{\tau}_x/\varepsilon^2} (\rho^2 + \left|Y_{\bar{\tau}_x}^{x,\varepsilon,m}\right|^2)^{1/\varepsilon^2}\right) \le \rho^{2/\varepsilon^2}.$$

But $\{\tau_x' < T, \tilde{\tau}_x = T\} \subseteq \{\left|Y_{\bar{\tau}_x}^{x,\varepsilon,m}\right| = \eta\}$ and we get

$$e^{-\tilde{C}T/\varepsilon^2} (\rho^2 + \eta^2)^{1/\varepsilon^2} P(\tau_x' < T, \tilde{\tau}_x = T) \le \rho^{2/\varepsilon^2},$$

from which we conclude that

$$\limsup_{\varepsilon \to 0} \sup_x \varepsilon^2 \log P(\tau_x' < T, \tilde{\tau}_x = T) \le \tilde{C}T + \log \frac{\rho^2}{\rho^2 + \eta^2}$$

and (2.81) follows at once. □

Remark Using a different method, Azencott extended the basic estimates of Freidlin and Wentzell further beyond the hypotheses of Theorem 2.29, allowing for explosions, and $a(\cdot)$ not necessarily elliptic. For a precise statement see Proposition 2.3 and Theorem 2.4 in [7]. In this more general situation one loses the simplified form of the action function as in (2.69). The above approximating sequences $X^{x,\varepsilon,m}$ can also be used under these more general conditions (see Section 5.6 in [85]). The situation is related to that mentioned in the remark following Theorem 1.19.

Coefficients $b(\cdot)$ or $\sigma(\cdot)$ which might be discontinuous along some curves arise naturally in some applications; see e.g. [28] for some results and discussion.

2.6 Exit from a domain

One of the classical applications of the large deviation estimates seen in Section 2.4 is to the problem of escape from a bounded domain D which under the action of the deterministic system is attracted to a fixed point or a limit cycle. Due to the presence of noise, the diffusion $X^{x,\varepsilon}$ will eventually escape from D, with probability one. The basic questions are:

(i) through which points on ∂D will it typically escape?
(ii) what is the typical exit time?
(iii) is there a typical path which provides the escape?

Answers to such questions have been given by Freidlin and Wentzell. Having in mind the metastability for these systems, to be discussed in Chapter 5, it is convenient first to study these basic problems. We do it in this section for the sake of completeness; the results are taken from Chapter 4 in [122]. They are explained in various degrees of generality in several of the already quoted monographs.

For simplicity we assume that $X^{x,\varepsilon}$ is given by (2.50), where the vector field $b(\cdot)$ satisfies a global Lipschitz condition (cf. (2.51)). In this section we also make the following assumptions.

Assumptions 2.30

(a) D is a bounded open domain in $\mathbb{R}^d$, with a (smooth) boundary ∂D of class C^2 and $\langle b(x), n(x)\rangle < 0$ for each $x \in \partial D$, where $n(x)$ is the outward unit normal vector to ∂D, at x.[2]
(b) $x_0 \in D$ is an asymptotically stable equilibrium point of the deterministic system, i.e. for each G neighbourhood of x_0 there is a neighbourhood $\tilde{G}$ of x_0 so that $\tilde{G} \subseteq G$ and if $x \in \tilde{G}$ then $X_t^{x,0} \to x_0$ as $t \to +\infty$, without exiting from G.
(c) The set $\overline{D} = D \cup \partial D$ is attracted to x_0. For any $x \in \overline{D}$, $X_t^{x,0} \in D$ for each $t > 0$ and $\lim_{t\to+\infty} X_t^{x,0} = x_0$.

[2] Domains D, G in this section and in Chapter 5 are taken as open.

Remarks (i) Assumptions 2.30 correspond to the *simplest* situation. For the study of metastability in Chapter 5, we need to consider domains that are not fully attracted. An interesting case, not discussed here, and which presents delicate features, is when D is attracted to a fixed point $x_0 \in D$ (or a cycle) but ∂D is not (see eg. [75]). The regularity on ∂D can be somewhat relaxed but we do not discuss such extensions either. The reader will notice that the condition on the vector field imposed in (a) can be relaxed provided (c) holds.

(ii) The restriction to constant diffusion coefficient stays for simplicity. This discussion could be extended without further difficulties to the situation in Section 2.5.

Notation To simplify the notation we shall eliminate the superscript x in $X^{x,\varepsilon}$ and will instead write P_x (E_x respectively) to indicate the condition $X_0^{x,\varepsilon} = x$, if $\varepsilon > 0$. (We hope no confusion will come from this, due to the presence of a second super index in our notation, indicating whether the noise is present or not.)

A crucial role in the treatment of the above basic questions is played by the *quasi-potential* with respect to x_0, $V(x_0, \cdot)$, defined as follows:

Definition 2.31 For $x, y \in \mathbb{R}^d$ set

$$V(x, y) = \inf\{I_T(\varphi) \colon \varphi(0) = x,\ \varphi(T) = y,\ T > 0\},$$

where $I_T(\cdot)$ is the rate function of Section 2.4, i.e.

$$I_T(\varphi) = \begin{cases} \frac{1}{2}\int_0^T |\dot\varphi(t) - b(\varphi(t))|^2\,dt & \text{if } \varphi \text{ is absolutely continuous,} \\ +\infty & \text{otherwise.} \end{cases}$$

Obviously $V(x, y) \geq 0$; in some limiting sense (described by the l.d.p.) $V(x, y)$ represents the 'cost' for the stochastically perturbed system to move from (nearby) x to y. It is very simple to see that under our assumptions $V(\cdot, \cdot)$ is jointly continuous. Indeed, from the definition we see that $V(x, z) \leq V(x, y) + V(y, z)$, for each x, y, z, and moreover we have the following.

Lemma 2.32 *If $b(\cdot)$ is continuous, then for each compact $C \subseteq \mathbb{R}^d$ we may take $L < +\infty$ so that for any $x, y \in C$ there exists φ, a C^∞ function, so that $\varphi(0) = x$, $\varphi(|x - y|) = y$ and $I_{|x-y|}(\varphi) \leq L\,|x - y|$; in particular $V(x, y) \leq L\,|x - y|$. The same L works over all $\mathbb{R}^d$ if $b(\cdot)$ is bounded.*

Proof Just take a linear interpolation $\varphi(t) = x + \frac{t}{|y-x|}(y - x)$. □

A first result concerning the basic questions tells us that the escape from D must happen near a point on ∂D where $V(x_0, \cdot)$ is minimal. (For simplicity we suppose that there is only one such point, but this is irrelevant.)

Theorem 2.33 (Freidlin and Wentzell) *Under the above conditions, i.e. Assumptions 2.30 and $b(\cdot)$ Lipschitz, and if there exits a unique $y_0 \in \partial D$ so that*

$$V(x_0, y_0) = \min_{y \in \partial D} V(x_0, y), \tag{2.84}$$

then, for each $\delta > 0$ and each $x \in D$

$$\lim_{\varepsilon \downarrow 0} P_x(\left|X^\varepsilon_{\tau_\varepsilon} - y_0\right| > \delta) = 0, \tag{2.85}$$

where $\tau_\varepsilon = \inf\{t : X^\varepsilon_t \notin D\}$ (with the convention that $\inf \emptyset = +\infty$).

The convergence in (2.85) is uniform on each compact set contained in D.

Notation (a) If $A \subseteq \mathbb{R}^d$ is open or closed, $\tau_\varepsilon(A)$ denotes the hitting time to A,

$$\tau_\varepsilon(A) = \inf\{t \geq 0 \colon X^\varepsilon_t \in A\},$$

with the convention that $\inf \emptyset = +\infty$. Thus $\tau_\varepsilon = \tau_\varepsilon(D^c)$, where $D^c = \mathbb{R}^d \setminus D$.

(b) $B_r(x) = \{y \in \mathbb{R}^d \colon \ |y - x| < r\}$; $\bar{B}_r(x) = \{y \in \mathbb{R}^d \colon \ |y - x| \leq r\}$; $S_r(x) = \{y \in \mathbb{R}^d \colon |y - x| = r\}$.

Before proving the previous theorem we verify the lemma below, which basically says that X^ε cannot stay wandering around in $\overline{D} \setminus B_r(x_0)$ for too long; this would have a high 'price' in terms of the rate function, due to our conditions on the deterministic system.

Lemma 2.34 *Under the conditions of Theorem 2.33, given $r > 0$ we have:*

(a) there exist $c > 0$, $T_0 > 0$ so that $I_T(\varphi) \geq c(T - T_0)$ for any φ such that $\varphi(t) \in \overline{D} \setminus B_r(x_0)$ for all $t \in [0, T]$;

(b) there exist $\tilde{c} > 0$, $T_0 > 0$ so that for any ε sufficiently small and any $x \in \overline{D} \setminus B_r(x_0)$

$$P_x(\tau_\varepsilon(D^c \cup B_r(x_0)) > T) \leq \exp\{-\varepsilon^{-2}\tilde{c}(T - T_0)\}.$$

Proof of Lemma 2.34 Of course (a) is relevant only for $T \geq T_0$. From Assumptions 2.30, the deterministic system $X^{x,0}$ reaches $B_{r/2}(x_0)$ in a finite time $t(x)$, for each $x \in \overline{D}$. Since the map $x \mapsto X^{x,0}_t$ is continuous, we see that $x \mapsto t(x)$ is upper semi-continuous, so that $T_0 := \sup_{x \in \overline{D}} t(x) < +\infty$. Consequently, the set

$$F = \{\varphi \in C([0, T_0], \mathbb{R}^d) \colon \varphi(t) \in \overline{D} \setminus B_r(x_0), \ \forall t \in [0, T_0]\}$$

is closed and does not contain any path of the deterministic system. This implies, by the lower semi-continuity of I_{T_0}, that $A := \inf_{\varphi \in F} I_{T_0}(\varphi) > 0$. From the additivity of $I_T(\varphi)$ in T, it follows that if $\varphi(t) \in \overline{D} \setminus B_r(x_0)$ for all $t \in [0, T]$, then $I_T(\varphi) \geq A[T/T_0] \geq c(T - T_0)$, with $c = A/T_0$, proving (a).

To prove (b), from Assumptions 2.30 we may take $0 < \alpha < r/2$ small enough, so that these assumptions continue to be valid for an α-neigbourhood of D, $D_{(\alpha)} := \left\{y \in \mathbb{R}^d : \varrho(y, D) < \alpha\right\}$, where $\varrho(y, D) = \inf\{|y - x| \colon x \in D\}$. As in (a), there exist $A > 0$, $T_0 > 0$ so that $\varphi([0, T_0]) \subseteq \overline{D_{(\alpha)}} \setminus B_{r/2}(x_0)$ implies $I_{T_0}(\varphi) > A$. In particular, if $I_{T_0}(\varphi) \leq A$, and $\psi([0, T_0]) \subseteq \overline{D} \setminus B_r(x_0)$ then we have $\varrho_{T_0}(\varphi, \psi) \geq \alpha$. Thus, given $h > 0$ there exists $\varepsilon_0 > 0$ so that

$$P_x\left(\tau_\varepsilon(D^c \cup B_r(x_0)) > T_0\right) \leq P_x\left(\varrho_{T_0}(X^\varepsilon, \mathbb{F}^x_{T_0}(A)) \geq \alpha\right) \leq \exp\{-\varepsilon^{-2}(A-h)\},$$

for all x, all $\varepsilon < \varepsilon_0$. Since this inequality holds for all x, using the Markov property we get (b) with $\tilde{c} = (A-h)/T_0$. □

Proof of Theorem 2.33 The proof of Theorem 2.33 may be, as originally presented by Freidlin and Wentzell, split into two basic points.

(I) For each $x \in D$ and each $r > 0$

$$\lim_{\varepsilon \to 0} P_x\left(\tau_\varepsilon < \tau_\varepsilon(B_r(x_0))\right) = 0 \tag{2.86}$$

(with uniform convergence on each compact $C \subseteq D$).

(II) Given $\delta > 0$ we may find $r > 0$ sufficiently small so that

$$\lim_{\varepsilon \to 0} \sup_{y \in S_r(x_0)} P_y(\left|X^\varepsilon_{\tau_\varepsilon} - y_0\right| > \delta) = 0. \tag{2.87}$$

□

The proof follows at once from (I) and (II) above, since the path continuity and strong Markov property imply that:

$$\begin{aligned} P_x(\left|X^\varepsilon_{\tau_\varepsilon} - y_0\right| > \delta) =& P_x(\left|X^\varepsilon_{\tau_\varepsilon} - y_0\right| > \delta,\ \tau_\varepsilon < \tau_\varepsilon(B_r(x_0))) \\ &+ P_x(\left|X^\varepsilon_{\tau_\varepsilon} - y_0\right| > \delta,\ \tau_\varepsilon \geq \tau_\varepsilon(B_r(x_0))) \\ \leq& P_x(\tau_\varepsilon < \tau_\varepsilon(B_r(x_0))) + \sup_{y \in S_r(x_0)} P_y(\left|X^\varepsilon_{\tau_\varepsilon} - y_0\right| > \delta). \end{aligned} \tag{2.88}$$

The proof of (I) follows from (2.53) and the fact that under our assumptions the deterministic system $X^{x,0}_t$ must visit $B_{r/2}(x_0)$ in a finite time $t(x)$, staying at some positive distance $\alpha(x)$ from D^c during all the time interval $[0, t(x)]$, so that $P_x\left(\tau_\varepsilon < \tau_\varepsilon(B_r(x_0))\right) \leq P_x(\varrho_{t(x)}(X^\varepsilon, X^{x,0}) > \alpha(x) \wedge r/2)$. (Also, if $C \subseteq D$ is compact, we may take $\sup_{x \in C} t(x) < +\infty$ and $\inf_{x \in C} \alpha(x) > 0$, from which we get the uniformity on C.)

To prove (II), if $r > 0$ and $\bar{B}_{2r}(x_0) \subseteq D$, write $S_u = S_u(x_0)$ for $u \leq 2r$, and consider the Markov chain imbedded in X^ε through the stopping times η_n, defined as follows:

$$\eta_0 = 0,$$

$$\sigma_0 = \inf\{t > 0\colon X^\varepsilon_t \in S_{2r}\},$$

and inductively

$$\eta_{i+1} = \inf\{t > \sigma_i\colon X^\varepsilon_t \in S_r \cup \partial D\},$$

$$\sigma_{i+1} = \inf\{t > \eta_{i+1}\colon X^\varepsilon_t \in S_{2r}\},$$

for $i \geq 0$, with the usual convention that $\inf \emptyset = +\infty$. Starting at any $x \in S_r$ we then consider the imbedded Markov chain $Z^\varepsilon_n = X^\varepsilon_{\eta_n}$, $n \geq 0$ (see Remark 2.35

below) whose state space is $S_r \cup \partial D$ and let $\nu = \min\{n \geq 1 : Z^\varepsilon_n \in \partial D\}$. In this way, (2.87) becomes

$$\lim_{\varepsilon \to 0} \sup_{x \in S_r} P_x(|Z^\varepsilon_\nu - y_0| > \delta) = 0. \tag{2.89}$$

Taking r sufficiently small (depending on δ) we check that (2.89) holds.

Remark 2.35 In principle we may have $\eta_m(\omega) = +\infty$ which would make the above definition of Z^ε_m meaningless. Nevertheless, except for a null set, in order to get $\eta_m = +\infty$ we need that $X^\varepsilon_{\eta_n} \in \partial D$ for some $n < m$, which allows the process never to come back to S_r, yielding $\sigma_{n+1} = \eta_{n+1} = \cdots = +\infty$. One solution would be to modify $b(\cdot)$ outside $\overline{D}$ in such a way that it keeps returning to D (and so to S_r). Since $b(\cdot)$ is unchanged in $\overline{D}$, the probabilities in the statement of Theorem 2.33 are not affected. Alternatively, one could add an extra point 'ς' to the state space and define $Z^\varepsilon_n = \varsigma$ if $\eta_n = +\infty$. Since Z^ε cannot go to ς without passing by ∂D, this does not interfere with the events of interest.

To verify (2.89), since τ_ε (and so ν) is a.s. finite and the event $\{\nu \geq n\} = \{Z^\varepsilon_j \in S_r, \forall\, j \leq n-1\}$ belongs to the σ-field generated by $Z^\varepsilon_1, \ldots, Z^\varepsilon_{n-1}$, the Markov property gives us, for each $x \in S_r$:

$$\begin{aligned} &P_x(|Z^\varepsilon_\nu - y_0| > \delta) \\ &= \sum_n E_x(\mathbf{1}_{(\nu \geq n)} P_{Z^\varepsilon_{n-1}}(Z^\varepsilon_1 \in \partial D \setminus B_\delta(y_0))) \\ &\leq \sum_n E_x(\mathbf{1}_{(\nu \geq n)} P_{Z^\varepsilon_{n-1}}(Z^\varepsilon_1 \in \partial D)) \sup_{z \in S_r} \frac{P_z(Z^\varepsilon_1 \in \partial D \setminus B_\delta(y_0))}{P_z(Z^\varepsilon_1 \in \partial D)} \\ &= \sup_{z \in S_r} \frac{P_z(Z^\varepsilon_1 \in \partial D \setminus B_\delta(y_0))}{P_z(Z^\varepsilon_1 \in \partial D)}. \end{aligned} \tag{2.90}$$

It suffices to prove that the last term on the right tends to zero provided $r > 0$ is properly chosen. For this let $\chi := \min_{y \in \partial D \setminus B_\delta(y_0)} V(x_0, y) - V(x_0, y_0)$, which is positive, by assumption. If L is given by Lemma 2.32, let us also require that $r \in (0, \chi/20L)$, and for such a choice let us prove that for ε sufficiently small:

$$\inf_{x \in S_r} P_x(Z^\varepsilon_1 \in \partial D) \geq \exp\{-\varepsilon^{-2}(V(x_0, y_0) + 4\chi/10)\}, \tag{2.91}$$

$$\sup_{x \in S_r} P_x(Z^\varepsilon_1 \in \partial D \setminus B_\delta(y_0)) \leq \exp\{-\varepsilon^{-2}(V(x_0, y_0) + 5\chi/10)\}. \tag{2.92}$$

The proof will then follow due to (2.90). For (2.91) it is enough to construct, for suitable $T < +\infty$ and each $x \in S_r$, a trajectory $\varphi^x \in C([0, T], \mathbb{R}^d)$, with $\varphi^x(0) = x$, $I_T(\varphi^x) \leq V(x_0, y_0) + 3\chi/10$ and such that whenever $\psi \in C([0, T], \mathbb{R}^d)$ and $\varrho_T(\psi, \varphi^x) < r$ then, after hitting S_{2r}, ψ leaves D before returning to S_r. In such a case the lower estimate (b)$'_u$ in Corollary 2.26 gives (2.91), provided ε is small enough. The construction of φ^x is done in four pieces: the main one is obtained from any trajectory $\tilde{\varphi} \in C([0, t_0], \mathbb{R}^d)$ which connects x_0 to y_0 and has action slightly above $V(x_0, y_0)$ (say less than $V(x_0, y_0) + \chi/10$). We call t_1, x_1 the last time and position where $\tilde{\varphi}$ hits S_{2r}. We use Lemma 2.32 and interpolate between x to x_0 and x_0 to x_1. From x_1 we follow $\tilde{\varphi}$ until hitting y_0, and then we make another interpolation to a point $\tilde{y} \notin D_{(r)}$, $|\tilde{y} - y_0| = r$ in order to exit from $D_{(r)}$. These interpolation pieces contribute very little; according to Lemma 2.32 their contribution is at most $4Lr < 2\chi/10$, and $T = t_0 - t_1 + 4r$, giving us equation (2.91). (Interpolating between x and x_0 and then to x_1 gave a constant time for all trajectories; this is not important and if instead we had trajectories in intervals $[0, \tilde{t}(x)]$ with $\sup_{x \in S_r} \tilde{t}(x) = T < +\infty$ we would continue by the deterministic path in the time interval $[\tilde{t}(x), T]$, to fit exactly in the form of (b)$'_u$ of Corollary 2.26.)

The argument for (2.92) is analogous, using instead (c)$'_u$ of Corollary 2.26. By path continuity and the strong Markov property:

$$\sup_{x \in S_r} P_x \left(Z_1^\varepsilon \in \partial D \setminus B_\delta(y_0) \right) \leq \sup_{y \in S_{2r}} P_y \left(X^\varepsilon_{\tau_\varepsilon(S_r \cup \partial D)} \in \partial D \setminus B_\delta(y_0) \right). \tag{2.93}$$

Now, given $t_0 > 0$ we decompose the event on the right-hand side of (2.93) according to $\tau_\varepsilon(S_r \cup \partial D) \leq t_0$ or $\tau_\varepsilon(S_r \cup \partial D) > t_0$. Lemma 2.34 tells us that the second part can be made negligible, i.e. we may take t_0 so that

$$\sup_{y \in S_{2r}} P_y(\tau_\varepsilon(S_r \cup \partial D) > t_0) \leq \exp\{-\varepsilon^{-2}(V(x_0, y_0) + \chi)\}, \tag{2.94}$$

for all ε sufficiently small. It remains to consider

$$\sup_{y \in S_{2r}} P_y(X^\varepsilon_{\tau_\varepsilon(S_r \cup \partial D)} \in \partial D \setminus B_\delta(y_0),\ \tau_\varepsilon(S_r \cup \partial D) \leq t_0).$$

Take $c_0 = V(x_0, y_0) + 6\chi/10$ and observe that if $F = \{\varphi \in C([0, t_0], \mathbb{R}^d) : \varphi(0) \in S_{2r},\ I_{t_0}(\varphi) \leq c_0\}$ then for each $\varphi \in F$ and $t \in [0, t_0]$, $\varphi(t)$ must be at distance at least r from $\partial D \setminus B_\delta(y_0)$. Indeed, if this were not the case there would exist $t_1 \leq t_0$ and $z \in \partial D \setminus B_\delta(y_0)$ so that $|\varphi(t_1) - z| \leq r$, and in such case we could, after interpolating x_0 to $\varphi(0)$, following φ up to $\varphi(t_1)$ and then interpolating $\varphi(t_1)$ to z, obtain a function ψ connecting x_0 to $\partial D \setminus B_\delta(y_0)$ in some finite time and with action function at most $V(x_0, y_0) + 3\chi/4$ contradicting the choice of χ. Thus,

$$\sup_{y \in S_{2r}} P_y(X^\varepsilon_{\tau_\varepsilon(S_r \cup \partial D)} \in \partial D \setminus B_\delta(y_0),\ \tau_\varepsilon(S_r \cup \partial D) \leq t_0)$$

$$\leq \sup_{y \in S_{2r}} P_y(\varrho_{t_0}(X^\varepsilon, \mathbb{F}^y_{t_0}(c_0)) \geq r).$$

Applying (c)$'_u$ of Corollary 2.26 and recalling (2.94), we get (2.92) for ε sufficiently small. This ends the proof of Theorem 2.33. □

Remark 2.36 The previous proof extends at once when the fixed point x_0 is replaced by an attracting limit cycle. The situation of several attractors in D is highly relevant for the problem of metastability and will be discussed partially in Chapter 5. Another comment refers to the uniqueness of minimizing y_0. The proof of Theorem 2.33 shows that if $C := \{y \in \partial D\colon V(x_0, y) = \min_{z\in\partial D} V(x_0, z)\}$ then $\lim_{\varepsilon\to 0} P_x(X^\varepsilon_{\tau_\varepsilon} \notin C_{(\delta)}) = 0$ where $C_{(\delta)}$ is the δ-neighbourhood of C. It does not say anything with respect to different points in C; this is beyond the 'rough' large deviation estimates used in the proof. Except for special situations, e.g. if symmetry provides uniform distribution in C, a more refined analysis is required.

Before proceeding with the basic questions raised above, let us discuss briefly the determination of $V(x_0, \cdot)$, as required for the application of Theorem 2.33. Since $V(x_0, \cdot)$ is defined through a variational problem, its differentiability would be useful at this point, due to the Jacobi equation. Though $V(x_0, \cdot)$ is Lipschitz, it is not necessarily differentiable. The simplest case, where no difficulty occurs, is that of a gradient system

$$b(x) = -\nabla U(x), \quad x \in \mathbb{R}^d, \tag{2.95}$$

for $U\colon \mathbb{R}^d \to \mathbb{R}$ which we assume to have continuous and bounded derivatives up to second order. In particular, $b(\cdot)$ satisfies (2.51). (Considering the process stopped upon its escape from D, it would suffice to have $U(\cdot)$ of class C^2, cf. proof of Theorem 2.24.) The characterization of $V(x_0, \cdot)$ in this case is very simple and the next proposition gives at least one reason for the name 'quasi-potential'.

Remark Under (2.95), proper growth conditions on $U(\cdot)$ at infinity guarantee the existence of a reversible probability measure for X^ε. In this case, the exit times can be analysed by exploiting more the reversibility instead of the l.d.p. This will be seen in Chapter 5.

Proposition 2.37 *Besides the assumptions of Theorem* 2.33, *let us suppose that* (2.95) *holds (everywhere), for some $U\colon \mathbb{R}^d \to \mathbb{R}$ of class C^2. Then, for each $y \in \overline{D}$*

$$V(x_0, y) = 2\,(U(y) - U(x_0)). \tag{2.96}$$

Moreover, given $T > -\infty$ and $y \in \overline{D}$, the reversed deterministic trajectory

$$\psi(t) = X^{y,0}_{T-t}, \quad -\infty < t \le T,$$

is the unique trajectory which attains the minimum

$$V(x_0, y) = \inf\left\{\int_{-\infty}^{T} |\dot\varphi(t) - b(\varphi(t))|^2\, dt\colon \varphi(-\infty) = x_0,\ \varphi(T) = y\right\}.$$

Proof Let φ be such that $I_T(\varphi) < +\infty$. Using the identity $|u - v|^2 = |u + v|^2 - 4\langle u, v\rangle$, $u, v \in \mathbb{R}^d$, and (2.95) we have:

$$\begin{aligned} I_T(\varphi) &= \frac{1}{2}\int_0^T |\dot\varphi(s) - b(\varphi(s))|^2\, ds \\ &= \frac{1}{2}\int_0^T |\dot\varphi(s) + b(\varphi(s))|^2\, ds - 2\int_0^T \langle\dot\varphi(s), b(\varphi(s))\rangle\, ds \\ &= \frac{1}{2}\int_0^T |\dot\varphi(s) + b(\varphi(s))|^2\, ds + 2\int_0^T \langle\dot\varphi(s), \nabla U(\varphi(s))\rangle\, ds \\ &= \frac{1}{2}\int_0^T |\dot\varphi(s) + b(\varphi(s))|^2\, ds + 2(U(\varphi(T)) - U(\varphi(0))). \end{aligned} \tag{2.97}$$

From (2.97) we immediately get that for any $y \in \mathbb{R}^d$:

$$V(x_0, y) \geq 2(U(y) - U(x_0)). \tag{2.98}$$

On the other hand, if $y \in \overline{D}$ and ψ is the time reversed orbit of $X^{y,0}$ such that $\psi(T) = y$ (i.e. $\psi(t) = X^{y,0}_{T-t}$, $0 \leq t \leq T$), then $\dot\psi(s) = -b(\psi(s))$ and the first term on the right-hand side of (2.97) vanishes, implying that $I_T(\psi) = 2(U(y) - U(X^{y,0}_T))$.

Letting $T \to +\infty$, we have $X^{y,0}_T \to x_0$ under Assumptions 2.30, and we get equality in (2.98), for each $y \in \overline{D}$. As for the uniqueness, (2.97) implies that

$$\int_{T_1}^T |\dot\varphi(s) - b(\varphi(s))|^2\, ds \geq 2(U(\varphi(T)) - U(\varphi(T_1))),$$

with equality if and only if φ is absolutely continuous in (T_1, T) and $\int_{T_1}^T |\dot\varphi(s) + b(\varphi(s))|^2\, ds = 0$, namely $\dot\varphi(t) = -b(\varphi(t))$ for $T_1 < t < T$. Since $b(\cdot)$ is Lipschitz the uniqueness follows, given $\varphi(T) = y$. Allowing T_1 to be taken as $-\infty$, we get the uniqueness of the extremal trajectory, given $\psi(T) = y$. □

Remark 2.38 The previous argument shows that the infimum in Definition 2.31 might not be attainable (along finite intervals). On the other hand, allowing semi-infinite trajectories ($\psi(t)\colon -\infty \leq t \leq T$), Freidlin and Wentzell prove that the infimum is indeed reached under the more general conditions of Theorem 2.33 (see [295]). Here $T > -\infty$ is arbitrary; nothing changes under the time shift.

Concerning the last statement of Proposition 2.37, its main importance is to go beyond the result of Theorem 2.33, giving not only $X^\varepsilon(\tau_\varepsilon)$ but also the 'final excursion' leading to ∂D. Typically, this will be near the extremal trajectory for $y = y_0$ (if this is unique up to translations in time). To turn this into a precise statement one needs to take a neighbourhood G_0 of x_0 such that the extremal trajectory leaves G_0 in a 'regular' fashion (without infinitely many oscillations around its boundary). Freidlin and Wentzell have proven that after its last visit to

G_0 the perturbed trajectory will follow closely the extremal one, until time τ_ε. For a precise statement see Theorem 2.3, Chapter 4 in [122]. In Chapter 6 we shall discuss a similar problem in another set-up.

Remark 2.39 Under the assumptions of the previous proposition, x_0 is the point of minimum of $U(\cdot)$ and its unique critical point in $\overline{D}$. For simplicity we have imposed too restrictive conditions. They may be relaxed to:

$$b(x) = -\nabla U(x) + \tilde{b}(x), \qquad x \in \overline{D},$$

where

(a) $\langle \nabla U(x), \tilde{b}(x) \rangle = 0$ for all $x \in \overline{D}$,

(b) $U(x) > U(x_0)$ and $\nabla U(x) \neq 0$ for all $x \in \overline{D} \setminus \{x_0\}$.

Under these last conditions, (2.96) still holds for each $y \in \overline{D}$ such that $U(y) \le \min_{z \in \partial D} U(z)$. Here it suffices that $U(\cdot)$ is of class C^1. When $U(\cdot)$ is of class C^2 the second statement of the previous proposition also holds. Fixing $\psi(T) = y$, the extremal trajectory is the unique solution of $\dot{\psi}(t) = \nabla U(\psi(t)) + \tilde{b}(\psi(t))$ for $t < T$, with $\psi(T) = y$.

The assumption of class C^2 guarantees the uniqueness of such ψ. The fact that $\lim_{t \to -\infty} \psi(t) = x_0$ follows quite easily from the assumption that $U(\cdot)$ does not have other critical points in $\overline{D}$, after observing that $\partial U(\psi(t))/\partial t = |\nabla U(\psi(t))|^2$. The statement follows once we verify that $\frac{1}{2}\int_t^T \left|\dot{\psi}(s) - b(\psi(s))\right|^2 ds = 2\,(U(y) - U(\psi(t)))$ and that $V(x_0, y) \ge 2(U(y) - U(x_0))$ for $y \in \overline{D}$ such that $U(y) \le \min_{z \in \partial D} U(z)$. To see this, use (a) to rewrite $I_T(\varphi)$ analogously to (2.97); if $\varphi(t) \in \overline{D}$ for all $0 \le t \le T$, one has

$$I_T(\varphi) = \frac{1}{2}\int_0^T \left|\dot{\varphi}(s) - \nabla U(\varphi(s)) - \tilde{b}(\varphi(s))\right|^2 ds + 2\int_0^T \langle \dot{\varphi}(s), \nabla U(\varphi(s)) \rangle ds.$$

The existence of the above orthogonal decomposition of $b(\cdot)$ is indeed equivalent to the continuous differentiability of $V(x_0, \cdot)$. (See Theorem 3.1 and further analysis in Chapter 4 of [122].)

Example To exemplify the content of the previous observations let us take a quadratic potential $U(x) = \frac{1}{2}\langle Bx, x \rangle$, where B is a $d \times d$ matrix with real entries. In this case, $\nabla U(x) = Ax$ where $A = (B + B^{\mathbb{T}})/2$, a symmetric real matrix. If $b(x) := -Ax$ we have a gradient system and if A is positive definite we can use Proposition 2.37 with $x_0 = 0$.

If we now let $\tilde{b}(x) = \frac{1}{2}(B^{\mathbb{T}} - B)x$ then $-\nabla U(x) + \tilde{b}(x) = -Bx$ and the condition $\langle \nabla U(x), \tilde{b}(x) \rangle = 0$ for all x is equivalent to $|B^{\mathbb{T}} x|^2 = |Bx|^2$ for all x. Thus we assume that $BB^{\mathbb{T}} = B^{\mathbb{T}}B$. If moreover $(B + B^{\mathbb{T}})/2$ is positive definite (eigenvalues of $-B$ with negative real part), the asymptotic stability of $x_0 = 0$ is ensured. We can apply the previous result for $b(x) := -Bx$ and $x_0 = 0$. In this case $X_t^{x,0} \to 0$, for any x. If the domain D satisfies the conditions in Theorem 2.33 and y_0 is the unique point of minimum of $U(\cdot)$ on ∂D, the 'final part' of the escape will be (modulo a time shift) near the solution of $\dot{\psi}(t) = B^{\mathbb{T}}\psi(t)$ for $t < 0$ with $\psi(0) = y_0$, with large probability.

Concerning the order of magnitude of τ_ε, the following is a basic result.

Theorem 2.40 (Freidlin and Wentzell) *If $b(\cdot)$ is Lipschitz, Assumptions* 2.30 *hold true, and $V_0 := \inf_{y\in\partial D} V(x_0, y)$, then for any $x \in D$:*
(*i*) $\lim_{\varepsilon\to 0} \varepsilon^2 \log E_x(\tau_\varepsilon) = V_0$;
(*ii*) $\lim_{\varepsilon\to 0} P_x\left(e^{(V_0-\zeta)/\varepsilon^2} < \tau_\varepsilon < e^{(V_0+\zeta)/\varepsilon^2}\right) = 1$, *for any $\zeta > 0$.*
Moreover, both convergences hold uniformly in x, on each compact subset of D.

Proof Before getting started with the proof, we may observe that $V_0 > 0$ under Assumptions 2.30. This can be seen from Remark 2.38, which we have not proven in general. It is useful to observe that for the upper bounds for τ_ε this will not play any role in the proof below. On the other hand, the lower bounds become quite simple if $V_0 = 0$ (see also Remark 2.41 below).

We first prove the following (uniform) upper estimate:

$$\limsup_{\varepsilon\to 0} \varepsilon^2 \sup_{x\in D} \log E_x(\tau_\varepsilon) \le V_0. \tag{2.99}$$

This will follow once we check that for any $\zeta > 0$ there exists a constant $T < +\infty$ and $\varepsilon_0 > 0$ so that for all $0 < \varepsilon < \varepsilon_0$

$$\min_{x\in D} P_x(\tau_\varepsilon < T) \ge \frac{1}{2} e^{-\frac{V_0+2\zeta/3}{\varepsilon^2}}. \tag{2.100}$$

Indeed, the Markov property and (2.100) imply that for $0 < \varepsilon < \varepsilon_0$ and all $x \in D$:

$$\begin{aligned} E_x(\tau_\varepsilon) &\le T \sum_{n=0}^{+\infty} P_x(\tau_\varepsilon \ge nT) \le T \sum_{n=0}^{+\infty} \left(\sup_{y\in D} P_y(\tau_\varepsilon \ge T) \right)^n \\ &\le \frac{T}{\min_{y\in D} P_y(\tau_\varepsilon < T)} \le 2T e^{\frac{V_0+2\zeta/3}{\varepsilon^2}}. \end{aligned}$$

Since $\zeta > 0$ is arbitrary we get (2.99). Now, to check (2.100) we recall two observations from the proof of Theorem 2.33.

(a) With the construction used in the proof of (2.91) we may find $r > 0$, $T_1 < +\infty$ with the property that for any $y \in \bar{B}_{2r}(x_0)$ there exists a function $\varphi^y \in C([0, T_1], \mathbb{R}^d)$ such that $\varphi^y(0) = y$, φ^y leaves the set $D_{(r)}$ before time T_1, and moreover $I_{T_1}(\varphi^y) \le V_0 + \zeta/2$. Thus, $\varepsilon_0 > 0$ may be taken so that for any $\varepsilon \in (0, \varepsilon_0]$ and any $y \in \bar{B}_{2r}(x_0)$:

$$P_y(\tau_\varepsilon < T_1) \ge P_y(\varrho_{T_1}(X^\varepsilon, \varphi^y) < r) \ge e^{-\frac{V_0+2\zeta/3}{\varepsilon^2}}, \tag{2.101}$$

where the first inequality is immediate and the second follows from (b)$'_u$ of Corollary 2.26.

(b) With $r > 0$ small chosen as in (a), we use Assumptions 2.30 to find $T_2 < +\infty$ so that for each $x \in \overline{D}$ the deterministic path $X_t^{x,0}$ arrives at $B_r(x_0)$ before T_2. Thus

$$P_x(\tau_\varepsilon(\bar{B}_{2r}(x_0)) < T_2) \geq P_x(\varrho_{T_2}(X^\varepsilon, X^{x,0}) \leq r),$$

which tends to one uniformly in x as $\varepsilon \to 0$, according to (2.53).

By the strong the Markov property, for any $x \in D$ we have:

$$P_x(\tau_\varepsilon < T_1 + T_2) \geq P_x(\tau_\varepsilon(\bar{B}_{2r}(x_0)) < T_2) \inf_{y \in \bar{B}_{2r}(x_0)} P_y(\tau_\varepsilon < T_1)$$

and the previous observations yield (2.100) with $T = T_1 + T_2$ and ε small enough. This completes the proof of (2.99), to which we apply the Markov inequality and see that

$$\lim_{\varepsilon \to 0} P_x\left(\tau_\varepsilon > e^{\frac{V_0+\zeta}{\varepsilon^2}}\right) = 0,$$

for any $\zeta > 0$ and $x \in D$, verifying the upper bound in the statement (ii).

Concerning the lower estimates, it suffices to prove that if $x \in D$ and $\zeta > 0$, then

$$\lim_{\varepsilon \to 0} P_x\left(\tau_\varepsilon < e^{\frac{V_0-\zeta}{\varepsilon^2}}\right) = 0. \tag{2.102}$$

Indeed, rewriting the Markov inequality $E_x(\tau_\varepsilon) \geq e^{\frac{V_0-\zeta}{\varepsilon^2}} P_x\left(\tau_\varepsilon \geq e^{\frac{V_0-\zeta}{\varepsilon^2}}\right)$ we see that

$$\liminf_{\varepsilon \to 0} \varepsilon^2 \log E_x(\tau_\varepsilon) \geq V_0 - \zeta,$$

which would conclude the proof.

To check (2.102) we first recall the imbedded Markov chain used in the proof of Theorem 2.33 and verify that given $0 < \zeta < V_0$, if $r > 0$ is small enough, one has

$$\sup_{x \in S_r} P_x(\nu = 1) \leq e^{-\frac{V_0-\zeta}{\varepsilon^2}}, \tag{2.103}$$

provided ε is sufficiently small. Indeed, using the strong Markov property at σ_0 we can write, for $0 < T < +\infty$:

$$\begin{aligned}\sup_{x \in S_r} P_x(\nu = 1) &= \sup_{y \in S_{2r}} P_y(\tau_\varepsilon = \tau_\varepsilon(S_r \cup \partial D)) \\ &\leq \sup_{y \in S_{2r}} P_y(\tau_\varepsilon = \tau_\varepsilon(S_r \cup \partial D) < T) \\ &\quad + \sup_{y \in S_{2r}} P_y(\tau_\varepsilon = \tau_\varepsilon(S_r \cup \partial D) \geq T).\end{aligned} \tag{2.104}$$

For the last term we use Lemma 2.34: if $r > 0$ and $\bar{B}_{2r}(x_0) \subseteq D$ we can take $T = T_r < +\infty$ so that

$$\sup_{y \in S_{2r}} P_y(\tau_\varepsilon = \tau_\varepsilon(S_r \cup \partial D) \geq T) \leq e^{-\frac{V_0}{\varepsilon^2}},$$

provided ε is small enough.

For the first term on the right-hand side of (2.104) we argue as in the proof of (2.92): given $\zeta > 0$ we may take $r > 0$ small enough so that the level set $\bigcup_{y \in S_{2r}} \mathbb{F}_T^y(V_0 - \zeta/2)$ is at positive distance from the set of trajectories which start at S_{2r} and hit ∂D before time T. From the upper estimate $(c)'_u$ of Corollary 2.26 we get that for such an $r > 0$, and ε sufficiently small:

$$\sup_{y \in S_{2r}} P_y(\tau_\varepsilon = \tau_\varepsilon(S_r \cup \partial D) < T) \leq e^{-\frac{V_0 - 2\zeta/3}{\varepsilon^2}},$$

which together with the last two inequalities gives (2.103) provided $r > 0$ is suitably small (depending on ζ) and then ε is small enough.

There is a uniform lower bound $t_r > 0$ (independent of $\varepsilon > 0$) for the time that the deterministic paths $X^{y,0}$ with $y \in S_{2r}$ need in order to reach $B_{3r/2}(x_0)$ and we can assume (since r is small) that all such paths keep distance larger than $r/2$ from D^c. Using the strong Markov property as before and (2.53), we get, for ε sufficiently small,

$$\inf_{y \in S_r} P_y(\eta_1 \geq t_r) \geq \inf_{y \in S_{2r}} P_y(\varrho_{t_r}(X^\varepsilon, X^{y,0}) < r/2) \geq \frac{2}{3}. \tag{2.105}$$

Given $\zeta > 0$ we take $r > 0$ as above. If $x \in D$ and $n_\varepsilon \geq 1$ we write

$$P_x\left(\tau_\varepsilon < e^{\frac{V_0 - 2\zeta}{\varepsilon^2}}\right) \leq P_x(\tau_\varepsilon \leq \tau_\varepsilon(S_r)) + \sup_{y \in S_r}\left(P_y(\nu \leq n_\varepsilon) + P_y(\eta_{n_\varepsilon} < e^{\frac{V_0 - 2\zeta}{\varepsilon^2}})\right),$$

where we have used the strong Markov property and the fact that $\{\nu \geq n_\varepsilon\} \subseteq \{\tau_\varepsilon \geq \eta_{n_\varepsilon}\}$.

We already know that the first term on the right-hand side of the above equation vanishes as $\varepsilon \to 0$. To take care of the second term, with a proper choice of n_ε, notice that (2.103) and the Markov property of Z^ε imply that

$$\inf_{x \in S_r} P_x(\nu > n) \geq (1 - e^{-\frac{V_0 - \zeta}{\varepsilon^2}})^n,$$

for all $n \geq 1$, with r and ε as in (2.103). Setting $n_\varepsilon = [\exp\{\varepsilon^{-2}(V_0 - 3\zeta/2)\}] + 1$, we have:

$$\sup_{y \in S_r} P_y(\nu \leq n_\varepsilon) \leq 1 - (1 - e^{-\frac{V_0 - \zeta}{\varepsilon^2}})^{n_\varepsilon},$$

which tends to zero as $\varepsilon \to 0$. It remains to check that for this choice of n_ε we have

$$\lim_{\varepsilon\to 0} \sup_{y\in S_r} P_y(\eta_{n_\varepsilon} < e^{\frac{V_0-2\zeta}{\varepsilon^2}}) = 0.$$

From the definition of n_ε we write $P_y(\eta_{n_\varepsilon} < e^{(V_0-2\zeta)/\varepsilon^2}) \le P_y(\eta_{n_\varepsilon}/n_\varepsilon < e^{-\zeta/2\varepsilon^2})$. We now check that this vanishes as $\varepsilon \to 0$. But $\eta_n \ge \sum_{j=1}^n t_r \mathbf{1}_{(\eta_j-\eta_{j-1}\ge t_r)}$, so that recalling (2.105) and using the strong Markov property we have, for ε sufficiently small and $n \ge 1$:

$$\sup_{y\in S_r} P_y\left(\frac{\eta_n}{n} < e^{-\frac{\zeta}{2\varepsilon^2}}\right) \le P\left(\frac{R_n}{n} < \frac{1}{t_r}e^{-\frac{\zeta}{2\varepsilon^2}}\right), \tag{2.106}$$

where R_n represents the number of successes in n independent Bernoulli trials with probability of success being $2/3$. Since $t_r^{-1}e^{-\zeta/2\varepsilon^2} \le 1/3$ for ε small enough, the conclusion follows at once and we get (2.102) with ζ replaced by 2ζ. This holds for $\zeta > 0$ arbitrarily small, thus completing the proof of (i) and (ii).

The statement on the uniformity follows from the proof and previous observations, according to Corollary 2.26. □

Remark 2.41 The asymptotic stability of x_0 implies that given $r > 0$ we may take $r_1 \in (0, r/2)$ so that for any $y \in B_{r_1}(x_0)$, $X_t^{y,0}$ converges to x_0, as $t \to +\infty$, without ever leaving $B_{r/2}(x_0)$. Using this, the rough estimate (2.53), and the Markov property, we see that $c_1 > 0$ can be taken so that

$$\sup_{y\in B_{r_1}(x_0)} P_y\big(\tau_\varepsilon(S_r(x_0)) < e^{\frac{c_1}{\varepsilon^2}}\big) \le e^{-\frac{c_1}{\varepsilon^2}}, \tag{2.107}$$

for all $\varepsilon > 0$ small enough. (In particular, this shows that $V_0 > 0$ in Theorem 2.40.)

To check (2.107), notice that we can also take $r_2 < r_1/2$ such that the deterministic orbits starting in $B_{r_2}(x_0)$ never leave $B_{r_1/2}(x_0)$. Now, we take t_1 finite so that $X_t^{y,0}$ hits $B_{r_2}(x_0)$ during $[0, t_1]$, for any $y \in B_{r_1}(x_0)$; in particular $X_{t_1}^{y,0} \in B_{r_1/2}(x_0)$. For $t_i = it_1$, $i = 0, 1, \ldots,$ we say that the interval $[t_i, t_{i+1}]$ is *regular* if: $X_t^{y,\varepsilon} \in B_{r_1}(x_0)$, for $t = t_i, t_{i+1}$ and the trajectory does not exit from $B_r(x_0)$ during $[t_i, t_{i+1}]$. Then, (2.53) implies that

$$\inf_{y\in B_{r_1}(x_0)} P_y([0,t_1] \text{ is regular}) \ge \inf_{y\in B_{r_1}(x_0)} P_y(\varrho_{t_1}(X^{y,\varepsilon}, X^{y,0}) \le r_1/2) \ge 1 - e^{-\frac{C}{\varepsilon^2}},$$

for suitable C positive (depending on r_1, t_1, κ, d) and $\varepsilon > 0$ small. Using the Markov property and repeating this for $i = 1, \ldots, [e^{C/2\varepsilon^2}]$ we get (2.107) if $c_1 < C/2$ and ε is small.

The study of the asymptotic distribution of $\tau_\varepsilon / E_x \tau_\varepsilon$ is postponed to Chapter 5.

The basic l.d.p. of Freidlin and Wentzell is useful in providing information on the asymptotic behaviour of the stationary measure. Clearly, proper conditions on the vector field $b(x)$ for $|x|$ large are needed to guarantee that X^ε has a (unique) stationary probability measure μ_ε (see [176]).

Just to get a feeling of the problem, let us recall the gradient case $b(x) = -\nabla U(x)$, with $U(\cdot)$ having a unique global minimum at x_0. Proper growth conditions on U (e.g. if U grows at least linearly at infinity) guarantee the ergodicity. In this case μ_ε is explicitly known, given by

$$\mu_\varepsilon(A) = \frac{\int_A e^{-\varepsilon^{-2} 2U(x)}\, dx}{\int_{\mathbb{R}^d} e^{-\varepsilon^{-2} 2U(x)}\, dx}.$$

We see that the Laplace–Varadhan method, cf. Theorem 1.18, is applicable and under the conditions of Theorem 2.40 we have, as $\varepsilon \to 0$:

$$\frac{1}{\varepsilon^2} \ln \mu_\varepsilon(A) \sim -2 \inf_{x \in A} (U(x) - U(x_0)).$$

Freidlin and Wentzell estimates can be applied to generalize this description to non-gradient cases, with the quasi-potential $V(x_0, y)$ playing the fundamental role; see Chapter 4 in [122].

3

Large deviations and statistical mechanics

Having in mind the discussion on metastability, we need a brief introduction to statistical mechanics. This will be done in Section 3.3. Considerations on the connection between large deviation theory and the statistical description of (equilibrium) thermodynamical systems appear naturally and are the content of Section 3.4, focusing on basic aspects rather than generality, with references to more advanced literature.

Apart from such motivations, the extension of the results of Chapter 1 beyond independent variables is natural in many contexts. We discuss the Gärtner–Ellis method briefly and apply it to finite Markov chains, as the simplest situation to start with. While discussing large deviations for (equilibrium) statistical mechanics models, it is important to stress the role of subadditivity and convexity, already illustrated in Section 1.5, where the Ruelle–Lanford method was considered in the context of i.i.d. variables. Some basic results are taken from Section 3 of Pfister's lecture notes [243], simplified for our situation.

3.1 Large deviations for dependent variables. Gärtner–Ellis theorem

Examining the proof of the Cramér theorem in Section 1.4, we are naturally led to allow a moderate dependence among the X_i variables. This extension is due to Gärtner [132] and Ellis [108]. To describe it briefly, we consider $\mathbb{R}^d$-valued variables W_n (replacing the previous $\bar{X}_n$), $n \geq 1$. Let μ_n be the law of W_n on $\mathbb{R}^d$ and $\hat{\mu}_n$ its moment generating function:

$$\hat{\mu}_n(\zeta) = Ee^{\langle \zeta, W_n \rangle}, \quad \zeta \in \mathbb{R}^d,$$

where $\langle \cdot, \cdot \rangle$ denotes the Euclidean inner product. The first basic assumptions involve the limiting behaviour of $\hat{\mu}_n$.

Assumptions 3.1 There exists $a_n \nearrow +\infty$ so that
(i) $\varphi(\zeta) := \lim_{n\to+\infty} a_n^{-1} \log \hat{\mu}_n(a_n\zeta)$ exists in $(-\infty, +\infty]$, for each ζ;
(ii) $0 \in \mathcal{D}_\varphi^\circ$, the interior of $\mathcal{D}_\varphi := \{\zeta \in \mathbb{R}^d : \varphi(\zeta) < +\infty\}$.

In the Cramér theorem, $a_n = n$ and $a_n^{-1} \log \hat{\mu}_n(a_n\zeta) = \log \hat{\mu}(\zeta) \in (-\infty, +\infty]$ for all n. As in that case we may consider the Fenchel–Legendre transform of φ:

$$\varphi^*(x) = \sup_{\zeta\in\mathbb{R}^d} (\langle\zeta, x\rangle - \varphi(\zeta)). \tag{3.1}$$

Basic properties of φ and φ^* to be needed below are the following.

Lemma 3.2 *Under Assumptions* 3.1 *we have*:
(a) $\varphi\colon \mathbb{R}^d \to (-\infty, +\infty]$ *is convex;*
(b) $\varphi^*\colon \mathbb{R}^d \to [0, +\infty]$ *is convex and satisfies Definition* 1.13;
(c) If $y = \nabla\varphi(\zeta)$ *for some* $\zeta \in \mathcal{D}_\varphi^\circ$, *then*

$$\varphi^*(y) = \langle\zeta, y\rangle - \varphi(\zeta) \tag{3.2}$$

and $y \in \mathcal{E}(\varphi, \varphi^*)$ *has* ζ *as exposing hyperplane, according to the next definition.*

Definition 3.3 A point $y \in \mathbb{R}^d$ is said to be φ^*-exposed if there exists $\zeta \in \mathbb{R}^d$ such that

$$\varphi^*(z) > \varphi^*(y) + \langle\zeta, z - y\rangle, \quad \text{for any } z \neq y.$$

In this case, we say that ζ is (determines) an exposing hyperplane and set

$$\mathcal{E}(\varphi, \varphi^*) := \{y \in \mathbb{R}^d : \exists\, \zeta \in \mathcal{D}_\varphi^\circ \text{ exposing hyperplane}\}.$$

Proof of Lemma 3.2 (a) The convexity of φ follows at once from that of $\log \hat{\mu}_n$, seen in Section 1.1.

(b) The lower semi-continuity, convexity and the fact that $\varphi^*(x) \geq -\varphi(0) = 0$ follow at once from the definition, just as in Section 1.1. We now check that under Assumptions 3.1 the function φ^* has bounded level sets: for that we start by taking $\delta > 0$ so that the closed Euclidean ball $\bar{B}_\delta(0) \subseteq \mathcal{D}_\varphi^\circ$. Being convex, φ is continuous on $\mathcal{D}_\varphi^\circ$ so that $\sup_{\zeta\in\bar{B}_\delta(0)} \varphi(\zeta) = a < +\infty$. Consequently (with $|\cdot|$ the Euclidean norm)

$$\varphi^*(x) \geq \sup_{\zeta\in\bar{B}_\delta(0)} (\langle\zeta, x\rangle - \varphi(\zeta)) \geq \delta|x| - a$$

from which the conclusion follows.

(c) The identity (3.2) follows as part (iii) of Proposition 1.22. To check the last statement we assume that $\varphi^*(x) \leq \varphi^*(y) + \langle\zeta, x - y\rangle$ for some $x \in \mathbb{R}^d$ and show that $x = y$. Indeed, together with (3.1) and (3.2), this assumption implies that for each $\tau \in \mathbb{R}^d$:

$$\varphi(\zeta) = \langle \zeta, y \rangle - \varphi^*(y) \le \langle \zeta, x \rangle - \varphi^*(x) \le -\langle \tau, x \rangle + \varphi(\zeta + \tau),$$

from which, replacing τ by $\delta\tau$ with $\delta > 0$ and letting $\delta \to 0$, we see that

$$\langle \tau, x \rangle \le \lim_{\delta \to 0} \frac{\varphi(\zeta + \delta\tau) - \varphi(\zeta)}{\delta} = \langle \tau, \nabla\varphi(\zeta) \rangle \quad \text{for each } \tau \in \mathbb{R}^d.$$

Consequently $x = \nabla\varphi(\zeta) = y$. □

We may now state the following basic theorem.

Theorem 3.4 (Gärtner–Ellis theorem) *Under Assumptions* 3.1 *we have:*

(*a*) $\limsup_{n\to+\infty} a_n^{-1} \log \mu_n(C) \le -\inf_{x \in C} \varphi^*(x)$ *for any closed set* $C \subseteq \mathbb{R}^d$;

(*b*) $\liminf_{n\to+\infty} a_n^{-1} \log \mu_n(G) \ge -\inf_{x \in G\cap\mathcal{E}} \varphi^*(x)$ *for any open set* $G \subseteq \mathbb{R}^d$, *where* $\mathcal{E} = \mathcal{E}(\varphi, \varphi^*)$.

(*c*) *If, besides Assumptions* 3.1, *we suppose*:

(*i*) φ *is lower semi-continuous on* $\mathbb{R}^d$ *and differentiable on* $\mathcal{D}_\varphi^\circ$,

(*ii*) *either* $\mathcal{D}_\varphi = \mathbb{R}^d$ *or* φ *is steep (at* $\partial\mathcal{D}_\varphi$*), i.e.* $\lim_{\zeta \in \mathcal{D}_\varphi^\circ, \zeta \to \partial\mathcal{D}_\varphi} |\nabla\varphi(\zeta)| = +\infty$,

then we may replace $G \cap \mathcal{E}$ *by* G *in item* (*b*) *above, yielding the validity of a l.d.p for* (μ_n), *with scaling* a_n *and good rate function* φ^*.

The greater involvement of convex analysis in part (c) is the delicate point. It is easy to give examples of probability measures μ on $\mathbb{R}$ so that $\log \hat{\mu}$ is not steep (see below). The proofs of the Cramér theorem in Sections 1.4 and 1.5 have avoided this.

Proof (a) This is quite similar to the proof of the upper bound in Theorem 1.23. Applying the Markov inequality, if C is a Borel set, $n \ge 1$ and $\zeta \in \mathbb{R}^d$ we have:

$$\mu_n(C) \le e^{-\inf_{y\in C}\langle \zeta, y\rangle} \hat{\mu}_n(\zeta).$$

Taking logarithms, dividing by a_n and minimizing in ζ, we can write:

$$\frac{1}{a_n} \log \mu_n(C) \le -\sup_{\zeta \in \mathbb{R}^d} \left(\inf_{y \in C} \langle \zeta, y \rangle - \frac{1}{a_n} \log \hat{\mu}_n(a_n \zeta) \right). \tag{3.3}$$

Let C be a compact set. The upper bound in (i) is trivial if $\inf_{x\in C} \varphi^*(x) = 0$. Otherwise, take $0 < a < \inf_{x \in C} \varphi^*(x)$. In particular, if $y \in C$ we may find $\zeta_y \in \mathbb{R}^d$ and $\delta > 0$ so that $\langle \zeta_y, y \rangle - \varphi(\zeta_y) = a + \delta$. This implies that $\varphi(\zeta_y) < +\infty$ and that $n_y \in \mathbb{N}$ may be taken so that

$$\langle \zeta_y, y \rangle - \frac{1}{a_n} \log \hat{\mu}_n(a_n \zeta_y) > a + \delta/2, \qquad \text{for each } n \ge n_y.$$

Thus, if $\delta_y|\zeta_y| < \delta/2$, (3.3) gives that

$$\frac{1}{a_n}\log \mu_n(B_{\delta_y}(y)) \le -a, \qquad \text{for each } n \ge n_y.$$

Since C is compact, we may cover it with finitely many balls $B_{\delta_{y_i}}(y_i)$ and, as in the proof of Theorem 1.23, we conclude that

$$\limsup_{n\to+\infty}\frac{1}{a_n}\log \mu_n(C) \le -a.$$

Since a may be taken arbitrarily close to $\inf_{y\in C}\varphi^*(y)$, the proof of (a) for compact sets is complete. It does not require $0 \in \mathcal{D}^\circ_\varphi$.

For the extension to closed sets it suffices to verify the exponential tightness (see remark (i) after Definition 1.17). This again goes as in Theorem 1.23: we first take $r > 0$ so that $\bar{B}_r(0) \subseteq \mathcal{D}_\varphi$ and apply the exponential Markov inequality to the coordinates $W_n^{(i)} = \langle \mathbf{e}_i, W_n\rangle$ (with $\mathbf{e}_1, \ldots, \mathbf{e}_d$ denoting the canonical unitary vectors), to write

$$\mu_n(\mathbb{R}^d \setminus [-k,k]^d) \le e^{-ra_nk}\sum_{i=1}^{d}(\hat{\mu}_n(ra_n\mathbf{e}_j) + \hat{\mu}_n(-ra_n\mathbf{e}_j)),$$

so that

$$\limsup_{n\to+\infty}\frac{1}{a_n}\log \mu_n(\mathbb{R}^d \setminus [-k,k]^d) \le -rk + \max_{i=1,\ldots,d}(\varphi(r\mathbf{e}_i)\vee\varphi(-r\mathbf{e}_i)).$$

Sending $k \to +\infty$ we conclude the exponential tightness and the proof of part (a).

(b) As in Proposition 1.15, it suffices to check that

$$\liminf_{n\to+\infty}\frac{1}{a_n}\log \mu_n(B_\delta(x)) \ge -\varphi^*(x)$$

for each $x \in \mathcal{E}$ and $\delta > 0$. Having fixed $x \in \mathcal{E}$ we may find $\zeta \in \mathcal{D}^\circ_\varphi$, exposing hyperplane for φ^* at x. In particular, $a_n\zeta \in \mathcal{D}_{\hat{\mu}_n}$ for large n (say $n \ge n_x$) in which case we may consider the new measures

$$\nu_n(dy) = \frac{1}{\hat{\mu}_n(a_n\zeta)}e^{a_n\langle\zeta,y\rangle}\mu_n(dy). \tag{3.4}$$

Proceeding as in the proof of Theorem 1.23, if $0 < \delta' < \delta$:

$$\begin{aligned}
\frac{1}{a_n}\log \mu_n(B_\delta(x)) &\ge \frac{1}{a_n}\log \mu_n(B_{\delta'}(x)) \\
&= \frac{1}{a_n}\log \hat{\mu}_n(a_n\zeta) + \frac{1}{a_n}\log\int_{B_{\delta'}(x)} e^{-a_n\langle\zeta,y\rangle}\nu_n(dy) \\
&\ge \frac{1}{a_n}\log \hat{\mu}_n(a_n\zeta) - \langle\zeta,x\rangle - \delta'|\zeta| + \frac{1}{a_n}\log \nu_n(B_{\delta'}(x))
\end{aligned}$$

so that

$$\liminf_{n\to+\infty} \frac{1}{a_n} \log \mu_n(B_\delta(x)) \geq \varphi(\zeta) - \langle \zeta, x\rangle + \lim_{\delta'\downarrow 0} \liminf_{n\to+\infty} \frac{1}{a_n} \log \nu_n(B_{\delta'}(x)).$$

Since $\varphi(\zeta) - \langle \zeta, x\rangle \geq -\varphi^*(x)$ it suffices to show that

$$\lim_{\delta'\downarrow 0} \liminf_{n\to+\infty} \frac{1}{a_n} \log \nu_n(B_{\delta'}(x)) = 0. \tag{3.5}$$

In the proof of Theorem 1.23 this followed from the weak law of large numbers. Here instead, we use part (a) and the property of $\mathcal{E}$. For this let $\hat{\nu}_n$ be the moment generating function of ν_n, so that from (3.4), and if $\tau \in \mathbb{R}^d$:

$$\lim_{n\to+\infty} \frac{1}{a_n} \log \hat{\nu}_n(a_n\,\tau) = \varphi(\zeta+\tau) - \varphi(\zeta) =: \psi(\tau).$$

Assumptions 3.1 hold for the modified measures, so that

$$\psi^*(x) = \sup_{\tau\in\mathbb{R}^d} (\langle \tau, x\rangle - \psi(\tau)) = \varphi^*(x) - \langle \zeta, x\rangle + \varphi(\zeta)$$

defines a good rate function. By part (a):

$$\limsup_{n\to+\infty} \frac{1}{a_n} \log \nu_n(\mathbb{R}^d \setminus B_{\delta'}(x)) \leq - \inf_{z\in\mathbb{R}^d\setminus B_{\delta'}(x)} \psi^*(z) = -\psi^*(y),$$

for some $y \in \mathbb{R}^d \setminus B_{\delta'}(x)$. But

$$\psi^*(y) = \varphi^*(y) - \langle \zeta, y\rangle + \varphi(\zeta) \geq (\varphi^*(y) - \langle \zeta, y\rangle) - (\varphi^*(x) - \langle \zeta, x\rangle) > 0,$$

since $x \in \mathcal{E}$ and ζ is an exposing hyperplane. Thus, for each $\delta' > 0$:

$$\limsup_{n\to+\infty} \frac{1}{a_n} \log \nu_n(\mathbb{R}^d \setminus B_{\delta'}(x)) < 0.$$

Since $\nu_n(\mathbb{R}^d) = 1$, we get $\lim_{n\to+\infty} \nu_n(B_{\delta'}(x)) = 1$ and (3.5) follows, proving (b).

For part (c) we need a result from convex analysis. First a basic and useful notion: if $C \subseteq \mathbb{R}^d$ is non-empty and convex, its relative interior, $\mathrm{ri}(C)$, is defined as

$$\mathrm{ri}(C) := \{x \in C\colon\ \forall y \in C,\ x - \epsilon(y-x) \in C \text{ for some } \epsilon > 0\}.$$

Clearly $C^\circ \subseteq \mathrm{ri}(C)$, but they might differ (e.g. for $C = \{x\}$, $\mathrm{ri}(C) = \{x\}$, $C^\circ = \emptyset$). $\mathrm{ri}(C)$ is the interior of C with respect to its affine hull, $\{\sum \alpha_i x_i : \forall i, x_i \in C, \sum \alpha_i = 1\}$. Basic properties include:

- C non-empty and convex $\Rightarrow \mathrm{ri}(C) \neq \emptyset$;
- $z \in \mathrm{ri}(C)$, $y \in C \Rightarrow \epsilon z + (1-\epsilon)y \in \mathrm{ri}(C)$, $\forall \epsilon \in (0,1]$;
- $g\colon \mathbb{R}^d \to (-\infty, +\infty]$ convex$\Rightarrow g$ continuous on $\mathrm{ri}(\mathcal{D}_g)$, where $\mathcal{D}_g := \{x\colon g(x) < +\infty\}$.

In our considerations $g = \varphi^*$ and $C = \mathcal{D}_{\varphi^*}$ which, under Assumptions 3.1, is a non-empty convex set. (Part (a) and the lower semi-continuity imply that $\varphi^*(x) = 0$ for some x.)

We need the following lemma whose proof we omit (see Corollary 26.4.1 in [255]):

Lemma 3.5 $\mathrm{ri}(\mathcal{D}_{\varphi^*}) \subseteq \mathcal{E}$, *under the assumptions in part* (c) *of Theorem* 3.4.

A comment on the proof of Lemma 3.5: given $y \in \mathrm{ri}(\mathcal{D}_{\varphi^*})$, one shows the existence of $\zeta \in \mathcal{D}_\varphi^\circ$ so that (3.2) holds and $y = \nabla\varphi(\zeta)$ (thus $y \in \mathcal{E}$ by Lemma 3.2). This has basically two steps: (1) show the existence of $\zeta \in \mathcal{D}_\varphi$ for which (3.2) holds; this suffices if $\mathcal{D}_\varphi = \mathbb{R}^d$; (2) using the steepness at $\partial\mathcal{D}_\varphi$ show that ζ may be taken in $\mathcal{D}_\varphi^\circ$. Together with the maximality condition in (3.2) this will imply that $y = \nabla\varphi(\zeta)$.

Proof of part (c). Due to part (b) and Lemma 3.5, it remains to prove that for any open set $G \subseteq \mathbb{R}^d$:

$$\inf_{x\in G\cap \mathrm{ri}(\mathcal{D}_{\varphi^*})} \varphi^*(x) \le \inf_{x\in G} \varphi^*(x).$$

We may suppose that $G \cap \mathcal{D}_{\varphi^*} \neq \emptyset$, otherwise the statement is trivial. But, if $z \in \mathrm{ri}(\mathcal{D}_{\varphi^*})$, and $y \in G \cap \mathcal{D}_{\varphi^*}$, then $\epsilon z + (1-\epsilon)y \in G \cap \mathrm{ri}(\mathcal{D}_{\varphi^*})$ for $\epsilon > 0$ sufficiently small, as previously observed (and since G is open). Letting $\epsilon \downarrow 0$ and using the convexity of φ^*:

$$\inf_{x\in G\cap \mathrm{ri}(\mathcal{D}_{\varphi^*})} \varphi^*(x) \le \lim_{\epsilon\downarrow 0} \varphi^*(\epsilon z + (1-\epsilon)y) \le \varphi^*(y),$$

from which the conclusion follows. □

Examples To relate Theorem 3.4 and Theorem 1.23 it is useful to provide examples of probability measures μ on $\mathbb{R}^d$ such that $0 \in \mathcal{D}_{\hat\mu}^\circ$, with $\log\hat\mu$ not steep. One such example would be μ with density $f(x) = ce^{-|x|}/(1+|x|^{d+2})$. See Chapter 2 of [85] for more examples and discussion. In the i.i.d. case we have seen that steepness can be avoided.

Comments (i) An interesting class of examples, suitable for application of Theorem 3.4, comes from empirical averages $\bar{X}_n$ of moderately interacting variables X_i, as for instance moving averages $X_i = \sum_j b_{i+j}\xi_j$, where the ξ_i are i.i.d. with a finite moment generating function, and $\sum_i |b_i| < +\infty$, $\sum_i b_i = 1$.

(ii) It is easy to see examples where the assumptions of part (c) in the above theorem fail but a l.d.p. holds, with φ^* as rate function. Here is a trivial one: let W_n be exponentially distributed with rate $a_n \uparrow +\infty$. Then $\varphi(\zeta) = +\infty\mathbf{1}_{\{\zeta\ge 1\}}$; $\varphi^*(x) = +\infty$ for $x < 0$ and $\varphi^*(x) = x$ otherwise. The verification of a l.d.p. with scaling a_n and rate function φ^* is direct, but the information coming from

Theorem 3.4 is rather trivial, since $\mathcal{E} = \{0\}$. For more interesting examples see e.g. Exercise 2.3.24 in [85] or the beginning of Section 3.4.

(iii) A classical application of Theorem 3.4 is related to the so-called 'moderate deviations': let $\bar{X}_n$ be the empirical averages of an i.i.d. $\mathbb{R}^d$-valued sequence (X_i). On their common law μ assume that it has an invertible covariance matrix A, $EX_1 = 0$, and $0 \in \mathcal{D}^\circ_{\hat{\mu}}$. Since $n^{1/2}\bar{X}_n$ converges in law to a centred d-dimensional Gaussian with covariance matrix A, if $W_n = n^\alpha X_n$ with $0 < \alpha < 1/2$ then $W_n \to 0$ in probability. Applying the Gärtner–Ellis theorem one gets a l.d.p. for (W_n), with rate function $I(x) = \frac{1}{2}\langle x, A^{-1}x\rangle$, and scaling $a_n = n^{1-2\alpha}$. The proof is direct application; a simple expansion of $\hat{\mu}$ around $\zeta = 0$ shows that $\varphi(\zeta) = \frac{1}{2}\langle A\zeta, \zeta\rangle$. For further discussion and applications see e.g. Section 3.7 in [85].

(iv) An earlier reference related to Theorem 3.4 is due to Sievers, cf. [278], see also [246]. In [132], Gärtner stated and proved the above theorem for the case $\mathcal{D}_\varphi = \mathbb{R}^d$ and applied such methods to prove a l.d.p. for the occupation measures of a class of Markov (diffusion) processes, thus obtaining an extension of the Sanov theorem. For a more general class of Markov processes, this problem was considered by Donsker and Varadhan in a fundamental series of articles [93–96] which settled the main results.

(v) Let $P(W_n = 1) = 1/2 = P(W_n = -1)$ for all n. Assumptions 3.1 hold for any $a_n \nearrow +\infty$, with

$$\varphi(\zeta) = |\zeta|, \quad \text{for any } \zeta; \qquad \varphi^*(x) = +\infty \mathbf{1}_{\{|x|>1\}}.$$

But $P(W_n \in (-1, 1)) = 0$, and the lower bound in part (c) fails, in this case. We see that a l.d.p. holds (for any such scaling a_n) with rate function $I(x) = +\infty \mathbf{1}_{\{|x| \neq 1\}}$. This trivial example shows that the lower bound in part (c) of Theorem 3.4 may fail without the differentiability assumption, even if $\mathcal{D}_\varphi = \mathbb{R}^d$. As in the example, this does not prevent a l.d.p. with a rate function $I \neq \varphi^*$, typically non-convex. Indeed, if a l.d.p. holds and Theorem 1.18 is applicable to each $F_\zeta(\cdot) := \langle \zeta, \cdot \rangle$ (e.g. W_n uniformly bounded), then

$$\varphi(\zeta) = \sup_{x \in \mathbb{R}^d} (\langle \zeta, x\rangle - I(x)) = I^*(\zeta),$$

so that $\varphi^* = (I^*)^*$. This is the closed convex hull of I, i.e. the largest convex and lower semi-continuous function, pointwise less than or equal to $I(\cdot)$ (see [255], p. 104).

Non-convex rate functions arise quite naturally in the context of mean field models, as we discuss later in Section 4.1.

(vi) Under Assumptions 3.1, $\varphi(\cdot)$ is differentiable at the origin if and only if $\varphi^*(\cdot)$ vanishes at the unique point $x_0 = \nabla\varphi(0)$. In such a case, by part (a) in Theorem 3.4, $P(|W_n - x_0| > \delta)$ decays exponentially in a_n, as $n \to +\infty$, for any $\delta > 0$. (For a converse see Theorem II.6.3 in [108].) In the context of statistical mechanics this is closely related to first order phase transitions, to be addressed in Section 3.4.

(vii) Various versions of the Gärtner–Ellis theorem in infinite dimensional topological vector spaces have been obtained by several authors (e.g. [12]); see Chapter 4 in [85] for discussion and references. In a very abstract set-up (leaving aside convexity) this brings us to the converse of the Varadhan lemma (cf. [39]) already mentioned in Section 1.2.

3.2 Large deviations for Markov chains

In order to capture the basic points with simplicity, we restrict the discussion to the easiest non-trivial situation: a set-up analogous to that in Section 1.3 replacing the assumption of i.i.d. by that of a finite ergodic Markov chain.

Let S be a finite or countable set. A transition (or stochastic) matrix q on S means any function $q\colon S \times S \to [0,1]$ such that $\sum_{y\in S} q(x,y) = 1$, for each $x \in S$. A sequence of random variables $(X_n : n \geq 1)$ on a given probability space $(\Omega, \mathcal{A}, P)$ is said to be a Markov chain with transition matrix q if for any $k \geq 1$

$$P(X_{k+1} = x_{k+1} | X_1 = x_1, \ldots, X_k = x_k) = q(x_k, x_{k+1}),$$
$$\text{whenever} \quad P(X_1 = x_1, \ldots, X_k = x_k) > 0.$$

Assuming such a property, the distribution of the variables (X_n) is determined once we give the initial distribution, i.e. the probabilities $P(X_1 = x)$ for each $x \in S$. P_x refers to the conditional probability given $X_1 = x$, and more generally if ν is a probability measure on S ($\nu_x = \nu\{x\}$):

$$P_\nu(X_1 = x_1, \ldots, X_k = x_k) = \nu_{x_1} q(x_1, x_2) \cdots q(x_{k-1}, x_k), \quad k \geq 1, \;\; x_1, \ldots, x_k \in S. \tag{3.6}$$

It might be convenient (though not necessary) to take the canonical construction of (X_n), i.e. $\Omega = S^{\mathbb{N}}$, $\mathcal{A} = \mathcal{B}(S^{\mathbb{N}})$ the usual product σ-field with X_n the nth coordinate and P_ν the unique probability on $\mathcal{A}$ that verifies (3.6), also called the law of the chain. (Since q is a transition matrix, (3.6) verifies the consistency condition (in k) and P_ν is the unique extension to $S^{\mathbb{N}}$, according to the Kolmogorov extension theorem, cf. [20].)

Definition 3.6 A probability measure μ on S is said to be stationary for the chain if

$$\sum_{x\in S} \mu_x q(x,y) = \mu_y, \qquad \text{for any } \; y \in S. \tag{3.7}$$

If μ is stationary we see that (3.7) holds for any power q^k ($k \in \mathbb{N}$) of the matrix q, from which we easily see that μ is stationary if and only if $P_\mu(X_k = x_1, \ldots, X_{k+m-1} = x_m)$ does not depend on k, for any $m \geq 1$ and $x_1, \ldots, x_m$ in S.

In other words, μ is stationary if and only if P_μ is invariant under the shift $\theta\colon (\omega_j) \mapsto (\omega_{j+1})$ on $S^{\mathbb{N}}$, that is, $P_\mu \in \mathcal{M}_1^\theta(S^{\mathbb{N}})$ in the notation of Section 1.3. In commonly used terminology we say that (X_n) is stationary under P_μ. (q^0 is the identity matrix).

Definition 3.7 The transition matrix q is called irreducible if for any pair (x, y) there exists $n = n(x, y)$ so that $q^n(x, y) > 0$. This definition applies as well to any matrix with non-negative entries, with S countable or finite.

It is a well known fact that if S is finite and q is an irreducible stochastic matrix, it admits a unique stationary probability measure μ and $\mu_x > 0$ for each x (see Theorem 3.10 below). Such chains are also called ergodic (since P_μ is ergodic under θ).

In what follows we assume:

Assumption 3.8 (X_n) is a Markov chain with values on $S = \{1, \dots, r\}$ and irreducible transition matrix $q = (q(x, y))_{x,y \in S}$.

An extension of the Sanov theorem for finite 'alphabets' goes as follows.

Theorem 3.9 *Under Assumption* 3.8, *let* $Q_n^{(1)}$ *be the law of the empirical measure* $L_n = L_n[X_1, \dots, X_n] := \frac{1}{n}\sum_{i=1}^n \delta_{X_i}$ *on* $\mathcal{M}_1(S)$, *where the initial distribution* ν *is omitted in the notation. For any* ν, *the sequence* $(Q_n^{(1)})$ *satisfies a l.d.p. with scaling* n *and rate function*

$$I^{(1)}(\alpha) = \sup_{u>0} \sum_{y=1}^r \alpha_y \log\left(\frac{u_y}{(uq)_y}\right), \qquad \alpha \in \mathcal{M}_1(S), \tag{3.8}$$

where $u > 0$ *means* $u\colon S \to (0, +\infty)$ *and* $(uq)_y := \sum_{x=1}^r u_x q(x, y)$, *for each* $y \in S$.

Remark We identify $\mathcal{M}_1(S)$ with the simplex $\Theta_r \subseteq \mathbb{R}^r$, and consider the same metric as in Section 1.3. Notice that if $u > 0$ and q is irreducible, then $uq > 0$ and (3.8) makes sense.

Before proving (3.8), let us compare it with the i.i.d. case, when the rate function had a closed expression $\mathcal{H}(\alpha|\mu)$, the relative entropy of α with respect to the underlying measure μ (cf. (1.49)). The present variational expression is not easily computable in general. We must understand how a more detailed description ('higher level') will come to a closed expression, (3.8) resulting by a contraction principle, as the variational form suggests. We shall discuss it in this simple situation.

Proof As the results for L_n in Section 1.3 can be obtained from the Cramér theorem on $\mathbb{R}^r$, one possibility consists in applying the Gärtner–Ellis theorem. With this in mind and keeping the notation from Section 1.3, we consider $W_n = L_n$. They take values on $\mathcal{M}_1(S)$, naturally identified with the simplex $\Theta_r \subseteq \mathbb{R}^r$, i.e. $L_n = (L_n(1), \dots, L_n(r))$ where $L_n(x) = \frac{1}{n}\sum_{k=1}^n \mathbf{1}_{\{X_k = x\}}$. According to

Assumptions 3.1 and letting $f: S \to \mathbb{R}^r$ given by $f(x) = \mathbf{e}_x$, the xth canonical unitary vector, we first need to compute, for any initial state $x = x_1$ in S and any $\zeta \in \mathbb{R}^r$:

$$\begin{aligned}\varphi_n(n\zeta) = E_x\left(e^{\langle n\zeta, L_n\rangle}\right) &= \sum_{x_2,\dots,x_n} \prod_{k=1}^{n-1} q(x_k, x_{k+1}) e^{\sum_{k=1}^n \langle \zeta, f(x_k)\rangle} \\ &= \sum_y e^{\zeta_x} q_\zeta^{n-1}(x, y)\end{aligned} \tag{3.9}$$

where the matrix q_ζ is defined as $q_\zeta(x, y) = q(x, y)e^{\zeta_y}$, and q_ζ^n denotes its nth power. The matrix q_ζ has non-negative entries and it is irreducible if so is q. Thus, we may use the following classical result. □

Theorem 3.10 (Perron–Frobenius) *Let $A = (A(x, y))_{x,y \in S}$ be a finite, irreducible matrix with non-negative entries. Its spectral radius $\mathbf{r}(A) := \max\{|u| : u$ is eigenvalue of $A\}$ satisfies:*

(a) $\mathbf{r}(A)$ is an eigenvalue of A which admits left and right eigenvectors for A with positive coordinates; the (left and right) eigenspaces corresponding to $\mathbf{r}(A)$ have dimension one;

(b) for every $x \in S$ and every $v \in \mathbb{R}^r$ with positive coordinates

$$\begin{aligned}&\lim_{n\to+\infty} \frac{1}{n} \log \sum_{y=1}^r A^n(x, y)\, v_y \\ &= \lim_{n\to+\infty} \frac{1}{n} \log \sum_{y=1}^r v_y\, A^n(y, x) = \log \mathbf{r}(A).\end{aligned}$$

Proof Part (a) is a classical result whose proof we omit (see e.g. [115] or [275]); (b) follows quite easily from (a). Indeed, if w is a right eigenvector corresponding to $\mathbf{r}(A)$ and has positive coordinates, we may find $c > 0$ such that $c^{-1} w_y \le v_y \le c w_y$ for all y. Thus

$$c^{-1} A^n(x, y) w_y \le A^n(x, y) v_y \le c A^n(x, y) w_y$$

and since $\sum_{y=1}^r A^n(x, y) w_y = (\mathbf{r}(A))^n w_x$ for each x, we get the first part of (b). Similarly for the second. □

If A is a finite irreducible stochastic matrix, it then follows that $\mathbf{r}(A) = 1$. The left eigenvector w with positive coordinates, normalized to $\sum_x w_x = 1$ corresponds to the unique stationary probability measure for the chain. For probabilistic proofs of this (based on couplings) see e.g. [201] and references therein.

Conclusion of the proof of Theorem 3.9 The Perron–Frobenius theorem tells us that $\lim n^{-1} \log \varphi(n\zeta) = \log \mathbf{r}(q_\zeta)$ for each $\zeta \in \mathbb{R}^r$, so that Assumptions 3.1 hold. In order to apply Theorem 3.4 it remains to know that $\zeta \mapsto \log \mathbf{r}(q_\zeta)$ is differentiable. By part (a) of Theorem 3.10, $\mathbf{r}(q_\zeta)$ is an eigenvalue of multiplicity one, that is, a simple root of the characteristic equation $\det(q_\zeta - z\mathbb{I}) = 0$. The differentiability follows by the implicit function theorem. (In fact, $\mathbf{r}(q_\zeta)$ is analytic in ζ, see e.g. Theorem 7.7.1 in [187].)

Applying Theorem 3.4 we get a l.d.p. for the distributions of L_n on $\mathbb{R}^r$ under each P_x, with scaling n and a good rate function, given by

$$I(\alpha) = \sup_{\zeta\in\mathbb{R}^r} \left(\langle\zeta,\alpha\rangle - \log \mathbf{r}(q_\zeta)\right). \tag{3.10}$$

But Θ_r is closed and $P_x(L_n \in \Theta_r) = 1$, so that $I(\alpha) = +\infty$ if $\alpha \notin \Theta_r$, by condition (b) of Definition 1.12. We need to verify that $I(\alpha) = I^{(1)}(\alpha)$, cf. (3.8), for $\alpha \in \Theta_r$. It suffices to show that given $u \in (0, +\infty)^r$ we can find $\zeta \in \mathbb{R}^r$ for which

$$\sum_{x=1}^r \alpha_x \log\left(\frac{u_x}{(uq)_x}\right) = \langle\zeta,\alpha\rangle - \log \mathbf{r}(q_\zeta) \tag{3.11}$$

and conversely. Given $u \in (0, +\infty)^r$, let $\zeta_x = \log(u_x/(uq)_x)$ for each x. The sum on the left-hand side of (3.11) coincides with $\langle\zeta,\alpha\rangle$; on the other side $u\,q_\zeta = u$ and thus $u\,q_\zeta^n = u$ for each n. Since $u \in (0, +\infty)^r$, we apply part (b) of Theorem 3.10 concluding that $\mathbf{r}(q_\zeta) = 1$ and that (3.11) holds. Conversely, given $\zeta \in \mathbb{R}^r$ we take $u \in (0, +\infty)^r$ such that $u\,q_\zeta = \mathbf{r}(q_\zeta)u$, according to part (a) of Theorem 3.10. Thus, $\mathbf{r}(q_\zeta) = (u\,q_\zeta)_x/u_x = (uq)_x\, e^{\zeta_x}/u_x$ for each x and

$$\log \mathbf{r}(q_\zeta) = \sum_x \alpha_x \log\left(\frac{(uq)_x}{u_x} e^{\zeta_x}\right) = \sum_x \alpha_x \log\left(\frac{(uq)_x}{u_x}\right) + \langle\zeta,\alpha\rangle$$

implying (3.11) and concluding the proof. □

Beyond the empirical measure; words of length 2

Let us go back to compare (3.8) and (1.49). The simplicity of (1.49) can be credited to the fact that for $X_1, \dots, X_n$ i.i.d., S-valued, their joint distribution $P(X_1 = x_1, \dots, X_n = x_n)$ is itself a function of the empirical measure $L_n[x_1, \dots, x_n] := n^{-1}\sum_{i=1}^n \delta_{x_i}$, which counts occurrences of each digit $x \in S$ in the sequence $(x_1, \dots, x_n)$. This reduced the large deviation estimation to a combinatorial one, bringing into play the entropy function. In the present situation,

$$P_\nu(X_1 = x_1, \dots, X_n = x_n) = \nu_{x_1} \prod_{i=1}^{n-1} q(x_i, x_{i+1}) = \nu_{x_1} \prod_{(i,j)\in S^2} (q(x,y))^{n_{x,y}} \tag{3.12}$$

where $n_{x,y}$ is the number of appearances of (x, y) as (x_i, x_{i+1}) for $i = 1, \dots, n-1$ and we use the convention that $0^0 = 1$. Since $n_{x,y}/(n-1) = \tilde{L}_n^{(2)}[x_1, \dots, x_n](x, y)$, with

$$\tilde{L}_n^{(2)}[x_1, \dots, x_n] := \frac{1}{n-1}\sum_{i=1}^{n-1} \delta_{(x_i,x_{i+1})}, \tag{3.13}$$

we rewrite (3.12) as (with $0 \log 0 = 0$)

$$P_\nu(X_1 = x_1, \ldots, X_n = x_n) = \nu_{x_1} e^{(n-1)\, g(\tilde{L}_n^{(2)}[x_1,\ldots,x_n])},$$

where $g(\alpha) = \sum_{x,y} \alpha(x, y) \log q(x, y)$ for each $\alpha \in \mathcal{M}_1(S^2)$.

Thus, instead of the occupation measure $L_n[x_1, \ldots, x_n]$, due to the one-step dependence given by the transition matrix, the main role (leaving aside the initial measure) is now played by the occupation measure for words of length 2, and we are led to study the l.d.p. for $\tilde{L}_n^{(2)} := \tilde{L}_n^{(2)}[X_1, \ldots, X_n]$.

Keeping the same notation as in Section 1.3, we also set $L_n^{(2)} := L_n^{(2)}[X_1, \ldots, X_n]$ as

$$L_n^{(2)}[x_1, \ldots, x_n] = \frac{1}{n} \sum_{i=1}^n \delta_{(x_i, x_{i+1})}, \qquad \text{with } x_{n+1} = x_1. \tag{3.14}$$

Remark Since $|L_n^{(2)} - \tilde{L}_n^{(2)}|_1 \le 2/n$, it does not matter which choice we make; each one has its conveniences. Due to the periodic extension $L_n^{(2)}$ takes values on $\hat{\mathcal{M}}_1(S^2)$, the set of probability measures on S^2 with equal marginals (both marginals give L_n). Since this is a closed subspace of $\mathcal{M}_1(S^2)$, the choice allows us to formulate the l.d.p. on $\hat{\mathcal{M}}_1(S^2)$.

Let us start with the following further simplification.

Extra assumption $q(x, y) > 0$ for each $x, y \in S$; $\nu_x > 0$ for each $x \in S$.

Under this assumption $P_\nu(X_1 = x_1, \ldots, X_n = x_n) > 0$ and (3.12) may be rewritten as:

$$P_\nu(X_1 = x_1, \ldots, X_n = x_n) = \frac{\nu_{x_1}}{q(x_n, x_1)} e^{n\, g(L_n^{(2)}[x_1,\ldots,x_n])}, \tag{3.15}$$

with the same g as above.

The previous observations show that the combinatorial argument discussed briefly in Section 1.3 provides an explicit form for the rate function for the distribution of $L_n^{(2)}$.

In fact, a useful shortcut reducing the present situation to the i.i.d. case is provided by the method of tilting. For that we must see that the distribution of $L_n^{(2)}$ for the Markov chain has a Radon–Nikodym derivative with respect to that in some i.i.d. case that fits into the Varadhan lemma (see remark (ii) following Theorem 1.18).

Let μ be the stationary probability measure ($\mu_x > 0$ for each x) and μ^n the product measure on S^n:

$$\mu^n(x_1, \ldots, x_n) = \mu_{x_1} \cdots \mu_{x_n} = e^{n\, h(L_n^{(2)}[x_1,\ldots,x_n])} \tag{3.16}$$

where $h(\alpha) = \sum_{x,y} \alpha(x, y) \log \mu_y$. Putting (3.15) and (3.16) together:

$$P_\nu(X_1 = x_1, \ldots, X_n = x_n) = \frac{\nu_{x_1}}{q(x_n, x_1)} e^{n\, F(L_n^{(2)}[x_1,\ldots,x_n])} \mu^n(x_1, \ldots, x_n) \tag{3.17}$$

where $F(\alpha) = g(\alpha) - h(\alpha) = \sum_{x,y} \alpha(x,y) \log q(x,y)/\mu_y$.

We have:

- the distributions of $L_n^{(2)}$ under μ^n satisfy a l.d.p. on $\hat{\mathcal{M}}_1(S^2)$; the (good) rate function is the relative entropy $\mathcal{H}(\alpha|\bar{\alpha} \times \mu)$, cf. (1.60), $\bar{\alpha}$ being the marginal of α;
- the factor $\nu_{x_1}/q(x_n, x_1)$ (not a function of $L_n^{(2)}$) is uniformly bounded away from zero and infinity; consequently it does not interfere in the logarithmic equivalence;
- the previous items together with (3.17) entitle us to use remark (ii) after Theorem 1.18 in Section 1.2, to get the following.

Theorem 3.11 *Let $Q_n^{(2)}$ be the distribution of $L_n^{(2)}$ on $\hat{\mathcal{M}}_1(S^2)$, under P_ν. Under the extra assumption the sequence $(Q_n^{(2)})$ satisfies a l.d.p. with scaling n and good rate function given by*

$$I^{(2)}(\alpha) = \mathcal{H}(\alpha|\bar{\alpha} \times \mu) - F(\alpha) = \sum_{x,y} \alpha(x,y) \log\left(\frac{\alpha(x,y)}{\bar{\alpha}_x q(x,y)}\right), \quad \alpha \in \hat{\mathcal{M}}_1(S^2). \tag{3.18}$$

In other words, $I^{(2)}(\alpha) = \mathcal{H}(\alpha|\bar{\alpha} \times q)$ where $(\bar{\alpha} \times q)(x,y) := \bar{\alpha}_x q(x,y)$.

Proof From the previous observations, if $\tilde{Q}_n^{(2)}$ denotes the distribution of $L_n^{(2)}$ under the product measure μ^n, positive constants c_1, c_2 may be found so that for any Borel set A in $\hat{\mathcal{M}}_1(S^2)$

$$c_1 \int_A e^{n F(\alpha)}\, \tilde{Q}_n^{(2)}(d\alpha) \le Q_n^{(2)}(A) \le c_2 \int_A e^{n F(\alpha)}\, \tilde{Q}_n^{(2)}(d\alpha).$$

The result then follows at once from the i.i.d. case ((1.60) in Section 1.3) and Varadhan lemma, cf. remark (ii) after Theorem 1.18 in Section 1.2 (here $1/c_2 \le Z_n \le 1/c_1$). □

Remarks (i) In (3.18), $(\bar{\alpha} \times q)(x,y) = 0 \Rightarrow \alpha(x,y) = 0$ under the extra assumption; such a term is omitted, with the usual convention $0 \log 0 = 0 \log(0/0) = 0$.

(ii) If $\alpha \in \hat{\mathcal{M}}_1(S^2)$ we have $I^{(2)}(\alpha) = 0$ if and only if $\alpha(x,y) = \bar{\alpha}_x q(x,y)$ for each x, y. Since both marginals of α coincide we see that $\bar{\alpha}_y = \sum_x \bar{\alpha}_x q(x,y)$, for each y, i.e. $\bar{\alpha} = \mu$. Thus:

$$I^{(2)}(\alpha) = 0 \quad \Leftrightarrow \quad \alpha(x,y) = \mu_x q(x,y), \qquad \text{for all } x, y \in S.$$

This implies the exponential convergence of $L_n^{(2)}$.

(iii) *Another proof of Theorem* 3.11. The variables $Y_i = (X_i, X_{i+1})$, $i = 1, 2, \ldots$ constitute a Markov chain on S^2, with transition matrix:

$$q_{(2)}((x,y),(x',y')) = \delta_{y,x'}\, q(x',y'), \qquad (x,y),\ (x',y') \in S^2, \tag{3.19}$$

which is irreducible under the extra assumption of $q(x, y) > 0$ for each (x, y). On the other hand $\tilde{L}_n^{(2)}[X_1, \ldots, X_n] = L_{n-1}[Y_1, \ldots, Y_{n-1}]$. Theorem 3.9 applied to the chain (Y_n) yields a l.d.p. for the distribution of $\tilde{L}_n^{(2)}$ under P_ν for any initial measure ν. It has scaling n and rate function given by:

$$\tilde{I}^{(2)}(\alpha) = \sup_{u>0} \sum_{(x,y)\in S^2} \alpha(x, y) \log\left(\frac{u(x, y)}{(uq_{(2)})(x, y)}\right), \quad \alpha \in \mathcal{M}_1(S^2).$$

Since $|\tilde{L}_n^{(2)} - L_n^{(2)}|_1 \le 2/n$, applying Theorem 3.11 and the uniqueness of the rate function, we see that $\tilde{I}^{(2)}(\alpha) = +\infty$ if $\alpha \notin \hat{\mathcal{M}}_1(S^2)$ and $\tilde{I}^{(2)}(\alpha) = I^{(2)}(\alpha) = \mathcal{H}(\alpha|\bar{\alpha} \times q)$, cf. (3.18).

Instead, one may verify the last identity directly, yielding an alternative proof for Theorem 3.11 (see e.g. Section 3.1 in [85].)

(iv) *Another proof of Theorem* 3.9. Due to the contraction principle and the uniqueness of the rate function, Theorems 3.9 and 3.11 provide the following identity, for $\alpha \in \mathcal{M}_1(S)$:

$$\inf\{I^{(2)}(\eta) \colon \eta \in \hat{\mathcal{M}}_1(S^2), \bar{\eta} = \alpha\} = \sup_{u>0} \sum_{x=1}^{r} \alpha_x \log\left(\frac{u_x}{(uq)_x}\right). \tag{3.20}$$

Nevertheless, a direct proof of (3.20) is not hard, so that Theorem 3.9 may be recovered from Theorem 3.11 (proven so far under the extra assumption). We now outline it.

We first let $\alpha \in \mathcal{M}_1(S)$ with $\alpha_x > 0$ for each x. Thus $g(u) := \sum_{x=1}^{r} \alpha_x \log\,(u_x/(uq)_x)$ is well defined and smooth on $(0, +\infty)^r$, attaining its supremum at a critical point, as we see from its form. We check that u' is critical if and only if

$$\sum_{x=1}^{r} \alpha_x \frac{u'_y\, q(y, x)}{(u'q)_x} = \alpha_y, \qquad \text{for all } y,$$

i.e. α is the stationary measure for the transition matrix $\bar{q}_{u'}(x, y) = u'_y q(y, x)/(u'q)_x$. Setting

$$\eta'(x, y) := \alpha_y \bar{q}_{u'}(y, x), \qquad \text{for any } (x, y) \in S^2,$$

simple computations, using $\eta'(x, y)/\alpha_x q(x, y) = \alpha_y u'_x/\alpha_x (u'q)_y$ and (3.18), show that:

- $\eta' \in \hat{\mathcal{M}}_1(S^2)$, it has marginal α, and $I^{(2)}(\eta') = \sum_x \alpha_x \log\,(u'_x/(u'q)_x)$;
- $I^{(2)}(\eta) = I^{(2)}(\eta') + \mathcal{H}(\eta|\eta')$ whenever $\eta \in \hat{\mathcal{M}}_1(S^2)$ has marginal α.

Since $\mathcal{H}(\eta|\eta') \ge 0$, with equality if and only if $\eta = \eta'$, we verify (3.20) directly, for such α. For general α, replace S by $\{x \colon \alpha_x > 0\}$ in the definition of $g(\cdot)$, and all goes just the same.

(v) The rate function in (3.20) may also be expressed as follows:

$$\inf\{I^{(2)}(\eta)\colon \eta \in \hat{\mathcal{M}}_1(S^2), \bar{\eta} = \alpha\} = \sup_{u>0} \sum_{x=1}^{r} \alpha_x \log \left(\frac{u_x}{(qu)_x} \right).$$

This is checked similarly as for (3.20): if $\alpha_x > 0$ for each x, one considers the function $h(u) := \sum_{x=1}^{r} \alpha_x \log (u_x/(qu)_x)$; its supremum on $(0, +\infty)^r$ is again attained at a critical point $\tilde{u} > 0$, and this solves

$$\sum_{x=1}^{r} \alpha_x \frac{q(x, y)\tilde{u}_y}{(q\tilde{u})_x} = \alpha_y, \qquad \forall y \in \mathbb{R}^d,$$

i.e. α is the stationary measure for the transition matrix $\tilde{q}_{\tilde{u}}(x, y) := q(x, y)\tilde{u}_y/(q\tilde{u})_x$. Let $\tilde{\eta}(x, y) = \alpha_x \tilde{q}_{\tilde{u}}(x, y)$. $\tilde{\eta}$ has marginals α and simple computations as before show that $I^{(2)}(\tilde{\eta}) = \sum_x \alpha_x \log (\tilde{u}_x/(q\tilde{u})_x)$, as well as that $I^{(2)}(\eta) = I^{(2)}(\tilde{\eta}) + \mathcal{H}(\eta|\tilde{\eta})$ for any $\eta \in \hat{\mathcal{M}}_1(S^2)$ with marginal α. The rest goes as before, including the passage to general $\alpha \in \mathcal{M}_1(S)$ (for further discussion see [108] or [157]).

(vi) *Theorem* 3.11 *holds under Assumption* 3.8, *for any initial measure.* To achieve this elimination of the extra assumption, one needs a little modification of the previous proofs.

The proof based on the Gärtner–Ellis theorem, cf. previous remark (iii), applies to any initial measure. To relax the extra assumption of $q(\cdot, \cdot) > 0$ to the irreducibility condition, it suffices to replace S^2 by $\tilde{S}_{(2)} = \{(x, y) \in S^2 \colon q(x, y) > 0\}$; (Y_n) may be thought of as taking values on $\tilde{S}_{(2)}$, on which the transition matrix given by (3.19) is irreducible.

One may also extend the proof of Theorem 3.11 based on the tilting argument, but we omit such considerations.

Empirical process

For any fixed $k \geq 2$, let $L_n^{(k)} = L_n^{(k)}[X_1, \ldots, X_n]$ be the empirical measure corresponding to words of length k, i.e.

$$L_n^{(k)}[x_1, \ldots, x_n] = \frac{1}{n} \sum_{i=1}^{n} \delta_{(x_i, \ldots, x_{i+k-1})}, \quad \text{where} \quad x_{n+j} = x_j \text{ for } 1 \leq j \leq k-1.$$

Let $(Q_n^{(k)})$ $((\tilde{Q}_n^{(k)}))$ denote the distributions of $L_n^{(k)}$ under P_ν (under μ^n, respectively). These are probability measures on the space $\hat{\mathcal{M}}_1(S^k)$, defined in (1.57). Under the conditions of Theorem 3.11, for any Borel set A in $\hat{\mathcal{M}}_1(S^k)$, we have:

$$c_1 \int_A e^{n\, F_k(\alpha)}\, \tilde{Q}_n^{(k)}(d\alpha) \leq Q_n^{(k)}(A) \leq c_2 \int_A e^{n\, F_k(\alpha)}\, \tilde{Q}_n^{(k)}(d\alpha)$$

where $F_k(\alpha) = \sum_{(x_1, \ldots, x_k)} \alpha(x_1, \ldots, x_k) \log (q(x_{k-1}, x_k)/\mu_{x_k})$, and c_1, c_2 are as in Theorem 3.11. Thus, a l.d.p. for the family $(Q_n^{(k)})$ follows from that in the i.i.d.

case, using the Varadhan lemma. The scaling is n and, recalling (1.60), the rate function becomes:

$$\begin{aligned} I^{(k)}(\alpha) &= \sum_{(x_1,\dots,x_k)\in S^k} \alpha(x_1,\dots,x_k) \log\left(\frac{\alpha(x_1,\dots,x_k)}{\bar\alpha(x_1,\dots,x_{k-1})\,q(x_{k-1},x_k)}\right) \\ &= \mathcal{H}(\alpha|\bar\alpha\times q), \end{aligned} \tag{3.21}$$

where $(\bar\alpha\times q)\,(x_1,\dots,x_k) := \bar\alpha(x_1,\dots,x_{k-1})\,q(x_{k-1},x_k)$ and

$$\bar\alpha(x_1,\dots,x_{k-1}) := \sum_j \alpha(x_1,\dots,x_{k-1},j) = \pi_{k-1}\alpha(x_1,\dots,x_{k-1}).$$

(This may be extended to the situation of Assumption 3.8, as previously discussed.)

Given the previous discussion, and proceeding as in Section 1.3 one may extend the previous results to a l.d.p. for the full empirical process. Given $X_1,\dots,X_n$ a realization of the Markov chain, let $X^{(n)}\in S^{\mathbb{N}}$ be obtained by repeating $X_1,\dots,X_n$ periodically over and over, and define

$$R_n = \frac{1}{n}\sum_{j=1}^n \delta_{\theta_j(X^{(n)})}, \tag{3.22}$$

where θ_j denotes the shift by j to the left, given by $(\theta_j y)_i = y_{i+j}$ for $i\in\mathbb{N}$, $y\in S^{\mathbb{N}}$.

The variable R_n takes values on $\mathcal{M}_1^\theta(S^{\mathbb{N}})$, the set of shift invariant probability measures on $S^{\mathbb{N}}$. Let Q_n denote its law under P_ν. Under Assumption 3.8, the same approximation argument used in the proof of Theorem 1.21 yields a l.d.p. for (Q_n), with scaling n and rate function given by:

$$I^{(\mathbb{N})}(\alpha) = \sup_{k\ge 2} I^{(k)}(\pi_k\alpha) = \sup_{k\ge 1}\mathcal{H}(\pi_k\alpha|\pi_{k-1}\alpha\times q), \qquad \alpha\in\mathcal{M}_1^\theta(S^{\mathbb{N}}), \tag{3.23}$$

where $(\pi_{k-1}\alpha\times q)(x_1,\dots,x_k) = (\pi_{k-1}\alpha)(x_1,\dots,x_{k-1})\,q(x_{k-1},x_k)$.

Comments (i) An earlier reference related to Theorem 3.9 is [218]. Miller obtained tail estimates for $n^{-1}\sum_{i=1}^n X_i$, when (X_n) is a finite, irreducible and aperiodic Markov chain. Combined with ideas from [21] in the i.i.d. case, the relation with the Perron–Frobenius theorem is made clear and applied.

(ii) Our discussion has been restricted to the simplest situation. On more general spaces S there are, roughly speaking, two types of results: under a uniform ergodicity assumption one might get uniform estimates, analogous to those in the finite case; non-uniform estimates are possible under less restrictive conditions. This theory has been settled in the series of articles by Donsker and Varadhan [93–96], and extended in many directions by several authors. A few basic references include the articles of Iscoe, Ney and Nummelin [169], de Acosta [1], Ellis [109],

Ellis and Wyner [110], Bolthausen and Schmock [25]. For a detailed account and further references see [85].

(iii) Donsker and Varadhan theory also applies to continuous time Markov processes. In Section 3.4 we shall discuss extensions related to statistical mechanical models.

3.3 A brief introduction to equilibrium statistical mechanics

This brief discussion follows closely ideas in the text of Gallavotti [129], to which the reader may refer for a more in depth analysis.

Statistical mechanics was introduced between the end of nineteenth century and the beginning of twentieth century, especially by the work of Boltzmann and Gibbs. It aims to explain the macroscopic behaviour of thermodynamic systems at equilibrium in terms of the interaction between their microscopic, elementary constituents.

The object of the analysis of statistical mechanics is, typically, a system of N identical point particles enclosed in a box $\Lambda \subseteq \mathbb{R}^3$ (for instance a cube) with volume $|\Lambda| \equiv V$. The number N is very large, typically of the order of Avogadro's number, namely $N \sim 10^{23}$. The microscopic states are specified by the values of positions and momenta, identified as the points $x \equiv (\mathrm{q}, \mathrm{p}) = \{\mathrm{q}_i, \mathrm{p}_i\}_{i=1,\dots,N}$ of the *phase space* $\Gamma := \Lambda^N \times \mathbb{R}^{3N}$. Assuming the interaction to be among pairs, the Hamiltonian of the system is taken of the form:[1]

$$H(x) = \sum_{i=1}^{N} \frac{|\mathrm{p}_i|^2}{2\mathrm{m}} + \frac{1}{2} \sum_{\substack{i,j=1 \\ i \neq j}}^{N} \phi(|\mathrm{q}_i - \mathrm{q}_j|), \tag{3.24}$$

where m represents the mass of the particles, and ϕ satisfies some conditions described below. Such systems are here called a one-component system of point particles (OCSPP).

The time evolution is described by Hamilton equations inside Λ, and elastic reflections with the walls of the container. It preserves the total energy H. It is possible to show (see [208]) that under quite general regularity conditions on the pair potential ϕ and on the boundaries of Λ, the above described problem of motion admits a unique solution, global in time, for λ-almost all initial conditions in Γ, where λ denotes the Lebesgue measure on $\mathbb{R}^{6N}$. Such evolution, denoted by X_t, will then determine a measurable flow on (Γ, λ). As proven in [208], Liouville theorem generalizes to this situation, i.e. the Lebesgue measure restricted to Γ is invariant under X_t (see [5]). As a corollary, if μ is a measure of the form $\mu(dx) = \rho(x)\lambda(dx)$, where $\rho(x) = \rho(X_t x)$ (λ-a.e.) (λ almost everywhere), then it is invariant under X_t.

[1] Due to the factor 1/2, the last term in (3.24) is the sum of $\phi(|\mathrm{q}_i - \mathrm{q}_j|)$ over unordered pairs.

Given E, $0 < DE \leq E$, the set

$$\Gamma_E := \{x \in \Gamma : E - DE \leq H(x) \leq E\}, \tag{3.25}$$

as well as the restricted normalized Lebesgue measure

$$d\lambda_E(x) := (\lambda(\Gamma_E))^{-1} \mathbf{1}_{\Gamma_E}(x) d\lambda(x), \tag{3.26}$$

are invariant under the time evolution, so that we may consider X_t as a measurable flow on (Γ_E, λ_E). The quantity DE is assumed to be 'macroscopically small' so that Γ_E represents a 'thick' (but not too much) version of the manifold of constant energy which, in turn, is given by $\tilde{\Gamma}_E := \{x \in \Gamma : H(x) = E\}$.

On the other hand, since $\tilde{\Gamma}_E$ is also invariant under time evolution, the procedure in [208] constructs a probability measure μ_E on (the Borel sets of) $\tilde{\Gamma}_E$ which is invariant under X_t. More precisely, let

$$d\mu_E(x) := \frac{d\sigma(x)}{|\nabla H(x)|} \Big/ \int_{\tilde{\Gamma}_E} \frac{d\sigma(y)}{|\nabla H(y)|} \tag{3.27}$$

where $d\sigma(x)$ represents the uniform measure on the surface $\tilde{\Gamma}_E$. The invariance of μ_E, at least formally, follows from the decomposition of λ:

$$d\lambda = d\sigma d\ell = d\sigma dE / |\nabla H(x)|$$

where $d\ell$ represents length on the direction orthogonal to $\tilde{\Gamma}_E$, and since we already know that $d\lambda$ and dE are invariant under X_t. For a proof of the invariance see e.g. [208] or [250].

A basic idea of equilibrium statistical mechanics consists in associating to any given thermodynamic equilibrium state an invariant measure for the flow X_t on the space of microscopic states. The important concept is that of *ensemble*, which is a family (μ_α) of invariant measures parametrized by the 'intensive' external parameters $\alpha = (\alpha_1, \ldots, \alpha_n)$. For the OCSPP one has $n = 2$; for instance, the energy per particle u, and the specific volume v, are the relevant intensive quantities for a large system which is thermally and mechanically isolated (cf. (3.28) below). From thermodynamics it is known that thermodynamic fluids consisting of an OCSPP are described completely by three thermodynamic parameters, such as the entropy per particle s, together with u and v, or by the pressure P, together with v and the absolute temperature T. But only two of them are independent, due to a link $s = s(u, v)$ or $\mathrm{P} = \mathrm{P}(v, \mathrm{T})$, called the *equation of state*.

In his original formulation, Boltzmann worked with a discrete system, but apart from this difference, the family of invariant measures defined by (3.26) corresponds to what he called *ergode*, and Gibbs later named *microcanonical ensemble*. One might prefer to take the microcanonical ensemble defined on the surfaces $\tilde{\Gamma}_E$; in such a case, since $\lambda(\tilde{\Gamma}_E) = 0$, one would take the orthogonal projection μ_E, or in more probabilistic terms, the regular conditional distribution given $H = E$, defined by (3.27). The choice of the 'thick' manifold avoids some technical problems. The important parameters, as $N \to +\infty$, are

$$u := \frac{E}{N} \quad \text{and} \quad v := \frac{V}{N}, \tag{3.28}$$

which represent, in this case, the external control parameters α. (That is, in the above limit, the dependence on DE is negligible, provided it is kept macroscopically finite, $DE = NDu$ with $0 < Du \leq u$, see (3.41).)

A thermodynamic transformation consists in a (continuous) sequence of equilibrium states corresponding to different values of the external control parameters. Which kind of parameters we should use is suggested by the specific experimental situation; for instance u, v are suitable for describing a system without thermal and mechanical exchanges whereas v and T are the correct parameters for a system without mechanical exchanges in contact with a reservoir at temperature T.

Boltzmann postulated that the microcanonical ensemble is the correct system to describe thermodynamic equilibrium states of isolated systems. A possible justification for the introduction of the microcanonical ensemble is the so-called 'ergodic hypothesis'. It says that the *dynamical system* $(\tilde{\Gamma}_E, \mu_E, X_t)$ is *ergodic* (see for instance [6]) in the sense that Borel subsets of $\tilde{\Gamma}_E$ which are invariant under X_t must have measure zero or one.

Given an *observable*, that is a function $f \in \mathrm{L}^1(\tilde{\Gamma}_E, \mu_E)$, its (asymptotic) *time average* is given by:

$$\bar{f}_{time}(x) = \lim_{t' \to +\infty} \frac{1}{t'} \int_0^{t'} f(X_t x) dt; \tag{3.29}$$

its μ_E-a.s. existence is guaranteed by the Birkhoff theorem (see [6]). Clearly, the ergodicity assumption implies that $\bar{f}_{time}$ is μ_E-a.e. constant on $\tilde{\Gamma}_E$, thus coinciding with the *phase average*

$$\bar{f} := \int_{\tilde{\Gamma}_E} f(x) \mu_E(dx). \tag{3.30}$$

In this framework, the ergodic hypothesis asserts that the time averages can be replaced by phase averages $\bar{f}$ with respect to the microcanonical measure. This point of view poses various problems: first of all many simple physical systems, like the free system with no interaction, any system of coupled harmonic oscillators, or more generally any integrable system (see [128]), are not ergodic; general results on ergodicity of significant systems with many degrees of freedom are still missing (see however interesting results in [40]). Moreover, ergodicity provides only a weak approach to equilibrium (in the sense that time averages tend to forget the initial state and to converge to equilibrium averages). It is a strong condition since it involves the consideration of *any* L^1 observable f, thus including functions which are sensitive to the details of microscopic configurations. On the basis of Poincaré recurrence theorem (see [6]) one could estimate the typical time needed for the time averages to be reasonably well approximated by phase averages: if D is a domain in $\tilde{\Gamma}_E$, the time average of its indicator function $\mathbf{1}_D$, i.e. the empirical or occupation measure of D, is expected to stabilize to its limiting value, $\mu_E(D)$, only after many visits to D.

For a realistic situation, such as one mole of hydrogen under normal conditions, such an estimate for the return time to a reasonably small cell of the phase space is much larger than the age of the Universe! (see [129]).

On the other hand we are not really interested in the approach to equilibrium of *any* L^1 function; rather, we are interested in some global functions, such as the mean energy, i.e. observables given by (empirical) spatial averages; their behaviour is usually much better than that of a generic function, as one can foresee. For the much simpler situation of lattice spin systems, we shall discuss this in relation to large deviation theory.

Comment The consideration of a system of point particles in the continuum simplifies the treatment. The original formulation by Boltzmann involved a discrete system, with the volume Γ divided in small cubes, or *cells*, Δ, of volume h^{3N}, where $h > 0$ would be a parameter. Considering such a discrete system one would have to be more careful when determining its evolution, since the cells Δ are not preserved by the dynamics. The discrete time evolution corresponds to $X_{n\tau}, n \geq 1$, where τ is a suitable small time. The scales involved are chosen so that an approximation for the time evolution consists of permutations of the cells. (See Section 1.3 in [129] for a discussion on this delicate point and for the formulation in terms of the discrete model.)

3.3.1 Orthodicity. Ensembles

Concerning the possibility of interpreting an ensemble as a thermodynamic system, Boltzmann introduced the fundamental notion of *orthodicity*. Before discussing this here, we need some definitions and notation:

(1) $v = V/N$, the 'specific volume';
(2) $\mathcal{T}(x) := N^{-1} \sum_{i=1}^{N} |\mathrm{p}_i|^2/2\mathrm{m}$, the 'kinetic energy per particle';
(3) $\mathcal{U}(x) := N^{-1} H(x)$, the 'total energy per particle';
(4) $\mathcal{P}(x)$, the momentum transferred to the walls of the container per unit surface and per unit time, starting from x.

We should make a comment about item (4). Let $t' > 0$; observing the system during an interval of duration t', one might try to measure the force necessary to exert on the walls to prevent them from moving due to the collisions of the particles with them. Dividing this quantity by t' and by the total surface $|\partial\Lambda|$ we would have an experimental quantity with the above description, and which would depend on t' and on the microscopic state x at the time the measurement started. Since $2\langle \mathrm{p}_i, \mathrm{n}\rangle$ is the contribution of a single collision, with incoming momentum p_i, whose component with respect to n (the outer normal to $\partial\Lambda$) is $\langle \mathrm{p}_i, \mathrm{n}\rangle > 0$, we would be led to consider

$$\mathcal{P}(x,t') = \frac{1}{t'} \sum_{0\le t\le t'} \frac{1}{|\partial\Lambda|} \sum_{\substack{\mathrm{q}_i\in X_t x,\\ \mathrm{q}_i\in\partial\Lambda}} 2|\langle \mathrm{p}_i,\ \mathrm{n}\rangle|.$$

Arguments in [208] prove that the above sum is well defined and finite for a set of full measure under μ_E (see also [248]). Moreover, due to the invariance of the measure it follows that the integral of $\mathcal{P}(\cdot, t')$ with respect to μ_E does not depend on t' (thus t' is omitted from the notation, cf. (3.33)).

Given an ensemble $\{\mu_\alpha\}$ consider the phase averages of the above observables:

$$\mathrm{T}(\alpha) := \int \mathcal{T}(x)\mu_\alpha(dx), \tag{3.31}$$

$$u(\alpha) := \int \mathcal{U}(x)\mu_\alpha(dx), \tag{3.32}$$

$$\mathrm{P}(\alpha) := \int \mathcal{P}(x)\mu_\alpha(dx). \tag{3.33}$$

We say that a family μ_α of invariant measures on the phase space Γ is an *orthodic ensemble* if

$$\frac{du + \mathrm{P}dv}{\mathrm{T}} \equiv \frac{\sum_i \frac{\partial u}{\partial\alpha_i} d\alpha_i + \mathrm{P}(\alpha)\sum_i \frac{\partial v}{\partial \alpha_i} d\alpha_i}{\mathrm{T}(\alpha)} \tag{3.34}$$

is an exact differential, that is, if there exists a function $s = s(\alpha)$ such that

$$ds \equiv \frac{du + \mathrm{P}dv}{\mathrm{T}}. \tag{3.35}$$

The importance of such a concept is that an orthodic ensemble gives rise to a microscopic model of thermodynamics, provided $\mathrm{T}(\alpha)$ is identified with the absolute temperature (multiplied by a universal constant) and the function $s(\alpha)$ with the thermodynamic entropy.

In this way, we recover the first and second principles of thermodynamics. In addition, for a specific thermodynamic fluid, whose potential energy function is known, we are also able to find, at least in principle, the equation of state: $s = s(u, v)$. Once we know $s(u, v)$, using thermodynamic relations it is possible to derive the more familiar expression $\mathrm{P} = \mathrm{P}(v, \mathrm{T})$. Indeed from the above definition of the form ds we have:

$$\frac{\partial s(u,v)}{\partial u} = \frac{1}{\mathrm{T}(u,v)}, \qquad \frac{\partial s(u,v)}{\partial v} = \frac{\mathrm{P}(u,v)}{\mathrm{T}(u,v)}. \tag{3.36}$$

Solving the first equation in (3.36) in u, and then inserting its expression in terms of v and T in the second equation, we get the desired relation $\mathrm{P} = \mathrm{P}(v, \mathrm{T})$.

Since we want to give a description of macroscopic systems containing an enormous number of microscopic constituents, we are indeed interested in the asymptotic regime called the *thermodynamic limit*. In the present situation, this amounts to considering a sequence of regions $\Lambda = \Lambda_N$ invading, in a regular way, the whole space $\mathbb{R}^3$, with $V \equiv |\Lambda| \to +\infty$ in such a way that $V/N \to v$ and $E/N \to u$, as $N \to +\infty$, where v and u are constants. Here, 'regular' means a regular ratio between surface and volume (*van Hove* regular, cf. Definition 3.19), as for instance a sequence of growing cubes or spheres.

In order for our OCSPP to exhibit correct thermodynamic behaviour (see [258]) we have to make suitable hypotheses on the potential. These hypotheses are very different, in nature, with respect to those needed to guarantee the existence of a global solution of the Hamilton equations and rather are related to the *extensive* behaviour of the energy.

In this connection, let us now give some definitions.

Definition 3.12 The potential energy associated with the configuration $(\mathrm{q}_1, \dots, \mathrm{q}_N) \in \mathbb{R}^{3N}$ is of the form

$$\Phi(\mathrm{q}_1, \dots, \mathrm{q}_N) := \frac{1}{2} \sum_{\substack{i,j=1 \\ i \neq j}}^{N} \phi(|\mathrm{q}_i - \mathrm{q}_j|), \tag{3.37}$$

and in such a situation the potential ϕ is called:

(i) *stable*, if there exists a real constant B such that

$$\Phi(\mathrm{q}_1, \dots, \mathrm{q}_N) \geq -BN, \quad \text{for all } (\mathrm{q}_1, \dots, \mathrm{q}_N) \in \mathbb{R}^{3N}; \tag{3.38}$$

(ii) *tempered*, if there exist $c > 0$, $r_0 > 0$, $\epsilon > 0$ such that

$$|\phi(r)| \leq \frac{c}{r^{3+\epsilon}}, \quad \text{for all } r > r_0; \tag{3.39}$$

(iii) *short range*, if there exists $r_0 < +\infty$ such that $\phi(r) = 0$ for $r > r_0$;

(iv) *hard core*, if there exists $r_1 > 0$ such that $\phi(r) = +\infty$ for $r \leq r_1$, i.e the region $|\mathrm{q}_i - \mathrm{q}_j| \leq r_1$ is not accessible, cf. (3.37).

Examples (a) If there exist $c < +\infty$ and $0 < r_1 < r_0 < +\infty$, so that $\phi(r) = +\infty$ for $r \leq r_1$; $\phi(r) = 0$ for $r \geq r_0$, and ϕ is continuous on $(r_1, r_0]$, with $\phi(r) \geq -c$ for all such r, then ϕ is stable.

(b) Hard core may be replaced by sufficient divergence at the origin: if there exist $c < +\infty$ and $0 < r_1 < r_0 < +\infty$ such that $\phi(r)$ is positive, continuous and decreasing on $(0, r_1)$, with $\phi(r) \sim r^{-(3+\epsilon)}$ at the origin, with $\epsilon > 0$, and $\phi(r) = -c$ if $r_1 \leq r \leq r_0$, then ϕ is stable. It is easy to check (3.38). (See Section 3.2 in [258]; cf. also Chapter 9 in [250].)

Remark Assumptions of stability and temperedness play a very important role in the existence of the thermodynamic limit; their violation could produce catastrophic situations. Coalescence is one example, with particles staying too close to each other if, for instance, stability is lost due to an attractive interaction near the origin ((3.38) fails if ϕ is continuous and $\phi(0) < 0$). On the other hand, forces with too long range (attractive or repulsive) breaking the temperedness condition, might also cause the particles not to occupy the volume properly. (For a discussion see Section 4.1 of [129].)

In order to obtain a microscopic model of thermodynamics, it suffices that an ensemble become orthodic in the thermodynamic limit. If we find two different orthodic ensembles the problem of their equivalence is immediately posed.

To conclude the above brief discussion, we must say that ergodicity by no means implies orthodicity. At this level, we simply leave ergodicity as a justification behind the assumption of the microcanonical ensemble as the basis of statistical mechanics but, rather, try to prove its orthodicity.

Let us use different notation for the choice of microcanonical ensemble given by (3.26) to make it coherent with (3.36): this is the family of measures $\tilde{\mu}_{\Lambda,u,v}$ given by

$$\tilde{\mu}_{\Lambda,u,v}(f) = \frac{1}{\mathcal{N}^{\Lambda}_{u,v,Du}} \int_{Nu-NDu \le H(\mathrm{q},\mathrm{p}) \le Nu} f(\mathrm{q},\mathrm{p}) d\mathrm{q} d\mathrm{p}, \tag{3.40}$$

where $\Lambda = \Lambda_N$ as before, $d\mathrm{q}d\mathrm{p} = d\mathrm{q}_1 \cdots d\mathrm{q}_N d\mathrm{p}_1 \cdots d\mathrm{p}_N$; $Nu = E$ is the total energy, $DE = NDu$ and the normalization factor $\mathcal{N}^{\Lambda}_{u,v,Du}$, given by the volume of the region Γ_E in (3.25), is called the *microcanonical partition function.*

It can be shown, for instance under the hypotheses of short range and hard core potential, that the following limit exists:

$$s(u,v) = \lim_{N\to+\infty} \frac{1}{N} \log\left(\frac{1}{N!}\mathcal{N}^{\Lambda_N}_{u,v,Du}\right), \tag{3.41}$$

does not depend on Du, and that the microcaconical ensemble is orthodic in the thermodynamic limit, with

$$\lim_{N\to+\infty} \frac{du + \mathrm{P}dv}{\mathrm{T}} = \frac{\partial s}{\partial u} du + \frac{\partial s}{\partial v} dv. \tag{3.42}$$

For a proof of this see Section 2.3 of [129]. In other words, in the thermodynamic limit, the logarithm of the microcanonical partition function per unit particle converges to the thermodynamic entropy per unit particle. From (3.42) we get the relations (3.36) directly, where T, P are the thermodynamic limits of the quantities defined in (3.31), (3.33), in the case of a microcanonical ensemble.

Remark The factor $1/N!$ in (3.41) is due to the indistinguishability of the particles. To understand why the dependence on Du disappears in the above limit in (3.41), let us look at the simplest possible situation, i.e. the ideal gas ($\Phi \equiv 0$). Using polar coordinates in $(\mathrm{p}_1, \ldots, \mathrm{p}_N)$ to compute the volume of the region Γ_E we see that:

$$\frac{1}{N!}\mathcal{N}^{\Lambda}_{u,v,Du} = \frac{|\Lambda|^N}{N!}\left((2\mathrm{m}Nu)^{3N/2} - (2\mathrm{m}N(u-Du))^{3N/2}\right)\frac{\pi^{d/2}}{\Gamma(3N/2)}$$

where $\Gamma(x)$ denotes the usual gamma function of Euler ($\Gamma(k) = (k-1)!$, if k is a positive integer). Using the Stirling formula, cf. (1.5), we easily get (3.41).

For a proof of (3.41) in the interacting case see Section 2.3 of [129]: if the Λ are cubes of side 2^n (with $2^{3n}/N \to v$) a subadditivity argument is employed, taking into account the assumptions on the potential; one then sees that the limit is the same under van Hove regularity. The argument is analogous to that sketched for Proposition 3.13.

We now want to introduce the *canonical ensemble*, which turns out to be orthodic even in finite volume. It is suited to describing a thermodynamic system with fixed volume, temperature, and number of particles. We may think of it as useful for treating a large subsystem A of a very large isolated system B. It is the family of measures $\mu_{\Lambda,\beta,v}$ given by:

$$\mu_{\Lambda,\beta,v}(f) := \frac{1}{Z(\Lambda,\beta,N)} \int_\Gamma f(\mathrm{q},\mathrm{p}) e^{-\beta H(\mathrm{q},\mathrm{p})} d\mathrm{q} d\mathrm{p}, \tag{3.43}$$

where $\Gamma = \Lambda^N \times \mathbb{R}^{3N}$, $\beta \geq 0$, and

$$Z(\Lambda,\beta,N) := \int_\Gamma e^{-\beta H(\mathrm{q},\mathrm{p})} d\mathrm{q} d\mathrm{p} \tag{3.44}$$

is called the canonical partition function.

The description of the canonical ensemble, as that of a large subsystem A of a very large isolated system B, i.e. immersed in a *heat bath*, may be made rigorous as a limit theorem. As an exercise consider the ideal gas ($\Phi \equiv 0$); the essential point is to prove that if $(Y_1^{(k)}, \dots, Y_k^{(k)})$ are uniformly distributed on the k-dimensional sphere $\{\sum_{i=1}^k y_i^2 = ck\}$, where $0 < c < +\infty$, then for each fixed $j \in \mathbb{N}$, $(Y_1^{(k)}, \dots, Y_j^{(k)})$ converge in law to a vector of i.i.d. coordinates, distributed as centred Gaussian with variance c, as $k \to +\infty$; see Section 2.1 of [215].

Remark Except for the non-interacting case, $\Phi \equiv 0$, the partition function given by (3.44) depends on the shape of Λ, not only on (β, V, N). The expression (3.47) for the pressure $\mathrm{P}(\Lambda,\beta,N)$ will keep such a dependence. This tells us that the correct correspondence with thermodynamics is expected only as $N \to +\infty$. Moreover, it is important that in this limit, and if $\Lambda = \Lambda_N$ invades $\mathbb{R}^3$ in a regular way, this dependence becomes negligible, for a reasonable class of potentials Φ.

The orthodicity in finite volume follows from the identities:

$$\mathrm{T}(N,\beta,v) := \mu_{\Lambda,\beta,v}\left(\frac{1}{N}\sum_{i=1}^N \frac{|\mathrm{p}_i|^2}{2\mathrm{m}}\right) = \frac{3}{2\beta} =: \frac{3}{2}k\mathrm{T} \tag{3.45}$$

where k is a universal constant and T is the absolute temperature,

$$u(\Lambda, \beta, v) = -\frac{1}{N}\frac{\partial \log Z(\Lambda, \beta, N)}{\partial \beta}, \tag{3.46}$$

$$\mathrm{P}(\Lambda, \beta, v) = \frac{1}{\beta}\frac{\partial \log Z(\Lambda, \beta, N)}{\partial V}. \tag{3.47}$$

Indeed, in this case we define:

$$s(\Lambda, \beta, v) := k\left(\beta\, u(\Lambda, \beta, v) + \frac{1}{N}\log Z(\Lambda, \beta, N)\right)$$

and simple verification shows (3.35). As for the above identities, (3.45) follows from simple calculations, since the momenta are independent Gaussian random variables under $\mu_{\Lambda,\beta,v}$, and this shows that the average kinetic energy depends only on β; (3.46) is trivial; a proof of (3.47) is sketched briefly below (see [215]).

Considering the particular case of an ideal gas ($\Phi \equiv 0$) we identify k with the Boltzmann constant (given by $k = R/\mathcal{N}$, R being the ideal gas constant and $\mathcal{N}$ Avogadro's number), so that

$$\beta = \frac{1}{k\mathrm{T}}. \tag{3.48}$$

Identifying s with the entropy, (3.46), (3.47), and (3.48) imply the following relation of the Helmholtz free energy per particle $f(\beta, v, N)$ with the partition function:

$$f := u - \mathrm{T}s = -\frac{1}{\beta N}\log Z(\Lambda, \beta, N). \tag{3.49}$$

Sketch of the proof of (3.47) Let us start by computing the variation of $\log Z$ with a small variation in Λ, assuming $\partial\Lambda$ to be smooth. We may write $Z = Z^{\mathrm{p}} Z^{\mathrm{q}}$ where Z^{p} is the integral on the momenta, and setting $d\mathrm{q} = d\mathrm{q}_1 \cdots d\mathrm{q}_N$,

$$Z^{\mathrm{p}} = (2\pi \mathrm{m}/\beta)^{3N/2} \qquad Z^{\mathrm{q}} = \int_{\Lambda^N} e^{-\beta\Phi(\mathrm{q}_1,\ldots,\mathrm{q}_N)} d\mathrm{q}.$$

We easily get:

$$\begin{aligned}
\Delta \log Z &= \frac{1}{Z^{\mathrm{q}}}\left[\int_{(\Lambda\cup\Delta\Lambda)^N} e^{-\beta\Phi(\mathrm{q}_1,\ldots,\mathrm{q}_N)} d\mathrm{q} - \int_{\Lambda^N} e^{-\beta\Phi(\mathrm{q}_1,\ldots,\mathrm{q}_N)} d\mathrm{q}\right] + o(|\Delta\Lambda|) \\
&= \frac{1}{Z^{\mathrm{q}}}\left[N\int_{\Delta\Lambda} d\mathrm{q}_1 \int_{(\Lambda)^{N-1}} e^{-\beta\Phi(\mathrm{q}_1,\ldots\mathrm{q}_N)} d\mathrm{q}_2, \ldots, d\mathrm{q}_N\right] + o(|\Delta\Lambda|) \\
&=: \int_{\Delta\Lambda} h(\mathrm{q}_1) d\mathrm{q}_1 + o(|\Delta\Lambda|)
\end{aligned}$$

(the last integral is the probability of finding a particle in $\Delta\Lambda$). Thus,

$$\Delta \log Z = \frac{|\Delta\Lambda|}{|\partial\Lambda|}\int_{\partial\Lambda} h(\mathrm{q})\, d\sigma + o(|\Delta\Lambda|),$$

where $d\sigma$ refers to the orthogonal projection of the Lebesgue measure on $\partial\Lambda$; we then conclude that

$$\lim_{\Delta V\to 0}\frac{\Delta \log Z}{\Delta V}=\frac{1}{|\partial\Lambda|}\int_{\partial\Lambda}h(\mathrm{q})d\sigma. \tag{3.50}$$

We now relate the right-hand side of (3.50) to the average of $\mathcal{P}$ with respect to the canonical measure $\mu_{\Lambda,\beta,v}$, cf. (3.33). At this level we make some approximations: if Λ is a cube, we neglect what happens at the corners, so that we have a sum over a finite number of flat walls, and in the small time interval Δt we neglect, for those particles which are about to collide with the wall, the interactions with other particles, i.e. their velocities are assumed to be constant over Δt. For a particle at $\mathrm{q}\in\Lambda$ which may collide with one flat wall during Δt, let $p=\langle \mathrm{p},\mathrm{n}\rangle$, where n is the outer normal, r the distance to $\partial\Lambda$. Summing over the various flat walls we are led to approximate $\mathrm{P}(\Lambda,\beta,v)$ by

$$\frac{1}{|\partial\Lambda|}\frac{1}{|\Delta t|}\int_{\Lambda}h(\mathrm{q})d\mathrm{q}\int_{\mathrm{m}r/|\Delta t|}^{+\infty}\frac{e^{-\frac{\beta p^2}{2\mathrm{m}}}}{\sqrt{2\pi\mathrm{m}/\beta}}2p\,dp$$

for $|\Delta t|$ small. Using simple calculations on Gaussian distributions, letting $|\Delta t|\to 0$, and recalling (3.50) we would obtain (3.47). (If $\partial\Lambda$ is smooth, we may do a similar approximation.) □

Another ensemble, in which the number of particles is not fixed, called a *grand canonical ensemble*, can also be proven to be orthodic in the thermodynamic limit (see [248]).

To introduce the grand canonical ensemble we have to generalize some previously introduced notions. The grand canonical ensemble can be seen as describing a large subsystem of a very large system, with which is possible not only an exchange of energy but also an exchange of particles. The new, enlarged phase space is

$$\mathcal{X}_\Lambda=\cup_{n\ge 0}\mathcal{X}_\Lambda^n,$$

where, for $n\ge 1$

$$\mathcal{X}_\Lambda^n := \Lambda^n\times\mathbb{R}^{3n}$$

is the phase space of a system containing exactly n particles and $\mathcal{X}_\Lambda^0$ just contains one isolated point representing the empty state.

The grand canonical measure $\nu=\nu_{\Lambda,\beta,\vartheta}$ is given by:

$$\nu(\mathcal{X}^0):=\frac{1}{\Xi_\Lambda(\vartheta,\beta)}, \tag{3.51}$$

$$\nu|_{\mathcal{X}_\Lambda^n}:=\frac{1}{\Xi_\Lambda(\vartheta,\beta)}\frac{e^{\beta\vartheta n}}{n!}e^{-\beta H((\mathrm{q})_n,(\mathrm{p})_n)}d(\mathrm{q})_n d(\mathrm{p})_n, \tag{3.52}$$

where the parameter ϑ is called the *chemical potential* and we use $(\mathrm{q})_n$, $(\mathrm{p})_n$ to denote $(\mathrm{q}_1, \ldots, \mathrm{q}_n)$, $(\mathrm{p}_1, \ldots, \mathrm{p}_n)$, respectively. The quantity:

$$\Xi_\Lambda(\vartheta, \beta) := \sum_{n=0}^{\infty} \frac{e^{\beta\vartheta n}}{n!} \int_{\Lambda^n \times \mathbb{R}^{3n}} e^{-\beta H((\mathrm{q})_n, (\mathrm{p})_n)} d(\mathrm{q})_n d(\mathrm{p})_n \tag{3.53}$$

is called the *grand canonical partition function*. As in (3.41), the factor $1/n!$ is introduced to take into account the indistinguishability of particles. By integrating over the momenta we get:

$$\Xi_\Lambda(\vartheta, \beta) := 1 + \sum_{n \geq 1} \frac{z^n}{n!} \int_{\Lambda^n} e^{-\beta\Phi((\mathrm{q})_n)} d(\mathrm{q})_n, \tag{3.54}$$

where

$$z := e^{\beta\vartheta} \left(\frac{2\pi \mathrm{m}}{\beta}\right)^{3/2} \tag{3.55}$$

is called the *fugacity*.

$\Xi_\Lambda(\vartheta, \beta)$ is defined via an infinite series and, even in a finite volume, we need some conditions on the interaction potential in order for this series to be convergent. It is clear that the stability condition (3.38) suffices for the convergence.

In the grand canonical ensemble the number of particles n is not fixed, but rather is a random variable with mean

$$\langle n \rangle = \frac{1}{\Xi_\Lambda(\vartheta, \beta)} \sum_{n=0}^{\infty} \frac{n e^{\beta\vartheta n}}{n!} \int_{\Lambda^n \times \mathbb{R}^{3n}} e^{-\beta H((\mathrm{q})_n, (\mathrm{p})_n)} d(\mathrm{q})_n d(\mathrm{p})_n. \tag{3.56}$$

It is possible to show (see [248]) that the grand canonical ensemble is orthodic in the thermodynamic limit, which in this case consists in taking $\Lambda \nearrow \mathbb{R}^3$ in a regular way.

Let us consider the *grand canonical pressure* P^{gc}, that is the ensemble average of the momentum $\mathcal{P}(x)$ transferred to the walls of the container per unit surface and per unit time by the system in the microscopic state x. Under suitable conditions on the interactions ((3.38) and (3.39) are sufficient), and on the sequence of regions $\Lambda \nearrow \mathbb{R}^3$ (for instance a sequence of increasing spheres) it is possible to show that the following limits exist, and so define the grand canonical pressure in the thermodynamic limit:

$$\mathrm{P}^{gc}(\vartheta, \beta) := \lim_{\Lambda \nearrow \mathbb{R}^3} \nu_{\Lambda,\beta,\vartheta}(\mathcal{P}) = \lim_{\Lambda \nearrow \mathbb{R}^3} \frac{1}{\beta|\Lambda|} \log \Xi_\Lambda(\vartheta, \beta). \tag{3.57}$$

For a proof of the existence of the limit in the right-hand side of (3.57) see Section 3.4 of [258]. To get a partial justification of (3.57) consider for a moment the free case (ideal gas) corresponding to a vanishing interaction potential: $\Phi \equiv 0$. In this simple situation the measure $\nu = \nu_{\Lambda,\beta,\vartheta}$ corresponds to a Poisson system of points, with constant intensity on Λ and independent Gaussian momenta, i.e. n is distributed as a Poisson random variable with average $z|\Lambda|$; given n, the positions are i.i.d. and uniformly distributed on Λ, and the momenta are i.i.d. Gaussian variables on $\mathbb{R}^3$, with zero average and covariance matrix $\frac{\mathrm{m}}{\beta}\mathbb{I}$, where $\mathbb{I}$ denotes the identity matrix. In particular,

$$z = \frac{1}{|\Lambda|} \log \Xi_\Lambda(\vartheta, \beta), \qquad \rho := \frac{\langle n \rangle}{|\Lambda|} = z$$

so that

$$\frac{1}{\beta|\Lambda|} \log \Xi_\Lambda(\vartheta, \beta) = k\mathrm{T}\rho = k\mathrm{T}\frac{n_g \mathcal{N}}{V}, \tag{3.58}$$

where V is the volume, $\mathcal{N}$ is Avogadro's number, n_g is the number of moles so that $n_g\mathcal{N}$ is the number of molecules; but, as we saw before, $k\mathcal{N} = R$, the perfect gas constant and we know from the perfect gas equation that $n_g R\mathrm{T}/V = \mathrm{P}$, the pressure.

Let us now discuss briefly the problem of the equivalence between canonical and grand canonical ensembles, in the thermodynamic limit. Similarly, one could discuss the equivalence between canonical and microcanonical ensembles, which turns out to be more complicated; we refer to [129], as well as [215].

In order to perform the thermodynamic limit in the canonical ensemble, just as in the microcanonical case, we need to take into account the indistinguishability of particles, so that:

$$Z_n^c(\Lambda, \beta) := \frac{1}{n!} Z(\Lambda, \beta, n), \tag{3.59}$$

where $Z(\Lambda, \beta, n)$ is the previously defined canonical partition function. (We multiply and divide by $n!$ the right-hand side of (3.43).) There exists a thermodynamic limit of the canonical free energy per particle

$$f_c(\rho, \beta) := - \lim_{N,V \to +\infty, \frac{N}{V} = \rho} \frac{1}{\beta N} \log Z_N^c(\Lambda, \beta) \tag{3.60}$$

and letting $V = |\Lambda|$, the volume of $\Lambda = \Lambda_N$, we may write

$$\Xi_\Lambda(\vartheta, \beta) = \sum_{n=0}^{\infty} e^{\beta\vartheta n} Z_n^c(\Lambda, \beta) = \sum_{n=0}^{\infty} e^{V\beta\rho_n(\vartheta - f_c(\rho_n, \beta) + \epsilon(n))}, \tag{3.61}$$

where $\rho_n = n/V$ is the density, and $\epsilon(n)$ is an infinitesimal quantity as $n \to +\infty$.

Before going on with the equivalence of ensembles let us discuss some purely thermodynamic relations between some thermodynamic functions, obtained via Legendre transform.

Given a specific thermodynamic system let $f = f(\rho, \mathrm{T})$ denote the Helmholtz free energy per particle expressed in terms of the density ρ and the temperature

T: $f = u - \mathrm{T}s$ with u the internal energy per particle, and s the entropy per particle. Writing $v = 1/\rho$ for the specific volume, the first and second principles of thermodynamics imply

$$df = du - \mathrm{T}ds - sd\mathrm{T} = du - (du + \mathrm{P}dv) - sd\mathrm{T}, \tag{3.62}$$

from which we get the following relation between the pressure P and the Helmholtz free energy f:

$$\mathrm{P} = -\frac{\partial f}{\partial v}. \tag{3.63}$$

Let

$$g(\rho) = g(\rho, \mathrm{T}) := \rho f(\rho, \mathrm{T}) \tag{3.64}$$

be the Helmholtz free energy per unit volume, where the temperature T is assumed to be fixed once and for all, and omitted from the notation. As we now check, the pressure P may be expressed as the Fenchel–Legendre transform of $g(\cdot)$:

$$\tilde{\mathrm{P}}(\vartheta) = g^*(\vartheta) := \sup_\rho (\vartheta\rho - g(\rho)), \tag{3.65}$$

using φ^* to denote the Fenchel–Legendre transform of a function φ.

Let us first suppose that g is of class C^2, with $g''(\rho) > 0$. In this case, g is strictly convex, and we see at once that for each ϑ in the range of g' the supremum in (3.65) is attained at a unique $\rho = \rho_0(\vartheta)$, the unique solution of

$$\vartheta = g'(\rho). \tag{3.66}$$

(This solution is unique since $g'(\cdot)$ is assumed to be strictly increasing.) Thus

$$\tilde{\mathrm{P}}(\vartheta) = \vartheta\rho_0(\vartheta) - g(\rho_0(\vartheta)) \tag{3.67}$$

and from (3.66):

$$\tilde{\mathrm{P}}(g'(\rho)) = g'(\rho)\rho - g(\rho) = -\left.\frac{\partial f}{\partial v}\right|_{v=\frac{1}{\rho}} \equiv \mathrm{P}(\rho), \tag{3.68}$$

or equivalently $\mathrm{P}(\rho_0(\vartheta)) = \tilde{\mathrm{P}}(\vartheta)$, whenever ϑ belongs to the range of g'.

Remark. With an abuse of notation, in this case we may write P in place of $\tilde{\mathrm{P}}$.

The meaning of the Legendre conjugated variable ϑ can be easily understood. In the present case

$$\vartheta = g'(\rho) = f - \frac{\partial f}{\partial v}v = f + \mathrm{P}v = u - \mathrm{T}s + \mathrm{P}v, \tag{3.69}$$

usually called the *chemical potential* or still Gibbs free energy.

Before proceeding to the general case, let us recall the geometric interpretation of the Fenchel–Legendre transform, as seen in Chapter 1: considering plane coordinates (ρ, η) (ρ horizontal, η vertical) $g^*(\vartheta)$ represents the maximal vertical (signed) distance between the straight line $\eta = \vartheta\rho$ and the graph of the function $\rho \mapsto g(\rho)$, as in Figure 1.2. (If $g\colon \mathcal{D}_g \to \mathbb{R}$ is convex, we may set $g(\rho) = +\infty$ for $\rho \notin \mathcal{D}_g$.)

Equation (3.66) asserts that if ϑ belongs to the range of g' the vertical distance is maximal where the slope of the tangent to the graph of g is exactly ϑ. If $g'' > 0$ everywhere, there is exactly one such point. Being the pointwise supremum of linear functions, g^* is always convex. Moreover, if g is of class C^2 and strictly convex, we easily see that:

(1) its Legendre transform $\tilde{\mathrm{P}} = g^*$ is C^2 and strictly convex;
(2) $(g^*)^* = g$, i.e.

$$g(\rho) = \sup_{\vartheta} \left(\vartheta\rho - \tilde{\mathrm{P}}(\vartheta)\right). \tag{3.70}$$

Remark. As mentioned in Chapter 1 and earlier in this chapter, $(g^*)^* = g$ if and only if g is convex and lower semi-continuous; generally $(g^*)^*$ gives the largest lower semi-continuous, convex function whose graph is below that of g. (Of course, one may also define g^* for non-convex g.)

Let us now consider the case when g, though convex, is not strictly convex, i.e. its graph contains a straight segment. As we shall see, the physical meaning for this is the presence of a first order phase transition. Indeed, if there exist constants $a > 0, b \in \mathbb{R}, 0 < \rho_1 < \rho_2$ such that

$$g(\rho) = -a + b\rho, \qquad \text{for all} \quad \rho \in [\rho_1, \rho_2], \tag{3.71}$$

from (3.63), after recalling that $f(\rho) = \frac{1}{\rho} g(\rho) = v g(\frac{1}{v})$, we get:

$$\mathrm{P}(\rho) = a, \qquad \text{if}\ \rho \in [\rho_1, \rho_2], \tag{3.72}$$

namely, the graph of P has a 'plateau' in the interval $[\rho_1, \rho_2]$, which is characteristic behaviour for a liquid/gas phase transition.

In other terms, let us suppose that besides being linear in $[\rho_1, \rho_2]$, g is C^2 and strictly convex outside $[\rho_1, \rho_2]$. Let $\bar{\vartheta} = g'(\rho)$ for $\rho \in (\rho_1, \rho_2)$. We easily solve the previous geometrical problem on the plane (η, ρ), and we get:

(1) $\max_\rho(\bar{\vartheta}\rho - g(\rho))$ is achieved in the whole interval $[\rho_1, \rho_2]$;
(2) $\vartheta \mapsto \tilde{\mathrm{P}}(\vartheta)$ is C^2 and strictly convex on $\mathcal{D}^\circ_{\tilde{\mathrm{P}}} \setminus \{\bar{\vartheta}\}$ (since it is convex, it is continuous on $\mathcal{D}^\circ_{\tilde{\mathrm{P}}}$);
(3) $\tilde{\mathrm{P}}$ is not differentiable at $\bar{\vartheta}$; its right and left derivatives (which exist by the convexity) are different, i.e. the graph of $\tilde{\mathrm{P}}(\cdot)$ has a 'kink' at $\bar{\vartheta}$:

$$\frac{d\tilde{\mathrm{P}}}{d\vartheta^-}(\bar{\vartheta}) = \lim_{\vartheta \to \bar{\vartheta}^-} \tilde{\mathrm{P}}'(\vartheta) = \rho_1, \qquad \frac{d\tilde{\mathrm{P}}}{d\vartheta^+}(\bar{\vartheta}) = \lim_{\vartheta \to \bar{\vartheta}^+} \tilde{\mathrm{P}}'(\vartheta) = \rho_2. \tag{3.73}$$

Conversely, given a convex function $\tilde{\mathrm{P}} : \mathbb{R}^+ \to \mathbb{R}^+$ with a 'kink' at $\bar{\vartheta}$ but C^2 and strictly convex otherwise, we may again solve, on the plane (ϑ, η), the geometrical problem of finding, for a given $\rho > 0$, the maximum of the vertical distance between the straight line $\eta = \rho\vartheta$ and the graph of the function $\tilde{P}(\vartheta)$. If ρ_1 and ρ_2 are, respectively, the left and right derivatives of $\tilde{P}$ at $\bar{\vartheta}$, it is easily seen, by simple geometric argument, that for $\rho < \rho_1$, $\rho > \rho_2$ in the range of $\tilde{P}'$ the maximum is achieved in a unique point $\vartheta = \vartheta_0(\rho)$ with $\vartheta_0(\rho)$ being invertible outside $[\rho_1, \rho_2]$, whereas for *every* $\rho \in [\rho_1, \rho_2]$ the maximum is constantly achieved at $\bar{\vartheta}$, and it varies linearly in ρ for $\rho \in [\rho_1, \rho_2]$.

From the above discussion it is clear that even in the presence of a first order phase transition g and P are related by Legendre transform but in this case $g(\rho)$ is not strictly convex, as it contains an affine branch describing coexistence of liquid and vapour; $\tilde{P}(\vartheta)$ is not differentiable and many different densities (representing different fractions of our fluid being in the liquid phase) correspond to the same value $\bar{\vartheta}$ of the chemical potential where $\tilde{P}(\vartheta)$ is singular.

After the above, purely thermodynamical considerations, let us now go back to the problem of equivalence of canonical and grand canonical ensembles.

We shall see in Proposition 3.13, that under quite general conditions on the interaction potential, the canonical free energy (thermodynamic limit):

$$g_c(\rho, \beta) = \lim_{N,V\to+\infty,\, \frac{N}{V}=\rho} -\frac{1}{\beta V} \log Z_N^c(\Lambda_N, \beta), \tag{3.74}$$

is convex in ρ.

Let us suppose that, as a function of ρ (fixed β), g_c is of class C^2 and strictly convex; in such a case, for any ϑ in the range of g_c', the quantity $\vartheta\rho - g_c(\rho, \beta)$ reaches its absolute maximum at a unique value $\rho_0 = \rho_0(\vartheta, \beta)$; as in the proof of Theorem 1.18,

$$\Xi_\Lambda(\vartheta, \beta) = \sum_{n=0}^{\infty} e^{n\beta\vartheta} Z_n^c(\beta, V) = \sum_{n=0}^{\infty} e^{V\beta\left(\vartheta\rho_n - g_c(\rho_n, \beta) + o(n)\right)},$$

should be well approximated, for large V, by the term corresponding to the maximum of $\vartheta\rho - g_c(\rho, \beta)$, in the sense that

$$\frac{1}{V} \log \Xi_\Lambda(\vartheta, \beta) \sim \beta \max_\rho (\vartheta\rho - g_c(\rho, \beta)) \quad \text{as } V \to +\infty. \tag{3.75}$$

The value $\rho_0(\vartheta, \beta)$ is the unique solution of $\vartheta = \frac{\partial g_c}{\partial \rho}(\rho, \beta)$, as in (3.66), so that

$$\vartheta = f_c(\rho_0, \beta) + \rho_0 \left.\frac{\partial f_c}{\partial \rho}\right|_{\rho=\rho_0} = f_c(\rho_0, \beta) - \rho_0 \left.\frac{\partial f_c}{\partial v}\right|_{v=\frac{1}{\rho_0}} \frac{1}{\rho_0^2}.$$

The parameter ϑ has the meaning of *chemical* potential computed for $\rho = \rho_0(\vartheta, \beta)$ and $\mathrm{T} = 1/k\beta$, as we see after writing the last equation as

$$\vartheta = f_c(\rho_0(\vartheta, \beta), \beta) + \mathrm{P}_c(\rho_0(\vartheta, \beta), \beta) \frac{1}{\rho_0(\vartheta, \beta)}, \tag{3.76}$$

that is, $\vartheta = u - \mathrm{T}s + \mathrm{P}v$. Here we are using the thermodynamic limit for the canonical ensemble; the smoothness assumption allows us to say that P_c is obtained from f_c via (3.63).

Under strict convexity of $g_c(\cdot, \beta)$, the maximizer $\rho_0(\vartheta, \beta)$ has the meaning of 'grand canonical density', in the sense that

$$\rho_{g\,c}(\vartheta, \beta) \equiv \frac{\sum_{n=0}^{\infty} \frac{n}{V} \exp(V(\beta\vartheta\rho_n - \beta g_c(\rho_n, \beta) + o(n)))}{\sum_{n=0}^{\infty} \exp(V(\beta\vartheta\rho_n - \beta g_c(\rho_n, \beta) + o(n)))} \simeq \rho_0(\vartheta, \beta) \quad (3.77)$$

using the above mentioned approximation as $V \to +\infty$. The equivalence of the canonical and grand canonical ensembles, as far as the pressure is concerned, is expressed by the relation:

$$\mathrm{P}_c(\rho_0(\vartheta, \beta), \beta) \equiv \mathrm{P}^{g\,c}(\vartheta, \beta), \quad (3.78)$$

that is, the grand canonical pressure coincides with the thermodynamic limit of the canonical one, as previously defined. To see this, in the case of strict convex $g_c(\cdot, \beta)$, we recall the expression (3.57) for the grand canonical pressure, and (3.75) to write:

$$\mathrm{P}^{g\,c}(\vartheta, \beta) \simeq \vartheta\rho_0(\vartheta, \beta) - \rho_0(\vartheta, \beta) f_c(\rho(\vartheta, \beta), \beta)$$

which together with (3.76) leads to (3.78). It is clear that with a similar procedure we can verify the complete equivalence; more precisely, one can show that all grand canonical thermodynamic functions, when expressed in terms of the parameters (ϑ, β), coincide with the corresponding canonical quantities computed for $(\rho_0(\vartheta, \beta), \beta)$.

The above brief discussion can be made completely rigorous, showing that a complete and satisfactory equivalence between canonical and grand canonical ensembles is possible in the regime of uniqueness of thermodynamic phase, corresponding to strict convexity of the Helmholtz free energy. In contrast, this equivalence breaks down for values of thermodynamic parameters corresponding to first order phase transitions: when $\tilde{\mathrm{P}}(\vartheta)$ has a kink at $\bar{\vartheta}$ there are many possible canonical ensembles corresponding to the same value of the chemical potential ϑ, those with $\rho \in [\rho_1, \rho_2]$. A complete equivalence can be recovered by introducing the 'extended grand canonical ensembles' (see [129]) corresponding to assigning particular boundary conditions outside our box Λ.

A satisfactory description of the phenomenon of phase transitions requires the introduction of more sophisticated concepts like the Gibbs measure. Before proceeding to the introduction of the theory of Gibbs measures in the simplified set-up of lattice gases, let us sketch a proof of the convexity of the Helmholtz free energy in the context of OCSPP.

Proposition 3.13 *Consider an OCSPP with the potential energy as in Definition* 3.12, *with ϕ stable and tempered. Let $\{\Lambda_n\}$ be the sequence of cubes centred* *at the origin with edge 2^n. Then the following limit exists*

$$g_c(\rho, \beta) = \lim_{N\to+\infty, n\to+\infty, \frac{N}{|\Lambda_n|}=\rho} -\frac{1}{\beta|\Lambda_n|} \log Z_N^c(\Lambda_n, \beta) \tag{3.79}$$

and defines a convex function of the density ρ. Moreover, the same limit in (3.79) *is obtained if Λ_n is replaced by a van Hove sequence $\tilde{\Lambda}_N$ (cf. Definition* 3.19) *invading $\mathbb{R}^3$, with $N/|\tilde{\Lambda}_N| \to \rho$.*

Sketch of the proof We sketch the proof for ϕ as in example (a), after Definition 3.12. For the full proof, and for more generality, see [117, 258]. Recalling (3.44), (3.59) and integrating over the momenta:

$$Z_N^c(\Lambda, \beta) = (2\pi \mathrm{m}/\beta)^{3N/2}\, \tilde{Z}_N^c(\Lambda, \beta), \tag{3.80}$$

where

$$\tilde{Z}_N^c(\Lambda, \beta) := \frac{1}{N!}\int_{\Lambda^N} e^{-\beta\Phi((\mathrm{q})_N)} d(\mathrm{q})_N. \tag{3.81}$$

The stability condition (3.38) implies that $\tilde{Z}_N^c(\Lambda, \beta) \le e^{\beta BN}|\Lambda|^N/N!$, and since $\log N! \ge N\log N - N$ we see that

$$\frac{1}{|\Lambda|}\log \tilde{Z}_N^c(\Lambda, \beta) \le \rho(B\beta + 1 - \log \rho)$$

if $N/|\Lambda| = \rho \in (0, +\infty)$. (The left-hand side above could be $-\infty$; indeed, if ρ is large, due to the hard core condition one has $\tilde{Z}_N^c(\Lambda, \beta) = 0$ and the limit in (3.79) will be $+\infty$.)

We now consider Λ_n (Λ_n') the cube centred at the origin, with side 2^n ($2^n - r_0$, respectively), with $N_n = \rho|\Lambda_n| = \rho 2^{3n}$. Thus,

$$\frac{1}{|\Lambda_n|}\log \tilde{Z}_{N_n}^c(\Lambda_n', \beta) \le \frac{1}{|\Lambda_n|}\log \tilde{Z}_{N_n}^c(\Lambda_n, \beta).$$

On the other hand, since $2^{n+1} - r_0 = 2(2^n - r_0) + r_0$ we may decompose Λ_{n+1}' as the union of 2^3 cubes which are translates of Λ_n', separated from each other by 'corridors' of width r_0 (see Figure 3.1); the short range condition and translation invariance give:

$$\tilde{Z}_{N_{n+1}}^c(\Lambda_{n+1}', \beta) \ge \tilde{Z}_{N_n}^c(\Lambda_n', \beta)^8, \tag{3.82}$$

as follows by restricting to the configurations with N_n particles in each of the eight translates of Λ_n', since $N_{n+1} = 8N_n$. But $|\Lambda_{n+1}| = 8|\Lambda_n|$, and (3.82) gives that

$$\frac{1}{|\Lambda_n|}\log \tilde{Z}_{N_n}^c(\Lambda_n', \beta)$$

is non-decreasing in n. From the previous observations, we see promptly that the limit in (3.79) exists for such a sequence Λ_n'. The same limit as for the previous special cubes is obtained along any van Hove sequence Λ invading $\mathbb{R}^3$, provided

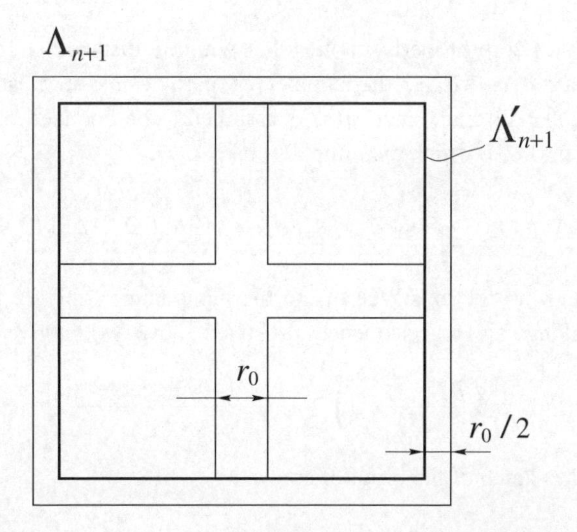

Figure 3.1

$N/|\Lambda| = \rho$. This involves covering Λ with the special cubes Λ_n of the previous case, properly controlling the interaction among them, and taking into account Definition 3.19. This is similar to, but more involved than, the argument below. We refer to [117, 258], or [286] for all this.

Already assuming the existence of the limit in (3.79), we now sketch a proof of the convexity in ρ. We may consider the cubes Λ, of side L, and $N = \rho|\Lambda|$. Let us suppose we divide Λ into two equal parallelepipeds Δ_1, Δ_2 with sides $L, L, L/2$; given ρ_1, ρ_2 two possible densities, so that $N_1 = \rho_1|\Delta_1|$, $N_2 = \rho_2|\Delta_2|$, are integers, and $N = N_1 + N_2$; in particular $\rho = (\rho_1 + \rho_2)/2$, since $|\Delta_1| = |\Delta_2| = |\Lambda|/2$.

From (3.80), (3.81) we get:

$$\tilde{Z}^c_N(\Lambda, \beta) \geq \frac{1}{N_1!N_2!} \times \int_{\Delta_1^{N_1}} d(\mathrm{q})_{N_1} \int_{\Delta_2^{N_2}} d(\mathrm{q}')_{N_2} e^{-\beta\Phi((\mathrm{q})_{N_1})} e^{-\beta\Phi((\mathrm{q}')_{N_2})} e^{-\beta W((\mathrm{q})_{N_1},(\mathrm{q}')_{N_2})} \tag{3.83}$$

where $W((\mathrm{q})_{N_1}, (\mathrm{q}')_{N_2})$ is the interaction between the N_1 particles in Δ_1 and the N_2 particles in Δ_2 given by

$$W((\mathrm{q})_{N_1}, (\mathrm{q}')_{N_2}) := \sum_{i=1}^{N_1}\sum_{j=1}^{N_2} \phi(|\mathrm{q}_i - \mathrm{q}'_j|). \tag{3.84}$$

From the short range and hard core properties of the interaction we easily get, for L sufficiently large,

$$Z^c_N(\Lambda, \beta) \geq Z^c_{N_1}(\Delta_1, \beta) Z^c_{N_2}(\Delta_2, \beta) e^{-\mathrm{const}|\partial\Lambda|}. \tag{3.85}$$

Indeed, the hard core properly implies that within a distance r_1 from the common boundary $\partial\Delta_1 \cap \partial\Delta_2$, the number of particles is at most of order of const $\times|\partial\Delta_1 \cap \partial\Delta_2|$ and, since $\phi(r)$ is assumed to be bounded from below on $r > r_1$, we get (3.85), which then implies

$$\frac{1}{|\Lambda|} \log Z_N^c(\Lambda, \beta) \geq \frac{1}{|\Lambda|} \log Z_{N_1}^c(\Delta_1, \beta_1) + \frac{1}{|\Lambda|} \log Z_{N_2}^c(\Delta_2, \beta) + O\left(\frac{|\partial\Lambda|}{|\Lambda|}\right).$$

This, together with the existence of the thermodynamic limit for the canonical free energy along van Hove sequences, as stated above, would give

$$g_c\left(\frac{\rho_1 + \rho_2}{2}, \beta\right) \leq \frac{g_c(\rho_1, \beta) + g_c(\rho_2, \beta)}{2}$$

concluding the sketch of the proof of convexity. □

Besides showing the usefulness of convexity, the preceding discussion had a similarity with arguments used in large deviation analysis. These are indeed closely related, and large deviations are important in the connection between thermodynamics and the statistical behaviour of macroscopic observables of the system. On the one hand, we have argued the equivalence between the grand canonical and canonical ensembles under strict convexity of the free energy per particle. On the other hand, it seems natural that such equivalence should be closely connected with a concentration on the distribution of the observed density of particles around the grand canonical average. This can be rephrased in terms of a large deviation statement, if the rate function has a unique point of minimum, leading to the previously mentioned *exponential convergence*. The point is then to see that the free energy per particle yields a rate function for the density of particles. An analogous argument applies to the full equivalence between the canonical and the microcanonical, where we need the exponential convergence of $N^{-1}H((\mathrm{q})_N, (\mathrm{p})_N)$. All this applies for a reasonable class of interaction potentials Φ, which includes those in example (a), following Definition 3.12.

In Section 3.4 we shall look at this again in the simpler situation of lattice spin systems.

3.3.2 Lattice systems: Gibbs measures, thermodynamic limit

Statistical mechanics not only provides a microscopic model for thermodynamics and allows (at least in principle) deduction of the equation of state from the microscopic Hamiltonian; it also gives the possibility to describe the 'equilibrium fluctuations', and to compute non-thermodynamical quantities like correlation functions.

One of the most interesting achievements of equilibrium statistical mechanics is the microscopic description of phase transitions.

Lattice gases and lattice spin systems are extremely simplified statistical mechanical models, which however exhibit many interesting properties. In this context we wish to outline the previously described programme.

To define the lattice gas we partition $\mathbb{R}^d$ into cubes of unitary side-length, we suppose that in each unit cube there is at most one particle, and that the interaction between particles does not change significantly inside a unit cube. This corresponds to choosing the unit of length of the same order of magnitude as the range of a hard core exclusion, and sufficiently shorter than the range of intermolecular forces.

In the continuous case the grand canonical phase space was:

$$\cup_{n=0}^{\infty} \Lambda^n \times \mathbb{R}^{dn}$$

moreover the momenta did not play any role since it was possible to integrate them out, cf. (3.54). Indistinguishability of particles implies that two configurations obtained from each other by a permutation have to be identified; in other words a configuration is given once the number and the positions of the particles are given, independently of the order. In the simplified lattice gas model, beyond assuming that in each unit cube there is at most one particle, we give up specifying its position inside the cube; thus a configuration is given just by assigning the 'occupation number' (0 or 1) in each unit cube. The centre of each cube belongs to the lattice $\mathbb{Z}^d$.

Notation In this and the next section $\Lambda \subset\subset \mathbb{Z}^d$ means that Λ is a non-empty finite subset of $\mathbb{Z}^d$, in which case $|\Lambda| = \sharp\Lambda$ denotes its cardinality. For $i \in \mathbb{Z}^d$, $|i|$ denotes the Euclidean norm in $\mathbb{R}^d$. We use O to denote the origin in $\mathbb{Z}^d$.

A configuration in $\Lambda \subset\subset \mathbb{Z}^d$ is a function on Λ with values in $\{0, 1\}$. We may write it as $n_\Lambda = \{n_i\}_{i\in\Lambda}$, where $n_i = 0$ or 1 is the occupation number at the site $i \in \Lambda$, and we might identify n_Λ with the set $\{i : n_i = 1\}$, i.e. the set of cubes where a particle is present.

The states[2] of the lattice gas are thus identified with probability measures ρ on the finite set $\mathcal{X}_\Lambda = \{0, 1\}^\Lambda$. Let us assume the interaction between particles at sites i and j in $\mathbb{Z}^d$ to be of the form $\phi_{i,j} = \phi(|i - j|)$, where $\phi\colon \mathbb{R}^+ \to \mathbb{R}$ satisfies

$$\sum_{i\in\mathbb{Z}^d} |\phi(|i|)| < \infty.$$

The role of the Hamiltonian is now played by the (potential) energy, given by (cf. (3.37)):

$$H_\Lambda(n_\Lambda) = \frac{1}{2} \sum_{\substack{i,j\in\Lambda \\ i\neq j}} \phi(|i - j|) n_i n_j,$$

[2] We also use *state* to designate a configuration, a possible value of a stochastic process. No confusion should arise.

and the grand canonical partition function (see (3.54)) becomes:

$$\Xi_\Lambda(\beta,\vartheta) = \sum_{n_\Lambda \in X_\Lambda} \exp\beta\Big(\vartheta\sum_{i\in\Lambda} n_i - \frac{1}{2}\sum_{\substack{i,j\in\Lambda\\ i\neq j}} \phi(|i-j|)n_i n_j\Big).$$

Another interesting case, isomorphic to the lattice gas, is that of Ising-like spin systems, which may be seen as a simplified model for describing magnetic systems: one thinks of a magnetic moment sitting at each site of $\mathbb{Z}^d$ and taking only two possible orientations: up and down [127].

To each site $i \in \mathbb{Z}^d$ we associate a spin variable σ_i taking values in $\{-1,+1\}$. Thus, for $\Lambda \subseteq \mathbb{Z}^d$ non-empty, a configuration in Λ is a function $\sigma_\Lambda : \Lambda \to \{-1,+1\}$.

$$\mathcal{X}_\Lambda = \{-1,+1\}^\Lambda$$

is the configuration space for Ising spin systems. If $\Lambda \subset\subset \mathbb{Z}^d$, the energy associated with a configuration $\sigma_\Lambda \in \mathcal{X}_\Lambda$ is set as

$$H_\Lambda(\sigma_\Lambda) = -\frac{1}{2}\sum_{\substack{i,j\in\Lambda\\ i\neq j}} J(|i-j|)\sigma_i\sigma_j - \sum_{i\in\Lambda} h_i\sigma_i, \tag{3.86}$$

where h_i represents the local magnetic field and $J(\cdot)$ describes the interaction among spins (assumed to be translationally invariant and 'isotropic').

The isomorphism with the lattice gas comes from the identification $n_i = (1+\sigma_i)/2$. Indeed:

$$\begin{aligned}
&\vartheta\sum_{i\in\Lambda} n_i - \frac{1}{2}\sum_{\substack{i,j\in\Lambda\\ i\neq j}} \phi(|i-j|)n_i n_j \\
&= \vartheta\sum_{i\in\Lambda}\frac{1+\sigma_i}{2} - \frac{1}{2}\sum_{\substack{i,j\in\Lambda\\ i\neq j}}\phi(|i-j|)\frac{1+\sigma_i}{2}\frac{1+\sigma_j}{2} \\
&= \frac{\vartheta}{2}|\Lambda| - \frac{1}{2}\sum_{\substack{i,j\in\Lambda\\ i\neq j}}\frac{1}{4}\phi(|i-j|) + \frac{\vartheta}{2}\sum_{i\in\Lambda}\sigma_i - \frac{1}{4}\sum_{i\in\Lambda}\sigma_i\sum_{j\in\Lambda\setminus\{i\}}\phi(|i-j|) \\
&\quad - \frac{1}{2}\sum_{\substack{i,j\in\Lambda\\ i\neq j}}\frac{1}{4}\phi(|i-j|)\sigma_i\sigma_j,
\end{aligned}$$

from which we get

$$h_i = \frac{\vartheta}{2} - \frac{1}{4}\sum_{j\in\Lambda\setminus\{i\}}\phi(|i-j|), \qquad J(|i-j|) = -\frac{1}{4}\phi(|i-j|).$$

We want now to develop a formalism which will allow us to describe infinite volume thermodynamic equilibrium and, in particular, to give mathematically precise definitions concerning phase transitions. The physical interest of introducing the thermodynamic limit (whose importance was already clear when we introduced orthodic ensembles and their equivalence) comes mainly from the requirement that thermodynamic properties are independent of the size of the sample, provided it is macroscopic.

Moreover, it is possible to show that in a finite volume the thermodynamic functions (pressure, free energy etc.) are analytic in the thermodynamic parameters and then, from this point of view, we cannot give an acceptable definition of phase transition without passing to the thermodynamic limit.

The *thermodynamic limit of the state* corresponds to the infinite volume equilibrium measure; it is related to the study of the statistical properties (e.g. the correlation functions) of our thermodynamic system far away from the boundaries of the container. In other words we are interested in the study of observables which are localized well inside the bulk, trying to show that their behaviour is weakly dependent on the size of the box, provided it is sufficiently large.

We shall now give some definitions using the language of spin systems. The lattice gas is, as we saw before, isomorphic to an Ising spin system.

The space of infinite volume configurations is:

$$\mathcal{X} = \mathcal{X}_{\mathbb{Z}^d} := \times_{i \in \mathbb{Z}^d} \mathcal{X}_i$$

where each $\mathcal{X}_i$ is a copy of the same $\mathcal{X}_O = S$, a finite set which represents the configuration space of a single spin. $\mathcal{X}_O = \{-1, +1\}$ in the particular case of the Ising spin systems. If $\Lambda \subseteq \mathbb{Z}^d$ is non-empty, let $\mathcal{X}_\Lambda = \times_{i \in \Lambda} \mathcal{X}_i = S^\Lambda$, endowed with the product topology, where on $\mathcal{X}_O$ all sets are assumed to be open (discrete topology). The family of (*finite dimensional*) *cylinder sets*, as defined below, constitutes a basis of open sets.

Definition 3.14 Given $\Delta \subset\subset \mathbb{Z}^d$, let π_Δ denote the projection on $\mathcal{X}_\Delta$: $\pi_\Delta(\sigma) = \sigma|_\Delta$, the restriction of σ to Δ. The sets of the form

$$\pi_\Delta^{-1}(A) = \{\sigma \in \mathcal{X} : \sigma|_\Delta \in A\}, \tag{3.87}$$

with $A \subseteq \mathcal{X}_\Delta$ and $\Delta \subset\subset \mathbb{Z}^d$ are called (finite dimensional) cylinder sets. Given a cylinder C, its support, denoted by $\Delta(C)$, is the smallest Δ for which $C = \pi_\Delta^{-1}(A)$ for some $A \subseteq \mathcal{X}_\Delta$. A function f on $\mathcal{X}$ is called cylinder if and only if there exist $\Delta \subset\subset \mathbb{Z}^d$ and a function g on $\mathcal{X}_\Delta$, so that $f(\sigma) = g(\sigma|_\Delta)$ for each σ; its support $\Delta(f)$ is defined accordingly. The set of real valued cylinder functions on $\mathcal{X}$ is denoted by $\mathcal{F}_{loc}$.

Notation A change of notation with respect to Section 1.3: for brevity, on that occasion we wrote π_k for $\pi_{\{1,\ldots,k\}}$.

The cylinder functions are evidently continuous, and an extension of (1.64) provides a metric for the product topology on $\mathcal{X}$. This is a compact space. Therefore, any continuous function must be uniformly continuous, and we see that

$f: \mathcal{X} \to \mathbb{R}$ is continuous if and only if

$$\lim_{n\to+\infty} \sup_{\sigma|_{\Lambda_n}=\sigma'|_{\Lambda_n}} |f(\sigma) - f(\sigma')| = 0$$

where $\Lambda_n = \{-n, \ldots, n\}^d$. Taking e.g. $f_n(\sigma) := f(\sigma \circ_{\Lambda_n} \tau)$ for a fixed $\tau \in \mathcal{X}$, cf. (3.88) below, we see that $\mathcal{F}_{loc}$ is dense in $C(\mathcal{X})$, the space of continuous (real valued) functions with the sup-norm $\|\cdot\|_\infty$. (See also [253].)

Notation As in Section 1.3, $\mathcal{M}_1(\mathcal{X})$ denotes the space of probability measures on $\mathcal{B} = \mathcal{B}(\mathcal{X})$, the Borel σ-field of $\mathcal{X}$, which in the present case coincides with the σ-field generated by the cylinder sets. If $\Lambda \subseteq \mathbb{Z}^d$ is non-empty, $\mathcal{B}_\Lambda = \mathcal{B}(\mathcal{X}_\Lambda)$ denotes both the Borel σ-field of $\mathcal{X}_\Lambda$ for the product topology, and the σ-field on $\mathcal{X}$ generated by the projection $\pi_\Lambda : \mathcal{X} \to \mathcal{X}_\Lambda$, i.e. $\{\pi_\Lambda^{-1}(B) : B \in \mathcal{B}(\mathcal{X}_\Lambda)\}$. (For finite Λ, $\mathcal{B}_\Lambda = \{A : A \subseteq \mathcal{X}_\Lambda\}$; it also denotes the class of cylinder sets with support in Λ, cf. Definition 3.14.)

$\mathcal{M}_1(\mathcal{X})$ is endowed with the usual weak topology induced by the (bounded) continuous functions, i.e. the coarsest topology which renders continuous the maps $\mu \mapsto \int_\mathcal{X} f d\mu$, for $f \in C(\mathcal{X})$. ($\mu(f) = \langle f, \mu\rangle = \int_\mathcal{X} f d\mu$ if no confusion arises.)

Remark 3.15 Take $\mathcal{X}_O$ finite, with discrete topology.

(a) $\mu^{(n)} \to \mu$ weakly if and only if $\mu^{(n)}(C) \to \mu(C)$ for any cylinder set C. Indeed, one implication follows from the inclusion $\mathcal{F}_{loc} \subseteq C(\mathcal{X})$; the converse follows from the density of $\mathcal{F}_{loc}$ in $C(\mathcal{X})$, since $|\langle f, \mu\rangle| \le \|f\|_\infty$ for $\mu \in \mathcal{M}_1(\mathcal{X})$.

(b) A metric generating this topology on $\mathcal{M}_1(\mathcal{X})$ is given by:

$$\varrho(\mu, \mu') = \sum_{k\ge 0} 2^{-k} |\pi_{\Lambda_k}\mu - \pi_{\Lambda_k}\mu'|_1,$$

with $\Lambda_k = \{-k, \ldots, k\}^d$, $\pi_\Lambda\mu$ the marginal of $\sigma|_\Lambda$ under μ, i.e. $\pi_\Lambda\mu(A) := \mu(\pi_\Lambda^{-1}(A))$ for any $A \subseteq \mathcal{X}_\Lambda$ and, if Λ is finite, $|\alpha - \alpha'|_1 = \sum_{\tilde\sigma \in \mathcal{X}_\Lambda} |\alpha(\tilde\sigma) - \alpha'(\tilde\sigma)|$ on $\mathcal{M}_1(\mathcal{X}_\Lambda)$, as in Section 1.3.

(c) From the compactness of $\mathcal{X}$ we know that $\mathcal{M}_1(\mathcal{X})$ is also compact; if a sequence $\mu^{(n)}$ in $\mathcal{M}_1(\mathcal{X})$ is such that $\mu^{(n)}(C)$ converges, for each cylinder set C, then this limit, call it $\mu(C)$, determines a unique $\mu \in \mathcal{M}_1(\mathcal{X})$, and $\mu^{(n)} \to \mu$. In particular, the class of cylinder sets (or cylinder functions) form a *convergence determining* class, cf. [20]. This follows as in Section 1.3, just replacing S^k by $\mathcal{X}_{\Lambda_k}$, with Λ_k as above.

(d) According to the Riesz theorem, the dual of the Banach space $C(\mathcal{X})$ (sup-norm) is $\mathcal{M}(\mathcal{X})$, the space of finite signed-measures on $(\mathcal{X}, \mathcal{B}(\mathcal{X}))$. The topology under consideration is the restriction of the so-called w^*-topology to $\mathcal{M}_1(\mathcal{X})$.

Extending the notation in Section 1.3 we consider the translation operators on $\mathcal{X}$: $(\theta_i\sigma)_j = \sigma_{i+j}$ for each $i, j \in \mathbb{Z}^d$. If f is a function on $\mathcal{X}$, let $\theta_i f(\sigma) = f(\theta_i\sigma)$

for each σ; if μ is a measure on $(\mathcal{X}, \mathcal{B})$, $\theta_i \mu$ will denote the measure induced by μ under θ_i, i.e. $\theta_i \mu(f) = \mu(\theta_i f)$ for each bounded Borel measurable $f : \mathcal{X} \to \mathbb{R}$.

We say that a measure ν on $\mathcal{X}$ is translationally invariant if $\nu = \theta_i \nu$ for any $i \in \mathbb{Z}^d$. The set of translationally invariant probability measures on $(\mathcal{X}, \mathcal{B})$ is denoted by $\mathcal{M}_1^\theta(\mathcal{X})$.

Remark Let $\nu \in \mathcal{M}_1(\mathcal{X})$. Then, $\nu \in \mathcal{M}_1^\theta(\mathcal{X})$ if and only if $\nu(\theta_i f) = \nu(f)$ for all $f \in \mathcal{F}_{loc}$ and all $i \in \mathbb{Z}^d$.

We now give a summary of useful definitions concerning Gibbs measures on $\mathcal{X}$. For simplicity we restrict ourselves to the case of finite range interactions, commenting briefly on extensions. Ising spin systems with finite range pairwise interaction are important examples.

Notation (a) $\Lambda + j := \{i + j : i \in \Lambda\}$, if $j \in \mathbb{Z}^d$, $\Lambda \subseteq \mathbb{Z}^d$. (b) If $\Lambda \subseteq \mathbb{Z}^d$ and $\sigma, \tau \in \mathcal{X}$, $\sigma \circ_\Lambda \tau$ denotes the configuration which agrees with σ in Λ and with τ in $\Lambda^c := \mathbb{Z}^d \setminus \Lambda$:

$$(\sigma \circ_\Lambda \tau)_x = \begin{cases} \sigma_x & \forall\, x \in \Lambda \\ \tau_x & \forall\, x \in \Lambda^c. \end{cases} \tag{3.88}$$

Definition 3.16 Extending (3.86) we consider energy functions given by a family Φ, of interaction potentials $\Phi_\Delta : \mathcal{X} \to \mathbb{R}$, $\Delta \subset\subset \mathbb{Z}^d$, verifying:

(i) Φ_Δ is $\mathcal{B}_\Delta$-measurable;

(ii) (translation invariance) $\Phi_{\Delta+i}(\sigma) = \Phi_\Delta(\theta_i \sigma)$ for each σ and i;

(iii) (finite range) these exists $r_0 < +\infty$ so that $\Phi_\Delta \equiv 0$ if the diameter of Δ exceeds r_0 (i.e. $\Phi_\Delta \equiv 0$ if $|i - j| > r_0$ for some i and j in Δ).

$\mathbb{B}_\theta^0$ denotes the class of such interactions Φ.

Given $\Phi \in \mathbb{B}_\theta^0$ and $\Lambda \subset\subset \mathbb{Z}^d$, the energy associated with a configuration $\sigma \in \mathcal{X}_\Lambda$ with boundary condition $\tau \in \mathcal{X}_{\Lambda^c}$, is given by:

$$H_\Lambda^\tau(\sigma) = \sum_{\Delta :\, \Delta \cap \Lambda \neq \emptyset} \Phi_\Delta(\sigma \circ_\Lambda \tau). \tag{3.89}$$

Finite range Ising spin systems with isotropic pairwise interaction are important particular cases: $\mathcal{X}_O = \{-1, +1\}$, $\Phi_\Delta = 0$ if $|\Delta| > 2$, $\Phi_{\{i\}} = -h\sigma_i$ and $\Phi_{\{i,j\}} = -J(|i - j|)\sigma_i \sigma_j$, where $h \in \mathbb{R}$ and $J : \mathbb{N} \to \mathbb{R}$ satisfies

$$\text{there exists } r_0 < +\infty \text{ such that } J(n) = 0 \text{ for all } n > r_0. \tag{3.90}$$

In this case (3.89) becomes:

$$H_\Lambda^\tau(\sigma) = -h \sum_{i \in \Lambda} \sigma_i - \frac{1}{2} \sum_{\substack{i,j \in \Lambda \\ i \neq j}} J(|i - j|)\sigma_i \sigma_j - \sum_{i \in \Lambda,\, j \in \partial_{r_0} \Lambda} J(|i - j|)\sigma_i \tau_j \tag{3.91}$$

where $\partial_r \Lambda = \{i \notin \Lambda : \exists j \in \Lambda \text{ such that } |i - j| \leq r\}$.

We have now taken a constant magnetic field h to satisfy condition (ii) of Definition 3.16; the isotropy condition could be relaxed; the case $J(\cdot) \geq 0$ corresponds to ferromagnetic interactions. Due to the hypothesis of finite range interaction, $H_\Lambda^\tau(\sigma)$ depends on τ only through $\tau|_{\partial_{r_0}\Lambda}$.

Definition 3.17 Let Φ be as above and $\Lambda \subset\subset \mathbb{Z}^d$. The finite volume *Gibbs measure* in Λ, at inverse temperature $\beta > 0$, and with boundary condition $\tau \in \mathcal{X}_{\Lambda^c}$ is given by:

$$\mu_{\Lambda,\beta}^\tau(\sigma) = \frac{\exp(-\beta H_\Lambda^\tau(\sigma))}{Z_\Lambda^\tau(\beta)}, \qquad \sigma \in \mathcal{X}_\Lambda, \tag{3.92}$$

with $H_\Lambda^\tau(\cdot)$ given by (3.89) and

$$Z_\Lambda^\tau(\beta) = \sum_{\sigma \in \mathcal{X}_\Lambda} \exp(-\beta H_\Lambda^\tau(\sigma)). \tag{3.93}$$

($Z_\Lambda^\tau(\beta)$ is usually called a partition function and $\mu_{\Lambda,\beta}^\tau$ is a probability measure on $\mathcal{X}_\Lambda$.)

Notation The physical interpretation is $\beta = (k\mathrm{T})^{-1}$, where k is the Boltzmann constant, and T the temperature. Φ is usually omitted from the notation; if required to make it explicit we may write e.g. $Z_\Lambda^\tau(\beta, \Phi)$ or $Z_\Lambda^\tau(\Phi)$. We often omit β as well, writing μ_Λ^τ for brevity.

The finite volume Gibbs measure μ_Λ^τ may be seen as a measure on $\mathcal{X}$, with the configuration outside Λ 'frozen' as $\tau|_{\Lambda^c}$: if $f \in \mathcal{F}_{loc}$, its finite volume Gibbs expectation is set as

$$\mu_\Lambda^\tau(f) := \sum_{\sigma \in \mathcal{X}_\Lambda} \mu_\Lambda^\tau(\sigma) f(\sigma \circ_\Lambda \tau). \tag{3.94}$$

We may also define a finite volume Gibbs measure in $\Lambda \subset\subset \mathbb{Z}^d$ with boundary conditions distributed according to an a priori measure λ on $\mathcal{X}_{\Lambda^c}$. Again we can define it formally as a measure on $\mathcal{X}$.

Definition 3.18 Given a probability measure λ on $(\mathcal{X}, \mathcal{B})$ and $\Lambda \subset\subset \mathbb{Z}^d$, we call a finite volume Gibbs measure in Λ with boundary condition λ, the measure μ_Λ^λ defined by

$$\mu_\Lambda^\lambda(f) = \int_{\mathcal{X}} \mu_\Lambda^\tau(f)\, \lambda(d\tau) \tag{3.95}$$

for any $f \in \mathcal{F}_{loc}$, according to (3.94).

Due to the finite range assumption, $\tau \mapsto \mu_\Lambda^\tau(f) \in \mathcal{F}_{loc}$ for $f \in \mathcal{F}_{loc}$ and Λ finite, and the integral in (3.95) reduces to a finite sum; it determines a probability measure on $\mathcal{X}$. The distribution of $\sigma|_{\Lambda^c}$ under μ_Λ^λ coincides with that under λ, and the conditional distribution on $\mathcal{X}_\Lambda$ given $\sigma|_{\Lambda^c} = \tau|_{\Lambda^c}$ is given by μ_Λ^τ (on $\mathcal{X}_\Lambda$), which depends only on $\tau|_{\partial r_0 \Lambda}$.

Consider now a sequence of domains regularly invading the whole $\mathbb{Z}^d$. By 'regular' we mean a van Hove sequence, as defined below.

Let $a \in \mathbb{N}^d$. We set $\Lambda(a) = \{x \in \mathbb{Z}^d \colon \forall i, 0 \le x_i \le a_i\}$ and $\Lambda(a,k) = \Lambda(a) + ka$, the translate $\Lambda(a)$ by the vector ka, with $k \in \mathbb{Z}$. Given $\Lambda \subset\subset \mathbb{Z}^d$ we set:

$$\begin{aligned} N_a^+(\Lambda) &= \sharp\{k \in \mathbb{Z} \colon \Lambda(a,k) \cap \Lambda \neq \emptyset\}, \\ N_a^-(\Lambda) &= \sharp\{k \in \mathbb{Z} : \Lambda(a,k) \subseteq \Lambda\}. \end{aligned} \tag{3.96}$$

Definition 3.19 An increasing sequence of finite sets (Λ_n) whose union is $\mathbb{Z}^d$ is said to invade $\mathbb{Z}^d$ in the sense of van Hove, and we write $\Lambda_n \nearrow \mathbb{Z}^d$, if for all a:

$$\lim_n N_a^-(\Lambda_n) = +\infty, \qquad \lim_n \frac{N_a^-(\Lambda_n)}{N_a^+(\Lambda_n)} = 1. \tag{3.97}$$

A basic example, usually used for the sake of concreteness, is $\Lambda_n := \{-n, \dots, n\}^d$.

According to part (c) of Remark 3.15, if $\lim_{n\to+\infty} \mu^\tau_{\Lambda_n}(C)$ exists for any cylinder set C, then it determines a probability measure μ^τ on $(\mathcal{X}, \mathcal{B})$, and $\mu^\tau_{\Lambda_n} \to \mu^\tau$ weakly.

We use $H_\Lambda(\sigma)$ to denote the energy associated with the configuration $\sigma \in \mathcal{X}_\Lambda$ with *free* (or empty) boundary conditions, obtained by summing in (3.89) only among $\emptyset \neq \Delta \subseteq \Lambda$. $Z_{\Lambda,\beta}$ and $\mu_{\Lambda,\beta}$ denote the corresponding partition function and Gibbs measure. (For the Ising spin system this is obtained, formally, by taking $\tau_j \equiv 0$ on the right-hand of (3.91).)

Let $\Lambda \subset\subset \mathbb{Z}^d$. The interaction between the spins in Λ and those in its complement is denoted by

$$W_{\Lambda,\Lambda^c}(\sigma) := \sum_{\Delta\cap\Lambda\neq\emptyset,\, \Delta\cap\Lambda^c\neq\emptyset} \Phi_\Delta(\sigma). \tag{3.98}$$

We may prefer to write $W_{\Lambda,\Lambda^c}(\sigma) = W_{\Lambda,\Lambda^c}(\sigma_\Lambda, \sigma_{\Lambda^c})$, with $\sigma = (\sigma_\Lambda, \sigma_{\Lambda^c})$ and $\sigma_\Delta = \sigma|_\Delta$ for $\Delta \subseteq \mathbb{Z}^d$ non-empty. For any $\Lambda \subset\subset \mathbb{Z}^d$:

$$H^\tau_\Lambda(\sigma) = H_\Lambda(\sigma) + W_{\Lambda,\Lambda^c}(\sigma \circ_\Lambda \tau).$$

The finite volume Gibbs measures determine what is called the local Gibbsian specifications: for $\Delta \subset\subset \mathbb{Z}^d$ this is given by

$$q_\Delta(\sigma_\Delta \mid \sigma_{\Delta^c}) := \mu^{\sigma_{\Delta^c}}_{\Delta,\beta}(\sigma_\Delta) = \frac{\exp\left[-\beta(H_\Delta(\sigma_\Delta) + W_{\Delta,\Delta^c}(\sigma_\Delta, \sigma_{\Delta^c}))\right]}{Z^{\sigma_{\Delta^c}}_\Delta(\beta)} \tag{3.99}$$

with $\sigma_\Delta \in \mathcal{X}_\Delta$, $\sigma_{\Delta^c} \in \mathcal{X}_{\Delta^c}$, and cf. (3.93). Due to the finite range assumption this depends on σ_{Δ^c} only through $\sigma_{\partial_{r_0}\Delta}$, and we thus write $q_\Delta(\sigma_\Delta \mid \sigma_{\partial_{r_0}\Delta})$.

The $q_\Delta(\cdot \mid \cdot)$ has to be seen as a function defined on $\mathcal{X}_{\partial_{r_0}\Delta}$, with values in $\mathcal{M}_1(\mathcal{X}_\Delta)$, or as a probability kernel on $\mathcal{X}_{\partial_{r_0}\Delta} \times \mathcal{X}_\Delta$.

To see how the various finite volume Gibbs measures are related we compute, given $\Lambda \subset\subset \mathbb{Z}^d$ and Δ such that $\Delta \cup \partial_{r_0}\Delta \subseteq \Lambda$, the conditional distribution for the configuration in Δ given the configuration in $\Lambda \setminus \Delta$, under any measure μ_Λ^τ. For this, extending the previous notation, if $\Delta_1 \subseteq \Lambda$ is non-empty, $\pi_{\Delta_1}\mu_\Lambda^\tau$ denotes the marginal of $\sigma|_{\Delta_1}$ under μ_Λ^τ, i.e. since $\mathcal{X}_\Lambda = \mathcal{X}_{\Delta_1} \times \mathcal{X}_{\Lambda\setminus\Delta_1}$:

$$\pi_{\Delta_1}\mu_\Lambda^\tau(\sigma_{\Delta_1}) = \sum_{\sigma_{\Lambda\setminus\Delta_1}} \mu_\Lambda^\tau(\sigma_{\Delta_1}, \sigma_{\Lambda\setminus\Delta_1})$$

We easily check that if $\Delta \cup \partial_{r_0}\Delta \subseteq \Lambda$, for each $\tau \in \mathcal{X}_{\Lambda^c}$:

$$\mu_\Lambda^\tau(\sigma_\Delta \mid \sigma_{\Lambda\setminus\Delta}) := \frac{\mu_\Lambda^\tau(\sigma_\Lambda)}{\pi_{\Lambda\setminus\Delta}\,\mu_\Lambda^\tau(\sigma_{\Lambda\setminus\Delta})} = q_\Delta(\sigma_\Delta \mid \sigma_{\partial_{r_0}\Delta}), \tag{3.100}$$

i.e. $q_\Delta(\sigma_\Delta \mid \sigma_{\partial_{r_0}\Delta})$ is the Gibbs conditional probability (under μ_Λ^τ) to find the configuration σ_Δ in Δ given the configuration $\sigma_{\Lambda\setminus\Delta}$ outside; it depends only on $\sigma_{\partial_{r_0}\Delta}$ due to assumption (iii) of Definition 3.16.

When $\Delta \subseteq \Lambda$ but $\Delta \cup \partial_{r_0}\Delta \nsubseteq \Lambda$, we get the following expression for the conditional probability:

$$\mu_\Lambda^\tau(\sigma_\Delta \mid \sigma_{\Lambda\setminus\Delta}) := \frac{\mu_\Lambda^\tau(\sigma_\Lambda)}{\pi_{\Lambda\setminus\Delta}\,\mu_\Lambda^\tau(\sigma_{\Lambda\setminus\Delta})} = q_\Delta(\sigma_\Delta \mid (\sigma \circ_\Lambda \tau)|_{\partial_{r_0}\Delta}),$$

as can be seen from simple calculations.

Let us suppose from now on that $\Delta \cup \partial_{r_0}\Delta \subseteq \Lambda$. In this case, we immediately see that for every finite volume Gibbs measure in Λ, with boundary conditions distributed according to a measure λ in $\mathcal{M}_1(\mathcal{X})$, we can write:

$$\mu_\Lambda^\lambda(\sigma_\Lambda) = q_\Delta(\sigma_\Delta \mid \sigma_{\partial_{r_0}\Delta})\, \pi_{\Lambda\setminus\Delta}\mu_\Lambda^\lambda(\sigma_{\Lambda\setminus\Delta}).$$

(Fixed boundary conditions $\tau \in \mathcal{X}$ correspond to λ a Dirac mass on τ.) If $\Delta \cup \partial_{r_0}\Delta \subseteq \tilde{\Delta} \subseteq \Lambda$, summing over the configurations in $\Lambda \setminus \tilde{\Delta}$ on the previous equation, we get:

$$\pi_{\tilde{\Delta}}\,\mu_\Lambda^\lambda(\sigma_{\tilde{\Delta}}) = q_\Delta(\sigma_\Delta \mid \sigma_{\partial_{r_0}\Delta})\pi_{\tilde{\Delta}\setminus\Delta}\mu_\Lambda^\lambda(\sigma_{\tilde{\Delta}\setminus\Delta}).$$

Passing to the limit, if $\Lambda_n \nearrow \mathbb{Z}^d$ in the sense of van Hove, and supposing $\mu_{\Lambda_n}^\lambda \to \mu^\lambda$, we get, for any finite $\tilde{\Delta} \supseteq \Delta \cup \partial_{r_0}\Delta$:

$$\pi_{\tilde{\Delta}}\mu^\lambda(\sigma_{\tilde{\Delta}}) = q_\Delta(\sigma_\Delta \mid \sigma_{\partial_{r_0}\Delta})\,\pi_{\tilde{\Delta}\setminus\Delta}\mu^\lambda(\sigma_{\tilde{\Delta}\setminus\Delta}).$$

The above relation is verified for any such limiting Gibbs measure, obtained by following any sequence of boundary conditions.

Definition 3.20 A probability measure μ on $(\mathcal{X}, \mathcal{B})$ satisfies the Dobrushin–Lanford–Ruelle (DLR) equations with respect to the interaction Φ as in Definition 3.16, at inverse temperature β, if for every $\tilde{\Delta} \subset\subset \mathbb{Z}^d$, $\Delta \cup \partial_{r_0}\Delta \subseteq \tilde{\Delta}$, and every configuration $\sigma_{\tilde{\Delta}}$:

$$\pi_{\tilde{\Delta}}\mu(\sigma_{\tilde{\Delta}}) = q_\Delta(\sigma_\Delta \mid \sigma_{\partial_{r_0}\Delta})\,\pi_{\tilde{\Delta}\setminus\Delta}\mu(\sigma_{\tilde{\Delta}\setminus\Delta}), \tag{3.101}$$

where $q_\Delta(\cdot \mid \cdot)$ is given by (3.99).

The meaning of the above equations (also called *equilibrium equations*) is as follows: an infinite volume state described by a measure μ is an equilibrium state for a system with interaction Φ, if in every finite volume Δ the system is at thermal equilibrium with the exterior (in the sense of Gibbs grand canonical prescription $q_\Delta(\cdot \mid \cdot)$) and this, in turn, is distributed according to μ. In other words, $q_\Delta(\cdot \mid \cdot)$ is the conditional distribution on $\mathcal{B}_\Delta$ given $\mathcal{B}_{\Delta^c}$ under μ.

We now analyse briefly the structure of DLR measures, or DLR states, as the measures verifying Definition 3.20 are usually called.

Let $\mathcal{G}_\Lambda$ denote the set of Gibbs measures in the finite volume Λ:

$$\mathcal{G}_\Lambda = \{\mu \in \mathcal{M}_1(\mathcal{X})\colon \mu(d\sigma) = \mu_\Lambda^\lambda(d\sigma), \ \text{ for some } \lambda \in \mathcal{M}_1(\mathcal{X})\}.$$

The following proposition summarizes the basic result.

Proposition 3.21 *Let $\Lambda_n \nearrow \mathbb{Z}^d$ in the sense of van Hove.* Then:

$$\mathcal{G} := \cap_n \mathcal{G}_{\Lambda_n}$$

is non-empty, convex and compact. $\mathcal{G}$ coincides with the set of all DLR measures.

Sketch of the proof We first remark that $\mathcal{G}_{\Lambda_n} \subseteq \mathcal{G}_{\Lambda_m}$, for $n > m$. It is easy to see that $\mathcal{G}_\Lambda$ is a closed subset of $\mathcal{M}_1(\mathcal{X})$ which, in turn, is weakly compact as we have already remarked. From this we immediately get that $\mathcal{G}$ is non-empty and compact. Its convexity is immediate. It is also clear that every measure in $\mathcal{G}$ satisfies DLR equations: for each $\Delta \subset\subset \mathbb{Z}^d$ we can find n so large that $\Delta \cup \partial_{r_0}\Delta \subseteq \Lambda_n$, from which (3.101) is easily concluded. On the other hand, if μ is DLR, then for any n it is of the form $\mu = \mu_{\Lambda_n}^\lambda$, with $\lambda = \mu$. □

From the above proposition it follows in particular that, given a regular sequence $\Lambda_n \nearrow \mathbb{Z}^d$ as $n \to +\infty$ and a random sequence of boundary conditions λ_n, due to compactness we can always find limiting points of $\{\mu_{\Lambda_n}^{\lambda_n}\}$. In general we find different limits for different sequences of boundary conditions.

In this way we can formulate rigorously the notion of absence or presence of phase transitions for a given lattice spin system driven by interactions as in Definition 3.16.

Remark 3.22 The previous discussion can be extended properly if (iii) of Definition 3.16 is relaxed to (iii)′ below, with the right-hand side of (3.99) defining a continuous function of σ_{Δ^c}. The class of such interactions is denoted by $\mathbb{B}_\theta$.

(iii)′ $|||\Phi||| := \sum_{\Delta\colon O \in \Delta} \|\Phi_\Delta\|_\infty < +\infty$ (absolute summability).

Definition 3.23 Given a family of interaction potentials $\Phi \in \mathbb{B}_\theta$ and inverse temperature β, we say that there is an absence of phase transition (for such Φ and β) if $\sharp\mathcal{G} = 1$, i.e. if we have uniqueness of the DLR measure.

Non-uniqueness ($\sharp\mathcal{G} > 1$) means that different equilibrium states are possible with the same values of $\{\Phi_\Delta\}$ and β; it then corresponds to a phase transition.

Remark Given Φ and β, it is clear that non-uniqueness can possibly be detected by showing that for a regular sequence $\Lambda_n \nearrow \mathbb{Z}^d$ as $n \to +\infty$, we can find different sequences λ_n, λ'_n such that for a given $f \in \mathcal{F}_{loc}$ we have:

$$\lim_{n\to+\infty} \mu_{\Lambda_n}^{\lambda_n}(f) \neq \lim_{n\to+\infty} \mu_{\Lambda_n}^{\lambda'_n}(f). \tag{3.102}$$

We shall look at this in the next section in the context of the standard Ising system.

For fixed Φ and β as above, we let $\mathcal{G}_\theta$ denote the convex set of DLR translationally invariant measures, i.e. $\mathcal{G}_\theta = \mathcal{G} \cap \mathcal{M}_1^\theta(\mathcal{X})$. Due to property (ii) of Definition 3.16, we see that if $\mu \in \mathcal{G}$ and $i \in \mathbb{Z}^d$, then $\theta_i\mu \in \mathcal{G}$; together with the convexity and compactness of $\mathcal{G}$ this implies that $\mathcal{G}_\theta \neq \emptyset$ (taking accumulation points of $|\Lambda_n|^{-1}\sum_{i\in\Lambda_n}\theta_i\mu$). The set of its *extremal* points (μ is extremal in a convex set $\mathcal{E}$ if $\mu = a\mu_1 + (1-a)\mu_2$ with $\mu_1, \mu_2 \in \mathcal{E}$, $0 < a < 1$ implies $\mu_1 = \mu_2 = \mu$) coincides with the set of *ergodic* DLR measures, where a measure μ in $\mathcal{M}_1^\theta(\mathcal{X})$ is called ergodic if any translationally invariant subset $B \in \mathcal{B}$ verifies $\mu(B) \in \{0, 1\}$ (see [134], Chapter 14).

The ergodic DLR measures are then associated with the *pure phases* of the system.

Notation If required to indicate the interaction Φ and the value of β we write $\mathcal{G}(\beta, \Phi)$.

We now recall very briefly some notions and results concerning the variational principles of thermodynamics from the point of view of rigorous statistical mechanics.

Having fixed Φ and $\beta > 0$, start with the simple observation that if $\Lambda \subset\subset \mathbb{Z}^d$, the Gibbs measure $\mu_{\Lambda,\beta}$ is the unique point of maximum of the function

$$\Pi_\Lambda(\alpha) := \frac{1}{|\Lambda|}\{\mathrm{T}\mathcal{S}(\alpha) - \sum_{\sigma\in\mathcal{X}_\Lambda} H_\Lambda(\sigma)\alpha(\sigma)\}, \qquad \alpha \in \mathcal{M}_1(\mathcal{X}_\Lambda)$$

where $\mathcal{S}(\alpha) = -k\sum_{\sigma\in\mathcal{X}_\Lambda}\alpha(\sigma)\log\alpha(\sigma)$. (*Notation.* $\mathcal{S} = k\mathcal{H}$ in Section 1.3.)

As seen in Section 1.3, this is easily checked by Lagrange multipliers and the concavity of Π_Λ. Also $\Pi_\Lambda(\mu_{\Lambda,\beta}) = (\beta|\Lambda|)^{-1}\log Z_{\Lambda,\beta}$.

In the context of translation invariant measures, there is an analogous (nontrivial) characterization at infinite volume.

Definition 3.24 The specific (or mean) entropy is the functional s on $\mathcal{M}_1^\theta(\mathcal{X})$, defined by:

$$s(\alpha) := \lim_{\Lambda \nearrow \mathbb{Z}^d} -\frac{k}{|\Lambda|} \sum_{\sigma_\Lambda \in \mathcal{X}_\Lambda} \pi_\Lambda \alpha(\sigma_\Lambda) \log \pi_\Lambda \alpha(\sigma_\Lambda), \quad \alpha \in \mathcal{M}_1^\theta(\mathcal{X}). \tag{3.103}$$

(k is the Boltzmann constant and $\Lambda \nearrow \mathbb{Z}^d$ in the van Hove sense, cf. Definition 3.19.)

The existence of the above limit, indeed independent of the particular sequence $\Lambda_n \nearrow \mathbb{Z}^d$ (van Hove), and that it defines an upper semi-continuous functional on $\mathcal{M}_1^\theta(\mathcal{X})$ is a multidimensional extension of (1.68) in Section 1.3. It will be proven in the next section, but taking only $\Lambda = \Lambda_n = \{-n, \dots, n\}^d$, for simplicity.

Definition 3.25 Let $\Phi \in \mathbb{B}_\theta$. The specific (or mean) energy u is a functional on $\mathcal{M}_1^\theta(\mathcal{X})$, defined by:[3] ($\langle f, \alpha \rangle = \int f d\alpha$ cf. previous notation)

$$u(\alpha) := \langle A_\Phi, \alpha \rangle, \quad \text{where} \quad A_\Phi(\sigma) := \sum_{\Delta:\, O \in \Delta} \frac{1}{|\Delta|} \Phi_\Delta(\sigma).$$

Example. $A_{J,h}(\sigma) = -h\sigma_O - \frac{1}{2}\sum_{j \neq O} J(|j|)\sigma_O\sigma_j$ for the Ising system in (3.91). Translationally invariant DLR measures are characterized through the following variational principle, due to Lanford and Ruelle.

Theorem 3.26 *For $\Phi \in \mathbb{B}_\theta$ and $\beta > 0$, let Π be the functional on $\mathcal{M}_1^\theta(\mathcal{X})$ defined by:*

$$\Pi(\alpha) := \mathrm{T}s(\alpha) - u(\alpha), \qquad \alpha \in \mathcal{M}_1^\theta(\mathcal{X}). \tag{3.104}$$

Π attains a maximum in $\mathcal{M}_1^\theta(\mathcal{X})$, and the points of maximum are precisely the elements of $\mathcal{G}_\theta$. ($\mathrm{T} = (k\beta)^{-1}$ *represents the temperature.*)

The proof is postponed to the next section, due to the important connection with large deviations. (*Notation.* If required to indicate Φ and β we write $\Pi_{\beta,\Phi}$.)

To relate Π and Π_Λ above, we summarize a few observations to be used afterwards.

Remark 3.27 If $\Phi \in \mathbb{B}_\theta^0$, we have $A_\Phi \in \mathcal{F}_{loc}$. Using a multidimensional version of the Birkhoff ergodic theorem (cf. [134], Chapter 14), we see that if $\alpha \in \mathcal{M}_1^\theta(\mathcal{X})$ and $\Lambda_n \nearrow \mathbb{Z}^d$ (van Hove), then

$$U(\sigma) := \lim_n \frac{1}{|\Lambda_n|} \sum_{i \in \Lambda_n} A_\Phi(\theta_i \sigma) \qquad \text{exists } \alpha \text{ a.s. in } \sigma.$$

[3] For this definition we need less restrictive conditions on Φ.

(The null set might depend on the sequence (Λ_n).) Also $\langle U, \alpha\rangle = u(\alpha)$. On the other hand,

$$\sum_{i\in\Lambda} A_\Phi(\theta_i\sigma) = \sum_{i\in\Lambda}\sum_{\Delta:\, i\in\Delta} \frac{1}{|\Delta|}\Phi_\Delta(\sigma) = H_\Lambda(\sigma) + \sum_{i\in\Lambda}\sum_{\Delta:\, i\in\Delta\not\subseteq\Lambda} \frac{1}{|\Delta|}\Phi_\Delta(\sigma),$$

so that for any $\tau \in \mathcal{X}$, cf. (3.98), and with $\partial_r^- \Lambda := \{i \in \Lambda : \exists j \notin \Lambda \text{ such that } |i - j| \le r\}$:

$$\left| H_\Lambda^\tau(\sigma) - \sum_{i\in\Lambda} A_\Phi(\theta_i\sigma)\right| = \left| W_{\Lambda,\Lambda^c}(\sigma \circ_\Lambda \tau) - \sum_{i\in\Lambda}\sum_{\Delta:\, i\in\Delta\not\subseteq\Lambda} \frac{1}{|\Delta|}\Phi_\Delta(\sigma)\right| \le 2|\partial_{r_0}^-\Lambda| \sum_{\Delta:\, O\in\Delta} \|\Phi_\Delta\|_\infty.$$

But $|\partial_r^-\Lambda|/|\Lambda| \to 0$ if $\Lambda \nearrow \mathbb{Z}^d$ (van Hove), so that

$$\lim_{\Lambda\nearrow\mathbb{Z}^d} \sup_{\sigma,\tau} \frac{|H_\Lambda^\tau(\sigma) - \sum_{i\in\Lambda} A_\Phi(\theta_i\sigma)|}{|\Lambda|} = 0. \tag{3.105}$$

In particular, $u(\alpha) = \lim_{\Lambda\nearrow\mathbb{Z}^d} |\Lambda|^{-1}\langle H_\Lambda^\tau, \alpha\rangle$, uniformly in τ, for any $\alpha \in \mathcal{M}_1^\theta(\mathcal{X})$.

The considerations extend to $\Phi \in \mathbb{B}_\theta$: $A_\Phi \in C(\mathcal{X})$[4] and:

$$\frac{|W_{\Lambda,\Lambda^c}(\sigma \circ_\Lambda \tau)|}{|\Lambda|} \vee \frac{1}{|\Lambda|}\sum_{i\in\Lambda}\sum_{\Delta:\, i\in\Delta\not\subseteq\Lambda} \frac{1}{|\Delta|}|\Phi_\Delta(\sigma)| \le \sum_{\Delta:\, O\in\Delta} \frac{|\{i\in\Lambda : \Delta + i \not\subseteq \Lambda\}|}{|\Lambda|}\|\Phi_\Delta\|_\infty, \tag{3.106}$$

which tends to zero as $\Lambda \nearrow \mathbb{Z}^d$ (van Hove) under condition (iii)′ of Remark 3.22 ($a \vee b = \max\{a, b\}$).

In the lattice gas interpretation the maximum of Π represents the thermodynamic limit of the grand canonical pressure. In the magnetic (spin) interpretation $-\Pi$ represents a sort of grand canonical free energy. This variational principle is the grand canonical version of the extended second principle of thermodynamics, which says that among the states of given volume, temperature and chemical potential, the thermodynamic equilibrium state is the one maximizing the pressure. The more familiar 'canonical' formulation, valid in the case of a fixed number of particles, says that among the states of given volume and temperature the thermodynamic equilibrium state is the one minimizing the free energy. We shall come back to this in the next section, after a brief discussion on large deviations, since there is a close connection to be analysed.

[4] For $\Phi \in \mathbb{B}_\theta$, A_Φ has bounded oscillations, i.e. $\sum_{i\in\mathbb{Z}^d} \sup_{\sigma_j=\sigma_j',\forall j\neq i} |A_\Phi(\sigma) - A_\Phi(\sigma')| < +\infty$.

Remark 3.28 When $\Lambda \subset\subset \mathbb{Z}^d$, the maximum of Π_Λ on $\mathcal{M}_1(\mathcal{X}_\Lambda)$ is given by $F_\Lambda(\beta, \Phi) := (\beta|\Lambda|)^{-1} \log Z_{\Lambda,\beta} = \Pi_\Lambda(\mu_{\Lambda,\beta})$. The interpretation of the previous paragraph is based on the existence of the thermodynamic limit

$$\lim_{\Lambda \nearrow \mathbb{Z}^d} \frac{1}{\beta|\Lambda|} \log Z_{\Lambda,\beta} =: F(\beta, \Phi),$$

which may be proven as Proposition 3.13, and gives the maximum value of $\Pi_{\beta,\Phi}$ in Theorem 3.26. We prove this in the next section (Lemma 3.42), with $\Lambda_n = \{-n, \dots, n\}^d$ for simplicity. For $\Phi \in \mathbb{B}_\theta$ the limit is insensitive to boundary conditions. (See the remark following Lemma 3.42.)

Comment Theorem 3.26 can be proven for interactions $\Phi \in \mathbb{B}_\theta$, meaning that for this class of interactions not only do we have well defined thermodynamical quantities such as pressure and free energy, but also there is equivalence between the set of translationally invariant Gibbs measures and the solutions of the associated variational principle in $\mathcal{M}_1^\theta(\mathcal{X})$. However, to extend the main results valid for short range interactions to the many body, long range case, the stronger norm

$$\text{(iii)}'' \qquad |||\Phi|||_\alpha := \sum_{O \in \Delta} e^{\alpha|\Delta|} \|\Phi_\Delta\|_\infty < +\infty \tag{3.107}$$

for some $\alpha > 0$ is needed. In particular, as far as the description of the high temperature phase is concerned, Dobrushin and Martirosyan [91] have shown that in an arbitrary neighbourhood of zero for the norm in (iii)$'$ of Remark 3.22 there are potentials giving rise to phase transitions, which is not the case with the norm $||| \cdot |||_\alpha$ in (3.107), with $\alpha > 0$.

The standard Ising model: low temperature phase transition

We now look in more detail at the *ferromagnetic* Ising model, with $\mathcal{X}_O = \{-1, +1\}$ and nearest neighbour (n.n.) interaction, i.e. $J(n) = J\mathbf{1}_{\{n=1\}}$ in (3.91) with $J > 0$ a fixed number. The existence of a phase transition for suitable values of temperature and magnetic field is one of the basic questions to be considered.

Given $\Lambda \subset\subset \mathbb{Z}^d$, the energy associated with a configuration $\sigma \in \mathcal{X}_\Lambda$, with boundary condition $\tau \in \mathcal{X}_{\Lambda^c}$ is now given by:

$$H_\Lambda^\tau(\sigma) = -\frac{J}{2} \sum_{\substack{i,j \in \Lambda \\ |i-j|=1}} \sigma_i \sigma_j - h \sum_{i \in \Lambda} \sigma_i - J \sum_{\substack{i \in \Lambda, j \notin \Lambda \\ |i-j|=1}} \sigma_i \tau_j. \tag{3.108}$$

Since the interaction has range one, the boundary condition τ is involved only through its restriction to the external boundary $\partial_1 \Lambda$, which we just denote as $\partial \Lambda$:

$$\partial\Lambda := \{i \notin \Lambda \colon \exists j \in \Lambda \text{ with } |i - j| = 1\}.$$

Of particular interest are the boundary conditions $+$ and $-$, corresponding to $\tau_i = +1$ for all $i \in \Lambda^c$ and $\tau_i = -1$ for all $i \in \Lambda^c$, respectively. Given $\Delta \subset\subset \mathbb{Z}^d$

let

$$\chi_\Delta(\sigma) := \prod_{i\in\Delta} \sigma_i. \tag{3.109}$$

Notice that $\mathbf{1}_{\{\sigma_i=a\}} = (1+a\sigma_i)/2$ for $a \in \{-1,+1\}$ and thus any function $f \in \mathcal{F}_{loc}$ may be written as a linear combination of such χ_Δ with $\Delta \subseteq \Delta(f)$. Consequently, the finite volume Gibbs measure, μ_Λ^τ defined in (3.92), (3.93) is characterized fully by the complete set of finite volume spin expectations,

$$\langle \chi_\Delta \rangle_\Lambda^\tau := \mu_\Lambda^\tau(\chi_\Delta), \tag{3.110}$$

with Δ varying on the subsets of Λ.

Moreover, all possible limiting Gibbs measures of $\mu_{\Lambda_n}^{\tau_n}$ along a sequence of boundary conditions τ_n and $\Lambda_n \nearrow \mathbb{Z}^d$ in the van Hove sense, are characterized through the thermodynamic limits of $\mu_{\Lambda_n}^{\tau_n}(\chi_\Delta)$ corresponding to the same sequence of boundary conditions.

Notation We shall often use the notation $\langle\cdot\rangle_{\Lambda,\beta,h}^\tau$, or $\mu_{\Lambda,\beta,h}^\tau$ to show dependence on β, h. Having fixed J we omit it from the notation, for simplicity.

For a fixed pair h, β of values of the magnetic field and of the inverse temperature, if, given an increasing sequence of cubes Λ, there exist two sequences $\tau_{\Lambda^c}^1, \tau_{\Lambda^c}^2$ of boundary conditions and a finite set Δ of sites in $\mathbb{Z}^d$ such that

$$\liminf_{\Lambda\nearrow\mathbb{Z}^d} \langle \chi_\Delta \rangle_{\Lambda,\beta,h}^{\tau_{\Lambda^c}^1} > \limsup_{\Lambda\nearrow\mathbb{Z}^d} \langle \chi_\Delta \rangle_{\Lambda,\beta,h}^{\tau_{\Lambda^c}^2},$$

then we would conclude that $\sharp\mathcal{G} \geq 2$, so establishing the existence of a phase transition.

Notation When $\Delta = \{i\}$ we shall replace $\chi_{\{i\}}$ simply by σ_i, provided no confusion may arise, i.e. $\langle\sigma_i\rangle_{\Lambda,\beta,h}^\tau = \mu_{\Lambda,\beta,h}^\tau(\chi_{\{i\}}) = \mu_{\Lambda,\beta,h}(\sigma_i = +1) - \mu_{\Lambda,\beta,h}(\sigma_i = -1)$.

Observe that if we denote by

$$m_\Lambda^\tau(\beta, h) := \frac{1}{|\Lambda|}\sum_{i\in\Lambda} \langle\sigma_i\rangle_{\Lambda,\beta,h}^\tau \tag{3.111}$$

the finite volume average magnetization and by

$$F_\Lambda^\tau(\beta, h) := \frac{1}{\beta|\Lambda|} \log Z_\Lambda^\tau(\beta, h), \tag{3.112}$$

the finite volume grand canonical (negative) Gibbs free energy,[5] cf. (3.93), then:

$$m_\Lambda^\tau(\beta, h) = \frac{\partial}{\partial h} F_\Lambda^\tau(\beta, h). \tag{3.113}$$

[5] *Notation.* $F_\Lambda^\tau(\beta, h) = F_\Lambda^\tau(\beta, \Phi)$ cf. Remark 3.28, when Φ is the interaction associated with a n.n. Ising model. This notation emphasizes the dependence of β, h, for a fixed J.

The convexity of $F_\Lambda^\tau(\beta,\cdot)$ is a basic important fact to be used later; taking logarithms it is seen to be equivalent to $Z_\Lambda^\tau(\beta, ah_1+(1-a)h_2) \le (Z_\Lambda^\tau(\beta,h_1))^a\, Z_\Lambda^\tau(\beta,h_1)^{1-a}$ for each $h_1, h_2 \in \mathbb{R}$, $0<a<1$, which follows at once from the Hölder inequality. (We may also check the convexity of $\beta \mapsto \beta F_\Lambda^\tau(\beta,h)$; also $\mathrm{T} \mapsto F_\Lambda^\tau(1/k\mathrm{T}, h)$ is convex; see e.g. [250], Section D2.)

Theorem 3.29 *Let $d \ge 2$ and $h=0$. There exists $\beta_0 > 0$ such that if $\beta > \beta_0$, and Λ is a sequence of volumes invading the whole $\mathbb{Z}^d$ in the sense of van Hove, cf. Definition* 3.19 (*e.g. a sequence of cubes centred at the origin, with side increasing to infinity*), *then we have*:

$$\liminf_{\Lambda\nearrow\mathbb{Z}^d} \langle\sigma_O\rangle^+_{\Lambda,\beta,0} > \limsup_{\Lambda\nearrow\mathbb{Z}^d} \langle\sigma_O\rangle^-_{\Lambda,\beta,0}. \tag{3.114}$$

As previously remarked:

$$\lim_{\Lambda\nearrow\mathbb{Z}^d} F_\Lambda^\tau(\beta,h) =: F^\tau(\beta,h) = F(\beta,h) \tag{3.115}$$

exists and does not depend on the sequence of boundary conditions. (See also Lemma IV.6.2 in [108], or Theorem. 8.3.1 in [250]. We shall come back to the proof in the next section.)

A phase transition in the sense specified by the above theorem can be interpreted as instability of the finite volume Gibbs measure (in particular of the local magnetization $\langle\sigma_O\rangle_\Lambda^\tau$) with respect to the boundary conditions, *uniformly in the volume*. By varying just the interaction of the boundaries of the container with the system, we vary the situation in the bulk: there is a 'transmission of information' at arbitrarily large distance. Indeed, in our specific case, we speak of 'long range order'. However, this instability is not felt 'directly' by the thermodynamic function $F(\beta,h)$, as (3.115) shows. But we have seen that, at least at finite volume, the average magnetization is equal to the derivative of the free energy with respect to h, cf. (3.113). Thus, at infinite volume and β large, we expect a discontinuity of this derivative at $h=0$ (first order phase transition), to be proven below. Such non-analytic behaviour of thermodynamic functions is another way to 'detect' a phase transition, beyond non-uniqueness of the Gibbs measure. The relation between such notions in the frame of Pirogov–Sinai theory is the object of Friedli's recent thesis, where a result of Isakov [168] for the Ising model is suitably extended, cf. [123, 124] also Section 4.1.3. Finally, let us recall that the Hamiltonian for $h=0$ is spin-flip invariant. The low temperature phase transition for the Ising model can also be regarded as a manifestation of a 'spontaneous symmetry breaking'. It is known that this is not always the case: in the general context of Pirogov–Sinai theory (see [244, 280]) there are many examples of first order phase transitions which are not at all associated with any spontaneous symmetry breaking. In such cases the proof of the existence of a phase transition is much more difficult; determination of the coexistence line is one of the main problems, as it cannot be found simply via symmetry considerations.

Remark To show rigorously the existence of a phase transition in the sense that $\sharp\mathcal{G} \geq 2$, it suffices to have (3.114). However, as a consequence of the ferromagnetic interaction (see Theorem 3.34 below) it can be seen that the expectations $\langle\sigma_O\rangle^+_{\Lambda,\beta,h}$ and $\langle\sigma_O\rangle^-_{\Lambda,\beta,h}$ are monotonic in Λ, so that limits exist in (3.114).

Proposition 3.30 *Let us consider $d \geq 2$ and $h = 0$. There exists $\beta_0 > 0$ so that for each $\beta > \beta_0$ a positive constant $c(\beta)$ may be found in a way that*

$$\langle\sigma_O\rangle^+_{\Lambda,\beta,0} > c(\beta) \qquad \textit{whenever} \quad O \in \Lambda \subset\subset \mathbb{Z}^d.$$

Proof of Theorem 3.29 We just observe that $H^+_\Lambda(\sigma) = H^-_\Lambda(-\sigma)$ for each $\sigma \in \mathcal{X}_\Lambda$, and use the map $\sigma \mapsto -\sigma$ to get $Z^+_\Lambda(\beta, 0) = Z^-_\Lambda(\beta, 0)$ so that $\langle\sigma_O\rangle^+_{\Lambda,\beta,0} = -\langle\sigma_O\rangle^-_{\Lambda,\beta,0}$, from which the result follows at once from Proposition 3.30. □

Proof of Proposition 3.30 The argument goes back to Peierls [144, 236], being based on a geometrical description of configurations in terms of contours. For simplicity we only consider $d = 2$ and use a special property of ferromagnetic interactions, cf. Remark 3.33 below, to extend the result to $d \geq 2$. Again for simplicity we suppose that Λ is a square. The boundary conditions are taken as $\tau_i = +1$ for all $i \in \partial\Lambda$. Given a generic configuration $\sigma \in \mathcal{X}_\Lambda$ consider all sites in $\Lambda \cup \partial\Lambda$ and draw a unit segment orthogonal to any 'broken edge', i.e. to any unit segment joining a pair of n.n. sites with opposite spins in σ. In this way we obtain, for any $\sigma \in \mathcal{X}_\Lambda$, a polygon lying on the dual lattice $(\mathbb{Z}^2)^* := \mathbb{Z}^2 + (1/2, 1/2)$. At each site of $(\mathbb{Z}^2)^*$ an even number of unit segments of this polygon meet: 0, 2 or 4. When this number is 4, we use some convention to 'round off' the corners (e.g. 'cutting' slightly the corner in the north-east, south-west direction, see Figure 3.2). In this way we obtain a decomposition of our polygon into a set $\Gamma(\sigma) = \{\gamma_1, \ldots, \gamma_k\}$ of closed self-avoiding and pairwise mutually avoiding contours lying on $(\mathbb{Z}^2)^*$. ($\Gamma(\sigma) = \emptyset$ for $\sigma = +\underline{1}$, i.e. $\sigma_i = +1$ for all $i \in \Lambda$.)

Given Λ with + boundary conditions there is a one-to-one correspondence between the set $\mathcal{X}_\Lambda$ of spin configurations and the set G_Λ of collections of closed self-avoiding and pairwise mutually avoiding contours; indeed the sign of the spin at x is determined uniquely by the number of enclosed contours that we have to cross, starting from x, to reach $\partial\Lambda$; moreover, if given $\sigma \neq +\underline{1}$, we have $\Gamma(\sigma) = \{\gamma_1, \ldots, \gamma_k\}$, then

$$H^+_\Lambda(\sigma) = -J\mathcal{N}_\Lambda + 2J\sum_{i=1}^k |\gamma_i|, \tag{3.116}$$

where we have denoted by $|\gamma|$ the perimeter of the closed contour γ corresponding to the number of 'broken edges' in that configuration σ such that $\Gamma(\sigma) \equiv \{\gamma\}$;

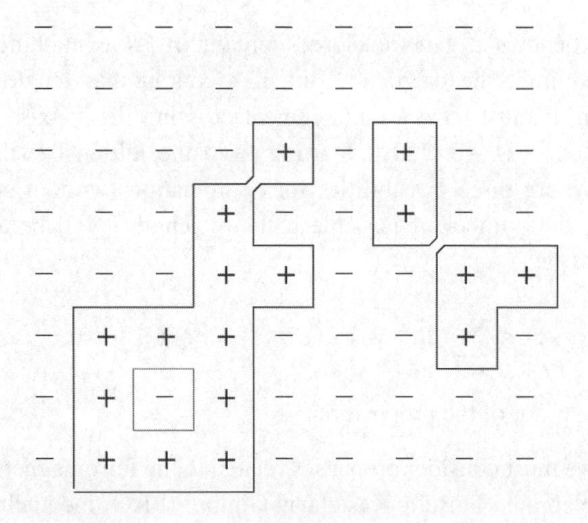

Figure 3.2

$\mathcal{N}_\Lambda$ is a constant given by the number of nearest neighbour edges in Λ and between Λ and $\partial\Lambda$. We may look at the contours as 'local excitations' of the ground state $+\underline{1}$.

The system may be described via an 'ensemble of contours' with space of configurations G_Λ, where the image of the original Gibbs measure on $\mathcal{X}_\Lambda$ is given by ρ, with

$$\rho(\Gamma) = \frac{e^{-2\beta J \sum_{\gamma\in\Gamma}|\gamma|}}{\sum_{\Gamma'\in G_\Lambda} e^{-2\beta J \sum_{\gamma'\in\Gamma'}|\gamma'|}}.$$

We write:

$$\rho_1(\gamma) := \rho(\{\Gamma \in G_\Lambda : \Gamma \ni \gamma\} = \frac{\sum_{\Gamma'\in G_\Lambda:\Gamma'\ni\gamma} e^{-2\beta J \sum_{\gamma'\in\Gamma'}|\gamma'|}}{\sum_{\Gamma''\in G_\Lambda} e^{-2\beta J \sum_{\gamma''\in\Gamma''}|\gamma''|}}$$

and use the notation: 'Γ' comp γ' to express that γ is not in the collection $\Gamma' = \{\gamma_1', \ldots, \gamma_k'\}$, but $\gamma_1', \ldots, \gamma_k'$ are compatible with its presence, in the sense that no γ_i' intersects γ. In this way, we have:

$$\rho_1(\gamma) = e^{-2\beta J|\gamma|} \frac{\sum_{\Gamma'\in G_\Lambda:\Gamma' \text{ comp } \gamma} e^{-2\beta J \sum_{\gamma'\in\Gamma'}|\gamma'|}}{\sum_{\Gamma''\in G_\Lambda} e^{-2\beta J \sum_{\gamma''\in\Gamma''}|\gamma''|}} \le e^{-2\beta J|\gamma|}.$$

Since $\langle\sigma_O\rangle^+_{\Lambda,\beta,0} = 1 - 2\mu^+_{\Lambda,\beta,0}(\sigma_O = -1)$, our goal is to obtain a proper upper bound for $\mu^+_{\Lambda,\beta,0}(\sigma_O = -1)$. We get

$$\mu^+_{\Lambda,\beta,0}(\sigma_O = -1) \le \sum_{\gamma\circ O} \rho_1(\gamma),$$

where $\gamma \circ O$ denotes a generic closed contour in $\mathbb{Z}^d$ containing, in its interior (i.e. encircling), the origin O. But, if a contour has length l and encircles the origin, it must have a unit segment crossing the x-axis somewhere in $\{(-l-1/2, 0), \dots, (l+1/2, 0)\}$. Starting from the leftmost such segment, at each step there are three possibilities for continuation (without coming back). Consequently, the number of possible paths of length l will be at most $(2l+2)3^{l-1}$, and we get:

$$\mu^+_{\Lambda,\beta,0}(\sigma_O = -1) \le \sum_{l=4}^{\infty}(2l+2)3^{l-1}e^{-2\beta J l} =: \epsilon(\beta) \to 0, \quad \text{as } \beta \to \infty,$$

concluding the proof of the proposition. □

At this point we must consider properties related to the ferromagnetic interaction, such as the celebrated Fortuin–Kasteleyn–Ginibre (FKG) inequalities. To introduce them we need the following.

Definition 3.31 For $\Lambda \subseteq \mathbb{Z}^d$ non-empty, we consider the following partial order on $\mathcal{X}_\Lambda$: given $\sigma, \sigma' \in \mathcal{X}_\Lambda$ we write

$$\sigma \prec \sigma' \quad \text{if} \quad \sigma_i \le \sigma'_i \quad \text{for all } i \in \Lambda. \tag{3.117}$$

Identifying a configuration in $\mathcal{X}_\Lambda$ with the subset of $+1$ spins, this relation is nothing but the inclusion. A function $f : \mathcal{X}_\Lambda \to \mathbb{R}$ is said to be increasing if $f(\sigma) \le f(\sigma')$ whenever $\sigma \prec \sigma'$.

Theorem 3.32 (Fortuin, Kasteleyn, Ginibre) *Let $J : \Lambda \times \Lambda \to \mathbb{R}$ be an interaction and $\underline{h} : \Lambda \to \mathbb{R}$ a non-homogeneous magnetic field. For $\Lambda \subset\subset \mathbb{Z}^d$, consider the Hamiltonian*

$$H_{\{J\},\underline{h},\Lambda}(\sigma) = -\sum_{i,j\in\Lambda} J(i,j)\sigma_i\sigma_j - \sum_{i\in\Lambda} h_i\sigma_i$$

and the associated finite volume Gibbs measure $\mu_{\{J\},\underline{h},\Lambda}$, given by:

$$\mu_{\{J\},\underline{h},\Lambda}(\sigma) = \frac{1}{Z_{\{J\},\underline{h},\Lambda}} e^{-H_{\{J\},\underline{h},\Lambda}(\sigma)} \quad \forall \sigma \in \mathcal{X}_\Lambda,$$

where $Z_{\{J\},\underline{h},\Lambda} = \sum_{\sigma\in\mathcal{X}_\Lambda} e^{-H_{\{J\},\underline{h},\Lambda}(\sigma)}$ is the normalizing constant. If $J(\,\cdot\,) \ge 0$ in $\Lambda \times \Lambda$, then for any $\underline{h}$ and for any pair of increasing functions $f, g : \mathcal{X}_\Lambda \to \mathbb{R}$, we have:

$$\mu_{\{J\},\underline{h},\Lambda}(fg) \ge \mu_{\{J\},\underline{h},\Lambda}(f)\mu_{\{J\},\underline{h},\Lambda}(g). \tag{3.118}$$

For the proof see e.g. [120], [108], Chapter IV, or [199], Chapter III. (For product measures, i.e. $J \equiv 0$, this was first proven by Harris, see [148].)

In the notation of the previous theorem the inverse temperature β has been absorbed in the functions J and $\underline{h}$; this last possibly includes the 'molecular field' generated by an external configuration. Inequality (3.118) is called the FKG inequality; any measure on $\mathcal{X}_\Lambda$ ($\mathcal{X}$) which satisfies it for any pair f, g of increasing (cylinder and increasing, respectively.) is said to be a FKG measure.

Remark 3.33 Another important inequality in the context of ferromagnetic interactions was obtained by Griffiths [145] for pair interactions and extended by Kelly and Sherman [174] to more general ferromagnetic spin systems. In the context of Theorem 3.32 it implies that if $h_i \geq 0$ and $J_{i,j} \geq 0$ for all i, j, then the expected value of a spin is non-decreasing in each $J_{i,j}$. In particular, one gets that the expected magnetization in the statement of Proposition 3.30 for $d \geq 2$ is not smaller than that in the case $d = 2$, for the same value of β. For the precise statement and a proof see e.g. Theorem 1.21, Chapter IV of [199]. On the other hand, for a more general treatment of the Peierls argument for $d \geq 2$ see e.g. [258].

Remark In the next two theorems, by $\Lambda \nearrow \mathbb{Z}^d$ we mean a sequence of growing cubes centred at the origin ($\Lambda_n = \{-n, \ldots, n\}^d$). However, it will appear clear that, at least for ferromagnetic interactions, most of the results indeed hold for any van Hove sequence of volumes containing the origin.

Theorem 3.34

(i) *The following limits exist*:

$$\lim_{\Lambda \nearrow \mathbb{Z}^d} \langle \sigma_O \rangle^+_{\Lambda,\beta,h} =: \langle \sigma_O \rangle^+_{\beta,h};$$

$$\lim_{\Lambda \nearrow \mathbb{Z}^d} \langle \sigma_O \rangle^-_{\Lambda,\beta,h} =: \langle \sigma_O \rangle^-_{\beta,h}.$$

(ii) *For each* $h_0 \in \mathbb{R}$:

$$\lim_{h \to h_0^+} \langle \sigma_O \rangle^+_{\beta,h} = \langle \sigma_O \rangle^+_{\beta,h_0};$$

$$\lim_{h \to h_0^-} \langle \sigma_O \rangle^-_{\beta,h} = \langle \sigma_O \rangle^-_{\beta,h_0}.$$

(iii) *The infinite volume (negative) free energy* $F(\beta, h)$, *cf.* (3.115), *is differentiable in* h, *for* $h \neq 0$. *We have*

$$\langle \sigma_O \rangle^+_{\beta,h} = \langle \sigma_O \rangle^-_{\beta,h} = \frac{\partial}{\partial h} F(\beta, h), \quad \textit{for all } h \neq 0.$$

(iv) *The left and right derivatives of* $F(\beta, \cdot)$ *always exist. Also*

$$\frac{\partial}{\partial h^+} F(\beta, h)|_{h=0} = \langle \sigma_O \rangle^+_{\beta,0}, \qquad \frac{\partial}{\partial h^-} F(\beta, h)|_{h=0} = \langle \sigma_O \rangle^-_{\beta,0}.$$

Proof In the context of Theorem 3.32, $\langle \sigma_O \rangle_{\{J\},\underline{h},\Lambda}$ is non-decreasing as a function of any individual h_i; to see this, it is sufficient to differentiate with respect to h_i, and to apply Theorem 3.32. Moreover,

$$\langle\sigma_O\rangle^+_{\Lambda',\beta,h'} \le \langle\sigma_O\rangle^+_{\Lambda,\beta,h} \tag{3.119}$$

provided $\Lambda \subseteq \Lambda'$ and $h' \le h$. Indeed, in the frame of the previous theorem, the measure $\mu^+_{\Lambda,\beta,h}$ can be obtained from $\mu^+_{\Lambda',\beta,h}$ by setting as $+\infty$ the magnetic fields h_i with $i \in \partial\Lambda$, and taking the marginal on $\mathcal{X}_\Lambda$. Part (i) follows at once from (3.119). Also

$$\langle\sigma_O\rangle^+_{\beta,h} \le \langle\sigma_O\rangle^+_{\Lambda,\beta,h},$$

so that

$$\limsup_{h\to h_0^+}\langle\sigma_O\rangle^+_{\beta,h} \le \limsup_{h\to h_0^+}\langle\sigma_O\rangle^+_{\Lambda,\beta,h} = \langle\sigma_O\rangle^+_{\Lambda,\beta,h_0}.$$

Passing to the limit $\Lambda \nearrow \mathbb{Z}^d$ we get

$$\limsup_{h\to h_0^+}\langle\sigma_O\rangle^+_{\beta,h} \le \langle\sigma_O\rangle^+_{\beta,h_0}.$$

On the other hand, making $\Lambda' = \Lambda \nearrow \mathbb{Z}^d$ in (3.119) we get:

$$\langle\sigma_O\rangle^+_{\beta,h_0} \le \liminf_{h\to h_0^+}\langle\sigma_O\rangle^+_{\beta,h}.$$

The argument for $\langle\sigma_O\rangle^-_{\Lambda,\beta,h}$ is fully analogous, mutatis mutandis, and we get part (ii).

For part (iii), we take for granted the existence of the thermodynamic limit $F(\beta,h)$, cf. (3.115) (independent on the boundary conditions). Its convexity in h follows at once from that of $F_\Lambda(\beta,h)$. At this point we omit the proof that the function $F(\beta,h)$ is continuously differentiable in h, for $h \neq 0$ (see, for instance [108], Theorem IV.5.1). Actually, Lee and Yang have shown it to be a real analytic funtion of h, for $h \neq 0$ (see [196]). We then apply a standard fact based on convexity: let $\{f_n; n = 1, 2, \ldots\}$ be a sequence of convex functions on an open interval $G \subseteq \mathbb{R}$, such that $f(\zeta) := \lim_{n\to+\infty} f_n(\zeta)$ exists for every $\zeta \in G$. If each f_n and f are differentiable at some point $\zeta_0 \in G$, then $f_n'(\zeta_0)$ converges to $f'(\zeta_0)$ as $n \to +\infty$ (see e.g. [108], Lemma IV6.3).

Applying this for $h \neq 0$, given any sequence τ of boundary conditions, the van Hove limit

$$\lim_{\Lambda\nearrow\mathbb{Z}^d} m^\tau_\Lambda(\beta,h) =: m(\beta,h) \tag{3.120}$$

exists, is independent of τ, and equals $\partial F(\beta,h)/\partial h$. Moreover, by the convexity in h we know that the right and left derivatives

$$\frac{\partial}{\partial h^+}F(\beta,h), \qquad \frac{\partial}{\partial h^-}F(\beta,h) \tag{3.121}$$

always exist and are right- and left-continuous, respectively. Therefore, the previous relations imply:

$$\frac{\partial}{\partial h^+} F(\beta, h)\,|_{h=0} = \lim_{h\to 0^+} m(\beta, h) =: m^*(\beta)$$

$$\frac{\partial}{\partial h^-} F(\beta, h)\,|_{h=0} = \lim_{h\to 0^-} m(\beta, h) =: -m^*(\beta)$$

with the quantity $m^*(\beta)$ being called 'spontaneous magnetization'.

Taking into account the already proven part (ii) and the above expressions for the partial derivatives of F, parts (iii) and (iv) will follow once we check that for $h \neq 0$

$$\langle \sigma_O \rangle^+_{\beta,h} = m(\beta, h) = \langle \sigma_O \rangle^-_{\beta,h}. \tag{3.122}$$

For this, we start by observing that for every $i \in \mathbb{Z}^d$:

$$\langle \sigma_O \rangle^+_{\beta,h} = \langle \sigma_i \rangle^+_{\beta,h}. \tag{3.123}$$

Indeed, having fixed $i \in \mathbb{Z}^d$, by the translation invariance of the interaction we have,

$$\langle \sigma_O \rangle^+_{\Lambda,\beta,h} = \langle \sigma_i \rangle^+_{\Lambda+i,\beta,h}, \quad \text{for all } \Lambda \subset\subset \mathbb{Z}^d. \tag{3.124}$$

Considering an increasing sequence of cubes $\{\bar{\Lambda}_n\}$ centred at the origin such that

$$\bar{\Lambda}_1 \subseteq \bar{\Lambda}_2 + i \subseteq \bar{\Lambda}_3 \subseteq \bar{\Lambda}_4 + i \subseteq \dots$$

by monotonicity, the limits of $\langle \sigma_O \rangle^+_{\Lambda,\beta,h}$ along the even and odd subsequences of $\Lambda = \bar{\Lambda}_n$ exist, coincide and are equal to the limit along any sequence of growing cubes. From this and (3.124) we immediately get (3.123). To check (3.122), we now take $\epsilon > 0$ and let Δ_ϵ be a cube centred at the origin, such that

$$\langle \sigma_O \rangle^+_{\Delta_\epsilon,\beta,h} \le \langle \sigma_O \rangle^+_{\beta,h} + \epsilon. \tag{3.125}$$

Given a large cube Λ which contains Δ_ϵ let $A_\epsilon(\Lambda) = \{i \in \Lambda \colon \Delta_\epsilon + i \subseteq \Lambda\}$. If $i \in A_\epsilon(\Lambda)$, recalling (3.123) and (3.119), we may write:

$$\langle \sigma_O \rangle^+_{\beta,h} = \langle \sigma_i \rangle^+_{\beta,h} \le \langle \sigma_i \rangle^+_{\Lambda,\beta,h} \le \langle \sigma_i \rangle^+_{\Delta_\epsilon+i,\beta,h} = \langle \sigma_O \rangle^+_{\Delta_\epsilon,\beta,h} \le \langle \sigma_O \rangle^+_{\beta,h} + \epsilon,$$

and we obtain

$$\langle \sigma_O \rangle^+_{\beta,h} \le \frac{1}{|A_\epsilon(\Lambda)|} \sum_{i\in A_\epsilon(\Lambda)} \langle \sigma_i \rangle^+_{\Lambda,\beta,h} \le \langle \sigma_O \rangle^+_{\beta,h} + \epsilon. \tag{3.126}$$

A similar argument holds for minus boundary conditions. But $|\Lambda|/|A_\epsilon(\Lambda)| \to 1$ as $\Lambda \nearrow \mathbb{Z}^d$ for each $\epsilon > 0$, implying (3.122) and concluding the proof. □

The previous theorem tells us that in the present case a discontinuity of the first derivative of $F(\beta, \cdot)$ occurs if and only if the spontaneous magnetization is positive, which implies non-uniqueness of the Gibbs measure. In the same context, the non-uniqueness of the Gibbs measure also implies spontaneous magnetization, as stated below.

Theorem 3.35 *For each $\beta > 0$, $h \in \mathbb{R}$, the weak limits in $\mathcal{M}_1(\mathcal{X})$*

$$\mu^+_{\beta,h} = \lim_{\Lambda \nearrow \mathbb{Z}^d} \mu^+_{\Lambda,\beta,h}, \qquad \mu^-_{\beta,h} = \lim_{\Lambda \nearrow \mathbb{Z}^d} \mu^-_{\Lambda,\beta,h}$$

exist, are translationally invariant and ergodic. Moreover,

$$\sharp\mathcal{G} = 1 \Leftrightarrow \mu^+_{\beta,h} = \mu^-_{\beta,h} \Leftrightarrow \frac{\partial}{\partial h} F(\beta, h) \text{ exists.}$$

Proof (Sketch) We sketch the proof through the following observations. It exemplifies the usefulness of the FKG property. (For more details see e.g. Theorem IV.6.5 in [108].)

(a) The argument leading to (3.119), based on Theorem 3.32, gives in fact:

$$\mu^+_{\Lambda',\beta,h'}(f) \leq \mu^+_{\Lambda,\beta,h}(f), \qquad \text{if,} \quad h' \leq h, \ \Lambda \subseteq \Lambda' \subset\subset \mathbb{Z}^d \tag{3.127}$$

for each increasing cylinder function f with $\Delta(f) \subseteq \Lambda$. In particular, having fixed h we have the existence of the limit of

$$\lim_{\Lambda \nearrow \mathbb{Z}^d} \mu^+_{\Lambda,\beta,h}(f)$$

for any increasing sequence of finite volumes $\Lambda \nearrow \mathbb{Z}^d$; the limit must be independent of the given sequence.

(b) Any function $f \in \mathcal{F}_{loc}$ may be written as a difference of two increasing cylinder functions. To see this just observe that if $\Delta \subset\subset \mathbb{Z}^d$ the functions

$$f_\Delta := \mathbf{1}_{[\sigma_i = 1, \, \forall i \in \Delta]}, \quad g_\Delta := \mathbf{1}_{[\exists i \in \Delta, \, \sigma_i = 1]}, \tag{3.128}$$

are increasing, as well as the product $f_\Delta \, g_{\tilde{\Delta}}$, for $\Delta, \tilde{\Delta}$ disjoint finite sets, and

$$\mathbf{1}_{[\sigma_i = 1, \, \forall i \in \Delta; \, \sigma_j = -1, \, \forall j \in \tilde{\Delta}]} = f_\Delta - f_\Delta \, g_{\tilde{\Delta}}.$$

Thus, if $\mu, \nu \in \mathcal{M}_1(\mathcal{X})$ and $\mu(f) = \nu(f)$ for each increasing cylinder function f, then $\mu = \nu$.

(c) From the last observation and the compactness of $\mathcal{M}_1(\mathcal{X})$, we see that a sequence ν_n in $\mathcal{M}_1(\mathcal{X})$ is weakly convergent if and only if $\nu_n(f)$ converges, for each increasing cylinder function f. Together with (a) and (b) this gives the existence of the limit $\mu^+_{\beta,h} = \lim_{\Lambda \nearrow \mathbb{Z}^d} \mu^+_{\Lambda,\beta,h}$. The same applies to the minus boundary condition.

(d) For the translational invariance of $\mu^+_{\beta,h}$ and $\mu^-_{\beta,h}$, the argument leading to (3.123) tells us that

$$\mu^+_{\beta,h}(f) = \mu^+_{\beta,h}(\theta_i f)$$

for any increasing $f \in \mathcal{F}_{loc}$ and we then apply observation (b) above (the same for $\mu^-_{\beta,h}$). As for the ergodicity, (3.129) below implies that $\mu^+_{\beta,h}$ and $\mu^-_{\beta,h}$ are extremal in $\mathcal{G}$ and a fortiori in $\mathcal{G}_\theta$, from which their ergodicity follows, as previously remarked.

(e) We want to show that $\sharp\mathcal{G} = 1$ when the spontaneous magnetization vanishes. Applying again Theorem 3.32, as in (a), we see that for any $\Lambda \subset\subset \mathbb{Z}^d$, $h \in \mathbb{R}$, $\tau \in \mathcal{X}_{\Lambda^c}$, and any increasing cylinder function f with $\Delta(f) \subseteq \Lambda$,

$$\mu^-_{\Lambda,\beta,h}(f) \le \mu^\tau_{\Lambda,\beta,h}(f) \le \mu^+_{\Lambda,\beta,h}(f). \tag{3.129}$$

Thus, if we show that $\mu^+_{\beta,h}(f_\Delta) = \mu^-_{\beta,h}(f_\Delta)$ for each finite set Δ and f_Δ as in (3.128), and recall the proof of (b), we can conclude that $\sharp\mathcal{G} = 1$. But, we check that

$$0 \le \langle f_\Delta\rangle^+_{\beta,h} - \langle f_\Delta\rangle^-_{\beta,h} \le |\Delta|\big(\langle\sigma_O\rangle^+_{\beta,h} - \langle\sigma_O\rangle^-_{\beta,h}\big). \tag{3.130}$$

The equivalences in the statement of the theorem follow at once from the previous observations, (3.130) and Theorem 3.34.

To check (3.130) we notice that the function $g(\sigma) := \sum_{i\in\Delta}\sigma_i - f_\Delta(\sigma)$ is increasing. Applying (3.129) to $f = f_\Delta$ and to $f = g$ we get:

$$0 \le \langle f_\Delta\rangle^+_{\Lambda,\beta,h} - \langle f_\Delta\rangle^-_{\Lambda,\beta,h} \le \sum_{i\in\Delta}\big(\langle\sigma_i\rangle^+_{\Lambda,\beta,h} - \langle\sigma_i\rangle^-_{\Lambda,\beta,h}\big);$$

then we take the limit as $\Lambda \nearrow \mathbb{Z}^d$ and use $\langle\sigma_i\rangle^+_{\beta,h} = \langle\sigma_O\rangle^+_{\beta,h}$. □

Remark 3.36 (Stochastic order) The importance of the FKG property in the above classical arguments is clear through the inequalities (3.119), (3.127) and (3.129).

The order $\prec$ in $\mathcal{X}$ induces an order relation in $\mathcal{M}_1(\mathcal{X})$, usually called stochastic order: $\mu \prec \nu$ iff $\mu(f) \le \nu(f)$ for each increasing continuous $f : \mathcal{X} \to \mathbb{R}$. We see that $\mu \prec \nu$ iff $\mu(f) \le \nu(f)$ for each increasing $f \in \mathcal{F}_{loc}$ and that indeed $\prec$ defines a partial order on $\mathcal{M}_1(\mathcal{X})$. (For this recall (b) in the proof of Theorem 3.35.) We rephrase (3.127) and (3.129) as:

- $\mu^+_{\Lambda,\beta,h}$ decreases and $\mu^-_{\Lambda,\beta,h}$ increases, as Λ increases;
- both $\mu^+_{\Lambda,\beta,h}$ and $\mu^-_{\Lambda,\beta,h}$ increase in h;
- $\mu^-_{\Lambda,\beta,h} \prec \mu^\tau_{\Lambda,\beta,h} \prec \mu^+_{\Lambda,\beta,h}$, in particular, $\mu^-_{\beta,h} \prec \mu \prec \mu^+_{\beta,h}$ for all $\mu \in \mathcal{G}(\beta,h)$ and $\sharp\mathcal{G}(\beta,h) = 1$ iff $\mu^-_{\beta,h} = \mu^+_{\beta,h}$.

Using the compactness of $\mathcal{M}_1(\mathcal{X})$, the argument in the previous proof shows that any monotone sequence in $\mathcal{M}_1(\mathcal{X})$ is weakly convergent.

The stochastic order will appear later in the context of stochastic dynamics associated with the Ising model (Chapter 7) and in the analysis of the Harris contact process (Chapter 4), where more details will be provided and the relation with coupling properties will be stated.

Comments For simplicity we have stated Theorem 3.35 in the n.n. case; it holds for a larger class of Ising ferromagnets. See for instance [258], p.113, [260], [192].

For the n.n. Ising ferromagnet (and a large class of ferromagnetic spin systems), besides the existence of $\partial F(\beta, h)/\partial h$, at $h \neq 0$, for any $\beta > 0$, it is possible to show that for small enough β, the function $F(\beta, \cdot)$ is differentiable also at $h = 0$. According to Theorem 3.35 this is equivalent to $\sharp\mathcal{G} = 1$ for such values of β. A probabilistic approach to prove uniqueness is based on conditions introduced by Dobrushin and further elaborated by Dobrushin and Shlosman (see [90], Chapter 8 of [134], or Theorem 3.1, Chapter IV of [199]). (See also [214].)

The Griffiths inequality, mentioned earlier, implies that $m^*(\beta)(= \mu^+_{\beta,0}(\sigma_O))$ increases in β, for an Ising ferromagnet. As a consequence, for any $d \geq 2$, there exists $\beta_c \in (0, +\infty)$ so that $m^*(\beta) = 0$ if $\beta < \beta_c$, and $m^*(\beta) > 0$ for $\beta > \beta_c$. Also, for fixed β, $\mu^+_{\beta,0}(\sigma_O)$ increases in the dimension d, cf. Remark 3.33.

If $d = 2$ the value of β_c is known exactly, through the famous Onsager relation $\sinh 2\beta_c J = 1$, as well as $m^*(\beta) = (1 - (\sinh 2\beta_c J)^{-4})^{1/8}$ if $\beta \geq \beta_c$, showing that at β_c there is a unique Gibbs measure in this case. For a discussion and references on various methods (combinatorial, algebraic, etc.) used in the derivation of this relation and further results on the two-dimensional Ising ferromagnet see Section 7.2 in [129] or Section 6.2 in [134] and the corresponding bibliographical notes therein. (When $d = 1$ the model becomes a simple Markov chain. From the Perron–Frobenius theorem we see that for any $\beta \geq 0$, $F(\beta, \cdot)$ is analytic in $\mathbb{R}$; in this case $\sharp\mathcal{G} = 1$ and $\beta_c = +\infty$, as can be seen also by other standard arguments.)

3.4 Large deviations for Gibbs measures

Considering the standard (ferromagnetic) Ising model on $\mathcal{X} = \{-1, +1\}^{\mathbb{Z}^d}$, it is natural to ask about the validity of a l.d.p. for the laws of the empirical magnetization

$$W_n = \frac{1}{|\Lambda_n|} \sum_{i \in \Lambda_n} \sigma_i,$$

under $\mu^\tau_{\Lambda_n,\beta,h}$, and its relation to spontaneous magnetization, if $d \geq 2$ ($\Lambda_n = \{-n, \ldots, n\}^d$).

We first examine the answer given by Theorem 3.4: recalling Assumptions 3.1, and assuming the validity of (3.115), we have that

$$\varphi_{\beta,h}(\zeta) := \lim_{n\to+\infty} \frac{1}{|\Lambda_n|} \log \mu^{\tau}_{\Lambda_n,\beta,h}(e^{\zeta \sum_{i\in\Lambda_n}\sigma_i})$$

exists, does not depend on τ, for any β, h, and:

$$\varphi_{\beta,h}(\zeta) = \beta\,(F(\beta, h + \zeta/\beta) - F(\beta, h))$$

where $F(\beta, h)$ is the thermodynamic limit in (3.115). Consequently, $\varphi_{\beta,h}$ is differentiable at $\zeta \neq -\beta h$ and its differentiability at $\zeta = -\beta h$ coincides with the existence of $\partial F(\beta, \tilde{h})/\partial \tilde{h}\mid_{\tilde{h}=0}$. In particular, if $\beta < \beta_c$, all hypotheses of Theorem 3.4 are verified for any value of h; it provides a l.d.p. for W_n, with scaling $|\Lambda_n|$ and rate function given by the Fenchel–Legendre transform $I_{\beta,h} = \varphi^*_{\beta,h}$.

Moreover, if $h \neq 0$ or $\beta < \beta_c$ we have

$$\varphi^*_{\beta,h}(m) = 0 \Longleftrightarrow m = m(\beta, h),$$

implying that $\mu^{\tau}_{\Lambda_n,\beta,h}(|W_n - m(\beta, h)| \geq \delta)$ decays exponentially in $|\Lambda_n|$, for any $\delta > 0$ (uniformly in τ), according to comment (vi) at the end of Section 3.1. The same holds under the unique infinite volume Gibbs measure.

The l.d.p. for the sequence (W_n) (with scaling $|\Lambda_n|$) holds for any β, h (rate function $I_{\beta,h} = \varphi^*_{\beta,h}$), even when part (c) of Theorem 3.4 is not applicable due to lack of differentiability of the free energy. This follows from Theorem 3.39 below[6] and the contraction principle, and it was first proven by Lanford [189] for a much more general class of interactions. (We shall come back to this.) The l.d.p. is also valid for the distribution of W_n under any $\mu \in \mathcal{G}_\theta(\beta, h)$.

To obtain the validity of a l.d.p. (valid in more general situations) we apply the Ruelle–Lanford method, starting with a probabilistic construction of the rate function, based on the entropy, instead of starting from $\varphi_{\beta,h}$.

From the properties of the free energy and the geometric description of the Fenchel–Legendre transform, for $\beta > \beta_c$ we have:

$$I_{\beta,0}(m) = 0 \Longleftrightarrow -m^*(\beta) \leq m \leq m^*(\beta).$$

That is, for $h = 0$, $\beta > \beta_c$, the volume scaling $|\Lambda_n|$ is too rough for the description of large fluctuations of the empirical magnetization within $[-m^*(\beta), m^*(\beta)]$; their probability is *not* exponentially small in the *volume*. The mechanism which generates this has to be understood in terms of coexistence of the $-$ and $+$ phases; the *cost* would come only from their separation, which suggests a *surface* scaling. Bounds showing the surface to be the right scaling were first derived by Schonmann [266]. This has led to a more refined study of large deviations (*surface* scaling), an object of intensive research. The two-dimensional case was studied

[6] see also remark (i) following Lemma 3.43.

initially by Dobrushin, Kotecky and Shlosman [92], and by Pfister [242]; these two articles settle not only precise bounds but also the relation to phase separation, when β is large enough. Ioffe [165, 166] and Ioffe and Schonmann [167] extended the analysis to all $\beta > \beta_c$. (In [270] the authors consider the situation when the magnetic field vanishes suitably as the volume tends to infinity, see also [182].) For $d \geq 3$, results have been obtained more recently by Pisztora [245], Bodineau [23], Cerf and Pisztora [57, 58], and by Bodineau, Ioffe and Velenik [24]. This analysis is too far beyond the scope of this text. For related results in the context of Kac interactions see e.g. [17].

3.4.1 Large deviation principle for the empirical field

While treating finite Markov chains we have seen that the consideration of large deviation properties of empirical averages, brings us naturally to the empirical distribution of words (of length 2 in that case). Longer memory transitions would bring in words of larger length, this being a good reason for the study of the full empirical process. Let us now extend this type of discussion to random fields, considering the associated empirical fields.

Let us consider a spin system described by (3.89) with $\Phi \in \mathbb{B}^0_\theta$. The partition function $Z^\tau_\Lambda(\beta)$ is a sum on all $\sharp S^{|\Lambda|}$ configurations $\sigma \in \mathcal{X}_\Lambda$. To think in terms of a probability average we give the same weight to each such configuration, writing $\sharp S^{-|\Lambda|} Z^\tau_\Lambda(\beta)$ as an expectation with respect to the uniform distribution on $\mathcal{X}_\Lambda$. Let ν be the product measure on $\mathcal{X}$ under which the coordinates σ_i are uniformly distributed on S. Using the notation of Definition 3.25 and (3.105) in Remark 3.27, we have:

$$\frac{1}{|\Lambda_n|} \log Z^\tau_{\Lambda_n}(\beta)$$
$$= \log \sharp S + \frac{1}{|\Lambda_n|} \log \int_{\mathcal{X}} \exp\left\{ - \sum_{i \in \Lambda_n} \beta A_\Phi(\theta_i \sigma) + \psi_n(\sigma, \tau) \right\} d\nu$$

with $|\psi_n(\sigma, \tau)| \leq 2\beta C |\partial^-_{r_0} \Lambda_n|$, and $C = \sum_{\Delta:\, O \in \Delta} \|\Phi_\Delta\|_\infty$. In terms of the empirical field

$$\tilde{R}_n(\sigma) := \frac{1}{|\Lambda_n|} \sum_{i \in \Lambda_n} \delta_{\theta_i \sigma}, \qquad \sigma \in \mathcal{X}, \tag{3.131}$$

we may write (uniformly in τ)

$$\left| \frac{1}{|\Lambda_n|} \log Z^\tau_{\Lambda_n}(\beta) - \log \sharp S - \frac{1}{|\Lambda_n|} \log \int_{\mathcal{X}} e^{-|\Lambda_n| \langle \beta A_\Phi, \tilde{R}_n \rangle} d\nu \right| \leq 2\beta C \frac{|\partial^-_{r_0} \Lambda_n|}{|\Lambda_n|}$$

which tends to zero as $n \to +\infty$.

Recalling the Varadhan lemma (Theorem 1.18), and since $\alpha \in \mathcal{M}_1(\mathcal{X}) \mapsto \langle \beta A_\Phi, \alpha \rangle$ is continuous (and so bounded) we see that a l.d.p. for the laws of the empirical field $\tilde{R}_n$ under the product measure ν would provide, cf. notation in Remark 3.28:

$$\beta F(\beta, \Phi) = \log \sharp S + \sup\{-\beta \langle A_\Phi, \alpha \rangle - I^{\mathbb{Z}^d}(\alpha) : \alpha \in \mathcal{M}_1(\mathcal{X})\}.$$

The 'reference' measure ν enters through $I^{\mathbb{Z}^d}(\alpha) - \log \sharp S$; an identification of the rate function will clarify this point, cf. Lemma 3.42 below and the remark following it.

As seen in Remark 3.27, the previous considerations apply to any interaction $\Phi \in \mathbb{B}_\theta$, with $C|\partial_{r_0}^- \Lambda_n|/|\Lambda_n|$ replaced by the right-hand side of (3.106), with $\Lambda = \Lambda_n$. In the notation of Lemma 3.42 we have

$$\beta F(\beta, \Phi) = \log \sharp S + \varphi(-\beta A_\Phi \mid \nu). \tag{3.132}$$

Remark 3.37 As seen from the above discussion and the experience with Markov chains in previous sections, once a l.d.p. for the laws of $\tilde{R}_n$ under the product measure ν is established, it provides, through tilting (remark (ii) following Theorem 1.18), a l.d.p. under Gibbs measures for interactions $\Phi \in \mathbb{B}_\theta$. If $I(\cdot) = I^{\mathbb{Z}^d}(\cdot)$ is the rate function for the l.d.p. under ν, one gets a l.d.p. for the laws of $\tilde{R}_n$ under the Gibbs measure $\mu^\tau_{\Lambda_n,\beta}$, with rate function given by

$$I_\Phi(\alpha) := I(\alpha) + \langle \beta A_\Phi, \alpha \rangle - \inf_{\tilde{\alpha} \in \mathcal{M}_1(\mathcal{X})} \{I(\tilde{\alpha}) + \langle \beta A_\Phi, \tilde{\alpha} \rangle\}.$$

Due to (3.105), the estimates in Definition 1.12 will then hold uniformly in τ. That is, for any Borel set A in $\mathcal{M}_1(\mathcal{X})$:

$$\liminf_{n \to +\infty} \inf_\tau \frac{1}{|\Lambda_n|} \log \mu^\tau_{\Lambda_n}\{\sigma : \tilde{R}_n(\sigma) \in A\} \geq - \inf_{\alpha \in A^\circ} I_\Phi(\alpha),$$

$$\limsup_{n \to +\infty} \sup_\tau \frac{1}{|\Lambda_n|} \log \mu^\tau_{\Lambda_n}\{\sigma : \tilde{R}_n(\sigma) \in A\} \leq - \inf_{\alpha \in \overline{A}} I_\Phi(\alpha).$$

Remark Under the measures μ_{Λ_n} (empty boundary conditions) the l.d.p. holds for a wider class of potentials (with the same rate function). Indeed, in this case, only the second term on the left-hand side of (3.106) matters; it is seen to tend to zero as $n \to +\infty$ if (iii)′ in Remark 3.22 is replaced by $\sum_{O \in \Delta} \frac{1}{|\Delta|} \|\Phi_\Delta\|_\infty < +\infty$.

In spite of the reduction indicated above, we consider directly a wider class of measures. This reveals better the main ingredients. We keep $\mathcal{X} = S^{\mathbb{Z}^d}$, with S finite. The space $\mathcal{M}_1(\mathcal{X})$ is endowed with the metric ϱ, cf. Remark 3.15, associated with the weak convergence, and the closed subspace $\mathcal{M}_1^\theta(\mathcal{X})$ of the translation invariant probability measures in $(\mathcal{X}, \mathcal{B}(\mathcal{X}))$ has an important role. A particularly suitable property was considered in Pfister's lecture notes [243] (cf. also [198] and [88] for more restrictive conditions).

Definition 3.38 A probability measure μ on $(\mathcal{X}, \mathcal{B}(\mathcal{X}))$ is said to be asymptotically decoupled (on the volume scale) if there exist functions $g\colon \mathbb{N} \to \mathbb{N}$, $c\colon \mathbb{N} \to [0, +\infty)$ such that $g(n)/n \to 0$, $c(n)/|\Lambda_n| \to 0$ as $n \to +\infty$, and $\forall i \in \mathbb{Z}^d$, $\forall n \in \mathbb{N}$, $\forall A \in \mathcal{B}_{\Lambda_n+i}$, $\forall B \in \mathcal{B}_{(\Lambda_{n+g(n)}+i)^c}$ we have

$$e^{-c(n)}\mu(A)\mu(B) \le \mu(A \cap B) \le e^{c(n)}\mu(A)\mu(B). \tag{3.133}$$

If only the left (right) inequality in (3.133) is assumed, one says that μ is asymptotically decoupled from below (above, respectively).

Remark If $\mu \in \mathcal{M}_1^\theta(\mathcal{X})$ it suffices to verify the condition in (3.133) for $i = O$.

Examples (1) Product probability measures on $\mathcal{X}$; $g(n) = c(n) = 0$ for each n.
(2) DLR measures μ associated with an interaction $\Phi \in \mathbb{B}_\theta^0$, cf. Definition 3.20. $g(n) = 0$, $c(n) = 4C\beta|\partial_{r_0}^- \Lambda_n|$, with $C = \sum_{\Delta\colon \{O\}\subsetneq\Delta} \|\Phi_\Delta\|_\infty$. To check this, observe that for any such μ and $\Lambda \subset\subset \mathbb{Z}^d$, $q_\Lambda(\sigma_\Lambda|\sigma_{\partial_{r_0}\Lambda})$, cf. (3.99), provides a version of the regular conditional distribution on $\mathcal{B}_\Lambda$ given $\mathcal{B}_{\Lambda^c}$, and

$$e^{-4\beta C|\partial_{r_0}^-\Lambda|}\pi_\Lambda\mu(\sigma_\Lambda) \le q_\Lambda(\sigma_\Lambda|\tau) \le e^{4\beta C|\partial_{r_0}^-\Lambda|}\pi_\Lambda\mu(\sigma_\Lambda), \quad \text{for all} \quad \tau \in \mathcal{X}_{\Lambda^c}, \tag{3.134}$$

implying (3.133).

(3) If the finite range condition in the previous example is relaxed to $\sum_{\Delta\colon O\in\Delta} \|\Phi_\Delta\|_\infty < +\infty$, still $g(n) = 0$, with $c(n) = 4\beta \sum_{\Delta\colon O\in\Delta} |\{i \in \Lambda_n\colon \Delta + i \not\subseteq \Lambda_n\}| \|\Phi_\Delta\|_\infty$, cf. the previous arguments and (3.106) (see [198] for more details).

(4) Stationary irreducible and aperiodic finite Markov chains (taking $d = 1$). See [243] for further examples in a more general context.

Theorems 1.21 may be extended as follows due to Lewis, Pfister and Sullivan.

Theorem 3.39 *Let $\tilde{R}_n$ denote the empirical field defined by* (3.131), *with $\Lambda_n = \{-n, \ldots, n\}^d$. If $\mu \in \mathcal{M}_1^\theta(\mathcal{X})$ is asymptotically decoupled and Q_n denotes the law of $\tilde{R}_n$ under μ, then (Q_n) satisfies a l.d.p. on $\mathcal{M}_1(\mathcal{X})$, with scaling $|\Lambda_n|$ and rate function*

$$I^{\mathbb{Z}^d}(\alpha) = \begin{cases} h(\alpha|\mu) := \lim_{k\to+\infty} \frac{1}{|\Lambda_k|}\mathcal{H}(\pi_{\Lambda_k}\alpha|\pi_{\Lambda_k}\mu) & \text{if } \alpha \in \mathcal{M}_1^\theta(\mathcal{X}) \\ +\infty & \text{if } \alpha \notin \mathcal{M}_1^\theta(\mathcal{X}), \end{cases}$$

where $\mathcal{H}(\pi_{\Lambda_k}\alpha|\pi_{\Lambda_k}\mu)$ is the relative entropy defined as in (1.99). (*The existence of the limit on the right-hand side is part of the statement.*)[7]

Remark The l.d.p. also holds for the periodic empirical field as in Section 1.3: given $\sigma \in \mathcal{X}_{\Lambda_n}$ we let $\sigma^{(n)}$ be its periodic extension to all $\mathbb{Z}^d$, analogously to

[7] In our case the space is compact. Being lower semi-continuous, $I^{\mathbb{Z}^d}$ is a good rate function. $h(\cdot|\mu)$ has compact level sets in more general situations (see [243]).

(3.22) and

$$R_n(\sigma) := \frac{1}{|\Lambda_n|} \sum_{i \in \Lambda_n} \delta_{\theta_i \sigma^{(n)}}.$$

For each $k \geq 1$ and $\sigma \in \mathcal{X}$ we have $|\pi_{\Lambda_k} R_n(\pi_{\Lambda_n}\sigma) - \pi_{\Lambda_k} \tilde{R}_n(\sigma)|_1 \leq 2|\{i \in \Lambda_n \colon \Lambda_k + i \not\subseteq \Lambda_n\}|/|\Lambda_n|$, in the notation of part (b) of Remark 3.15. Recalling the definition of $\varrho(\mu, \mu')$, we see that the validity of the l.d.p. for one such field implies it for the other. The periodic field has the advantage that $R_n(\sigma) \in \mathcal{M}_1^\theta(\mathcal{X})$, for each n, σ. On the other hand, $\langle f, \tilde{R}_n(\sigma)\rangle = |\Lambda_n|^{-1} \sum_{i \in \Lambda_n} f(\theta_i \sigma)$ gives exactly the spatial average. (*Notation.* Sometimes we write $R_n(\sigma) = R_n(\pi_{\Lambda_n}\sigma)$, for $\sigma \in \mathcal{X}$.)

Our discussion of Theorem 3.39 follows Section 3 in [243], taking into account the simplifications due to our assumption of finite S. Such l.d.p. for Gibbs measures have been proven initially by Comets [65], Follmer and Orey [118], and Olla [234], extending Lanford's arguments in [189]. Several extensions and different proofs have been presented, as in [88, 135, 198, 243], where the connection with thermodynamics is carefully analysed in a more general context and with different viewpoints.

The following basic estimate is taken from [243]; it captures the essentials of the subaddivity argument already used in Lemma 1.29 and in Proposition 3.13.

Lemma 3.40 *Let $\mu \in \mathcal{M}_1^\theta(\mathcal{X})$ be asymptotically decoupled from below, $f \colon \mathcal{X} \to \mathbb{R}^p$ be a cylinder function ($p \geq 1$), and $\psi \colon \mathbb{R}^p \to [0, +\infty)$ be a convex function. If one sets*

$$\mathcal{E}_\Lambda(a) = \left\{\sigma \in \mathcal{X} \colon \psi\Big(\frac{1}{|\Lambda|} \sum_{i \in \Lambda} f(\theta_i \sigma)\Big) < a\right\},$$

where $\Lambda \subset\subset \mathbb{Z}^d$ and $a > 0$, then given $a > 0$, $0 < a' < a$, $\delta \in (0, 1)$ we may find M (depending on all previous parameters) so that for each $m \geq M$ there exists $N = N(m)$ (also depending on all previous parameters) so that for $m \geq M, n \geq N$ we have:

$$\frac{1}{|\Lambda_n|} \log \mu(\mathcal{E}_{\Lambda_n}(a)) \geq (1 - \delta) \frac{1}{|\Lambda_m|} \log \mu(\mathcal{E}_{\Lambda_m}(a')) - \frac{c(m)}{|\Lambda_m|},$$

where $\Lambda_n = \{-n, \ldots, n\}^d$. (Notation. If required to indicate ψ, we write $\mathcal{E}_\Lambda(\psi, a)$.)

Proof Let us fix $r \in \mathbb{N}$ so that $f \in \mathcal{B}_{\Lambda_r}$. Given $m \geq 1$, we set r' as the smallest integer larger than or equal to $g(m + r)/2$, and $m' = m + r + r'$. If $n \geq m$ we take ℓ, q integers determined by $2n + 1 = \ell(2m' + 1) + q$ where $0 \leq q \leq 2m'$. Partitioning $\{-n, \ldots, n\}$ into $\ell + 1$ successive disjoint blocks, $B_1, \ldots, B_\ell, B_{\ell+1}$, the first ℓ with cardinality $2m' + 1$ and the last one with cardinality q, we consider the ℓ^d translates of $\Lambda_{m'}$: $B_{l_1} \times \cdots \times B_{l_d}$ with $l_1, \ldots, l_d \in \{1, \ldots, \ell\}$, relabelled as $\tilde{\Lambda}(l)$, $l = 1, \ldots, \ell^d$. Thus, for suitable a_l, $l = 1, \ldots, \ell^d$,

$$\tilde{\Lambda}(l) = a_l + \Lambda_{m'}, \qquad \text{and we let, if } q \neq 0,$$

$$\tilde{\Lambda}(\ell^d + 1) = \Lambda_n \setminus \cup_{l=1}^{\ell^d} \tilde{\Lambda}(l) = \{x \in \Lambda_n : \max_{l=1,\ldots,d} x_l \in \{n - q + 1, \ldots, n\}\}.$$

In order to use the decoupling property we set

$$\Lambda(l) = a_l + \Lambda_m, \qquad l = 1, \ldots, \ell^d,$$

$$\Lambda(\ell^d + 1) = \tilde{\Lambda}(\ell^d + 1) \cup \cup_{l=1}^{\ell^d} (\tilde{\Lambda}(l) \setminus \Lambda(l)).$$

The picture is similar to that in Figure 3.1, but now the $\Lambda(l)$ (ℓ^d such sets) are separated by corridors of width $2(r + r')$ which might tend to $+\infty$ with m, but $(r + r')/m \to 0$ by the asymptotic decoupling property; there is also a remaining part $\tilde{\Lambda}(\ell^d + 1)$.

Simple computations show that

$$\frac{|\tilde{\Lambda}(\ell^d + 1)|}{|\Lambda_n|} \leq \frac{2dm'(2n + 1)^{d-1}}{(2n + 1)^d} \leq \frac{2dm'}{2n + 1} \tag{3.135}$$

and that

$$\frac{|\Lambda(\ell^d + 1)|}{|\Lambda_n|} \leq \frac{2dm'}{2n + 1} + \frac{d(r + r')}{m'}. \tag{3.136}$$

Let $b = \max_\sigma |\psi(f(\sigma))|$. Then $b < \infty$ (S is finite), and since $g(m)/m \to 0$ as $m \to +\infty$, we may take M large enough in such a way that if $m \geq M$ there exists $N = N(m)$ finite so that

$$\frac{|\Lambda(\ell^d + 1)|}{|\Lambda_n|} \leq \delta \wedge (a - a')/b, \qquad \text{if} \quad m \geq M, n \geq N(m).$$

(Indeed, first take m large so that the second term on the right-hand side of (3.136) is small; then choose n large, depending on m for the first term to be small.)

In particular, using the convexity we have that for such m and n:

$$\psi\left(\frac{1}{|\Lambda_n|} \sum_{x \in \Lambda_n} f(\theta_x \sigma)\right) \leq \sum_{l=1}^{\ell^d} \frac{|\Lambda(l)|}{|\Lambda_n|} \psi\left(\frac{1}{|\Lambda(l)|} \sum_{x \in \Lambda(l)} f(\theta_x \sigma)\right) + (a - a').$$

so that

$$\mathcal{E}_{\Lambda_n}(a) \supseteq \cap_{l=1}^{\ell^d} \mathcal{E}_{\Lambda(l)}(a').$$

From the lower bound in (3.133) and the translation invariance of μ, we get:

$$\mu(\mathcal{E}_{\Lambda_n}(a)) \geq e^{-\ell^d c(m)} \mu(\mathcal{E}_{\Lambda_m}(a'))^{\ell^d}.$$

Taking logarithms, dividing by $|\Lambda_n|$, and since $1 - \delta \leq \ell^d |\Lambda_m| / |\Lambda_n| \leq 1$ for such m, n, we easily get the conclusion. □

Corollary 3.41 *Let $\mu \in \mathcal{M}_1^\theta(\mathcal{X})$ be asymptotically decoupled from below.*
(a) For each $k \geq 0$ and $\alpha \in \mathcal{M}_1(\mathcal{X}_{\Lambda_k})$ we may define:

$$
\begin{aligned}
-I^{(k)}(\alpha) &= \lim_{\varepsilon \downarrow 0} \limsup_{n \to +\infty} \frac{1}{|\Lambda_n|} \log \mu(|\pi_{\Lambda_k} \tilde{R}_n - \alpha|_1 < \varepsilon) \\
&= \lim_{\varepsilon \downarrow 0} \liminf_{n \to +\infty} \frac{1}{|\Lambda_n|} \log \mu(|\pi_{\Lambda_k} \tilde{R}_n - \alpha|_1 < \varepsilon).
\end{aligned}
\tag{3.137}
$$

$I^{(k)}$ is lower semi-continuous and convex on $\mathcal{M}_1(\mathcal{X}_{\Lambda_k})$. The laws of $\pi_{\Lambda_k} \tilde{R}_n$ under μ satisfy a l.d.p. with rate function $I^{(k)}$ (scaling $|\Lambda_n|$).
(b) Set $I^{\mathbb{Z}^d}(\alpha) := \sup_{k \geq 1} I^{(k)}(\pi_{\Lambda_k} \alpha)$. $I^{\mathbb{Z}^d} : \mathcal{M}_1(\mathcal{X}) \to [0, +\infty]$ is a lower semi-continuous and convex function. For each $\alpha \in \mathcal{M}_1(\mathcal{X})$ we have

$$
\begin{aligned}
-I^{\mathbb{Z}^d}(\alpha) &= \lim_{\varepsilon \downarrow 0} \limsup_{n \to +\infty} \frac{1}{|\Lambda_n|} \log \mu(\varrho(\tilde{R}_n, \alpha) < \varepsilon) \\
&= \lim_{\varepsilon \downarrow 0} \liminf_{n \to +\infty} \frac{1}{|\Lambda_n|} \log \mu(\varrho(\tilde{R}_n, \alpha) < \varepsilon).
\end{aligned}
\tag{3.138}
$$

The laws of $\tilde{R}_n$ under μ satisfy a l.d.p. on $\mathcal{M}_1(\mathcal{X})$, with rate function $I^{\mathbb{Z}^d}$ (scaling $|\Lambda_n|$). Moreover, $I^{\mathbb{Z}^d}(\alpha) = +\infty$ if $\alpha \notin \mathcal{M}_1^\theta(\mathcal{X})$.

Proof For each fixed k, we apply Lemma 3.40 to get the first statement. Indeed, enumerating $\mathcal{X}_{\Lambda_k}$ as $\{\tilde{\sigma}_1, \ldots, \tilde{\sigma}_{\tilde{k}}\}$, we consider the cylinder function $f = (f_1, \ldots, f_{\tilde{k}})$ on $\mathcal{X}$, where $f_l(\sigma) = \mathbf{1}_{[\pi_{\Lambda_k} \sigma = \tilde{\sigma}_l]}$ for each l, and the convex function ψ_α, on $\mathbb{R}^{\tilde{k}}$ given by $\psi_\alpha(x_1, \ldots, x_{\tilde{k}}) = \sum_{l=1}^{\tilde{k}} |x_l - \alpha\{\tilde{\sigma}_l\}| = |x - \alpha|_1$.

Applying Lemma 3.40 to the events $\mathcal{E}_{\Lambda_n}(\varepsilon)$ associated with such f and ψ_α we get that if $0 < \varepsilon' < \varepsilon$

$$
\liminf_{n \to +\infty} \frac{1}{|\Lambda_n|} \log \mu(|\pi_{\Lambda_k} \tilde{R}_n - \alpha|_1 < \varepsilon) \geq \limsup_{n \to +\infty} \frac{1}{|\Lambda_n|} \log \mu(|\pi_{\Lambda_k} \tilde{R}_n - \alpha|_1 < \varepsilon')
$$

from which (3.137) follows at once. It is easy to deduce the lower semi-continuity of $I^{(k)}$ from (3.137). Indeed, if $u < I^{(k)}(\alpha)$ we may take $\varepsilon > 0$ so that

$$
\limsup_{n \to +\infty} \frac{1}{|\Lambda_n|} \log \mu(|\pi_{\Lambda_k} \tilde{R}_n - \alpha|_1 < \varepsilon) \leq -u.
$$

Now, if $|\tilde{\alpha} - \alpha|_1 < \varepsilon/2$, it follows that for each $\varepsilon' \leq \varepsilon/2$

$$
\limsup_{n \to +\infty} \frac{1}{|\Lambda_n|} \log \mu(|\pi_{\Lambda_k} \tilde{R}_n - \tilde{\alpha}|_1 < \varepsilon') \leq -u;
$$

letting $\varepsilon' \to 0$ we get $I^{(k)}(\tilde{\alpha}) \geq u$ for all such $\tilde{\alpha}$ which shows the lower semi-continuity.

The lower semi-continuity of $I^{\mathbb{Z}^d}$ follows from that of $I^{(k)}$ and the continuity of $\alpha \mapsto \pi_{\Lambda_k} \alpha$. Recalling the definition of ϱ, cf. Remark 3.15, identities (3.138) are easily derived from those in (3.137) as in the proof of Theorem 1.21.

As for the validity of the l.d.p. in parts (a) and (b), the lower bound follows at once from the identities (3.137) and (3.138), respectively. For the upper bound we use the compactness of the spaces $\mathcal{M}_1(\mathcal{X}_{\Lambda_k})$ and $\mathcal{M}_1(\mathcal{X})$, for S finite. The details are the same as in the proof of Theorem 1.21 in Section 1.3, and are thus omitted here.

Concerning the convexity, it suffices to check that for each k and each $\alpha, \tilde{\alpha} \in \mathcal{M}_1(\mathcal{X}_{\Lambda_k})$ we have $I^{(k)}((\alpha+\tilde{\alpha})/2) \le (I^{(k)}(\alpha) + I^{(k)}(\tilde{\alpha}))/2$. For this one may consider a modification of Lemma 3.40, after observing that

$$
\begin{aligned}
\psi_{\frac{\alpha+\tilde{\alpha}}{2}}\left(\frac{1}{|\Lambda_n|}\sum_{x\in\Lambda_n} f(\theta_x\sigma)\right) \le{} & \sum_{i=1}^{[\ell^d/2]} \frac{|\Lambda(i)|}{|\Lambda_n|}\psi_\alpha\left(\frac{1}{|\Lambda(i)|}\sum_{x\in\Lambda(i)} f(\theta_x\sigma)\right) \\
&+ \sum_{i=[\ell^d/2]+1}^{\ell^d} \frac{|\Lambda(i)|}{|\Lambda_n|}\psi_{\tilde{\alpha}}\left(\frac{1}{|\Lambda(i)|}\sum_{x\in\Lambda(i)} f(\theta_x\sigma)\right) \\
&+ \frac{|\Lambda(\ell^d+1)|}{|\Lambda_n|} b + \left|\frac{1}{2} - [\ell^d/2]\frac{|\Lambda_m|}{|\Lambda_n|}\right| \\
&+ \left|\frac{1}{2} - (\ell^d - [\ell^d/2])\frac{|\Lambda_m|}{|\Lambda_n|}\right|
\end{aligned}
$$

so that the proof is modified taking ψ_α on the cubes $\Lambda(i)$, $1 \le i \le [\ell^d/2]$, and $\psi_{\tilde{\alpha}}$ if $[\ell^d/2]+1 \le i \le \ell^d$. As in the proof of Lemma 3.40 we may take M so that if $m \ge M$ and $n \ge N(m)$:

$$
\mathcal{E}_{\Lambda_n}(\psi_{\frac{\alpha+\tilde{\alpha}}{2}}, a) \supseteq \cap_{i=1}^{[\ell^d/2]} \mathcal{E}_{\Lambda(i)}(\psi_\alpha, a') \cap \cap_{i=[\ell^d/2]+1}^{\ell^d} \mathcal{E}_{\Lambda(i)}(\psi_{\tilde{\alpha}}, a'),
$$

the notation $\mathcal{E}_\Lambda(\psi, a)$ being as indicated in Lemma 3.40. The conclusion then follows from the argument of Lemma 3.40, and (3.137).

To prove the last statement, let $\alpha \in \mathcal{M}_1(\mathcal{X}) \setminus \mathcal{M}_1^\theta(\mathcal{X})$. This entails $\varrho(\alpha, \theta_j\alpha) =: a > 0$, for some $j \in \mathbb{Z}^d$. On the other hand, for each $k \ge 1$ we see that $\sup_\sigma |\pi_{\Lambda_k}\tilde{R}_n(\sigma) - \pi_{\Lambda_k}\theta_j\tilde{R}_n(\sigma)|_1 \to 0$ as $n \to +\infty$, so that

$$
\lim_{n\to+\infty} \sup_\sigma \varrho(\tilde{R}_n(\sigma), \theta_j\tilde{R}_n(\sigma)) = 0
$$

and therefore

$$
[\varrho(\tilde{R}_n, \alpha) < a/3] \cap [\varrho(\theta_j\tilde{R}_n, \theta_j\alpha) < a/3] = \emptyset, \qquad \text{if} \quad n \text{ is large enough.}
$$

We thus have an open set $G \subseteq \mathcal{M}_1(\mathcal{X})$, with $\alpha \in G$ and $\{\sigma : \tilde{R}_n(\sigma) \in G\} = \emptyset$ for n large. In particular,

$$
-I^{\mathbb{Z}^d}(\alpha) \le \liminf_{n\to+\infty} \frac{1}{|\Lambda_n|} \log \mu(\tilde{R}_n \in G) = -\infty.
$$

□

Lemma 3.42 *Let $\mu \in \mathcal{M}_1^\theta(\mathcal{X})$ be asymptotically decoupled from below. If $f \in C(\mathcal{X})$ we set*

$$\varphi_n(f|\mu) = \frac{1}{|\Lambda_n|} \log \int_{\mathcal{X}} \exp\left\{\sum_{i\in\Lambda_n} f(\theta_i\sigma)\right\} \mu(d\sigma).$$

The limit $\varphi(f|\mu) := \lim_{n\to+\infty} \varphi_n(f|\mu)$ exists and defines a convex function on the Banach space $C(\mathcal{X})$ (with the sup norm $\|\cdot\|_\infty$). *Moreover,*

$$\varphi(f|\mu) = \sup_{\alpha\in\mathcal{M}_1(\mathcal{X})} (\langle f, \alpha\rangle - I^{\mathbb{Z}^d}(\alpha)), \tag{3.139}$$

where $I^{\mathbb{Z}^d}$ is given in Corollary 3.41 *and $\langle f, \alpha\rangle = \int f d\alpha$. The map $f \mapsto \varphi(f|\mu)$ is lower semi-continuous for the weak topology on $C(\mathcal{X})$.*

Proof Notice that

$$\varphi_n(f|\mu) = \frac{1}{|\Lambda_n|} \log \int_{\mathcal{X}} \exp\{|\Lambda_n|\langle f, \tilde{R}_n\rangle\} d\mu$$

and since the map $\alpha \in \mathcal{M}_1(\mathcal{X}) \mapsto \langle f, \alpha\rangle$ is continuous (bounded), the existence of the limit and expression (3.139) follow from Corollary 3.41 and the Varadhan lemma. The convexity of $\varphi_n(\cdot|\mu)$ can be seen from the Hölder inequality (as for $F_\Lambda(\beta, \cdot)$ in (3.112)) and it yields that of $\varphi(\cdot|\mu)$. The last statement follows immediately from (3.139), since (by definition) the map $f \in C(\mathcal{X}) \mapsto \langle f, \alpha\rangle$ is weakly continuous, for each $\alpha \in \mathcal{M}_1(\mathcal{X})$, and it suffices to consider the supremum among α so that $I^{\mathbb{Z}^d}(\alpha) < +\infty$. (Also $|\varphi(f \mid u) - \varphi(g \mid \mu)| \le ||f - g||_\infty$.) □

Remark The existence of the limit $\varphi(f \mid \mu)$ may be obtained directly with an argument as in Lemma 3.40, for μ asymptotically decoupled either from below or from above. The same applies for convexity and lower semi-continuity (see Proposition 3.1 in [243]).

Duality The dual of $C(\mathcal{X})$ is $\mathcal{M}(\mathcal{X})$, the space of all finite signed-measures on $(\mathcal{X}, \mathcal{B})$. If we extend $I^{\mathbb{Z}^d}$, by setting $I^{\mathbb{Z}^d}(\alpha) = +\infty$ if $\alpha \in \mathcal{M}(\mathcal{X}) \setminus \mathcal{M}_1(\mathcal{X})$, it remains convex and lower semi-continuous for the w^*-topology, since $\mathcal{M}_1(\mathcal{X})$ is convex and closed in $\mathcal{M}(\mathcal{X})$; obviously (3.139) does not change if the supremum is taken on all $\mathcal{M}(\mathcal{X})$. We are then in a situation to apply a duality theorem on Fenchel–Legendre transforms (cf. [38], Theorem 3.10), to conclude that for each $\alpha \in \mathcal{M}(\mathcal{X})$:

$$I^{\mathbb{Z}^d}(\alpha) = \varphi^*(\alpha|\mu) := \sup_{f\in C(\mathcal{X})} (\langle f, \alpha\rangle - \varphi(f|\mu)), \tag{3.140}$$

where we used that $C(\mathcal{X})$ is the dual of $\mathcal{M}(\mathcal{X})$ (w^*-topology), cf. Lemma 1.38. We omit the proof of the duality theorem. See [38] or Lemma 4.5.8 in [85]. (The Hahn–Banach theorem is used, since it involves separation through hyperplanes.)

Remark In the general situation treated in [198] and [243], the role of $C(\mathcal{X})$ as space of observables is taken by the $\|\cdot\|_\infty$-closure of the set of bounded, measurable cylinder functions. The argumentation based on Lemma 3.40 leads to a l.d.p. in its dual, generally larger than $\mathcal{M}(\mathcal{X})$. A delicate point there is to prove that the rate function takes the value $+\infty$ outside $\mathcal{M}(\mathcal{X})$ (see [243], Lemma 3.4).

The final step for the identification of $I^{\mathbb{Z}^d}$ is given in the next lemma which completes the proof of Theorem 3.39.

Lemma 3.43 *Let $\mu \in \mathcal{M}_1^\theta(\mathcal{X})$ be asymptotically decoupled from above. Then the limit $h(\alpha|\mu) := \lim_{k\to+\infty} |\Lambda_k|^{-1}\mathcal{H}(\pi_{\Lambda_k}\alpha|\pi_{\Lambda_k}\mu)$ exists and defines an affine, lower semi-continuous function on $\mathcal{M}_1^\theta(\mathcal{X})$. Moreover, if $\alpha \in \mathcal{M}_1^\theta(\mathcal{X})$ we have $\varphi^*(\alpha|\mu) = h(\alpha|\mu)$.*

Notation For $\alpha \in \mathcal{M}_1(\mathcal{X})$, $\alpha_{|\mathcal{B}_\Lambda}$ denotes its restriction to $\mathcal{B}_\Lambda$, and we write $\mathcal{H}_\Lambda(\alpha|\mu) := \mathcal{H}(\alpha_{|\mathcal{B}_\Lambda}|\mu_{|\mathcal{B}_\Lambda}) = \mathcal{H}(\pi_\Lambda\alpha|\pi_\Lambda\mu)$, cf. (1.99). We see that

$$\mathcal{H}_\Lambda(\alpha|\mu) \le \mathcal{H}_{\Lambda'}(\alpha|\mu), \qquad \text{if} \quad \Lambda \subseteq \Lambda'. \tag{3.141}$$

Indeed, if $\pi_{\Lambda'}\alpha \ll \pi_{\Lambda'}\mu$ and $f_{\Lambda'}$ is the Radon–Nikodym derivative, the μ-conditional expectation of $f_{\Lambda'}$ given $\mathcal{B}_\Lambda$ is $f_\Lambda = d\pi_\Lambda\alpha/d\pi_\Lambda\mu$. Recalling (1.99) (that is (1.49) in the finite case), the convexity of $x \mapsto x\log x$ on $[0,+\infty)$, and the Jensen inequality for conditional expectations, we get (3.141). In the finite case, the lower semi-continuity and convexity of $\alpha \mapsto \mathcal{H}(\pi_\Lambda\alpha|\pi_\Lambda\mu)$ were seen in Section 1.3 (cf. also Section 1.6).

Proof We first check that if $\mu \in \mathcal{M}_1^\theta(\mathcal{X})$ verifies the upper bound in (3.133), the specific relative entropy $h(\alpha|\mu)$ is well defined and has the above stated properties. The proof is taken from [243] (cf. also [88]). Let us recall the decomposition and the notation in the proof of Lemma 3.40, for $r = 0$. The validity of the second inequality in (3.133) implies that

$$\pi_{\Gamma_n}\mu \ll \times_{l=1}^{\ell^d}\pi_{\Lambda(l)}\mu, \qquad \text{where} \quad \Gamma_n = \cup_{l=1}^{\ell^d}\Lambda(l)$$

and a version of the Radon–Nikodym derivative can be taken bounded from above by $e^{\ell^d c(m)}$. As a consequence we have:

$$\mathcal{H}_{\Gamma_n}(\alpha|\times_{l=1}^{\ell^d}\pi_{\Lambda(l)}\mu) \le \mathcal{H}_{\Gamma_n}(\alpha|\mu) + \ell^d c(m).$$

Recalling (1.61) and since $\mathcal{H}_{\Lambda(l)}(\alpha|\mu) = \mathcal{H}_{\Lambda_m}(\alpha|\mu)$ for $l = 1,\dots,\ell^d$, due to the translation invariance of α and μ, we may write

$$\ell^d\mathcal{H}_{\Lambda_m}(\alpha|\mu) \le \mathcal{H}_{\Gamma_n}(\alpha|\mu) + \ell^d c(m) \le \mathcal{H}_{\Lambda_n}(\alpha|\mu) + \ell^d c(m),$$

where the last inequality follows from (3.141) since $\Gamma_n \subseteq \Lambda_n$. Therefore, for n, m as in the given decomposition:

$$\mathcal{H}_{\Lambda_n}(\alpha|\mu) \ge \ell^d\mathcal{H}_{\Lambda_m}(\alpha|\mu) - \ell^d c(m).$$

Given $\delta \in (0, 1)$, we can find $M = M(\delta)$, and for $m \geq M$, $N = N(m, \delta)$ so that $m \geq M, n \geq N(m)$ implies $\ell^d |\Lambda_m|/|\Lambda_n| \geq 1 - \delta$, and so:

$$\frac{1}{|\Lambda_n|}\mathcal{H}_{\Lambda_n}(\alpha|\mu) \geq (1-\delta)\frac{1}{|\Lambda_m|}\mathcal{H}_{\Lambda_m}(\alpha|\mu) - \frac{c(m)}{|\Lambda_m|}, \quad \text{if } m \geq M, n \geq N(m). \tag{3.142}$$

Sending first n, and then m, to infinity, we get the existence of the limit defining $h(\alpha|\mu)$. Moreover, (3.142) implies that

$$h(\alpha|\mu) \geq (1-\delta)\frac{1}{|\Lambda_m|}\mathcal{H}_{\Lambda_m}(\alpha|\mu) - \frac{c(m)}{|\Lambda_m|}, \quad \text{if } m \geq M,$$

and the lower semi-continuity of $h(\cdot|\mu)$ follows from that of $\mathcal{H}_{\Lambda_m}(\cdot|\mu)$. From the convexity of $\mathcal{H}_\Lambda(\cdot|\mu)$ that of $h(\cdot|\mu)$ also follows; for the concavity we check that if $p_l > 0$, $\sum_{l=1}^k p_l = 1$ and $\alpha_l \in \mathcal{M}_1(\mathcal{X})$, then

$$\sum_{l=1}^k p_l \mathcal{H}_\Lambda(\alpha_l|\mu) \leq \mathcal{H}_\Lambda\left(\sum_{l=1}^k p_l\alpha_l|\mu\right) + \sum_{l=1}^k p_l \log\frac{1}{p_l}.$$

It remains to check that $\varphi^*(\cdot|\mu) = h(\cdot|\mu)$ on $\mathcal{M}_1^\theta(\mathcal{X})$. Recalling (1.55), if $f \in \mathcal{F}_{loc}$ with $\Delta(f) \subseteq \Lambda_k$ we have

$$\left\langle \sum_{i\in\Lambda_n} \theta_i f, \alpha \right\rangle - \log \int_{\mathcal{X}} \exp\left\{\sum_{i\in\Lambda_n} \theta_i f\right\} d\mu \leq \mathcal{H}_{\Lambda_{n+k}}(\alpha|\mu).$$

Dividing by $|\Lambda_n|$, using that $\alpha \in \mathcal{M}_1^\theta(\mathcal{X})$, letting $n \to +\infty$ and taking the supremum leads to

$$\sup_{f\in\mathcal{F}_{loc}} (\langle f, \alpha\rangle - \varphi(f|\mu)) \leq h(\alpha|\mu).$$

Due to the continuity of $f \mapsto \langle f, \alpha\rangle - \varphi(f|\mu)$ on $C(\mathcal{X})$ and the density of $\mathcal{F}_{loc}$, the supremum does not change if f varies on $C(\mathcal{X})$, leading to $\varphi^*(\alpha|\mu) \leq h(\alpha|\mu)$ on $\mathcal{M}_1^\theta(\mathcal{X})$. The converse inequality can be obtained with a reasoning similar to Lemma 3.40, partitioning Λ_n into $\tilde{\Lambda}(l)$, $l = 1, \ldots, \ell^d + 1$, taking $r = 0$ in that proof, so that $\tilde{\Lambda}(l) = \Lambda_{m'} + a_l$, for $l = 1, \ldots, \ell^d$, with $m' = m + r'$, and r' is the smallest integer larger than or equal to $g(m)/2$. In this notation, if f is bounded and measurable we write

$$\left|\sum_{i\in\Lambda_n}\theta_i f - \sum_{i\in\Lambda_{m'}}\sum_{l=1}^{\ell^d}\theta_{a_l+i} f\right| = \left|\sum_{i\in\tilde{\Lambda}(\ell^d+1)} \theta_i f\right| \leq |\tilde{\Lambda}(\ell^d+1)|\, \|f\|_\infty.$$

From (3.135), given $m \geq 1$, $\delta \in (0, 1)$, there exists N so that $|\tilde{\Lambda}(\ell^d + 1)| \|f\|_\infty/|\Lambda_n| \leq \delta$ for $n \geq N$, and consequently

$$\varphi_n(f|\mu) \leq \frac{1}{|\Lambda_n|}\log\int_{\mathcal{X}}\exp\left\{\sum_{i\in\Lambda_{m'}}\sum_{l=1}^{\ell^d}\theta_{a_l+i} f\right\} d\mu + \delta. \tag{3.143}$$

By the Jensen inequality and the translation invariance of μ we have

$$\int_{\mathcal{X}} \exp\left\{\sum_{i\in\Lambda_{m'}}\sum_{l=1}^{\ell^d}\theta_{a_l+i}f\right\}d\mu \le \int_{\mathcal{X}} \exp\left\{|\Lambda_{m'}|\sum_{l=1}^{\ell^d}\theta_{a_l}f\right\}d\mu,$$

and if $f \in \mathcal{F}_{loc}$, with $\Delta(f) \subseteq \Lambda_m$, the upper bound in (3.133) implies that

$$\int_{\mathcal{X}} \exp\left\{|\Lambda_{m'}|\sum_{l=1}^{\ell^d}\theta_{a_l}f\right\}d\mu \le \left(\int_{\mathcal{X}} e^{|\Lambda_{m'}|f}\,d\mu\right)^{\ell^d} e^{\ell^d c(m)},$$

also by the translation invariance. Inserting into (3.143), we get

$$\langle f,\alpha\rangle - \varphi_n(f|\mu) \ge \frac{1}{|\Lambda_{m'}|}\left(\langle|\Lambda_{m'}|f,\alpha\rangle - \log\int_{\mathcal{X}} e^{|\Lambda_{m'}|f}\,d\mu\right) - \frac{c(m)}{|\Lambda_{m'}|} - \delta,$$

for all such f and large n, so that using (1.55) again:

$$\varphi^*(\alpha|\mu) \ge \frac{1}{|\Lambda_{m'}|}\mathcal{H}_{\Lambda_m}(\alpha|\mu) - \frac{c(m)}{|\Lambda_{m'}|} - \delta$$

from which the lemma follows upon letting $m \to +\infty$ and $\delta \to 0$. □

Remarks (i) For interactions $\Phi \in \mathbb{B}_\theta$, the l.d.p. also holds for the laws of $\tilde{R}_n$ under the Gibbs measures $\mu^\tau_{\Lambda_n}$, uniformly in τ, cf. Remark 3.37. In particular, on $\mathcal{M}_1^\theta(\mathcal{X})$, $h(\cdot|\mu)$ is the same for any $\mu \in \mathcal{G}_\theta$, and $h(\tilde{\mu}|\mu) = 0$ if $\tilde{\mu}, \mu \in \mathcal{G}_\theta$. More precisely, under such conditions, as in (3.134) and example (3) right after it, we have:

$$\frac{1}{|\Lambda_n|}\mathcal{H}_{\Lambda_n}(\mu^\tau_{\Lambda_n} \mid \mu^{\tau'}_{\Lambda_n}) \le 4\beta \sum_{\Delta:\, O\in\Delta} \frac{|\{i \in \Lambda_n : \Delta + i \not\subseteq \Lambda_n\}|}{|\Lambda_n|}\|\Phi_\Delta\|_\infty,$$

$$\forall\, \tau, \tau' \in \mathcal{X},$$

which tends to zero for Λ_n as before. Also $\varphi(\cdot|\mu) = \varphi(\cdot|\tilde{\mu})$ on $C(\mathcal{X})$ if $\mu, \tilde{\mu} \in \mathcal{G}_\theta$.

(ii) The identification $I^{\mathbb{Z}^d}(\cdot) = h(\cdot|\mu)$ in Theorem 3.39 is restricted to $\mathcal{M}_1^\theta(\mathcal{X})$; non-translation invariant measures α for which $h(\alpha|\mu) < +\infty$ are easily constructed.

(iii) For examples of translation invariant probability measures μ on $\{-1,+1\}^{\mathbb{Z}}$ for which the limits in Lemmas 3.42 and 3.43 fail to exist see e.g. [111], p. 1092, and references therein.

Gibbs variational principle As argued in the beginning of this section, by applying Theorem 3.39 to ν, the product of uniform probabilities on S, we see that if $\Phi \in \mathbb{B}_\theta$, the thermodynamic limit $F(\beta,\Phi)$ is given by:

$$\begin{aligned} F(\beta,\Phi) &= \sup\{-\langle A_\Phi,\alpha\rangle - \beta^{-1}h(\alpha|\nu) + \beta^{-1}\log \sharp S : \alpha \in \mathcal{M}_1^\theta(\mathcal{X})\} \\ &= \sup\{\mathrm{T}s(\alpha) - u(\alpha) : \alpha \in \mathcal{M}_1^\theta(\mathcal{X})\}, \end{aligned} \tag{3.144}$$

since $h(\cdot\,|\nu) = \log \sharp S - k^{-1}s(\cdot)$, and $\mathrm{T} = 1/k\beta$ is the temperature (here k is the Boltzmann constant).

The Gibbs variational principle (Theorem 3.26) characterizes the measures in $\mathcal{G}_\theta$ as those elements in $\mathcal{M}_1^\theta(\mathcal{X})$ where the the supremum in (3.144) is attained.

Proof of Theorem 3.26 (Sketch) From Remark 3.37 and the discussion that precedes it, we have the relation between the rate functions under $\mu \in \mathcal{G}_\theta(\beta, \Phi)$ and under ν, in Theorem 3.39. Consequently, if $\alpha \in \mathcal{M}_1^\theta(\mathcal{X})$:

$$h(\alpha|\mu) = h(\alpha|\nu) + \beta\langle A_\Phi, \alpha\rangle - \inf\{h(\tilde{\alpha}|\nu) + \beta\langle A_\Phi, \tilde{\alpha}\rangle : \tilde{\alpha} \in \mathcal{M}_1^\theta(\mathcal{X})\} \tag{3.145}$$

from which we see that any $\mu \in \mathcal{G}_\theta$ minimizes the function $u(\cdot) - \mathrm{T}s(\cdot)$ over $\mathcal{M}_1^\theta(\mathcal{X})$.

The converse is the essential point of this celebrated theorem. We sketch a proof assuming $\Phi \in \mathbb{B}_\theta^0$, cf. Definition 3.16. This follows [88] and uses an argument due to Preston, which holds more generally, cf. Theorem 7.1 in [247].

Let $\lambda \in \mathcal{M}_1^\theta(\mathcal{X})$ be a point of minimum of $u(\cdot) - \mathrm{T}s(\cdot)$ over $\mathcal{M}_1^\theta(\mathcal{X})$. We must determine that λ verifies the DLR equation (3.101), at inverse temperature β.

From the previous observation (see also remark (i) above), we know that

$$\lim_{n\to+\infty} \frac{1}{|\Lambda_n|}\mathcal{H}_{\Lambda_n}(\lambda \mid \mu^\lambda_{\Lambda_n}) = 0,$$

and our goal is to verify the validity of (3.101), i.e. $\lambda = \mu^\lambda_\Delta$, for any $\Delta \subset\subset \mathbb{Z}^d$, cf. Definition 3.18, or equivalently, that $\mathcal{H}_{\tilde{\Delta}}(\lambda \mid \mu^\lambda_\Delta) = 0$, for any $\tilde{\Delta} \subset\subset \mathbb{Z}^d$. That is, from the asymptotics of $\mathcal{H}_{\Lambda_n}(\lambda \mid \mu^\lambda_{\Lambda_n})$ one indeed concludes precise information on each $\Delta \subset\subset \mathbb{Z}^d$.

Since $\mathcal{H}_{\tilde{\Delta}}(\lambda \mid \mu^\lambda_\Delta)$ is non-decreasing in $\tilde{\Delta}$ it suffices to prove that $\mathcal{H}_{\Lambda_m}(\lambda \mid \mu^\lambda_\Delta) = 0$, for all m large enough. But, if the interaction has range r_0, for $\Delta \cup \partial_{r_0}\Delta \subseteq \tilde{\Delta} \subseteq \Lambda$, we have:

$$\mathcal{H}_{\tilde{\Delta}}(\lambda \mid \mu^\lambda_\Lambda) - \mathcal{H}_{\tilde{\Delta}\setminus\Delta}(\lambda \mid \mu^\lambda_\Lambda) = \mathcal{H}_{\tilde{\Delta}}(\lambda \mid \mu^\lambda_\Delta), \tag{3.146}$$

as we can verify after recalling (3.99) in the present context.

Given m large so that $\Delta \cup \partial_{r_0}\Delta \subseteq \Lambda_m$, $\ell \geq 1$ odd and n so that $2n+1 = \ell(2m+1)$ we decompose Λ_n into ℓ^d disjoint translates of Λ_m: $\tilde{\Delta}_l = \Lambda_m + a(l)$, $l = 1, \ldots, \ell^d$, and we let $\Delta_l = \Delta + a(l)$, $G_l = \cup_{k=1}^l \tilde{\Delta}_k$ for $l = 1, \ldots, \ell^d$. Thus:

$$\mathcal{H}_{\Lambda_n}(\lambda \mid \mu^\lambda_{\Lambda_n}) = \sum_{i=1}^{\ell^d} [\mathcal{H}_{G_i}(\lambda \mid \mu^\lambda_{\Lambda_n}) - \mathcal{H}_{G_{i-1}}(\lambda \mid \mu^\lambda_{\Lambda_n})],$$

with the convention that $\mathcal{H}_{G_0}(\cdot \mid \cdot) = 0$, and by (3.146) we have

$$\begin{aligned}
&\mathcal{H}_{G_i}(\lambda \mid \mu^{\lambda}_{\Lambda_n}) - \mathcal{H}_{G_{i-1}}(\lambda \mid \mu^{\lambda}_{\Lambda_n})\\
&\geq \mathcal{H}_{G_i}(\lambda \mid \mu^{\lambda}_{\Lambda_n}) - \mathcal{H}_{G_i \setminus \Delta_i}(\lambda \mid \mu^{\lambda}_{\Lambda_n})\\
&= \mathcal{H}_{G_i}(\lambda \mid \mu^{\lambda}_{\Delta_i}) \geq \mathcal{H}_{\tilde{\Delta}_i}(\lambda \mid \mu^{\lambda}_{\Delta_i}) = \mathcal{H}_{\Lambda_m}(\lambda|\mu^{\lambda}_{\Delta})
\end{aligned}$$

using the translation invariance of λ and of the interaction. Consequently

$$\mathcal{H}_{\Lambda_m}(\lambda|\mu^{\lambda}_{\Delta}) \leq \frac{1}{\ell^d}\mathcal{H}_{\Lambda_n}(\lambda \mid \mu^{\lambda}_{\Lambda_n}),$$

which tends to zero as $\ell \to +\infty$, concluding the proof in the finite range case. □

For the full proof when $\Phi \in \mathbb{B}_\theta$ (and in a more general situation) see [247] or [134], Section 15.4; cf. also [243]; for continuous systems see [136].

Comment The convex duality given by (3.139) and (3.140), with $I^{\mathbb{Z}^d}(\alpha) = h(\alpha|\mu)$ if $\alpha \in \mathcal{M}_1^\theta(\mathcal{X})$ and $+\infty$ otherwise, suggests a more general form of the variational principle in $\mathcal{M}_1^\theta(\mathcal{X})$ associated with the definition of equilibrium states proposed by Ruelle. ($C(\mathcal{X})$ is too big for this; see [111].) When $\mu = \nu$ is the product measure as before, and $f = -A_\Phi$, with $\Phi \in \mathbb{B}_\theta$, Theorem 3.26 characterizes the points of maximum in (3.139) as the translation invariant DLR measures for Φ. Recall:

- if $\Phi \in \mathbb{B}_\theta$, A_Φ has bounded oscillations ($\sum_{i\in\mathbb{Z}^d} \sup_{\sigma_j=\sigma'_j, \forall j\neq i} |A_\Phi(\sigma) - A_\Phi(\sigma')| < +\infty$);
- the control on the boundary terms was needed in the verification of the l.d.p.

A very general situation has been considered in [198] and [243]: the class of observables is exactly the space of bounded oscillation functions. With a stronger form of asymptotic decoupling on the reference measure μ, suitable control of boundary terms may be introduced, leading to an extension of the Gibbs variational principle and a quite general construction on the relation thermodynamics/large deviations for lattice systems (see [243], Chapter 4).

3.4.2 Lanford theory. Equivalence of ensembles

In the following brief discussion we sometimes absorb β into Φ (i.e. $\beta = 1$) and write $F(\Phi)$ for the (negative) free energy: $F(\Phi) = \log \sharp S + \varphi(-A_\Phi|\nu)$, with ν the product measure as before and $\Phi \in \mathbb{B}_\theta$.

Large deviations for empirical averages (level 1)

Let us first recall the l.d.p. for the empirical magnetization W_n of the standard Ising model, discussed at the beginning of this section. Its validity follows at once from the contraction principle and Theorem 3.39, since the map $\alpha \in \mathcal{M}_1^\theta(\mathcal{X}) \mapsto$

$\langle\sigma_O\rangle_\alpha$ is continuous in our case (finite S) and W_n is the image of the empirical field $\tilde{R}_n$. As a consequence, the (lower semi-continuous) rate function is given by:

$$I_{\beta,h}(m) := \inf\{h(\alpha|\mu) : \alpha \in \mathcal{M}_1^\theta(\mathcal{X}), \langle\sigma_O\rangle_\alpha = m\}$$

($\inf_\emptyset = +\infty$). Since $\alpha \in \mathcal{M}_1^\theta(\mathcal{X}) \mapsto \langle\sigma_O\rangle_\alpha$ is affine, the convexity of $h(\cdot|\mu)$ implies that $I_{\beta,h}$ is convex. On the other hand, by the Varadhan lemma we see that $\varphi_{\beta,h}$ is the Fenchel–Legendre transform of $I_{\beta,h}$. Thus, by convex duality we obtain $I_{\beta,h} = \varphi^*_{\beta,h}$.

We see that $I_{\beta,h}(m) = 0$ if and only if there exists $\mu \in \mathcal{G}_\theta(\beta, h)$ with magnetization m; in particular $I_{\beta,0}(m) = 0$ if and only if $m \in [-m^*(\beta), m^*(\beta)]$, as mentioned earlier.

The contraction principle furnishes a 'restricted variational principle'. The representation as $\varphi^*_{\beta,h}$ in the above example recalls a 'general method' in the derivation of large deviation estimates: perturb the measure suitably so as to make the given situation 'typical' and then compute the cost of the perturbation, optimizing it. (In this example the representation suggests perturbation with an extra magnetic field.)

From the l.d.p. for the empirical field we have at our disposal a large class of macroscopic observables to which the previous argument applies. This is the key of the so-called Lanford theory, providing a relation to the thermodynamic formalism, as proposed in [189] and settled in [65, 118, 234]. More general formulations appear in [135, 198, 243].

For a brief description, let us first fix $\Phi \in \mathbb{B}_\theta$, $\beta = 1$, to which we apply the l.d.p. for the empirical fields (see Remark 3.37 too). Now let $\Psi \in \mathbb{B}_\theta$. Since the specific energy $\alpha \mapsto \langle A_\Psi, \alpha\rangle$ is continuous, the contraction principle provides a l.d.p. for the distributions of $\langle A_\Psi, R_n\rangle$, where R_n is the periodic empirical field in Λ_n (cf. remark following Theorem 3.39). The previous argument provides, for $B \in \mathcal{B}(\mathbb{R})$:

$$\begin{aligned}\liminf_{n\to+\infty} \inf_\tau \frac{1}{|\Lambda_n|} \log \mu^\tau_{\Phi,\Lambda_n}\{\sigma : \langle A_\Psi, R_n(\sigma)\rangle \in B\} &\geq -\inf_{x\in B^\circ} \bar{I}_{\Phi,\Psi}(x),\\ \limsup_{n\to+\infty} \sup_\tau \frac{1}{|\Lambda_n|} \log \mu^\tau_{\Phi,\Lambda_n}\{\sigma : \langle A_\Psi, R_n(\sigma)\rangle \in B\} &\leq -\inf_{x\in \bar{B}} \bar{I}_{\Phi,\Psi}(x),\end{aligned} \tag{3.147}$$

where, for $\mu \in \mathcal{G}_\theta(\Phi)$, $x \in \mathbb{R}$,

$$\bar{I}_{\Phi,\Psi}(x) = \inf\{h(\alpha|\mu) : \alpha \in \mathcal{M}_1^\theta(\mathcal{X}), \langle A_\Psi, \alpha\rangle = x\} = \sup_{\zeta\in\mathbb{R}}\{\zeta x - \varphi(\zeta A_\Psi|\mu)\}. \tag{3.148}$$

It follows that $\bar{I}_{\Phi,\Psi}(x) = 0$ if and only $\langle A_\Psi, \alpha\rangle = x$ for some $\alpha \in \mathcal{G}_\theta(\Phi)$. Also,

$$\mathcal{D}_{\bar{I}_{\Phi,\Psi}} := \{x : \bar{I}_{\Phi,\Psi}(x) < +\infty\} = \{\langle A_\Psi, \alpha\rangle : \alpha \in \mathcal{M}_1^\theta(\mathcal{X}), h(\alpha|\mu) < +\infty\},$$

which is a bounded interval, since A_Ψ is bounded. Since $h(\cdot|\mu)$ is lower semi-continuous on the compact space $\mathcal{M}_1^\theta(\mathcal{X})$, the infimum in (3.148) is attained whenever $\bar{I}_{\Phi,\Psi}(x) < +\infty$. Notice that $\varphi(\zeta A_\Psi|\mu) = F(\Phi - \zeta\Psi) - F(\Phi)$ in (3.148). Concerning the supremum in (3.148) being attained, we quote the following (see Lemma 4.9 in [135]).

Lemma 3.44 *Let $\Phi, \Psi \in \mathbb{B}_\theta$, $x, \zeta \in \mathbb{R}$ and $\mu \in \mathcal{G}_\theta(\Phi)$. We write $\bar{I} = \bar{I}_{\Phi,\Psi}$. The following are equivalent:*

(a) $x = \langle A_\Psi, \alpha\rangle$ *for some* $\alpha \in \mathcal{G}_\theta(\Phi - \zeta\Psi)$;
(b) $\zeta x - \varphi(\zeta A_\Psi|\mu) = \bar{I}(x)$;
(c) $x(\zeta' - \zeta) \le F(\Phi - \zeta'\Psi) - F(\Phi - \zeta\Psi)$ *for all* $\zeta' \in \mathbb{R}$;
(d) $\bar{I}(x) + \zeta y \le \bar{I}(x + y)$ *for each* $y \in \mathbb{R}$;
(e) $x \in \mathcal{D}_{\bar{I}}$, *and* $\emptyset \ne \{\alpha \in \mathcal{M}_1^\theta(\mathcal{X}) : \langle A_\Psi, \alpha\rangle = x, h(\alpha|\mu) = \bar{I}(x)\} \subseteq \mathcal{G}_\theta(\Phi - \zeta\Psi)$.

If $x \in \mathcal{D}_{\bar{I}}$ and $\zeta, \tilde{\zeta}$ such that (a)–(e) hold for both pairs (x, ζ) and $(x, \tilde{\zeta})$, then $\mathcal{G}(\Phi - \zeta\Psi) = \mathcal{G}(\Phi - \tilde{\zeta}\Psi)$.

Proof We omit the proof of the last statement. As for the equivalences, we easily verify that (b) ⇔ (c) and that (b) ⇒ (d). (e) ⇒ (a) is trivial. We now check that (a) ⇒ (b) and (d) ⇒ (e). For (a) ⇒ (b), taking any $\mu \in \mathcal{G}_\theta(\Phi)$, $\tilde{\mu} \in \mathcal{G}_\theta(\Phi - \zeta\Psi)$, from (3.144) and (3.145) we write, for any $\alpha \in \mathcal{M}_1^\theta(\mathcal{X})$:

$$h(\alpha|\mu) = h(\alpha|\nu) + \langle A_\Phi, \alpha\rangle + F(\Phi) - \log \sharp S,$$

$$h(\alpha|\tilde{\mu}) = h(\alpha|\nu) + \langle A_\Phi, \alpha\rangle - \zeta\langle A_\Psi, \alpha\rangle + F(\Phi - \zeta\Psi) - \log \sharp S$$

so that

$$h(\alpha|\mu) = h(\alpha|\tilde{\mu}) + \zeta\langle A_\Psi, \alpha\rangle - F(\Phi - \zeta\Psi) + F(\Phi). \tag{3.149}$$

Under (a) we have $h(\alpha|\tilde{\mu}) = 0$, and $\zeta x - F(\Phi - \zeta\Psi) + F(\Phi) = h(\alpha|\mu) \ge \bar{I}(x)$, implying (b), since the reverse inequality is trivial. Let us now check (d) ⇒ (e). Since $\inf_{y\in\mathbb{R}} \bar{I}(y) = 0$, we see right away that under (d) we have $x \in \mathcal{D}_{\bar{I}}$ and the set in (e) is non-empty. Let $\alpha \in \mathcal{M}_1^\theta(\mathcal{X})$, with $\langle A_\Psi, \alpha\rangle = x$, $h(\alpha|\mu) = \bar{I}(x)$, and let ζ be as in (*d*). Taking $\tilde{\mu} \in \mathcal{G}_\theta(\Phi - \zeta\Psi)$ let $y = \langle A_\psi, \tilde{\mu}\rangle - x$. From (d), and since $x + y = \langle A_\psi, \tilde{\mu}\rangle$, we have

$$h(\alpha|\mu) + \zeta\langle A_\Psi, \tilde{\mu} - \alpha\rangle = \bar{I}(x) + \zeta y \le \bar{I}(x + y) \le h(\tilde{\mu}|\mu).$$

Using (3.149) we get $h(\alpha|\tilde{\mu}) \le h(\tilde{\mu}|\tilde{\mu}) = 0$, and (the proof of) Theorem 3.26 implies that $\alpha \in \mathcal{G}_\theta(\Phi - \zeta\Psi)$, concluding the proof of the equivalences. □

Remark Items (c) and (d) have a geometric interpretation. It can be seen that if $x \in D^\circ_{I_{\Phi,\Psi}}$, the conditions are verified for some $\zeta \in \mathbb{R}$.

Notice that for $\sigma \in \mathcal{X}_{\Lambda_n}$ we have

$$\langle A_\Psi, R_n(\sigma)\rangle = \frac{1}{|\Lambda_n|} H^{\Psi,per}_{\Lambda_n}(\sigma), \tag{3.150}$$

where $H_\Lambda^{\Psi,per}(\sigma)$ is the Hamiltonian for the interaction Ψ with periodic boundary conditions. (For $\Psi \in \mathbb{B}_\theta$, we may as well consider Hamiltonians with fixed (or free) boundary conditions not affecting the previous l.d.p., cf. Remark 3.27.) If $\Phi = 0$ and Ψ has range 0 we are back to Cramér theorem for simple random variables.

One may as well consider the joint distribution of finitely many such observables, $A_{\Psi^{(1)}}, \ldots, A_{\Psi^{(k)}}$. The convex analysis is more intricate, as we could guess from Section 3.1. Lemma 3.44 extends; on the relative interior $\mathrm{ri}(\mathcal{D}_{\bar{I}_{\Phi,\Psi}})$ we still have the same good geometric behaviour (see e.g. [135, 243]). We do not pursue this, but look at a particular example in the next subsection.

Infinite dimensional situations may also be considered, as in [135], since $\mathbb{B}_\theta$ is a Banach space with the norm $|||\cdot|||$, cf. Remark 3.22. If $\mathbb{V}$ is a closed subspace, and $\mathbb{V}^*$ its dual, the previous observable is replaced by the restriction to $\mathbb{V}$, of the specific energy function $u_{\mathbb{V}}\colon \mathcal{M}_1^\theta(\mathcal{X}) \to \mathbb{V}^*$, which is continuous if $\mathbb{V}^*$ has the standard weak* topology. In other words, let $u_{\mathbb{V}}(\alpha)(\Psi) = \langle A_\Psi, \alpha\rangle$, for $\Psi \in \mathbb{V}$. The l.d.p. for the laws of $u_{\mathbb{V}}(R_n)$ under the measures $\mu^\tau_{\Phi,\Lambda_n}$ is obtained from Theorem 3.39 (see Remark 3.37, with $I_\Phi(\alpha) = h(\alpha|\mu)$, $\mu \in \mathcal{G}_\theta(\Phi)$). It holds uniformly in τ, and the rate function $\bar{I}_{\Phi,\mathbb{V}}$ is represented as a Fenchel–Legendre transform of $\Psi \in \mathbb{V} \mapsto F(\Phi - \Psi) - F(\Phi)$:

$$\begin{aligned}\bar{I}_{\Phi,\mathbb{V}}(x) &= \inf\{h(\alpha|\mu)\colon \alpha \in \mathcal{M}_1^\theta(\mathcal{X}), u_{\mathbb{V}}(\alpha) = x\} \\ &= \sup_{\Psi\in\mathbb{V}} (x(\Psi) - (F(\Phi-\Psi) - F(\Phi))), \quad x \in \mathbb{V}^*.\end{aligned}$$

Such considerations provide an interpretation to tangent hyperplanes to the convex map $\Psi \in \mathbb{V} \mapsto F(\Phi - \Psi)$. For the full proof see [135].

The example $\mathbb{V} = \{\Psi \in \mathbb{B}_\theta : \Psi_\Delta = 0,\ \text{unless } |\Delta| = 1\}$ appears in the analysis of empirical measures, discussed below. (Since we have S finite, this $\mathbb{V}$ is finite-dimensional, but it works more generally.)

Large deviations for empirical measures (level 2)

The map $\pi_0 = \pi_{\Lambda_0}\colon \mathcal{M}_1(\mathcal{X}) \to \mathcal{M}_1(S)$, which associates with each $\alpha \in \mathcal{M}_1(\mathcal{X})$ its marginal at the origin, is continuous. When $\mu \in \mathcal{M}_1^\theta(\mathcal{X})$ verifies the assumption of Theorem 3.39, the contraction principle (Theorem 1.19) yields the following expression for $I^{(0)}$, the rate function for the laws of the empirical measures $L_n(\sigma) = \pi_0(R_n(\sigma)) = |\Lambda_n|^{-1}\sum_{i\in\Lambda_n}\delta_{\sigma_i}$ under μ, cf. Corollary 3.41:[8]

$$I^{(0)}(\eta) = \inf\{h(\alpha|\mu)\colon \alpha \in \mathcal{M}_1^\theta(\mathcal{X}),\ \pi_0\alpha = \eta\}.$$

(As before, the infimum is attained.) A detailed description of this variational problem was provided by Föllmer and Orey in [118], when $\mu \in \mathcal{G}_\theta(\Phi)$ for some

[8] Notational mismatch: $I^{(0)}$ here corresponds to $I^{(1)}$ in Section 1.3, convenient in Corollary 3.41.

$\Phi \in \mathbb{B}_\theta$. On the other hand, recall that if $g\colon S \to \mathbb{R}$ and

$$\varphi^{(0)}(g|\mu) = \varphi(g \circ \pi_0 \mid \mu) = \lim_{n\to+\infty} \frac{1}{|\Lambda_n|} \log \int_{\mathcal{X}} \exp\left(\sum_{i\in\Lambda_n} g(\sigma_i)\right) d\mu$$

then, by the Varadhan lemma,

$$\varphi^{(0)}(g|\mu) = \sup_{\eta\in\mathcal{M}_1(S)} (\langle g, \eta\rangle - I^{(0)}(\eta)).$$

By duality, since $I^{(0)}$ is convex and lower semi-continuous:

$$I^{(0)}(\eta) = \sup_{g\colon S\to\mathbb{R}} (\langle g, \eta\rangle - \varphi^{(0)}(g|\mu)). \tag{3.151}$$

(Letting $r = \sharp S$, g and η are elements of $\mathbb{R}^r$, η varying in Θ_r, as in Section 3.2. $I^{(0)}(\eta) = +\infty$, if $\eta \notin \Theta_r$. Compare with (3.9) and (3.10).)

Given $g\colon S \to \mathbb{R}$ we perturb Φ with a self interaction: let $\Phi^g_{\{i\}}(\sigma) = \Phi_{\{i\}}(\sigma) - g(\sigma_i)$ for each $i \in \mathbb{Z}^d$, $\Phi^g_\Delta = \Phi_\Delta$ otherwise. It is easy to check that $F(\Phi^g) - F(\Phi) = \varphi^{(0)}(g|\mu)$, and from (3.145) we have that if $\mu_g \in \mathcal{G}_\theta(\Phi^g)$, and $\alpha \in \mathcal{M}_1^\theta(\mathcal{X})$, then:

$$h(\alpha|\mu) = h(\alpha|\mu_g) + \langle g, \pi_0\alpha\rangle - \varphi^{(0)}(g|\mu). \tag{3.152}$$

In particular, $h(\mu_g|\mu) = \langle g, \pi_0\mu_g\rangle - \varphi^{(0)}(g|\mu)$ if $\mu_g \in \mathcal{G}_\theta(\Phi^g)$.

Remark. The above definition still makes sense if $g(x) = -\infty$ for some (but not all) $x \in S$; $\mathcal{G}_\theta(\Phi^g) \neq \emptyset$, and clearly $\pi_0\mu_g\{x\colon g(x) = -\infty\} = 0$. Equation (3.152) still holds, the terms being finite if $\pi_0\alpha\{x\colon g(x) = -\infty\} = 0$.

As suggested from (3.151) and (3.152) one has (see [118], Theorem 5.2) the following.

(a) If $\eta \in \mathcal{M}_1(S)$ and $I^{(0)}(\eta) < +\infty$, there exists $g\colon S \to [-\infty, +\infty)$, with $\eta\{x\colon g(x) = -\infty\} = 0$, and $I^{(0)}(\eta) = \langle g, \eta\rangle - \varphi^{(0)}(g|\mu)$.

(b) If $\alpha \in \mathcal{M}_1^\theta(\mathcal{X})$ satisfies $\pi_0\alpha = \eta$ and $h(\alpha|\mu) = I^{(0)}(\eta)$, then indeed $\alpha \in \mathcal{G}(\Phi^g)$. Conversely if $\alpha \in \mathcal{G}_\theta(\Phi^g)$ and $\pi_0\alpha = \eta$, we have $h(\alpha|\mu) = I^{(0)}(\eta)$.

Sketch of proof Subtracting a constant reduces to $g \in (-\infty, 0]^r$ in (3.151); one may extract an approximating sequence, which converges to some $g \in [-\infty, 0]^r$; thus $I^{(0)}(\eta) \le \langle g, \eta\rangle - \varphi^{(0)}(g|\mu)$, and $\eta\{x\colon g(x) = -\infty\} = 0$. To get (a) and the first statement in (b), recall (3.152), and that $h(\alpha|\mu_g) \ge 0$ with equality if and only if $\alpha \in \mathcal{G}_\theta(\Phi^g)$. The second statement in (b) follows from (3.152) with $\alpha = \mu_g$.

When $\Phi = 0$ we recover the Sanov theorem for finite alphabets, seen in Section 1.3; when Φ is a n.n. interaction, μ represents a (stationary) Markov random field and we have the spatial analogue of the analysis in Section 3.2 for finite Markov chains.

Boltzmann principle of equivalence of ensembles

In the present context, the product measure $\nu_\Lambda = \nu_O^\Lambda$ (ν_O uniform on S) plays the role of the Liouville measure on $\mathcal{X}_\Lambda$. Write $\nu = \nu_{\mathbb{Z}^d}$. Given $\Psi \in \mathbb{B}_\theta$ and Λ_n as before, an approximate microcanonical distribution on $\mathcal{X}_{\Lambda_n}$, at the specific energy u, may be given by cf. (3.150):

$$\tilde{\mu}_{\Lambda_n,u,\delta}(\cdot) := \nu_{\Lambda_n}\left\{\cdot \mid \frac{1}{|\Lambda_n|} H_{\Lambda_n}^{\Psi,per} \in [u-\delta, u+\delta]\right\}.$$

A natural description of microcanonical/grand canonical equivalence in the absence of phase transitions could be: if $\beta = \beta(u)$ is such that $\langle A_\Psi, \mu\rangle = u$ and $\{\mu\} = \mathcal{G}_\theta(\beta\Psi)$ then the measures $\tilde{\mu}_{\Lambda_n,u,\delta}$ converge to μ, as $n \to +\infty$ and then $\delta \downarrow 0$. For non-trivial[9] Ψ, it can be seen (with the help of Theorem 3.26) that if such β exists, it is unique. The question on the existence of such β is treated by Lemma 3.44, taking $\Phi = 0$ there; in particular, the answer is positive if $u \in \mathcal{D}^\circ_{\bar{I}_\Psi}$, where $\bar{I}_\Psi = \bar{I}_{0,\Psi}$ in the notation there.

Remark In the present formulation $\beta \in \mathbb{R}$; the inverse temperature is $|\beta|$, with interaction Φ or $-\Phi$ according to the sign of β.

Similar behaviour is expected for microcanonical type measures with 'thick' energy shells, before sending δ to 0; β depends on the shell.

Allowing for phase transitions, this type of argument gives that all the limiting points of these microcanonical measures belong to $\mathcal{G}_\theta(\beta\Psi)$, with β as above; this inclusion might be strict as for the two-dimensional Ising model (cf. Example 3.2 in [135]).

To give a brief indication of the role of the l.d.p. in this discussion we state the following theorem, taken from [135].

Theorem 3.45 *Let $\Psi \in \mathbb{B}_\theta$, and $C \subseteq \mathbb{R}$ an interval such that $C^\circ \cap \mathcal{D}_{\bar{I}_\Psi} \neq \emptyset$, and that $C_{min} \cap \mathcal{D}^\circ_{\bar{I}_\Psi} \neq \emptyset$, where*

$$C_{min} := \{x \in \overline{C} \colon \bar{I}_\Psi(x) = \inf_{y \in \overline{C}} \bar{I}_\Psi(y)\}.$$

There exists $\beta \in \mathbb{R}$ so that the (non-empty) set of accumulation points of

$$\tilde{\mu}_{\Lambda_n,C} := \nu_{\Lambda_n}\left\{\cdot \mid \frac{1}{|\Lambda_n|} H_{\Lambda_n}^{\Psi,per} \in C\right\} \tag{3.153}$$

is contained in $\mathcal{G}_\theta(\beta\Phi) \cap \{\alpha \in \mathcal{M}_1^\theta(\mathcal{X}) \colon \langle A_\Phi, \alpha\rangle \in C_{min}\}$. For any such β, we have $-\beta x - \varphi(-\beta A_\Psi|\nu) = \bar{I}_\Psi(x)$ in (3.148) for any $x \in C_{min}$.

[9] Non-trivial here means that Ψ is not equivalent to 0, i.e. $\langle A_\Psi, \alpha\rangle \neq 0$ for some $\alpha \in \mathcal{M}_1^\theta(\mathcal{X})$.

Proof (Sketch) For simplicity we consider only the case that $C_{min} = \{u\}$ with $u \in C^\circ \cap \mathcal{D}^\circ_{\bar{I}_\Psi}$. Recalling (3.150), the event on which we condition is

$$\mathcal{Y}_n = \mathcal{Y}_n^C := \{\sigma \in \mathcal{X}_{\Lambda_n} : \langle A_\Psi, R_n(\sigma)\rangle \in C\},$$

and from the lower bound in (3.147), for $\Phi = 0$:

$$\liminf_{n\to+\infty} \frac{1}{|\Lambda_n|} \log \nu_{\Lambda_n}(\mathcal{Y}_n) \geq -\inf_{x \in C^\circ} \bar{I}_\Psi(x) = -\bar{I}_\Psi(u) > -\infty. \tag{3.154}$$

In particular, $\nu_{\Lambda_n}(\mathcal{Y}_n) > 0$, and $\tilde{\mu}_{\Lambda_n,C}$ is well defined for large n. (The existence of accumulation points is trivial in our case due to compactness.) Also, from the definition of periodic field R_n, if $f : \mathcal{X}_{\Lambda_n} \to \mathbb{R}$ we see that

$$\int_{\mathcal{Y}_n} f \, d\nu_{\Lambda_n} = \int_{\mathcal{Y}_n} \langle f, R_n\rangle d\nu_{\Lambda_n}.$$

Thus, it suffices to prove that there exists β, so that if G is open and $\mathcal{G}_\theta(\beta\Phi) \cap \{\alpha \in \mathcal{M}_1^\theta(\mathcal{X}) : \langle A_\Phi, \alpha\rangle \in C_{min}\} \subseteq G$, then

$$\limsup_{n\to+\infty} \nu(R_n \notin G \mid \langle A_\Psi, R_n\rangle \in C) = 0, \tag{3.155}$$

where, for brevity, $R_n(\sigma) = R_n(\pi_{\Lambda_n}\sigma)$. Now, if G is open and $\{\alpha \in \mathcal{M}_1^\theta(\mathcal{X}) : \langle A_\Phi, \alpha\rangle \in C_{min},\ h(\alpha|\nu) = \bar{I}_\Psi(u)\} \subseteq G$, Theorem 3.39 tells us that

$$\limsup_{n\to+\infty} \frac{1}{|\Lambda_n|} \log \nu(R_n \notin G, \langle A_\Psi, R_n\rangle \in C)$$
$$\leq -\inf\{h(\alpha|\nu) : \alpha \in \mathcal{M}_1^\theta(\mathcal{X}) \setminus G, \langle A_\Psi, \alpha\rangle \in \overline{C}\},$$

which is strictly smaller than $-\bar{I}_\Psi(u)$. Together with (3.154), this yields (3.155) for such G. It remains to prove the existence of $\beta \in \mathbb{R}$ so that

$$\{\alpha \in \mathcal{M}_1^\theta(\mathcal{X}) : \langle A_\Phi, \alpha\rangle \in C_{min},\ h(\alpha|\nu) = \bar{I}_\Psi(u)\}$$
$$= \mathcal{G}_\theta(\beta\Phi) \cap \{\alpha \in \mathcal{M}_1^\theta(\mathcal{X}) : \langle A_\Phi, \alpha\rangle \in C_{min}\},$$

which follows if for some $\beta \in \mathbb{R}$,

$$\{\alpha \in \mathcal{M}_1^\theta(\mathcal{X}) : \langle A_\Phi, \alpha\rangle \in C_{min},\ h(\alpha|\nu) = \bar{I}_\Psi(u)\} \subseteq \mathcal{G}_\theta(\beta\Psi).$$

When $C_{min} = \{u\}$ this, as well as the final statement, follows at once from Lemma 3.44 with $\Phi = 0$, since $u \in \mathcal{D}^\circ_{\bar{I}_\Psi}$. (It is here that we used this extra assumption.) For the full proof see [135], p. 1871. It uses the last statement in Lemma 3.44, whose proof we omitted. □

Remarks (a) Less is required for the convergence of $\tilde{\mu}_{n,u,\delta}$, as $n \to +\infty$ and then $\delta \downarrow 0$, cf. [88]. This is so, because for the previous upper estimate we have

$$-\inf\{h(\alpha|\nu) : \alpha \in \mathcal{M}_1^\theta(\mathcal{X}) \setminus G, \langle A_\Psi, \alpha\rangle \in [u-\delta, u+\delta]\},$$

which tends to $-\inf\{h(\alpha|\nu) : \alpha \in \mathcal{M}_1^\theta(\mathcal{X}) \setminus G, \langle A_\Psi, \alpha\rangle = u\}$, as $\delta \downarrow 0$. The rest is the same, cf. Lemma 3.44. (Conversely, this result may be deduced from Theorem 3.45, see [135].) For a further analysis for the two-dimensional Ising model in the phase transition region see [88].

(b) Theorem 3.45 relates conditioning (microcanonical) and tilting (grand canonical). For a general treatment, see [198], and [243]. If we do not take the Hamiltonian with *periodic* boundary conditions, the $\tilde{\mu}_{\Lambda_n,C}$ should be replaced by the average of $\theta_i \tilde{\mu}_{\Lambda_n,C}$, $i \in \Lambda_n$ (see the given references).

(c) Replacing ν_{Λ_n} by a periodic Gibbs measure for an interaction $\Phi \in \mathbb{B}_\theta$ one is able to study canonical/grand canonical equivalences.

4

Metastability. General description. Curie–Weiss model. Contact process

4.1 The van der Waals–Maxwell theory

Metastability is a relevant phenomenon for thermodynamic systems close to a first order phase transition. Examples are supercooled vapours and liquids, supersaturated vapours and solutions, as well as ferromagnets in the part of the hysteresis loop where the magnetization is opposite to the external magnetic field. A metastable state occurs when some thermodynamic parameter such as the temperature, pressure or magnetic field is changed from a value giving rise to a stable state with a unique phase, say X, to one for which at least part of the system should be in some new equilibrium phase Y. Then, in particular experimental situations, instead of undergoing the phase transition, the system goes over continuously into a 'false' equilibrium state with a unique phase X', far from Y but actually close to the initial equilibrium phase X. It is this apparent equilibrium situation that is called a 'metastable state'. Its properties are very similar to those of the stable equilibrium state; for example for a supersaturated vapour one can determine the pressure experimentally as a function of the temperature and the specific volume. We speak of the 'metastable branch' of the isothermal curve.

The distinguishing feature of metastability is that, eventually, either via an external perturbation or via a spontaneous fluctuation, a nucleus of the new phase appears, starting an irreversible process which leads to the stable equilibrium state Y, where the phase transition has taken place.

Let us first illustrate metastability in more detail with the example of liquid–vapour phase transition; subsequently we shall describe magnetic systems.

Consider an isothermal compression of a gas at a temperature T below its critical temperature T_c. (see Figure 4.1). Suppose we start with a specific volume v slightly larger than the value v_g corresponding to the condensation point at temperature T (the specific volume v is given by $v = V/N$, where V is the volume and N is the number of molecules). If our sample is reasonably free of impurities

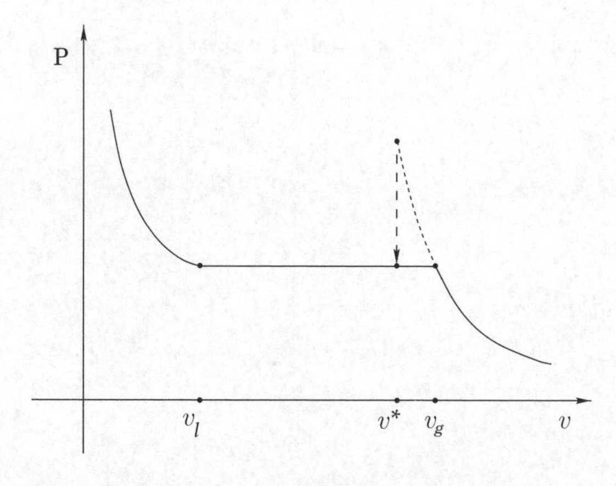

Figure 4.1

and if we proceed in the compression slowly by avoiding significant density gradients, we can obtain a pure gaseous phase, with $v < v_g$, commonly called a 'supersaturated vapour'.

It is observed experimentally that, by continuing in the isothermal compression, we end up with a value v^* of the specific volume for which the gas becomes unstable; then we start observing the appearance and growth of the new liquid phase until the segregation between liquid and gas is completed, and a fraction $\alpha = (1/v^* - 1/v_g)(1/v_l - 1/v_g)^{-1}$ of the fluid is finally in the liquid phase with specific volume v_l.

Now, in the final stable equilibrium, the liquid and gaseous phases are segregated and coexist at the saturated vapour pressure.

For $v \in (v^*, v_g)$ the properties of the (supersaturated) vapour are quite similar to those of the normal gas with $v > v_g$. Apparently we have an equilibrium pure gaseous phase. The main difference is that even a relatively small disturbance can induce the 'nucleation' of the liquid phase. An impurity or any external perturbation inducing a localized density change can produce a 'nucleus of condensation' around which the liquid phase starts growing by absorbing molecules from the surrounding supersaturated vapour. If v is very near to (but still larger than) v^*, even a spontaneous thermal fluctuation can be responsible for the formation of a condensation nucleus. In this last case we speak of 'homogeneous nucleation' since these density fluctuations take place everywhere in the bulk. Again the final stable state corresponds to a fraction $\alpha = (1/v - 1/v_g)(1/v_l - 1/v_g)^{-1}$ of liquid coexisting with the saturated gas. This behaviour is typical of a conservative evolution which preserves the number of molecules.

Another relevant example of phase transition giving rise to metastability phenomena is that which takes place for ferromagnetic systems at temperature T

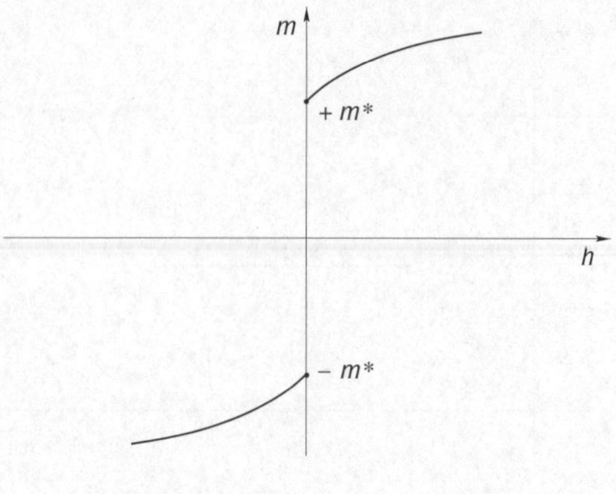

Figure 4.2

below the Curie point T_c. As we shall see, this case is naturally described in a non-conservative context.

Figure 4.2 shows, for $T < T_c$, a typical graph of magnetization m versus external magnetic field h for an ideal ferromagnet. We observe the phenomenon of spontaneous magnetization:

$$\lim_{h \to 0^+} m(h) > 0, \qquad \lim_{h \to 0^-} m(h) < 0. \tag{4.1}$$

For $T > T_c$ there is a *paramagnetic* behaviour:

$$\lim_{h \to 0} m(h) = 0. \tag{4.2}$$

As we have seen in Section 3.3 the existence of a spontaneous magnetization can be interpreted as coexistence, for the same value of the magnetic field $h = 0$, of two different phases with opposite magnetization. What is observed for a real ferromagnet is the *hysteresis loop* shown in Figure 4.3. For $T < T_c$, suppose we change the value of the magnetic field from small negative to zero, and then to small positive values. Our magnet persists, apparently in equilibrium, in a state with magnetization opposite to the external magnetic field h. This situation corresponds to a metastable state since the 'true' equilibrium state would have magnetization parallel to the field. We can continue increasing h up to a limiting value h^*, called the *coercive field*, at which we observe a decay to the stable state with positive magnetization. In a similar way we can cover the rest of the loop. The static and dynamic properties of the system along the metastable arc are similar to those we have already described in the case of fluids. The main difference here is that we do not have the global constraint that we had before, of fixed total number of molecules; now the corresponding quantity, i.e. the total number of 'up' spins, is not at all fixed. From the point of view of equilibrium statistical mechanics the situation described for a fluid corresponds to the *canonical*

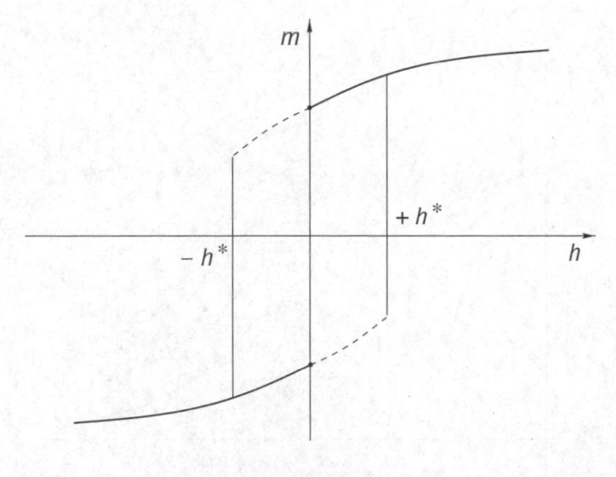

Figure 4.3

ensemble whereas the situation for a magnet corresponds to the *grand canonical ensemble*.

The specific features that a theory of metastability have to explain are both static and dynamic in nature. We would like to find the 'equation of state' of metastable states. For example in the case of liquid–vapour phase transition we would like to determine the value of $v^* = v^*(\mathrm{T})$ and the arc of the isotherm describing the supersaturated vapour between v^* and v_g. On the other hand, the most relevant dynamical quantity is the 'lifetime' of the metastable state. Its correct definition is also part of the problem. We expect the lifetime to become very small as v tends to v^*, whereas we expect that it tends to infinity as v goes to v_g.

A first possible explanation of metastability can be found in the framework of the classical van der Waals–Maxwell theory.

The equation of state of real gases introduced by van der Waals is based on semiphenomenological grounds. His approach is neither purely thermodynamical nor is it based on a genuine microscopical point of view. The van der Waals equation of state for one mole of fluid is:

$$\left(\mathrm{P} + \frac{a}{v^2}\right)(v - b) = k\mathrm{T}, \tag{4.3}$$

where P is the pressure, T is the absolute temperature, k is Boltzmann's constant and $a \geq 0$, $b \geq 0$ are parameters depending on the particular fluid.

This equation of state is simply obtained from the 'perfect gas law' $\mathrm{P}v = k\mathrm{T}$, by introducing the corrective terms a/v^2 and $-b$. The first term takes into account the intermolecular attraction, and the term $-b$ takes into account the intrinsic volume occupied by molecules, seen as extended instead of point-like objects.

For a more detailed treatment of the van der Waals equation we refer, for instance, to [163].

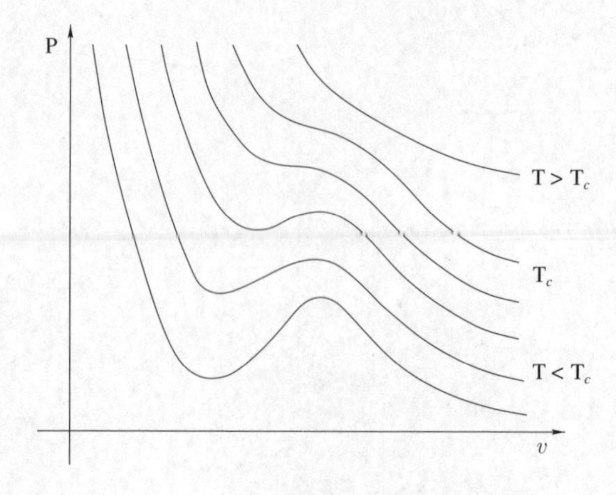

Figure 4.4

In Figure 4.4 we draw some typical van der Waals isotherms. For $\mathrm{T} = \mathrm{T}_c := 8a/27kb$, the graph of the pressure P as a function of v has a horizontal inflection point; moreover, for every $\mathrm{T} \geq \mathrm{T}_c$ the pressure is monotonic in v, whereas for $\mathrm{T} < \mathrm{T}_c$ there is an interval of non-monotonicity corresponding to negative compressibility. This unphysical feature of the van der Waals equation can be corrected by the help of the *Maxwell's equal area rule*. Maxwell's idea is that below T_c the hypothesis of spatial homogeneity, implicit in the heuristics which is at the basis of van der Waals equation, has to be left behind and the more realistic hypothesis of phase segregation has to be included in the theory. The new Maxwell heuristics goes as follows: as a consequence of the second principle of thermodynamics we have that the Helmholtz free energy per unit particle, $F = F(v, \mathrm{T})$, is given by:

$$F(v, \mathrm{T}) = -\int_{isotherm} \mathrm{P}(v, \mathrm{T}) dv \Leftrightarrow \mathrm{P}(v, \mathrm{T}) = -\frac{\partial F(v, \mathrm{T})}{\partial v} \tag{4.4}$$

$$\Rightarrow F(v, \mathrm{T}) = -k\mathrm{T} \log(v - b) - \frac{a}{v}. \tag{4.5}$$

In Figure 4.5, for a given $\mathrm{T} < \mathrm{T}_c$ we draw simultaneously the graphs of $\mathrm{P}(v)$ and $F(v)$. The non-monotonicity of $\mathrm{P}(v)$ has, as a counterpart, the existence of a non-convexity interval for $F(v)$. Consider the *convex envelope* of F, denoted by $CE(F)$, defined as the maximal convex function which is everywhere less than or equal to F. In our case $CE(F)$ is obtained from F by means of the 'common tangent' construction as shown in Figure 4.5. The idea is that the straight segment between the points A and B represents coexistence of two phases. Indeed, the states A and B have the same temperature (as they lie on the same isotherm) and pressure (by (4.4), as they have a common tangent); so they can describe coexisting phases with different specific volumes: the liquid and the vapour. The

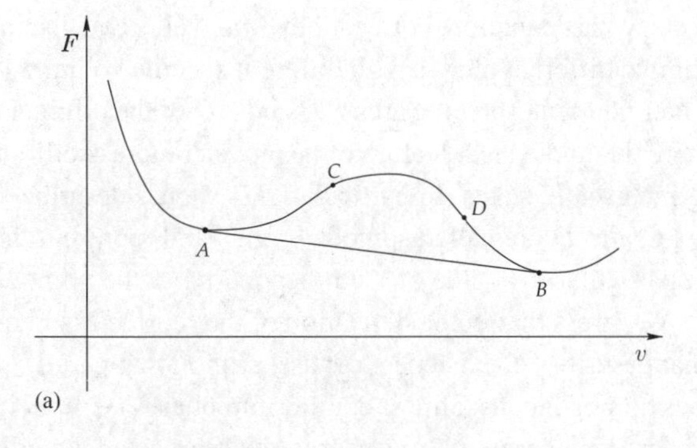

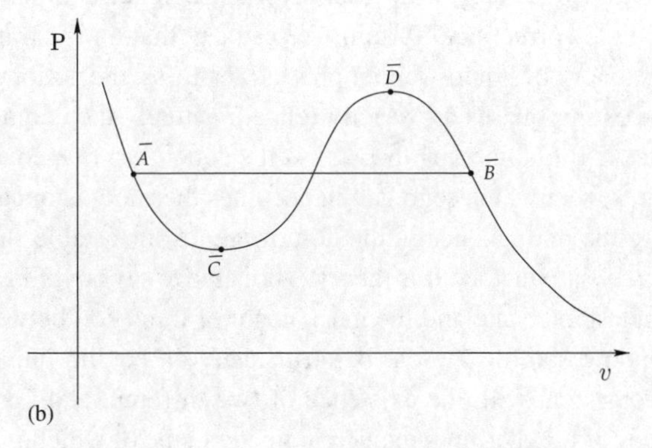

Figure 4.5

points in the straight segment describe a situation in which part of the system is in state A and part in state B, since in this case the free energy is a convex combination of the corresponding free energies. In the graph of $\mathrm{P}(v)$, the segment between A and B corresponds to the horizontal segment between $\bar{A}$ and $\bar{B}$. The common tangent condition

$$\frac{F(v_B) - F(v_A)}{v_B - v_A} = \left.\frac{\partial F(v)}{\partial v}\right|_{v=v_A} = \left.\frac{\partial F(v)}{\partial v}\right|_{v=v_B}, \tag{4.6}$$

becomes

$$\bar{\mathrm{P}}(v_B - v_A) = \int_{v_A}^{v_B} \mathrm{P}(v)dv, \tag{4.7}$$

where $\bar{\mathrm{P}} := -\partial F(v)/\partial v|_{v=v_{\bar{A}}} = -\partial F(v)/\partial v|_{v=v_{\bar{B}}}$ is the common value of the pressure at the points of the horizontal segment joining $\bar{A}$ to $\bar{B}$. The meaning of (4.7) is that the segment $\bar{A}\bar{B}$ cuts the van der Waals isotherm in such a way that the areas of the two regions $\Gamma_{\bar{A}}, \Gamma_{\bar{B}}$ between the segment $\bar{A}\bar{B}$ and the van der Waals isotherm are equal.

Since van der Waals equation, corrected by Maxwell's equal area rule, predicts coexistence of two different phases with different specific volumes (the liquid and the gas), it is natural to interpret the arcs $\bar{B}\bar{D}$ and $\bar{A}\bar{C}$ of the original van der Waals isotherm (where the unphysical feature of negative compressibility is still absent) as describing metastable states. In particular $\bar{B}\bar{D}$ should describe supersaturated vapour. Points $\bar{C}$ and $\bar{D}$ are called spinodal points and represent the threshold of instability; in particular the value of v corresponding to the point $\bar{D}$ can be identified with v^*. We stress that in the derivation of the 'correct' isotherm containing the plateau that describes coexistence, the assumption that part of 'van der Waals loop' describes homogeneous, almost equilibrium phases is implicit. These, however, have higher free energy with respect to the segregated situation which then turns out to be the correct equilibrium. We can say that, in the framework of the van der Waals theory of liquid–vapour phase transitions, the notions of metastable states and coexisting phases are strictly related: actually they come out together. The endpoints $\bar{A}$, $\bar{B}$ obtained by the Maxwell's equal area rule emerge in the theory as the points from which segregation becomes thermodynamically convenient with respect to the permanence in the homogeneous metastable states. Although qualitatively very satisfactory, this theory is not able to say anything about the lifetime of the metastable state and the mechanism of transition between metastable and corresponding stable states. It is intrinsically an equilibrium theory which, however, is consistent with the existence of two different equilibrium states for the same values of the thermodynamic parameters, one of which is the true stable equilibrium. An infinite free energy barrier separates the two equilibria, so that the infinite lifetime not only is not excluded by this theory but, rather, it appears as a natural conjecture in the line of thought leading to the van der Waals–Maxwell theory. We do not expect that this feature persist in more complete and realistic approaches.

We want to mention that the metastable branch $\bar{B}\bar{D}$ is obtained as an analytic continuation of the isotherm beyond the condensation point $\bar{B}$; similarly for the branch $\bar{A}\bar{C}$. This feature also depends on the peculiarity of van der Waals theory.

As we said before, the van der Waals theory is semiphenomenological in nature; however it is possible to give a partial justification of this theory by means of a very simple model, in the framework of equilibrium statistical mechanics.

Consider N particles in a volume V. Assume that the 'allowed volume' is $\tilde{V} = V - bN, 0 < b < v \equiv V/N$. This does not mean that we introduce an authentic hard core repulsion between particles; rather, it is a caricature of the true model which happens to be very difficult to describe. The subtraction of the term bN is meant to take into account in a very simple way, according to a mean field philosophy, the total volume occupied by the molecules.

The interaction between any pair of particles is not decaying with the interparticle distance: it is given by the *constant value* $-2a/V, a > 0$ (the denominator V is needed to get an extensive behaviour for the total energy).

The canonical free energy is given by

$$F_N(v, \mathrm{T}) = -k\mathrm{T}\frac{1}{N}\log Z_N^c(\mathrm{T}, v), \tag{4.8}$$

where k is the Boltzmann constant, T is the absolute temperature, and $Z_N^c(\mathrm{T}, v)$ is the canonical partition function given by:

$$Z_N^c(\mathrm{T}, v) = \frac{1}{N!}\int_{\tilde{V}^N} \exp\left(-\frac{H_N}{k\mathrm{T}}\right) dx_1 \dots dx_N, \tag{4.9}$$

$$H_N = -\frac{a}{V}(N^2 - N). \tag{4.10}$$

From (4.8)–(4.10) we get:

$$F_N(v, \mathrm{T}) = -\frac{a}{v} - k\mathrm{T}\log(v - b) + \mathrm{const} + o_N(1),$$

namely the previous expression (4.5) (except for the irrelevant constant and an infinitesimal term $o_N(1) \to 0$ as $N \to +\infty$).

Thus, from a microscopic point of view, the van der Waals equation can be regarded as a *mean field theory*; the limiting assumption of 'mean field interaction' amounts to supposing that every molecule interacts with the same intensity with all other molecules or, almost equivalently, that it feels the mean field generated by them. The absence of spatial dependence and consequently the fact that the range of the interaction coincides with the size of the box containing the fluid is responsible for the non-convexity of the free energy. Indeed, as we have seen in Section 3.3, for 'short range interactions' it can be proven rigorously, in the framework of equilibrium statistical mechanics, that the free energy is a convex function of the specific volume.

We shall see later that the extreme hypothesis of constant potential can be corrected via the introduction of the so-called Kac potentials (see (4.20)).

4.1.1 The Curie–Weiss theory

To describe ferromagnetic systems below the Curie temperature T_c, we may use an approach similar to that of the van der Waals theory for fluids; again metastable states are intrinsically connected with the description of the phase transition. To clarify this point let us deduce the Curie–Weiss theory of ferromagnets in the framework of equilibrium statistical mechanics of lattice spin systems.

Consider a system of N spins: $(\sigma_i)_{i=1,\dots,N}$ where $\sigma_i \in \{-1, +1\}$. A *configuration* σ is an element of $\mathcal{X}_N := \{-1, +1\}^N$. The energy associated with a configuration σ is:

$$H_N(\sigma) := -\frac{J}{2N}\sum_{i,j=1}^{N} \sigma_i\sigma_j - h\sum_{i=1}^{N}\sigma_i, \tag{4.11}$$

which may be written as

$$H_N(\sigma) = -N\left(\frac{J}{2}(m_N(\sigma))^2 + h\,m_N(\sigma)\right),$$

where $m_N(\sigma) = N^{-1}\sum_{i=1}^N \sigma_i$. In particular, $H_N(\sigma)$ depends on σ only through $m_N(\sigma)$, implying the extreme simplicity of the model. We define the (grand canonical) free energy:

$$F_N(\beta, h) = -\frac{1}{\beta N}\log Z_N(\beta, h), \tag{4.12}$$

where $\beta = 1/k\mathrm{T}$ (T denotes temperature, k the Boltzmann constant) and $Z_N(\beta, h)$ is the (grand canonical) partition function

$$Z_N(\beta, h) := \sum_{\sigma\in\mathcal{X}_N} e^{-\beta H_N(\sigma)} = \sum_{\sigma\in\mathcal{X}_N} e^{\beta N\left(\frac{J}{2}(m_N(\sigma))^2 + h\,m_N(\sigma)\right)}. \tag{4.13}$$

We want to compute the limit of $F_N(\beta, h)$ as $N \to +\infty$. Let ϖ_N denote the distribution of $m_N(\sigma)$, with σ uniformly distributed in $\mathcal{X}_N$: $\varpi_N(m) = 2^{-N}\binom{N}{N((m+1)/2)}$, $m \in \mathcal{Y}_N$, where

$$\mathcal{Y}_N := \{-1, -1+2/N, -1+4/N, \dots, 1-4/N, 1-2/N, 1\}. \tag{4.14}$$

Then (4.13) reads as:

$$2^{-N} Z_N(\beta, h) = \int_{\mathcal{Y}_N} e^{\beta N\left(\frac{J}{2}m^2 + h\,m\right)}\varpi_N(dm).$$

As was seen in Section 1.1 (cf. (1.8)), the sequence (ϖ_N) satisfies a l.d.p. with scaling N and rate function given by $\mathcal{E}(m) + \log 2$, where

$$\mathcal{E}(m) = \frac{1+m}{2}\log\left(\frac{1+m}{2}\right) + \frac{1-m}{2}\log\left(\frac{1-m}{2}\right), \tag{4.15}$$

for $m \in [-1, 1]$, and by $+\infty$ for $|m| > 1$. Applying Theorem 1.18, (1.41) gives

$$\lim_{N\to+\infty} F_N(\beta, h) = \min_{m\in[-1,1]} f_{\beta,h}(m),$$

where $f_{\beta,h}(m)$ is the canonical free energy,[1] given by:

$$\beta f_{\beta,h}(m) := \mathcal{E}(m) - \beta\frac{J}{2}m^2 - \beta hm. \tag{4.16}$$

[1] Omitting the fixed parameter J from the notation.

To find the minima of such function, we first find its critical points m^*, i.e. we solve:

$$\beta J m^* + \beta h = \mathcal{E}'(m^*).$$

Since $\mathcal{E}'(m) = \frac{1}{2}\log((1+m)/(1-m))$, the previous equation is equivalent to:

$$m^* = \tanh \beta(Jm^* + h). \tag{4.17}$$

(1) If $\beta J \le 1$, (4.17) has a unique solution $m(\beta, h)$; $m(\beta, 0) = 0$, whereas $m(\beta, h)$ has the same sign as h, for $h \ne 0$. $f_{\beta,h}(\cdot)$ has a minimum at $m(\beta, h)$.

(2) If $\beta J > 1$, there exists $h_0 = h_0(\beta, J)$ (the coercive field) such that for $|h| > h_0$ (4.17) has a single solution $m(\beta, h)$, the minimum point of $f_{\beta,h}(\cdot)$. $m(\beta, h)$ has the same sign as h. At $|h| = h_0$, (4.17) has two solutions; one gives the unique minimum point of $f_{\beta,h}(\cdot)$ and has the same sign as h; the other is an inflection point.

(3) If $\beta J > 1$ and $|h| < h_0$, the canonical free energy $f_{\beta,h}(\cdot)$ has a double well structure.
   - When $h = 0$, the double well is symmetric. Equation (4.17) has three solutions: $0, \pm m^*(\beta)$ $(m^*(\beta) > 0)$. Zero gives a local maximum and $\pm m^*(\beta)$ are the points of absolute minimum.
   - When $0 < |h| < h_0$, (4.17) still has three solutions and $f_{\beta,h}(\cdot)$ keeps a double well structure, with two points of local minima $m^-(\beta, h) < 0 < m^+(\beta, h)$. Now there is a deeper well. The point of absolute minimum has the same sign as h (i.e. the magnetization is parallel to the external field). The third solution of (4.17), denoted by $m^0(\beta, h)$, gives a local maximum and has sign opposite to h. In this case the absolute minimum is interpreted as the stable equilibrium state, whereas the local minimum, with magnetization opposite to the field, is interpreted as the metastable state.

Remark The previous argument provides a l.d.p. for the distributions of $m_N(\sigma)$ under the Gibbs measures corresponding to Curie–Weiss interaction: for $\sigma \in \mathcal{X}_N$ we set

$$\mu_{N,\beta,h}(\sigma) = \frac{1}{Z_N(\beta,h)} e^{-\beta H_N(\sigma)} = \frac{1}{Z_N(\beta,h)} e^{\beta N\left(\frac{J}{2}(m_N(\sigma))^2 + h\, m_N(\sigma)\right)}.$$

Thus, if g is measurable on $\mathbb{R}$:

$$\int_{\mathcal{X}_N} g(m_N(\sigma))d\mu_{N,\beta,h}(\sigma) = \frac{\int_{[-1,+1]} g(m) e^{N\beta(\frac{J}{2}m^2 + hm)}\varpi_N(dm)}{\int_{[-1,+1]} e^{N\beta(\frac{J}{2}m^2+hm)}\varpi_N(dm)} \tag{4.18}$$

where ϖ_N is the distribution of $m_N(\sigma)$ with σ uniformly distributed on $\mathcal{X}_N$.

We are thus in condition to apply the *tilting* argument (remark (ii) after Theorem 1.18): the validity of a l.d.p. for the sequence ϖ_N yields at once a l.d.p. for the distribution of $m_N(\sigma)$ under $\mu_{N,\beta,h}$, hereby denoted by $\nu_{N,\beta,h}(m)$. It has scaling N and the rate function is

$$I_{\beta,h}(m) = \beta f_{\beta,h}(m) - \min_{m' \in [-1,1]} \beta f_{\beta,h}(m'). \tag{4.19}$$

From previous observations, there is a unique root of $I_{\beta,h}(m) = 0$, if $\beta \leq 1/J$ or $h \neq 0$. As remarked in Section 3.1, this provides exponential convergence of $m_N(\sigma)$ to the unique minimum. When $h = 0$ and $\beta > 1/J$ there are two points $\pm m^*(\beta)$ at which $I_{\beta,h}(\cdot)$ vanishes and $m_N(\sigma)$ fails to have a deterministic limit; in fact, it converges in law to $(\delta_{m^*(\beta)} + \delta_{-m^*(\beta)})/2$, as $N \to +\infty$.

We deduce that this simple model predicts the correct qualitative behaviour of a ferromagnet at $h = 0$. From (4.16) we also see that $\lim_{h\to 0} m(\beta, h) = 0 = m(\beta, 0)$ when $\beta J \leq 1$, whereas $\lim_{h\to 0\pm} m^+(\beta, h) = \pm m^*(\beta)$ when $\beta J > 1$, describing the spontaneous magnetization.

Again we see that in the framework of a mean field theory the existence of metastable states emerges naturally. They are intrinsically related to the mean field theory of phase transitions.

Let us now summarize the main ideas about metastability arising in the framework of the mean field approach to phase transitions:

- metastability can be described via the equilibrium Gibbsian formalism;
- it is associated to the existence of local minima of the free energy; indeed both in van der Waals and in Curie–Weiss models, metastability arises from the (unphysical) non-convexity of the canonical free energy.
- the metastable branch can be obtained via an analytic continuation beyond the condensation point.

We shall see that for realistic systems with short range interactions the above features fail, at least in a naïve form, and we shall need new ideas to describe metastability in that case.

4.1.2 Kac potentials

Now, in connection with the above discussion, we want to introduce the modern version of mean field theory based on the so-called Kac potentials (see [172, 193]). It is a 'correct' theory at least in the sense that it gives rise to the correct behaviour of the thermodynamic functions. Contrary to the 'naïve' mean field theory we shall recover the right convexity properties of the free energy, and the unphysical feature of thermodynamical instability of the system will be eliminated.

Consider a continuous system of particles in $\mathbb{R}^d$ whose interaction is given by a two-body potential as in Definition 3.12, with:

$$\phi(r) = q(r) + w_\gamma(r), \qquad w_\gamma(r) = \gamma^d w(\gamma r), \qquad r \geq 0, \tag{4.20}$$

where w and q are suitable regular functions. For instance, we assume that there exist $r_0 > 0$, $D > 0$, $\varepsilon > 0$ such that:

$$q(r) = \infty \quad \text{if} \quad r \leq r_0 \qquad \text{(hard core condition)} \tag{4.21}$$

$$|q(r)| \leq Dr^{-d-\varepsilon} \quad \text{if} \quad r \geq r_0 \qquad \text{(fast decay at infinity)} \tag{4.22}$$

$$-Dr^{-d-\varepsilon} \leq w(r) \leq 0 \quad \text{(fast decay and attractiveness)}, \tag{4.23}$$

and

$$\int_{\mathbb{R}^d} w(|\mathbf{r}|)d\mathbf{r} = \alpha.$$

For such systems, the so-called van der Waals limit (see [193]) consists in first taking the thermodynamic limit and, subsequently, the limit $\gamma \to 0$, sending to infinity the range of the attractive potential.

Let $f(\rho, \gamma)$ be the thermodynamic limit for the Helmholtz free energy:

$$f(\rho, \gamma) := -\lim_{\Lambda \nearrow \mathbb{R}^d} \frac{1}{\beta|\Lambda|} \log Z(\Lambda, \beta, \rho). \tag{4.24}$$

Here $Z(\Lambda, \beta, \rho)$ is the canonical partition function for $n = \rho|\Lambda|$ particles enclosed in the region Λ:

$$Z(\Lambda, \beta, \rho) = \frac{1}{n!} \int_{\Lambda^n} \exp\left(-\beta \sum_{i<j} \phi(|r_i - r_j|)\right) dr_1 \dots dr_n, \tag{4.25}$$

where by $\Lambda \nearrow \mathbb{R}^d$ we mean the limit when Λ invades the whole $\mathbb{R}^d$ in the van Hove sense (see Section 3.3); to fix ideas we may take a sequence of cubes centred at the origin. The above limit in (4.24) exists by virtue of the assumptions in (4.21)–(4.23) (see [117, 258]). (*Notation*. Here f corresponds to what we denoted by g in Section 3.3, cf. (3.79).)

The following theorem, due to Lebowitz and Penrose [193], describes the behaviour in the van der Waals limit. It extends the initial result by Kac, Uhlenbeck and Hemmer [172].

Theorem 4.1 (Lebowitz–Penrose)

$$f(\rho, 0^+) := \lim_{\gamma \to 0^+} f(\rho, \gamma) = CE\left(f_0(\rho) + \frac{1}{2}\alpha\rho^2\right), \tag{4.26}$$

where $f_0(\rho)$ is the canonical free energy of the 'reference system' described by the potential $q(r)$, namely the quantity defined as in (4.24), (4.25) *with $q(r)$ in place of $\phi(r)$. (Notation. $CE(g)$ indicates, as before, the convex envelope of the real valued function g, i.e. the maximal convex function nowhere exceeding g.)*

The proof of the above theorem can be found in [193]. (See also [249].) We remark that the existence of a first order phase transition, which is detected by the appearance of a straight segment in the graph of the canonical free energy as a function of the density ρ, can only be a consequence of the non-convexity of the free energy of the corresponding naïve model, namely:

$$f^{naïve}(\rho, \beta) = f_0(\rho) + \frac{1}{2}\alpha\rho^2.$$

Thus, in the 'true' theory the metastable branch disappears and Maxwell's equal area rule is justified on rigorous grounds.

Similar results hold for lattice spin systems described by the energy function:

$$H_\gamma(\sigma_\Lambda) = -\frac{1}{2}\sum_{i \neq j} J_\gamma(i, j)\sigma_i\sigma_j \; - \; h\sum_i \sigma_i, \qquad \sigma_i \in \{-1, +1\},$$

$$J_\gamma(i, j) := \gamma^d J(\gamma|i - j|) \geq 0, \qquad \int_{\mathbb{R}^d} J(|\mathbf{r}|)d\mathbf{r} \; = \; 1. \tag{4.27}$$

Similarly to the continuous case, we may define the canonical free energy $f(m, \gamma)$. The following theorem holds:

Theorem 4.2

$$f(m, 0^+) := \lim_{\gamma \to 0+} f(m, \gamma) \; = \; CE\left(-\frac{1}{2}m^2 - hm + \frac{1}{\beta}\mathcal{E}(m)\right).$$

The proof can be found in [286] (for more general potentials). We remark that for $h \neq 0$ the absolute minimum of $f(m, 0^+)$ is reached for a unique value of the magnetization $m = m(h, \beta)$, whereas for $h = 0$ and $\beta > \beta_c \equiv 1$ the graph of $f(m, 0^+)$ shows a horizontal plateau corresponding to the segment between $-m^*(\beta)$ and $+m^*(\beta)$.

The above theorems show that the correct version of mean field theory, given by Kac potentials, both for continuous systems of particles and lattice spin systems, gives rise to a convex canonical free energy. So, a physically acceptable approach rules out the possibility of including, in a naïve way, metastability in the equilibrium mean field theory.

4.1.3 Metastability and the Gibbsian approach

We want now to discuss some general rigorous results in equilibrium statistical mechanics related to the impossibility of including metastability in the usual Gibbsian equilibrium theory, at least when interactions decay sufficiently fast at large distance.

In their fundamental paper [190], Lanford and Ruelle state a general result on 'absence of metastability', valid under general hypotheses on the decay of the interaction at large distances. They prove that a translationally invariant Gibbs (DLR) measure necessarily satisfies the variational principle of thermodynamics. (For the case of lattice spin systems see Theorem 3.26.) Thus, a state which is metastable in the sense that (i) it is an equilibrium state, namely a translationally invariant Gibbs measure, and (ii) it does not correspond to an absolute minimum of the free energy functional, cannot exist. In other words, for a short range system, it is impossible to describe metastable states in the framework of usual equilibrium Gibbsian formalism, for instance as local minima for the free energy functional.

In this connection let us describe briefly an approach to metastability due to Sewell.

In [276] the author develops a theory aiming to include metastability in a generalized Gibbsian set-up. He characterizes equilibrium and metastable states in terms of a variety of stability conditions.

(I) *Equilibrium states.* They minimize the free energy density of the system with respect to global variations of the state. In this sense they are globally thermodynamically stable.

(II) *Ideal metastable states or metastable states of the 'highest grade'.* They do not minimize the free energy density globally but, rather, they minimize it only with respect to modifications of the state confined to bounded regions of the space. In this sense they are locally thermodynamically stable. Since global and local thermodynamic stability coincide for 'realistic' short range systems, ideal metastable states are possible only for special models with suitable long range or many-body forces.

(III) *Normal (i.e. non-ideal) metastable states.* They minimize the restriction of the free energy density to some reduced state space. Normal metastable states have finite but large lifetime, as we have to overstep a large barrier to remove the state from this reduced state space. These states are essentially those that can be included in the approach developed by Penrose and Lebowitz, based on restricted ensembles, which we describe in Section 4.1.5.

The possibility of continuing the free energy analytically beyond the condensation point of a first order phase transition remained as open problem until the fundamental result of Isakov (see [168]) which we quote now. Consider the standard Ising model in dimension $d \geq 2$, at low enough temperature (see Section 3.3); let J be the ferromagnetic coupling constant, h the external magnetic field, and β the inverse temperature. By $F(\beta, h)$ we denote the thermodynamic limit (along a van Hove sequence) of the grand canonical free energy:

$$F(\beta, h) = \lim_{\Lambda \nearrow \mathbb{Z}^d} -\frac{1}{\beta|\Lambda|} \log Z_\Lambda(\beta, h),$$

where $Z_\Lambda(\beta, h)$ is the grand canonical partition function. If β is large enough we can find $n_0(\beta)$ finite, so that:

$$\lim_{h\to 0^+} \frac{1}{n!}\frac{\partial^n F}{\partial h^n} = (n!)^{\frac{1}{d-1}}(2\beta)^n(2(d-1)(2\beta J + C\xi))^{-\frac{nd}{d-1}},$$

for all $n > n_0(\beta)$, where $C \in \mathbb{R}$ depends only on the dimension d and $|\xi| \le 1$.

This implies that the free energy (and then the magnetization $m = m(\beta, h)$) has an essential singularity in $h = 0$ (at large β) forbidding any real analytic continuation. Isakov's result has been extended recently to the whole class of two-phase lattice models within the framework of Pirogov–Sinai theory.[2] In this set-up Friedli and Pfister showed that the validity of the Peierls condition implies non-analyticity at the transition point (see [123, 124]).

However, a C^∞ extrapolation beyond the condensation point can be associated with metastability in a dynamical context. (see [271]).

There is a point of view on metastability, going back to Langer (see [191]), based on the analytic continuation of the free energy beyond the condensation point. It is believed that, in the case of ferromagnetic, short range systems, because of the essential singularity at the condensation point, the free energy acquires an imaginary part in the metastable domain. This imaginary part is interpreted as a decay rate from the metastable situation. In [228] some very simplified models are considered, whose free energy can be continued analytically and to which one can assign a stochastic dynamics and deduce a decay rate. It is found that the imaginary part of the continued free energy is proportional to the decay rate (see also [227].) See [133] for a counterexample.

In [239], Penrose applied the above point of view to classical nucleation theory. He considered the so-called droplet model and showed that the corresponding thermodynamical function $P(z)$, representing the pressure as a function of the fugacity, can be continued analytically and that the imaginary part is proportional to the rate of nucleation of the corresponding metastable state, as given by Becker–Döring theory; see Section 4.1.4 below.

Remark Since the van der Waals theory emerges as the limit of finite range Kac potentials, it is important to understand how the described lack of analytic continuation fits with the limiting behaviour as $\gamma \downarrow 0$. In the context of Theorem 4.2, with $d \ge 2$ and $h = 0$, in [47] and [30] it is proven that, if $\beta > 1$, then there exists $\gamma_\beta > 0$ so that the system with scaling γ presents phase transition for $\gamma < \gamma_\beta$ and the canonical free energy has a plateau in $[-m^*_\gamma(\beta), m^*_\gamma(\beta)]$. Using an interaction given by a step function, it is proven in [125] that there exist β_0 and γ_0 so that if $\beta > \beta_0$ and $\gamma \in (0, \gamma_0)$ the canonical free energy has no analytic continuation

[2] See [244] or [280] for a precise description of Pirogov–Sinai theory.

through $\pm m^*_\gamma(\beta)$. In the grand canonical ensemble, the estimates in [125] show how the analytic continuation is recovered in the limit.

We can conclude this critical analysis by saying that metastability has to be seen as a genuine dynamical phenomenon.

In any case, both static and dynamic properties have to be explained. In particular, one has to give a definition, based on dynamic grounds, for the lifetime of the metastable situation, and to describe the mechanism of transition.

Mainly, this amounts to characterizing the 'nucleation pattern'; as we shall see, the crucial point is to find shape and size of the 'critical nucleus'.

4.1.4 Classical theory of nucleation

The so-called 'classical theory of nucleation' can be considered as a first attempt to formulate a dynamical approach to metastability and nucleation. It goes back to the 1930s, being closely associated with the names of Becker and Döring. It can be considered as a dynamical version of the so-called droplet model, which we now introduce briefly.

Consider the standard Ising model in the lattice gas language (see Section 3.3.2); the corresponding grand canonical partition function in the cube $\Lambda \subset\subset \mathbb{Z}^d$ (empty boundary conditions) at fugacity z and inverse temperature β, is given by

$$\Xi = \Xi(\beta, z) := \sum_{\xi \in \{0,1\}^\Lambda} z^{N(\xi)} e^{-\beta V(\xi)} \tag{4.28}$$

where $N(\xi)$ denotes the number of occupied sites in the configuration ξ and $V(\xi)$ is the energy associated with ξ: $V(\xi) = -Up(\xi)$ where U is a positive constant (attractive interaction) and $p(\xi) = \sharp\{$non-ordered pairs i, j such that $|i - j| = 1, \xi(i) = \xi(j) = 1\}$. In terms of previous notation $z = e^{\beta\vartheta}$, where ϑ is the chemical potential ($\sharp$ denotes cardinality).

Given any configuration $\xi \in \{0, 1\}^\Lambda$, we partition the corresponding set of occupied sites into maximal connected (by nearest neighbours) components c called 'clusters'. We can write

$$N(\xi) = \sum_i N(c_i), \qquad V(\xi) = \sum_i V(c_i), \tag{4.29}$$

where $c_1, c_2, \ldots$ are the clusters constituting ξ, and $N(c_i)$, $V(c_i)$ are the number of particles and the energy, respectively, of the cluster c_i. Thus, we write

$$\Xi = 1 + \sum_{\{c\}} \prod_i z^{N(c_i)} e^{-\beta V(c_i)}, \tag{4.30}$$

where $\{c\}$ denotes a possible arrangement of clusters such that no two clusters overlap or use neighbouring sites; the term 1 represents the contribution of the empty configuration.

We want now to introduce an approximation to the partition function, which indeed gives rise to a rigorous upper bound, and can be taken to define the droplet model.

This approximation mainly consists in neglecting the hard core exclusion between clusters, allowing them to touch by nearest neighbours and even to overlap; it is reasonable to introduce this crude approximation when the parameters z and β give rise to a very rarefied system of clusters. This happens, in particular, for sufficiently small z. Let us first rewrite the true partition function as

$$\Xi = \sum_{n=0}^{n_{max}(\Lambda)} \frac{1}{n!} \sum_{\substack{\text{compatible} \\ c_1,\dots,c_n}} \prod_{i=1}^{n} z^{N(c_i)} e^{-\beta V(c_i)}, \tag{4.31}$$

where the second sum is over ordered n-tuples of pairwise compatible (i.e. neither overlapping nor touching by nearest neighbours) clusters; $n_{max}(\Lambda)$ is the maximum number of possible clusters in Λ.

We denote by $\bar{\Xi}$ the partition function corresponding to a system containing an arbitrary number of non-interacting clusters (it is also permitted to repeat clusters):

$$\bar{\Xi} = \sum_{n=0}^{\infty} \frac{1}{n!} \left(\sum_{c \subseteq \Lambda} z^{N(c)} e^{-\beta V(c)} \right)^n = e^{A_\Lambda}, \tag{4.32}$$

where $A_\Lambda = \sum_{x \in \Lambda} A(\Lambda, x)$, $A(\Lambda, x) = \sum_{c \subseteq \Lambda : x \in c} \frac{1}{N(c)} z^{N(c)} e^{-\beta V(c)}$ and the sums are taken over connected sets c only.

To obtain the droplet model we make a further approximation, reasonable for the majority of sites x (far from the boundary $\partial\Lambda$). It consists in substituting $A(\Lambda, x)$ by its thermodynamic limit

$$\sum_{c \subset\subset \mathbb{Z}^d :\, x \in c} \frac{1}{N(c)} z^{N(c)} e^{-\beta V(c)} = \sum_{l=1}^{+\infty} z^l Q_l,$$

where (as before O denotes the origin in $\mathbb{Z}^d$)

$$Q_l = \sum_{\substack{O \in c \subset\subset \mathbb{Z}^d : \\ N(c) = l}} \frac{1}{l} e^{-\beta V(c)}. \tag{4.33}$$

We suppose for the moment β, z such that the corresponding series is absolutely convergent, and write

$$\Xi^{DM} := \exp\left(|\Lambda| \sum_{l=1}^{\infty} z^l Q_l \right) \tag{4.34}$$

for the partition function of the simplified model.

Thus, the 'droplet model' represents a free system of polymers with the same 'activity function' $z^{N(c)}e^{-\beta V(c)}$ as in the Ising model. Therefore we can take a Poisson distribution with average $|\Lambda| z^l Q_l$ for the number of clusters with l particles, so that the total average number of clusters in Λ is $|\Lambda| \sum_{l \geq 1} z^l Q_l$. There is no interaction among clusters. The average number of clusters with l particles, per unit volume, becomes

$$m_l^{DM} = z^l Q_l. \tag{4.35}$$

One can argue that (4.35) is a good approximation for the true expression of m_l when the clusters are very diluted, in the thermodynamic limit $\Lambda \nearrow \mathbb{Z}^d$. Moreover $m_l \leq m_l^{DM}$ in any case (see [241]). As an example, let us consider the two-dimensional case at very low temperature. In this case only clusters without 'holes' and of square shape will contribute significantly and from (4.33) one gets (see e.g. Appendix 2 in [241]):

$$e^{-\beta U} \exp \beta\big(2Ul - 2U\sqrt{l}\big) \leq Q_l \leq \frac{9}{8} 3^{2l} \exp \beta\big(2Ul - 2U\sqrt{l}\big), \tag{4.36}$$

so that $\sum_{l=1}^{\infty} z^l Q_l$ is clearly convergent for $z < \frac{1}{9} \exp\{-2U\beta\}$. On the other hand we can identify $z_S = \exp\{-2U\beta\}$ as the asymptotic value, for large β, of the fugacity corresponding to saturated vapour; recall for this the correspondance between the parameters J, h (coupling constant and magnetic field) of the spin representation of the Ising model and U, ϑ (binding energy and chemical potential) of the lattice gas representation: $4J = U,\ h = U + \vartheta/2$, so that $h = 0$ corresponds to $\vartheta \equiv (\log z)/\beta = -2U$.

Let us now introduce the dynamical model constituting the basis of classical nucleation theory. The object of our study is a system of extended clusters. The main assumptions in this treatment are that clusters are statistically independent and that processes of fragmentation or coalescence involving only clusters containing more than one particle are ignored. The state of the system is just given in terms of the variables $\{m_l\}_{l=1,2,\ldots}$ having the meaning of specific 'occupation numbers' of the sizes $l = 1, 2, \ldots$; $m_l(t)$ represents the number of clusters with l particles, per unit volume, at time t. It is assumed that the dynamics is particle conserving; therefore the only way an l-particle cluster can grow is by absorbing a 1-particle cluster. The number of times this happens per unit volume and per unit time is assumed to be proportional to the density m_l of l-particle clusters and also to the density m_1 of 1-particle clusters; it is therefore taken of the form $a_l m_l m_1$ with a_l suitable constants. On the other hand the only way an $(l+1)$-particle cluster can shrink is by emitting a 1-particle cluster; the number of times this happens per unit volume and per unit time is assumed to be $b_{l+1} m_{l+1}$ with b_{l+1} suitable constants. The net rate at which the l-particle clusters are being converted to $l+1$-particle clusters is therefore

$$J_l(t) = a_l m_l(t) m_1(t) - b_{l+1} m_{l+1}(t), \quad l = 1, 2, \ldots \tag{4.37}$$

and the net rate of change of m_l is given by

$$\frac{dm_l(t)}{dt} = J_{l-1}(t) - J_l(t), \quad l = 2, 3, \ldots; \tag{4.38}$$

this equation does not apply for $l = 1$ and some additional hypothesis has to be introduced. A first model, called A corresponds to the assumption appearing in the original paper by Becker and Döring (see [15, 241]), that m_1 does not change significantly with time, so that one in fact assumes $dm_1/dt = 0$. With this assumption equations (4.38), valid for $l = 2, 3, \ldots$, become linear, constant coefficients and therefore mathematically quite simple to treat. Model A corresponds to a system coupled to a reservoir that can emit or absorb only 1-particle clusters. A more realistic model, called B, was introduced in [241]: the condition $dm_1/dt = 0$ is replaced by the assumption that $\sum_{l \geq 1} l m_l$ does not change with time. Thus, $dm_1(t)/dt = 0$ in model A and, recalling (4.38), $dm_1(t)/dt = -2J_1(t) - \sum_{l \geq 2} J_l(t)$ in model B.

For both models the equilibrium solution corresponding to $J_l = 0$ for each l is given by:

$$m_l^{eq} = c_l (m_1^{eq})^l \tag{4.39}$$

with m_1^{eq} = constant and

$$c_1 = 1, \qquad c_l = \frac{a_1 \ldots a_{l-1}}{b_2 \ldots b_l}, \qquad \text{if} \quad l \geq 2. \tag{4.40}$$

From (4.37) we have

$$\frac{b_{l+1}}{a_l} = \frac{m_l^{eq} m_1^{eq}}{m_{l+1}^{eq}}. \tag{4.41}$$

To get an expression as (4.35) for m_l^{eq}, since $Q_1 = 1$ in (4.33) we choose:

$$a_l = 1, \ l \geq 1, \qquad b_l = \frac{Q_{l-1}}{Q_l}, \qquad l \geq 2. \tag{4.42}$$

Let us now go to the treatment of metastability in the framework of the classical theory of nucleation for the model A. In the droplet model the metastable situation corresponds to values of the parameters z, β for which the series $\sum_{N=1}^{\infty} z^N Q_N$ is divergent; in particular for the example of the two-dimensional case at very low temperature, the metastable regime corresponds to a fugacity slightly larger than the saturated vapour value: $z > z_S$. The equilibrium solution given by (4.39), (4.40) does not make sense anymore. Indeed in the real Ising model, when $z > z_S$, the typical configuration will contain a very large droplet of the liquid phase and the basic assumptions of the droplet model (small and sparse clusters) fail.

The basic idea is to assume that every time the number of particles in a cluster passes a specified large value u, the cluster is removed from the system, replacing (4.38) by

$$\frac{dm_l(t)}{dt} = J_{l-1} - J_l, \quad l = 2, 3, \ldots, u; \quad m_l(t) = 0 \text{ if } l \geq u. \tag{4.43}$$

Metastability is associated with a steady state solution of (4.43) in which J_l is independent of l. The common value J is interpreted as nucleation rate and can be calculated as follows. Let

$$\begin{aligned} \Phi_1 &:= m_1, \quad \Phi_l := \frac{m_l}{(m_1)^{l-1}c_l} = \frac{m_l b_2 \ldots b_l}{(m_1)^{l-1}a_1 \ldots a_{l-1}}, \ l = 2, 3, \ldots, \\ R_l &:= \frac{1}{(m_1)^l c_l a_l} = \frac{1}{(m_1)^l} \frac{b_2 \ldots b_l}{a_1 \ldots a_l} \end{aligned} \tag{4.44}$$

so that (4.37) becomes $J_l R_l = \Phi_l - \Phi_{l+1}$, for $l \geq 1$. Summing this for $l = 1, \ldots, u$ and using $\Phi_1 = m_1$, $\Phi_{u+1} = 0$ (since $m_{u+1} = 0$) we obtain

$$J = \frac{m_1}{\sum_{l=1}^{u} R_l}. \tag{4.45}$$

In the two-dimensional case, at very large β and the coefficients given by (4.42) we see that, once u is large enough, J will not depend much on its choice in the above modified evolution (setting $m_l(t) = 0$ if $l > u$). For this notice that $R_l \approx \exp\{-\beta[(\vartheta + 2U)l - 2U\sqrt{l}]\}$ as $\beta \to +\infty$, cf. (4.36) and writing $z = e^{\beta\vartheta}$. The maximum of this last expression is achieved at $l^* = (U/(2U + \vartheta))^2$, assuming this is an integer (otherwise at one of the two closest integers). As $\beta \to +\infty$ we then get a reasonable estimate of the nucleation rate J in this very simplified model.

Such a simple treatment is not possible in the case of model B which, more realistically than model A, is only globally conservative and takes into account the depletion of monomers as large clusters are formed. For model B, a steady state solution with constant non-vanishing J_l, like the one we have seen above for model A, does not exist. Penrose, in [238], shows the existence of a class of solutions of the kinetic equations of model B which have (in a well defined sense) very long lifetimes. In particular, an initial condition $m_l(0)$ such that for a suitable value $l = u$, $m_l(0) \sim m_l^{eq}$ for $l \leq u$ and $m_l(0) \sim 0$ for $l > u$, gives rise to a solution which varies extremely slowly with time. In the two-dimensional case at very low temperature it turns out that the correct choice is $u = l^*$.

Let us now discuss the above approach critically.

The main assumption of the classical theory of nucleation is that the clusters (droplets) are very sparse and non-interacting. This is extrapolated from the equilibrium situation at very low temperature. Moreover, because of the identity (4.41), only the equilibrium values of m_l are relevant to define the dynamics. An average over the shapes of the droplets is implicit; this, at very low temperature, is equivalent to assuming that the shape of the clusters is approximately square.

In the 'true problem', as we shall see, the decay from metastability is a real non-equilibrium phenomenon and there is no reason, a priori, to deduce simplifying hypotheses from the equilibrium situation. It involves a system of interacting droplets, with given sizes and shapes and which can coalesce. In particular, the assumption that coalescence and fragmentation involving extended clusters can be neglected deserves a dynamical justification.

We shall see that in the same asymptotic regime of very small temperature it is now possible to study, from a microscopic point of view, the same aspects (geometry of droplets, coalescence) that are neglected programmatically in the classical theory of nucleation.

4.1.5 The point of view of the evolution of the ensembles

We now describe briefly the first rigorous approach to metastability where dynamic and static properties have been considered together. For reasons that will appear clear in what follows we shall call it the point of view of the 'evolution of the ensembles'. It was introduced by Penrose and Lebowitz in the paper [240]. These authors characterize metastable states on the basis of the following three conditions.

(1) Only one thermodynamic phase is present.
(2) The lifetime is very large in the sense that a system that starts in this state is likely to take a long time to get out.
(3) The decay from the metastable to the stable state is an irreversible process in the sense that the return time to the metastable state is much longer than the decay time.

The main idea introduced in [240] is that of 'restricted ensemble'. An initial state is considered, given by a probability measure μ on the phase space Γ; a law governing the time evolution is supposed to be assigned. For example, for a continuous system of classical particles it is natural to assume a deterministic time evolution given by the Liouville equation. This initial measure, describing a metastable state, has to be chosen according to the above three criteria. The idea is to chose μ according to a suitable (and substantial) modification of the canonical Gibbs prescription: zero probability is given to the class of configurations producing phase segregation; in other words, μ is Gibbsian but conditioned to a suitable subset R of Γ. For instance, in the case of a supersaturated vapour the density is constrained to an interval of values, forbidding the appearance of the liquid phase.

Let us now summarize the Lebowitz–Penrose theory for a continuous system of particles.

The phase space is $\Gamma := \Lambda^N \times \mathbb{R}^N$ and a generic point in Γ is denoted by

$$\xi = \mathrm{q}_1, \dots, \mathrm{q}_N, \ \mathrm{p}_1, \dots, \mathrm{p}_N,$$

with q_i and p_i denoting position and momentum, respectively, of the ith particle. Given a Borel subset R of Γ we set :

$$\mu_R(d\xi) = \frac{\exp\{-\beta H(\xi)\}\mathbf{1}_R(\xi)d\xi}{\int_R \exp\{-\beta H(\zeta)\}d\zeta},$$

where $H(\xi)$ is the Hamiltonian and $\mathbf{1}_R$ denotes the indicator function of the set R ($\mathbf{1}_R(\xi) = 1$ if $\xi \in R$, and 0 otherwise). Let $P_t^{\mu_R}$ denote the evolution at time t of the initial measure μ_R, along the Liouville equation. For $t \geq 0$ we write p_t for the probability of being outside R at time t, starting from μ_R at $t = 0$, that is:

$$p_t := P_t^{\mu_R}(\Gamma \setminus R).$$

It is shown in [240] that, as a simple consequence of the Gibbsian expression, p_t increases in t, while

$$\frac{dp_t}{dt} \leq \left.\frac{dp_t}{dt}\right|_{t=0},$$

so the rate of increase of p_t is maximal at $t = 0$. It is then reasonable to define the 'escape rate' λ as:

$$\lambda := \left.\frac{dp_t}{dt}\right|_{t=0}. \tag{4.46}$$

In their approach Penrose and Lebowitz identify the lifetime of the metastable state as the inverse of the escape rate λ. This quantity is computable in terms of the initial state and the generator of the dynamics; it is an 'incipient' dynamical quantity, related to the initial tendency of the system to leave R. It is small if there is a 'bottleneck' effect in the flow of probability outside R. Using the restricted ensemble prescription, in order to verify the three metastability criteria, one has to chose the set R so that:

(i) μ_R describes a pure phase,

(ii) λ is very small,

(iii) under the equilibrium measure in Λ, the measure of the set R is exponentially small, i.e. there exists a positive constant c such that

$$\mu_{eq}(R) = \frac{\int_R \exp\{-\beta H(\xi)\}d\xi}{\int_\Gamma \exp\{-\beta H(\xi)\}d\xi} \leq \exp(-c|\Lambda|)$$

where $|\Lambda|$ stands for the Lebesgue measure of Λ.

Let us now give a short exposition of the application, considered in [240], of the above approach to the case of Kac potentials given by (4.20), in the vapour–liquid transition. The authors choose a grand canonical restricted ensemble and discuss the equivalence at the level of restricted ensembles. So, instead of being a subset of the canonical phase space Γ, with fixed number of particles, R will be a subset of the grand canonical phase space

$$\mathcal{X}_\Lambda = \cup_{n \geq 0}(\Lambda^n \times \mathbb{R}^{3n}).$$

In fact R will not impose any restriction on the momenta which here play no significant role and, for many purposes, could be integrated out.

Conditions (4.21), (4.22) and (4.23) for the validity of Theorem 4.1 are supposed to hold. We also suppose that the convex envelope $CE(f_0(\rho) + \frac{1}{2}\alpha\rho^2)$ is strictly smaller than $f_0(\rho) + \frac{1}{2}\alpha\rho^2$, in the interval (ρ_g, ρ_l) $(0 < \rho_g < \rho_l)$. To be concrete we treat the supersaturated vapour, corresponding to a density close to ρ_g. To define the set R the following choices are made in [240]: choose ρ_-, ρ_+ such that for each ρ in $[\rho_-, \rho_+]$, we have

(i) $f_0(\rho) + \frac{1}{2}\alpha\rho^2 > f(\rho, 0^+)$,
(ii) $f_0''(\rho) + \alpha > 0$.

The set R will be defined so that in the corresponding grand canonical restricted ensemble the global average density will belong to the interval (ρ_-, ρ_+). The restricted grand canonical chemical potential ϑ will be related to the density ρ by the formula:

$$\vartheta = \frac{d(f_0(\rho) + \frac{1}{2}\alpha\rho^2)}{d\rho}.$$

For simplicity we assume Λ to be a cube which we imagine divided into cubic cells $\omega_1, \ldots, \omega_M$ of equal volume $|\omega|$ $(|\Lambda| = M|\omega|)$. Let n_i denote the number of particles in the volume ω_i. R is then defined as the set of all configurations verifying

$$\rho_-|\omega| \leq n_i \leq \rho_+|\omega|, \qquad i = 1, \ldots, M. \tag{4.47}$$

To guarantee the requirements the following asymptotic regime is considered:

$$r_0^d \ll |\omega| \ll \gamma^{-d} \ll |\Lambda|.$$

The constraint $r_0^d \ll |\omega|$ means that the linear size of ω is much larger than the range of the hard core interaction between particles. It ensures that $|\omega|^{1/d}$ is a 'mesoscopic' scale (the number of particles in ω is typically very large) and the validity of a law of large numbers with respect to the restricted (conditioned) measure. The constraint $|\omega| \ll \gamma^{-d}$ means that the range γ^{-1} of the Kac potential is so large that on the scale $|\omega|^{1/d}$ the long range part of the interaction is almost constant, giving mean field behaviour on this scale. Finally, the constraint $\gamma^{-d} \ll |\Lambda|$ says that the volume is much larger than the range of the interaction, which is important to achieve the correct thermodynamic behaviour (right convexity properties of the thermodynamic functions).

The choice in [240] involves making both γ^{-1} and $|\omega|$ tend to infinity with $|\Lambda|$, with

$$\log|\Lambda| \ll |\omega| \ll \gamma^{-d} \ll |\Lambda|. \tag{4.48}$$

With the above choices Penrose and Lebowitz carry out their programme; in particular, as far as the 'static properties' of the metastable state are concerned, they show that the equation of state is exactly that predicted by the van der Waals–Maxwell theory: the 'metastable branch' is obtained by continuing, beyond the condensation point, the van der Waals–Maxwell isotherm. This metastable branch is a piece of the original van der Waals curve corrected with the Maxwell equal area rule: precisely the part of the curve between the condensation point and the 'spinodal point' corresponding to the point $\bar{D}$ in Figure 4.5(b) where thermodynamic instability takes place (inflection point in Figure 4.5 (a)).

The main dynamical result is that the lifetime λ^{-1} tends to infinity in the joint triple limit in which $|\omega|, \gamma^{-d}, |\Lambda|$ tend to infinity keeping (4.48). They indeed find the physically interesting behaviour:

$$\lambda \leq \tilde{c}\ |\Lambda| \exp(-c|\omega|), \tag{4.49}$$

with $\tilde{c}$ and c suitable positive constants; $c|\omega|$ is a lower bound to the 'activation free energy' necessary to get out from R in a single ω_i and the pre-factor $|\Lambda|$ accounts for the choice of the ω_i in Λ. It turns out that the escape rate λ is directly related to the probability, in the restricted ensemble, to be at the limit of the constraints that specify R, in (4.47).

Let us now turn to a brief description of the Lebowitz–Penrose approach applied to the case of the short range Ising model. This was developed by Capocaccia, Cassandro and Olivieri in [44].

In the following we use the definitions and notation of Section 3.3.2.

We first introduce a class of stochastic Markovian dynamics for our system: the so-called Glauber dynamics, also called stochastic Ising models or kinetic Ising models. Using the equivalence between the spin system and lattice gas formulations of the Ising model these can be seen as particular interacting particle systems in the sense of Liggett's book [199], and they can be defined directly in an infinite volume as well as in a finite volume Λ with given boundary conditions outside Λ.

A Glauber dynamics is a *single spin-flip* dynamics which is *reversible* with respect to the Gibbs measure corresponding to a given interaction. More precisely, given an Ising spin system with coupling constants $\underline{J} \equiv \{J_{|x-y|}\}_{x \neq y \in \mathbb{Z}^d}$ and external magnetic field h, a Glauber dynamics for this Ising system is a Markov process on the state space $\mathcal{X} := \{-1, +1\}^{\mathbb{Z}^d}$ whose generator L (see [199], p. 15) acts on a generic cylinder function (i.e. a function depending on finitely many coordinates) f as

$$(Lf)(\sigma) = \sum_{x \in \mathbb{Z}^d} c(x, \sigma)(f(\sigma^{(x)}) - f(\sigma)), \tag{4.50}$$

where $\sigma^{(x)}$ is the configuration obtained from σ by flipping the spin at the site x, namely:

$$\sigma^{(x)}(y) = \begin{cases} \sigma(y) & \text{if } y \neq x \\ -\sigma(x) & \text{if } y = x, \end{cases} \tag{4.51}$$

and $c(x, \sigma)$ is called the rate of flip of the spin at the site x when the system is in the configuration $\sigma \in \mathcal{X}$. It is known (see [199], p. 27) that under suitable regularity conditions on the rates $c(x, \sigma)$ the generator is well defined and it generates a unique Markov semigroup.

The main assumption that we make on the rates $c(x, \sigma)$ is the reversibility property with respect to the Gibbs measures $\mu = \mu_{\underline{J},h}$ for our Ising system. This means that the generator L is self-adjoint in $\mathrm{L}^2(\mathcal{X}, d\mu)$, and in particular μ is a stationary measure for the generated Markov process (see [199] p. 91). An equivalent explicit condition for reversibility is the so-called detailed balance:

$$c(x, \sigma) = c(x, \sigma^{(x)}) \exp(-\beta \Delta_x H(\sigma)), \tag{4.52}$$

where

$$\Delta_x H(\sigma) := \sigma(x) \left(\sum_{y \in \mathbb{Z}^d} 2J_{|x-y|}\sigma(y) + 2h \right) \tag{4.53}$$

which formally equals $H(\sigma^{(x)}) - H(\sigma)$.

A finite volume version of this dynamics in Λ with boundary condition τ outside Λ is easily defined in a similar way. It is a continuous time Markov chain on the state space $\mathcal{X}_\Lambda = \{-1, +1\}^\Lambda$ with generator L_Λ^τ given by:

$$(L_\Lambda^\tau f)(\sigma) = \sum_{x \in \Lambda} c_\Lambda^\tau(x, \sigma)(f(\sigma^{(x)}) - f(\sigma)), \tag{4.54}$$

where the reversibility condition becomes

$$c_\Lambda^\tau(x, \sigma) = c_\Lambda^\tau(x, \sigma^{(x)}) \exp(-\beta \Delta_x H_\Lambda^\tau(\sigma)) \tag{4.55}$$

where $\Delta_x H_\Lambda^\tau(\sigma) = H_\Lambda^\tau(\sigma^{(x)}) - H_\Lambda^\tau(\sigma)$ and $H_\Lambda^\tau(\sigma)$ given by (3.91) or (3.108). An interesting example is the so-called Metropolis dynamics:

$$c(x, \sigma) = \exp\{-\beta(\Delta_x H(\sigma))^+\}$$

where $a^+ = \max\{a, 0\}$ is the positive part of a.

In [44] the authors study the problem of metastability for the standard n.n. Ising model, cf. (3.108), when the time evolution is given by a Glauber dynamics. The system is supposed to be enclosed in an arbitrarily large finite cube Λ with homogeneous minus boundary condition; the magnetic field h is positive and sufficiently small; the temperature $\mathrm{T} = \beta^{-1}$ is sufficiently low.

In this case the subset R is defined as the set of all configurations giving rise to suitably small clusters of pluses. To be precise, consider for any spin configuration $\sigma \in \mathcal{X}_\Lambda$ the set $C(\sigma)$ given as the union of all closed unit cubes in $\mathbb{R}^d$ centred at the sites x where the spin $\sigma_x = +1$. For any σ the 'sea of minuses' is well defined: it is the subset of Λ given by the totality of sites x for which $\sigma_x = -1$ and there exists a (nearest neighbour connected) path of minuses in Λ connecting x with the boundary $\partial\Lambda$. Let $\gamma_1, \ldots, \gamma_k$ denote the maximal connected components of the boundary of $C(\sigma)$ all made by unit segments touching the sea of minuses. Any such γ can be seen as a polyhedron made by unit faces. For $d = 2$ it is a closed polygon. They will be called 'outer contours'.

Given a generic configuration σ, we denote by $\Gamma(\sigma)$ the set of all its outer contours. If γ is an outer contour we denote by $\Theta(\gamma)$ the region it encloses, and by $|\Theta(\gamma)|$ its volume (area in $d = 2$). For $c > 0$ we set

$$R \equiv R_c := \{\sigma \in \mathcal{X}_\Lambda : \max_{\gamma \in \Gamma(\sigma)} |\Theta(\gamma)| \leq c^2\}, \tag{4.56}$$

and having in mind the conditions of the Lebowitz–Penrose approach we would like to find an 'optimal' choice of the parameter c (i.e. with largest lifetime). In this connection, and considering only the two-dimensional case, treated in [44], let us introduce a simple heuristic argument leading, on the basis of purely static, energetic considerations, to the notion of 'critical droplet'. This notion plays a fundamental role not only in the present point of view of evolution of ensembles, but also as we shall see, in other more complete dynamical approaches to metastability.

First of all let us take a very low temperature; it seems reasonable to assume that, under these conditions, the typical shape of a plus droplet in the sea of minuses will be almost square; moreover the free energy (both bulk and surface) will almost coincide with the energy. Let $E(L)$ denote the energy formation of a square droplet of pluses $Q(L)$ with side L (and, say, with lower left corner at the origin):

$$E(L) := H(\underline{\sigma}^{(L)}) - H(-\underline{1}),$$

where by $-\underline{1}$ we denote the configuration with all minuses and by $\underline{\sigma}^{(L)}$ we denote the configuration where the pluses are exactly the sites in $Q(L)$. We have

$$E(L) = 8JL - 2hL^2 \tag{4.57}$$

and observe at once that the maximum value for $E(L)$ is attained at the integer(s) closest to $L^* = 2J/h$.

From an energetical point of view droplets with side L smaller than L^* will have a tendency to shrink whereas droplets with L larger than L^* will have a tendency to grow. The 'activation energy' necessary to nucleate the stable phase locally is given by:

$$E(L^*) = 8JL^* - 2h(L^*)^2 = \frac{8J^2}{h}. \tag{4.58}$$

This heuristic argument is at the basis of the rigorous result of [44], which provides, at low values of temperature, a range of c where $R = R_c$ given by (4.56) verifies the prescriptions.

More precisely, for β large enough the authors apply the Minlos and Sinai [219] analysis of contours to show that if $c \le 4(\beta J - \log 6)/\beta h$ the restricted measure defines a (negatively magnetized) pure phase: in the thermodynamic limit the limiting free energy is infinitely differentiable in β and h; the central limit is valid for the usual empirical observables. Moreover, $\mu(R_c) \to 0$ in the thermodynamic limit. Concerning the escape rate, for large β, and suitable restrictions on J and h, they obtain lower and upper bounds on the optimal c and the minimal escape rate λ, which leads to

$$\frac{\lambda}{|\Lambda|} \approx \exp\left(-\frac{8\beta J^2}{h}\right) \quad \text{as} \quad \beta \to +\infty, \tag{4.59}$$

in agreement with the heuristic argument and the droplet model.

Remark The precise restrictions and bounds obtained in [44] have some dependence on the class of rates used (e.g. $J > \sqrt{2}h \ge AJe^{-2\beta J}$, with A a suitable absolute constant is good for a class of rates.) It is important to have the minimal escape rate per unit volume in (4.59) much smaller than the minimal value of the single flip rate; for the Metropolis dynamics this means $4J^2/h > 4J + h$, compatible with the restrictions in [44].

Again the factor $|\Lambda|$ appears in the escape rate λ, cf. (4.59). This is due to the arbitrary choice of the location where nucleation takes place. In contrast to the previous case of long range Kac potential considered in [240], here in the short range situation, the lifetime stays finite and the 'escape rate per unit volume' can only be made very small by properly choosing the parameters β and h. This is related to the phenomenon of 'homogeneous nucleation'.

We shall see that in more complete dynamical approaches, the determination of the typical shapes of the droplets during nucleation is in general very delicate. Generally, the reasoning leading to the assumption of a shape solving the isoperimetric problem is not justified and leads to incorrect results.

Analogous results were obtained by Cassandro and Olivieri in the framework of the Widom–Rowlinson model; this is a classical continuous model with two kinds of particles, A and B, in which there is a hard core repulsion of range r between unlike particles and no interaction between those of the same type. The activities z_A, z_B and the range r are the only parameters. For this model Ruelle, in

[259], proved rigorously the existence of a phase transition when $z_A = z_B$ is sufficiently large. The proof is based on an analysis in terms of contours similar to the Peierls contours for the Ising model, seen in Section 3.3.2. In [46], a suitable restricted ensemble for the Widom–Rowlinson model was introduced when the activities z_A, z_B are large and close to each other; the set R was defined similarly to that given in [44] for the Ising model. The main difference is that in [46] the natural deterministic time evolution, given by the Liouville equation, was used. The main result there tells that the relaxation time per unit volume can be made arbitrarily large by taking the activities sufficiently large and close to each other.

4.2 The pathwise approach: basic description

In this section we start the discussion of a different approach to the dynamical phenomenon of metastability.

The situation, as in the Lebowitz–Penrose approach, could be that of a stochastic process with a unique stationary measure and which, for suitable initial conditions, behaves for 'a very long time' as if it were described by another stationary measure. Finally, and unpredictably, it undergoes the proper transition.

In contrast to the Lebowitz–Penrose approach, where the evolution of the ensembles is analysed, the basic idea of this approach is to single out typical trajectories and to look at the statistics along such typical paths. Time averages constitute natural statistics along trajectories, and the proposal is to characterize metastability through their following behaviour: stability–sharp transition–stability. This is made precise below, and the 'unpredictability' manifests itself through an approximate exponential random time for the transition. The exponential is a quite wide distribution (the average coincides with the standard deviation) and while looking at the evolution of ensembles (i.e. the expectation in the probability space) we usually do not see the sharp behaviour; it is simply smoothed out. It becomes quite difficult to distinguish, in the framework of the evolution of ensembles, between the case when single trajectories do behave as above (metastability) or when they simply evolve smoothly, and very slowly, towards the equilibrium, which has nothing to do with metastability. The pathwise approach, introduced in [48], by Cassandro, Galves, Olivieri and Vares is one proposal for this clarification.

The basic objects we look at are the empirical averages along typical trajectories. The main ingredient is the existence of two time scales: one, much shorter, on which apparent stability is achieved. This may be called a 'thermalization' process. On the longer time scale one finally observes 'tunnelling', that is, escape from the metastable situation. In order to make this precise, we analyse a limiting situation and set it in terms of convergence to a jump process, in a manner to be made rigorous later. As a limit theorem this misses an exact prescription for 'the metastable state'.

In the rest of this chapter, and in the next one, we illustrate the approach, through three examples.

(a) *The Curie–Weiss chain.* This is a very simple birth and death chain, connected to the Curie–Weiss model which we discussed in the previous section. Due to its extreme simplicity, the expected 'tunnelling' time is simply computable, and everything is treated by completely elementary methods. It is discussed in detail just as a very simple (rather trivial and particular) illustration.

(b) *The Harris contact process on a finite set.* Considered in the limiting situation as this set becomes infinite, this example has several interesting aspects, including a non-trivial spatial structure. We do not discuss all details but try to give an idea of various procedures and methods. Ideas from percolation theory are essential. The contact process is a very well studied interacting particle system on which a vast literature exists (see [199, 200]). Concerning its metastable behaviour some open problems will be mentioned at the end of the chapter.

(c) *Itô processes in the Freidlin–Wentzell regime.* The analysis of metastability for this class of processes is obtained as an application of large deviation estimates and of Freidlin and Wentzell theory. The results of Chapter 2 need to be extended in order to treat 'tunnelling' problems. Besides large deviations, coupling methods constitute an extra ingredient, used to verify the already mentioned unpredictability. Particular attention is devoted to the gradient case, for which an alternative proof is outlined, exploiting the explicit knowledge of the stationary (reversible) measure and coupling methods. These models are discussed extensively in Chapter 5.

In Chapters 6 and 7 we get to the study of metastability for a class of systems that includes the short range Ising models in the low temperature regime. This is one of the main topics of our analysis and is discussed extensively. Examples, extensions and open problems are further discussed in Chapter 7.

4.3 The Curie–Weiss Markov chain

We now turn to the study of a dynamical model for the Curie–Weiss theory, introduced in Section 4.1.1. Our main interest is the verification of its metastable behaviour in terms of the just described 'pathwise approach'. It is an extremely simplified and unphysical model, but it serves as an initial illustration.

Since the energy function in (4.11) depends on the configuration σ only through the average of the spins $m_N(\sigma)$ (the magnetization), it is natural to forget the individual spins σ_i and think only in terms of $m_N(\sigma)$. Thus we change from the configurational space to $\mathcal{Y}_N$, cf. (4.14), and consider the distribution of $m_N(\sigma)$

under the Gibbs measure $\mu_{N,\beta,h}$, denoted by $\nu_{N,\beta,h}$:

$$\nu_{N,\beta,h}(m) = \frac{1}{Z_N}\binom{N}{N(\frac{m+1}{2})} e^{\beta N(\frac{m^2}{2}+hm)} = \frac{1}{Z_N} e^{-N\beta f_{N,\beta,h}(m)}, \tag{4.60}$$

with

$$f_{N,\beta,h}(m) = -\left(\frac{m^2}{2}+hm\right) - \frac{1}{N\beta}\log\binom{N}{N(\frac{m+1}{2})}, \tag{4.61}$$

for each $m \in \mathcal{Y}_N$. (To avoid heavy indexing, the dependence on β and h is sometimes dropped from the notation, provided no confusion arises.)

As very simple stochastic dynamics associated with the Curie–Weiss model we may take any Markov chain on $\mathcal{Y}_N$ having ν_N as unique invariant measure. That is, a Markov chain $(\xi_N(n) : n \geq 0)$ on $\mathcal{Y}_N$ whose transition matrix $(Q_N(m,\tilde{m}))$ satisfies:

$$\sum_{m\in\mathcal{Y}_N} \nu_N(m)\, Q_N(m,\tilde{m}) = \nu_N(\tilde{m}), \quad \text{for each} \quad \tilde{m} \in \mathcal{Y}_N. \tag{4.62}$$

Among these, birth and death chains refer to those situations where $Q_N(m,\tilde{m}) = 0$ if $|m-\tilde{m}| > 2/N$, $m,\tilde{m} \in \mathcal{Y}_N$. (The number of up spins, given by $N(1+m)/2$, increases or decreases by one at each step.) In connection with our model, birth and death chains appear naturally from a specific class of spin-flip Markov dynamics on the configurations σ; see Remark 4.4 below.

For birth and death chains, we easily check that (4.62) simplifies to

$$\nu_N(m)\, Q_N(m,\tilde{m}) = \nu_N(\tilde{m})\, Q_N(\tilde{m},m), \quad \text{for} \quad m,\tilde{m} \in \mathcal{Y}_N, \;\; |m-\tilde{m}| = 2/N. \tag{4.63}$$

Remark According to the usual definition

$$Q(m,\tilde{m}) = P(\xi(j+1) = \tilde{m} \,|\, \xi(0) = m_0, \dots, \xi(j-1) = m_{j-1}, \xi(j) = m),$$

whenever $P(\xi(0) = m_0, \dots, \xi(j-1) = m_{j-1}, \xi(j) = m) > 0$ and $j \geq 0$. If ν is a probability on $\mathcal{Y}_N$, we use P_ν to denote the law of the chain when $\xi(0)$ is distributed according to ν, i.e. the unique probability on the Borel σ-field of the sequence space $(\mathcal{Y}_N)^{\mathbb{Z}_+}$ verifying

$$P_\nu\{\xi : \xi(0) = m_0, \dots, \xi(k) = m_k\} = \nu(m_0) Q_N(m_0,m_1) \dots Q_N(m_{k-1},m_k),$$

for any $k \geq 1, m_0, \dots, m_k \in \mathcal{Y}_N$.

Equation (4.62) is usually called the 'balance condition'. It says that starting with $\xi_N(0)$ distributed according to ν_N the chain is 'time stationary', i.e.

$$\begin{aligned} P_{\nu_N}(\xi_N(0) = m_0, \xi_N(1) = m_1, \dots, \xi_N(k) = m_k) &= P_{\nu_N}(\xi_N(j) \\ = m_0, \xi_N(j+1) = m_1, \dots, \xi_N(j+k) = m_k), \end{aligned} \tag{4.64}$$

for each $j, k \geq 0$, each $m_0, \ldots, m_k \in \mathcal{Y}_N$. (In this case the measure P_ν may be uniquely extended to $(\mathcal{Y}_N)^{\mathbb{Z}}$, preserving (4.64).)

On the other hand, the validity of $\nu_N(m)Q_N(m, \tilde{m}) = \nu_N(\tilde{m})Q_N(\tilde{m}, m)$ for each pair of different points $m, \tilde{m}$ is usually called a 'detailed balance'. For general Markov chains this is stronger than (4.62) and gives the time reversibility of the measure P_{ν_N}:

$$P_{\nu_N}(\xi_N(0) = m_0, \ldots, \xi_N(k) = m_k) = P_{\nu_N}(\xi_N(0) = m_k, \ldots, \xi_N(k) = m_0) \tag{4.65}$$

for each $k \geq 0$, $m_0, \ldots, m_k \in \mathcal{Y}_N$. (We also say that the process $\xi_N(\cdot)$ is reversible under ν_N, or still that ν_N is reversible for the given process.)

Reversibility is an important notion. (It was already mentioned in the previous section and will return in Section 5.2 and in Chapter 6.) It is also a special one, in the sense that 'most' processes simply do not exhibit it. Birth and death chains are the simplest examples of reversible chains; as previously mentioned, in this special situation (4.62) indeed implies (4.63), i.e. stationary birth and death chains are also reversible.

In the particular case under consideration, we may rewrite (4.63) as:

$$Q_N(m, \tilde{m})\sqrt{\frac{\nu_N(m)}{\nu_M(\tilde{m})}} = Q_N(\tilde{m}, m)\sqrt{\frac{\nu_N(\tilde{m})}{\nu_M(m)}}, \quad \text{for} \quad m, \tilde{m} \in \mathcal{Y}_N,$$

$$|m - \tilde{m}| = 2/N,$$

and, recalling (4.60),

$$Q_N(m, \tilde{m}) = \sqrt{\frac{\nu_N(\tilde{m})}{\nu_M(m)}} g_N(m, \tilde{m}) = e^{-\frac{N}{2}\beta(f_{N,\beta,h}(\tilde{m}) - f_{N,\beta,h}(m))} g_N(m, \tilde{m}), \tag{4.66}$$

where $g_N(m, \tilde{m}) = g_N(\tilde{m}, m) > 0$ provided $|m - \tilde{m}| = 2/N$. This last is chosen in a way that we have a transition probability matrix. Here are some common examples.

Examples (1)

$$Q_N(m, \tilde{m}) = \frac{\upsilon}{\sqrt{N}} e^{-\frac{N\beta}{2}(f_{N,\beta,h}(\tilde{m}) - f_{N,\beta,h}(m))},$$

for $m, \tilde{m} \in \mathcal{Y}_N$, $|m - \tilde{m}| = 2/N$, and

$$\begin{aligned} Q_N(1, 1) &= 1 - Q_N(1, 1 - 2/N); \\ Q_N(-1, -1) &= 1 - Q_N(-1, -1 + 2/N) \\ Q_N(m, m) &= 1 - Q_N(m, m + 2/N) \\ &\quad - Q_N(m, m - 2/N) \quad \text{if } m \in \mathcal{Y}_N \setminus \{-1, +1\}, \end{aligned} \tag{4.67}$$

where $\upsilon > 0$ is fixed constant (independent of N), sufficiently small so that $Q_N(m, m) \geq 0$ for each m. (And consequently we get a stochastic matrix.)

(2)

$$Q_N(m, \tilde{m}) = \frac{1}{2} \frac{e^{-\frac{N\beta}{2}(f_{N,\beta,h}(\tilde{m}) - f_{N,\beta,h}(m))}}{e^{-\frac{N\beta}{2}(f_{N,\beta,h}(\tilde{m}) - f_{N,\beta,h}(m))} + e^{\frac{N\beta}{2}(f_{N,\beta,h}(\tilde{m}) - f_{N,\beta,h}(m))}},$$

if $m, \tilde{m} \in \mathcal{Y}_N$ with $|m - \tilde{m}| = 2/N$, and the diagonal terms $Q_N(m, m)$ follow (4.67).

(3) (Metropolis)

$$Q_N(m, \tilde{m}) = \frac{1}{2} e^{-N\beta(f_{N,\beta,h}(\tilde{m}) - f_{N,\beta,h}(m))^+} \tag{4.68}$$

if $m, \tilde{m} \in \mathcal{Y}_N$ and $|m - \tilde{m}| = 2/N$, and the diagonal terms $Q_N(m, m)$ are defined according to (4.67). ($x^+ = \max\{x, 0\}$.)

The results below apply to a large class of examples, including those above. Different choices of the function $g_N(\cdot, \cdot)$ in (4.66) might make the 'flow' much slower between certain pairs $m, \tilde{m}$ (two small $g_N(m, \tilde{m})$). Of course, this is reflected in the thermalization and tunnelling times, estimated in Corollary 4.9 and Theorem 4.10.

Remark 4.3 Examples (2) and (3) have the following property: given a small $\delta > 0$ we can take a positive constant a_δ (independent of N) for which $Q_N(m, m + 2/N) \ge a_\delta$ for each $m \in [-1, 1 - \delta] \cap \mathcal{Y}_N$, and $Q_N(m, m - 2/N) \ge a_\delta$ if $m \in [-1 + \delta, 1] \cap \mathcal{Y}_N$. In Example (1) this lower bound is of order $N^{-1/2}$. The differences among these cases are much smaller than the relevant scale for metastability and will not even appear in the rougher statement needed for Theorem 4.10 below. They are seen in Corollary 4.9.

Remark 4.4 We have chosen to define a Markov chain directly on $\mathcal{Y}_N$, but one could take some microscopic type dynamics $\sigma(t)$ satisfying the detailed balance condition with respect to $\mu_{N,\beta,h}(\sigma) = e^{-\beta H_N(\sigma)}/Z_N$, the Gibbs measure on $\mathcal{X}_N$ at inverse temperature β. For chains where only one spin may flip at each step, the transition probability $q_N(\sigma, \tilde{\sigma})$ vanishes unless $\tilde{\sigma} = \sigma^{(i)}$, for some $i = 1, \dots, N$, where $\sigma^{(i)}$ denotes the configuration obtained from σ by flipping the spin at site i, when we must have, as in (4.66):

$$q_N(\sigma, \sigma^{(i)}) = \frac{\sqrt{\mu_{N,\beta,h}(\sigma^{(i)})}}{\sqrt{\mu_{N,\beta,h}(\sigma)}} s_N(\sigma, \sigma^{(i)}) = e^{-\frac{\beta}{2}(H_N(\sigma^{(i)}) - H_N(\sigma))} s_N(\sigma, \sigma^{(i)}),$$

where $s_N(\sigma, \sigma^{(i)}) = s_N(\sigma^{(i)}, \sigma) > 0$. Since $H_N(\sigma)$ is a function of $m_N(\sigma)$ ($H_N(\sigma) = H_N(m_N(\sigma))$ with little abuse of notation) one is naturally led to consider examples where $s_N(\sigma, \sigma^{(i)})$ depends on $(\sigma, \sigma^{(i)})$ through $(m_N(\sigma), m_N(\sigma^{(i)}))$. In this situation $m_N(\sigma(t))$ will again be a Markov chain, for which ν_N is reversible. One such natural example was used by Griffiths, Weng and Langer in [146]: at each time unit a site $i \in \{1, \ldots, N\}$ is chosen at random and the corresponding spin is flipped with probability

$$\tilde{\upsilon}\frac{\sqrt{\mu_{N,\beta,h}(\sigma^{(i)})}}{\sqrt{\mu_{N,\beta,h}(\sigma)}} = \tilde{\upsilon}e^{-\frac{\beta}{2}(H_N(\sigma^{(i)})-H_N(\sigma))};$$

otherwise, nothing happens. $\tilde{\upsilon} > 0$ is taken independent of N and sufficiently small to make these numbers less than one. The evolution induced on $m_N(\sigma(t))$ is then a Markov chain on $\mathcal{Y}_N$ whose transition probabilities are given below.

Example (4) (Griffiths, Weng and Langer)

$$Q_N(m, m + \frac{2}{N}) = \tilde{\upsilon}\frac{1-m}{2}e^{-\frac{\beta}{2}(H_N(m+\frac{2}{N})-H_N(m))},$$

$$Q_N(m, m - \frac{2}{N}) = \tilde{\upsilon}\frac{1+m}{2}e^{-\frac{\beta}{2}(H_N(m-\frac{2}{N})-H_N(m))},$$

if $m, m + 2/N \in \mathcal{Y}_N$ or $m, m - 2/N \in \mathcal{Y}_N$, respectively; the diagonal terms $Q_N(m, m)$ follow rules as in (4.67). As in Examples (2) and (3), the above $Q_N(m, \tilde{m})$, for $|m - \tilde{m}| = 2/N$ are uniformly bounded away from zero. They differ from those in Example (1) by the factor $\frac{\tilde{\upsilon}}{2\upsilon}\sqrt{N(1-(m \wedge \tilde{m}))(1+(m \vee \tilde{m}))}$.

Remark It might be preferable to work with continuous time Markov chains, as already introduced in Section 4.1.5, but there is no significant difference.

Before stating a theorem with the pathwise description of metastable behaviour for such chains, we need to recall a few estimates from Section 4.1.1. We fix $J = 1$.

Recalling the Gibbs free energy $f_{\beta,h}(m) = -(m^2/2 + hm) + \beta^{-1}\mathcal{E}(m)$, where $\mathcal{E}(m)$ is given by (4.15), and the Stirling formula (1.5), we can write, for $m \in \mathcal{Y}_N \setminus \{-1, +1\}$:

$$\begin{aligned}\beta f_{N,\beta,h}(m) = \beta f_{\beta,h}(m) &+ \frac{1}{N}\left(\log\frac{\sqrt{1-m^2}}{2} + \log\sqrt{2N\pi}\right) \\ &+ \frac{1}{N}\left(\epsilon_{N(\frac{1+m}{2})} + \epsilon_{N(\frac{1-m}{2})} - \epsilon_N\right),\end{aligned} \tag{4.69}$$

with $1/(12k+1) < \epsilon_k < 1/12k$. Also $f_{N,\beta,h}(-1) = f_{\beta,h}(-1)$ and $f_{N,\beta,h}(1) = f_{\beta,h}(1)$ for each N. For large N, (4.69) provides a good approximation for the transition probabilities in terms of the derivative of $\beta f_{\beta,h}(m)$. The approximation is uniform on each compact interval contained in $(-1, +1)$.

If $\beta > 1$ and $0 < h < \beta^{-1}\log(\sqrt{\beta} - \sqrt{\beta - 1}) + \sqrt{(\beta-1)/\beta} =: h_0(\beta)$, the graph of $f_{\beta,h}(\cdot)$ has a double well structure, with absolute minimum at $m^+(\beta, h) > 0$, a local minimum at $m^-(\beta, h) < 0$, and a local maximum at $m^0(\beta, h)$; the critical points are the solutions of equation (4.17) with $J = 1$.

From now on we fix β and h under the above conditions. The double well is suggestive of metastable behaviour and our next task is to prove it. The Curie–Weiss chain behaves similarly to a one-dimensional Itô process in the Freidlin–Wentzell regime.

Notation Since the parameters β and h are fixed we write simply m^-, m^0, m^+. The same is done for $f(m)$ and $f_N(m)$ (instead of $f_{\beta,h}(m)$ and $f_{N,\beta,h}(m)$, respectively).

Remark For finite birth and death chains one gets explicit expressions for expected hitting times by working out difference equations ([115], Chapter XIV). They were used in [48] to provide upper and lower bounds on the two time scales (thermalization and tunnelling). Here we have modified the proofs a little, so that possible extensions become more clear. Nevertheless, in this section we use strongly the one dimensionality of the process and its property of point recurrence. Generalizations to situations where this is not true, as in higher dimensional Itô processes, are treated in Chapter 5 with the help of suitable couplings, and the theory of Freidlin and Wentzell. Higher dimensional analogues of the Curie–Weiss chains appear naturally; the Curie–Weiss model under the action of a random magnetic field is an example. Their metastable behaviour has been described in [216] and [31], using analytical and probabilistic tools. Reversibility plays a fundamental role and allows quite precise estimates. This is outlined briefly at the end of this chapter, together with indications to (simpler) arguments that lead to rougher estimates.

Since m^-, m^0 and m^+ might not belong to $\mathcal{Y}_N$, it is convenient to fix $i_-(N)$, $i_0(N)$, $i_+(N)$ integers so that

$$m_a(N) := -1 + \frac{2i_a(N)}{N} \le m^a < -1 + \frac{2(i_a(N)+1)}{N} \tag{4.70}$$

for $a = -, +$ or 0. We fix $\varepsilon > 0$ so that for all N sufficiently large, $-1 + \varepsilon < m_-(N) < m_0(N) < m_+(N) < 1 - \varepsilon$, which we may do since $-1 < m^- < m^0 < m^+ < +1$. Also, let $i_*(N)$ be defined as the integer satisfying

$$m_*(N) := -1 + \frac{2i_*(N)}{N} \le m^0 + \frac{1}{N^{1/4}} < -1 + \frac{2(i_*(N)+1)}{N}. \tag{4.71}$$

Notation An upper index j in $\xi_N^j(\cdot)$ indicates the initial condition $\xi_N^j(0) = -1 + 2j/N$, i.e. j is the number of positive spins at time 0. This might also be indicated as a subindex in the probability distribution: $P_j(\xi_N \in A) = P(\xi_N \in A \mid \xi_N(0) = -1 + 2j/N) = P(\xi_N^j \in A)$, for $j \in \{0, \ldots, N\}$ and A a Borel set in the product space $(\mathcal{Y}_N)^{\mathbb{Z}_+}$. E_j will refer to the expectation with respect to P_j.

Definition 4.5 If $i, j \in \{0, 1, \ldots, N\}$ we set

$$T_N^{i,j} = \min \left\{ n \geq 1 : \xi_N^i(n) = -1 + \frac{2j}{N} \right\},$$

with $T_N^{i,j}$ taken as $+\infty$ if the above set is empty. For $i \neq j$ $(i = j)$, $T_N^{i,j}$ is the time of the first visit (revisit, respectively) to $-1 + 2j/N$, starting at $-1 + 2i/N$. We set $T_N = T_N^{i_-(N), i_*(N)}$.

Notation According to the previous notation $P_i(T_N^j = k) = P(T_N^{i,j} = k)$.

T_N represents the 'tunnelling' time for this simple model. Having arrived at $m_*(N)$ the process will, with overwhelming probability, reach $m_+(N)$ before going back to $m_0(N)$, and in a relatively short time, for N large. (This is proven below.)

The theorem below gives the (rough) order of magnitude of T_N and its asymptotic unpredictability, or loss of memory.

Theorem 4.6 *Let* $\Delta = \beta\left(f(m^0) - f(m^-)\right)$. *For any of the previously described examples we have:*

(*a*) $\lim_{N \to +\infty} P(e^{N(\Delta - \delta)} < T_N < e^{N(\Delta + \delta)}) = 1$, *for each* $\delta > 0$.

(*b*) T_N / ET_N *converges in law to a unit mean exponential random variable, as* $N \to +\infty$.

While proving (a) above, we shall indeed get a more precise estimate, cf. Corollary 4.9 below.

Theorem 4.6 involves the behaviour of the process ξ_N starting to the left of $m_*(N)$, until it hits $m_*(N)$. Thus, we may as well consider a new chain $\bar{\xi}_N$ on

$$\bar{\mathcal{Y}}_N = \left\{ -1, -1 + \frac{2}{N}, \ldots, -1 + \frac{2i_*(N)}{N} \right\}, \tag{4.72}$$

with transition probabilities:

$$\begin{aligned} \bar{Q}_N(m, \tilde{m}) &= Q_N(m, \tilde{m}) \quad \text{if} \quad m, \tilde{m} \in \bar{\mathcal{Y}}_N, \\ \bar{Q}_N(m_*(N), m_*(N)) &= 1 - \bar{Q}_N(m_*(N), m_*(N) - 2/N). \end{aligned} \tag{4.73}$$

In other words, $\bar{\xi}_N$ has the same transition probabilities as ξ_N to the left to $m_*(N)$, where a reflecting barrier has been introduced. We have a trivial coupling of $\bar{\xi}_N$ and ξ_N up to T_N: starting in $\bar{\mathcal{Y}}_N$, the two processes move together until for the first time ξ_N jumps out of $\bar{\mathcal{Y}}_N$. Recalling (4.73), at this step $\bar{\xi}_N$ remains at $m_*(N)$, and from then on they behave independently. This describes a well defined Markov chain on the product $\mathcal{Y}_N \times \bar{\mathcal{Y}}_N$, whose transition probability matrix can be easily written.

We check that $\bar{\nu}_N$, defined as ν_N conditioned to $\bar{\mathcal{Y}}_N$, is reversible for $\bar{\xi}_N$. That is, setting

$$\bar{\nu}_N(m) = \frac{\nu_N(m)}{\sum_{m' \in \bar{\mathcal{Y}}_N} \nu_N(m')}, \quad \text{for} \quad m \in \bar{\mathcal{Y}}_N, \tag{4.74}$$

we have $\bar{\nu}_N(m)\,\bar{Q}_N(m, m') = \bar{\nu}_N(m')\,\bar{Q}_N(m', m)$, for any $m, m' \in \bar{\mathcal{Y}}_N$.

For the proof of Theorem 4.6 we may replace ξ_N by $\bar{\xi}_N$. The hitting times for the chain $\bar{\xi}_N$ are denoted by $\bar{T}_N^{i,j}$ and defined analogously to $T_N^{i,j}$, if $i, j \le i_*(N)$. One has $\bar{T}_N := \bar{T}_N^{i_-(N), i_*(N)} = T_N$ with probability one in the above mentioned coupling.

We start by the following simple observation which gives a lower bound for $\bar{T}_N$, and is an immediate consequence of the stationarity of $\bar{\nu}_N$.

Proposition 4.7 *There exists a positive constant c_1, so that for all N sufficiently large and each T positive integer*

$$P\left(\bar{T}_N \le T\right) \le c_1\, T\, e^{-N\Delta}. \tag{4.75}$$

Proof Since all the jumps have size $2/N$, $P\left(\bar{T}_N \le T\right) \le P_{i_-(N)}(\bar{T}_N^{i_0(N)} \le T)$. Moreover,

$$\begin{aligned}
P_{i_-(N)}\left(\bar{T}_N^{i_0(N)} \le T\right) &\le \sum_{n=1}^{T} P_{i_-(N)}\left(\bar{\xi}_N(n) = m_0(N)\right) \\
&= \sum_{n=1}^{T} \frac{1}{\bar{\nu}_N(m_-(N))} P_{\bar{\nu}_N}\left(\bar{\xi}_N(0) = m_-(N), \bar{\xi}_N(n) = m_0(N)\right) \\
&\le \sum_{n=1}^{T} \frac{1}{\bar{\nu}_N(m_-(N))} P_{\bar{\nu}_N}\left(\bar{\xi}_N(n) = m_0(N)\right) \\
&= T\, \frac{\bar{\nu}_N(m_0(N))}{\bar{\nu}_N(m_-(N))},
\end{aligned}$$

using the stationarity of $\bar{\nu}_N$ at the last step. Recalling the expression for $\bar{\nu}_N$ we may write:

$$P\left(\bar{T}_N \le T\right) \le T\, e^{-N\beta(f_N(m_0(N)) - f_N(m_-(N)))}.$$

From (4.69) and the observation following (4.70), a constant $c > 0$ may be taken so that for all N sufficiently large:

$$N\beta(f_N(m_0(N)) - f_N(m_-(N))) \ge N\beta(f(m_0(N)) - f(m_-(N))) - c.$$

The proof follows by observing that we may also take $\tilde{c} > 0$ so that for all N large

$$\beta(f(m_0(N)) - f(m_-(N))) \ge \Delta - \tilde{c}/N.$$

□

Remark. For higher dimensions (or if jumps are not only between nearest neighbours) we should keep $\mathrm{T}\,\bar{\nu}_N(m_*(N))/\bar{\nu}_N(m_-(N))$ for the upper bound, including a factor $e^{cN^{3/4}}$ on the right-hand side of (4.75). This still fits to the lower bound in part (a) of Theorem 4.6.

We now need an upper bound for the tunnelling time T_N. For finite birth and death chains, closed expressions for the expected hitting times are elementary and easily obtained, as summarized below.

Lemma 4.8 *Let $(\xi(n) : n \geq 0)$ be a Markov chain with values on the finite set $\{0, 1, \ldots, M\}$ and transition probabilities $Q(i, j) = P(\xi(n+1) = j \mid \xi(n) = i)$, given by:*

$$Q(i,j) = \begin{cases} p_i & \text{if } j = i+1 \\ q_i & \text{if } j = i-1 \\ r_i & \text{if } j = i, \end{cases}$$

for $1 \leq i \leq M-1$, where $p_i > 0,\ q_i > 0,\ r_i \geq 0,\ p_i + q_i + r_i = 1$, and

$$Q(0,1) = p_0 = 1 - Q(0,0), \quad Q(M, M-1) = q_M = 1 - Q(M,M), \tag{4.76}$$

where $0 < p_0, q_M < 1$. (We write $q_0 = 1 - p_0,\ p_M = 1 - q_M$.)

Let also

$$\tau_j = \min\{n \geq 0 : \xi(n) = j\}. \tag{4.77}$$

(i.e. $P_j(\tau_j = 0) = 1$, and under $P_i, i \neq j,\ \ \tau_j$ is the time of first visit to the state j.) Then, for any $0 \leq i < j \leq M$:

$$E_i\left(\tau_j\right) = \sum_{\ell=i}^{j-1} \sum_{r=0}^{\ell} \left\{ \frac{1}{p_r} \prod_{s=r+1}^{\ell} \frac{q_s}{p_s} \right\} = \sum_{\ell=i}^{j-1} \sum_{r=0}^{\ell} \left\{ \frac{\nu_r}{\nu_\ell} \frac{1}{p_\ell} \right\}, \tag{4.78}$$

$$E_j\left(\tau_i\right) = \sum_{\ell=i+1}^{j} \sum_{r=\ell}^{N} \left\{ \frac{1}{q_r} \prod_{s=\ell}^{r-1} \frac{p_s}{q_s} \right\} = \sum_{\ell=i+1}^{j} \sum_{r=\ell}^{N} \left\{ \frac{\nu_r}{\nu_\ell} \frac{1}{q_\ell} \right\}, \tag{4.79}$$

with the convention that $\prod_{s=\ell}^{\ell-1} = 1$, and where $\nu = (\nu_i)$ denotes the stationary (reversible) measure *with $\nu_i = \nu(i)$.*

Proof This is an elementary application of the method of difference equations, as in Chapter XIV of [115]. The second equality of each of the above equations follows at once from the detailed balance ($\nu_i p_i = \nu_{i+1} q_{i+1}$) and so we prove only the first.

We prove only (4.78); it is the same as (4.79) after renaming the states.

Let us fix j, and for $i \leq j$ we set $u_i = E_i\left(\tau_j\right)$ so that $u_j = 0$, according to (4.77). Now let $1 \leq i \leq j-1$, and conditioning on the position after the first step we may write:

$$u_i = p_i(u_{i+1}+1) + q_i(u_{i-1}+1) + r_i(u_i+1).$$

Recalling that $p_i + q_i + r_i = 1$ we rewrite this as

$$(p_i + q_i)u_i = p_i\, u_{i+1} + q_i\, u_{i-1} + 1,$$

and letting $d_i = u_i - u_{i+1}$, for $1 \le i \le j-1$, we have

$$p_i\, d_i = q_i\, d_{i-1} + 1.$$

Due to (4.76), and since $u_j = 0$ we also have $d_{j-1} = u_{j-1}$, and $d_0 = 1/p_0$. Thus, for each $i \le j-1$ and using the convention $\prod_{s=r+1}^{r} = 1$, we get:

$$d_i = \sum_{r=0}^{i} \frac{1}{p_r} \prod_{s=r+1}^{i} \frac{q_s}{p_s}.$$

Summing up it follows that

$$u_i = \sum_{\ell=i}^{j-1} d_\ell = \sum_{\ell=i}^{j-1}\sum_{r=0}^{\ell} \frac{1}{p_r} \prod_{s=r+1}^{\ell} \frac{q_s}{p_s},$$

as claimed. □

Corollary 4.9

(a) *There are positive constants c_1, c_2 (independent of N) such that for each N:*

$$c_1\, N^{\kappa+1} e^{N\Delta} \le E\bar{T}_N \le c_2\, N^{\kappa+1}\, e^{N\Delta}, \tag{4.80}$$

where $\Delta = \beta\big(f(m^0) - f(m^-)\big)$, $\kappa = 1/2$ in Example (1), $\kappa = 0$ in Examples (2), (3) and (4) above.

(b) *There are positive constants c_3, c_4 (independent of N) such that for each N*

$$\sup_{i \le i_*(N)} E_i\, \bar{T}_N^{i_-(N)} \le c_3\, e^{c_4 N^{3/4}}. \tag{4.81}$$

Proof (a) The estimate is derived from Lemma 4.8. We just need a few computations. Observe first that for each of the previous examples, positive constants $c, \tilde{c}$ can be taken in a way that $cN^{-\kappa} \le p_\ell \le \tilde{c}N^{-\kappa}$ if $i_-(N) \le \ell < i_*(N)$. Thus, applying the previous lemma and recalling (4.60), we have to provide lower and upper estimates for:

$$\sum_{\ell=i_-(N)}^{i_*(N)-1} \sum_{r=0}^{\ell} e^{N\beta\left\{f_N\left(-1+\frac{2\ell}{N}\right) - f_N\left(-1+\frac{2r}{N}\right)\right\}}.$$

According to (4.69), this can be bounded from below and from above by suitable constants multiplying:

$$e^{N\Delta} \sum_{\ell=i_-(N)}^{i_*(N)-1} e^{-N\beta\left\{f\left(-1+\frac{2i_0(N)}{N}\right)-f\left(-1+\frac{2\ell}{N}\right)\right\}} \sum_{r=0}^{\ell} e^{-N\beta\left\{f\left(-1+\frac{2r}{N}\right)-f\left(-1+\frac{2i_-(N)}{N}\right)\right\}}.$$

The double sum can be bounded, both from below and from above, by suitable constants times N, as can be seen upon approximation of the sums by Gaussian integrals, due to the shape of $f(\cdot)$, and since the maximum at m^0 as well as the minimum at m^- are quadratic.

As for part (b), since we do not care about the precise value of constant c_4, the analysis is less precise and the estimate takes the same form in all the given examples; we simply take the maximum value of the ratio ν_r/ν_ℓ and multiply by the number of terms. □

Remark. Applying these last rough bounds also in part (a) would yield $c_1\, N^\kappa e^{N\Delta} \le E(\bar{T}_N) \le c_2\, N^{\kappa+2}\, e^{N\Delta}$, for suitable positive constants c_1, c_2. This would suffice for Theorem 4.6 as well as Theorem 4.10 below.

Proof of Theorem 4.6. As already observed we may consider the Markov chain $\bar{\xi}_N$, and $\bar{T}_N = \bar{T}_N^{i_-(N),i_*(N)}$. For $\varphi_N > 0$ we write:

$$P\left(\left(\frac{1}{\varphi_N} e^{N\Delta} < \bar{T}_N < e^{N\Delta}\, N^{\kappa+1}\varphi_N\right)^c\right)$$
$$= P\left(\bar{T}_N \le \frac{1}{\varphi_N}\, e^{N\Delta}\right) + P(\bar{T}_N \ge e^{N\Delta}\, N^{\kappa+1}\varphi_N).$$

Now, if $\varphi_N \to +\infty$, the first term tends to zero by Proposition 4.7; for the second term we apply Corollary 4.9 and the Markov inequality, proving part (a).

To check part (b) we set $\gamma_N := \min\{n \ge 1 \colon P(\bar{T}_N \le n) \ge 1 - e^{-1}\}$. Notice that $P(\bar{T}_N = 1) = 0$ for N large, and so

$$P(\bar{T}_N < \gamma_N) < 1 - e^{-1} \le P(\bar{T}_N \le \gamma_N). \tag{4.82}$$

From part (a) it follows at once that $N^{-1} \log \gamma_N \to \Delta$, as $N \to +\infty$. According to part (b) of Corollary 4.9 we may take $\alpha_N > 0$ so that

$$\begin{aligned} &\text{(i)} \quad \alpha_N/\gamma_N \to 0, \\ &\text{(ii)} \quad \varepsilon_N := \sup_{i \le i_*(N)} E_i\, \bar{T}_N^{i_-(N)}/\alpha_N \to 0, \end{aligned} \tag{4.83}$$

as $N \to +\infty$. (Possible choices include $\alpha_N = e^{N\varsigma}$, with $3/4 < \varsigma < 1$.)

Remark Consideration of the restricted chain $\bar{\xi}_N$ in (4.83) (ii) is necessary. This is not true for ξ_N (cf. Corollary 4.15 below).

The asymptotic loss of memory of $\bar{T}_N/\gamma_N$ is an easy consequence of the faster recurrence of $\bar{\xi}_N$ on $m_-(N)$. To see this, we let, for $s \ge 0$:

$$\bar{t}_{N,s} = \inf\{n \ge s\gamma_N : \bar{\xi}_N(n) = m_-(N)\}. \tag{4.84}$$

(With the usual convention that $\inf \emptyset = +\infty$.)

Applying the Markov property at time $s\gamma_N$, together with (4.83) (ii) and the Markov inequality, we have:

$$P(\bar{T}_N > s\gamma_N,\, \bar{t}_{N,s} > s\gamma_N + \alpha_N) \le P(\bar{T}_N > s\gamma_N)\, \varepsilon_N. \tag{4.85}$$

On the other hand, if $s, t > 0$ and $t\gamma_N > \alpha_N$, by the strong Markov property of $\bar{\xi}_N$ (applied at $\bar{t}_{N,s}$), we can write:

$$\begin{aligned} P(\bar{T}_N >(s+t)\gamma_N,\, \bar{t}_{N,s} \le s\gamma_N + \alpha_N) &= P(\bar{T}_N > s\gamma_N,\, \bar{t}_{N,s} \le s\gamma_N + \alpha_N) \\ &\quad \times P(\bar{T}_N > (s+t)\gamma_N \mid \bar{T}_N > s\gamma_N,\, \bar{t}_{N,s} \le s\gamma_N + \alpha_N) \\ &\le P(\bar{T}_N > s\gamma_N) P(\bar{T}_N > t\gamma_N - \alpha_N). \end{aligned} \tag{4.86}$$

As for the lower bound, arguing similarly, and if $t\gamma_N > \alpha_N$:

$$\begin{aligned} &P(\bar{T}_N > (s+t)\gamma_N,\, \bar{t}_{N,s} \le s\gamma_N + \alpha_N) \\ &\quad = \sum_{s\gamma_N \le k \le s\gamma_N + \alpha_N} P(\bar{T}_N > \bar{t}_{N,s} = k) P(\bar{T}_N > (t+s)\gamma_N - k) \\ &\quad \ge P(\bar{T}_N > s\gamma_N + \alpha_N,\, \bar{t}_{N,s} \le s\gamma_N + \alpha_N) P(\bar{T}_N > t\gamma_N) \\ &\quad \ge (P(\bar{T}_N > s\gamma_N + \alpha_N) - \varepsilon_N) P(\bar{T}_N > t\gamma_N), \end{aligned} \tag{4.87}$$

using (4.83) (ii) at the last inequality. From (4.85), (4.86), and (4.87) we get:

$$\begin{aligned} P(\bar{T}_N > (s+t)\gamma_N) &\le P(\bar{T}_N > s\gamma_N) \big(\varepsilon_N + P(\bar{T}_N > t\gamma_N - \alpha_N)\big) \\ P(\bar{T}_N > (s+t)\gamma_N) &\ge (P(\bar{T}_N > s\gamma_N + \alpha_N) - \varepsilon_N) P(\bar{T}_N > t\gamma_N). \end{aligned} \tag{4.88}$$

Let N_0 be such that $\alpha_N \le \gamma_N$ for all $N \ge N_0$. Thus for any integer $k \ge 2$ we write

$$P(\bar{T}_N > (k+2)\gamma_N) \le P(\bar{T}_N > k\gamma_N)(\varepsilon_N + P(\bar{T}_N > \gamma_N)).$$

Taking $r := 2e^{-1}$ we may assume that $\varepsilon_N + P(\bar{T}_N > \gamma_N) \le \varepsilon_N + e^{-1} \le r < 1$, for each $N \ge N_0$. Thus, for each $N \ge N_0$ and each $k \ge 3$ we get

$$P\big(\bar{T}_N > k\gamma_N\big) \le r^{[k/2]}, \tag{4.89}$$

which immediately implies the tightness of the family $\big\{\bar{T}_N/\gamma_N : N \ge 2\big\}$ on $[0, +\infty)$. In view of (4.83) and (4.88), any random variable τ that is a distributional limit of a subsequence $\bar{T}_{N'}/\gamma_{N'}$ satisfies

$$P(\tau > t + s) = P(\tau > t) P(\tau > s), \tag{4.90}$$

for any $s, t > 0$ continuity points of its distribution. From the density of such points, and the right continuity of the distribution function, (4.90) holds for all $s, t \geq 0$. Thus, $P(\tau > t) = e^{-at}$, for all $t > 0$, where $a = -\log P(\tau > 1) \in (0, +\infty]$. The case $a = +\infty$ corresponds to τ identically null, being excluded by (4.82), since

$$\begin{aligned} P(\tau < 1) &\leq \liminf_{N' \to +\infty} P(\bar{T}_{N'}/\gamma_{N'} < 1) \\ &\leq 1 - e^{-1} \leq \limsup_{N' \to +\infty} P(\bar{T}_{N'}/\gamma_{N'} < 1) < P(\tau < 1). \end{aligned}$$

We conclude that $a = 1$. (See Lemma 4.34 for a more general statement.)

To conclude the proof of Theorem 4.6 we write

$$\frac{1}{\gamma_N} E\bar{T}_N = \frac{1}{\gamma_N} \int_0^{+\infty} P(\bar{T}_N > s) ds = \int_0^{+\infty} P(\bar{T}_N > \gamma_N s)\, ds.$$

Due to (4.89), we may apply the dominated convergence theorem. Thus

$$\lim_{N \to +\infty} \frac{E\bar{T}_N}{\gamma_N} = \int_0^{+\infty} \lim_{N \to +\infty} P(\bar{T}_N > s\gamma_N) ds = \int_0^{+\infty} e^{-s} ds = 1. \qquad \square$$

We now discuss the other ingredient of the pathwise approach to metastability, i.e. the 'thermalization' around a metastable state. As mentioned briefly before, this property is formalized here through a limit theorem ($N \to +\infty$ in the present example). A precise definition of the metastable state is missing; what appears can be thought as an approximation only. In the present example, we could think of $\bar{\nu}_N$, the Curie–Weiss measure conditioned to $\bar{\mathcal{Y}}_N$, as a suitable approximation of the metastable state. Thought of as probability measures on the interval $[-1, +1]$, and considering the usual w^*-topology, we have $\lim_{N \to +\infty} \bar{\nu}_N = \delta_{m^-}$, the Dirac point measure at m^-. This is the 'state' which appears in our limit theorem. It is possible to perform a more detailed analysis, so that better approximations are involved.

Notation In the rest of this section $\mathcal{M}_1 = \mathcal{M}_1([-1, 1])$ will denote the space of probability measures on the Borel sets of the interval $[-1, 1]$, endowed with the usual w^*- topology. For $x \geq 0$, $[x]$ denotes the integer part of x, as usual.

For $N \geq 1$, $R \geq 1$ integers, we consider the following empirical process of moving time averages of length R:

$$A_{N,R}(s) := \frac{1}{R} \sum_{k=[s]+1}^{[s]+R} \delta_{\xi_N(k)}, \tag{4.91}$$

where δ_x denotes the Dirac point measure at x. It takes values on $\mathcal{M}_1$. If B is a Borel subset of $[-1, 1]$, we write $A_{N,R}(s, B)$ for the measure given to the set B; it counts the fraction of time spent in B during $[s] + 1, \dots, [s] + R$. If g is a Borel real function on $[-1, 1]$, $A_{N,R}(s, g)$ will denote the integral of g with respect to the measure given by (4.91).

We may now state the following.

Theorem 4.10 *There exist $R_N \to +\infty$ so that $R_N/\gamma_N \to 0$ for γ_N defined by* (4.82), *and such that for each B open interval containing m^- and each $\delta > 0$*

$$\lim_{N\to+\infty} P_{i_-(N)}\left(\inf_{s<T_N-R_N} A_{N,R_N}(s,B) > 1-\delta,\ T_N > R_N\right) = 1. \tag{4.92}$$

Possible choices include $R_N = [e^{\alpha N}]$ where $0 < \alpha < \Delta$.

This theorem involves the behaviour of the chain ξ_N up to time T_N. Using the previously described coupling we may replace ξ_N by $\bar{\xi}_N$ in its proof. The basic estimate is contained in the next lemma, which refers to $\bar{\xi}_N$ (and is not true for ξ_N).

Lemma 4.11 *Let $R_N = e^{\alpha N}$ with $0 < \alpha < \Delta$. Given $\delta \in (0,1)$ and $\rho > 0$, there exists $c_\delta > 0$ and $N_0 = N_0(\rho,\delta,\alpha)$ so that for all $N \geq N_0$ and integer $R \geq R_N$*

$$\sup_{i\leq i_*(N)} P_i\left(\frac{1}{R}\sum_{k=1}^{R} \mathbf{1}_\rho(\bar{\xi}_N(k)) \leq 1-\delta\right) \leq \exp\left(-c_\delta\,\frac{R}{R_N}\right), \tag{4.93}$$

where $\mathbf{1}_\rho = \mathbf{1}_{(m^- -\rho,m^-+\rho)}$.

Proof Of course, it suffices to consider ρ small. Fix α' such that $0 < \alpha' < \alpha \wedge \inf_{|m-m^-|=\rho} \beta\left(f(m) - f(m^-)\right)$. Let $R'_N = [e^{N\alpha'}]$ and

$$\tilde{\epsilon}_N := \sup_{i\leq i_*(N)} P_i\left(\bar{T}_N^{i_-(N)} > \sqrt{R'_N}\right) + P_{i_-(N)}\left(\psi_N < R'_N\right), \tag{4.94}$$

where $\psi_N := \min\{k \geq 1 : |\bar{\xi}_N(k) - m^-| \geq \rho\}$. By part (b) of Corollary 4.9, and arguing as in Proposition 4.7, we see that

$$\lim_{N\to+\infty} \tilde{\epsilon}_N = 0. \tag{4.95}$$

We fix N_0 so that $R'_N/R_N \leq \delta/2$, $1/\sqrt{R'_N} \leq \delta/4$ and $\tilde{\epsilon}_N \leq \delta/5$ for all $N \geq N_0$.

If $N \geq N_0$ and $R \geq R_N$, due to $R'_N/R_N \leq \delta/2$ we may write

$$P_i\left(\frac{1}{R}\sum_{k=1}^{R} \mathbf{1}_\rho(\bar{\xi}_N(k)) \leq 1-\delta\right) \leq P_i\left(\frac{1}{k_N R'_N}\sum_{k=1}^{k_N R'_N} \mathbf{1}_\rho(\bar{\xi}_N(k)) \leq 1-\frac{\delta}{2}\right), \tag{4.96}$$

where $k_N = [R/R'_N]$.

For each $k = 1, \ldots, k_N$ let us say that the time interval $[(k-1)R'_N, kR'_N)$ is *good* if the process $\bar{\xi}_N(\cdot)$ hits $m_-(N)$ before time $(k-1)R'_N + \sqrt{R'_N}$ and spends the rest of this time interval inside $(m^- - \rho, m^- + \rho)$. Otherwise, it is called *bad*. Let $Y_{N,k}$ be the indicator function of the event [the time interval $[(k-1)R'_N, kR'_N)$ is *bad*]. Thus for any $i \leq i_*(N)$:

$$\max_{k=1,\ldots,k_N} P_i\left(Y_{N,k} = 1 \mid Y_{N,1} = y_1, \ldots, Y_{N,k-1} = y_{k-1}\right) \leq \tilde{\epsilon}_N, \tag{4.97}$$

for any choice of $y_1, \ldots, y_{k-1} \in \{0, 1\}$. Since $\tilde{\epsilon}_N \le \delta/5$, performing successive conditioning and applying (4.97) we get, for $\lambda > 0$:

$$E_i\left(e^{\lambda \sum_{k=1}^{k_N} Y_{N,k}}\right) \le \left(1 + \frac{\delta}{5}(e^{\lambda} - 1)\right)^{k_N}.$$

Using (4.96) and (4.97) we see that

$$P_i\left(\frac{1}{R}\sum_{k=1}^{R} \mathbf{1}_\rho(\xi_N(k)) \le 1 - \delta\right) \le P_i\left(\frac{1}{k_N}\sum_{k=1}^{k_N} Y_{N,k} \ge \frac{\delta}{4}\right) \le e^{-k_N c_\delta}, \quad (4.98)$$

for all $N \ge N_0$, from which the lemma follows. At the last inequality we have used the exponential Markov inequality and the preceding observation with $\lambda = \lambda(\delta) > 0$ small enough so that $1 + \delta(e^{\lambda} - 1)/5 < e^{\lambda\delta/4}$. (Alternatively, we may apply the upper bound in Theorem 1.1 and Remark 4.12 below.) □

Remark 4.12 Let $0 < p < 1$ and let $Y_1, Y_2, \ldots$ be random variables taking values on $\{0, 1\}$, such that $P(Y_1 = 1) \le p$, and for each $k \ge 1$:

$$p(y_1, \ldots, y_k) := P(Y_{k+1} = 1 \mid Y_1 = y_1, \ldots, Y_k = y_k) \le p$$

for any choice of $y_1, \ldots, y_{k-1} \in \{0, 1\}$ such that $P(Y_1 = y_1, \ldots, Y_k = y_k) > 0$. (If $P(Y_1 = y_1, \ldots, Y_k = y_k) = 0$, we set $p(y_1, \ldots, y_k) = 0$.) On a suitable probability space we take $U_1, U_2, \ldots$ i.i.d. uniformly distributed on the interval $(0, 1)$ and set $Z_k = \mathbf{1}_{[U_k \le p]}$. The variables $Z_1, Z_2, \ldots$ are i.i.d. Bernoulli with parameter p. Setting $\tilde{Y}_1 = \mathbf{1}_{[U_1 \le P(Y_1=1)]}$, and recursively $\tilde{Y}_{k+1} = \mathbf{1}_{[U_{k+1} \le p(\tilde{Y}_1, \ldots, \tilde{Y}_k)]}$, we see at once that $\tilde{Y}_k \le Z_k$ a.s. for each k, and that the random sequence $(\tilde{Y}_1, \tilde{Y}_2, \ldots)$ has the same distribution as $(Y_1, Y_2, \ldots)$.

Proof of Theorem 4.10 Fix $R_N = [e^{\alpha N}]$ with $0 < \alpha < \Delta$. It suffices to prove (4.92) for $B = (m^- - \rho, m^- + \rho)$ with $\rho > 0$ small. Let $\mathbf{1}_\rho$ be the indicator function of B, as in Lemma 4.11. Using the previously mentioned coupling of the chains ξ_N and $\bar{\xi}_N$, which move together up to $T_N = \bar{T}_N$, the probability in (4.92) is rewritten as

$$P_{i_-(N)}\left(\min_{s=0,\ldots,[\bar{T}_N - R_N]} \frac{1}{R_N}\sum_{k=s+1}^{s+R_N} \mathbf{1}_\rho(\bar{\xi}_N(k)) > 1 - \delta,\ \bar{T}_N > R_N\right),$$

which is clearly bounded from below by $1 - P_{i_-(N)}(\bar{T}_N \le R_N) - P_{i_-(N)}(F_N)$ where

$$F_N = \bigcup_{\substack{\ell \in \mathbb{Z} \\ 0 \le \ell \le K_N}} \left[K_N \ge 1, \frac{1}{R_N}\sum_{k=\ell R_N+1}^{(\ell+1)R_N} \mathbf{1}_\rho(\bar{\xi}_N(k)) \le 1 - \delta/2\right]$$

and $K_N = [\bar{T}_N / R_N]$. Proposition 4.7 implies that $P_{i_-(N)}(\bar{T}_N \le R_N)$ tends to zero as $N \to +\infty$. On the other hand, fixing $0 < a < 1$ and taking any $k_N \in \mathbb{N}$ we have

$$P_{i_-(N)}(F_N) \le P_{i_-(N)}(K_N \ge k_N) + k_N \max_{i \le i_*(N)} P_i\left(\frac{1}{R_N}\sum_{k=1}^{R_N} \mathbf{1}_\rho(\bar\xi_N(k)) \le 1-\delta/2\right)$$
$$\le P_{i_-(N)}(\bar T_N \ge k_N R_N) + k_N\, e^{-c'_\delta R_N^{1-a}},$$

for all N sufficiently large ($N \ge N_0(\alpha,\rho,\delta,a)$), where we have used Lemma 4.11 with $[R_N^a]$ replacing R_N and $\delta/2$ instead of δ ($c'_\delta = c_{\delta/2}$ in that lemma). Thus, for all N large

$$P_{i_-(N)}(F_N) \le P(\bar T_N \ge k_N R_N) + k_N\, e^{-c'_\delta R_N^{1-a}}. \tag{4.99}$$

Choosing for instance $a = 1/2$, $k_N = \gamma_N/\sqrt{R_N}$, and applying Theorem 4.6 we see that $P_{i_-(N)}(F_N)$ tends to 0 as $N \to +\infty$, concluding the proof for the above choice of R_N. □

Remark 4.13 For the previous result we do not use fully the estimates in part (a) of Corollary 4.9; the rougher estimates discussed at the end of that proof would suffice. We may as well consider other starting points near $m_-(N)$. (cf. also Corollary 4.15 below.)
The statement of Theorem 4.10 could be strengthened in some directions. The main one involves the consideration of (4.93) with ρ and δ suitably tending to zero with N (slowly enough). This involves a modification in (4.94) and on the possible choices for R_N, using a more precise analysis.

Let R_N be as in Theorem 4.6. To describe the behaviour of the process ξ_N after time T_N, we first observe that, if starting at $m_*(N)$, ξ_N hits $m_+(N)$ before hitting $m_0(N)$ and in a time much shorter than R_N, with overwhelming probability, as $N \to +\infty$. For this we notice the following elementary fact.

Lemma 4.14 *In the notation and hypotheses of Lemma* 4.8 *let*

$$u_i = P_i(\tau_M < \tau_0), \qquad i = 1, \dots, M-1.$$

For each i,

$$u_i = \frac{\sum_{k=1}^{i}\prod_{j=1}^{k-1}\frac{q_j}{p_j}}{\sum_{k=1}^{M}\prod_{j=1}^{k-1}\frac{q_j}{p_j}} = \frac{\sum_{k=1}^{i}\frac{1}{\nu_{k-1}p_{k-1}}}{\sum_{k=1}^{M}\frac{1}{\nu_{k-1}p_{k-1}}}.$$

We omit the proof. Simply apply the method of difference equations as before.

Corollary 4.15 *Let* $\alpha > 0$. *For the process* ξ_N *starting at* $m_*(N)$, *the probability to hit* $m_+(N)$ *in a time shorter than* $e^{\alpha N}$ *and before hitting* $m_0(N)$, *tends to one as* $N \to +\infty$.

Proof The statement on the hitting probability follows from the last lemma applied to the Curie–Weiss chain. The rest is a consequence of arguments analogous to those of Lemma 4.8 and Corollary 4.9. □

Starting now from $m_+(N)$ we have a behaviour analogous to that described by Theorems 4.6 and 4.10, but for the deeper well of $\beta f(\cdot)$. That is:

(i) Let R_N be as in Theorem 4.10, then for any $\delta > 0$ and B open interval containing m^+ one has

$$\lim_{N\to+\infty} P_{i_+(N)}\left(\sup_{s<T_N^{i_0(N)}-R_N} A_{N,R_N}(s,B) \le 1-\delta\right) = 0;$$

(ii) Since $f(m^+) < f(m^-)$, we can take $\tilde{\Delta} > \Delta$ so that $\lim_{N\to+\infty} P_{i_+(N)}(T_N^{i_0(N)} \le e^{\tilde{\Delta}N}) = 0$.

For each N, the chain ξ_N performs infinitely many visits to both $m_-(N)$ and $m_+(N)$, but the return time to $m_0(N)$ after hitting $m_+(N)$ becomes increasingly larger than γ_N, as $N \to +\infty$; we typically do not see it on the time scale γ_N.

Under $P_{i_-(N)}$, the $\mathcal{M}_1$-valued process $A_N(s) := A_{N,R_N}(s\,\gamma_N)$ behaves asymptotically ($N \to +\infty$) as a Markov jump process which performs a unique jump from δ_{m^-} to δ_{m^+}, at a unit mean exponential random time. To make this formal as a limit in law in the Skorohod space $D([0,+\infty), \mathcal{M}_1)$ we need a small modification of $A_N(\cdot)$ around time T_N/γ_N, replacing it by $\tilde{A}_N(\cdot)$, given by:

$$\tilde{A}_N(s) = \begin{cases} A_{N,R_N}(T_N - \gamma_N) & \text{if } \dfrac{T_n - R_N}{\gamma_N} \le s < \dfrac{T_N}{\gamma_N}, \\ A_{N,R_N}(s\,\gamma_N) & \text{otherwise.} \end{cases}$$

4.4 The Harris contact process

We now discuss another example sharing the same features of metastability as a dynamical phenomenon, observed on the basis of pathwise behaviour. This is the Harris contact process restricted to $\{-N,\dots,N\}^d$, in the limit $N \to +\infty$.

Introduced by Harris in [149], and conceived with a biological interpretation, the contact process on $\mathbb{Z}^d$ is a continuous time Markov process taking values on $\mathcal{P}(\mathbb{Z}^d)$, the set of all subsets of $\mathbb{Z}^d$. It is described informally as follows: particles are distributed in $\mathbb{Z}^d$ in such a way that each site is either empty or occupied by at most one particle. The evolution is Markovian: each particle disappears after an exponential waiting time of rate one, independently of all the rest; at any time, each particle has the possibility to create a new particle at each empty neighbouring site, with rate λ, also independently of everything else.

Notation As before, sites $x, y \in \mathbb{Z}^d$ are called nearest neighbours (n.n.), or simply neighbours, if $|x - y| = 1$, with $|x - y|$ denoting the Euclidean distance in $\mathbb{R}^d$.

In the biological interpretation, occupied sites correspond to infected individuals, while empty sites correspond to healthy individuals. The recovery rate is one, λ is the rate of propagation of the infection (in each direction), and the process can be thought of as a very simplified mathematical model for the spread of an infection. The addition of enriching ingredients leads to various extensions.

Notation Identifying each $\xi \subseteq \mathbb{Z}^d$ with its indicator function $\mathbf{1}_\xi$, we may think of the process as taking values on $\mathcal{X} := \{0, 1\}^{\mathbb{Z}^d}$, shifting freely between $\mathcal{P}(\mathbb{Z}^d)$ and $\mathcal{X}$. In this notation, $\xi(x, t) = 1$ or 0, according to whether $x \in \xi(t)$ or not, with $t \geq 0$ denoting the time. (With respect to Section 3.3, the change from $\{-1, 1\}$ to $\{0, 1\}$ just follows convention.)

The need for a precise definition and construction of infinite systems was already mentioned in Section 4.1.5. Based on the informal description, if the initial set is infinite the process should leave it 'instantaneously'; infinitely many changes should happen during any finite interval $[0, \varepsilon)$, in contrast to the case of a Markov chain. Since the rates are bounded and the propagation happens only through neighbours, we expect that, with overwhelming probability, the evolution at any given finite collection of sites, during a fixed finite time interval, does not depend on what happens 'too far away'. This allows a formal mathematical construction of the infinite system as limit of finite systems.

One possible method of construction involves semigroup theory, a classical analytical tool based on the Hille–Yosida theorem. How to apply it to infinite particle systems is demonstrated in detail in Chapter I of Liggett's book [199], to which we refer. Another general procedure is based on the solution of martingale problems, as developed by Holley and Stroock, analogous to the construction of diffusions by Stroock and Varadhan. (see Chapter I of [199].)

As for a large class of systems, the contact process allows a very useful and more explicit construction. It involves a random graph in the space-time diagram $\mathbb{Z}^d \times \mathbb{R}_+$ and several properties are seen from it more readily. Such a graphical construction is due to Harris and we now describe it in the case of the contact process. (see [149] or [143].)

For each $x \in \mathbb{Z}^d$ consider $(\tau_n^x)_{n\in\mathbb{N}}$, $(\tau_n^{x,x+\mathbf{e}_i})_{n\in\mathbb{N}}$, $(\tau_n^{x,x-\mathbf{e}_i})_{n\in\mathbb{N}}$ for $i = 1, \ldots, d$, the arrival times of $2d + 1$ independent Poisson processes, where (τ_n^x) has rate 1 and the others have rate $\lambda > 0$, which is the parameter of the model. ($\mathbf{e}_i$ denotes the canonical unitary vector in the ith direction.) All such Poisson processes, also as x varies on $\mathbb{Z}^d$, are taken independently. For each $x, y \in \mathbb{Z}^d$ with $|x - y| = 1$, and $n \geq 1$, we draw arrows in $\mathbb{Z}^d \times \mathbb{R}_+$, from $(x, \tau_n^{x,y})$ to $(y, \tau_n^{x,y})$. Then, we put down a cross sign ($\times$) at each of the points (x, τ_n^x), $n \geq 1$. A segment linking (x, s) to (x, t) is called a time segment and has the orientation from (x, s) to (x, t) if $s < t$. Given two points (x, s) and (y, t) in $\mathbb{Z}^d \times \mathbb{R}_+$, with $s < t$, we say that there is a path from (x, s) to (y, t) if there is a connected chain of oriented time segments and arrows in the constructed random graph, which links (x, s) to

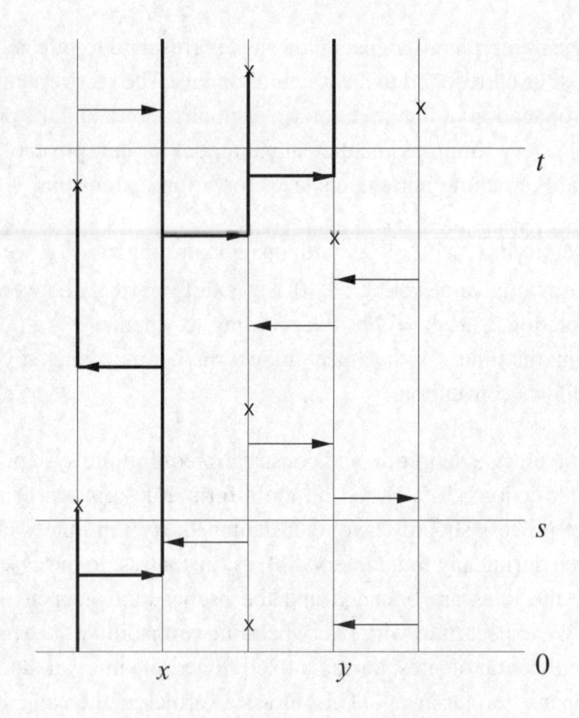

Figure 4.6

(y, t), without going through any cross sign, and following the orientation of time segments and arrows (see Figure 4.6.)

Let $A \in \mathcal{P}(\mathbb{Z}^d)$. The basic contact process with parameter λ and starting at A, denoted by $(\xi^A(t)\colon t \geq 0)$, is defined as follows: $\xi^A(0) = A$, and if $t > 0$

$$\xi^A(t) = \{y \in \mathbb{Z}^d : \text{there is a path from } (x, 0) \text{ to } (y, t) \text{ for some } x \in A\}. \tag{4.100}$$

(We say that A is the initial configuration.)

Remark As in Section 3.3, we endow $\mathcal{X} = \{0, 1\}^{\mathbb{Z}^d}$ with the usual product topology, given the discrete topology on $\{0, 1\}$. (Recalling the identification with $\mathcal{P}(\mathbb{Z}^d)$, $\xi_n \to \xi$ as $n \to +\infty$ if and only if, given any finite set $F \subseteq \mathbb{Z}^d$, there exists $n_0 = n_0(F)$ such that $\xi_n \cap F = \xi \cap F$ for all $n \geq n_0$.) Based on a system of independent Poisson processes (τ_n^x), $(\tau_n^{x,y})$ (also called 'marks'), one can verify that the graphical construction yields, almost surely, trajectories $t \mapsto \xi^A(t)$ in the Skorohod space $D([0, +\infty), \mathcal{X})$ of paths on $\mathcal{X}$ which are right continuous, with left limits. If P_A denotes the law of $\xi^A(\cdot)$ on $D([0, +\infty), \mathcal{X})$, then it can be seen that $A \mapsto P_A(B)$ is a Borel measurable function, for each Borel set B in the Skorohod space.

The Markov property, as well as several other basic properties, are readily seen from the graphical construction. Let $s > 0$ and consider the 'marks' in $\mathbb{Z}^d \times [s, +\infty)$. This is still a Poisson system just as before, independent of the 'marks' in $\mathbb{Z}^d \times [0, s)$. (Notice that the probability of finding a 'mark' at a fixed time s is zero.) Therefore, if for $A \subseteq \mathbb{Z}^d$ we set:

$$ {}_{(s)}\xi^A(t) = \{y \in \mathbb{Z}^d : \text{there is a path from } (x, s) \text{ to } (y, s+t) \text{ for some } x \in A\}, \tag{4.101}$$

then ${}_{(s)}\xi^A(\cdot)$ is a basic contact process starting from A. Moreover, given s:

$$\xi^A(s + \cdot) = {}_{(s)}\xi^{\xi^A(s)}(\cdot) \quad \text{a.s.,} \tag{4.102}$$

and we immediately get the Markov property.

The same random graph may be used to construct several contact processes, thus providing a natural coupling of them. This includes the restriction of the contact process to various volumes $\Lambda \subseteq \mathbb{Z}^d$ (cubes, slabs or cylinders, for instance): we follow the previous definition but consider only the paths contained entirely in $\Lambda \times \mathbb{R}_+$, i.e. we use only the Poisson processes $(\tau_n^x)_{n\in\mathbb{N}}$, $(\tau_n^{x,z})_{n\in\mathbb{N}}$, with $x, z \in \Lambda$, $|x - z| = 1$. Such restricted processes are denoted by $\xi_\Lambda^A(t)$. In the particular case $\Lambda = \Lambda_N^d := \Lambda_N \times \cdots \times \Lambda_N$ with $\Lambda_N = \{-N, \ldots, N\}$, $N \in \mathbb{N}$, we write $\xi_N^A(t)$. Other restricted processes will appear in the proofs. (The reason for the superindex d in the above notation, in contrast to that in Section 3.3, comes from Section 4.5.)

Notation If $A \subseteq \mathbb{Z}^d$ we usually write $\xi_\Lambda^A(t)$ for $\xi_\Lambda^{A\cap\Lambda}(t)$; $\mathcal{X}_\Lambda = \{0, 1\}^\Lambda$.

Remark 4.16 $\xi_N^A(\cdot)$ is a continuous time Markov chain with values on the finite set $\mathcal{P}(\Lambda_N)$ (identified with $\mathcal{X}_{\Lambda_N}$). From the graphical construction one easily verifies the following finite volume approximation: given $t_0 < +\infty$ and a finite set $B \subseteq \mathbb{Z}^d$, we have

$$\lim_{N\to+\infty} \sup_{0\le t\le t_0} P(\xi^A(t) \cap B \neq \xi_N^A(t) \cap B) = 0,$$

for any $A \subseteq \mathbb{Z}^d$. On the other hand, as $t \to +\infty$, the infinite system $\xi^A(t)$ behaves quite differently from $\xi_N^A(t)$. This is part of the discussion below.

The main advantage of the previous construction is that $\xi_N^A(\cdot)$, $\xi^B(\cdot)$, with $N \in \mathbb{N}$, and $A, B \in \mathcal{P}(\mathbb{Z}^d)$ are all given as functions of the same 'marks'. They are all coupled in a natural way. The following monotonicity relations hold automatically (a.s.) for all $t > 0$:

$$\begin{aligned} &\text{(i)} \quad \xi_N^A(t) \subseteq \xi^A(t), \quad \text{if} \quad A \subseteq \Lambda_N^d, \\ &\text{(ii)} \quad \xi_N^A(t) \subseteq \xi_N^B(t), \quad \text{if} \quad A \subseteq B \subseteq \Lambda_N^d, \\ &\text{(iii)} \quad \xi^A(t) \subseteq \xi^B(t), \quad \text{if} \quad A \subseteq B \subseteq \mathbb{Z}^d. \end{aligned} \tag{4.103}$$

(Items (ii) and (iii) correspond to what is called attractiveness.)

Notation When the initial configuration is the maximal one we omit the superscript: $\xi_N(t) = \xi_N^{\Lambda_N^d}(t)$, $\xi(t) = \xi^{\mathbb{Z}^d}(t)$.

Concerning the ergodic behaviour of the contact process, for $\xi_N^A(\cdot)$ the situation is trivial: it is a finite state Markov chain, the empty set $\emptyset$ is a trap (i.e. absorbing), and it is reached from any initial state. That is, letting

$$T_N^A := \inf\{t > 0 \colon \xi_N^A(t) = \emptyset\}, \tag{4.104}$$

then $T_N^A \leq T_N^{\Lambda_N^d} < +\infty$ a.s. In particular, the unique invariant measure is $\delta_\emptyset$, the Dirac point-mass at $\emptyset$. The process is ergodic, in the sense that for any A, $\xi_N^A(t)$ converges in law to $\delta_\emptyset$, as $t \to \infty$.

The infinite system shows different behaviour, which was one of the basic reasons for the great interest it raised. Of course, $\emptyset$ remains a trap, but for λ large enough (depending on d) a non-trivial invariant measure $\mu^{(\lambda)} \neq \delta_\emptyset$ appears as the limit in law of $\xi(t)$, when $t \to +\infty$. In this sense the system exhibits a dynamical phase transition.

For N large the chain $\xi_N(\cdot)$ may be seen as a suitable truncation (perturbation) of the infinite system and $\mu^{(\lambda)}$ gives place to a metastable state. Our next goal is to make this precise; keeping it in mind we need to recall some basic facts on the above mentioned dynamical phase transition. This is done briefly in the next subsection. It is just a summary, omitting most of the proofs and discussing briefly those arguments that will be used further. A basic reference for the study of contact process is Chapter VI of [199]. For an account of more recent developments see [200].

4.4.1 Basic properties

Monotonicity and attractiveness

Consider $\mathcal{P}(\mathbb{Z}^d)$ with the partial order given by inclusion. In the identification with $\mathcal{X}$, it corresponds to the coordinatewise order: $\xi \prec \xi'$ means that $\xi(x) \leq \xi'(x)$ for all $x \in \mathbb{Z}^d$. This induces a partial order on the space of probability measures on $\mathcal{X}$, as already mentioned in Section 3.3: if μ_1 and μ_2 are probability measures on $(\mathcal{X}, \mathcal{B}(\mathcal{X}))$ we say that μ_1 is stochastically smaller than μ_2, and write $\mu_1 \prec \mu_2$, if and only if $\mu_1(f) \leq \mu_2(f)$ for each continuous and increasing function $f \colon \mathcal{X} \to \mathbb{R}$. Here f is called increasing if $f(\xi) \leq f(\xi')$ whenever $\xi \prec \xi'$. (*Notation.* $\mu(f) = \int f d\mu$ as before.)

The following classical result gives an equivalent coupling property.

Proposition 4.17 *Let μ_1, μ_2 be probability measures on $\mathcal{X} = \{0, 1\}^{\mathbb{Z}^d}$. The following are equivalent*:

(*i*) *there exists a measure* $\bar{\mu}$ *on* $\mathcal{X} \times \mathcal{X}$ *with marginals* μ_1 *and* μ_2, *i.e.*

$$\bar{\mu}(A \times \mathcal{X}) = \mu_1(A), \quad \bar{\mu}(\mathcal{X} \times A) = \mu_2(A) \quad \textit{for all} \quad A \in \mathcal{B}(\mathcal{X}), \tag{4.105}$$

having the property that

$$\bar{\mu}\{(\xi, \xi') : \xi \prec \xi'\} = 1. \tag{4.106}$$

(*ii*) $\mu_1 \prec \mu_2$.

The implication (i) ⇒ (ii) is immediate: if $\bar{\mu}$ verifies (4.105) and (4.106), then for any f increasing and integrable one has

$$\mu_1(f) = \int_{\mathcal{X}\times\mathcal{X}} f(\xi)\mathrm{d}\bar{\mu}(\xi, \xi') \le \int_{\mathcal{X}\times\mathcal{X}} f(\xi')\mathrm{d}\bar{\mu}(\xi, \xi') = \mu_2(f).$$

The important part is the implication (ii) ⇒ (i). For a proof and further discussion see e.g. Theorem 2.4, Chapter II of [199] or the references therein.

Recalling (4.103), we see right away that the use of the same 'marks' in the graphical construction of the contact process provides examples of a direct verification of the coupling property (i) in the previous proposition. In this way the proof of this proposition is not essential for the following discussion. In general, however, the construction of a suitable coupling is not a simple task, from which comes the usefulness and importance of this classical result.

It is proper to observe that $\prec$ indeed defines a partial order on the space of probability measures on $(\mathcal{X}, \mathcal{B}(\mathcal{X}))$, which as before is denoted by $\mathcal{M}_1(\mathcal{X})$. Other properties being trivial we check that $\mu_1 \prec \mu_2$, $\mu_2 \prec \mu_1 \Rightarrow \mu_1 = \mu_2$. This follows at once from Proposition 4.17, but may be seen directly from the definition, since $\mu_1(f) = \mu_2(f)$ for each increasing cylinder function f implies that $\mu_1 = \mu_2$. Indeed, any cylinder function can be written as the difference of two increasing cylinder functions, cf. observation (b) in the proof of Theorem 3.35. (see also Remark 3.36).

Going back to the contact process, we see from (4.103)(iii) that whenever $A \subseteq B$, the law of $\xi^A(t)$ is stochastically smaller than that of $\xi^B(t)$. The same holds for the process restricted to Λ_N or more general volumes. The property of order preservation is usually called attractiveness or stochastic monotonicity, cf. [162]. It will also be important in Chapter 7, when studying stochastic Ising models.

Remark If $(\Omega, \mathcal{A}, P)$ is a probability space on which the Poisson processes of 'marks' were constructed and μ is a probability measure on $(\mathcal{X}, \mathcal{B}(\mathcal{X}))$, the contact process $\xi^\mu(t)$ with initial distribution μ may be constructed on the space $\tilde{\Omega} = \mathcal{X} \times \Omega$ with the product σ-algebra, replacing P by $\mu \times P$. That is, we first choose an initial A according to μ and then run the evolution $\xi^A(\cdot)$, according to the given 'marks'. In this way $P(\xi^\mu(\cdot) \in B) = \int P(\xi^\eta(\cdot) \in B)\mu(d\eta)$, where B is a Borel set on the path space. Having two initial measures μ_1 and μ_2 coupled cf. (4.106), we may use the same 'marks' to couple the processes (on $\bar{\Omega} = \mathcal{X} \times \mathcal{X} \times \Omega$, $\bar{P} = \bar{\mu} \times P$) in a way that with probability one $\xi^{\mu_1}(t) \subseteq \xi^{\mu_2}(t)$, for all $t \geq 0$. By Proposition 4.17 we see that if $\mu_1 \prec \mu_2$, then the law of $\xi^{\mu_1}(t)$ is stochastically smaller than that of $\xi^{\mu_2}(t)$.

From (4.102) and (4.103) we see that the laws of $\xi(t)$ decrease stochastically in t. In particular, by observation (c) in the proof of Theorem 3.35 we get the existence of a probability $\mu^{(\lambda)}$ on $\mathcal{X}$ to which the laws of $\xi(t)$ converge, as $t \to +\infty$. The attractiveness implies that any weak limit of the laws of $\xi^A(t_n)$, as $t_n \to +\infty$, is stochastically smaller than $\mu^{(\lambda)}$.[3] Consequently, the following are equivalent:

(i) $\mu^{(\lambda)} = \delta_\emptyset$;

(ii) $\xi^A(t)$ converges in law to $\delta_\emptyset$ for any initial A (ergodicity);

(iii) $\lim_{t\to+\infty} P(x \in \xi(t)) = 0 \quad$ for any $\quad x \in \mathbb{Z}^d$.

This is a general feature of attractive processes on $\{0, 1\}^{\mathbb{Z}^d}$: there is always a maximal (and a minimal) invariant measure and ergodicity is equivalent to their coincidence. (For a detailed discussion of attractive systems see Chapter III of [199].)

Remark 4.18 As we have observed, the graphical construction yields, for any $s, t \geq 0$, a coupling of $\xi(s+t)$ and $\xi(t)$ concentrated on $\{(\eta, \eta') : \eta \subseteq \eta'\}$ (cf. (4.102) and (4.103)). We may let $s \to +\infty$ and use compactness to get a measure $\bar{\mu}_t$ concentrated on the upper triangle $\{(\eta, \eta') : \eta \subseteq \eta'\}$, whose first marginal is $\mu^{(\lambda)}$ and the second gives the law of $\xi(t)$. (That is, $\mu^{(\lambda)}$ is stochastically smaller than the law of $\xi(t)$ and $\bar{\mu}_t$ is a coupling as described in Proposition 4.17.)

Harris inequality

An important property related to the partial order $\prec$ has already appeared in Section 3.3: we say that a probability measure μ on $\mathcal{X}$ is FKG[4], or that it has positive correlations, if $\mu(fg) \geq \mu(f)\,\mu(g)$ for all functions $f, g \colon \mathcal{X} \to \mathbb{R}$, increasing and continuous. In statistical mechanics such measures are associated with ferromagnetic (attractive) interactions.

To verify the FKG property it suffices to check the correlation inequality for each f and g increasing cylinder functions. In other words, μ is FKG if and only if all its finite dimensional projections are so (adopting the analogous definition on $\mathcal{X}_\Lambda$, with $\Lambda \subseteq \mathbb{Z}^d$).

The product measures on $\mathcal{X}$ are basic important examples of FKG measures, as shown by the classical FKG inequality, due to Harris, in [148] ($J \equiv 0$ in Theorem 3.32). A more general argument goes as follows: (i) the product of two FKG measures gives an FKG measure, with the coordinatewise order on the product

[3] For a direct coupling construction notice that $\{(\eta, \eta') \in \mathcal{X} \times \mathcal{X} : \eta \prec \eta'\}$ is a closed set in $\mathcal{X} \times \mathcal{X}$.
[4] for Fortuin, Kasteleyn and Ginibre, who first proved Theorem 3.32.

space; (ii) the FKG property holds trivially if the space is totally ordered, as $\{0, 1\}$. (see [199], p.78.)

Without quite explicit knowledge of the measure, it is generally hard to verify the FKG property. Therefore, it is interesting to know whether a given stochastic dynamics preserves the FKG property, that is, if the initial measure is FKG, then so is the law of the process at any time $t > 0$. We would typically apply this to any fixed configuration, or more generally, to any product measure at time zero.

The following relation of the FKG property and attractive dynamics was set by Harris in [150]: an attractive continuous time Markov chain on a finite partially ordered state space preserves the FKG property if and only if it allows jumps only between comparable states (for proofs see also Theorem 2.14, Chapter II of [199], or [67]).

We see at once that such a property applies to the contact process on a finite volume Λ, or to any attractive chain on $\mathcal{X}_\Lambda = \{0, 1\}^\Lambda$ for which only one spin changes at each jump. Recalling Remark 4.16, this extends to the infinite volume contact process. It implies that if $t \geq 0$ and f, g are increasing cylinder functions, then

$$E(f(\xi(t))g(\xi(t))) \geq E(f(\xi(t)))E(g(\xi(t))).$$

Letting $t \to +\infty$ in the previous inequality, we see that $\mu^{(\lambda)}$ has positive correlations. Moreover, applying the Markov property and the attractiveness, we see that also the law of $(\xi(t_1), \dots, \xi(t_k))$ is FKG on the product space of k copies of $\mathcal{X}$, for any $k \geq 1, t_1, \dots, t_k \geq 0$ (see Corollary 2.21, Chapter II of [199]).

In some proofs below we need a variation of the previous statement. For Λ and $\tilde{\Lambda}$ finite, consider the pair $(\xi_\Lambda(t), \xi_{\tilde{\Lambda}}(t))$ constructed on the same basic graph. This is an attractive process with respect to the coordinatewise order on $\mathcal{X}_\Lambda \times \mathcal{X}_{\tilde{\Lambda}}$. We easily see that only jumps between comparable configurations are allowed. Applying the previous result and the approximation through finite volumes (Remark 4.16), we conclude that for any $\Lambda, \tilde{\Lambda} \subseteq \mathbb{Z}^d$, if f and g are increasing and continuous, then

$$E\left(f(\xi_\Lambda(t))g(\xi_{\tilde{\Lambda}}(t))\right) \geq E\big(f(\xi_\Lambda(t))\big)E\left(g(\xi_{\tilde{\Lambda}}(t))\right). \tag{4.107}$$

As in the previous paragraph, this extends to functions on the time evolution.

An alternative way to verify (4.107) and its extension along the time evolution is to set it in terms of the percolation graph on which all variants of the contact process were constructed: roughly speaking, 'increasing' should mean that it increases as more $\tau_n^{x,y}$ 'marks' are included, and decreases as more $\times$ 'marks' are included.

Consider first a discrete approximation based on the following random graph: for $\delta > 0$ small, $x \in \mathbb{Z}^d, k \geq 0$, arrows $(x, k\delta) \to (x, (k+1)\delta)$ are set with probability $1 - \delta$ and arrows $(x, k\delta) \to (y, (k+1)\delta)$ are set with probability $\lambda\delta$ if $|y - x| = 1$, with full independence among different arrows. Using the FKG property of the product measure, and then letting $\delta \to 0$, it is not hard to get (4.107). (See Corollary B18 and p. 11 in [200], or [100] for a proof of the Harris inequality along such lines.)

Self-duality

Self-duality is a quite special property, which in the case of the contact process can be stated as follows: if $A,\ B \subseteq \mathbb{Z}^d$ and $t > 0$,

$$P(\xi^A(t) \cap B \neq \emptyset) \;=\; P(\xi^B(t) \cap A \neq \emptyset). \tag{4.108}$$

The above relation is particularly useful if exactly one among the sets A or B is finite. For finite B, $\xi^B(\cdot)$ is a continuous time Markov chain with values on the (countable) space of finite subsets of $\mathbb{Z}^d$; (4.108) describes the law of the infinite system $\xi^A(t)$ in terms of such chains.

Given the graphical construction, (4.108) follows from the time reversal property for a system of independent Poisson processes, as we now check. Fix t positive, and consider the restriction of the random graph to $\mathbb{Z}^d \times [0, t]$. If we now reverse the direction of all arrows and of the time, keeping the same $\times$ 'marks', we get another graph, through which we define for $0 \le s \le t$:

$$\hat{\xi}^{A,t}(s) = \{x \in \mathbb{Z}^d\colon\ \exists \text{ a reversed path from some } (y, t),\ y \in A, \text{ to } (x, t-s)\}. \tag{4.109}$$

By a simple property of Poisson processes, the law of $(\hat{\xi}^{A,t}(s)\colon 0 \le s \le t)$ is the same as that of $(\xi^A(s)\colon 0 \le s \le t)$. On the other hand, the construction implies at once that $\xi^B(t) \cap A \neq \emptyset$ if and only if $\hat{\xi}^{A,t}(t) \cap B \neq \emptyset$, proving (4.108). (Notice that t is fixed in the construction of the reversed graph. The dependence of $\hat{\xi}^{A,t}(\cdot)$ on t is quite irregular.)

Taking $A = \mathbb{Z}^d$ and B finite in (4.108) we obtain:

$$P(\xi(t) \cap B \neq \emptyset) = P(\xi^B(t) \neq \emptyset) = P(T^B > t),$$

where $T^B := \inf\{t > 0\colon \xi^B(t) = \emptyset\}$ (setting $\inf \emptyset = +\infty$). In particular, letting $t \to +\infty$ we get

$$\mu^{(\lambda)}\{\eta\colon \eta \cap B \neq \emptyset\} = P(T^B = +\infty), \tag{4.110}$$

and taking $B = \{0\}$ in the last equation, $\rho_\lambda := \mu^{(\lambda)}\{\eta\colon 0 \in \eta\} = P(T^{\{0\}} = +\infty)$.

Dynamical phase transition

Let θ_x denote the translation operators on $\mathcal{X}$, as introduced in Chapter 3: $\theta_x\eta(y) = \eta(y + x)$, for each $x,\ y \in \mathbb{Z}^d$. (Treating η as a subset of $\mathbb{Z}^d$, $\theta_x\eta$ is its translation by $-x$; equivalently, the observer is shifted by x.) If f is a function on $\mathcal{X}$ we let $\theta_x f(\eta) = f(\theta_x\eta)$, for each η, and if μ is a measure on $\mathcal{X}$, $\theta_x\mu$ will denote the measure induced by μ under θ_x: $\theta_x\mu(f) = \mu(\theta_x f)$ for each bounded measurable $f\colon \mathcal{X} \to \mathbb{R}$.

From the random graph construction and the translation invariance of the Poisson system of 'marks', we see that $\mu^{(\lambda)}$ is translation invariant, i.e. $\theta_x \mu^{(\lambda)} = \mu^{(\lambda)}$ for each x. Thus, $\mu^{(\lambda)} = \delta_\emptyset$ if and only if $\rho_\lambda = 0$.

Another property which follows at once from the graphical construction is the monotonicity in λ. Namely, if $\lambda < \lambda'$ then the law of $\xi^A(t)$ with rate λ is stochastically smaller than that with rate λ'. To see this, we may simply start with the random graph corresponding to λ', and construct a new graph by keeping the same $\times$ 'marks' while for the arrows each one is kept (disregarded) with probability λ/λ' $(1 - \lambda/\lambda'$, respectively) independently of all the rest. The process constructed on the new graph is clearly smaller than that on the initial graph and corresponds to a contact process with rate λ. In particular, if $\lambda < \lambda'$ we have $\mu^{(\lambda)} \prec \mu^{(\lambda')}$. Setting

$$\lambda_c^d = \sup\{\lambda \colon \rho_\lambda = 0\}, \tag{4.111}$$

the previous discussion tells us:

(i) if $\lambda < \lambda_c^d$, then $\mu^{(\lambda)} = \delta_\emptyset$ and $P(T^B = +\infty) = 0$, for any finite set B;
(ii) if $\lambda > \lambda_c^d$, then $\mu^{(\lambda)} \neq \delta_\emptyset$ and $P(T^{\{0\}} = +\infty) > 0$.

A priori $\lambda_c^d \in [0, +\infty]$, i.e. one of the above alternatives might be empty. An immediate comparison of the cardinality of $\xi^{\{0\}}(t)$ with a branching process whose birth rate is $2\lambda d$, and whose death rate is one, shows at once that $\lambda_c^d \geq 1/2d$. This is too crude a comparison (at least for d small) and others, less trivial but still simple, yield better estimates from below. One such example is $\lambda_c^1 \geq 1$, obtained by comparing the diameter of $\xi^{\{0\}}(t)$ (for $d = 1$) with a random walk which increases by one with rate 2λ and decreases by one with rate 2, whenever it is larger than or equal to 2; this walk has a negative drift if $\lambda < 1$, which would force the set $\xi^{\{0\}}$ to be reduced to a point infinitely many times a.s. on the set $\{T^{\{0\}} = +\infty\}$; due to the positive death rate this forces a.s. extinction. Exploiting the duality in a more elaborate fashion, one gets $\lambda_c^d \geq 1/(2d - 1)$, cf. Example 5.11, Chapter III of [199]; see also p. 289 of [199] for a better lower bound when $d = 1$.

The dynamical phase transition corresponds to $\lambda_c^d < +\infty$; this is harder to prove, and constitutes an important classical result, initially proven by Harris in [149]. For any value of d it is known that $\lambda_c^d \leq 2/d$, as proven by Holley and Liggett. (see Corollary 4.4, Chapter VI of [199]). For better approximations of λ_c^d see p. 128 of [200] and references therein.

The behaviour at λ_c^d has been a challenging problem. It was settled by Bezuidenhout and Grimmett, using dynamical renormalization techniques. Their results imply that $\mu^{(\lambda_c^d)} = \delta_\emptyset$. (see [18].)

Another basic property of $\mu^{(\lambda)}$ is its ergodicity with respect to the translations θ_x, i.e. translation invariant events have probability zero or one. Proofs of this, valid for more general translation invariant attractive systems can be found in [98] or in Proposition 2.16, Chapter III of [199]. The starting point is the ergodicity of the law of each $\xi(t)$. For finite range interactions (not necessarily attractive) this was proven by Holley in [161]. Both proofs exploit the attractiveness to conclude the ergodicity of the limiting measure (see also Theorem 4.15, Chapter I of [199], and references therein).

The measures $\mu^{(\lambda)}$ exhibit exponential decay of correlations. This is closely related to the speed of convergence of the dynamics as $t \to +\infty$, and it plays an important role in the control of the thermalization process in metastability. Therefore we need a brief discussion of some basic results on the ergodic behaviour of the contact processes, before entering into the discussion on metastability.

Ergodic behaviour

When $\lambda > \lambda_c^d$, the first natural question concerns the full description of the invariant measures and convergence to equilibrium. Due to peculiar properties of the random graph on $\mathbb{Z} \times [0, +\infty)$, involving basically crossing of paths, the one-dimensional case is special, allowing a much simpler treatment with the help of the so-called edge processes. For this reason the basic problems concerning the ergodic behaviour were settled first for $d = 1$.

To keep the discussion of the main points as simple as possible, in this section our treatment of metastability is restricted to the case $d = 1$. In the next section we discuss briefly the extension to $d > 1$.

Let $d = 1$, $\lambda > \lambda_c^1$. The characterization of $\delta_\emptyset$ and $\mu^{(\lambda)}$ as the unique extremal invariant measures constitutes a fundamental result which holds for more general one-dimensional attractive and nearest neighbour systems, cf. Theorem 3.13 in Chapter III of [199].

In the case of a one-dimensional contact process, more complete information on the ergodic behaviour is summarized by the following results due to Durrett ([98]), with earlier weaker versions by Harris ([151]) and Griffeath [141, 142].

(a) Complete convergence theorem *If* $\lambda > \lambda_c^1$, $A \subseteq \mathbb{Z}$, *and* $f: \mathcal{P}(\mathbb{Z}) \to \mathbb{R}$ is *continuous, then*

$$\lim_{t\to\infty} E(f(\xi^A(t))) = P(T^A < \infty)\, f(\emptyset) + P(T^A = +\infty)\, \mu^{(\lambda)}(f).$$

In other words, $\xi^A(t)$ *converges in law to* $P(T^A < \infty)\,\delta_\emptyset + P(T^A = +\infty)\,\mu^{(\lambda)}$.

(b) Pointwise ergodic theorem *If* $\lambda > \lambda_c^1$, $A \subseteq \mathbb{Z}$, *and* $f: \mathcal{P}(\mathbb{Z}) \to \mathbb{R}$ *is continuous, then*

$$\frac{1}{t}\int_0^t f(\xi^A(s))ds \to f(\emptyset)\,\mathbf{1}_{[T^A<\infty]} + \mu^{(\lambda)}(f)\,\mathbf{1}_{[T^A=+\infty]} \quad \text{a.s.}$$

For the full proofs see Theorem 2.28 and Theorem 2.33, Chapter VI of [199] (cf. also [98]). We sketch the basic arguments for (a), since they will be used extensively later.

Proof of (a) (Sketch) Considering $A = \{0\}$, the statement becomes equivalent to:

$$\lim_{t\to+\infty} P(\xi^{\{0\}}(t) \cap B \neq \emptyset, T^{\{0\}} > t) = \mu^{(\lambda)}\{\xi : \xi \cap B \neq \emptyset\} P(T^{\{0\}} = +\infty), \tag{4.112}$$

for any finite set B.

The proof sketched below works only for $d = 1$ and uses heavily the 'edge' processes: $r_t := \max \xi^{\mathbb{Z}\cap(-\infty,0]}(t)$ and $\ell_t := \min \xi^{\mathbb{Z}\cap[0,+\infty)}(t)$. ($\xi^{\mathbb{Z}\cap(-\infty,0]}(t)$ and $\xi^{\mathbb{Z}\cap[0,+\infty)}(t)$ are a.s. non-empty, and r_t and ℓ_t are well defined, integer valued.) Since the random graph on which the process is constructed lies on the plane, we can take advantage of its nearest neighbour character (horizontal arrows only among nearest neighbours) and use crossing properties of paths to see that $\{T^{\{0\}} > t\} = \{\ell_s \le r_s, \forall s \le t\}$, and that on this event $\xi^{\{0\}}(t) \cap [\ell_t, r_t] = \xi(t) \cap [\ell_t, r_t]$.

A crucial point is the existence, for $\lambda > \lambda_c^1$, of a constant $\alpha(\lambda) > 0$ so that $\lim_{t\to+\infty} r_t/t = \alpha(\lambda) = -\lim_{t\to+\infty} \ell_t/t$ a.s., i.e. a linear growth condition. The Kingman–Liggett subadditive ergodic theorem (Theorem 2.6, Chapter VI of [199]) is used for verification of the a.s. (and L^1) convergence. Further analysis is needed to determine that $\alpha(\lambda) > 0$ for $\lambda > \lambda_c^1$. (See Theorem 2.19 and Theorem 2.27, Chapter VI of [199] for the proof of the convergence and that $\lambda_c^1 = \inf\{\lambda : \alpha(\lambda) > 0\}$.)

Given these ingredients and having fixed a finite set B, we have

$$\xi^{\{0\}}(t) \cap B = \xi(t) \cap B \quad \text{on} \quad \{[\ell_t, r_t] \supseteq B, T^{\{0\}} > t\},$$

and the indicator function of this last event tends a.s. to that of $\{T^{\{0\}} = +\infty\}$. Therefore one needs to argue that the conditional distribution of $\xi(t)$ given $\{T^{\{0\}} = +\infty\}$ tends to $\mu^{(\lambda)}$. If not for the conditioning, this would be just the definition of $\mu^{(\lambda)}$. Using the attractiveness we see that the conditioning does not spoil the limit: fix s and take $t > s$, so that by (4.102) and (4.103) (iii) we have $P(\xi(t) \cap B \neq \emptyset, \mathcal{E}_s) \le P(\xi(t-s) \cap B \neq \emptyset) P(\mathcal{E}_s)$, where $\mathcal{E}_s = \{T^{\{0\}} > s\}$ or its complement. Thus $\limsup_{t\to+\infty} P(\xi(t) \cap B \neq \emptyset, \mathcal{E}_s) \le \mu^{(\lambda)}\{\xi : \xi \cap B \neq \emptyset\} P(\mathcal{E}_s)$ for any s, in each of the two cases. But the sum on the left-hand side (for $\mathcal{E}_s$ and its complement) tends to $\mu^{(\lambda)}\{\xi : \xi \cap B \neq \emptyset\}$ entailing that $\lim_{t\to+\infty} P(\xi(t) \cap B \neq \emptyset, T^{\{0\}} > s) = \mu^{(\lambda)}\{\xi : \xi \cap B \neq \emptyset\} P(T^{\{0\}} > s)$, for each s. Since $P(s < T^{\{0\}} < +\infty)$ tends to zero as $s \to +\infty$ it is simple to conclude the announced convergence of the conditional distribution.

The extension to any A finite is simple; the conclusion then follows by attractiveness and since $P(T^A = +\infty)$ tends to 1 as $\sharp A$ tends to infinity, cf. Remark 4.19 below. □

The proof of the pointwise ergodic theorem combines an argument analogous to the one we have just sketched with the standard reasoning based on the ergodic theorem for the stationary process $\xi^{\mu^{(\lambda)}}$; we omit it.

Extensions of these results to $d \geq 2$ require other tools. The validity of the complete convergence theorem for $d \geq 2$ and any $\lambda > \lambda_c^d$ is a consequence of the results in [18]. Earlier results, due to Andjel, apply to λ large enough, cf. [4] and references therein.

Remark 4.19 Let $d = 1$ and $\lambda > \lambda_c^1$. Among the sets A of a given finite cardinality n, the survival probability $P(T^A = +\infty)$ is minimized if A is a 'block' $\{1, \ldots, n\}$. In fact, if $A = \{x_1, \ldots, x_n\}$ and $B = \{y_1, \ldots, y_n\}$ are such that $0 < x_{i+1} - x_i \leq y_{i+1} - y_i$ for each $i = 1, \ldots, n-1$, then

$$P(T^A = +\infty) \leq P(T^B = +\infty). \tag{4.113}$$

Since the death rate is constant, (4.113) should be a consequence of the way $\xi^A(t)$ grows, through empty neighbours of occupied sites. It may be verified through the following coupling argument, due to Liggett (cf. Theorem 1.9, Chapter VI; of [199]): it is possible to couple ξ^A and ξ^B in such a way that for any $t < T^A$ not only $\sharp\xi^B(t) \geq \sharp\xi^A(t)$, but also $\xi^B(t)$ is more spread out, as initially. That is, if $\varphi_t : \xi^A(t) \to \xi^B(t)$ associates to the ith element of $\xi^A(t)$ (usual order) the ith one in $\xi^B(t)$, then $|\varphi_t(x) - \varphi_t(y)| \geq |x - y|$. For this, set $\varphi_0(x_i) = y_i$ for each i; having defined the joint evolution up to a given time t, with the desired property, let the sites $x \in \xi^A(t)$ and $\varphi_t(x)$ use the same exponential death 'clocks'; those elements of $\xi^B(t)$ which are not in the image of φ_t have independent death 'clocks'. Consider the collection of independent exponential 'clocks' which determine the birth on the empty neighbours of $\xi^A(t)$: if the first change is a birth at $x \pm 1 \notin \xi^A(t)$ (due to an arrow from $x \in \xi^A(t)$), we create simultaneously a particle in ξ^B, at $\varphi_t(x) \pm 1$ (which is possible!); births at empty neighbours of unpaired sites in ξ^B occur independently. At each birth time s we update φ_s so that the ith element of $\xi^A(s)$ (usual order) continues to correspond to the ith element in $\xi^B(s)$, for each $i \leq \sharp\xi^A(s)$.

4.4.2 The metastable behaviour of the finite contact process for $d = 1$

We shall now make precise the general description given at the beginning of Section 4.4, for $\lambda > \lambda_c^1$. For large N and far from the border of Λ_N, the process $\xi_N(\cdot)$ behaves for a very long time as if $\mu^{(\lambda)}$ (projected on $\mathcal{X}_{\Lambda_N}$) were its unique invariant measure until a large fluctuation brings it to $\delta_\emptyset$. As in Section 4.3, the basic characteristics are:

(a) asymptotic unpredictability of T_N;
(b) thermalization.

Having in mind the simplest possible situation, we restrict the consideration to the one-dimensional case, where special features might be used, as we have already seen. These features were exploited in [48, 104, 265] where the results have been proven. The higher dimensional situation, treated in [220, 279], are discussed briefly in the next section.

Notation Having fixed $d = 1$, we omit the corresponding superscript from now on.

As auxiliary tool in the next proofs we use the restricted processes $\xi_{[-N,\infty)}(\cdot)$ and $\xi_{(-\infty,N]}(\cdot)$, defined just before Remark 4.16, i.e. using only the paths contained in $\{-N, -N+1, \ldots\} \times \mathbb{R}_+$, and $\{\ldots N-1, N\} \times \mathbb{R}_+$ respectively. That is, in the case of $\xi_{[-N,\infty)}(\cdot)$, a particle at $-N$ can only create a new particle at $-N+1$. The properties of attractiveness and duality are valid for them.

Clearly, $\xi_{[-N,\infty)}$, $\theta_N \xi_{[0,\infty)}$, and $\mathcal{R}\xi_{(-\infty,N]}$ all have the same law, where $\mathcal{R}\xi(y) = \xi(-y)$ for each y. We let $\mu^{(\lambda)}_{[-N,\infty)}$ ($\mu^{(\lambda)}_{(-\infty,N]}$) be the maximal invariant measure in each case (obtained from $\mu^{(\lambda)}_{[0,\infty)}$ through translation and reflection along the origin).

The contact process on $\{0, 1, \ldots\}$ also presents a dynamical phase transition, i.e. $\lambda_c^+ := \inf\{\lambda \colon \mu^{(\lambda)}_{[0,\infty)} \neq \delta_\emptyset\} < +\infty$; clearly $\lambda_c^+ \geq \lambda_c$. Using the renormalization method introduced in [101] one sees that $\lambda_c^+ = \lambda_c$ (see Remark 4.30 below).

We need the following simple fact: $\mu^{(\lambda)}_{[0,\infty)}(\{\emptyset\}) = 0$ whenever $\mu^{(\lambda)}_{[0,\infty)} \neq \delta_\emptyset$. To check this, if $\mu^{(\lambda)}_{[0,\infty)} \neq \delta_\emptyset$ we write $\mu^{(\lambda)}_{[0,\infty)} = (1-u)\mu + u\delta_\emptyset$, where μ is $\mu^{(\lambda)}_{[0,\infty)}$ conditioned on $\mathcal{X} \setminus \{\emptyset\}$ and $u = \mu^{(\lambda)}_{[0,\infty)}(\{\emptyset\})$. This gives $\mu^{(\lambda)}_{[0,\infty)} \prec \mu$ and, since $\emptyset$ is a trap, μ is also invariant. From the maximality of $\mu^{(\lambda)}_{[0,\infty)}$ we conclude that $\mu^{(\lambda)}_{[0,\infty)} = \mu$, i.e. $u = 0$.

Notation (i) As before we omit the superscript indicating the initial configuration if this is the maximal one, e.g. $T_N = T_N^{\Lambda_N}$.

(ii) As in Section 3.3, the support of a cylinder function g on $\mathcal{X}$, is defined as the smallest finite set A such that $g(\xi)$ depends only on the coordinates in A. Thinking of ξ as a subset of $\mathbb{Z}$ this means $g(\xi) = g(\xi \cap A)$. The support of g is denoted by $\Delta(g)$.

The asymptotic unpredictability of T_N may be stated as follows.

Theorem 4.20 *Let $\lambda > \lambda_c$ and define β_N through the relation*

$$P(T_N > \beta_N) = e^{-1}. \tag{4.114}$$

Then:

(*i*) *T_N/β_N converges in law to a unit mean exponential random variable, as $N \to +\infty$;*

(*ii*) $\lim_{N\to+\infty} \frac{E(T_N)}{\beta_N} = 1$.

Remark The distribution function of T_N is easily seen to be continuous and strictly increasing, so that β_N is well defined through (4.114). Since T_N is at least as large as the maximum of $2N+1$ i.i.d. unit mean exponential random variables, β_N grows at least as $\log N$, for any λ. (For $\lambda > \lambda_c$, β_N grows exponentially in N, cf. Remark 4.33 below.)

The speed of convergence in the previous theorem is an important point. We discuss it in Remark 4.31 below.

To formalize the thermalization property we consider the following $\mathcal{M}_1(\mathcal{X})$-valued processes, describing moving time averages of length $R > 0$:

$$A_R^N(s, \cdot) = \frac{1}{R} \int_s^{s+R} \delta_{\xi_N(u)}(\cdot)\, du, \tag{4.115}$$

where δ_ξ denotes the Dirac point measure at $\xi \in \mathcal{X}$ and $s \geq 0$.

Notation $A_R^N(s, f)$ stands for the integral of f with respect to $A_R^N(s, \cdot)$. We then have the following

Theorem 4.21 *If $\lambda > \lambda_c$, there exist positive numbers R_N so that*

(i) $R_N \to +\infty$, $R_N/\beta_N \to 0$, *as* $N \to +\infty$;

(ii) *for each $\varepsilon > 0$ and cylinder function f*

$$\lim_{N\to+\infty} P\left(\sup_{s<T_N-2R_N} |A_{R_N}^N(s, f) - \mu^{(\lambda)}(f)| \leq \varepsilon,\ T_N > 2R_N \right) = 1. \tag{4.116}$$

Indeed, for each such ε and f there exists a finite number $L = L(\varepsilon, f)$ so that

$$\lim_{N\to+\infty} P\left(\sup_{s<T_N-2R_N} \max_{i\in I_{f,N}(L)} |A_{R_N}^N(s, \theta_i f) - \mu^{(\lambda)}(f)| \leq \varepsilon,\ T_N > 2R_N \right) = 1, \tag{4.117}$$

where $I_{f,N}(L) = \{i \in \mathbb{Z} : \Delta(\theta_i f) \subseteq \{-N+L, \ldots, N-L\}\}$.

Analogously to the results in Section 4.3, the last two theorems imply that a slight modification of $A_{R_N}^N(\cdot\beta_N)$ converges in law to a Markov jump process with a unique jump from $\mu^{(\lambda)}$ to $\delta_\emptyset$.

Proof of Theorem 4.20 For part (i) it suffices to prove that

$$\lim_{N\to+\infty} (P(T_N > \beta_N(t+s)) - P(T_N > \beta_N t) P(T_N > \beta_N s)) = 0 \tag{4.118}$$

for each $s, t > 0$. (By (4.114), this implies at once that $P(T_N > \beta_N t)$ converges to e^{-t} for any $t \geq 0$ rational; the extension to all $t \geq 0$ follows by monotonicity.) The proof below does not depend on any estimate of β_N, and we verify directly that

$$v_N := \sup_{s,t\geq 0} \left| \frac{P(T_N > s+t)}{P(T_N > s)} - P(T_N > t) \right| \to 0. \tag{4.119}$$

By the Markov property and attractiveness, cf. (4.103)(ii),

$$P(T_N > t+s) = \sum_{A \neq \emptyset} P(T_N > s, \xi_N(s) = A)\, P(T_N^A > t)$$
$$\leq P(T_N > s) P(T_N > t).$$

On the other hand, if $\mathbb{A} \subseteq \mathcal{P}(\Lambda_N)$:

$$P(T_N > t+s) \geq \inf_{A \in \mathbb{A}} P(T_N^A > t) P(T_N > s, \xi_N(s) \in \mathbb{A}),$$

so that for each $s, t > 0$

$$\begin{aligned} 0 &\leq P(T_N > t) P(T_N > s) - P(T_N > s+t) \\ &\leq \left\{ P(T_N > t) - \inf_{A \in \mathbb{A}} P(T_N^A > t) \right\} P(T_N > s) + P(T_N > s,\ \xi_N(s) \notin \mathbb{A}) \\ &\leq \sup_{A \in \mathbb{A}} P(T_N \neq T_N^A)\, P(T_N > s) + P(T_N > s,\ \xi_N(s) \notin \mathbb{A}), \end{aligned} \tag{4.120}$$

where at the last step we are using the coupling provided by the graphical construction. The right-hand side of (4.120) does not depend on t and the goal is to determine suitable sets $\mathbb{A}$, for which it yields (4.119). For the first term to be small, it seems good to take the elements of $\mathbb{A}$ somehow 'thick'; they should not be too 'thick' if we want the second term to vanish as well. A good choice is

$$\mathbb{A}_b = \left\{ A \colon \frac{|A \cap [-b, -1]|}{b} \geq \rho,\ \frac{|A \cap [1, b]|}{b} \geq \rho \right\},$$

where $\rho \in (0, \rho_\lambda)$ is fixed, $b > 0$ is an integer to be chosen and, to simplify the notation, $|\cdot| = \sharp(\cdot)$ denotes the cardinality. Clearly, we take $N \geq b$.

We start by the second term on the right-hand side of (4.120). If $L \geq 1$, $N \geq b + L$ we write:

$$\begin{aligned} &P(T_N > s, \xi_N(s) \notin \mathbb{A}_b) \\ &\leq P(\xi_N(s) \notin \mathbb{A}_b, T_N > s, \min \xi_N(s) < -N + L, \max \xi_N(s) > N - L) \\ &\quad + P(\min \xi_N(s) \geq -N + L, T_N > s) + P(\max \xi_N(s) \leq N - L, T_N > s). \end{aligned}$$

Using the nearest neighbour character of the graph we have,

$$\{T_N > s\} \subseteq \{\xi_N(s) \cap [\min \xi_N(s), \max \xi_N(s)] = \xi(s) \cap [\min \xi_N(s), \max \xi_N(s)]\},$$

$$\{T_N > s\} \subseteq \{\min \xi_{[-N,+\infty)}(s) \ = \ \min \xi_N(s)\,,\ \max \xi_{(-\infty,N]}(s) \ = \ \max \xi_N(s)\},$$

and since $b \leq N - L < N$:

$$\begin{aligned} P(T_N > s, \xi_N(s) \notin \mathbb{A}_b) \leq\ & P(T_N > s, \xi(s) \notin \mathbb{A}_b) \\ & + P(T_N > s, \min \xi_{[-N,+\infty)}(s) \geq -N + L) \\ & + P(T_N > s, \max \xi_{(-\infty,N]}(s) \leq N - L). \end{aligned}$$

We may apply the Harris inequality to each of above intersections ($\{T_N > s\}$ is increasing and the other event is decreasing, in each case), so that the right-hand side on the previous expression is bounded from above by

$$P(T_N > s)\big\{P(\xi(s) \notin \mathbb{A}_b) + P(\min \xi_{[-N,+\infty)}(s) \geq -N + L) \\ + P(\max \xi_{(-\infty,N]}(s) \leq N - L)\big\}$$

which, by attractiveness, is less than or equal to

$$P(T_N > s)\left(\mu^{(\lambda)}(\mathbb{A}_b^c) + 2\mu^{(\lambda)}_{[0,+\infty)}\{\eta\colon \eta \cap \{0, \ldots, L-1\} = \emptyset\}\right),$$

provided $L \geq 1$ and $N \geq b + L$.

We now turn to the first term on the right-hand side of (4.120). A minor modification of Remark 4.19 gives that if $A \subseteq \{0, 1, \ldots\}$ and $|A| \geq L$, then

$$P(T^A_{[0,+\infty)} = +\infty) \geq P(T^{\{0,\ldots,L-1\}}_{[0,+\infty)} = +\infty),$$

which by self-duality coincides with $\mu^{(\lambda)}_{[0,+\infty)}\{\eta\colon \eta \cap \{0, \ldots, L-1\} \neq \emptyset\}$. Consequently, and using the translation invariance of the graph, if $b \geq L/\rho$ we have

$$P\left(T^{A\cap[-b,-1]}_{[-N,+\infty)} = +\infty\right) \geq \mu^{(\lambda)}_{[0,+\infty)}\{\eta\colon \eta \cap \{0, \ldots, L-1\} \neq \emptyset\}, \tag{4.121}$$

for each $A \in \mathbb{A}_b$ and $N > b$. The same applies to $P(T^{A\cap[1,b]}_{(-\infty,N]} = +\infty)$.

For $A \in \mathbb{A}_b$ and $N > b$ we set:

$$G_A := \{T^{A\cap[-b,-1]}_{[-N,+\infty)} = +\infty = T^{A\cap[1,b]}_{(-\infty,N]}\},$$

$$U_A := \inf\{u > 0\colon\ N \in \xi^{A\cap[-b,-1]}_{[-N,+\infty)}(u)\},$$

$$V_A := \inf\{v > 0\colon\ -N \in \xi^{A\cap[1,b]}_{(-\infty,N]}(v)\},$$

with the usual convention that $\inf \emptyset = +\infty$. We easily see that

$$P\left(T^{A\cap[-b,-1]}_{[-N,+\infty)} = +\infty = U_A\right) = P\left(T^{A\cap[1,b]}_{(-\infty,N]} = +\infty = V_A\right) = 0,$$

as well as $G_A \subseteq \{\max(U_A, V_A) < T^A_N \leq T_N\}$ a.s. On the other hand, again by the nearest neighbour character of the random graph, $\xi_N(u) = \xi^A_N(u)$ for $u > \max(U_A, V_A)$. Therefore, $P(T_N \neq T^A_N) \leq P(G^c_A)$, and from (4.121) we get for $b \geq L/\rho$ and $N > b$,

$$\sup_{A\in\mathbb{A}_b} P(T_N \neq T^A_N) \leq 2\mu^{(\lambda)}_{[0,+\infty)}\ \{\eta\colon \eta \cap \{0, \ldots, L-1\} = \emptyset\}.$$

We now put together the two estimates. If $0 < \rho < \rho_\lambda$ has been fixed, $L \geq 1$ and $b \geq L/\rho$ are integers, $N \geq b + L$, and υ_N is given by (4.119), then

$$\upsilon_N \leq \mu^{(\lambda)}(\mathbb{A}^c_b) + 4\mu^{(\lambda)}_{[0,+\infty)}\{\eta\colon \eta \cap \{0, \ldots, L-1\} = \emptyset\}.$$

It remains to show that this can be made small. As noticed before, $\mu^{(\lambda)}_{[0,+\infty)}(\emptyset) = 0$ for each $\lambda > \lambda_c$, and so $L = L(\varepsilon)$ may be taken so that

$$\mu^{(\lambda)}_{[0,+\infty)}\{\eta\colon \eta \cap \{0, \ldots, L(\varepsilon) - 1\} = \emptyset\} \leq \varepsilon/8.$$

On the other hand, since $\mu^{(\lambda)}$ is ergodic with density $\rho_\lambda > \rho$, we may take integer $b(\varepsilon) \ge L(\varepsilon)/\rho$ so that $\mu^{(\lambda)}(\mathbb{A}_b^c) \le \varepsilon/2$ for any $b \ge b(\varepsilon)$. Consequently, if $N > N(\varepsilon) := b(\varepsilon) + L(\varepsilon)$ the right-hand side of (4.119) is bounded from above by ε, concluding the proof of part (i).

Part (ii) is very simple in this case, as in Theorem 4.6. By (4.114) and the first inequality in (4.120) we have $P(T_N > s\beta_N) \le e^{-[s]}$, for each $s \ge 0$, which allows us to apply the dominated convergence theorem to the identity

$$\frac{E(T_N)}{\beta_N} = \frac{1}{\beta_N}\int_0^{+\infty} P(T_N > s)ds = \int_0^{+\infty} P(T_N > s\beta_N)ds,$$

yielding the conclusion, due to part (i). □

Comments Theorem 4.20 was originally proven in [48] for $\lambda > \lambda_c(o)$, the critical parameter when the infection propagates only to the right. The present proof is essentially that given by Schonmann in [265], where a suitable choice of the set $\mathbb{A}_b$ allows comparison with semi-infinite processes. Using the renormalization procedure introduced in [101], Durrett and Schonmann provided a shorter and elegant proof of Theorem 4.20 (see Remark 4.32 below). The previous proof works at once for $\lambda > \lambda_c^+$, the critical parameter of the semi-infinite process. On the other hand, the verification that $\lambda_c^+ = \lambda_c$ relies on such renormalization techniques (see Remark 4.30).

We now address the proof of Theorem 4.21. Explicit conditions on R_N are obtained from the exponential convergence to $\mu^{(\lambda)}$, stated as Theorem 4.23 below. This result was obtained by Durrett and Griffeath in [101], and the proof uses the just mentioned renormalization procedure. Postponing a brief discussion of this procedure, we first recall its main corollary for our present purposes, in the form of a large deviation estimate for the edge process.

Proposition 4.22 *If $\lambda > \lambda_c$ and $a < \alpha(\lambda)$, then $\lim_{t\to+\infty} t^{-1}\log P(r_t < at)$ exists and is negative.*

The discussion of the previous proposition is deferred to the end of this section. Following Theorem 5 in [101] or Theorem 3.23, Chapter VI of [199], we first see how it applies to the proof of exponential convergence, which will be crucial in our proof of Theorem 4.21.

Theorem 4.23 (Durrett and Griffeath) *Assume $\lambda > \lambda_c$. Then:*

(*i*) *there exist positive constants c, γ so that for any $t \ge 0$ and any finite set $B \subseteq \mathbb{Z}$*

$$P(t < T^B < +\infty) \le ce^{-\gamma t}; \tag{4.122}$$

(ii) *for any cylinder function* $f : \mathcal{P}(\mathbb{Z}) \to \mathbb{R}$ *there exists a positive constant* c_f *such that*

$$|Ef(\xi^{\mathbb{Z}}(t)) - \mu^{(\lambda)}(f)| \le c_f e^{-\gamma t}, \tag{4.123}$$

with γ *as in part (i).*

Proof We first check (4.122) for $B = \{0\}$. With the notation introduced in the discussion of the complete convergence theorem, $\{\xi^{\{0\}}(t) \neq \emptyset\} = \{\ell_s \le r_s, \forall s \le t\}$, and we can write

$$\begin{aligned} P(t < T^{\{0\}} < +\infty) &\le P(\ell_s > r_s \text{ for some } s > t) \\ &\le 2P(r_s < 0 \text{ for some } s > t), \end{aligned}$$

also using the symmetry between ℓ_s and r_s. From Proposition 4.22, given $0 < a < \alpha(\lambda)$ we may take $c, \gamma > 0$ so that $P(r_m < am) \le ce^{-\gamma m}$ for all $m \ge 1$. In this case $P(r_m < am$ for some integer $m \ge n) \le ce^{-\gamma n}/(1 - e^{-\gamma})$, taking care of the process along integer times. But r_t increases only by jumps of size one, which happen with rate λ, so that $P\,(r_s < 0$ for some $s \in [m-1, m), r_m > am) \le P(\sup_{s\in[m-1,m)}(r_m - r_s) > am) \le P(X > am)$, where X is a Poisson random variable with average λ. Using the exponential Markov inequality we see that $P(X > am) \le \tilde{c}e^{-\gamma m}$ for all m, and suitable $\tilde{c} = \tilde{c}(a, \gamma, \lambda) < +\infty$. Summing up for $m \ge t$, and putting together with the previous estimate we get (4.122), for $B = \{0\}$. That this implies (4.123) follows from the following observations. By self-duality, $Ef(\xi^{\mathbb{Z}}(t)) - \mu^{(\lambda)}(f) = P(t < T^B < +\infty)$ when $f = g_B = \mathbf{1}_{\{\xi :\, \xi\cap B\neq\emptyset\}}$. Since any cylinder function may be written as a finite linear combination of the g_B, (ii) follows at once from a weaker version of (i) (replacing c by $c|B|$), itself an immediate consequence of the case $B = \{0\}$. Indeed, according to Remark 4.18:

$$\begin{aligned} 0 &\le E\big(g_B(\xi^{\mathbb{Z}}(t))\big) - \mu^{(\lambda)}(g_B) = \bar{\mu}_t(\{(\eta, \eta') \colon \eta \cap B = \emptyset, \eta' \cap B \neq \emptyset\}) \\ &\le \sum_{x\in B} \bar{\mu}_t\{(\eta, \eta') : x \in \eta', x \notin \eta\} = \sum_{x\in B} \Big(P\{x \in \xi^{\mathbb{Z}}(t)\} - \mu^{(\lambda)}\{x \in \xi\}\Big). \end{aligned}$$

The restart argument used in Section 12 of [99] shows (i) with $c, \gamma > 0$ depending only on λ once it holds (with possibly different c, γ) for $B = \{0\}$: let $\tau_0 = 0$, $x_0 = \sup B$ and $\tau_1 = T^{\{x_0\}}$. If $\tau_1 < +\infty$ but $\xi^B(\tau_1) \neq \emptyset$ set $x_1 = \sup \xi^B(\tau_1)$; otherwise set $x_1 = 0$; repeat the procedure considering $\eta^1_t =_{(\tau_1)} \xi^{\{x_1\}}(t - \tau_1)$ for $t > \tau_1$, cf. (4.101), and $\tau_2 = \inf\{t > \tau_1 \colon \eta^1_t = \emptyset\}$, and defining recursively x_k, τ_{k+1} when $k \le K = \sup\{k \colon \tau_k < +\infty\}$. From the construction one has that $\{t < T^B < +\infty\} \subseteq \{\tau_K > t\}$. We see that $P(K \ge k) = (1 - \rho_\lambda)^k$ and conditioned on $K \ge k, \tau_i - \tau_{i-1}, i = 1, \dots, k$ are independent and distributed as $T^{\{0\}}$ conditioned to be finite, which has an exponential tail, and easily conclude that $P(\tau_K > t) \le ce^{-\gamma t}$ for suitable $c, \gamma > 0$, depending only on λ. (Of course, (4.122) becomes trivial for B infinite or empty.) □

We shall in fact need the following consequence of Theorem 4.23, proven in [265].

Corollary 4.24 *Let γ be as in Theorem* 4.23. *Given $f, g : \mathcal{P}(\mathbb{Z}) \to \mathbb{R}$ cylinder functions, a positive constant $c_{f,g}$ may be taken so that*

$$|E\,(f(\xi(s))g(\xi(r))) - E\,(f(\xi(s)))\; E\,(g(\xi(r)))\,| \le c_{f,g}\; e^{-\gamma|r-s|}.$$

Proof We may assume $0 < r < s$. Taking linear combinations, it suffices to consider $f = g_A$, $g = g_B$ with A and B finite, as in the previous proof. By self-duality, this can be phrased in terms of part (i) of Theorem 4.23. To see this, one considers the inverted graph structure introduced in the beginning of this section, letting, as in (4.109):

$$\hat{\xi}^{B,s}(u) = \{x \in \mathbb{Z}\colon \exists \text{ a reversed path from some } (y,s),\, y \in B, \text{ to } (x, s-u)\},$$

for $0 \le u \le s$, and

$$\hat{\xi}^{A,r}(u) = \{x \in \mathbb{Z}\colon \exists \text{ a reversed path from some } (y,r),\, y \in A, \text{ to } (x, r-u)\},$$

for $0 \le u \le r$. Then:

(a) $\text{law}(\hat{\xi}^{B,s}(u) : 0 \le u \le s) = \text{law}(\xi^B(u) : 0 \le u \le s)$,
(b) $\text{law}(\hat{\xi}^{A,r}(u) : 0 \le u \le r) = \text{law}\big(\xi^A(u) : 0 \le u \le r\big)$,
(c) the process $(\hat{\xi}^{A,r}(u) : 0 \le u \le r)$ depends only on the portion of the initial graph which is contained in $\mathbb{Z} \times [0, r]$,
(d) $g_B(\xi(s)) = \mathbf{1}_{\tilde{B}}$ and $g_A(\xi(r)) = \mathbf{1}_{\tilde{A}}$, where $\tilde{B} = \{\hat{\xi}^{B,s}(s) \ne \emptyset\}$ and $\tilde{A} = \{\hat{\xi}^{A,r}(r) \ne \emptyset\}$.

Due to (c) and the independence properties of Poisson processes the events $\tilde{A}$ and $H := \{\hat{\xi}^{B,s}(s-r) \ne \emptyset\}$ are independent, and $\tilde{B} \subseteq H$. From this we easily obtain:

$$\begin{aligned} E\big(g_B(\xi(s))g_A(\xi(r))\big) - E\big(g_B(\xi(s))\big)\, E\big(g_A(\xi(r))\big) &= P(\tilde{A} \cap \tilde{B}) - P(\tilde{A})P(\tilde{B}) \\ &= P(\tilde{A})P(H \cap \tilde{B}^c) - P(\tilde{A} \cap H \cap \tilde{B}^c), \end{aligned}$$

so that

$$|E\big(g_B(\xi(s))g_A(\xi(r))\big) - E\big(g_B(\xi(s))\big)\, E\big(g_A(\xi(r))\big)| \le P(H \cap \tilde{B}^c).$$

By (a), $P(H \cap \tilde{B}^c) = P(s - r < T^B < s)$, which we bound by $P(s - r < T^B < +\infty)$ and apply Theorem 4.23 to conclude the proof. □

Remark 4.25 Theorem 4.23 and Corollary 4.24 also apply to $\xi_{[0,+\infty)}$.

Proof of Theorem 4.21. We start with R_N satisfying condition (i) in the statement. By Theorem 4.20 we see that $P(T_N > 2R_N) \to 1$ as $N \to +\infty$. Replacing ε by 2ε in (4.117) and setting $K_N \colon = \max\{\ell \in \mathbb{Z}_+ : \ell R_N < T_N\}$, it suffices to have that for each f as in the statement and each $\varepsilon > 0$:

$$\lim_{N\to+\infty} P\left(\max_{\ell\in\mathbb{Z},0\le\ell<K_N} \max_{i\in I_{f,N}(L)} |A^N_{R_N}(\ell R_N, \theta_i f) - \mu^{(\lambda)}(f)| > \varepsilon,\, K_N \ge 1\right) = 0.$$

Notice that $K_N < +\infty$ a.s. If $m_N \in \mathbb{N}$ we write

$$\begin{aligned} P\Big(&\max_{\ell\in\mathbb{Z},0\le\ell<K_N} \max_{i\in I_{f,N}(L)} | A^N_{R_N}(\ell R_N, \theta_i f) - \mu^{(\lambda)}(f)| > \varepsilon,\, K_N \ge 1\Big) \\ &\le P(K_N > m_N) \\ &\quad + \sum_{j=1}^{m_N} P\left(\bigcup_{\ell=0}^{j-1} \bigcup_{i\in I_{f,N}(L)} B^N_{\ell,i},\, K_N = j\right) \\ &\le P(K_N > m_N) \\ &\quad + (2N+1)\sum_{j=1}^{m_N} j \max_{0\le\ell<j} \max_{i\in I_{f,N}(L)} P(B^N_{\ell,i},\, K_N = j), \end{aligned}$$

where $B^N_{\ell,i} = \{| A^N_{R_N}(\ell R_N, \theta_i f) - \mu^{(\lambda)}(f)| > \varepsilon\}$. Due to the definition of $I_{f,N}(L)$, if we set $C_{N,L} := \{\eta \colon \eta \cap [-N, -N+L] = \emptyset\}$, and $\tilde{C}_{N,L} := \{\eta \colon \eta \cap [N-L, N] = \emptyset\}$, the nearest neighbour character of the graph implies that

$$\left\{\xi_N(t) \notin C_{N,L} \cup \tilde{C}_{N,L}\right\} \subseteq \left\{\theta_i f(\xi_N(t)) = \theta_i f(\xi(t)),\, \forall i \in I_{f,N}(L)\right\},$$

and so

$$|\theta_i f(\xi_N(t)) - \theta_i f(\xi(t))| \le 2||f||_\infty \big(\mathbf{1}_{C_{N,L}}(\xi_N(t)) + \mathbf{1}_{\tilde{C}_{N,L}}(\xi_N(t))\big).$$

Moreover, on the event $\{t < T_N\}$ we have:

$$\begin{aligned} \mathbf{1}_{C_{N,L}}(\xi_N(t)) &= \mathbf{1}_{C_{N,L}}(\xi_{[-N,+\infty)}(t)), \\ \mathbf{1}_{\tilde{C}_{N,L}}(\xi_N(t)) &= \mathbf{1}_{\tilde{C}_{N,L}}(\xi_{(-\infty,N]}(t)), \end{aligned}$$

so that

$$\begin{aligned} &\max_{0\le\ell<j} \max_{i\in I_{f,N}(L)} P(B^N_{\ell,i},\, K_N = j) \\ &\quad\le \max_{0\le\ell<j} \max_{i\in I_{f,N}(L)} P\left(\left| \frac{1}{R_N} \int_{\ell R_N}^{(\ell+1)R_N} \theta_i f(\xi(t))dt - \mu^{(\lambda)}(f)\right| > \frac{\varepsilon}{2}\right) \\ &\quad + \max_{0\le\ell<j} 2P\left(2||f||_\infty \frac{1}{R_N} \int_{\ell R_N}^{(\ell+1)R_N} \mathbf{1}_{C_{N,L}}(\xi_{[-N,+\infty)}(t))dt > \frac{\varepsilon}{4}\right), \end{aligned} \tag{4.124}$$

where we used that $\xi_{[-N,+\infty)}(t)$ and $\mathcal{R}\xi_{(-\infty,N]}(t)$ have the same law. The probability on the first term of the right-hand side of (4.124) does not depend on i and we get

$$\begin{aligned} P\Big(&\max_{0\le\ell<K_N} \max_{i\in I_{f,N}(L)} \big| A^N_{R_N}(\ell R_N, \theta_i f) - \mu^{(\lambda)}(f)\big| > \varepsilon,\, K_N \ge 1\Big) \\ &\le P(K_N > m_N) + (2N+1)m_N^2 \Big\{ \max_{\ell\ge0} P(\Gamma^N_\ell) + 2\max_{\ell\ge0} P(\tilde{\Gamma}^N_\ell)\Big\}, \end{aligned} \tag{4.125}$$

where

$$\Gamma_\ell^N = \left[\left| \frac{1}{R_N} \int_{\ell R_N}^{(\ell+1)R_N} f(\xi(t))dt - \mu^{(\lambda)}(f) \right| > \frac{\varepsilon}{2} \right]$$

and

$$\tilde{\Gamma}_\ell^N = \left[\frac{1}{R_N} \int_{\ell R_N}^{(\ell+1)R_N} \mathbf{1}_{C_{N,L}}(\xi_{[-N,+\infty)}(t))dt > \frac{\varepsilon}{8||f||_\infty} \right]$$

(if $f = 0$ we set $\tilde{\Gamma}_\ell^N = \emptyset$).

The events $\tilde{\Gamma}_\ell^N$ (Γ_ℓ^N) depend on f and L (f, respectively). L is properly chosen, depending on f and ε. R_N has to be found verifying condition (i) in the statement and so that the expression on the right-hand side of (4.125) tends to zero, as $N \to +\infty$, for each f and ε.

Due to the spatial shift invariance of the random graph

$$P(\tilde{\Gamma}_\ell^N) = P\left(\frac{1}{R_N} \int_{\ell R_N}^{(\ell+1)R_N} \mathbf{1}_{C_{0,L}}(\xi_{[0,+\infty)}(s))ds > \frac{\varepsilon}{8||f||_\infty} \right),$$

and since $\mathbf{1}_{C_{0,L}}$ is a decreasing function, from the monotonicity of the process, we have

$$E\left(\frac{1}{R} \int_{\ell R}^{(\ell+1)R} \mathbf{1}_{C_{0,L}}(\xi_{[0,+\infty)}(s))ds \right) \le \mu^{(\lambda)}_{[0,+\infty)}\{\eta \colon \eta \cap \{0, \dots, L\} = \emptyset\},$$

uniformly on $R > 0$ and $\ell \ge 0$. As seen in the previous proof, we may take L depending on $\varepsilon/||f||_\infty$, so that

$$\mu^{(\lambda)}_{[0,+\infty)}\{\eta \colon \eta \cap \{0, \dots, L\} = \emptyset\} \le \frac{\varepsilon}{16||f||_\infty}.$$

With such fixed $L = L(\varepsilon, f)$ we set

$$Y_\ell^R = \frac{1}{R} \int_{\ell R}^{(\ell+1)R} \mathbf{1}_{C_{0,L}}(\xi_{[0,+\infty)}(s))ds,$$

so that from the above observations we have

$$\begin{aligned} \max_{\ell \ge 0} P(\tilde{\Gamma}_\ell^N) &\le \max_{\ell \ge 0} P\left(|Y_\ell^{R_N} - EY_\ell^{R_N}| \ge \frac{\varepsilon}{16||f||_\infty} \right) \\ &\le \left(\frac{16||f||_\infty}{\varepsilon} \right)^2 \max_{\ell \ge 0} \mathrm{Var}(Y_\ell^{R_N}) \\ &\le \left(\frac{16||f||_\infty}{\varepsilon} \right)^2 \frac{2}{R_N^2} \int_0^{R_N} ds \int_0^s du\, c_L e^{-\gamma_L(s-u)} \le \frac{C(L, f, \varepsilon)}{R_N}, \end{aligned} \tag{4.126}$$

where c_L and γ_L come from Remark 4.25 applied to $f = g = \mathbf{1}_{C_{0,L}}$, and $C(L, f, \varepsilon) = (16||f||_\infty/\varepsilon)^2 2c_L/\gamma_L$.

For $\max_{\ell\geq 0} P(\Gamma_\ell^N)$ the strategy is very similar. Let

$$X_\ell^R = \frac{1}{R}\int_{\ell R}^{(\ell+1)R} f(\xi(s))ds,$$

so that

$$EX_\ell^R = \frac{1}{R}\int_0^R Ef(\xi(\ell R+u))du.$$

Using Corollary 4.24 we see that:

$$\text{Var}(X_\ell^R) = \frac{2}{R^2}\int_0^R ds \int_0^s du \,\text{cov}\big(f(\xi(\ell R+u)),\, f(\xi(\ell R+s))\big) \leq \frac{C(f)}{R},$$

for a suitable positive constant $C(f)$ (depending on λ) simultaneously on ℓ. On the other hand, as $t \to +\infty$ the law of $\xi(t)$ converges to $\mu^{(\lambda)}$ so that EX_ℓ^R tends to $\mu^{(\lambda)}(f)$ uniformly in ℓ, as $R \to +\infty$. Given $\varepsilon > 0$ we take $R = R(\varepsilon, f)$ so that for $R \geq R(\varepsilon, f)$

$$\sup_{\ell\geq 0} |EX_\ell^R - \mu^{(\lambda)}(f)| \leq \frac{\varepsilon}{4}.$$

Thus, if N is large enough ($N \geq N_0(\varepsilon, f)$), applying the previous observations and the Chebychev inequality we get:

$$\begin{aligned}\max_{\ell\geq 0} P(\Gamma_\ell^N) &\leq \max_{\ell\geq 0} P\left(|X_\ell^{R_N} - \mu^{(\lambda)}(f)| \geq \frac{\varepsilon}{2}\right)\\ &\leq \max_{\ell\geq 0} P\left(|X_\ell^{R_N} - EX_\ell^{R_N}| \geq \frac{\varepsilon}{4}\right) \qquad (4.127)\\ &\leq \frac{16}{\varepsilon^2}\max_{\ell\geq 0} \text{Var}(X_\ell^{R_N}) \leq \frac{16C(f)}{\varepsilon^2 R_N}.\end{aligned}$$

Putting together (4.125), (4.126) and (4.127), and due to Theorem 4.20, it suffices to find $R_N > 0$ and m_N so that, as $N \to +\infty$: (a) $R_N/\beta_N \to 0$, (b) $m_N R_N/\beta_N \to +\infty$, and (c) $N{m_N}^2/R_N \to 0$. This is equivalent to:

$$1 \ll \frac{\beta_N}{R_N} \ll m_N \ll \left(\frac{R_N}{N}\right)^{1/2} \ll \left(\frac{\beta_N}{N}\right)^{1/2}, \qquad (4.128)$$

where $a_N \ll b_N$ stands for $\lim_{N\to+\infty} a_N/b_N = 0$. In particular we need $N \ll \beta_N$. This is a very rough lower bound for β_N. A simple proof (given in [265]) is written in the next lemma. We assume it for the moment and conclude the argument, for which we take R_N such that $N^{1/3}\beta_N^{2/3} \ll R_N \ll \beta_N$. In this case $\beta_N/R_N \ll \left(R_N/N\right)^{1/2}$ and we may take m_N 'in between'. For example, set $R_N = N^{1-a}\beta_N^a$ with $2/3 < a < 1$ and $m_N = \left(\beta_N/N\right)^{a'}$ with $1 - a < a' < a/2$, so that (4.128) holds. □

Lemma 4.26 $\lim_{N\to+\infty} N/\beta_N = 0$ *under the conditions of Theorem* 4.20.

Proof We again consider the processes $\xi_N(\cdot)$ and $\xi(\cdot)$ constructed on the same random graph, and let $B_N := \{T^{[-N/2,N/2]\cap\mathbb{Z}} = +\infty\}$, as well as

$$S_N := \inf\left\{t > 0 \colon \{-N, N\} \cap \xi^{[-N/2,N/2]\cap\mathbb{Z}}(t) \neq \emptyset\right\},$$

which is a.s. finite on B_N. Due to the common graph, $\xi_N^{[-N/2,N/2]\cap\mathbb{Z}}(t) = \xi^{[-N/2,N/2]\cap\mathbb{Z}}(t)$ if $t < S_N$. Thus, $B_N \subseteq [S_N \leq T_N]$.

Since the rightmost particle in $\xi^{[-N/2,N/2]\cap\mathbb{Z}}(t)$ is bounded from above by $N/2 + n_t$ where n_t is a Poisson process with rate λ (analogously for the leftmost particle), we have

$$P(T_N \leq aN) \leq 2P(\tilde{S}_N \leq aN) + P(B_N^c),$$

where $\tilde{S}_N$ is a sum of $[N/2]$ i.i.d. exponential random variables with rate λ.

By self-duality, cf. (4.110), $P(B_N) = \mu^{(\lambda)}\{\xi : \xi \cap [-N/2, N/2] \neq \emptyset\}$, so that $P(B_N^c)$ tends to zero as $N \to +\infty$. Also $P(\tilde{S}_N \leq aN)$ tends to zero if $2a < 1/\lambda$, by the weak law of large numbers. Thus $\lim_{N\to+\infty} P(T_N \leq aN) = 0$ if $2a < 1/\lambda$. Using Theorem 4.20 we easily deduce that $N \ll \beta_N$. □

Though sufficient for the conclusion of the proof of Theorem 4.21, the previous argument is just too simple to provide a reasonable lower bound for β_N, cf. Remark 4.33 below.

Comment Equation (4.116) was initially proven in [48] without recource to the exponential convergence of $\xi(t)$ and, consequently, with a much poorer control on R_N. The proof we discussed is essentially contained in [265].

The following corollary on the thermalization along spatial Cesaro averages is a simple consequence of (4.117).

Corollary 4.27 *Under the conditions of Theorem* 4.21, *for any* $\varepsilon > 0$ *and cylinder function* f *one has*

$$P\left(\sup_{0\leq s<T_N-2R_N}\left|A_{R_N}^N(s, \bar{f}) - \mu^{(\lambda)}(f)\right| \leq \varepsilon, T_N > 2R_N\right) \to 1,$$

as $N \to +\infty$, *where* R_N *is as in Theorem* 4.21, $I_{f,N} = \{i : \Delta(\theta_i f) \subseteq \Lambda_N\}$, *and*

$$\bar{f} = \frac{1}{|I_{f,N}|}\sum_{i\in I_{f,N}} \theta_i f\,.$$

Proof Omitted (see [265]). □

Taking $f(\eta) = \eta(0)$ in the previous corollary we have the thermalization property for the empirical spatial density $|\xi_N(\cdot)|/(2N+1)$; similarly for empirical pair correlation functions.

Problem In spite of the reasonably good description of the thermalization property, contained in Theorem 4.21 and Corollary 4.27, nothing is said about the final stage, i.e. the typical behaviour of $\xi_N(\cdot)$ in the time interval $[T_N - R_N, T_N]$. A description of 'how' the process disappears is still lacking and it is an interesting problem.

A renormalization procedure For the sake of completeness we give a brief description of the renormalization procedure leading to the proof of Proposition 4.22.

Such arguments were first applied to percolation in the works of Kesten [175], Russo [261] and Seymour and Welsh [277]. The application to the one-dimensional contact process appeared in [101], being extended by Gray to a quite general class of attractive one-dimensional systems, cf. [140]. The method has been shown to be useful in many applications and further developments. The basic point is that by changing the scale one is able to compare the original system to an oriented one-dependent percolation model where the density of 'open' sites is arbitrarily close to 1. The use of contour arguments becomes possible and provides very useful information for the original process at any supercritical parameter λ.

In this short discussion, aiming to point out the basic scheme, we follow [199] and [99], where the construction of [101] has been presented with modifications proposed in [140]. For full details we refer to any of these texts.

Definition 4.28 Let $S = \{(i, j) \in \mathbb{Z} \times \mathbb{Z}_+ : i + j \text{ is even}\}$. Sites (i, j), $(i', j') \in S$ are declared neighbours[5] iff $|(i, j) - (i', j')|_1 := |i - i'| + |j - j'| = 2$. A one-dependent oriented percolation model on S, with parameter p, refers to any set of random variables $\{U_{(i,j)} : (i, j) \in S\}$ such that:

(i) $P(U_{(i,j)} = 1) = p$ and $P(U_{(i,j)} = 0) = 1 - p$ for each (i, j);

(ii) if $F \subseteq S$ and $|(i, j) - (i', j')|_1 > 2$ for each $(i, j), (i', j') \in F$ distinct, then the random variables $\{U_{(i,j)} : (i, j) \in F\}$ are independent.

An open oriented path in this model is a sequence of sites $(i_1, j_1), \ldots, (i_m, j_m)$ with $j_{k+1} = j_k + 1$, $|i_{k+1} - i_k| = 1$, if $1 \le k \le m - 1$, such that $U_{(i_k, j_k)} = 1$ if $1 \le k \le m$. (That is, site (i, j) is 'open' if $U_{(i,j)} = 1$; otherwise it is closed.)

The attribute 'one-dependent' means that $U_{(i,j)}$, $(i, j) \in F$ are independent when F does not contain a pair of neighbours. (If $j \ge 2$, (i, j) has eight neighbours in S.)

One says that percolation from the origin occurs if there exists an infinite open oriented path starting at $(0, 0)$, i.e. if $C_{(0,0)} := \{(i, j) : \exists \text{ open oriented path from } (0, 0) \text{ to } (i, j)\}$ is infinite. The interpretation of a fluid flowing along channels through open sites is a natural one, being the reason for the expression

[5] The minimal $|\cdot|_1$ distance between distinct points in S is 2.

'(i, j) is wet' if $(i, j) \in C_{(0,0)}$. The occurrence of percolation means that 'the fluid reaches infinitely many sites'.

In other words, S is made into a graph with oriented edges connecting each (i, j) to each of $(i \pm 1, j + 1)$, on which one considers site percolation with the described dependence.

Remark 4.29 In the set-up of Definition 4.28 we need the following applications of Peierls type contour arguments (p close to one), which are crucial for Proposition 4.22. We follow Theorem 3.19 and Theorem 3.21, Chapter VI of [199], cf. also Sections 9 and 10 in [99].

(i) $P(C_{(0,0)}$ is infinite$) > 0$ if p is sufficiently near to 1.

Proof If $U_{(0,0)} = 1$ and $C_{(0,0)}$ is finite, we may consider a 'closed' contour which blocks percolation. (Contours are defined in the dual lattice of S. Since we have an oriented site model, the situation is a little different from that in Proposition 3.30.) Equivalently, we may think of the region $D := \cup_{(i,j)\in C_{(0,0)}} D_{(i,j)} \subseteq \mathbb{R}^2$, i.e. each site (i, j) is replaced by the diamond $D_{(i,j)} := \{(x, y) \in \mathbb{R}^2 \colon |x - i| + |y - j| \leq 1\}$. Non-occurrence of percolation implies that $\mathbb{R}^2 \setminus D$ has a unique infinite component. Its boundary Γ consists of an even number (at least four) of 'segments', i.e. any of the four segments forming the boundary of a diamond $D_{(i,j)}$. Let g be a fixed realization of Γ with $2n$ 'segments', each of which belongs to the boundary of a unique 'wet' $D_{(i,j)}$. Exactly n 'segments' are the upper (left or right) boundary of the corresponding 'wet' $D_{(i,j)}$, implying that $U_{(i-1,j+1)} = 0$ (upper left) or $U_{(i+1,j+1)} = 0$ (upper right). We see that g determines a set of at least $n/2$ different sites on which $U_{\cdot} = 0$ (the same site corresponds to at most two different upper 'segments'); since each site has at most eight different neighbours, we extract (in a deterministic procedure, using a given fixed total order in S) a set F with cardinality at least $n/18$, containing no pairs of neighbours, with $U_{(i,j)} = 0$ for each $(i, j) \in F$. Thus, $P(\Gamma = g \mid U_{(0,0)} = 1) \leq (1 - p)^{n/18}$ and since there are at most 3^{2n-2} different possible γ with $2n$ 'segments', we get $P(C_{(0,0)}$ is finite$) \leq (1 - p) + p \sum_{n \geq 2} 3^{2n-2}(1 - p)^{n/18} < 1$, for p close to one. □

(ii) Set $\tilde{r}_n := \max\{j \colon \exists$ open oriented path in S from $(m, 0)$ to (j, n) for some $m \leq 0\}$. If $\tilde{a} < 1$, $P(\tilde{r}_n \leq \tilde{a}n) \leq ce^{-\tilde{c}n}$ for suitable $c, \tilde{c} > 0$ and all $n \geq 1$, provided p is close enough to 1.

Proof Let us consider

$$W = \{(i, j) \in S \colon \exists \text{ open oriented path from } (m, 0) \text{ to } (i, j) \text{ for some } m \leq 0\},$$

and again it is convenient to replace W by $V = \cup_{(i,j)\in W} D_{(i,j)} \subseteq \mathbb{R}^2$. Given $n \geq 1$, with probability one, the set $\mathbb{R} \times [0, n] \setminus V$ has only one infinite component, whose boundary we denote by Γ. The number of 'segments' in Γ is of the form $n + 2m$ for some $m \geq 0$. Each such 'segment' is contained in the boundary of exactly one $D_{(i,j)}$, with $(i, j) \in W$, and is classified according to its position in this diamond, as upper left, upper right, lower left and lower right. For any fixed m, and calling n_{ul}, n_{ur}, n_{ll}, and n_{lr} the number of segments of the corresponding type, one has:

$$n_{lr} + n_{ll} + n_{ul} + n_{ur} = n + 2m,$$

$$n_{lr} + n_{ll} - n_{ul} - n_{ur} = \tilde{r}_n,$$

and so $2(n_{ul} + n_{ur}) = n - \tilde{r}_n + 2m$, implying that $\{\tilde{r}_n < \tilde{a}n\} \subseteq \{n_{ul} + n_{ur} > (2m + n(1 - \tilde{a}))/2\}$. Taking into account the same observations as in the previous remark, and since the number of possible different realizations of Γ with $n + 2m$ segments is at most 3^{n+2m}, one ends up with $P(\tilde{r}_n < \tilde{a}n) \le \sum_{m \ge 0} 3^{n+2m} (1-p)^{(2m+n(1-\tilde{a}))/36} < 3^{-n+1}$, provided $1 - p < 3^{-72/(1-\tilde{a})}$. □

Remark The reduction from $2n$ to $n/18$ on the number of 'segments' (bonds in the dual lattice) contributing to the estimate comes from the orientation, the fact that we have a site model, and that we cannot use nearest neighbour sites.

It remains to see how a rescaling procedure relates the supercritical contact process to a one-dependent percolation model, with p close to one, as demonstrated in [101] and [140]. This brief discussion follows Section 3, Chapter VI of [199]. One again uses the existence and positivity of an asymptotic drift, $\alpha(\lambda)$, for the 'edge' process r_t, if $\lambda > \lambda_c$, as previously discussed. Let $\alpha = \alpha(\lambda)$, $0 < \beta < \alpha/3$, and choose a parameter $M > 0$ such that $M\beta/2$ and $M\alpha$ are integers. For $(i, j) \in S$, let us define the event $\mathcal{E}_{(i,j)}$ as the set of those realizations in the graphical construction such that for each of given parallelograms $R_{(i,j)}$ and $L_{(i,j)}$ there is a path lying completely within the given parallelogram and connecting its bottom edge to the top edge. $R_{(i,j)}$ and $L_{(i,j)}$ are translates by $(Mi(\alpha - \beta), Mj)$ of $R_{(0,0)}$ and $L_{(0,0)}$, respectively, where (see Figure 4.7):

$$R_{(0,0)} = \{(x, t) \in \mathbb{Z} \times [0, M(1 + \beta/\alpha)] \colon \alpha t - 3M\beta/2 \le x \le \alpha t - M\beta/2\},$$

$$L_{(0,0)} = \{(x, t) \colon (-x, t) \in R_{(0,0)}\}.$$

Clearly $P(\mathcal{E}_{(i,j)})$ is the same for each $(i, j) \in S$. The event $\mathcal{E}_{(i,j)}$ depends only on the portion of the graph inside the rectangle

$$A_{(i,j)} := [Mi(\alpha - \beta) - M(\alpha + \beta/2), Mi(\alpha - \beta) + M(\alpha + \beta/2)] \times [Mj, M(j + 1 + \beta/\alpha)].$$

Since $0 < \beta < \alpha/3$, $A_{(i,j)} \cap A_{(i',j')} = \emptyset$ whenever $|(i, j) - (i', j')|_1 > 2$, implying condition (ii) in Definition 4.28 for variables $U_{(i,j)} := \mathbf{1}_{\mathcal{E}_{(i,j)}}$. To escape away from criticality as previously mentioned, one verifies that for fixed $\alpha = \alpha(\lambda)$ and $0 < \beta < \alpha/3$

$$\lim_{M \to +\infty} P(\mathcal{E}_{(0,0)}) = 1. \tag{4.129}$$

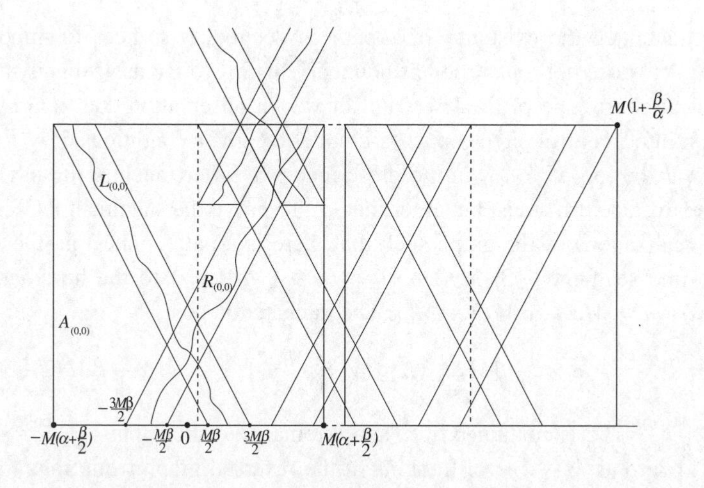

Figure 4.7

The parallelograms $L_{(i,j)}$, $R_{(i,j)}$ for $(i, j) \in S$ were constructed to 'match' properly, cf. Figure 4.7, in such a way that the existence of an open oriented path $(i_1, j_1), \ldots, (i_m, j_m)$ in S implies the existence of a path in the graph of the contact process from the bottom of $R_{(i_1,j_1)}$ to the top of $R_{(i_m,j_m)}$ and of $L_{(i_m,j_m)}$. In particular, percolation from $(0, 0)$ guarantees the survival of the contact process starting from $A := [-3M\beta/2, 3M\beta/2] \cap \mathbb{Z}$, and Remark 4.29 provides an upper bound for $P(T^A < +\infty)$, as one checks straightforwardly. It remains to prove (4.129).

Proof of (4.129) For fixed α and β as above, let $\mathcal{E}$ be the set of configurations exhibiting a path which lies inside $R_{(0,0)}$ and connects the bottom of $R_{(0,0)}$ to its top. It suffices to show that $\lim_{M\to+\infty} P(\mathcal{E}) = 1$. For $n \in \mathbb{Z}$, let r_t^n denote the right edge of the contact process starting from $\xi(0) = (-\infty, n] \cap \mathbb{Z}$, so that $r_t^0 = r_t$ and the processes $(r_t^n : t \geq 0)$ and $(n + r_t : t \geq 0)$ have the same law. From this and the a.s. convergence of r_t/t to α, as $t \to +\infty$, we see that taking e.g. $M_0 = [M\beta/3] + 1$ and $t_0 = M(1 + \beta/\alpha)$, one gets

$$\lim_{M\to+\infty} P\left\{r_t^{-M\beta/2-M_0} < \alpha t - \frac{\beta M}{2}, \forall t \leq t_0\right\} = 1,$$

and

$$\lim_{M\to+\infty} P\left\{r_{t_0}^{-M\beta/2-M_0} > M\alpha - \frac{\beta M}{2} + M_0\right\} = 1.$$

On the intersection of these two events the graph of $r_t^{-M\beta/2-M_0}$, $0 \leq t \leq t_0$ does not get to the right of $R_{(0,0)}$ and reaches its top at least M_0 sites to the right of the left extreme.

To guarantee the existence of a path as desired, it suffices to impose that $r^{-M\beta/2-M_0}$ does not touch the left boundary of $R_{(0,0)}$. By itself this is the more complicated estimate of the lower tail of r_t. An observation due to Gray transforms it into a control on the upper tail: under the given conditions, if $r^{-M\beta/2-M_0}$ arrives at the top of $R_{(0,0)}$ in the prescribed way after touching its left boundary, the average drift must be larger than α. To prove the statement let $\mathcal{D}_n$ be the set of realizations of the graph such that there is a path connecting the vertical space-time segment $\{-3\beta M/2+n\} \times [(n-1)/\alpha, n/\alpha)$ to the horizontal strip $\{(m, t_0)\colon m \geq M\alpha - \beta M/2 + M_0\}$. The intersection of

$$\{r_t^{-M\beta/2-M_0} < \alpha t - \beta M/2, \forall t \leq t_0,\ r_{t_0}^{-M\beta/2-M_0} > M\alpha - \beta M/2 + M_0\},$$

with $\cap_{n=1}^{M(\alpha+\beta)} \mathcal{D}_n^c$ is contained in $\mathcal{E}$, and it remains to show that $\sum_{n=1}^{M(\alpha+\beta)} P(\mathcal{D}_n)$ tends to zero as $M \to +\infty$. Looking at the reversed graph from time t_0 we see that

$$\begin{aligned} P(\mathcal{D}_{M(\alpha+\beta)-n}) &= P(r_s > M_0 + n \text{ for some } s \in [n/\alpha, (n+1)/\alpha)) \\ &\leq P(r_{n/a} > 3M_0/4 + n) + P(\sup_{s\in[n/\alpha,(n+1)/\alpha)} r_s - r_{n/a} > M_0/4). \end{aligned}$$

By the attractiveness and (4.102) $r_{t+u} \leq r_u + r_{u,t}$, where $(r_{u,t}\colon t \geq 0)$ is independent of r_u, and has the same law as $(r_t\colon t \geq 0)$. Using this and the comparison of r_t with a Poisson process with rate λ, we see immediately that the second term of the above decomposition decreases exponentially in M, uniformly in n. It remains to prove that

$$\lim_{M\to+\infty} \sum_{n=1}^{M(\alpha+\beta)} P(r_{n/\alpha} > 3M_0/4 + n) = 0. \tag{4.130}$$

Since Er_t/t tends to α as $t \to +\infty$, given $\varepsilon > 0$ to be chosen later, we can take $t = t_\varepsilon$ so that $Er_t < (\alpha + \varepsilon)t$. Previous observation tells us that the variable r_{kt} is stochastically smaller than a sum of k i.i.d. random variables distributed as r_t. But r_t is dominated by a Poisson random variable, so that $E(e^{\theta r_t}) < +\infty$ for $\theta > 0$, and we may use Theorem 1.1 to see that if $t = t_\varepsilon$ is picked as above, there exists $\gamma_\varepsilon > 0$ so that

$$P(r_{kt} \geq kt(\alpha + \varepsilon)) \leq e^{-\gamma_\varepsilon k},$$

for all $k \geq 1$. Choosing ε so that $\varepsilon(1 + \beta/\alpha) < \beta/4$ we see that given $\delta > 0$ we can take n_δ (independent of M) so that the sum for $n \geq n_\delta$ in (4.130) is smaller than δ. Taking M large we control the remaining terms. □

Proof of Proposition 4.22 The existence of the limit is a consequence of the previous observation that $r_{t+s} \leq r_s + r_{s,t}$, which entails

$$P(r_{t+s} < a(t+s)) \geq P(r_s < as)P(r_t < at).$$

We must check that the limit is strictly negative. With α, β, and M as in the previous construction and $\tilde{r}_n$ cf. Remark 4.29, we have $r_{nM} \geq (\alpha - \beta)M\tilde{r}_n - 3\beta M/2$. It suffices to take $\beta < \alpha/3$ so that $a < \alpha - \beta$, in which case we take any $a/(\alpha - \beta) < \tilde{a} < 1$ and M large enough for the renormalized model to fit into item (ii) of Remark 4.29, according to (4.129). For example, $\beta < \alpha/3 \wedge (\alpha - a)/2$ with the choice $\tilde{a} \in (2a/(\alpha + a), 1)$ works. □

The usefulness of the renormalization procedure is that the choice of M large enough allows us to use simple contour methods successfully. As one sees, it leads to poor control on the limit in Proposition 4.22, and consequently on the value γ in Theorem 4.23.

Remark 4.30 The previous construction allows us to see that if $\lambda > \lambda_c$ the restricted process $\xi^{\{0\}}_{[0,+\infty)}$ also has positive probability of survival, i.e. $\lambda_c^+ = \lambda_c$. (For large p the related one-dependent oriented site percolation model has positive probability of survival even if restricted to $\{(i, j) \in S : i \geq 0\}$.)

Remark 4.31 (Rate of convergence in Theorem 4.20) A general and simple argument on the stability of exponential laws (Lemma 4.34 below) gives:

$$\sup_{t \geq 0} |P(T_N/ET_N > t) - e^{-t}| \leq 2\upsilon_N,$$

with υ_N defined by (4.119). Having fixed $\rho \in (0, \rho_\lambda)$, if $L_N = [\rho(N-1)/(\rho + 1)]$ and $b_N = [(N-1)/(\rho + 1)] + 1$, the proof of Theorem 4.20 gives us:

$$2\upsilon_N \leq 2\mu^{(\lambda)}(\mathbb{A}^c_{b_N}) + 8\mu^{(\lambda)}_{[0,+\infty)}\{\eta : \eta \cap \{0, \ldots, L_N - 1\} = \emptyset\}. \tag{4.131}$$

($[x]$ denotes the integer part of $x \in \mathbb{R}_+$.) Both terms on the right-hand side of (4.131) refer to large deviation probabilities (under $\mu^{(\lambda)}$ and $\mu^{(\lambda)}_{[0,+\infty)}$).

(a) Adapting the renormalization procedure to the restricted process $\xi_{[0,+\infty)}$, the term $\mu^{(\lambda)}_{[0,+\infty)}\{\eta : \eta \cap A = \emptyset\} = P(T^A_{[0,+\infty)} < +\infty)$, with $A = \{0, \ldots, L - 1\}$, may be estimated with the help of the contour argument. This allows us to see that the second term on the right-hand side of (4.131) tends to zero exponentially in L_N (see [99] for more details).

(b) The term $\mu^{(\lambda)}(\mathbb{A}^c_b)$ involves a more elaborate estimate, since $\rho > 0$. But, in the proof of Theorem 4.20 we have freedom to choose $\rho \in (0, \rho_\lambda)$ small enough, in which case the exponential decay of $\mu^{(\lambda)}(\mathbb{A}^c_b)$ follows from the result used in (a) (for the unrestricted process). Indeed, as observed in [194]:

$$\begin{aligned}\mu^{(\lambda)}(\mathbb{A}^c_b) &\leq 2\mu^{(\lambda)}(\eta : \exists J \subseteq [1, b] \cap \mathbb{Z},\ |J| \geq [(1-\rho)b],\ \eta \cap J = \emptyset) \\ &\leq 2\binom{b}{[(1-\rho)b]} \tilde{c} e^{-\tilde{\gamma}[(1-\rho)b]},\end{aligned}$$

with $\tilde{\gamma} > 0$. Applying the Stirling formula (1.7), we see that $\limsup_{b \to +\infty} b^{-1} \log \mu^{(\lambda)}(\mathbb{A}^c_b) \leq -(1-\rho)\log(1-\rho) - \rho \log \rho - (1-\rho)\tilde{\gamma} < 0$ provided $\rho > 0$ is small (see Remark 4.33 below).

Collecting the previous observations and recalling the above choices of L_N, b_N, we conclude that given $\lambda > \lambda_c$, there exist $\bar{c} = \bar{c}(\lambda), \bar{\gamma} = \bar{\gamma}(\lambda) \in (0, +\infty)$ so that for each N

$$\sup_{t \geq 0} |P(T_N / ET_N > t) - e^{-t}| \leq \bar{c} e^{-\bar{\gamma} N}.$$

Remark 4.32 (A shorter proof of (i) of Theorem 4.20) (See [104].) If $0 \leq u < v$, consider the following event in the graphical construction

$$\mathcal{E}(N, u, v) := \{\exists \text{ a path contained in } \Lambda_N \times [u, v] \text{ connecting } \Lambda_N \times \{u\} \text{ to } \Lambda_N \times \{v\}\}.$$

By the construction and the time homogeneity of the Poisson point processes we have $P(\mathcal{E}(N, u, v)) = P(T_N > v - u)$ for any $N \geq 1$ integer and any $0 < u < v$. In particular, if $s, t > 0$ we may write

$$P(T_N > \beta_N s) P(T_N > \beta_N t) = P(\mathcal{E}(N, 0, \beta_N s))\, P(\mathcal{E}(N, \beta_N s, \beta_N (s + t))).$$

From Lemma 4.26 we know that $\beta_N / N \to +\infty$. If $\alpha = \alpha(\lambda)$ as before and $\beta_N s \geq 4N/\alpha$, we consider the following three parallelograms, where $L = 2N/0.9\alpha$ (see Figure 4.8):

- Q_0 has vertices $(N, \beta_N s - 0.5L)$, $(N + 0.1\alpha L, \beta_N s - 0.5L)$, $(-N, \beta_N s + 0.5L)$, and $(-N - 0.1\alpha L, \beta_N s + 0.5L)$;
- Q_1 has vertices $(-N - 0.1\alpha L, \beta_N s + 0.3L)$, $(-N, \beta_N s + 0.3L)$, $(N + 0.1\alpha L, \beta_N s + 1.3L)$, and $(N, \beta_N s + 1.3L)$;
- Q_{-1} has vertices $(-N - 0.1\alpha L, \beta_N s - 1.3L)$, $(-N, \beta_N s - 1.3L)$, $(N + 0.1\alpha L, \beta_N s - 0.3L)$, and $(N, \beta_N s - 0.3L)$.

Let $\mathcal{E}(N, s)$ denote the event that each of these three parallelograms contains a path (entirely contained in the parallelogram) connecting its bottom edge to the top edge. Since the graph lies on the plane and horizontal arrows connect nearest neighbour sites we see that on $\mathcal{E}(N, s)$, and if $\beta_N t \geq 4N/\alpha$, any couple of paths as prescribed in the definition of $\mathcal{E}(N, 0, \beta_N s)$ and $\mathcal{E}(N, \beta_N s, \beta_N (s + t))$ can be concatenated. Moreover, we see that

$$\mathcal{E}(N, 0, \beta_N s) \cap \mathcal{E}(N, \beta_N s, \beta_N (s + t)) \cap \mathcal{E}(N, s) \subseteq \mathcal{E}(N, 0, \beta_N (t + s)), \tag{4.132}$$

for any $t \geq 0$. The events entering (4.132) are increasing, so we may apply the Harris inequality to see that

$$P(T_N > \beta_N s) P(T_N > \beta_N t) P(\mathcal{E}(N, s)) \leq P(T_N > \beta_N (t + s)). \tag{4.133}$$

Thus, if $t \geq 0, s \geq 4N/(\alpha \beta_N)$, we may write

$$\begin{aligned} P(T_N > \beta_N (t + s)) &\leq P(T_N > \beta_N s) P(T_N > \beta_N t) \\ &\leq P(T_N > \beta_N (t + s)) / P(\mathcal{E}(N, s)). \end{aligned}$$

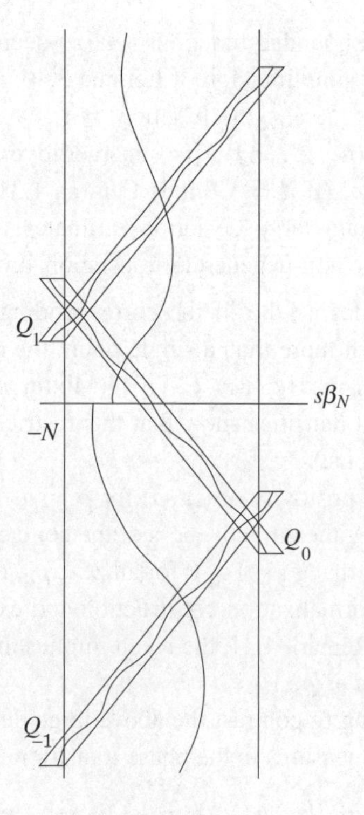

Figure 4.8

As in the proof of (4.129) we see that $P(\mathcal{E}(N, s))$ (it does not depend on the given s) tends to 1 as $N \to +\infty$. We obtain (4.118) for any $s, t > 0$, which proves (i) in Theorem 4.20.

The argument shows that the ratio $P(T_N > \beta_N(s+t))/(P(T_N > \beta_N s)\, P(T_N > \beta_N t))$ tends to one uniformly for $s \geq 0, t \geq 4N/(\alpha\beta_N)$, as $N \to +\infty$. In particular (compare with (4.119)):

$$\tilde{\upsilon}_N := \sup_{s \geq 0, t \geq 4N/\alpha\beta_N} \left| \frac{P(T_N/\beta_N > s+t)}{P(T_N/\beta_N > s)} - P(T_N/\beta_N > t) \right| \to 0,$$

from which a proper estimate of the rate of convergence follows, cf. Lemma 4.34.

Remark 4.33 (Large deviations. Order of magnitude of β_N) Let $d \geq 1$, $\lambda > \lambda_c^d$. Since $\mu^{(\lambda)}$ is FKG, translation invariant and non-degenerate ($0 < \rho_\lambda < 1$), the limits

$$-\varphi_+(\rho) := \lim_{n \to +\infty} \frac{1}{|\Lambda_n^d|} \log \mu^{(\lambda)}\{\eta \colon |\eta \cap \Lambda_n^d| \geq \rho |\Lambda_n^d|\},$$

$$-\varphi_-(\rho) := \lim_{n \to +\infty} \frac{1}{|\Lambda_n^d|} \log \mu^{(\lambda)}\{\eta \colon |\eta \cap \Lambda_n^d| \leq \rho |\Lambda_n^d|\}, \quad 0 \leq \rho \leq 1 \tag{4.134}$$

exist and define convex bounded functions ($\varphi_-(\rho)$ decreases, $\varphi_+(\rho)$ increases in ρ) This follows from a simplification of Lemma 3.40: if $\Lambda(n_1, \ldots, n_d)$ is a parallelogram of sides $n_1, \ldots, n_d$, the function $f(n_1, \ldots, n_d) := \log \mu^{(\lambda)}(\eta\colon |\eta \cap \Lambda(n_1, \ldots, n_d)| \geq \rho |\Lambda(n_1, \ldots, n_d)|)$ is superadditive in each n_i. Similarly if $\geq \rho$ is replaced by $\leq \rho$. (If $d = 1$ this is Lemma 1.31.) The argument is elaborated in [194] to obtain large deviation estimates for the empirical density $|\Lambda_n^d|^{-1} \sum_{i \in \Lambda_n^d} \eta_i$ under non-degenerate translation invariant FKG measures on $\mathcal{X} = \{0, 1\}^{\mathbb{Z}^d}$. It provides a l.d.p. if the corresponding functions $\varphi_\pm(\rho)$ do not vanish simultaneously at more than a single point, the rate function given by the convex function $\varphi(\rho) = \max\{\varphi_+(\rho), \varphi_-(\rho)\}$. (Mixtures of two Bernoulli product measures with different densities show that this restriction cannot be eliminated without further conditions.)

Applied to $\mu^{(\lambda)}$, the positivity of $\varphi_+(\rho)$ for $\rho > \rho_\lambda$ follows from a quite general result in [194] using the attractiveness of the contact process. Extra difficulty is involved in verifying that $\varphi_-(\rho) > 0$ for all $\rho < \rho_\lambda$. For $d = 1$ this was proven in [105], using the renormalization construction and exploiting the planar graph duality. (In relation to Remark 4.31, the result implies that $\mu^{(\lambda)}(\mathbb{A}_b^c)$ decays exponentially in b, for any $\rho < \rho_\lambda$.)

Remark It is interesting to contrast the above large deviation behaviour of $\mu^{(\lambda)}$ with that of the Gibbs measures in the phase transition region.

$P(T^{\Lambda_N^d} < +\infty) = \mu^{(\lambda)}\{\eta\colon \eta \cap \Lambda_N^d = \emptyset\}$ by self-duality, and the previous argument tells us that

$$\lim_{N \to +\infty} \frac{1}{N^d} \log P(T^{\Lambda_N^d} < +\infty) = -\tilde{\gamma}(\lambda),$$

where $\tilde{\gamma}(\lambda) = 2^d \varphi_-(0) \leq 2^d |\log(1 - \rho_\lambda)| \leq 2^d \log(1 + 2d\lambda)$, as one sees at once. That $\tilde{\gamma}(\lambda) > 0$ already follows from the argument used in Remark 4.33, if $d = 1, \lambda > \lambda_c$. In this case Durrett and Liu [102] proved that for any $\varepsilon > 0$, $P(T_N > e^{(\tilde{\gamma}(\lambda)+\varepsilon)N})$ tends to zero as N tends to infinity; Durrett and Schonmann [104] have shown that $P(T_N < e^{(\tilde{\gamma}(\lambda)-\varepsilon)N})$ vanishes as $N \to +\infty$, concluding that

$$\lim_{N \to +\infty} \frac{1}{N} \log T_N = \tilde{\gamma}(\lambda) \quad \text{in probability.}$$

Now using Theorem 4.20 we see that $\lim_{N \to +\infty} N^{-1} \log \beta_N = \tilde{\gamma}(\lambda)$, if $d = 1, \lambda > \lambda_c$.

Several proofs mentioned above require that $d = 1$. In the next section we comment briefly on the case $d \geq 2$. From the methods outlined there one should be able to give a full answer to the l.d.p. mentioned above.

One should contrast the previously described behaviour of T_N for $\lambda > \lambda_c$ with that in the sub-critical case: if $\lambda < \lambda_c^d$, $T_N / \log N \to d/\gamma_-(\lambda)$ in probability,

where

$$\gamma_-(\lambda) = -\lim_{t\to+\infty} \frac{1}{t}\log P(T^{\{0\}} > t) \in (0, +\infty)$$

(see [102] or Theorem 3.3, p. 72 in [200]). For $\lambda = \lambda_c, d = 1$, Durrett, Schonmann and Tanaka [106] have proven that $P(aN \le T_N \le bN^4)$ tends to 1 as $N \to +\infty$, if $a, b > 0$. The exact asymptotics of T_N at $\lambda = \lambda_c^d$ with $d \ge 2$ is not known, to our knowledge.

We conclude the section with a lemma related to the stability of exponential laws. Part (a) is taken from Chapter 1 of [8] and was used in Remark 4.31.

Lemma 4.34 *Let τ be a non-negative and non-degenerate random variable. Set $F(t) = P(\tau \le t)$, $\bar{F}(t) = 1 - F(t)$, and*

$$h(s,t) = \frac{\bar{F}(s+t)}{\bar{F}(s)} - \bar{F}(t), \quad s, t \ge 0, \tag{4.135}$$

with the convention that $h(s,t) = 0$ if $\bar{F}(s) = 0$.

(*a*) *If*

$$\sup_{s\ge 0, t\ge 0} |h(s,t)| \le u < 1, \tag{4.136}$$

then:

(*i*) *τ is integrable (in fact it has all positive moments)*;

(*ii*) *$\sup_{t\ge 0} |\bar{F}(t) - e^{-\lambda t}| \le 2u$, where $\lambda = (E\tau)^{-1}$.*

(*b*) *If we replace* (4.136) *by*

$$\sup_{s\ge 0, t\ge u} |h(s,t)| \le u < 1, \tag{4.137}$$

then item (*i*) *still holds*; *if $0 < \lambda u < \log((1+\sqrt{5})/2)$ and $G_\lambda(t) = (1 - e^{-\lambda t})^+$ we have $\varrho(F, G_\lambda) \le 2u \vee F(u)$, where*

$$\varrho(F, G) = \inf\{\delta > 0\colon F(x-\delta) - \delta \le G(x) \le F(x+\delta) + \delta, \ \forall x \in \mathbb{R}\}$$

is the Lévy distance, as in Section 1.6.

(*c*) *If $\tau_n, n \ge 1$ is a sequence as in* (*b*), *with $u_n \to 0$ and $P(\tau_n > 1) \le e^{-1} \le P(\tau_n \ge 1)$, then $E\tau_n \to 1$.* (Notation. *λ in this lemma has nothing to do with its use elsewhere in this section.*)

Proof Under (4.137), take $t_0 > u$ such that $\bar{F}(t_0) + u < 1$. Since $\bar{F}(s+t) = \bar{F}(s)(\bar{F}(t) + h(s,t))$ for all $s, t \ge 0$, we have $F(mt_0) \le \bar{F}(t_0)(\bar{F}(t_0) + u)^{m-1}$ for all $m \ge 1$, implying (i). For (ii) we may suppose $u < 1/2$ since otherwise the conclusion is trivial. Integrating in s the relation $\bar{F}(s+t) = \bar{F}(s)(\bar{F}(t) + h(s,t))$ we write

$$\int_0^{+\infty} \bar{F}(s+t)ds = \frac{1}{\lambda}\bar{F}(t) + \frac{1}{\lambda}\vartheta(t)$$

where $\vartheta(t) = \lambda \int_0^{+\infty} \bar{F}(s)h(s,t)ds$, for $t \geq 0$. Under (4.136) we have $|\vartheta(t)| \leq u$ for all $t \geq 0$. Calling $\psi(t) = \int_0^{+\infty} \bar{F}(s+t)ds = \int_t^{+\infty} \bar{F}(s)ds$ we have $\psi(0) = \lambda^{-1}$ and $\psi'(t) = -\bar{F}(t)$ for all $t \geq 0$, from which we see that $\lambda\psi(t) + \psi'(t) = \vartheta(t)$ so that

$$\psi(t) = \frac{1}{\lambda}e^{-\lambda t} + \int_0^t e^{-\lambda(t-s)}\vartheta(s)ds,$$

and we get the representation

$$\bar{F}(t) = e^{-\lambda t} + \lambda \int_0^t e^{-\lambda(t-s)}\vartheta(s)ds - \vartheta(t), \quad t \geq 0. \tag{4.138}$$

Item (ii) follows at once from this. As for part (b), we need to verify the estimate on the Lévy distance. The remaining cases being trivial, it suffices to check that for $z = 2u \vee F(u)$:

$$\bar{F}(t+z) \leq e^{-\lambda t} + z, \ \forall t \geq 0 \quad \text{and} \quad \bar{F}(t-z) \geq e^{-\lambda t} - z, \ \forall t \geq z.$$

For the first inequality, using (4.138) and the assumption we have for any $v \geq u$ and $t \geq 0$:

$$\begin{aligned}\bar{F}(t+v) &\leq e^{-\lambda(t+v)} + \int_0^u \lambda e^{-\lambda(t+v-s)}ds + 2u \\ &= e^{-\lambda(t+v)} + \int_{t+v-u}^{t+v} \lambda e^{-\lambda s}ds + 2u = e^{-\lambda(t+v-u)} + 2u \leq e^{-\lambda t} + 2u.\end{aligned}$$

For the second inequality, we split into two cases: if $t - z \leq u$, we just write $\bar{F}(t-z) \geq \bar{F}(u) = 1 - F(u) \geq e^{-\lambda t} - z$. If $t - z > u$, we use (4.138) to write

$$\begin{aligned}\bar{F}(t-z) \geq \bar{F}(t-2u) &\geq e^{-\lambda(t-2u)} - \int_0^u \lambda e^{-\lambda(t-2u-s)}ds - 2u \\ &= 2e^{-\lambda(t-2u)} - e^{-\lambda(t-3u)} - 2u \geq e^{-\lambda t} - 2u,\end{aligned}$$

where at the first inequality we used that $z \geq 2u$ and at the last one we used that $2e^{2\lambda u} - e^{3\lambda u} \geq 1$ if $0 < \lambda u < \log((1+\sqrt{5})/2)$.

For part (c), since $\varrho(F_n, G_{\lambda_n}) \leq 2u_n$, we have $\bar{F}_n(1+2u_n) - 2u_n \leq e^{-\lambda_n} \leq \bar{F}_n(1-2u_n) + 2u_n$. Recalling the assumption that $\bar{F}_n(1) \leq e^{-1} \leq \lim_{s\uparrow 1} \bar{F}_n(s)$, we see that $|\lambda_n - 1| \to 0$. (Alternatively, we could argue as in the proof of Theorem 4.6.) □

Remark 4.35 Part (a) of the previous lemma is suitable for application in Theorem 4.20 (cf. (4.119)) and part (b) is directly applicable to its shorter proof, cf. Remark 4.32. The lemma can be modified to apply under the estimates given in the proof of Theorems 4.6 or 5.21. For instance, in (4.88) the loss of memory controlled as

$$(\bar{F}_n(s+u_n) - u_n)\bar{F}_n(t) \leq \bar{F}_n(s+t) \leq \bar{F}_n(s)(\bar{F}_n(t-u_n) + u_n),$$

uniformly on $s \geq 0, t \geq u_n$ (see (4.88)) where τ_n are non-negative random variables as in part (c) above. This may be handled similarly to the case in (b), with a modified upper bound for the Lévy distance. Details are omitted.

4.5 Notes and comments

Metastability for higher dimensional contact process

Several arguments in the last section (and in the quoted proofs) required that $d = 1$. Due to the planarity of the graphical construction, one could take advantage of path intersection properties. Together with the subadditive ergodic theorem, this was a key fact in the proofs of basic theorems used in the description of metastability. One example is the linear growth of $\xi_t^{\{0\}}$ given its survival, used in Durrett–Griffeath's renormalization procedure.

The replacement of a priori fixed blocks in the renormalization procedure by a more flexible dynamical construction represented a breakthrough in the analysis of percolative systems. This is due to Barsky, Grimmett and Newman [14], who considered high dimensional (non-oriented) percolation and to Bezuidenhout and Grimmett [18], who constructed a variant of the procedure for the higher dimensional contact process.

Taking advantage of the flexibility of the construction, Bezuidenhout and Grimmett proved that if the contact process survives on $\mathbb{Z}^d \times [0, +\infty)$ with positive probability, the same holds on a sufficiently deep space-time slab $\Lambda_K^{d-1} \times \mathbb{Z} \times [0, +\infty)$, where $\Lambda_K = \{-K, \ldots, K\}$ as before.

The statement is based on the validity of a finite volume condition: if $\lambda > \lambda_c^d$, $\varepsilon > 0$, and M is any given (large) number, one can find r, L, T so that starting from $\Lambda_r^d \times \{0\}$, the (space-time) process restricted to $B_{L,T} = \Lambda_L^d \times [0, T]$ not only survives up to time T with probability at least $1 - \varepsilon$, but also produces at least M infected points on $\Lambda_L^d \times \{T\}$ and at least M well separated infected points on each side of $B_{L,T}$, i.e. points connected to $\Lambda_r^d \times \{0\}$ through a path contained in $B_{L,T}$. (In fact, one can find such many points on each orthant of the top and of the sides of $B_{L,T}$.) Having many well separated points one is able to grow again a suitable translate of $\Lambda_r^d \times \{0\}$, using only properly restricted paths. (The FKG inequality plays an important role in the verification of this property.)

Using this kind of estimate and successive restarting, one is able to compare the process on a slab $\Lambda_{2L}^{d-1} \times \mathbb{Z} \times [0, +\infty)$ to a suitable one-dependent oriented bond percolation on $S = \{(u, v) \in \mathbb{Z} \times \mathbb{Z}_+, u + v \text{ is even}\}$, each bond $(u, v) \to (u \pm 1, v + 1)$ being open with probability very near to one. For this, if $k \in \mathbb{N}$, one considers the sets

$$\mathcal{L}^{\pm} = \Lambda_{2L}^{d-1} \times \left\{(y, t) \in \mathbb{Z} \times \mathbb{R} \colon 0 \leq t \leq (2k+2)T,\ -5L \pm \frac{L}{2T} t \leq y \leq 5L \pm \frac{L}{2T} t\right\},$$

and for each $x \in \Lambda_{2L}^{d-1} \times [-2L, 2L]$ and $t \in [0, 2T]$, let $\mathcal{E}^{\pm}(x, t)$ be the event that $(x, t) + \Lambda_r^d \times \{0\}$ is connected inside $\mathcal{L}^{\pm}$ to each point in some translate $(y, s) + \Lambda_r^d \times \{0\}$, with $(y, s) \in \Lambda_L^{d-1} \times ((\pm k - 2)L, (\pm k + 2)L] \times [2kT, (2k + 2)T]$. The basic estimate iterated k times tells us that given $\delta > 0$, $k \in \mathbb{N}$, $k \geq 10$, we can take r, L and T so that $P(\mathcal{E}^{\pm}(x, t)) \geq 1 - \delta$, for all such (x, t). Using 'seeds' (fully infected translates of initial $\Lambda_r^d \times \{0\}$) located in $(ukL\mathbf{e}_d, 2vkT) + \Lambda_L^{d-1} \times \Lambda_{2L} \times [0, 2T]$, with $(u, v) \in S$, one gets the previously mentioned comparison, and the survival (with positive probability) in the space-time region

$$\mathcal{L} = \cup_{(u,v)\in S}\{(ukL\mathbf{e}_d, 2vkT) + (\mathcal{L}^+ \cup \mathcal{L}^-)\}$$

($\mathbf{e}_d$ denotes the canonical unit vector in $\mathbb{R}^d$).

Without the possibility of discussing this dynamical renormalization procedure, we refer to [18]. (See also [100, 200], where modified versions are presented.) An important feature is that once the basic growth condition involves only a finite volume, it is continuous in λ. This implies one of the important conclusions in [18]: the critical contact process dies out.

In terms of applications to metastability, ideas and methods from [18] were used by Mountford [220] and Simonis [279] to obtain multidimensional versions of Theorem 4.20 and Theorem 4.21, respectively, with $\xi_N(\cdot)$ the contact process on the space volume Λ_N^d.

Recalling Remark 4.32 and Lemma 4.34, and extending the previous notation in the obvious way, for the analogue of Theorem 4.20 it suffices to find $\tilde{\gamma}_N, \gamma_N$ tending to $+\infty$ so that, as $N \to +\infty$:

(a) $\gamma_N / \tilde{\gamma}_N \to 0$;

(b) $P(T_N < \tilde{\gamma}_N) \to 0$;

(c) $\sup_{A \subseteq \Lambda_N^d} P(\xi_N^A(\gamma_N) \neq \xi_N(\gamma_N),\ \xi_N^A(\gamma_N) \neq \emptyset) \to 0$.

For the sake of simplicity let $d = 2$. Given the comparison with a one-dependent percolation model as in [18], with given r, L, T, k and $1 - \delta$ the probability of each bond to be open, one considers $M = [(2N + 1)/4L]$ vertical strips $B_i = \{4Li, \ldots, 4L(i + 1) - 1\} \times \Lambda_N \subseteq \Lambda_N^2$ and the contact process restricted to each of them is compared to the percolation model restricted to $\{(u, v) \in S \colon u \in \{-cN, \ldots, cN\}\}$, for some $c > 0$. As in the case of one-dimensional supercritical contact process, this implies that condition (b) will be satisfied for $\tilde{\gamma}_N = e^{\tilde{c}N}$, for some $\tilde{c} > 0$.

With a beautiful combination of the results in [18], couplings and duality, Mountford proves in [220] that condition (c) is satisfied with $\gamma_N = 2(a_N + b_N)2kT$, where $a_N = (2N + 1)^4$ and $b_N = (2N + 1)^2$. For this he proves that $a < 1$ can be found so that

$$\sup_{A\subseteq\Lambda_N^2,\, x\in\Lambda_N^2} P(\xi_N^A(\gamma_N) \neq \emptyset, \hat{\xi}_N^{x,\gamma_N}(\gamma_N) \neq \emptyset, \xi_N^A(\gamma_N/2) \cap \hat{\xi}_N^{x,\gamma_N}(\gamma_N/2) \neq \emptyset) \leq a^N \tag{4.139}$$

for all N large, where $\hat{\xi}_N^{x,\gamma_N}$ is the dual process starting at $\{x\}$ at time γ_N (using the reversed graph restricted to $\Lambda_N^2 \times [0, \gamma_N]$). That this implies condition (c) above follows from duality and the graphical construcion. Indeed, (4.139) implies at once that

$$\sup_A \sum_{x \in \Lambda_N^2} P(\xi_N^A(\gamma_N) \neq \emptyset, \hat{\xi}_N^{x,\gamma_N}(\gamma_N) \neq \emptyset, \hat{\xi}_N^{x,\gamma_N}(\gamma_N) \cap A = \emptyset) \to 0,$$

from which one has condition (c) above, through duality. To prove (4.139), Mountford first shows that $a < 1$ may be found so that

$$\sup_A P(\xi_N^A(2kTa_N) \neq \emptyset, \xi_N^A(2kTa_N) \cap B_i = \emptyset, \text{for some } i) < a^N. \tag{4.140}$$

Applying this to the dual process starting at $\{x\}$ at time γ_N, one has

$$\sup_x P(\hat{\xi}_N^{x,\gamma_N}(2kTa_N) \neq \emptyset, \hat{\xi}_N^{x,\gamma_N}(2kTa_N) \cap B_i = \emptyset, \text{for some } i) < a^N. \tag{4.141}$$

Let $\mathcal{E}_A = \{\xi_N^A(2kTa_N) \cap B_i \neq \emptyset, \forall i\}$ and $\hat{\mathcal{E}}_x = \{\hat{\xi}_N^{x,\gamma_N}(2kTa_N) \cap B_i \neq \emptyset, \forall i\}$. We now look at both processes (forward and backward) on the time interval $[2kTa_N, \gamma_N/2]$ which depend on the portions of the graph restricted to $\Lambda_N^2 \times [2kTa_N, \gamma_N/2]$ and $[\gamma_N/2, \gamma_N/2 + 2kTb_N]$, and moreover we restrict them to each strip B_i, i.e. eliminate all the horizontal arrows not entirely contained in B_i, for some i. In this case we have evolutions which are conditionally independent given $\xi_N^A(2kTa_N)$, $\hat{\xi}^{x,a_N}(2kTa_N)$. Moreover, from the results in [18] one can prove the existence of $\tilde{\delta} = \tilde{\delta}(L, T, k, r) > 0$ so that for any $z, y \in B_i$, with probability at least $\tilde{\delta}$ there is a path connecting z to y in the graph restricted to $B_i \times [0, 2kTb_N]$. Applying this to both the direct and dual processes we conclude

$$\sup_{A \subseteq \Lambda_N^2, x \in \Lambda_N^2} P([\xi_N^A(\gamma_N/2) \cap \hat{\xi}^{x,\gamma_N}(\gamma_N/2) = \emptyset] \cap \mathcal{E}_A \cap \hat{\mathcal{E}}_x) \leq (1 - \tilde{\delta}^2)^M.$$

Together with (4.140), (4.141) this implies (4.139). For the full proof see [220].

Concerning the order of magnitude of $E(T_N)$, Chen [60] has shown the existence of positive constants c_1, c_2 so that $e^{c_1 N^d} \leq E(T_N) \leq e^{c_2 N^d}$ for all N if $\lambda > \lambda_c^d$. More recently, using the renormalization ideas of [18] together with contour and coupling methods, Mountford [221] has proven the existence of a positive constant $\gamma(\lambda)$ so that $\lim_{N \to +\infty} N^{-d} \log E(T_N) = \gamma(\lambda)$. For that (taking e.g. $d = 2$) he uses a special coupling between the contact process on $\Lambda_{2N} \times \Lambda_N$ and two independent irreducible chains on $\{0, 1\}^{\Lambda_N^2}$ similar to a contact process with regeneration, obtaining an 'approximate superadditivity' property which leads to the existence of the limit. Together with the asymptotic exponentiality, this implies that $N^{-d} \log T_N$ converges in probability to $\gamma(\lambda)$. More information on $\gamma(\lambda)$, cf. Remark 4.33, is still lacking.

Further examples of asymptotic behaviour (not only metastable) of finite systems when the infinite system is non-ergodic can be found in [68].

Local mean field model on the circle

To keep the discussion technically simple and still present the basic ideas we considered in Section 4.3 an extremely simplified model, which besides being *unphysical* has no geometric structure. One of the simplest extensions in the set-up of mean field spin models consists in replacing the Hamiltonian H_N in (4.11) by

$$H_N(\sigma) = -\sum_{i=1}^{N} h(i/N)\sigma(i) - \frac{1}{2N}\sum_{i,j=1}^{N} J((i-j)/N)\sigma(i)\sigma(j)$$

where J is a continuous symmetric function on $\mathbb{R}$ with period one. It describes the mean field interaction among spins, and the magnetic field variables are associated with a function h, continuous on unit circle $\mathbb{T} = \mathbb{R}(\text{mod}\,\mathbb{Z})$. $\sigma(i)$ can be interpreted as a spin variable at the (macroscopic) position $i/N \in \mathbb{T}$. As $N \to +\infty$, the laws of the empirical measures

$$\mu_N := \frac{1}{N}\sum_{i=1}^{N} \sigma_N(i)\delta_{\frac{i}{N}}$$

(δ_x denotes the Dirac point-mass at x) under the corresponding Gibbs measure at inverse temperature β satisfy a l.d.p. on $\mathcal{M}(\mathbb{T})$,[6] with scaling N and rate function given by $I(\mu) = \beta\mathcal{F}(\mu) - \inf_{\tilde{\mu}} \beta\mathcal{F}(\tilde{\mu})$, where:

$$\mathcal{F}(\mu) = -\int_{\mathbb{T}} m(r)(h(r) + \frac{1}{2}(J * m)(r))dr + \frac{1}{\beta}\int_{\mathbb{T}} \mathcal{E}(m(r))dr,$$

if μ is absolutely continuous with respect to the Lebesgue measure with density m in the unitary ball of $\mathrm{L}^\infty(\mathbb{T}, dr)$, $J * m(r) = \int_{\mathbb{T}} J(r-s)m(s)ds$, and $\mathcal{E}$ given by (4.15); otherwise $\mathcal{F}(\mu) = +\infty$. As for the empirical magnetization for the Curie–Weiss model, cf. (4.19), this result follows by a *tilting* argument from the non-interacting case (see also [107] or [108]).

If $J \geq 0$ (ferromagnetic interaction) the behaviour is similar to that of the Curie–Weiss model. Let $h = 0$. If $\beta \leq \beta_c := (\int_{\mathbb{T}} J(r)dr)^{-1}$, $m = 0$ is the unique point of minimum of $\mathcal{F}$; it becomes unstable at β_c, and for $\beta > \beta_c$ there are two global minima, given by the constant densities $\pm m^*$, where m^* is the unique positive solution of $\tanh \beta\beta_c^{-1}m^* = m^*$. The dynamical transition from one minimum to another was investigated by Comets in [66] through spin-flip dynamics reversible under the Gibbs measure at temperature β^{-1}. After obtaining a l.d.p. analogous to that of Freidlin and Wentzell (but in $\mathcal{M}(\mathbb{T})$) and a law of large

[6] The space of bounded signed measures on $(\mathbb{T}, \mathcal{B}(\mathbb{T}))$, with weak* topology.

numbers (the limiting equation has as stationary points exactly the points of minima of $\mathcal{F}$, cf. [177]), the analogue of Freidlin and Wentzell theory is developed: basic properties of the quasi-potential are studied and a proof (under some technical conditions) that typically the exit from attracting domains happens near the lowest saddle points is given. Extra technical difficulties appear in part due to the lack of continuity of the quasi-potential for the weak* topology. This has been done for a large class of rates and also in higher dimensions.

One particular example in [66] which brings in a nice geometric picture is the following: $h = 0$, $J(r) = 1 + 2br\cos(2\pi p)$ with $p \in \mathbb{Z} \setminus \{0\}$, and $0 < b \leq 1/2$. The description of the phase diagram becomes: for $\beta < 1 = \beta_c$ one has only the equilibrium point $m = 0$; for $\beta = \beta_c$ this becomes a saddle and the two new stable equilibrium points appear, the constants m^*, $-m^*$. For $\beta \geq \beta_p := 1/b$ there are also non-constant equilibria, which bifurcate from the zero solution and are obtained by all translations of a basic function $m_p(\cdot)$, described explicitly. For $\beta > \beta_p$ these are the lowest saddle points connecting the two constant equilibria, which can be understood as indication of the existence of the 'nucleation' phenomenon. (To our knowledge the full description of the extremal trajectories in this case, which would confirm this conjecture, has not yet been done. Simulations could also be helpful.) We refer to [66] for details, and in the next subsection discuss related works for the infinite volume version.

Glauber dynamics for Kac interaction

Several features of spin dynamics for mean field models in the set-up of Kac potentials have been investigated in the recent years, and questions related to metastability emerge naturally.

One may consider $\{-1, 1\}^{\mathbb{Z}^d}$-valued Markov processes, given by Glauber-type dynamics under a mean field interaction as in (4.27), with $J(\cdot) \geq 0$, symmetric, and with compact support. For convenience let $\int_{\mathbb{R}^d} J(r)dr = 1$. Following validation of the continuum[7] limit as $\gamma \to 0$, leading to the infinite volume evolution equation

$$\frac{\partial m_t}{\partial t} = -m_t + \tanh\{\beta(J * m_t + h)\}, \tag{4.142}$$

where $J * m(x) = \int J(x - y)m(y)dy$, β represents the inverse temperature, h is an external magnetic field, and $m_t(\cdot)$ corresponds to magnetization density, various aspects of the long time behaviour of the spin system (for $\gamma > 0$ small) have been studied (see [82] and references therein).

Carrying out the full pathwise description of metastability for Glauber-type dynamics presents non-trivial difficulties. De Masi, Olivieri and Presutti [83] made the first step, studying the spectral properties of operators obtained by the

[7] It involves space scales like γ^{-1}.

linearization of (4.142) around profiles in some given class, for $d = 1$. Recall that for $\beta > 1$ one has a mean field phase transition, in the sense described in Section 4.1; for $0 \leq h < h^*(\beta)$ there exist three stationary and spatially homogeneous solutions of (4.142), given by the roots of $m = \tanh(\beta m + \beta h)$, and written as $m^-(\beta, h) < m^0(\beta, h) \leq 0 < m^+(\beta, h)$. One knows that if $h = 0$, $m^+(\beta, 0) = -m^-(\beta, 0)$ are stable and correspond to the pure phases; $m^0(\beta, 0) = 0$ is unstable. For $h > 0$, $m^0(\beta, h) < 0$, $|m^-(\beta, h)| < m^+(\beta, h)$; in this case $m^-(\beta, h)$ corresponds to a metastable state, and $m^+(\beta, h)$ to the unique thermodynamically stable state; $m^0(\beta, h)$ remains unstable. The study of the above mentioned spectral properties is a basic step towards the understanding of the interface dynamics describing how one phase takes over, such as the instantons when $h = 0$, or the travelling fronts when $h > 0$.

The profiles of main interest for metastability are those where the region occupied by one of the phases is bounded. The initial problem is the existence of a critical droplet within the metastable phase, given by a suitable 'bump', i.e. a non-constant stationary solution $\tilde{m}$ of the equation (4.142), such that $\lim_{|x| \to +\infty} \tilde{m}(x) = m^-(\beta, h)$. In [84] it is proven that the linearization of (4.142) around the bump is described by an operator which has a single maximal eigenvalue larger than 1, with a positive eigenvector, and that the rest of its spectrum is contained in the open unit ball. Buttà, De Masi and Rosatelli [41] have proven the existence of a one-dimensional invariant manifold connecting the bump to the metastable and to the stable states. Having a droplet large enough of the $m^+(\beta, h)$ phase would lead to the stable phase, but otherwise the metastable state persists under small perturbations of the evolution (4.142). This suggests a similar tunnelling mechanism for the spin system, for small γ, but it needs further investigation since the times for which (4.142) is known to describe the behaviour of the spin system are shorter than those involved in the escape from the metastable phase.

The description of the typical profiles under the infinite volume Gibbs measure, for $d = 1$, was given in [50].

Random field Curie–Weiss models

In the domain of disordered systems (random interactions), it is natural to consider again the Curie–Weiss model as a good initial 'laboratory', in spite of its unphysical aspects. One of the simplest possibilities is to assume the presence of random magnetic fields, replacing the Hamiltonian H_N in (4.11) by (fixing $J = 1$):

$$H_N(\sigma) = -\frac{1}{2N} \sum_{i,j=1}^{N} \sigma_i \sigma_j - \sum_{i=1}^{N} h_i(\omega) \sigma_i \tag{4.143}$$

where $h_1, \dots, h_N$ are i.i.d. random variables on some probability space $(\Omega, \Sigma, \mathbb{P})$ with $\mathbb{P}(h_i = h) = 1/2 = \mathbb{P}(h_i = -h)$ for a fixed $h > 0$. A full description of the phase diagram of this model, from the point of view of equilibrium statistical mechanics, was given in [262]. The metastable behaviour of a class of stochastic dynamics related to this model has been analysed in [216] and more recently in [31], in a more general set-up. There are basically two types of results one searches for: a.s. in $\mathbb{P}$, or involving averages in the disorder (random fields). For the example above, since the random field takes only two possible values, for each realization of $(h_i)_i$ the Hamiltonian is a function of two variables: the average magnetization along the sites where $h_i = +h$, and along those where $h_i = -h$. Even in this simple situation, we see immediately that the very simple analysis based on the difference equations, used for the birth and death chains, can no longer be used: it must be suitably replaced by a more delicate study of the boundary value problem corresponding to the (random) generator of the present Markov chain. The problem can be treated with the estimation of the spectral gap of the generator corresponding to the process killed (or reflected) on the boundary of suitable domains of interest. In [216] the authors consider a continuous time spin-flip dynamics on $\mathcal{X}_N = \{-1, +1\}^N$ as described at the end of Remark 4.4, for H_N given by (4.143), reversible for the corresponding (random) Gibbs measure at inverse temperature β. Relevant macroscopic variables, corresponding to the $m_N(\sigma)$ in Section 4.3, are given by the pair $(m_N^+(\sigma), m_N^-(\sigma))$:

$$m_N^{\pm}(\sigma) = \frac{1}{N} \sum_{i:h_i=\pm h} \sigma_i.$$

Since the Hamiltonian can be written as

$$H_N(\sigma) = H_N(m_N^+(\sigma), m_N^-(\sigma)) = -N\left[\frac{1}{2}(m_N^+(\sigma) + m_N^-(\sigma))^2 \right.$$
$$\left. + h(m_N^+(\sigma) - m_N^-(\sigma))\right],$$

one may naturally consider spin-flip Markov dynamics such that the evolution induced by the map $\sigma \mapsto (m_N^+(\sigma), m_N^-(\sigma))$ is also Markovian, with values on a random subset of $[-1, +1]^2$:

$$\mathcal{Y}_N = \left\{\frac{-N_+}{N}, \frac{-N_+}{N} + \frac{2}{N}, \ldots, \frac{N_+}{N}\right\} \times \left\{\frac{-N_-}{N}, \frac{-N_-}{N} + \frac{2}{N}, \ldots, \frac{N_-}{N}\right\},$$

where N_+ (N_-) denote the number of spins where the field is positive (negative) respectively. One possible example is given by the spin-flip rates $c(i, \sigma) = e^{-\beta/2(H_N(\sigma^{(i)}) - H_N(\sigma))}$. The induced process $(m_N^+(\sigma(t)), m_N^-(\sigma(t)))$ is a Markov chain, reversible under the Gibbs measure on $\mathcal{Y}_N$:

$$\tilde{\mathcal{G}}_N^{\beta}(m^+, m^-) = \frac{e^{-\beta N f_N(m^+, m^-)}}{Z_N},$$

where

$$f_N(m^+, m^-) = -\frac{1}{2}(m^+ + m^-)^2 - h(m^+ - m^-)$$
$$- \frac{1}{\beta N} \log \left(\frac{N^+}{\frac{N^+}{2}(1 + \frac{Nm^+}{N^+})} \right) \left(\frac{N^-}{\frac{N^-}{2}(1 + \frac{Nm^-}{N^-})} \right).$$

The random variables $N^{\pm}/N$ tend $\mathbb{P}$ a.s. to $1/2$, implying the a.s. convergence of f_N, uniformly on the compact subsets of $(-1/2, 1/2)^2$, to the Gibbs canonical free energy[8] given by

$$f_{\beta,h}(m^+, m^-) = -\frac{1}{2}(m^+ + m^-)^2 - h(m^+ - m^-) + \frac{1}{2\beta}(\mathcal{E}(2m^+) + \mathcal{E}(2m^-)),$$

for $m^{\pm} \in [-1/2, 1/2]$, where $\mathcal{E}$ is given by (4.15). To obtain the points of minima of $f_{\beta,h}$, observe that its critical points (m^+, m^-) satisfy the pair of equations

$$m^{\pm} = \frac{1}{2} \tanh(\beta(m^+ + m^-) \pm \beta h).$$

The sum $m = m^+ + m^-$ must then satisfy the equation

$$m = \frac{1}{2}[\tanh(\beta(m + h)) + \tanh(\beta(m - h))]. \tag{4.144}$$

One checks that for $0 < \beta \leq 1$ the unique solution to (4.144) is $m = 0$; if $\beta > 1$ we can find $h_0(\beta) > 0$ so that if $0 < h < h_0(\beta)$ there are exactly three solutions of (4.144), $m = 0, \pm m^*(\beta, h)$. In this region the critical points of $f_{\beta,h}$ are three and are given by:

$$m_{(0)} = \left(\frac{1}{2} \tanh(\beta h), -\frac{1}{2} \tanh(\beta h) \right)$$
$$m_{(+)} = \left(\frac{1}{2} \tanh(\beta m^* + \beta h), \frac{1}{2} \tanh(\beta m^* - \beta h) \right)$$
$$m_{(-)} = \left(\frac{1}{2} \tanh(-\beta m^* + \beta h), -\frac{1}{2} \tanh(\beta m^* + \beta h) \right).$$

$m_{(+)}$ and $m_{(-)}$ are the points of minima with $f_{\beta,h}(m_{(+)}) = f_{\beta,h}(m_{(-)})$ and $m_{(0)}$ is a saddle point.

After suitably defining a kind of 'discrete basin of attraction' for $m_{(+)}$, the following results concerning metastability are proven in [216].

(a) For almost all realizations of $(h_i)_i$, if the magnetization process starts near $m_{(+)}$ and T_N denotes the escape time from the basin of $m_{(+)}$, there exist γ_N, dependent on the disorder, such that T_N/γ_N tends in law to a unit mean exponential random variable. The quantity γ_N may be given as the inverse of the (random)

[8] More precisely, $\lim_{\delta \to 0} \lim_{N \to +\infty} \inf\{f_N(m_1, m_2) \colon (m_1, m_2) \in \mathcal{Y}_N, |m_1 - m^+| + |m_2 - m^-| \leq \delta\} = f_{\beta,h}(m^+, m^-)$ $\mathbb{P}$ a.s.

spectral gap of the generator of the process killed upon leaving such a 'basin of attraction'. A first estimate for this gap is provided in [216]; it implies that γ_N is logarithmically equivalent to the non-random quantity $e^{\beta N \Delta f}$ as $N \to +\infty$, $\mathbb{P}-$ a.s., where $\Delta f = \Delta f_{\beta,h} = f_{\beta,h}(m_{(0)}) - f_{\beta,h}(m_{(+)})$. The same asymptotic properties apply to the spectral gaps of the generator of the full process (denoted below by $1/\tilde{\gamma}_N$), and for the generator of the process reflected upon meeting the 'boundary' of the 'basin of attraction'. More refined estimates on the time needed to reach the equilibrium were obtained in [119], providing a good insight on how a certain class of realizations of $(h_i)_i$ leads to faster approach to the equilibrium measure $\tilde{\mathcal{G}}_N$.

(b) For almost all realizations of $(h_i)_i$, and initial conditions as in (a) above, the distribution of $(m_N^+(\sigma(t\tilde{\gamma}_N)), m_N^-(\sigma(t\tilde{\gamma}_N)))$ is well approximated (for large N) by $e^{-t}\delta_{m_{(+)}} + (1 - e^{-t})\tilde{\mathcal{G}}_N^\beta$, the single time marginal of a measure-valued Markov jump process.

(c) As for the equilibrium Gibbs measure $\tilde{\mathcal{G}}_N^\beta(m^+, m^-)$, there is no $\mathbb{P}-$ almost sure convergence; they converge in law to a random mixture $\alpha\delta_{m_{(+)}} + (1 - \alpha)\delta_{m_{(-)}}$ where the random variable α takes the values 1 or 0 with probability 1/2 each. (For an analogous description in the infinite volume case (Kac model) see [51, 52].)

(d) The statements in (b) and (c) can also be proven in terms of the microscopic system, cf. [119]: under the conditions of (a), the distribution of $\sigma(t\tilde{\gamma}_N)$ is well approximated, for large N by $e^{-t}\mu^+ + (1 - e^{-t})\mathcal{G}_N^\beta$ or still by $e^{-t}\mu^+ + (1 - e^{-t})(\alpha_N\mu^+ + (1 - \alpha_N)\mu^-)$ where the measures $\mu^\pm$ are Bernoulli product measures with the average of σ_i given by $\tanh \beta(\pm m^* + h_i)$ (disorder dependent).

(e) The asymptotic distribution of the escape time τ_N in (a) when averaging also in the disorder was considered in [119], where it is proven that $(\log \tau_N - \beta N \Delta f)/N^{1/2}$ tends in distribution to a Gaussian random variable.

Analogous sharp estimates have been presented by Bovier *et al.* In [31], a large class of reversible Markov chains modelling low temperature stochastic dynamics for mean field interactions with disorder was studied. The authors consider discrete time Markov chains on some lattice imbedded in $\mathbb{R}^d$, reversible under a probability measure $\nu_N(x) = (1/Z_N)e^{-f_N(x)}$ where, analogously to the previous situations, f_N tends to a given landscape f as $N \to +\infty$. Regularity conditions on f_N and on f are assumed; the more important assumption involves the structure of the set local minima of f_N: it is assumed to remain with bounded cardinality and to approximate that of f. Freidlin and Wentzell theory is developed in this set-up. Using probabilistic techniques the authors obtain estimates on the spectral properties of the corresponding generators. (L^∞ estimates are obtained, instead of L^2); reversibility is used extensively and the probability of visiting a given point before returning to the starting point is estimated in terms of the landscape f_N. The use of a suitable family of paths that reduces to a one-dimensional situation, where the boundary value problem (difference equation) becomes simple, is important. The investigation of the long time behaviour of the process leads naturally to consideration of a certain tree structure on the set of points of minima for f_N. The basic estimate refers to the escape time of a given 'valley' of the landscape f_N, in terms of the value of $N(f_N(z) - f_N(x))$, where z corresponds to the lowest saddle and x is the minimum. Assuming, for simplicity, the transition probabilities between neighbouring sites to be uniformly bounded from below (in N), one has lower and upper estimates of order $cNe^{N(f_N(z)-f_N(x))}$. With this basic ingredient they can suitably control the Laplace transforms of the escape times and show their asymptotic exponentiality. Putting this together with the much smaller order of magnitude of expected return times at the minima, they get a picture of the long time behaviour, which due to the existence of several valleys, indeed involves various scales.

The situation described by [31] applies to Glauber dynamics for the random field Curie–Weiss model if the random field takes finitely many values. The symmetric case corresponds to the situation where the role of disorder becomes important with respect to symmetry breaking.

In Section 5.2 we sketch an alternative argument leading to estimates of tunnelling times in the context treated there. Analogous ideas can be applied to the Markov chains which appear naturally in the random field Curie–Weiss models, such as for instance Metropolis dynamics (multidimensional birth and death chains, in this case). The basic idea behind this proposal is to exploit more the explicit knowledge of the stationary (reversible) measure and to control the loss of memory in a suitable way. In Section 5.2 this relies on coupling methods, and for the models mentioned above it depends on rough estimates of the recurrence times. (This is simpler to perform in the case of a totally attracted domain, but it can also be extended to tunnelling times.)

5

Metastability. Models of Freidlin and Wentzell

Following the pathwise approach introduced in the previous chapter, we now discuss examples of metastable behaviour using the set-up of Chapter 2. In this context, we describe the role of the basic large deviation estimates and of the theory of Freidlin and Wentzell.

Let $X^{x,\varepsilon}$ be the Markov process obtained as the unique solution of the Itô equation

$$X_t^{x,\varepsilon} = x + \int_0^t b(X_s^{x,\varepsilon})\, ds + \varepsilon W_t, \quad t \geq 0, \tag{5.1}$$

where (W_t) is a standard d-dimensional Brownian motion, $x \in \mathbb{R}^d$ is the initial point, $\varepsilon > 0$, and the vector field $b\colon \mathbb{R}^d \to \mathbb{R}^d$ satisfies a global Lipshitz condition, cf. (2.51). According to Theorem 2.24, this is sufficient for the existence and pathwise uniqueness of the solution, for each x. The process $X^{x,\varepsilon}$ satisfies the strong Markov property, as remarked in Section 2.3. To simplify we shall in fact take $b(\cdot)$ continuously differentiable (class C^1) with bounded partial derivatives of first order.

The simplest picture to keep in mind is that of $b(x) = -\nabla U(x)$, where $U(\cdot)$ is a double well potential, as in Figure 5.1. When $U(\cdot)$ grows fast enough at infinity so that the integral $Z_\varepsilon := \int_{\mathbb{R}^d} \exp\{-\varepsilon^{-2} 2U(x)\}\, dx$ is finite, then a (unique) stationary probability measure exists; it is given by

$$\mu_\varepsilon(A) = \frac{1}{Z_\varepsilon} \int_A \exp\{-\varepsilon^{-2} 2U(x)\}\, dx. \tag{5.2}$$

As $\varepsilon \to 0$, the measures μ_ε converge to a Dirac point-mass at the bottom of the deepest well, called q. Nevertheless, starting in the shortest well, and for ε sufficiently small, the system exhibits characteristics associated with metastability:

(i) Due to the action of the field $b(\cdot)$ the process is attracted towards the bottom, called p; when very close to p the importance of noise is considerable, since the

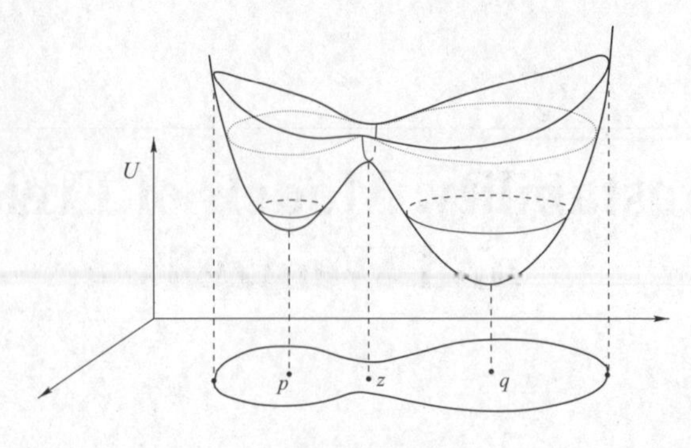

Figure 5.1

field becomes negligible. This allows the process to get away from the bottom and everything starts again, with a large number of attempts to escape from the given well, followed by a strong attraction towards its bottom. During this period the system may look as if it is in equilibrium with a stationary measure that is concentrated around p. We say that 'thermalization' occurs, as in Chapter 4.

(ii) Finally, after many frustrated attempts, the process succeeds in overcoming the barrier of the potential and reaching the deepest well. Due to the unsuccessful trials, we expect this 'tunnelling' time to show little memory.

(iii) Once in the deepest well, everything is repeated. The new barrier is higher so that the next 'tunnelling' time will happen on a much longer scale.

We want to make the above description rigorous in a more general situation. Due to the specific properties of the double well potential, we consider this case in detail in Section 5.2, exploiting its peculiarities and simplicity.

Concerning the proof of asymptotic exponentiality of the escape time, the situation here is more delicate than that in the Curie–Weiss chain. Among other factors, we can no longer use pointwise recurrence. A more careful analysis is needed to prove the loss of memory. Analytical tools were used by Day in [73], in the context of Section 2.6, and extended to our tunnelling problem in [131].

In [210], these analytical tools were replaced by more probabilistic methods, based on contraction properties of the stochastic map $x \mapsto X_T^{x,\varepsilon}$, at a suitable (large) time $T > 0$. These arguments were introduced in [209] and involve the consideration of renormalization schemes. One advantage is their applicability to some infinite dimensional situations. (We comment on that at the end of this chapter.)

The search for a convenient coupling of two solutions of (5.1) with different initial points appears naturally in this context. In [209] and [210] this is done with the use of a common noise. For our present situation it will be convenient to couple them differently. The next section is dedicated to explaining some details of this coupling, proposed by Lindvall and Rogers in [202]. We then come back to our basic problem.

The other ingredient is identification of the relevant time scales: that in which thermalization occurs and that of tunnelling. This type of problem is treated by Freidlin and Wentzell in Chapter 6 of [122]. Their results are recalled in Section 5.3, and applied to the discussion of metastability in Section 5.4.

5.1 The coupling method of Lindvall and Rogers

Coupling methods have often been used in Markov processes and many textbooks on stochastic processes include this useful tool. They have already appeared in Chapter 4. Generally, a coupling of two processes refers to any joint construction of them, i.e. on the same probability space. In a measure-theoretic sense, given two probability measures ν_1 and ν_2 on spaces Ω_1 and Ω_2, respectively, a coupling of ν_1 and ν_2 is any measure on the product space $\Omega_1 \times \Omega_2$ whose marginal distributions are ν_1 and ν_2, respectively. (General references include [201] and the recent monograph [287].)

The problem which we now have in mind concerns the loss of memory of the initial condition. Thus, one looks at two evolutions given by (5.1) starting from different points, and tries to construct a 'successful' coupling, i.e. one for which the two paths will meet in finite time, with probability one. If this happens, we can let them move together once they meet, being able to conclude that both exhibit the same asymptotic behaviour at large times. This kind of argument is used in several situations, for example for a proof of the uniqueness of the stationary probability measure for finite irreducible aperiodic Markov chains, in which case letting the processes evolve independently until they meet provides a successful coupling (see e.g. [201] and references therein).

For multidimensional diffusions we cannot expect independent coupling to be successful. (Two independent Brownian motions on $\mathbb{R}^2$ starting at different points do not meet, with probability one.) More adequate coupling methods are thus required. We consider one that was proposed by Lindvall and Rogers in [202]. It will be very useful to prove the asymptotic exponentiality of the tunnelling time for a large class of diffusions in the Freidlin and Wentzell regime. This includes the double well potential previously described. The coupling also furnishes upper estimates for the time scale on which thermalization occurs. The analysis includes the escape times considered in Section 2.6.

The coupling proposed by Lindvall and Rogers is based on the observation that if (W_t) is a d-dimensional standard Brownian motion and H is a $d \times d$ orthogonal matrix with real entries, then $W_t' := HW_t$ for $t \geq 0$ also describes a standard Brownian motion on $\mathbb{R}^d$. (We say that a matrix H is orthogonal if it is invertible and $H^{-1} = H^{\mathbb{T}}$, where $H^{\mathbb{T}}$ denotes the transpose matrix, i.e. $H^{\mathbb{T}}_{i,j} = H_{j,i}$ for $1 \leq i, j \leq d$.)

To see how this could be applied, let us start with the case when $b = 0$, $\sigma = \mathbb{I}$, the $d \times d$ identity matrix. We want a successful coupling of two standard

Brownian motions starting at different points. Let us fix $x \neq x'$ in $\mathbb{R}^d$ and consider the hyperplane orthogonal to the vector $x - x'$ and passing through the middle point $(x + x')/2$, i.e.

$$L_{x,x'} = \{y \in \mathbb{R}^d : \langle y - (x + x')/2, x - x' \rangle = 0\}.$$

Let (W_t) be a d-dimensional standard Brownian path starting at the origin, $X_t = x + W_t$, and let $\tilde{X}_t$ be the specular reflection of X_t through $L_{x,x'}$. Thus, the processes X and $\tilde{X}$ will meet for the first time at the (first) hitting time to $L_{x,x'}$ by X. Now, $\tilde{X}$ can be written as

$$\tilde{X}_t = x' + HW_t$$

where $H = H(x, x')$ is the matrix

$$H(x, x') = \mathbb{I} - 2\left(\frac{x - x'}{|x - x'|}\right)\left(\frac{x - x'}{|x - x'|}\right)^{\mathbb{T}} \tag{5.3}$$

with $\mathbb{I}$ denoting, as before, the $d \times d$ identity matrix, the vector $(x - x')/|x - x'|$ written as a $d \times 1$ matrix, $(\cdot)^{\mathbb{T}}$ denoting the transpose, and with the usual multiplication of matrices. (*Notation.* For $y \in \mathbb{R}^\ell$, $\ell \geq 1$, we let $|y|$ denote its Euclidean norm.)

To verify the last statement, notice that $H(x, x')(x - x') = -(x - x')$, and $H(x, x')v = v$, if v is orthogonal to $x - x'$. That is, the matrix $H(x, x')$ represents the specular reflection through the hyperplane orthogonal to $x - x'$ and passing through the origin, so that $x' + HW_t$ is the specular reflection of $x + W_t$ through $L_{x,x'}$.

Since $H(x, x')$ is a $d \times d$ orthogonal matrix we see that $\tilde{X}$ is a standard d-dimensional Brownian motion starting at x'. ($H(x, x')$ has determinant -1, corresponding to a reflection.) If we set

$$T_{x,x'} = \inf\{t > 0 \colon X_t \in L_{x,x'}\}$$

and

$$X'_t = \begin{cases} \tilde{X}_t & \text{if } t \leq T_{x,x'}, \\ X_t & \text{if } t > T_{x,x'}, \end{cases} \tag{5.4}$$

we can see that $T_{x,x'} < +\infty$ a.s., that is, we have a successful coupling. Indeed, we observe that $T_{x,x'}$ represents the (first) hitting time to the origin for a *one-dimensional* Brownian motion starting at $r = |x - x'|/2$.

The extension of the previous reflection method to the case of varying coefficients suggests a way to couple two solutions X^x, $X^{x'}$ of an Itô equation of the type:

$$dX_t = b(X_t)\,dt + \sigma(X_t)\,dW_t,$$

with distinct starting points x, x', provided the coefficients $b(\cdot)$ and $\sigma(\cdot)$ satisfy suitable conditions. Given a standard Brownian motion (W_t) on $\mathbb{R}^d$, starting at the origin, we may consider the (coupled) pair of Itô equations:

$$\begin{aligned} dX_t^x &= b(X_t^x)dt + \sigma(X_t^x)dW_t, \quad X_0^x = x, \\ dX_t^{x'} &= b(X_t^{x'})dt + \sigma(X_t^{x'})H(X_t^x, X_t^{x'})dW_t, \quad X_0^{x'} = x', \end{aligned} \tag{5.5}$$

where $H(\cdot, \cdot)$ takes values on the set of orthogonal matrices and satisfies some needed regularity conditions. A choice of the function H is considered good if it provides a successful coupling.

A possible set of regularity conditions on σ and b is that already used in Section 2.5:

(a) $\sigma(\cdot)$ and $b(\cdot)$ are globally Lipschitz;
(b) $\sigma(\cdot)$ and $b(\cdot)$ are bounded;
(c) $\sigma^{-1}(\cdot)$ exists and is bounded.

The function $H(\cdot, \cdot)$ in the construction that we adopt is given by:

$$H(u, v) = \mathbb{I} - 2y(u, v)(y(u, v))^{\mathbb{T}}$$

for $u \neq v$ in $\mathbb{R}^d$, where

$$y(u, v) = \frac{(\sigma(v))^{-1}(u - v)}{|(\sigma(v))^{-1}(u - v)|},$$

i.e. $H(u, v)$ corresponds to a reflection orthogonal to the vector $y(u, v)$. We set $H(u, u)$ as the identity. (This discontinuity on $H(\cdot, \cdot)$ is not a true difficulty, as we see next.)

With the above conditions on σ and b one can prove strong existence and uniqueness of the solution $(X^x, X^{x'})$ of equation (5.5) up to the hitting time to the diagonal. If $T_{x,x'} = \inf\{t > 0: X_t^x = X_t^{x'}\}$ is finite, then past $T_{x,x'}$ we let the two paths move together: we set $X^{x'}(t) = X^x(t)$, for $t > T_{x,x'}$, where X^x is the unique strong solution of the first equation in (5.5). This provides a strong solution of (5.5). It has the proper marginal distributions, given by the laws of the basic diffusion starting at x and x'. (We omit full details.)

Remark Under the Lipschitz condition, it is a standard procedure to relax the boundedness assumptions on σ, σ^{-1}, and b through the use of localization arguments, stopping the process when it hits the boundary of arbitrarily large balls. For our present purposes this is less relevant, since the questions we have in mind involve the process until it escapes from a fixed bounded domain. We omit a more technical discussion.

We need conditions that guarantee the a.s. finiteness of $T_{x,x'}$. This task becomes much easier in the case of a constant diffusion matrix (Example 5 in [202]). From now on we assume that $\sigma = \varepsilon\,\mathbb{I}$, where $\varepsilon > 0$. Let $(X^{x,\varepsilon}, X^{x',\varepsilon})$ be the corresponding solution of (5.5), as described above, with H given by (5.3). $T_{x,x'}^{\varepsilon}$ denotes the corresponding coupling time.

Let us consider $Y_t^\varepsilon = |X_t^{x,\varepsilon} - X_t^{x',\varepsilon}|$. Notice that

$$X_t^{x,\varepsilon} - X_t^{x',\varepsilon} = x - x' + \int_0^t \{b(X_s^{x,\varepsilon}) - b(X_s^{x',\varepsilon})\}ds \\ + \varepsilon \int_0^t (\mathbb{I} - H(X_s^{x,\varepsilon}, X_s^{x',\varepsilon}))dW_s. \tag{5.6}$$

We may use the (extended) Itô formula, cf. (2.35), with $f(u) = |u|$, to get that the one-dimensional process Y_t^ε satisfies, for $t < T_{x,x'}^\varepsilon$:

$$dY_t^\varepsilon = \left\langle b(X^{x,\varepsilon}) - b(X^{x',\varepsilon}), \frac{X_t^{x,\varepsilon} - X_t^{x',\varepsilon}}{|X_t^{x,\varepsilon} - X_t^{x',\varepsilon}|} \right\rangle dt \\ + \varepsilon \left\langle (\mathbb{I} - H(X^{x,\varepsilon}, X^{x',\varepsilon})) \frac{X_t^{x,\varepsilon} - X_t^{x',\varepsilon}}{|X_t^{x,\varepsilon} - X_t^{x',\varepsilon}|}, dW_t \right\rangle; \tag{5.7}$$

$$Y_0^\varepsilon = |x - x'|.$$

Remark Since the function $f(u) = |u|$ is not differentiable at the origin, we need in fact an extension of the usual Itô formula. We may use the stopping times $\eta_n = \inf\{t \geq 0 \colon |Y_t^\varepsilon| \leq 1/n\}$ and check the expression for the processes stopped at η_n, $Y_{t\wedge\eta_n}^\varepsilon$ and then let $n \to +\infty$ to get the extension of (2.35), applicable to Y_t^ε for $t < T_{x,x'}^\varepsilon$. Clearly, the right-hand side of (5.7) does not make sense if $X_t^{x,\varepsilon} = X_t^{x',\varepsilon}$. Since we are interested in the behaviour of Y^ε up to $T_{x,x'}^\varepsilon$, we shall not fill in all details of the localization procedure. To check (5.7) we apply this extension of Itô's formula (2.35) and observe that the term with second derivatives

$$\sum_{i,k} \frac{\partial^2 f}{\partial x_i \partial x_k}(X^{x,\varepsilon} - X^{x',\varepsilon}) \sum_j (\mathbb{I} - H(X_s^{x,\varepsilon}, X_s^{x',\varepsilon}))_{i,j} (\mathbb{I} - H(X_s^{x,\varepsilon}, X_s^{x',\varepsilon}))_{k,j} \tag{5.8}$$

vanishes, as a simple computation using (5.3) shows.

After some computations we also see that for $t < T_{x,x'}^\varepsilon$ the second term on the right-hand side of (5.7) equals

$$2\varepsilon \left\langle \frac{X_t^{x,\varepsilon} - X_t^{x',\varepsilon}}{|X_t^{x,\varepsilon} - X_t^{x',\varepsilon}|}, dW_t \right\rangle$$

which may be written as $2\varepsilon d\tilde{B}_t$, for a one-dimensional standard Brownian motion $\tilde{B}_t$, starting at the origin.

The control of the first term on the right-hand side of (5.7) is simplified if

$$\langle b(u) - b(v), u - v \rangle \leq 0 \tag{5.9}$$

for all $u \neq v$. In this case $T^{\varepsilon}_{x,x'}$ becomes bounded from above by

$$\tilde{S}^{\varepsilon}_{x,x'} = \inf\{t > 0\colon\ |x - x'| + 2\varepsilon \tilde{B}_t = 0\},$$

for a one-dimensional standard Brownian motion $\tilde{B}_t$ and we get:

$$\begin{aligned} P(\tilde{S}^{\varepsilon}_{x,x'} \le \varepsilon^{-3}) &= P(\inf_{s \le \varepsilon^{-3}} 2\varepsilon \tilde{B}_s \le -|x - x'|) \\ &= 2P(2\varepsilon \tilde{B}_{\varepsilon^{-3}} \le -|x - x'|) \\ &= 2P(\tilde{B}_1 \ge \frac{\varepsilon^{1/2}}{2}|x - x'|) \\ &\ge 1 - \frac{|x - x'|}{\sqrt{2\pi}}\varepsilon^{1/2}, \end{aligned} \tag{5.10}$$

where we have used the reflection principle (Theorem 2.14) in the second equality.

Observe that the coupling time has been estimated with such simplicity owing to two important points:

(i) the hypothesis of a constant diffusion matrix, which makes the terms involving second derivatives in Itô's formula, cf. (5.8), vanish;

(ii) condition (5.9) on the vector field $b(\cdot)$.

An example verifying (5.9) is $b(\cdot) = -\nabla U(\cdot)$, where U is convex.

Remark 5.1 As a basic application, consider the problem of proving the asymptotic exponentiality of the escape time τ_ε in the context of Section 2.6. Concerning (ii) above, it will suffice to guarantee the validity of (5.9) in a small neighbourhood of the stable fixed point x_0 (see the next paragraph). Let $b(\cdot) = -\nabla U(\cdot)$, where U is of class C^2 and x_0 is a hyperbolic point of minimum of $U(\cdot)$. Thus, the Hessian matrix $D^2U(x_0)$ (the matrix of second order partial derivatives of $U(\cdot)$ at x_0) is symmetric and positive definite. Let $\delta > 0$ be small. By the mean value theorem, we can write the left-hand side of (5.9), for $u, v \in B_\delta(x_0)$, as $\langle A(u, v)(u - v), u - v\rangle$, where $A(u, v)$ is a $d \times d$ matrix with entries $-(\partial^2 U/\partial x_i \partial x_j)(y(i, j))$ for suitable points $y(i, j)$ with $\max_{1\le k\le d} |y_k(i, j) - x_{0,k}| \le \delta$, so that $|y(i, j) - x_0| \le d^{1/2}\delta$. Since U is of class C^2 and the matrix $D^2U(x_0)$ is symmetric and positive definite, we see that (5.9) holds for $u, v \in B_\delta(x_0)$ and δ sufficiently small.

The arguments in Section 2.6 show that with overwhelming probability the process visits $B_\delta(x_0)$ and then takes a time of order at least $\exp\{c/\varepsilon^2\}$ for a positive constant $c = c_\delta$ to leave it. Since this is much larger than ε^{-3}, for ε small enough, (5.10) indicates the possibility of using the coupling technique to prove the asymptotic loss of memory of τ_ε. This is discussed in Section 5.4 in a more general situation.

The situation with item (i) above is delicate. The coupling method was discussed in [202] for nonconstant diffusion matrices, with conditions for a successful coupling when $\sigma(\cdot)$ does not vary too much. However, its applicability to our basic problem would require more precise estimates of the order of magnitude of the coupling time (its ε-dependence). This requires a more careful analysis that we shall not pursue in this text.

5.2 The case of a double well potential

We shall now make rigorous the description given at the beginning of this chapter in the simplest case of a vector field that derives from a double well potential. The goal is to prove the analogue of Theorems 4.10 and 4.6 in Chapter 4, summarizing the two basic ingredients: thermalization and unpredictability of the tunnelling times. More general situations are treated later in this chapter, using the full machinery of Freidlin and Wentzell.

Assumptions 5.2 Let $(X^{x,\varepsilon})$ be as in (5.1). We assume that for $x \in \mathbb{R}^d$

$$b(x) = -\nabla U(x),$$

where

(i) $U : \mathbb{R}^d \to \mathbb{R}$ is a function of class C^2;
(ii) $\liminf_{|x| \to +\infty} U(x)/|x| > 0$;
(iii) $U(\cdot)$ has exactly three critical points, denoted by p, q and z, all hyperbolic, i.e. the determinant of the Hessian matrix of U at these points is different from zero; p and q are assumed to be stable, i.e. they are points of local minimum of $U(\cdot)$; z is a saddle point;
(iv) $U(q) < U(p) < U(z)$;
(v) b is globally Lipschitz, i.e. condition (2.51) holds.

For the problems we have in mind condition (v) does not impose any extra restriction, as we shall see.

Remark 5.3 Under the previous assumptions it follows that the Hessian matrix $-D^2U(z)$ has only one positive eigenvalue, whose eigenspace is one-dimensional. The eigenvalues of $-D^2U(x)$, for $x = p, q$ are all negative.

Under Assumptions 5.2, the behaviour of the deterministic system $X_t^{x,0}$ is very simple. We have a decomposition

$$\mathbb{R}^d = \mathcal{W}_p^s \cup \mathcal{W}_q^s \cup \mathcal{W}_z^s,$$

where $\mathcal{W}_x^s$ denotes the (global) stable manifold of the fixed point x: $\mathcal{W}_x^s$ is invariant under the deterministic flow and attracted towards x (if $y \in \mathcal{W}_x^s$, then $X_t^{y,0} \in \mathcal{W}_x^s$ for all $t > 0$ and $\lim_{t \to +\infty} X_t^{y,0} = x$). For $x = p, q$ we shall write $\mathcal{W}_x^s = D_x$ to recall that in these cases $\mathcal{W}_x^s$ is an open domain in $\mathbb{R}^d$, also called the domain of attraction of x. The separatrix of D_p and D_q is $\mathcal{W}_z^s$, which is a manifold of codimension one, under the above assumptions.

The saddle point z also admits a non-empty unstable manifold, which we denote by $\mathcal{W}_z^u$. In this case, it is a one-dimensional manifold connecting p and q (see Remark 5.3).

We keep here the same notation introduced in Chapter 2. In particular, $\tau_\varepsilon^x(A)$ will denote the hitting time to the set A for the process $X^{x,\varepsilon}$. The starting point is omitted from the notation if it appears as a subscript in the probability measure. $B_r(x)$ ($\bar{B}_r(x)$) will denote the open (closed, respectively) Euclidean ball centred at x, with radius r; $S_r(x) = \{y \in \mathbb{R}^d : |y - x| = r\}$. Domains are taken open.

Definition 5.4 Let us fix $c > 0$ so that $\bar{B}_c(p) \subseteq D_p$ and $\bar{B}_c(q) \subseteq D_q$, and let us set

$$\tau_\varepsilon^x = \tau_\varepsilon^x(\bar{B}_c(q))$$

and

$$\bar{\tau}_\varepsilon^x = \inf\{t > \tau_\varepsilon^x : X_t^{x,\varepsilon} \in \bar{B}_c(p)\},$$

with the convention that $\inf(\emptyset) = +\infty$. Notice that τ_ε^x and $\bar{\tau}_\varepsilon^x$ are stopping times (for the canonical filtration associated with the given Brownian motion).

We now state the basic results on metastability for this example.

Theorem 5.5 *For each $\varepsilon > 0$ let β_ε be defined through the relation*

$$P_p(\tau_\varepsilon > \beta_\varepsilon) = e^{-1}. \tag{5.11}$$

Under Assumptions 5.2,

$$\lim_{\varepsilon\to 0} \varepsilon^2 \log \beta_\varepsilon = 2(U(z) - U(p)) =: \Delta, \tag{5.12}$$

and for each $x \in D_p$ and each $t > 0$,

$$\lim_{\varepsilon\to 0} P_x(\tau_\varepsilon > t\beta_\varepsilon) = e^{-t}. \tag{5.13}$$

Similarly, if $\bar{\beta}_\varepsilon$ is defined through the relation $P_q(\tau_\varepsilon(\bar{B}_c(p)) > \bar{\beta}_\varepsilon) = e^{-1}$, then for each $x \in D_p$ and each $t > 0$,

$$\lim_{\varepsilon\to 0} P_x(\bar{\tau}_\varepsilon - \tau_\varepsilon > t\bar{\beta}_\varepsilon) = e^{-t}.$$

Theorem 5.6 *Under the assumptions of Theorem* 5.5 *and with β_ε defined by* (5.11), *we can find $R_\varepsilon > 0$, with $\lim_{\varepsilon\to 0} R_\varepsilon = +\infty$, $\lim_{\varepsilon\to 0} R_\varepsilon/\beta_\varepsilon = 0$, such that the $\mathcal{M}_1(\mathbb{R}^d)$-valued processes ν_t^ε defined by*

$$\nu_t^\varepsilon(f) = \frac{1}{R_\varepsilon} \int_{t\beta_\varepsilon}^{t\beta_\varepsilon + R_\varepsilon} f(X_s^\varepsilon)ds, \quad f \in C_b(\mathbb{R}^d), \quad t > 0, \tag{5.14}$$

verify, for each $x \in D_p$, $f \in C_b(\mathbb{R}^d)$, *and* $t > 0$,

$$\lim_{\varepsilon\to 0} P_x\left(\sup_{0<s<(\tau_\varepsilon - 2R_\varepsilon)/\beta_\varepsilon} |\nu_s^\varepsilon(f) - f(p)| > \delta\right) = 0 \tag{5.15}$$

and

$$\lim_{\varepsilon\to 0} P_x\left(\sup_{\tau_\varepsilon/\beta_\varepsilon<s<(\bar{\tau}_\varepsilon - 2R_\varepsilon)/\beta_\varepsilon} |\nu_s^\varepsilon(f) - f(q)| > \delta\right) = 0. \tag{5.16}$$

Possible choices of R_ε *include* $R_\varepsilon = e^{\alpha/\varepsilon^2}$, *with* $0 < \alpha < \Delta$.

An immediate consequence of the last two theorems is that with a minor modification of the process (ν_t^ε) around $\tau_\varepsilon/\beta_\varepsilon$, we have convergence in law, in the Skorohod space of measure-valued trajectories. The limit is a jump process which performs a unique jump, from δ_p to δ_q, after a unit mean exponential random time.

Notation Given a measurable set $A \subseteq \mathbb{R}^d$ for which $\mu_\varepsilon(A) > 0$, we denote by $\mu_{\varepsilon,A}$ the corresponding restricted (conditioned) equilibrium measure

$$\mu_{\varepsilon,A}(dx) := \frac{1}{\mu_\varepsilon(A)} \mathbf{1}_A(x)\mu_\varepsilon(dx).$$

Proof of Theorem 5.5 We start by proving (5.12). The proof also provides the main point for the loss of memory involved in (5.13). Given $\delta > 0$ we shall prove that for any $x \in D_p$

$$\lim_{\varepsilon\to 0} P_x(e^{\frac{\Delta-\delta}{\varepsilon^2}} < \tau_\varepsilon < e^{\frac{\Delta+\delta}{\varepsilon^2}}) = 1 \tag{5.17}$$

where $\Delta = 2(U(z) - U(p))$. This implies (5.12), as easily checked.

Given a small positive number ϑ so that $\bar{B}_\vartheta(z) \cap (\bar{B}_c(p) \cup \bar{B}_c(q)) = \emptyset$, we may take a bounded domain $G = G(\vartheta)$ with boundary ∂G of class C^2, and such that:

(a) $\bar{B}_c(p) \cup \bar{B}_c(q) \subseteq G$, $z \in G$;

(b) $\langle b(x), n(x)\rangle < 0$ for all $x \in \partial G$, where $n(x)$ denotes the unit outward normal vector to ∂G at x;

(c) $G = G_p \cup G_q \cup G_z$ with $G_p \subseteq D_p$, $G_q \subseteq D_q$,

$$G_z \subseteq B_\vartheta(z). \tag{5.18}$$

Condition (b) guarantees that $G \cup \partial G$ is invariant under the deterministic flow $(X_t^{x,0})$, as time evolves positively. We also assume that

(d) $\min_{x\in\partial G} U(x) > U(z)$.

Since the potential $U(\cdot)$ is of class C^2, possible choices are $G = \{x : U(x) < \bar{u}\}$, where $\bar{u} = \bar{u}(\vartheta) > U(z)$ is suitably close to $U(z)$.

Having fixed any such auxiliary domain G, we may modify our problem slightly and restrict our attention to what happens inside G only. Due to

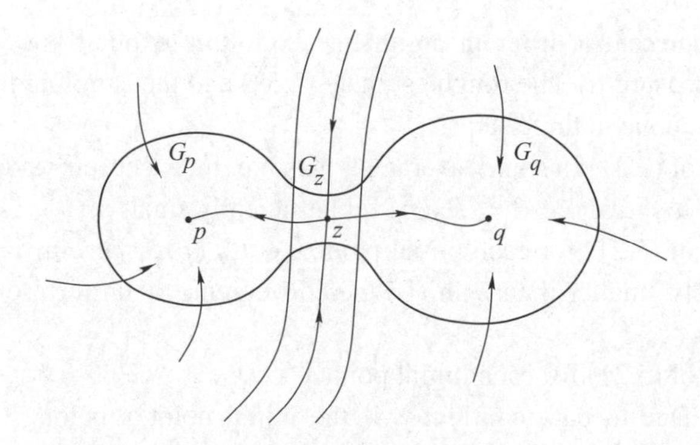

Figure 5.2

Assumptions 5.2, the $(d-1)$-dimensional manifold $\mathcal{W}_z^s$ is a separatrix for D_p and D_q. $\mathcal{W}_z^s \cap G$ divides G into two positively invariant domains $D_p \cap G$ and $D_q \cap G$. Our problem will be reduced to studying how the process $X^{x,\varepsilon}$ escapes from one of these domains to the other, once we show that this happens before the escape from G, with overwhelming probability. For convenience, in the proof of the upper bound for the exit time we shall avoid the presence of the critical point z in the boundary, for which reason we have singled out the small subdomain G_z, cf. (c) above. We call ∂^q and ∂^p the portions of the boundary of G_z contained in D_q and D_p respectively. In addition, we set

$$G' = G_p \cup \partial^p \cup \partial^q \cup G_z. \tag{5.19}$$

Notice that ∂^q and ∂^p may be taken transverse to $\mathcal{W}_z^u$.

As consequence of condition (d) above we have that

$$\min_{y \in \partial G' \setminus \partial^q} U(y) > U(z). \tag{5.20}$$

The picture to have in mind is that of Figure 5.2, drawn in the two-dimensional case.

We start by first proving the lower bound for the tunnelling time, i.e. that

$$\lim_{\varepsilon \to 0} P_x(\tau_\varepsilon < e^{\frac{\Delta-\delta}{\varepsilon^2}}) \to 0, \tag{5.21}$$

for any $x \in D_p$ and any $\delta > 0$.

Observe that (5.21) was essentially proven in Section 2.6, when completely attracted domains were considered and the upper bound in the l.d.p. was used. Indeed, given the continuity of the trajectories, and explicit knowledge of the quasi-potential inside D_p, cf. Proposition 2.37, the discussion in Section 2.6 gives us the validity of (5.21) for all $x \in B_c(p)$. The extension to all $x \in D_p$ follows as step (III) below.

We now present a different argument, exploiting explicit knowledge of the stationary measure μ_ε, the rough estimate (2.53) and the coupling methods. This proof will be done in three steps.

(I) Proof of (5.21) in some 'averaged' way, i.e. for x chosen randomly according to $\mu_{\varepsilon,B}$, where $B = B_\rho(p)$ for sufficiently small ρ.

(II) Proof of (5.21) for each initial point $x \in B_{\rho_1}(p)$, where ρ_1 is positive, but possibly smaller than ρ in (I); the convergence is uniform on such initial points.

(III) Proof of (5.21) for each initial point $x \in D_p$.

Step (I). Due to path continuity, if the initial point x belongs to G_p, τ_ε is greater than the exit time from $D_p \cap G$, for any choice of ϑ as above. From the stationarity of the measure μ_ε, the continuity of $U(\cdot)$ and the choice of G, we can take $\rho > 0$, $\vartheta' > 0$ small enough (depending on δ) so that for any T and ε positive:

$$
\begin{aligned}
P_{\mu_{\varepsilon,B_\rho(p)}}(X_T^\varepsilon \in \partial(D_p \cap G)_{(\vartheta')}) &\le \frac{\mu_\varepsilon(\partial(D_p \cap G)_{(\vartheta')})}{\mu_\varepsilon(B_\rho(p))} \\
&\le \frac{e^{-\frac{2}{\varepsilon^2}\inf_{x\in\partial(D_p\cap G)_{(\vartheta')}} U(x)}\lambda_d(\partial(D_p \cap G)_{(\vartheta')})}{e^{-\frac{2}{\varepsilon^2}\sup_{x\in B_\rho(p)} U(x)}\lambda_d(B_\rho(p))} \\
&\le \frac{\lambda_d(\partial(D_p \cap G)_{(\vartheta')})}{\lambda_d(B_\rho(p))} e^{-(\Delta-\delta/3)/\varepsilon^2}
\end{aligned}
\tag{5.22}
$$

where $A_{(r)} = \{x \in \mathbb{R}^d : |x - y| < r, \text{ for some } y \in A\}$ is the r-neighbourhood of A. Here we have used the continuity of U and that $\min_{x\in\partial(D_p\cap G)} U(x) = U(z)$, which follows from the choice of the set G, since $U(y) \ge U(z)$ for $y \in \mathcal{W}_z^s$. (λ_d denotes the d-dimensional Lebesgue measure.) Now, given $\delta > 0$ we set

$$T^- = T^-(\varepsilon, \delta) := e^{\frac{\Delta-\delta}{\varepsilon^2}}, \tag{5.23}$$

let $N = [T^-/\varepsilon^3]$ be the integer part of T^-/ε^3, $t_i = i\varepsilon^3$, $0 \le i \le N$, and $t_{N+1} = T^-$. Using the inequality in the reflection principle (Theorem 2.14) we can write

$$P\left(\max_{0\le i\le N}\ \sup_{t_i\le t\le t_{i+1}} |W_t - W_{t_i}| \ge 1\right) \le 2d(N+1)e^{-\frac{\alpha}{\varepsilon^3}} \tag{5.24}$$

for a suitable positive constant α.

Calling $\tilde\Omega$ the complement of the event in (5.24) and with the processes $X^{x,\varepsilon}$ constructed as in Section 2.4, it is easy to see that for ε sufficiently small and any initial x,

$$\tilde\Omega \cap (X_{t_i}^{x,\varepsilon} \in \overline{D_p \cap G},\ \forall 0 \le i \le N) \subseteq \left(\sup_{0\le i\le N}\ \sup_{t_i\le t\le t_{i+1}} |X_t^{x,\varepsilon} - X_{t_i}^{x,\varepsilon}| \le 2\varepsilon\right). \tag{5.25}$$

To check the last inclusion, recall that if $t_i \le t \le t_{i+1}$ we have

$$
\begin{aligned}
|X_t^{x,\varepsilon} - X_{t_i}^{x,\varepsilon}| &\le \left|\int_{t_i}^t b(X_s^{x,\varepsilon})ds\right| + \varepsilon|W_t - W_{t_i}| \\
&\le \int_{t_i}^t |b(X_s^{x,\varepsilon}) - b(X_{t_i}^{x,\varepsilon})|ds + |(t-t_i)b(X_{t_i}^{x,\varepsilon})| + \varepsilon|W_t - W_{t_i}|.
\end{aligned}
$$

Using the Lipschitz property given by (2.51) and the Gronwall inequality, we see that on the event $[X_{t_i}^{x,\varepsilon} \in \overline{D_p \cap G}] \cap \tilde{\Omega}$ and for $t_i \le t \le t_{i+1}$ we have:

$$
|X_t^{x,\varepsilon} - X_{t_i}^{x,\varepsilon}| \le e^{\kappa|t-t_i|}\left(\varepsilon + \varepsilon^3 \sup_{x \in \overline{D_p \cap G}} |b(x)|\right)
$$

and (5.25) follows at once, for $\varepsilon > 0$ small. From this and (5.24) we can write

$$
P_{\mu_{\varepsilon,B_\rho(p)}}(\tau_\varepsilon < T^-) \le 2d(N+1)e^{-\frac{\alpha}{\varepsilon^3}} + \sum_{i=1}^N P_{\mu_{\varepsilon,B_\rho(p)}}(X_{t_i}^\varepsilon \in \partial(D_p \cap G)_{(\vartheta')}),
$$

provided $\varepsilon > 0$ is small. Recalling (5.22) and (5.23) we conclude that

$$
\lim_{\varepsilon \to 0} P_{\mu_{\varepsilon,B_\rho(p)}}(\tau_\varepsilon < T^-) = 0, \tag{5.26}
$$

completing step (I).

Step (II). We now verify that the initial measure $\mu_{\varepsilon,B_\rho(p)}$ might be replaced by any initial point x close enough to p. For this we use the coupling method of Lindvall and Rogers, defined in Section 5.1. Pick $\rho > 0$ small enough to guarantee the validity of (5.9) for all $u \ne v \in B_\rho(p)$, as seen in Remark 5.1. We now recall Remark 2.41: due to the stability of p, we can take $\rho_1 > 0$ small enough and c_1 positive so that

$$
\sup_{y \in B_{\rho_1}(p)} P_y\left(\tau_\varepsilon(S_\rho(p)) < e^{\frac{c_1}{\varepsilon^2}}\right) \le e^{-\frac{c_1}{\varepsilon^2}}, \tag{5.27}
$$

where $S_\rho(x) = \{y \in \mathbb{R}^d : |y - x| = \rho\}$.

Due to the basic properties of the coupling discussed in Section 5.1 and as remarked at that occasion, for two trajectories starting at $x, x' \in B_{\rho_1}(p)$, we see that with probability tending to one as ε tends to zero, the two trajectories will meet before a time of order ε^{-3} and before any of them has escaped from $B_\rho(p)$. In particular,

$$
\lim_{\varepsilon \to 0} \sup_{x,x' \in B_{\rho_1}(p)} |P_x(\tau_\varepsilon > T^-) - P_{x'}(\tau_\varepsilon > T^-)| = 0. \tag{5.28}
$$

Together with (5.26) this implies (5.21), uniformly on $B_{\rho_1}(p)$, concluding step (II) (notice that in this step ρ_1 replaces the initial ρ and it might be much smaller).

Step (III). To conclude the proof it now suffices to exploit the closeness to the deterministic orbit and the strong Markov property, as already done in Chapter 2. More precisely, starting at $x \in D_p$ there is a time T_x, independent of ε, such that during the time interval $[0, T_x]$ the deterministic system reaches $B_{\rho_1/2}(p)$ and keeps at least a positive distance δ_x from ∂D_p. Using the simple estimate (2.53) we see that for small ε, with overwhelming probability, the stochastic system also reaches $B_{\rho_1}(p)$ before T_x and without exiting from D_p. We then use the strong Markov property at the first hitting time to $B_{\rho_1}(p)$ and the result of step (II), concluding the proof of (5.21). (The convergence is uniform for x in a compact subset of D_p.)

We now turn to the proof of the upper bound for τ_ε, namely that for any $\delta > 0$ and any $x \in D_p$:

$$\lim_{\varepsilon \to 0} P_x(\tau_\varepsilon > e^{\frac{\Delta+\delta}{\varepsilon^2}}) = 0. \tag{5.29}$$

Having fixed a domain G as before, and using the closeness to the deterministic orbit, it suffices to show that given $\delta > 0$ we can find ρ_1 positive such that

$$\lim_{\varepsilon \to 0} \sup_{x \in B_{\rho_1}(p)} P_x(\tau_\varepsilon(\partial^q) > T^+) = 0, \tag{5.30}$$

where $T^+ = T^+(\delta, \varepsilon) = e^{(\Delta+\delta)/\varepsilon^2}$, and we use the previous notation. Indeed, the orbits of the deterministic system issued from points in ∂^q reach $B_{c/2}(q)$ in a bounded time, and we apply the strong Markov property and (2.53).

To get (5.30) we want to find $T = T(\varepsilon)$ such that

$$\lim_{\varepsilon \to 0} T(\varepsilon)/e^{\frac{\Delta+\delta/2}{\varepsilon^2}} = 0$$

and for which we can show the existence of a suitable set of trajectories $\mathcal{E} = \mathcal{E}_{x,T}$ contained in $\{\varphi \in C([0, T], \mathbb{R}^d) : \varphi(0) = x\}$ verifying:

(i) $\inf_{x \in G'} P\left(X^{x,\varepsilon} \in \mathcal{E}_{x,T}\right) \geq \alpha_\varepsilon := T(\varepsilon)/e^{(\Delta+\delta/2)/\varepsilon^2}$;

(ii) every trajectory in $\mathcal{E}_{x,T}$ reaches ∂^q by time T.

Notice that (5.30) will follow once events $\mathcal{E}$ satisfying the above conditions (i) and (ii) are proven to exist. To see this, we split the time interval $[0, T^+]$ into subintervals of length $T(\varepsilon)$. Using the Markov property at the initial times of each subinterval we first obtain that

$$\sup_{x \in G'} P_x\left(\tau_\varepsilon(\partial G') > T^+\right) \leq (1 - \alpha_\varepsilon)^{[\frac{T^+}{T(\varepsilon)}]},$$

which tends to zero as $\varepsilon \to 0$. On the other hand, due to (5.20), the previous lower estimate yields $\delta_1 > 0$ (depending on G) and $\rho_1 > 0$ so that[1]

$$\lim_{\varepsilon \to 0} \sup_{x \in B_{\rho_1}(p)} P_x\left(\tau_\varepsilon(\partial G' \setminus \partial^q) < e^{\frac{\Delta+\delta_1}{\varepsilon^2}}\right) = 0.$$

In this case, (5.30) would follow if $\delta \leq \delta_1$. But the argument given below shows

[1] For instance, we can take $\rho_1 = c$.

that we can take δ arbitrarily small (independently of the fixed G) and construct events $\mathcal{E}_{x,T}$ as above.

In the framework of large deviation theory, a construction of such events $\mathcal{E}_{x,T}$, with T independent of ε (but depending on δ and G) follows from Theorem 2.25. The idea is to show that for each $x \in G'$ there is a trajectory φ^x starting at x, with rate function not larger than $\Delta + \delta/3$ and such that not only does φ^x exit from G' (through ∂^q) by time T, but also that any path sufficiently close to it in the metric $\varrho_T(\cdot, \cdot)$ will do the same. For this we let $\vartheta' \le \vartheta$ be so small that the rate function of the linear interpolation between any two points within a distance $2\vartheta'$ is at most $\delta/16$ and let $G(z, \vartheta')$ be the set of points in G_z at distance at most ϑ' from $\mathcal{W}^s_z$. From any point $x_0 \in G(z, \vartheta')$ we choose any of the points $x'_0 \in \mathcal{W}^s_z$ closest to it, and interpolate from x_0 to x'_0. We then follow the deterministic flow, arriving at some $x''_0 \in \bar{B}_{\vartheta'}(z)$, from where we interpolate to a point $x_1 \in \mathcal{W}^u_z \cap D_q$ with $|x_1 - z| = \vartheta'$. The deterministic path starting at a point $x_0 \in D_p \cap \partial G(z, \vartheta')$ hits $S_{\vartheta'}(p)$ in a certain bounded time t'_{x_0} and that starting at a point x_1 as above arrives at some point $x_3 \in G_q$ at a distance at least ϑ' from G_z in a certain bounded time t_{x_1}. The construction of the trajectory φ^x goes as follows: for $x \in G(z, \vartheta')$ we follow the above prescription (with $x_0 = x$) until we reach a point x_3 as above. For $x \in G' \backslash G(z, \vartheta')$ there are two different cases: for $x \in D_q$ we just follow the deterministic path until we reach a point x_3 as above; for $x \in D_p$ we first follow the deterministic path until we reach $S_{\vartheta'}(p)$ and then a little interpolation to a point $x_2 \in S_{\vartheta'}(p) \cap \mathcal{W}^u_z$ from which we can follow the reversal of the deterministic trajectory, reaching a point $x_0 \in D_p \cap S_{\vartheta'}(z)$, and then we follow the previous prescription. The total time t_x of such trajectories φ^x can be uniformly bounded in x, their rate function can be bounded from above by $\Delta + \delta/3$ and each path close enough to φ^x during time $[0, t_x]$ will reach ∂^q within t_x. Defining $\mathcal{E}_x$ as a small neighbourhood of φ^x, condition (ii) follows from the construction and (i) is a consequence of item $(b)'_u$ in Corollary 2.26. We get (5.30) as indicated above.

This concludes the proof of (5.17), from which (5.12) follows at once.

The proof of (5.13) is now almost evident. The loss of memory can be checked as a consequence of the coupling of Lindvall and Rogers, according to (5.28), together with the fact that if the process remains in G' for an exponentially long time, then it must visit $B_{\rho_1}(p)$ with overwhelming probability. With respect to the case in Section 2.6, this implies control on the time it stays around z. We provide details in the more general context of Section 5.4.

The second part of Theorem 5.5 is completely analogous to the first. □

The details for the proof of Theorem 5.6 are omitted, since they are similar to the corresponding ones in Chapter 4.

Remark Concerning the lower bound for τ_ε in (5.17), besides the proof based on the l.d.p., we have seen an alternative argument based on the explicit description of the invariant measure, on the closeness to the deterministic system (cf. (2.53)), and on coupling methods. It is not hard to see that the same can be done for the upper bound for the escape time τ_ε in Theorem 2.40 of Section 2.6, if $b = -\nabla U$ and U is a single well potential. Moreover, exploiting the reversibility of μ_ε (see next remark), one can also argue along these lines to describe the first excursion to ∂G for this particular choice of the vector field.

On the other hand, to carry out such an argument in order to prove the upper bound for the tunnelling time τ_ε in the context of Theorem 5.5 requires more involved details. One may properly choose the domain G' in the previous proof, with ϑ suitably small, and modify U outside G' by introducing another small well around a stable point $\tilde{q} \notin G'$, near to ∂^q. This can be done in such a way that the invariant measure of the modified process gives weight at least $\exp\{-\varepsilon^{-2}(\Delta + \delta/2)\}$ to a small neighbourhood around z. With the help of this auxiliary process we can construct events $\mathcal{E}_{x,T}$ ($T = T(\varepsilon)$ suitable) leading to the proof of (5.30), as above. The rest of the proof follows as before. The details of an argument along these lines are lengthy and we do not develop them here.

Remark Under Assumptions 5.2, the measure μ_ε given by (5.2) is reversible for X^ε, i.e. for any given $T > 0$, $(X_t^\varepsilon : 0 \le t \le T)$ and $(X_{T-t}^\varepsilon : 0 \le t \le T)$ have the same law under $P_{\mu_\varepsilon} := \int P_x(\cdot)\mu_\varepsilon(dx)$. This is expressable by the self-adjointness of the generator in the space $\mathrm{L}^2(\mathbb{R}^d, \mu_\varepsilon)$ and it can be checked after recalling (2.47) for $b = -\nabla U$ and $a = \varepsilon^2\mathbb{I}$. This concept appeared in Chapter 4 and it is further exploited in the next chapter, in the context of stochastic Ising models; we refer to Chapter I of [199].

5.3 General case. Asymptotic magnitude of the escape time

Having in mind extensions of Theorem 5.5, one is naturally led to study the escape time from a domain G that contains several asymptotically stable fixed points (or periodic orbits). This requires a more general method to deal with the transitions among small neighbourhoods of such points (or periodic orbits) inside G, up to the escape from G. The fundamental results are provided by Freidlin and Wentzell theory, and are described in Chapter 6 of [122]. In this section, for the sake of completeness, we recall some of their basic points leading to estimates of the order of magnitude of the escape time. Some details will be omitted, for which we refer to [122]. The other ingredient for the extension of Theorem 5.5 is a suitable application of the coupling discussed in Section 5.1. We need to adapt it, since condition (5.9) might not hold. This is done in Section 5.4, and it will enable us to go beyond the results of Freidlin and Wentzell, describing the asymptotic distribution of the escape time in the presence of a 'cycle' property.

We consider diffusion processes $X^{x,\varepsilon}$ on $\mathbb{R}^d$, given by Itô's equation (5.1). The vector field $b\colon \mathbb{R}^d \to \mathbb{R}^d$ is assumed to be of class C^1 and we assume as well that the entries of the Jacobian matrix $Db(x) = (\partial b^i/\partial x_j(x))$ are bounded.

In Section 2.6 we considered the case of a domain G positively invariant under the deterministic flow, and fully attracted to a unique point $x_0 \in G$. For investigation of the escape from G we were led to consider $V(x, y)$, the quasi-potential of Freidlin and Wentzell, cf. Definition 2.31, which represents the 'minimal cost' for the system to go from x to y, in terms of large deviations. The basic results concerning the escape from G were obtained through consideration of a discrete time Markov chain (Z_n^ε) that records the successive visits to $S_r(x_0) \cup \partial G$, after having passed by $S_{2r}(x_0)$, and stopping upon the first visit to ∂G, where $r > 0$ was taken small enough. In the limit of small noise, a crucial role was played by $V_0 = \min_{y\in\partial G} V(x_0, y)$ and by those points in ∂G where the minimum is attained.

If the domain G contains several stable equilibrium points (or attracting periodic orbits) for the deterministic system, the extension now involves the analysis of a Markov chain (Z_n^ε) taking values on the union of ∂G with the boundaries of small neighbourhoods of the different fixed points (or periodic orbits) inside G, until it reaches ∂G. The estimates should then involve the 'cost' of the given transitions under the restriction of not exiting from $G \cup \partial G$. This leads to the consideration of $V_G(x, y)$, defined for $x, y \in G \cup \partial G$:

$$V_G(x, y) = \inf\{I_T(\varphi)\colon \varphi(0) = x, \varphi(T) = y, \varphi(s) \in G \cup \partial G \\ \text{for all } s \in [0, T], T > 0\},$$

where I_T is the rate function introduced in Chapter 2.

Of course, $V_G(x, y) \ge V(x, y)$ for all $x, y \in G \cup \partial G$. We also see that if $x, y \in G$ and $V_G(x, y) \le \min_{z\in\partial G} V(x, z)$, then $V_G(x, y) = V(x, y)$.

Concerning the problems we have outlined, we need a quick review of essential points in the theory of Freidlin and Wentzell. We start with a summary of the basic properties of $V_G(\cdot, \cdot)$ that will be needed in the sequel. At this point we will be sketchy. The results in the proposition below are taken from Lemmas 1.2 and 1.4, Chapter 6 of [122], where they are proved more generally.

Proposition 5.7

(i) *V_G is finite and continuous in $G \times G$. If ∂G is of class C^2, then the continuity holds in $(G \cup \partial G) \times (G \cup \partial G)$.*

(ii) *For each $\gamma > 0$ and each K compact subset of G there exists $T_0 < +\infty$ such that for any $x, y \in K$ there is a path φ contained in $G \cup \partial G$ with $\varphi(0) = x$, $\varphi(s) = y$ for some $s \le T_0$ and such that*

$$I_s(\varphi) \le V_G(x, y) + \gamma.$$

(When ∂G is of class C^2, the above property holds for any compact $K \subseteq G \cup \partial G$.)

(iii) *Let G be a domain with ∂G a compact manifold of class C^2. For $\delta > 0$ we write $G_{(-\delta)} = \{y \in G : |y - x| > \delta, \ \forall x \in \partial G\}$ and, as before, $G_{(\delta)} = \{y \in \mathbb{R}^d : |y - x| < \delta, \text{ for some } x \in G\}$.*

Given $\gamma > 0$ and a compact set $K \subseteq G \cup \partial G$, we can find $\delta > 0$ such that

$$V_G(x, y) \le V_{G_{(\delta)}}(x, y) + \gamma, \tag{5.31}$$

$$V_G(x, y) \ge V_{G_{(-\delta)}}(x_{(-\delta)}, y_{(-\delta)}) - \gamma, \tag{5.32}$$

for any $x, y \in K$, where $z_{(-\delta)}$ denotes the element in $\overline{G_{(-\delta)}}$ closest to z, for any $z \in G \cup \partial G$.

Remark The assumptions on ∂G allow to define $z_{(-\delta)}$, provided $\delta > 0$ is small enough.

Sketch of the proof (Partial) (i) The finiteness of V_G is easily seen. For the continuity, it suffices to see that given $\gamma > 0$ and $x \in G \cup \partial G$, there exists $\delta > 0$ small enough so that if $y \in G \cup \partial G$ and $|x - y| \le \delta$, we can find a trajectory connecting x and y in a bounded time, entirely contained in $G \cup \partial G$, and with rate function at most γ. For $x \in G$ this follows at once from Lemma 2.32, since we can just make a small linear interpolation. A little more care is needed for the points at the boundary, but due to the smoothness of ∂G it works essentially in the same way, with a more delicate interpolation.

(ii) Using the compactness, for $\delta > 0$ sufficiently small (to be determined) we take a finite subset $\{x_1, \ldots, x_m\} \subseteq K$ such that $K \subseteq \cup_{i=1}^m B_\delta(x_i)$, and for each pair of points in this finite set we can find a trajectory as in the statement but with γ replaced by $\gamma/2$. As in (i), using Lemma 2.32, we can take δ so small such that $K_{(\delta)} \subseteq G$, and such that any two points in K at distance at most δ can be connected by a trajectory contained in $G \cup \partial G$, with rate function at most $\gamma/2$, in time at most δ (linear interpolation, simply). Putting these two together: given $x, y \in K$ we take x_i, x_j such that $|x - x_i| \vee |y - x_j| \le \delta$, and the trajectory is constructed by first connecting x to x_i, then x_i to x_j, and finally x_j to y according to the previous description.

When the boundary ∂G is smooth we can allow $K \subseteq G \cup \partial G$, and for that we would again have to modify the interpolation near ∂G. We omit details.

(iii) This is a little more involved. It suffices to prove one of the inequalities and the other is analogous. The main point in the proof of (5.32) is to show that, given $a < +\infty$ and $\gamma > 0$, there exists $\delta > 0$ so that if a path φ in $G \cup \partial G$ connects two points $x, y \in K \cap (G_{(-\delta)} \cup \partial(G_{(-\delta)}))$ in a time T, and $T + I_T(\varphi) \le a$, then we can replace it by a path in $G_{(-\delta)} \cup \partial(G_{(-\delta)})$ connecting x to y (possibly in a different time), without increasing the rate function by more than γ. The procedure may be obtained by first approximating φ by polygonal functions, which have to be perturbed during the time spent at distance at most δ from the boundary. Details are omitted; we refer to Lemma 1.4, Chapter 6 of [122]. □

Definition 5.8 Given a domain G which we assume, from now on, with ∂G of class C^2, and given $x, y \in G \cup \partial G$, we say that $x \sim_G y$ if $V_G(x, y) = V_G(y, x) = 0$. When $G = \mathbb{R}^d$ we write simply $\sim$.

Observe that $\sim_G$ is an equivalence relation in $G \cup \partial G$. Each equivalence class for $\sim_G$ is closed, due to the continuity of V_G. Since the rate function vanishes only on the trajectories of the deterministic system, we can check that if the equivalence class of x is non-trivial (strictly larger than $\{x\}$), then the deterministic trajectory (or orbit) $X_\cdot^{y,0}$ of any point y in this class must be entirely contained in the class. Thus, any non-trivial equivalence class for $\sim_G$ contains the ω-limit set of each of its points. (As usual, the ω-limit set of a point y is defined as the set of accumulation points of the positive orbit of the deterministic system, issued from y.) We now turn to the points of the Freidlin and Wentzell theory that provide the estimates of the escape times. For the moment we consider the set-up as in Section 2, Chapter 6 of [122], but restricted to $\mathbb{R}^d$ and taking $a(x) = \mathbb{I}$.

Assumptions 5.9

(a1) $G \subseteq \mathbb{R}^d$ is a bounded domain with ∂G of class C^2, containing a finite number of compact sets $K_1, \dots, K_l$ such that:

(a2) each K_i is an equivalence class for $\sim_G$;

(a3) if the ω-limit set of any point for the deterministic system $\dot{x} = b(x)$ lies entirely in $G \cup \partial G$, then it must be contained in some K_i, for $i = 1, \dots, l$.

Remark Notice that under the above assumptions we allow the existence of points in G for which the ω-limit set is not contained in G.

For the extension of the Markov chain approximation used in Section 2.6, we need to recall the following definitions, introduced by Freidlin and Wentzell:

$$\tilde{V}_G(K_i, K_j) = \inf\{I_T(\varphi) \colon \varphi(0) \in K_i, \varphi(T) \in K_j, \varphi(s) \in (G \cup \partial G) \setminus \cup_{\ell \neq i,j} K_\ell \text{ for all } 0 < s < T, T > 0\},$$

$$\tilde{V}_G(x, K_j) = \inf\{I_T(\varphi) \colon \varphi(0) = x, \varphi(T) \in K_j, \varphi(s) \in (G \cup \partial G) \setminus \cup_{\ell \neq j} K_\ell \text{ for all } 0 < s < T, T > 0\},$$

$$\tilde{V}_G(K_i, y) = \inf\{I_T(\varphi) \colon \varphi(0) \in K_i, \varphi(T) = y, \varphi(s) \in (G \cup \partial G) \setminus \cup_{\ell \neq i} K_\ell \text{ for all } 0 < s < T, T > 0\},$$

$$\tilde{V}_G(x, y) = \inf\{I_T(\varphi) \colon \varphi(0) = x, \varphi(T) = y, \varphi(s) \in (G \cup \partial G) \setminus \cup_{\ell} K_\ell \text{ for all } 0 < s < T, T > 0\},$$

where $i, j \in \{1, \dots, l\}$ and $x, y \in G \cup \partial G$. (By convention, the infimum over the empty set is taken as $+\infty$.) Moreover, it is still convenient to set:

$$\tilde{V}_G(K_i, \partial G) = \min_{y \in \partial G} \tilde{V}_G(K_i, y)$$

and analogously $\tilde{V}_G(x, \partial G) = \min_{y \in \partial G} \tilde{V}_G(x, y)$.

Definition 5.10 An equivalence class K for $\sim$ is said to be stable if $V(K, y) > 0$ for each $y \notin K$. Otherwise K is said to be unstable.

Notice that if x is a fixed point of the deterministic system and $K = \{x\}$, the notion coincides with the usual asymptotic (or Lyapunov) stability, defined in Section 2.6.

Having introduced the above quantities, Freidlin and Wentzell showed how the large deviation theory, together with the strong Markov property for the process $X^{x,\varepsilon}$, yield a Markov chain approximation similar to the one used in Section 2.6. This Markov chain is obtained by first taking a small positive number ρ such that the neighbourhoods $\partial G_{(\rho)}, (K_i)_{(\rho)}, i = 1, \dots, l$, are all at a positive distance from each other. Choosing $\rho_1 \in (0, \rho)$ and letting $w_i = (K_i)_{(\rho_1)}$, $w = \cup_{i=1}^l w_i$, $C = (G \cup \partial G) \setminus \cup_{i=1}^l (K_i)_{(\rho)}$, one considers the sequence of stopping times:

$$\eta_0 = 0, \quad \sigma_0 = \inf\{s \geq 0 \colon X_s^\varepsilon \in C\}, \tag{5.33}$$

and, for $n \geq 1$,

$$\begin{aligned} \eta_n &= \inf\{s \geq \sigma_{n-1} \colon X_s^\varepsilon \in \partial w \cup \partial G\}, \\ \sigma_n &= \inf\{s \geq \eta_n \colon X_s^\varepsilon \in C\}. \end{aligned} \tag{5.34}$$

These random times are well defined until the first hitting time to ∂G. After that we may simply stop the chain, i.e. if $N = \min\{n \colon X_{\eta_n}^\varepsilon \in \partial G\}$, then for $m \geq N$ we set $\sigma_m = \eta_m = \eta_N$, as in Section 2.6. Recalling the strong Markov property of the diffusion process X^ε, one is led to consider the imbedded Markov chain $Z_n^\varepsilon = X_{\eta_n}^\varepsilon$. The basic estimates are summarized in the following two lemmas due to Freidlin and Wentzell, given as Lemma 2.1 and Lemma 2.2, Chapter 6 of [122].

Lemma 5.11 (Freidlin and Wentzell) *For any $\gamma > 0$ there exists $\rho > 0$ such that for any $\rho_2 \in (0, \rho)$ there exists $\rho_1 \in (0, \rho_2)$ such that if $\delta_0 \in (0, \rho)$ and ε is sufficiently small, then for all $x \in (K_i)_{(\rho_2)}, i = 1, \dots, l$, and any $y \in \partial G$ one has (recalling $w_i = (K_i)_{(\rho_1)}$):*

$$\begin{aligned} e^{-\varepsilon^{-2}(\tilde{V}_G(K_i, K_j)+\gamma)} &\leq P(Z_1^\varepsilon \in \partial w_j | Z_0^\varepsilon = x) \leq e^{-\varepsilon^{-2}(\tilde{V}_G(K_i, K_j)-\gamma)}; \\ e^{-\varepsilon^{-2}(\tilde{V}_G(K_i, \partial G)+\gamma)} &\leq P(Z_1^\varepsilon \in \partial G | Z_0^\varepsilon = x) \leq e^{-\varepsilon^{-2}(\tilde{V}_G(K_i, \partial G)-\gamma)}; \\ e^{-\varepsilon^{-2}(\tilde{V}_G(K_i, y)+\gamma)} &\leq P(Z_1^\varepsilon \in \partial G \cap B_{\delta_0}(y) | Z_0^\varepsilon = x) \leq e^{-\varepsilon^{-2}(\tilde{V}_G(K_i, y)-\gamma)}. \end{aligned} \tag{5.35}$$

Lemma 5.12 (Freidlin and Wentzell) *For any $\gamma > 0$ there exists $\rho > 0$ such that for any $\rho_2 \in (0, \rho)$ there exists $\rho_1 \in (0, \rho_2)$ such that if $\delta_0 \in (0, \rho)$ and ε is sufficiently small, then for all $x \in G \setminus (\cup_{i=1}^{l} (K_i)_{(\rho_2)} \cup \partial G_{(\rho_2)})$ and all $y \in \partial G$ one has:*

$$\begin{aligned} e^{-\varepsilon^{-2}(\tilde{V}_G(x,K_j)+\gamma)} &\le P(Z_1^\varepsilon \in \partial w_j | Z_0^\varepsilon = x) \le e^{-\varepsilon^{-2}(\tilde{V}_G(x,K_j)-\gamma)}; \\ e^{-\varepsilon^{-2}(\tilde{V}_G(x,\partial G)+\gamma)} &\le P(Z_1^\varepsilon \in \partial G | Z_0^\varepsilon = x) \le e^{-\varepsilon^{-2}(\tilde{V}_G(x,\partial G)-\gamma)}; \\ e^{-\varepsilon^{-2}(\tilde{V}_G(x,y)+\gamma)} &\le P(Z_1^\varepsilon \in \partial G \cap B_{\delta_0}(y) | Z_0^\varepsilon = x) \le e^{-\varepsilon^{-2}(\tilde{V}_G(x,y)-\gamma)}. \end{aligned} \tag{5.36}$$

Proof of Lemma 5.11 (cf. [122]) Let us start with an auxiliary remark, complementing Proposition 5.7.

Remark. If K is a compact equivalence class, $\gamma > 0$ and $\rho > 0$ are given, then it is possible to find $T_0 < +\infty$ and for each $x, y \in K$ a trajectory connecting x to y, in a time $T \le T_0$, entirely contained in $K_{(\rho)}$ and with rate function less than γ. That is, a trajectory as described in Proposition 5.7 can be assumed to be entirely contained in $K_{(\rho)}$.

Proof of the remark. Observe that at the beginning of the proof of item (ii) in Proposition 5.7 we may start with trajectories that do not leave the neighbourhood $K_{(\rho)}$. Indeed, if a sequence of trajectories in $G \cup \partial G$ connects two points x, y in K and their rate function tends to zero, there must be one of these trajectories which does not leave the neighbourhood $K_{(\rho)}$; otherwise we would get (as an accumulation point) $z \notin K$, equivalent to x and y, in contradiction with the assumption that K is an equivalence class. We may then proceed as in the proof of (ii) of Proposition 5.7, checking the remark.

A brief reflection allows us to conclude that each of the estimates in (5.35) becomes trivial if the corresponding $\tilde{V}_G(\cdot, \cdot)$ is $+\infty$, that is, the transition probability must vanish in this case. Indeed, the non-existence of a trajectory with finite rate function that satisfies the restrictions involved in the definition of a term $\tilde{V}_G(\cdot, \cdot)$ in (5.35) implies the non-existence of any continuous path verifying the same restrictions. This clearly implies that the corresponding conditional probability vanishes. One may take $\bar{V} < +\infty$ as an upper bound for all finite values of the quantities $\tilde{V}_G(\cdot, \cdot)$ involved in (5.35).

The procedure for the proof is quite similar to that of Section 2.6, based on the large deviation principle. It suffices to prove the statements at the first and third lines of (5.35). The second line follows at once from the third one, by taking a finite cover of ∂G by open balls of small radius.

For each of the lower bounds on the conditional probability, one might exhibit a suitable tube of trajectories performing the corresponding transition and apply the lower bound in the l.d.p. Thus, we need a trajectory with properly bounded rate function such that any path sufficiently close to it in the supremum norm will perform the given transition.

Let us concentrate on the lower bound at the first line of (5.35).

We take $\rho \in (0, \gamma/10L)$, with L being as in Lemma 2.32, and such that the sets $(K_i)_{(\rho)}$, $i = 1, \dots, l$, $\partial G_{(\rho)}$ are all at distance at least ρ from each other. As in Proposition 5.7, we may take $\delta \in (0, \rho/2)$ so that:

$$\begin{aligned}
\tilde{V}_{G_{(\delta)}}(K_i, K_j) &\geq \tilde{V}_G(K_i, K_j) - 0.1\gamma, \\
\tilde{V}_{G_{(-\delta)}}(K_i, K_j) &\leq \tilde{V}_G(K_i, K_j) + 0.1\gamma, \\
\tilde{V}_{G_{(\delta)}}(K_i, y) &\geq \tilde{V}_G(K_i, y) - 0.1\gamma, \\
\tilde{V}_{G_{(-\delta)}}(K_i, y_{(-\delta)}) &\leq \tilde{V}_G(K_i, y) + 0.1\gamma.
\end{aligned} \tag{5.37}$$

We also fix any $\rho_2 \in (0, \rho)$. For ρ_1 sufficiently small, suitable $\bar{T} < +\infty$ and $\delta' > 0$, for each pair i, j such that $\tilde{V}_G(K_i, K_j) \leq \bar{V}$, and each $x \in (K_i)_{(\rho_2)}$ we want to find a trajectory φ^x with the following two properties:

(1) $\varphi^x(0) = x, \quad I_{\bar{T}}(\varphi^x) \leq \tilde{V}_G(K_i, K_j) + 0.7\gamma$;

(2) $[\varrho_{\bar{T}}(X^\varepsilon, \varphi^x) < \delta'] \subseteq [X^\varepsilon_{\eta_1} \in \partial w_j]$.

For the lower bound at the first line in (5.35) it suffices to check the existence of such a trajectory and to apply the basic lower estimate for large deviations. The main piece for the construction is given by a path φ^{K_i, K_j} that connects K_i to K_j without exiting from $G_{-\delta} \cup \partial(G_{(-\delta)})$, without visiting $\cup_{\ell \neq i,j} K_\ell$, and with rate function less than $\tilde{V}_G(K_i, K_j) + 0.2\gamma$ (using (5.37) and the definition of $\tilde{V}_{G_{(-\delta)}}$). Each of these paths φ^{K_i, K_j} involves a finite time, and since we have finitely many such trajectories, we may assume these times to be uniformly bounded. Now let $\rho_1 \in (0, \rho_2 \wedge \rho/2)$ such that the ρ_1 tube around each of the trajectories φ^{K_i, K_j} stays at a distance at least ρ_1 from $\cup_{\ell \neq i,j} K_\ell$. For each $x \in (K_i)_{(\rho_2)}$ we may define φ^x as follows.

Case $j \neq i$. (a) We first take an interpolation from x to a point $x' \in K_i$ (closest to x); this piece of trajectory stays at distance at least $\rho - \rho_2$ from C and has rate function less than 0.1γ, by Lemma 2.32. (b) Since K_i is a compact equivalence class we can use the remark at the beginning of the proof to connect x' to $\varphi^{K_i, K_j}(0)$ by a trajectory in $(K_i)_{\rho_2}$ and with rate function less than 0.1γ and bounded time. (c) From $\varphi^{K_i, K_j}(0)$ we follow φ^{K_i, K_j}, arriving at K_j as described above. (d) Since the sum of the times in (a), (b) and (c) is uniformly bounded by some $\bar{T} < +\infty$, we complete the trajectory up to time $\bar{T}$ just following the deterministic flow. This satisfies properties (1) and (2) above, for $\delta' < \delta \wedge \rho_1 \wedge (\rho - \rho_2)$.

Case $j = i$. (a) With $\delta' > 0$ as above we take φ^x connecting x to a point x' at distance $\rho + \delta'$ from K_i with rate function less than 0.3γ (using Lemma 2.32). (b) From x' we interpolate again to the closest point in K_i; the rate function of this piece is less than 0.2γ. (c) We may assume that φ^x stays at distance at least

$\delta' + \rho_1$ from $\cup_{\ell \neq i} K_\ell$ and the sum of the times in (a) and (b) is bounded from above by $\bar{T}$. We complete the trajectory up to time $\bar{T}$ with deterministic flow. Again, it is easy to see that for the choice of parameters as indicated, φ^x also has properties (1) and (2) in this case.

For the lower bound at the third line in (5.35), the argument is similar, using the compactness of ∂G to take a finite, δ_0-dense set in ∂G, $\{y_1, \ldots, y_k\}$, with ρ, ρ_2 as above ($\delta_0 \in (0, \rho)$ as in the statement). This allows us to work with a finite set of trajectories $\varphi^{K_i,s}$ connecting K_i to $(y_s)_{(-\delta)}$, without exiting from $G_{(-\delta)} \cup \partial(G_{(-\delta)})$, without visiting $\cup_{\ell \neq i} K_\ell$, and with rate function bounded from above by $\tilde{V}_G(K_i, \partial G) + 0.2\gamma$. The rest of the reasoning is similar with a proper $\delta' > 0$, ρ_1 and $\bar{T}$ adjusted to work for all cases at once (for details see [122]).

As for the upper estimates, and taking ρ as before, we first reduce to initial points $x \in \Gamma_i$, where $\Gamma_i := \partial((K_i)_{(\rho)})$, just using the strong Markov property, as in Section 2.6. Due to (5.37) and the choices of ρ, δ' as before (using Lemma 2.32), we see that any trajectory φ starting at Γ_i, reaching a δ'-neighbourhood of ∂w_j without leaving $(G_{(\delta)} \cup \partial(G_{(\delta)})) \setminus \cup_{\ell \neq i,j} K_\ell$ must have rate function at least $\tilde{V}_G(K_i, K_j) - 0.3\gamma$. Firstly, using Lemma 2.34 we find T_1 so that for all ε sufficiently small and all $x \in (G \cup \partial G) \setminus w$

$$P_x(\eta_1 > T_1) \le e^{-\varepsilon^{-2}\bar{V}}$$

and so, proceeding as in the proof of Theorem 2.40, we can write:

$$P_x(X^\varepsilon_{\eta_1} \in \partial w_j) \le e^{-\varepsilon^{-2}\bar{V}} + P_x\left(\varrho_{T_1}(X^\varepsilon, \mathbb{F}_x(\tilde{V}_G(K_i, K_j) - 0.3\gamma)) \ge \delta'\right).$$

The upper estimate for the conditional probability at the first line of (5.35) then follows from the upper bound in the l.d.p. The argument is similar for that at the third line of (5.35). (See [122], from where this proof is taken.) □

The proof of Lemma 5.12 is omitted; it follows arguments similar to those used for Lemma 5.11. As the proof shows, ρ may be taken arbitrarily small in these two lemmas.

The last two lemmas tell us that, for small ε, the transition probabilities of the Markov chain (Z^ε_n) are approximated, in some sense, by those of a finite valued Markov chain whose state space is the set of symbols $\mathcal{L} = \{K_1, \ldots, K_l, \partial G\}$. This is a key fact in the analysis of Freidlin and Wentzell concerning invariant measures as well as escape times.

In the context of finite state (irreducible) Markov chains (cf. Definition 3.7, Section 3.2), it might be convenient to express the exit time from a subset (as well as the position) in the language of graphs. The expressions in Lemma 4.8 are a very particular case, in the simple situation of birth and death chains.

Definition 5.13 Let $\mathcal{L}$ be a finite set and $\mathcal{Q}$ a non-empty proper subset. We denote by $\mathbb{G}(\mathcal{Q})$ the set of all maps $g : \mathcal{L} \setminus \mathcal{Q} \to \mathcal{L}$ with the property that g maps no non-empty subset of $\mathcal{L} \setminus \mathcal{Q}$ into itself. Equivalently, $\mathbb{G}(\mathcal{Q})$ corresponds to the set of graphs on $\mathcal{L}$ with arrows $i \to j, i \in \mathcal{L} \setminus \mathcal{Q}, j \in \mathcal{L}, j \neq i$, such that: (i) from each $i \in \mathcal{L} \setminus \mathcal{Q}$ exactly one arrow $i \to j$ is issued; (ii) for each $i \in \mathcal{L} \setminus \mathcal{Q}$ there is a chain of arrows starting at i and finishing at some point in $\mathcal{Q}$. If j is such point we say that g leads i to j. The set of graphs in $\mathbb{G}(\mathcal{Q})$ leading i to j is denoted by $\mathbb{G}_{i,j}(\mathcal{Q})$, for $i \in \mathcal{L} \setminus \mathcal{Q}$ and $j \in \mathcal{Q}$. (The set $\mathcal{L}$ is omitted from the above notation.)

Lemma 5.14 *Let $\mathcal{L} = \{1, \ldots, l\}$ and let us consider an irreducible discrete time Markov chain (Y_n) with values in $\mathcal{L}$. $(q(i, j))_{i,j}$ denotes the transition matrix. For $\emptyset \neq \mathcal{Q} \subsetneq \mathcal{L}$ and $g \in \mathbb{G}(\mathcal{Q})$, we set $\pi(g) = \prod_{(i \to j) \in g} q(i, j)$. Let also $\tau(\mathcal{Q}) = \min\{n : Y_n \in \mathcal{Q}\}$.*
Then, for any $i \in \mathcal{L} \setminus \mathcal{Q}$:

$$E(\tau(\mathcal{Q})|Y_0 = i) = \frac{\sum_{g \in \mathbb{G}(\mathcal{Q} \cup \{i\})} \pi(g) + \sum_{j \in \mathcal{L} \setminus \mathcal{Q},\, j \neq i} \sum_{g \in \mathbb{G}_{i,j}(\mathcal{Q} \cup \{j\})} \pi(g)}{\sum_{g \in \mathbb{G}(\mathcal{Q})} \pi(g)}, \tag{5.38}$$

$$P(Y_{\tau(\mathcal{Q})} = j | Y_0 = i) = \frac{\sum_{g \in \mathbb{G}_{i,j}(\mathcal{Q})} \pi(g)}{\sum_{g \in \mathbb{G}(\mathcal{Q})} \pi(g)}, \tag{5.39}$$

with the convention that $\mathbb{G}(\mathcal{L})$ reduces to the empty graph g, with $\pi(g) = 1$.

Remark The irreducibility assumption, cf. Definition 3.7, is equivalent to the existence, for each $j \in \mathcal{L}$, of a graph $g \in \mathbb{G}(\{j\})$ with $\pi(g) > 0$. In particular, the denominators in (5.38) and (5.39) are positive. (To verify the above identities only the positivity for the given $\mathcal{Q}$ is needed.) Under the irreducibility assumption we know that the chain has a unique stationary probability measure ν, given by

$$\nu_i = \frac{1}{Z} \sum_{g \in \mathbb{G}(\{i\})} \pi(g), \quad i \in \mathcal{L}, \tag{5.40}$$

where Z is the normalizing constant. This last identity follows by uniqueness and a direct verification of the balance equation $\nu_j = \sum_i \nu_i q(i, j)$, for each j. Since $\sum_i q(j, i) = 1$, we may write this equation as: $\nu_j \sum_{i \neq j} q(j, i) = \sum_{i \neq j} \nu_i q(i, j)$, for each j. Replacing ν according to (5.40) and expanding, we check it by inspection.

Observe that in the case of birth and death chains there is a tremendous simplification since only very simple graphs g have $\pi(g) \neq 0$; recall (4.78) and (4.79).

The approximation of (Z_n^ε) by a finite state Markov chain was described by Lemmas 5.11 and 5.12: the state space can be partitioned as $S = \cup_{i=1}^{l} \partial w_i \cup \partial G$ and the probability of a transition from a point in ∂w_i to ∂w_j or to ∂G is bounded from below and above, according to (5.35) and (5.36). To treat this chain, we need a suitable extension of Lemma 5.14. This is summarized below, following [122]. Letting $c = 1$ in the next lemma, one recovers Lemma 5.14.

Lemma 5.15 *Let (Y_n) be a Markov chain taking values on $S = \cup_{i=1}^{l} S_i$ with $S_i \cap S_j = \emptyset$ if $i \neq j$, with transition probabilities given by $Q(x, dy)$.*[2] *Let us assume the existence of $c \geq 1$ and non-negative numbers $q(i, j)$ so that*

$$c^{-1} q(i, j) \leq Q(x, S_j) \leq c q(i, j) \tag{5.41}$$

for each $x \in S_i$. For $\mathcal{Q}$ a non-empty proper subset of $\mathcal{L} := \{1, \dots, l\}$ let $\tau(\mathcal{Q}) = \min\{n\colon Y_n \in \cup_{j \in \mathcal{Q}} S_j\}$ and $\tilde{p}_{\mathcal{Q}}(x, S_j) = P(Y_{\tau(\mathcal{Q})} \in S_j | Y_0 = x)$. Let $r = \sharp(\mathcal{L} \setminus \mathcal{Q})$. Then for $x \in S_i$, $i \in \mathcal{L} \setminus \mathcal{Q}$:

$$c^{-4^r} M_{\mathcal{Q}}(i) \leq E(\tau(\mathcal{Q}) | Y_0 = x) \leq c^{4^r} M_{\mathcal{Q}}(i), \tag{5.42}$$

and if $j \in \mathcal{Q}$:

$$c^{-4^r} N_{\mathcal{Q}}(i, j) \leq \tilde{p}_{\mathcal{Q}}(x, S_j) \leq c^{4^r} N_{\mathcal{Q}}(i, j), \tag{5.43}$$

provided $\sum_{g \in \mathbb{G}(\mathcal{Q})} \pi(g)$ is positive, where $\pi(g)$ is as in Lemma 5.14 *and $M_{\mathcal{Q}}(i)$ and $N_{\mathcal{Q}}(i, j)$ are given by the expressions on the right-hand sides of* (5.38) *and* (5.39), *respectively.*

Notation The set of graphs appearing in the numerator of (5.38) will be denoted by $\mathbb{G}(i \nrightarrow \mathcal{Q})$; they have no loops, and if $\mathcal{L} \setminus \mathcal{Q}$ has r points (one of which is i, by assumption) then each of them consists of $r - 1$ arrows $u \to v$ with $u \in \mathcal{L} \setminus \mathcal{Q}$, $v \in \mathcal{L}$, and such that there is no chain of arrows connecting i to $\mathcal{Q}$; those in the first sum have no arrow starting at i, and those in the second sum have a chain starting at i and ending up at $j \in \mathcal{L} \setminus (\mathcal{Q} \cup \{i\})$, without reaching $\mathcal{Q}$. In particular, if $\mathcal{L} \setminus \mathcal{Q} = \{i\}$, the first term corresponds to the empty graph g, for which $\pi(g) = 1$ by convention, and the second term vanishes. In this case the numerator in (5.42) is set equal to one.

Proof To check (5.42) and (5.43) we follow closely [122]. Both inequalities will be proven by induction on $r = \sharp(\mathcal{L} \setminus \mathcal{Q})$.

(a) ***Proof of*** (5.43). When $r = 1$ the estimate refers to the first exit from S_i where $\mathcal{Q} = \mathcal{L} \setminus \{i\}$. We decompose the event so that the exit happens through S_j according to the number of jumps inside of S_i before the first exit. Using this decomposition and the Markov property we immediately get, for any $x \in S_i$:

$$\tilde{p}_{\mathcal{L} \setminus \{i\}}(x, S_j) = Q(x, S_j) + \sum_{n=1}^{\infty} \int_{S_i} Q(x, dy_1) \dots \int_{S_i} Q(y_{n-1}, dy_n) Q(y_n, S_j),$$

[2] That is, $Q(x, A)$ is a version of the regular conditional distribution $P(Y_{n+1} \in A \mid Y_n = x)$.

from which it follows that

$$\frac{\underline{A}_{i,j}}{1-\underline{A}_{i,i}} \le \tilde{p}_{\mathcal{L}\setminus\{i\}}(x, S_j) \le \frac{\bar{A}_{i,j}}{1-\bar{A}_{i,i}}$$

provided we write, for each $i, k \in \mathcal{L}$,

$$\underline{A}_{i,k} = \inf_{y\in S_i} Q(y, S_k), \qquad \bar{A}_{i,k} = \sup_{y\in S_i} Q(y, S_k).$$

By assumption $\frac{1}{c}q(i,k) \le \underline{A}_{i,k} \le \bar{A}_{i,k} \le cq(i,k)$, for each i, k. Moreover, the last assumption with $\mathcal{Q} = \mathcal{L}\setminus\{i\}$ becomes $\sum_{k\neq i} q(i,k) > 0$ from which we see that

$$1-\bar{A}_{i,i} = \inf_{y\in S_i}(1-Q(y,S_i)) = \inf_{y\in S_i}\sum_{k\neq i} Q(y, S_k) \ge \frac{1}{c}\sum_{k\neq i} q(i,k) > 0.$$

Similarly $1-\underline{A}_{i,i} \le c\sum_{k\neq i} q(i,k)$, and we get

$$\frac{1}{c^2}\frac{q(i,j)}{\sum_{k\neq i} q(i,k)} \le \tilde{p}_{\mathcal{L}\setminus\{i\}}(x, S_j) \le c^2\frac{q(i,j)}{\sum_{k\neq i} q(i,k)},$$

verifying (5.43) in this case (with c^2 instead of c^4).

We now proceed with the induction step. For this, let $\emptyset \neq \mathcal{Q} \subsetneq \mathcal{L}$ be such that $\sharp(\mathcal{L}\setminus\mathcal{Q}) = r+1$ and $\sum_{g\in\mathbb{G}(\mathcal{Q})}\pi(g) > 0$, having assumed the statement to be true for all subsets $\tilde{\mathcal{Q}}$ of $\mathcal{L}$ such that $\sum_{g\in\mathbb{G}(\tilde{\mathcal{Q}})}\pi(g) > 0$ and $1 \le \sharp(\mathcal{L}\setminus\tilde{\mathcal{Q}}) \le r$. Then let $i \in \mathcal{L}\setminus\mathcal{Q}$ and $j \in \mathcal{Q}$, and set $\tilde{\mathcal{Q}} = \mathcal{Q}\cup\{i\}$, $F = \cup_{k\in\mathcal{L}\setminus\tilde{\mathcal{Q}}} S_k$. From the induction hypothesis we have the estimates for $\tilde{p}_{\tilde{\mathcal{Q}}}(y,.)$, $y \in F$, and $\tilde{p}_{\mathcal{L}\setminus\{i\}}(x,.)$, $x \in S_i$. Thus, we decompose the event of interest $[Y_{\tau(\mathcal{Q})} \in S_j]$ according to the number of transitions from S_i to F before hitting $\mathcal{Q}$; for any given number of such transitions there are two ways for the process to reach $\cup_{k\in\mathcal{Q}} S_k$ through S_j: from F to S_j, or from F back to S_i and then to S_j. Using this decomposition and the strong Markov property we get:

$$\begin{aligned}\tilde{p}_{\mathcal{Q}}(x, S_j) = A(x) + \sum_{n\ge 1}\int_F \tilde{p}_{\mathcal{L}\setminus\{i\}}(x, dy_1)\int_{S_i}\tilde{p}_{\tilde{\mathcal{Q}}}(y_1, d\tilde{y}_1)\cdots \\ \times \int_F \tilde{p}_{\mathcal{L}\setminus\{i\}}(\tilde{y}_{n-1}, dy_n)\int_{S_i}\tilde{p}_{\tilde{\mathcal{Q}}}(y_n, d\tilde{y}_n)A(\tilde{y}_n),\end{aligned} \tag{5.44}$$

where

$$A(x) = \tilde{p}_{\mathcal{L}\setminus\{i\}}(x, S_j) + \int_F \tilde{p}_{\mathcal{L}\setminus\{i\}}(x, dy)\,\tilde{p}_{\tilde{\mathcal{Q}}}(y, S_j).$$

But, for any $x \in S_i$:

$$\underline{C}_{i,j} + \underline{D}_{i,j} \le A(x) \le \bar{C}_{i,j} + \bar{D}_{i,j},$$

where

$$\underline{C}_{i,j} = \inf_{x \in S_i} \tilde{p}_{\mathcal{L}\setminus\{i\}}(x, S_j), \qquad \bar{C}_{i,j} = \sup_{x \in S_i} \tilde{p}_{\mathcal{L}\setminus\{i\}}(x, S_j)$$

and

$$\underline{D}_{i,k} = \inf_{x \in S_i} \int_F \tilde{p}_{\mathcal{L}\setminus\{i\}}(x, dy)\tilde{p}_{\tilde{\mathcal{Q}}}(y, S_k), \quad \bar{D}_{i,k} = \sup_{x \in S_i} \int_F \tilde{p}_{\mathcal{L}\setminus\{i\}}(x, dy)\tilde{p}_{\tilde{\mathcal{Q}}}(y, S_k).$$

From (5.44) we then have, for each $x \in S_i$

$$\frac{\underline{C}_{i,j} + \underline{D}_{i,j}}{1 - \underline{D}_{i,i}} \le \tilde{p}_{\mathcal{Q}}(x, S_j) \le \frac{\bar{C}_{i,j} + \bar{D}_{i,j}}{1 - \bar{D}_{i,i}}.$$

The case $r = 1$ gives us:

$$c^{-4} \frac{q(i, j)}{\sum_{k \neq i} q(i, k)} \le \underline{C}_{i,j} \le \bar{C}_{i,j} \le c^4 \frac{q(i, j)}{\sum_{k \neq i} q(i, k)}. \tag{5.45}$$

Decomposing the integral in F to estimate $\bar{D}_{i,j}$, $\underline{D}_{i,j}$, we get, using the induction hypothesis again (the cases $\sharp(\mathcal{L} \setminus \mathcal{Q}) = 1$ or $\sharp(\mathcal{L} \setminus \mathcal{Q}) = r$):

$$\bar{D}_{i,j} \le \sum_{k \in \mathcal{L} \setminus \tilde{\mathcal{Q}}} c^4 \frac{q(i, k)}{\sum_{s \neq i} q(i, s)} c^{4^r} \frac{\sum_{g \in \mathbb{G}_{k,j}(\tilde{\mathcal{Q}})} \pi(g)}{\sum_{g \in \mathbb{G}(\tilde{\mathcal{Q}})} \pi(g)} \tag{5.46}$$

and the analogous lower bound for $\underline{D}_{i,j}$. Combining $q(i, k)$ and $\pi(g)$ in the numerator in the right-hand side of the previous equation we have a sum of $\pi(\tilde{g})$ on those graphs $\tilde{g} \in \mathbb{G}_{i,j}(\mathcal{Q})$ not containing the arrow $i \to j$. These last ones will be recovered when we add with $\bar{C}_{i,j}$, which will bring in the term $q(i, j) \sum_{g \in \mathbb{G}(\tilde{\mathcal{Q}})} \pi(g)$, after the reduction to the same denominator. Adding the upper bounds in (5.45) and (5.46) we reconstruct the sum over all $\mathbb{G}_{i,j}(\mathcal{Q})$, getting:

$$\bar{C}_{i,j} + \bar{D}_{i,j} \le c^{4+4^r} \frac{\sum_{g \in \mathbb{G}_{i,j}(\mathcal{Q})} \pi(g)}{\sum_{k \neq i} q(i, k) \sum_{g \in \mathbb{G}(\tilde{\mathcal{Q}})} \pi(g)} \tag{5.47}$$

and the analogous lower bound for $\underline{C}_{i,j} + \underline{D}_{i,j}$, with c^{4+4^r} replaced by c^{-4-4^r}. Similarly we estimate $1 - \bar{D}_{i,i}$ and $1 - \underline{D}_{i,i}$. As in the case $r = 1$, at this point we need the assumption that the denominator in (5.39) is positive, which implies that $\tau(\mathcal{Q})$ is finite with probability one (and a fortiori $\tau(\tilde{\mathcal{Q}}) < \infty$ a.s.). As a consequence:

$$
\begin{aligned}
1 - \bar{D}_{i,i} \\
&= \inf_{x \in S_i} \left(1 - \int_F \tilde{p}_{\mathcal{L}\setminus\{i\}}(x, dy)(1 - \tilde{p}_{\tilde{\mathcal{Q}}}(y, \cup_{k \in \mathcal{Q}} S_k) \right) \\
&= \inf_{x \in S_i} \left(\tilde{p}_{\mathcal{L}\setminus\{i\}}(x, \cup_{k \in \mathcal{Q}} S_k) + \int_F \tilde{p}_{\mathcal{L}\setminus\{i\}}(x, dy) \tilde{p}_{\tilde{\mathcal{Q}}}(y, \cup_{k \in \mathcal{Q}} S_k) \right) \\
&\geq c^{-4} \frac{\sum_{k \in \mathcal{Q}} q(i,k)}{\sum_{s \neq i} q(i,s)} + c^{-4} \frac{\sum_{s \notin \tilde{\mathcal{Q}}} q(i,s) \sum_{k \in \mathcal{Q}} c^{-4^r} \sum_{g \in \mathbb{G}_{s,k}(\tilde{\mathcal{Q}})} \pi(g)}{\sum_{s \neq i} q(i,s) \sum_{g \in \mathbb{G}(\tilde{\mathcal{Q}})} \pi(g)} \\
&\geq c^{-4-4^r} \frac{\sum_{k \in \mathcal{Q}} q(i,k) \sum_{g \in \mathbb{G}(\tilde{\mathcal{Q}})} \pi(g) + \sum_{s \notin \tilde{\mathcal{Q}}} q(i,s) \sum_{k \in \mathcal{Q}} \sum_{g \in \mathbb{G}_{s,k}(\tilde{\mathcal{Q}})} \pi(g)}{\sum_{k \neq i} q(i,k) \sum_{g \in \mathbb{G}(\tilde{\mathcal{Q}})} \pi(g)} \\
&= c^{-4-4^r} \frac{\sum_{g \in \mathbb{G}(\mathcal{Q})} \pi(g)}{\sum_{k \neq i} q(i,k) \sum_{g \in \mathbb{G}(\tilde{\mathcal{Q}})} \pi(g)}.
\end{aligned}
\tag{5.48}
$$

Putting together (5.47) and (5.48) we get:

$$
\frac{\bar{C}_{i,j} + \bar{D}_{i,j}}{1 - \bar{D}_{i,i}} \leq c^{2(4+4^r)} \frac{\sum_{g \in \mathbb{G}_{i,j}(\mathcal{Q})} \pi(g)}{\sum_{g \in \mathbb{G}(\mathcal{Q})} \pi(g)}
$$

and since $2(4 + 4^r) \leq 4^{r+1}$ we obtain the upper bound in the induction step. The lower bound is analogous, concluding the verification of (5.43).

(b) *Proof of* (5.42). We write $E_x(\cdot) = E(\cdot \mid Y_0 = x)$. The case $r = 1$, i.e. $\mathcal{Q} = \mathcal{L} \setminus \{i\}$ for some i, corresponds to the expected exit time from S_i, which is finite a.s., under the assumption. As in the proof of (5.43), and if $x \in S_i$, we have

$$
\begin{aligned}
E_x \tau(\mathcal{Q}) &= \sum_{n \geq 1} P(\tau(\mathcal{Q}) \geq n | Y_0 = x) \\
&= 1 + Q(x, S_i) + \sum_{n=1}^{\infty} \int_{S_i} Q(x, dy_1) \ldots \int_{S_i} Q(y_{n-1}, dy_n) Q(y_n, S_i),
\end{aligned}
$$

so that $1/(1 - \underline{A}_{i,i}) \leq E_x \tau(\mathcal{Q}) \leq 1/(1 - \bar{A}_{i,i})$, with $\underline{A}_{i,i}$ and $\bar{A}_{i,i}$ defined in the proof of (5.43). Inequalities (5.42) in the case $r = 1$ follow at once from the previous estimate. For the induction step, and using the same notation as in the proof of (5.43), we write:

$$
\begin{aligned}
E_x \tau(\mathcal{Q}) = R(x) + \sum_{n \geq 1} \int_F \tilde{p}_{\mathcal{L}\setminus\{i\}}(x, dy_1) \int_{S_i} \tilde{p}_{\tilde{\mathcal{Q}}}(y_1, d\tilde{y}_1) \cdots \\
\times \int_F \tilde{p}_{\mathcal{L}\setminus\{i\}}(\tilde{y}_{n-1}, dy_n) \int_{S_i} \tilde{p}_{\tilde{\mathcal{Q}}}(y_n, d\tilde{y}_n) R(\tilde{y}_n),
\end{aligned}
\tag{5.49}
$$

where

$$R(x) = E_x\, \tau(\mathcal{L} \setminus \{i\}) + \int_F \tilde{p}_{\mathcal{L}\setminus\{i\}}(x, dy)\, E_y\, \tau(\tilde{\mathcal{Q}}).$$

As before we get, for any $x \in S_i$, $i \notin \mathcal{Q}$:

$$\frac{\underline{G}_i + \underline{H}_i}{1 - \underline{D}_{i,i}} \leq E_x\, \tau(\mathcal{Q}) \leq \frac{\bar{G}_i + \bar{H}_i}{1 - \bar{D}_{i,i}}, \tag{5.50}$$

where

$$\underline{G}_i = \inf_{x \in S_i} E_x\, \tau(\mathcal{L} \setminus \{i\}), \qquad \bar{G}_i = \sup_{x \in S_i} E_x\, \tau(\mathcal{L} \setminus \{i\}),$$

$$\underline{H}_i = \inf_{x \in S_i} \int_F \tilde{p}_{\mathcal{L}\setminus\{i\}}(x, dy) E_y\, \tau(\tilde{\mathcal{Q}}), \qquad \bar{H}_i = \sup_{x \in S_i} \int_F \tilde{p}_{\mathcal{L}\setminus\{i\}}(x, dy) E_y\, \tau(\tilde{\mathcal{Q}}).$$

The case $r = 1$ gives us

$$c^{-4^r} \frac{1}{\sum_{k \neq i} q(i,k)} \leq \underline{G}_i \leq \bar{G}_i \leq c^{4^r} \frac{1}{\sum_{k \neq i} q(i,k)}.$$

The denominators in (5.50) have already been estimated in the previous proof. Thus, it remains to estimate $\underline{H}_i$, $\bar{H}_i$. We break the integral over F into a sum on the various S_k, $k \in \mathcal{L} \setminus \tilde{\mathcal{Q}}$, use (5.43) for $\tilde{p}_{\mathcal{L}\setminus\{i\}}(x, S_k)$ and the induction hypothesis to estimate $E_y(\tau(\tilde{\mathcal{Q}}))$, uniformly for $y \in S_k$. The procedure is quite similar to the previous proof and with some care we reconstruct the bounds in (5.42) with r replaced by $r + 1$. (For full details see Chapter 6 of [122].) □

Having recalled Freidlin and Wentzell estimates for the discrete time chain (Z_n^ε), one still needs to know how much time the continuous process X^ε spends in the corresponding w_i, during $[\eta_n, \sigma_n)$. In terms of the time scales involved here, this is simple if we take into account only the logarithm equivalence. It is described in Lemma 1.7 and Lemma 1.8, Chapter 6 of [122], which we recall briefly.

Lemma 5.16 (Freidlin and Wentzell) *(a) Let K be a compact set and N a given neigbourhood of K. For each $\gamma > 0$ there exists $\rho > 0$ so that for all $\varepsilon > 0$ small enough, we have*

$$\inf_{x \in \overline{K_{(\rho/2)}}} E_x \int_0^{\tau_\varepsilon(\mathbb{R}^d \setminus N)} \mathbf{1}_{K_{(\rho)}}(X_t^\varepsilon) dt > e^{-\gamma/\varepsilon^2}, \tag{5.51}$$

with $\overline{A}$ denoting the closure of A.

(b) Let K be a compact set contained in some equivalence class. For each $\gamma > 0$ there exists $\rho > 0$ so that for all $\varepsilon > 0$ small enough, we have

$$\sup_{x \in K_{(\rho)}} E_x \tau_\varepsilon(\mathbb{R}^d \setminus K_{(\rho)}) < e^{\gamma/\varepsilon^2}. \tag{5.52}$$

Sketch of the proof (a) Since K is compact and the deterministic flow is continuous, there exists $t_0 = t_0(\rho) > 0$ so that the orbits of the deterministic system starting from points in K remain in $K_{(2\rho/3)}$ for a time at least t_0. Moreover, ρ can be taken sufficiently small so that $K_{(\rho)} \subseteq N$, and such that any point $x \in \overline{K_{(\rho/2)}}$ can be connected to its closest point x' in K with rate function at most $\gamma/2$, without exiting $K_{(2\rho/3)}$, and in a (uniformly) bounded time $\bar{t}$. Let φ^x be a trajectory that makes this connection and then follows the deterministic flow. Then

$$P_x\left(\tau_\varepsilon(\mathbb{R}^d \setminus K_{(\rho)}) > t_0\right) \geq P_x\left(\varrho_{t_0}(X^\varepsilon, \varphi^x) < \rho/3\right)$$

and the conclusion follows from $(b)'_u$ of Corollary 2.26.

(b) This follows from Proposition 5.7 and the basic large deviation estimate, together with the Markov property. Indeed, let $\rho > 0$ be chosen small enough and let y be a point at distance 2ρ from the set K. If $x \in K_{(\rho)}$ we use linear interpolation to connect x to its closest point $x' \in K$, then we use item (ii) of Proposition 5.7 to connect x' to the point y' in K closest to y; from y' we use linear interpolation to connect to y. (When the closest point is not unique, we just make an arbitrary choice.) Let φ^x be the constructed trajectory. By taking ρ small enough the rate function of φ^x can be assumed to be less than $\gamma/2$ and the total time to be uniformly (in x) bounded from above by some $\bar{T}_0 < +\infty$. From $(b)'_u$ of Corollary 2.26 we then conclude that for $\varepsilon > 0$ small enough

$$\inf_{x \in K_{(\rho)}} P_x\left(\tau_\varepsilon(\mathbb{R}^d \setminus K_{(\rho)}) < \bar{T}_0\right) \geq e^{-\frac{2}{3}\gamma\varepsilon^{-2}}.$$

Using the Markov property, as in Section 2.6, we have

$$\sup_{x \in K_{(\rho)}} P_x\left(\tau_\varepsilon(\mathbb{R}^d \setminus K_{(\rho)}) \geq n\bar{T}_0\right) \leq (1 - e^{-\frac{2}{3}\gamma\varepsilon^{-2}})^n,$$

from which it follows that

$$E_x \tau_\varepsilon(\mathbb{R}^d \setminus K_{(\rho)}) \leq \bar{T}_0 e^{\frac{2}{3}\gamma\varepsilon^{-2}},$$

concluding the proof. □

We now have the basic ingredients for the estimates of the order of magnitude of $E_x\, \tau_\varepsilon(\mathbb{R}^d \setminus G)$, under Assumptions 5.9. First let us make a few simple observations.

Notation Since the quasi-potential $V_G(x, y)$ is constant on $K_i \times K_j$, let $V_G(K_i, K_j)$ denote the common value; similarly for $V_G(K_i, x)$ and $V_G(x, K_i)$. Also $V_G(K_i, \partial G) = \inf_{y \in \partial G} V_G(K_i, y)$.

Proposition 5.17
(i) *For each* $i \neq j$

$$V_G(K_i, K_j) = \min(\tilde{V}_G(K_i, K_{i_2}) + \cdots + \tilde{V}_G(K_{i_{k-1}}, K_j)), \qquad (5.53)$$

with the minimum taken among all finite sequences $(i_1, \dots, i_k)$ *in* $\{1, \dots, l\}$ *with* $i_1 = i$ *and* $i_k = j$. *(Clearly, in (5.53) it is enough to consider sequences with* $i_1, \dots, i_k$ *all distinct. In particular,* $2 \le k \le l$.*) An analogous observation applies to* $V_G(K_i, x)$ *and* $V_G(x, K_i)$.

(ii) If K_i *is an unstable class, there is a stable class* K_j *such that* $V_G(K_i, K_j) = 0$ *or* $V_G(K_i, \partial G) = 0$.

(iii) If $x \in G \setminus \cup_{i=1}^{l} K_i$, *there is a stable class* K_j *such that* $V_G(x, K_j) = 0$ *or* $V_G(x, \partial G) = 0$ *(ii) and (iii) have proper extensions if* $G = \mathbb{R}^d$).

Proof (i) Simple.

(ii) Though simple, it requires some verification: since K_i is unstable, we might find a point $y \in G \setminus K_i$ that can be reached from K_i through trajectories with arbitrarily small rate function. Starting now from the point y we are able to arrive at some other compact K_j (without exiting from G) or at ∂G, still with arbitrarily small rate function since, due to Assumptions 5.9, we can follow the deterministic flow until very close to some K_j or ∂G and then we make a small interpolation, as previously discussed. Since $y \notin K_i$ it is easy to see that $j \neq i$, and $V_G(K_i, K_j) = 0$. If K_j is stable we are done, if not we continue. Since we have finitely many classes, cf. Assumptions 5.9, the procedure will stop. The argument for (iii) is the same. □

Applying the results which we have summarized in the previous sequence of lemmas, we may now see how they lead to the evaluation of $\lim_{\varepsilon \to 0} \varepsilon^2 \log E_x \, \tau_\varepsilon(\mathbb{R}^d \setminus G)$. The important quantities are those entering (5.42) for the chain (Z_n^ε) defined right after (5.34) for suitably small ρ, ρ_1. In the denominator of (5.42) the crucial role is played by:

$$M_G := \min_{g \in \mathbb{G}(\partial G)} \sum_{(\alpha \to \beta) \in g} \tilde{V}_G(\alpha, \beta), \tag{5.54}$$

where we are considering graphs on the finite set of symbols $\{K_1, \dots, K_l, \partial G\}$. Due to the observation made in (ii) of Proposition 5.17 it is not hard to see that

$$M_G = \min_{g \in \mathbb{G}(\partial G)} \sum_{(\alpha \to \beta) \in g} V_G(\alpha, \beta). \tag{5.55}$$

In (5.55) the same minimum is obtained when we take $\{\partial G\} \cup \{K_i : K_i \text{ is stable}\}$ as the set of symbols. Some details are still missing, but we omit them (see Lemmas 5.1, 4.1 and 4.3, Chapter 6 of [122]).

As for the numerator, let us consider the graphs on the set $\{K_1, \dots K_l, x, \partial G\}$ and let

$$M_G(x) = \min_{g \in \mathbb{G}(x \nrightarrow \partial G)} \sum_{(\alpha \to \beta) \in g} \tilde{V}_G(\alpha, \beta), \tag{5.56}$$

cf. notational remark that follows Lemma 5.15. As before, we may replace $\tilde{V}_G(\alpha, \beta)$ by $V_G(\alpha, \beta)$ in (5.56), and in this case we may restrict the argument to the graphs on the subset obtained by eliminating all the unstable classes.

We now state and quickly review the following basic result.

Theorem 5.18 (Freidlin and Wentzell) *Under Assumptions* 5.9,

$$\lim_{\varepsilon \to 0} \varepsilon^2 \log E_x \, \tau_\varepsilon(\mathbb{R}^d \setminus G) = M_G - M_G(x), \tag{5.57}$$

for any $x \in G$. The convergence is uniform on each compact contained in G.

Proof (Sketch) Given $\gamma > 0$ we choose the parameters ρ, ρ_2, ρ_1 as in Lemmas 5.11, 5.12. We may assume that ρ is small enough so that the expected exit time from the ρ-neighbourhood of K_i is bounded from above by e^{γ/ε^2}, according to Lemma 5.16, part (b). We then consider the Markov chain (Z_n^ε) defined right after (5.34), up to its first exit from G. By an extension of item (b) of Lemma 2.34, we control the typical duration of the time intervals $[\sigma_n, \eta_{n+1})$ and we see that the important contribution comes from the expected number of jumps of the discrete time Markov chain (Z_n^ε) until it reaches ∂G. Estimating this according to Lemma 5.15 and the last observations leads to the conclusion. □

With respect to the asymptotic distribution of $X^\varepsilon_{\tau_\varepsilon(\mathbb{R}^d \setminus G)}$, we simply state the basic result within the theory of Freidlin and Wentzell, under the same Assumptions 5.9. For this description one may use a proper combination of Lemmas 5.11, 5.12 and inequalities (5.43) in Lemma 5.15, with the same type of arguments already used, but the variational problem is now more detailed than that involved in Theorem 5.18. Besides the quantity M_G defined in (5.54), one needs to define, for graphs on the set of symbols $\{K_1, \ldots, K_l, x, y, \partial G\}$ with $x \in G$ and $y \in \partial G$:

$$M_G(x, y) := \min_{g \in \mathbb{G}_{x,y}(\{y, \partial G\})} \sum_{(\alpha \to \beta) \in g} \tilde{V}_G(\alpha, \beta). \tag{5.58}$$

As before, the minimum is the same if the quantities $\tilde{V}_G(\alpha, \beta)$ are replaced by $V_G(\alpha, \beta)$. It can be proven that $\min_{y \in \partial G} M_G(x, y)$ does not depend on $x \in G$ and coincides with M_G.

A basic result of Freidlin and Wentzell gives the following information on $X^\varepsilon_{\tau_\varepsilon(\mathbb{R}^d \setminus G)}$.

Theorem 5.19 (Freidlin and Wentzell) *Under the same assumptions of Theorem* 5.18, *for any compact set $C \subseteq G$, $\gamma > 0$ and $\delta > 0$, there exist $\varepsilon_0 > 0$ and $\delta_0 \in (0, \delta)$ so that for any $x \in C$, $y \in \partial G$ and $\varepsilon \in (0, \varepsilon_0)$, one has:*

$$\begin{aligned} \exp\{-\varepsilon^{-2}(M_G(x, y) - M_G + \gamma)\} &\le P_x(X^\varepsilon_{\tau_\varepsilon(\mathbb{R}^d \setminus G)} \in B_{\delta_0}(y)) \\ &\le \exp\{-\varepsilon^{-2}(M_G(x, y) - M_G - \gamma)\}, \end{aligned} \tag{5.59}$$

with $M_G(x, y)$ and M_G defined by (5.58) *and* (5.55), *respectively.*

The previous theorem provides a suitable extension of Theorem 2.29. We shall not show the proof here and refer to Theorem 5.1, Chapter 6 of [122]. It tells us that in order to describe the exit place from G, starting sufficiently close to K_i (or at a point x for which the deterministic system comes to K_i without exiting from G) we need to identify the chains leading from K_i to ∂G on the graphs reaching the minimum in (5.58), and to consider each last step $K_j \to \partial G$. The escape will happen, typically, near the set of points in ∂G where $V_G(K_j, \cdot)$ is minimized, for one such j. We treat the analogous problem in the context of Chapter 6.

Recall that in the context of Section 2.6, the expectation of escape time is logarithmically equivalent to any of its quantiles, cf. Theorem 2.40, consistent with the asymptotic exponentiality in Section 5.2. It is easy to see that this is *no longer* true under the conditions of the previous theorem. This behaviour is closely related to the fact that the asymptotic value in (5.57) might depend on the initial point. To see this in a very simple situation, we might consider a one-dimensional example, with $b(x) = -(d/dx)U(x)$, where U is a smooth double well potential on $\mathbb{R}$, i.e. it has three critical points x_i, $i = 1, 2, 3$, with x_1, x_3 being local minima and x_2 a local maximum.

Let $G = (a, c)$ with $a < x_1 < x_2 < x_3 < c$. The previous conditions are satisfied with $K_i = \{x_i\}$ for $i = 1, 2, 3$. x_1 and x_3 are stable. We see immediately that $V_G(x, y) = V(x, y)$ for each $x, y \in G$. If $U(c) - U(x_3) = 1$, $U(x_2) - U(x_3) = 2$, $U(x_2) - U(x_1) = 6$, and $U(a) - U(x_1) = 5$ we see that $V(x_3, \partial G) = V(x_3, c) = 2$, $V(x_1, \partial G) = V(x_1, a) = 10$, $V(x_1, x_2) = 12$, and $V(x_3, x_2) = 4$. From Lemma 5.12 it follows at once that if the process starts to the right of x_2 it will, with overwhelming probability, exit G through c, and in a time of order approximately (logarithmic equivalence) e^{2/ε^2}. But, with probability of order approximately e^{-2/ε^2} it will arrive at x_1 before exiting from G, and then the typical time to exit will be of order at least e^{10/ε^2}. We see that the expected exit time, starting from x_3 is much greater than e^{2/ε^2}. We might easily compute $M_G = 12$, $M_G(x_1) = 2$, and $M_G(x_3) = 4$. With this trivial example we simply point out the difference with respect to the situation in Section 5.2.

Situations such as the previous example tell us that one cannot expect asymptotic unpredictability of the exit time to hold generally under Assumptions 5.9. The concept of cycles plays an important role.

5.4 Cycles and asymptotic unpredictability of the escape time

Our main purpose in this section is the extension of the results of Section 5.2, concerning the asymptotic exponentiality of escape times. A certain 'cycle property' plays a crucial role in this, here and in the context of the next chapter.

Basically, a cycle may be thought of as a non-empty subset $\mathcal{C}$ of the set of stable classes with the property that if the process starts near any of these classes, then it typically visits arbitrarily small neighbourhoods of any other class in $\mathcal{C}$ before arriving at those classes outside $\mathcal{C}$. (In the previous example $\{x_1, x_3\}$ is *not* a cycle!) For the moment, to keep notational consistency with the previous discussion, let G be a bounded domain containing the classes in $\mathcal{C}$ and phrase the cycle property with respect to the escape from G.

We now set the following conditions.

Assumptions 5.20

(b1) G is a bounded domain with boundary ∂G of class C^2.

(b2) In G there are finitely many compact sets $K_1, \ldots, K_l$, equivalence classes for $\sim$, such that if the ω-limit set of a point (for the deterministic system $\dot{x} = b(x)$) is entirely contained in $G \cup \partial G$, then it must be contained in some K_i, $i = 1, \ldots, l$. We also assume that ∂G does not intersect any ω-limit set for the deterministic system.

(b3) For some $1 \le r < l$, let $K_1, \ldots, K_r$ denote the stable classes (among $K_1, \ldots, K_l$). Suppose that each of them consists of a fixed point of the deterministic system, denoted by x_i, $i = 1, \ldots, r$. At least one of these critical points is supposed to be hyperbolic. Namely, we assume the existence of $i_0 \in \{1, \ldots, r\}$ such that all the eigenvalues of the Jacobian matrix of b at the point x_{i_0} have negative real part.

(b4) Setting $V = \max_{i,j \le r} V(K_i, K_j)$ and $\bar{V}_G = \min_{1 \le i \le r} \min_{y \in \partial G} V(K_i, y)$, we assume that $V < \bar{V}_G$ (the 'cycle' property).

From the stability (and since G is open) we may choose $\delta > 0$ so that $\bar{B}_\delta(x_i)$, the closed unit ball with centre x_i and radius δ, is contained in the basin of attraction of x_i, as well as in G, for $i = 1, \ldots, r$. We may assume δ to be such that all positive orbits of the deterministic system $\dot{x} = b(x)$ starting in this ball do not leave G, i.e. they converge to x_i without exiting G. Then let $\tau_\varepsilon = \tau_\varepsilon(\mathbb{R}^d \setminus G)$,

$$D_i = \bar{B}_\delta(x_i), \qquad D = \cup_{i=1}^r D_i. \tag{5.60}$$

Using the notation just introduced we may state and prove the following result.

Theorem 5.21 *Under Assumptions* 5.20 *we may define* β_ε *through the relation*

$$P_{x_{i_0}}(\tau_\varepsilon > \beta_\varepsilon) = e^{-1}. \tag{5.61}$$

Then:

(*i*) $\lim_{\varepsilon \to 0} P_x(\tau_\varepsilon > t\beta_\varepsilon) = e^{-t}$, *for each* $x \in D$ *and each* $t > 0$.

(*ii*) $\lim_{\varepsilon \to 0} E_x\, \tau_\varepsilon / \beta_\varepsilon = 1$, *for each* $x \in D$.

If G *is confining, in the sense that* $\langle b(y), n(y) \rangle < 0$ *for each* $y \in \partial G$, *where* $n(y)$ *indicates the outward unit normal vector to* ∂G *at the point* y, *then* (*i*) *and* (*ii*) *hold for any* $x \in G$.

The context of the previous theorem is very similar to that treated in [210]. The proof that we now discuss is based on the coupling method presented in

Section 5.1, together with the estimates discussed in Section 5.3, and it appears in [34]. Under the above assumptions, the results of Freidlin and Wentzell give us that

$$\liminf_{\varepsilon\to 0} \varepsilon^2 \log \beta_\varepsilon \geq \bar{V}_G, \tag{5.62}$$

and that for any $1 \leq i, j \leq r$, $h > 0$ and $x \in D_i$

$$\lim_{\varepsilon\to 0} P_x\left(\tau_\varepsilon(D_j) \leq e^{(V+h)/\varepsilon^2}\right) = 1, \tag{5.63}$$

where $\tau_\varepsilon(A)$ denotes the first hitting time to the set A, as before. Indeed, (5.62) follows from Lemmas 5.11, 5.15 and 5.16. Similarly for (5.63), which may also be seen as a consequence of Theorem 5.18 applied to $G \setminus D_j$.

Comparing (5.62) and (5.63) we immediately understand the 'cycle property' stated as the last condition (b4). They also suggest that any intermediate scale e^{a/ε^2} with $V < a < \bar{V}_G$ should be long enough to allow thermalization, but much shorter than the scale on which the escape from G happens. Nevertheless, for complete proof of the loss of memory described by the above theorem, we need more precise control, and this will be achieved with the help of coupling methods. Recalling Lemma 4.34 in Section 4.4, we see that the basic point is given by the next lemma.

Lemma 5.22 *There exists* $\delta_0 > 0$ *such that*

$$\lim_{\varepsilon\to 0} \sup_{x\in\bar{B}_{\delta_0}(x_{i_0})} \sup_{t\geq\varepsilon^{-3}/\beta_\varepsilon} |P_x(\tau_\varepsilon > t\beta_\varepsilon) - P_{x_{i_0}}(\tau_\varepsilon > t\beta_\varepsilon)| = 0. \tag{5.64}$$

As already announced, the basic idea is to present a joint realization of the two processes $X^{x,\varepsilon}$, $X^{x_{i_0},\varepsilon}$ in such a way that with overwhelming probability they should meet in a time much shorter than β_ε. Using the method of [202], we now need to work a little more than at the end of Section 5.1, since condition (5.9) might not hold, even if we restrict the processes to arbitrarily small neighbourhoods of x_{i_0}. Indeed, (5.9) tells us that the distance $|X^{u,0}(t) - X^{v,0}(t)|$ between the corresponding points in different deterministic orbits decreases with the time t. An example of a linear vector field $b(\cdot)$ in $\mathbb{R}^2$ not satisfying (5.9) is that given by the matrix

$$\begin{pmatrix} a & 1 \\ 0 & a \end{pmatrix}$$

for $a < 0$ with $|a|$ small, which has real eigenvalues.

For the suitable modification of that proof, let $J = Db(x_{i_0})$ be the Jacobian matrix of b at x_{i_0}. Since all eigenvalues of the matrix J have negative real part, it is well known that there exists a unique symmetric, positive definite matrix Λ such that

$$J^{\mathbb{T}}\Lambda + \Lambda J = -\mathbb{I} \tag{5.65}$$

where $J^{\mathbb{T}}$ denotes the transpose of J and $\mathbb{I}$ denotes the $d \times d$ identity matrix. Λ is called the Lyapunov matrix of J (see e.g. [13] p. 307 for a proof of this fact).

Let us then define the norm ν on $\mathbb{R}^d$ by

$$\nu(x) = (\langle x, \Lambda x \rangle)^{1/2}. \tag{5.66}$$

Since the vector field b has been assumed to be of class C^1, by the standard mean value theorem, if $x, x' \in B_{\delta_1}(x_{i_0})$, with $\delta_1 > 0$ small, we can write:

$$b(x) - b(x') = A(x, x')(x - x'),$$

where $A = A(x, x')$ is a $d \times d$ matrix with entries $(\partial b^u / \partial x_s)(y(u, s))$, $1 \leq u, s \leq d$, for suitable points $y(u, s)$ such that $\max_{1 \leq j \leq d} |y_j(u, s) - x_{i_0, j}| \leq \delta_1$. In particular, $y(u, s) \in \bar{B}_{(d)^{1/2}\delta_1}(x_{i_0})$. Since b is of class C^1, using (5.65) and the symmetry of Λ we can take $\delta_1 \in (0, \delta)$ small enough so that

$$\begin{aligned} \langle b(x) - b(x'), \Lambda(x - x') \rangle &= \langle A(x, x')(x - x'), \Lambda(x - x') \rangle \\ &= \frac{1}{2} \langle (A^{\mathbb{T}} \Lambda + \Lambda A)(x - x'), (x - x') \rangle \\ &\leq -\frac{1}{4} |x - x'|^2 \leq 0, \end{aligned} \tag{5.67}$$

for all $x, x' \in B_{\delta_1}(x_{i_0})$.

Proof of Lemma 5.22 Let δ_1 be chosen such that (5.67) is valid. We may now pick $\delta_0 \in (0, \delta_1)$ so that all positive orbits of the deterministic system issued from a point in $\bar{B}_{\delta_0}(x_{i_0})$ converge to x_{i_0} without leaving $\bar{B}_{\delta_1}(x_{i_0})$. For the coupling defined by (5.5) with $x' = x_{i_0}$ and the map H given by (5.3), we prove that with large probability the trajectories $X^{x,\varepsilon}$ and $X^{x_{i_0},\varepsilon}$ will meet before leaving $\bar{B}_{\delta_1}(x_{i_0})$ and within a time of order ε^{-3}. The uniformity over x in $\bar{B}_{\delta_0}(x_{i_0})$ in (5.64) follows from the proof.

Recalling (5.6) let us consider the one-dimensional process

$$Y_t^{\varepsilon} := \nu(X_t^{x,\varepsilon} - X_t^{x_{i_0},\varepsilon}),$$

where ν is given by (5.66). Applying (extended) Itô's formula, cf. (2.35), and as long as $t < S^{\varepsilon} := \inf\{s > 0 : X_s^{x,\varepsilon} = X_s^{x_{i_0},\varepsilon}\}$ we have:

$$\begin{aligned} dY_t^{\varepsilon} &= \left\langle b(X_t^{x,\varepsilon}) - b(X_t^{x_{i_0},\varepsilon}), \frac{\Lambda(X_t^{x,\varepsilon} - X_t^{x_{i_0},\varepsilon})}{\nu(X_t^{x,\varepsilon} - X_t^{x_{i_0},\varepsilon})} \right\rangle dt + \frac{\varepsilon^2}{2} A_t^{\varepsilon} dt \\ &\quad + \varepsilon \left\langle [\mathbb{I} - H(X_t^{x,\varepsilon}, X_t^{x_{i_0},\varepsilon})] \frac{\Lambda(X_t^{x,\varepsilon} - X_t^{x_{i_0},\varepsilon})}{\nu(X_t^{x,\varepsilon} - X_t^{x_{i_0},\varepsilon})}, dW_t \right\rangle, \quad Y_0^{\varepsilon} = \nu(x - x_{i_0}) \end{aligned} \tag{5.68}$$

(i.e. the stochastic differential equation is satisfied for the process up to the coupling time). The process A_t^ε above is that coming from the second derivatives in Itô's formula (2.35), and it is given by (5.8), with $f(u) = \nu(u)$.

Tedious but simple calculations show that indeed $A_t^\varepsilon = 0$ in the present case, analogously to (5.8) (see [34] for full details). Consequently, from (5.67) it follows that the drift part in (5.68) is less than or equal to zero, as long as $X_t^{x,\varepsilon}$ and $X_t^{x_{i_0},\varepsilon}$ remain in $B_{\delta_1}(x_{i_0})$, and up to S^ε.

On the other hand, and from the expression for H given by (5.3) we see that the martingale term on the right-hand side of (5.68) is

$$\begin{aligned} dM_t^\varepsilon &= \varepsilon \left\langle [\mathbb{I} - H(X_t^{x,\varepsilon}, X_t^{x_{i_0},\varepsilon})] \frac{\Lambda(X_t^{x,\varepsilon} - X_t^{x_{i_0},\varepsilon})}{\nu(X_t^{x,\varepsilon} - X_t^{x_{i_0},\varepsilon})}, dW_t \right\rangle \\ &= 2\varepsilon \left\langle \frac{\nu(X_t^{x,\varepsilon} - X_t^{x_{i_0},\varepsilon})}{|X_t^{x,\varepsilon} - X_t^{x_{i_0},\varepsilon}|} \frac{(X_t^{x,\varepsilon} - X_t^{x_{i_0},\varepsilon})}{|X_t^{x,\varepsilon} - X_t^{x_{i_0},\varepsilon}|}, dW_t \right\rangle, \end{aligned} \tag{5.69}$$

from where we see that the process Y_t^ε satisfies the following equation

$$\begin{aligned} Y_t^\varepsilon = Y_0 &+ \int_0^t \left\langle b(X_s^{x,\varepsilon}) - b(X_s^{x_{i_0},\varepsilon}), \frac{\Lambda(X_s^{x,\varepsilon} - X_s^{x_{i_0},\varepsilon})}{\nu(X_s^{x,\varepsilon} - X_s^{x_{i_0},\varepsilon})} \right\rangle ds \\ &+ 2\varepsilon \int_0^t \frac{\nu(X_s^{x,\varepsilon} - X_s^{x_{i_0},\varepsilon})}{|X_s^{x,\varepsilon} - X_s^{x_{i_0},\varepsilon}|} dB_s^0, \end{aligned}$$

with B^0 a standard one-dimensional Brownian motion ($dB_t^0 = \langle (X_t^{x,\varepsilon} - X_t^{x_{i_0},\varepsilon})/|X_t^{x,\varepsilon} - X_t^{x_{i_0},\varepsilon}|, dW_t \rangle$).

Recall that $S^\varepsilon = \inf\{t \ge 0\colon Y_t^\varepsilon = 0\}$ and let $T^\varepsilon(y) = \inf\{t \ge 0\colon X_t^{y,\varepsilon} \notin B_{\delta_1}(x_{i_0})\}$. Let us also consider the random time

$$\tilde{S}^\varepsilon := \inf \left\{ t \ge 0 : Y_0^\varepsilon + 2\varepsilon \int_0^t \frac{\nu(X_s^{x,\varepsilon} - X_s^{x_{i_0},\varepsilon})}{|X_s^{x,\varepsilon} - X_s^{x_{i_0},\varepsilon}|} dB_s^0 = 0 \right\}.$$

From (5.67) and the last remarks it then follows that:

$$\begin{aligned} P\Big(S^\varepsilon \le \varepsilon^{-3}\Big) &\ge P\Big(S^\varepsilon \le \varepsilon^{-3}, T^\varepsilon(x) \wedge T^\varepsilon(x_{i_0}) > \varepsilon^{-4}\Big) \\ &\ge P\Big(\tilde{S}^\varepsilon \le \varepsilon^{-3}, T^\varepsilon(x) \wedge T^\varepsilon(x_{i_0}) > \varepsilon^{-4}\Big) \\ &\ge P\Big(\tilde{S}^\varepsilon \le \varepsilon^{-3}\Big) - P\Big(T^\varepsilon(x) \wedge T^\varepsilon(x_{i_0}) \le \varepsilon^{-4}\Big), \end{aligned} \tag{5.70}$$

where $t \wedge s$ denotes the minimum between t and s. On the other hand from Lemma 5.11 there exist constants $c, c' > 0$ so that for ε sufficiently small we have:

$$P\Big(T^\varepsilon(x) \wedge T^\varepsilon(x_{i_0}) \le \varepsilon^{-4}\Big) \le 2 \sup_{y\in \bar{B}_{\delta_0}(x_{i_0})} P\Big(T^\varepsilon(y) \le \exp\{c'/\varepsilon^2\}\Big) \le \exp\{-c/\varepsilon^2\} \tag{5.71}$$

showing that the last term on the right-hand side of (5.70) is asymptotically negligible. It remains to control the first term, and for that let us observe that up to time S^ε, the compensator of the continuous martingale

$$M_t^\varepsilon = 2\varepsilon \int_0^t \frac{\nu(X_s^{x,\varepsilon} - X_s^{x_{i_0},\varepsilon})}{|X_s^{x,\varepsilon} - X_s^{x_{i_0},\varepsilon}|} dB_s^0$$

can be written as

$$\langle M^\varepsilon\rangle_t = 4\varepsilon^2 \int_0^t \frac{\nu^2(X_s^{x,\varepsilon} - X_s^{x_{i_0},\varepsilon})}{|X_s^{x,\varepsilon} - X_s^{x_{i_0},\varepsilon}|^2} ds \ge \varepsilon^2 k_1 t$$

for some constant k_1, by the uniform equivalence of the norms in $\mathbb{R}^d$. A classical result on time change for martingales (see [173], p. 174) allows us to write $M_t^\varepsilon = \tilde{W}_{\langle M^\varepsilon\rangle_t}$, where $\tilde{W}$ is a standard one-dimensional Brownian motion. This representation is achieved by setting $\tilde{W}_t = M^\varepsilon_{C_t^\varepsilon}$ where the process C_t^ε is the inverse of the compensator (cf. Theorem 4.6, Chapter 3 of [173]). Thus, we may conclude:

$$\begin{aligned} P\Big(\tilde{S}^\varepsilon \le \varepsilon^{-3}\Big) &= P\Big(\inf_{s\le \varepsilon^{-3}} M_s^\varepsilon \le -Y_0^\varepsilon\Big) \ge P\Big(\inf_{s\le k_1\varepsilon^{-1}} \tilde{W}_s \le -Y_0^\varepsilon\Big) \\ &= 1 - \int \mathbf{1}_{\{|y|\le \frac{Y_0^\varepsilon \varepsilon^{1/2}}{k_1^{1/2}}\}} \frac{e^{-y^2/2}}{\sqrt{2\pi}} dy \ge 1 - \frac{k_2\delta_0\varepsilon^{1/2}}{k_1^{1/2}}, \end{aligned} \tag{5.72}$$

where k_2 is a positive constant so that $\nu(v) \le k_2|v|$, for all $v \in \mathbb{R}^d$.

From (5.70), (5.71) and (5.72), we get at once that $\lim_{\varepsilon\to 0} P(S^\varepsilon \le \varepsilon^{-3}) = 1$, uniformly over x in $\bar{B}_{\delta_0}(x_{i_0})$. Recalling (5.62) we conclude the proof. □

Remark As in Section 5.1, we have overlooked some details in the previous arguments: since the expression on the right-hand side of (5.69) makes no sense if $X_s^{x,\varepsilon} = X_s^{x_{i_0},\varepsilon}$ for some $s < t$, one needs to be careful in writing it. But we are interested in the behaviour of the process Y^ε until it reaches zero, and we can use 'localization methods' through the introduction of suitable stopping times.

Lemma 5.22 provides the main ingredient for the asymptotic loss of memory summarized as follows (see also Lemma 4.34 in Section 4.4).

Lemma 5.23 *Under the assumptions of Theorem* 5.21, *and having defined* β_ε *through* (5.61), *let*

$$f_\varepsilon(t) = P_{x_{i_0}}(\tau_\varepsilon > t\beta_\varepsilon)$$

for $t > 0, \varepsilon > 0$. *Then, there exist positive numbers* δ_ε, *which tend to zero as* $\varepsilon \to 0$, *and such that for each* $s, t > 0$:

$$f_\varepsilon(s+\delta_\varepsilon)(f_\varepsilon(t) - \varphi(t,\varepsilon)) \le f_\varepsilon(t+s) \le f_\varepsilon(s)(f_\varepsilon(t-\delta_\varepsilon) + \varphi(t,\varepsilon)) \quad (5.73)$$

where $\varphi(t,\varepsilon)$ *is a function of* t *and* ε, *which tends to zero as* $\varepsilon \to 0$, *uniformly on* $t \ge t_0$, *for any fixed* $t_0 > 0$.

Proof We start with the following.

Claim. If $\tilde{\eta}_\varepsilon = e^{\alpha/\varepsilon^2}$ with $V < \alpha < \bar{V}_G$, then $\tilde{\eta}_\varepsilon/\beta_\varepsilon \to 0$ and

$$\lim_{\varepsilon\to 0}\sup_{x\in G} P_x(\tau_\varepsilon > \tilde{\eta}_\varepsilon, \tau_\varepsilon(B_{\delta_0}(x_{i_0})) > \tilde{\eta}_\varepsilon) = 0.$$

Proof of the claim. The first statement follows at once from (5.62) and the choice $\alpha < V_G$. Since $G \setminus D$ is bounded and all the stable classes in $G \cup \partial G$ are contained in the interior of D, the argument used for (b) in Lemma 2.34 implies that

$$\lim_{\varepsilon\to 0}\sup_{x\in G\setminus D} P_x(\tau_\varepsilon(G^c \cup D) > \tilde{\eta}_\varepsilon/2) = 0.$$

On the other hand, due to (5.63) and since $V < \alpha$, we have

$$\lim_{\varepsilon\to 0}\sup_{x\in D} P_x(\tau_\varepsilon(B_{\delta_0}(x_{i_0})) > \tilde{\eta}_\varepsilon/2) = 0.$$

Using the strong Markov property at $\tau_\varepsilon(D)$ we get:

$$\begin{aligned}&\sup_{x\in G} P_x(\tau_\varepsilon > \tilde{\eta}_\varepsilon,\ \tau_\varepsilon(B_{\delta_0}(x_{i_0})) > \tilde{\eta}_\varepsilon)\\ &\le \sup_{x\in G} P_x(\tau_\varepsilon > \tilde{\eta}_\varepsilon,\ \tau_\varepsilon(D) > \tilde{\eta}_\varepsilon/2) + \sup_{x\in D} P_x(\tau_\varepsilon(B_{\delta_0}(x_{i_0})) > \tilde{\eta}_\varepsilon/2).\end{aligned}$$

Both terms on the right-hand side of the last inequality tend to zero, proving the claim.

From now on, the proof of (5.73) follows by essentially the same argument used for the analogous results in Chapter 4. Indeed, let $s > 0$ and define:

$$R_\varepsilon^s = \inf\{u > s\beta_\varepsilon : X_u^\varepsilon \in B_{\delta_0}(x_{i_0})\}.$$

Using the Markov property at time $s\beta_\varepsilon$ we can write

$$\sup_{x\in G} P_x(\tau_\varepsilon > s\beta_\varepsilon + \tilde{\eta}_\varepsilon, R_\varepsilon^s > s\beta_\varepsilon + \tilde{\eta}_\varepsilon) \le \sup_{x\in G} P_x(\tau_\varepsilon > \tilde{\eta}_\varepsilon, \tau_\varepsilon(B_{\delta_0}(x_{i_0})) > \tilde{\eta}_\varepsilon)$$

which tends to zero by the claim. As $\tilde{\eta}_\varepsilon/\beta_\varepsilon \to 0$, we get that

$$\sup_{x\in G} P_x(\tau_\varepsilon > (s+t)\beta_\varepsilon, R_\varepsilon^s > s\beta_\varepsilon + \tilde{\eta}_\varepsilon) \to 0, \quad (5.74)$$

uniformly on $s \geq 0, t \geq t_0$, for any fixed $t_0 > 0$.

On the other hand, using the strong Markov property at R^s_ε, we see that

$$\begin{aligned} &P_{x_{i_0}}(\tau_\varepsilon > (s+t)\beta_\varepsilon, R^s_\varepsilon \leq s\beta_\varepsilon + \tilde{\eta}_\varepsilon) \\ &\leq P_{x_{i_0}}(\tau_\varepsilon > s\beta_\varepsilon) \sup_{y \in \bar{B}_{\delta_0}(x_{i_0})} P_y(\tau_\varepsilon > t\beta_\varepsilon - \tilde{\eta}_\varepsilon), \end{aligned} \tag{5.75}$$

as well as

$$\begin{aligned} &P_{x_{i_0}}(\tau_\varepsilon > (s+t)\beta_\varepsilon, R^s_\varepsilon \leq s\beta_\varepsilon + \tilde{\eta}_\varepsilon) \\ &\geq P_{x_{i_0}}(\tau_\varepsilon > s\beta_\varepsilon + \tilde{\eta}_\varepsilon, R^s_\varepsilon \leq s\beta_\varepsilon + \tilde{\eta}_\varepsilon) \inf_{y \in \bar{B}_{\delta_0}(x_{i_0})} P_y(\tau_\varepsilon > t\beta_\varepsilon). \end{aligned} \tag{5.76}$$

Lemma 5.23 follows at once from (5.74), (5.75) and (5.76), and Lemma 5.22. □

Remark The uniformity in t in (5.73) is indeed much stronger than what is written: it holds over $t \geq \tilde{\eta}_\varepsilon/\beta_\varepsilon$ as can be seen from Lemma 5.22 and (5.74). In particular, fixing $0 < \rho < \bar{V}_G - V$ it holds over $\{(s,t) \colon s \geq 0, t \geq \exp(-\rho/\varepsilon^2)\}$. Upper bounds for the error term $\varphi(t, \varepsilon)$ and for δ_ε are given in the proofs of Lemmas 5.22 and 5.23. They are not optimal, but provide bounds on the Lévy distance between the distribution of $\tau_\varepsilon/\beta_\varepsilon$ and the unit mean exponential distribution, cf. Lemma 4.34.

Notice that in the previous proof of Lemma 5.22 there is a dependence on the dimension d; the estimates get worse for large d, since it requires $(d)^{1/2}\delta_1$ to be small, used in our argument in order to guarantee the validity of (5.67).

Proof of Theorem 5.21 (i) For $x = x_{i_0}$ the proof follows at once from Lemma 5.23, as in part (b) of Theorem 4.6 (see also Lemma 4.34). Lemma 5.22 extends the result to any $x \in B_{\delta_0}(x_{i_0})$. For the general case we again use the claim at the beginning of the proof of Lemma 5.23 and the strong Markov property.

(ii) This is the same as in Theorem 4.6: let $x \in D$ and write

$$\frac{E_x \tau_\varepsilon}{\beta_\varepsilon} = \frac{1}{\beta_\varepsilon} \int_0^{+\infty} P_x(\tau_\varepsilon > t)\, dt = \int_0^{+\infty} P_x(\tau_\varepsilon > t\beta_\varepsilon)\, dt. \tag{5.77}$$

By part (i) and taking the limit $\varepsilon \to 0$, it suffices to justify interchanging this limit with the integral on the rightmost term of (5.77). For this we apply the dominated convergence theorem. Indeed, let

$$h_\varepsilon(t) = \sup_{y \in G} P_y(\tau_\varepsilon > t\beta_\varepsilon),$$

so that by the Markov property we have $h_\varepsilon(t+s) \leq h_\varepsilon(t) h_\varepsilon(s)$ and therefore

$$h_\varepsilon(2k) \leq (h_\varepsilon(2))^k. \tag{5.78}$$

On the other hand, using the strong Markov property at the hitting times $\tau_\varepsilon(D)$ and $\tau_\varepsilon(B_{\delta_0}(x_{i_0}))$, we can write

$$
\begin{aligned}
h_\varepsilon(2) \le & \sup_{x\in G} P_x(\tau_\varepsilon(D\cup G^c) > \beta_\varepsilon/2) \\
& + \sup_{x\in D} P_x(\tau_\varepsilon(B_{\delta_0}(x_{i_0})) > \beta_\varepsilon/2) \\
& + \sup_{x\in \bar{B}_{\delta_0}(x_{i_0})} P_x(\tau_\varepsilon > \beta_\varepsilon).
\end{aligned}
\tag{5.79}
$$

The first two terms on the right-hand side of (5.79) tend to zero, as we have noticed above, and the last one tends to e^{-1}. In particular we see that for ε small enough $h_\varepsilon(2) \le 2e^{-1} < 1$. From (5.78) we then see that the dominated convergence theorem can be applied to the right-hand side of (5.77) and (ii) follows, for any $x \in D$.

The statement about the confining case follows easily from the previous arguments. □

The notion of cycle is developed further in the next chapter. For comparison with the construction to be done there, we now discuss briefly a recursive procedure proposed by Freidlin and Wentzell to describe the full hierarchy of cycles. For this, one needs the following.

Assumptions 5.24 There are finitely many compact sets $K_1, \dots, K_l$, equivalence classes for $\sim$, such that each ω-limit set of the deterministic system is contained in some K_i. To fix the notation let us assume that $K_1, \dots, K_r$ are the stable classes $(1 \le r \le l)$ and set $\mathcal{L} = \{1, \dots, r\}$ (see the observation right after (5.55)).

Under these assumptions, and imposing some non-degeneracy conditions, the recursive procedure may be started at any $i_0 \in \mathcal{L}$. We *assume* that for any $i \in \mathcal{L}$ there is a *unique* $R(i) \in \mathcal{L} \setminus \{i\}$ such that $V(K_i, K_{R(i)}) = \min_{j\in\mathcal{L}\setminus\{i\}} V(K_i, K_j)$. Then let $i_1 = R(i_0)$, $i_2 = R(i_1), \dots,$ and let k be the order of the iteration when a repeated index first appears, i.e. the $i_1, \dots, i_{k-1}$ are all distinct and $i_k = i_n$, with $n \in \{0, \dots, k-1\}$; in this case, the cycles of rank one starting from $\{i_0\}$ will be $\{i_0\}, \dots, \{i_{n-1}\}, \{i_n \to i_{n+1} \to \dots \to i_{k-1} \to i_n\}$. (The cyclic order comes from the construction just given. Recall that i represents here the stable class K_i.) It is easy to see that the choice of different initial i_0 brings disjoint or coincident cycles with the same cyclic order.

This yields a partition of $\mathcal{L}$ into cycles (of rank one). Lemmas 5.11 and 5.15 tell us that if the process X^ε starts sufficiently near to some K_i in a given cycle, then for $\varepsilon > 0$ small, it will visit small neighbourhoods of any other class in this cycle before escaping to another cycle, with overwhelming probability. The time scales are given in terms of Lemma 5.15. Notice that the cyclic order in the above construction is the most probable, but *not* the unique order in which the system will typically visit the various classes within this cycle before escaping to another one. To understand how the system keeps moving in longer time scales, one needs to see how it moves from any cycle of rank one to another. This leads to a recursive process of constructing cycles of higher ranks.

For each cycle $\mathcal{C}_0^{(1)}$ of rank one, the tunnelling time to another rank one cycle is given in terms of the following quantity:

$$A(\mathcal{C}_0^{(1)}) = \min_{g \in \mathbb{G}(\mathcal{L} \setminus \mathcal{C}_0^{(1)})} \sum_{(i \to j) \in g} V(K_i, K_j)$$

as can be seen from Lemmas 5.11 and 5.15. *Assuming* that the minimum is attained at a *unique* graph $g_{\mathcal{C}_0^{(1)}}$, let $R_{\mathcal{C}_0^{(1)}}(j)$ be the terminal point for the chain of arrows in $g_{\mathcal{C}_0^{(1)}}$ that starts at j; we can see that this terminal point is the same for all $j \in \mathcal{C}_0^{(1)}$ and is denoted by $R_{\mathcal{C}_0^{(1)}}$. Then let $\mathcal{C}_1^{(1)}$ be the cycle of rank one to which $R_{\mathcal{C}_0^{(1)}}$ belongs. In this way one continues, defining $\mathcal{C}_k^{(1)}, k = 1, 2, \ldots$ until the first repetition appears, i.e. the first k such that $\mathcal{C}_k^{(1)} = \mathcal{C}_m^{(1)}$ for some $0 \le m < k$, and then consider the rank two cycles $\{\mathcal{C}_0^{(1)}\}, \{\mathcal{C}_1^{(1)}\}, \ldots, \{\mathcal{C}_{m-1}^{(1)}\}$, $\{\mathcal{C}_m^{(1)} \to \cdots \to \mathcal{C}_{k-1}^{(1)} \to \mathcal{C}_m^{(1)}\}$. The procedure continues, analogously to the case of rank one, since starting with two different rank one cycles we generate disjoint or coincident rank two cycles, and in this case the cyclic order is the same. Thus a partition of the set of rank one cycles is obtained.

Under the above assumptions (including the uniqueness of minimizers at each level) the procedure is recursive and it ends at a certain level with a unique cycle, due to the existence of finitely many classes, containing each possible ω-limit set.

The important property shared by cycles is that starting sufficiently close to any of its classes, the process visits small neighbourhoods of any class in the cycle before reaching a small neighbourhood of a class outside the cycle. (There is an implicit identification of a cycle $\mathcal{C}$ of any given rank with the classes belonging to those cycles of lower rank that form $\mathcal{C}$.) In particular, the situation mentioned at the end of Section 5.3 does not happen here, i.e. there is logarithmic equivalence between the expectation of the escape time and any of its quantiles, as in Theorem 2.6. This asymptotic value is given in Theorem 6.2, Chapter 6 of [122], written below. We omit the proof which is obtained from Theorem 5.18.

Theorem 5.25 (Freidlin and Wentzell) *Under the assumptions of the previous construction, let $\mathcal{C}$ be a cycle different from $\mathcal{L}$, and let*

$$\bar{V}(\mathcal{C}) = A(\mathcal{C}) - \min_{K_m \in \mathcal{C}} \min_{g \in \mathbb{G}_{\mathcal{C}}(\{m\})} \sum_{(i \to j) \in g} V(K_i, K_j),$$

where $\mathbb{G}_{\mathcal{C}}(\{m\})$ is given by Definition 5.13 *with $\mathcal{C}$ in the place of $\mathcal{L}$. If $\rho > 0$ is small enough, D_j is the ρ-neighbourhood of K_j and we write (for brevity)*

$$\tau_\varepsilon(\mathcal{L} \setminus \mathcal{C}) = \inf\{t > 0\colon X_t^\varepsilon \in \cup_{j\in\mathcal{L}\setminus\mathcal{C}} D_j\},$$

then:

$$\lim_{\varepsilon\to 0} \varepsilon^2 \log E_x\, \tau_\varepsilon(\mathcal{L} \setminus \mathcal{C}) = \bar{V}(\mathcal{C}),$$

and for $\gamma > 0$,

$$\lim_{\varepsilon\to 0} P_x \left(e^{\frac{\bar{V}(\mathcal{C})-\gamma}{\varepsilon^2}} < \tau_\varepsilon(\mathcal{L} \setminus \mathcal{C}) < e^{\frac{\bar{V}(\mathcal{C})+\gamma}{\varepsilon^2}} \right) = 1,$$

uniformly for all x in $\cup_{j\in\mathcal{C}} D_j$.

Under the previous assumptions, the full hierarchy of cycles allows us to study the asymptotic behaviour of $X^\varepsilon_{t(\varepsilon)}$ for $t(\varepsilon) \approx \exp\{a/\varepsilon^2\}$, as $\varepsilon \to 0$ (cf. Section 6, Chapter 6 of [122]).

Remark Under the assumptions of Theorem 5.21, by changing the field b outside G we can satisfy Assumptions 5.24. In this new system, we can see that condition (b4) of Assumptions 5.20 guarantees that the stable classes in G, i.e. $x_1, \ldots, x_r$, form a cycle (of some level) according to the last definition. The asymptotic exponentiality should extend to the escape times in Theorem 5.25. In the Markov chain set-up, this kind of problem is treated in the next chapter, with a further discussion on cycle decomposition.

5.5 Notes and comments

Theorems 5.5 and 5.6 were proven in [131] using an estimate due to Day [73, 74], which implies that

$$\sup_{x\in B_{\rho_1}(p)} \|\nu_x^\varepsilon - \nu_p^\varepsilon\| \le e^{-c/\varepsilon^2} \tag{5.80}$$

for some $c > 0$, where ν_x^ε denotes the law of $X^{x,\varepsilon}_{\tau_\varepsilon(S_\rho(p))}$, and $\|\cdot\|$ the total variation norm. Here one assumes $\bar{B}_\rho(p) \subseteq D_p$ and takes $\rho_1 \in (0, \rho)$ suitably small. From this estimate (obtained not only in the gradient case) Day first derived the asymptotic exponentiality of τ_ε in the context of Section 2.6. To prove (5.80) he shows that with probability exponentially close to one, the process visits an exponentially small ball around p before escaping from $B_\rho(p)$, and then uses a comparison with an Ornstein–Uhlenbeck process, whose forcing term is the Jacobian matrix of b at p.

This proof, as well as that given in Section 5.2, depends on a detailed analysis of the generator. The finite dimensionality of the state space is also crucial. A different approach to this problem was used by Martinelli and Scoppola in [209].

Under Assumptions 5.20 of the previous section, and if x_i is a stable fixed point with a positive Lyapunov exponent α_0, they show that once all paths are constructed using the same random noise, then all dynamics starting at any pair of points near $K_i = \{x_i\}$ will eventually become exponentially close, after some initial time layer which does not depend on ε. More precisely, if $\alpha < \alpha_0$ and $T > 0$ there exist $t_0, c_1, c_2, \varepsilon_0$ so that for any $0 < \varepsilon < \varepsilon_0$

$$P\left(\sup_{x \neq y \in \mathrm{B}_{i,T}} \frac{|X_t^{x,\varepsilon} - X_t^{y,\varepsilon}|}{|x-y|} \le c_2 e^{-\alpha t}, \forall t > t_0 \right) \ge 1 - e^{-c_1/\varepsilon^2} \tag{5.81}$$

where $\mathrm{B}_{i,T} = \{x \colon X_t^{x,0}$ reaches a small neighbourhood of K_i by time $T\}$.

If x and y belong to different basins of attraction of the deterministic system (as needed in the proof of Theorem 5.21), then c_2 or t_0 should be replaced by $c_2(\varepsilon)$ and $t_0(\varepsilon)$ respectively, cf. (5.63). The proof of (5.81) involves a multiscale type analysis, and uses techniques that are similar to those used by Martinelli and Scoppola in the study of Anderson localization for disordered systems.

This 'gluing' of trajectories was used successfully in [210] to prove Theorem 5.21 (under slightly more restrictive assumptions) as well as for an infinite dimensional analogue.

A related problem in the context of the Freidlin and Wentzell models refers to the validity of a l.d.p. for the family of empirical measures $\psi_\varepsilon := \frac{1}{\tau_\varepsilon} \int_0^{\tau_\varepsilon} \delta_{X_s^\varepsilon} ds$ (see [217]).

Non-linear heat equation with noise Interest in the analysis of stochastic partial differential equations has grown tremendously since the late 1970s. We mention a few motivations closer to our context. (a) The hydrodynamics of an incompressible viscous fluid is usually described through Navier–Stokes equations. They have an approximate character, and it becomes crucial to study their stability under random perturbations. (b) A broad class of partial differential equations appears in transport problems, chemical reactions subjected to diffusion, population genetics, etc. The motivation for stochastic stability is the same as in the previous example. (c) Interest in a class of evolutions related to models of quantum mechanics and quantum field theory, such as the stochastic quantization method.

The example that we now outline was introduced by Faris and Jona-Lasinio [113]. It fits (b) and (c) above quite well.

For a given $L > 0$ and $\varepsilon > 0$ one may consider the process $u^\varepsilon(r,t)$, described *formally* as the solution of the following stochastic partial differential equation:

$$\begin{aligned} \frac{\partial u^\varepsilon}{\partial t} &= \frac{\partial^2 u^\varepsilon}{\partial r^2} - V'(u^\varepsilon) + \varepsilon\, \xi(r,t), \\ u^\varepsilon(0,t) &= u^\varepsilon(L,t) = 0, \\ u^\varepsilon(r,0) &= \varphi(r), \end{aligned} \tag{5.82}$$

where $r \in [0, L]$, $V(u) = (\alpha/4)u^4 - (\beta/2)u^2$, $\alpha, \beta > 0$ and ξ is a white noise in space-time. A motivation for studying this model comes from a quantum mechanical double well unharmonic oscillator, but it may also be thought of as describing the motion of an elastic string (with fixed extremities) in a highly viscous and noisy environment.

We may think of u^ε (at least formally) as a small random perturbation of the following infinite dimensional deterministic system, with values on the space $C_D[0, L] := \{f \in C([0, L], \mathbb{R}) : f(0) = f(L) = 0\}$:

$$\begin{aligned} \frac{\partial u}{\partial t} &= -\frac{\delta S}{\delta u}, \\ u(\cdot, 0) &= u_0(\cdot), \end{aligned} \tag{5.83}$$

where the *potential* S is given by the following function

$$S(\psi) = \int_0^L \Big[\frac{1}{2}\Big(\frac{d\psi}{dr}\Big)^2 + V(\psi)\Big] dr \tag{5.84}$$

for $\psi \in C_D[0, L]$ absolutely continuous (S is thought of as $+\infty$ otherwise).

The qualitative analysis of the system given by (5.83) was initiated by Chafee and Infante in [59], where it is shown that for L large enough (fixed α, β), there are several critical points. Two of them are stable, called u_- and u_+, and there is a certain number of saddle points (depending on the value of L) that are given by instanton-type (or multi-instanton) functions verifying the Dirichlet boundary conditions. Further results on the analysis of systems of this type can be found in [152].

A rigorous meaning for (5.82) is provided in [113], where the validity of a l.d.p. for u^ε is also established, extending Freidlin and Wentzell estimates. For this one considers the space of continuous paths taking values in $C_D[0, L]$ endowed with the supremum norm. Using the large deviation principle, they obtained upper and lower estimates for the probability of passing from a neighbourhood of u_- to a neighbourhood of u_+. The rate function on $C_D[0, L]$ is given by

$$I_T(\psi) = \frac{1}{2}\int_0^T \int_0^L \left[\frac{\partial \psi}{\partial t}(r, t) - \frac{\partial^2 \psi}{\partial r^2}(r, t) + V'(\psi(r, t))\right]^2 drdt.$$

The variational problem of minimizing the value of I among the paths which connect u_- to u_+ is also solved in [113]; the solution suggests that escape from the domain of attraction of u_- happens close to the lowest saddle (i.e. with minimal value of S). This was proven by Brassesco in [32]. In [210] the authors used the ideas initially developed in [209] to extend Theorem 5.5, and Brassesco [32] completed the proof of the metastability picture, following the pathwise approach described in Chapter 4.

The procedure developed in [209] shows that the trajectories become very close as the time goes to infinity. To get the asymptotic exponentiality in the context of Theorem 5.21 the basic ingredient is a control on the difference between the values of τ_ε^x and τ_ε^y for two different points x, y (near one of the attractors, say). Since τ_ε^x is exponentially large (in ε^{-2}), we conclude that at this given time, the distance between the two different evolutions will be superexponentially small, due to (5.81). Taking into account the boundedness and regularity of $b(\cdot)$ one has the desired control.

On the other hand, for the infinite dimensional system described by (5.82), we notice that the drift term, given by the right-hand side of (5.83), is no longer continuous and bounded. This is so due to possible irregular behaviour in the space variable (if we consider the state space with the supremum norm). Thus, in spite of the validity of an analogue of (5.81), as proven in [209], the loss of memory of τ_ε requires a more careful analysis. Nevertheless, using the previous argument, we have that at the time τ_ε for one of the trajectories, both of them will be in a small neighbourhood of u_+, which is an attractor. This allows us to conclude that the corresponding τ_ε of the other trajectory will not differ much, from which we can conclude the asymptotic exponentiality.

The argument of the previous paragraph does not apply if we consider the escape time from a domain fully contained in the basin of attraction of u_-, analogous to that in Section 2.6. (In this case the drift is against the escape.) One possibility consists in finding a more suitable coupling. For the specific example this was done by Brassesco [33], adapting a coupling introduced by Mueller [222].

Motivated by the analysis of the quantum mechanical unharmonic oscillator, the full picture would indeed require the replacement of the bounded space interval $[0, L]$ by the whole real line. Cassandro, Olivieri and Picco (see [49]) studied a model similar to (5.82) with non-symmetric $V(\cdot)$, in the double limit, $\varepsilon \to 0$, $L \to +\infty$, obtaining large deviation estimates and estimating the probability of tunnelling. Their results allow us to take $L = L_\varepsilon$ tending slowly to infinity, corresponding in some sense to the mean field theory of Lebowitz and Penrose for ferromagnetic spin systems under a Kac interaction potential. This model presents a rich and complicated spatial structure which is nevertheless necessary for proper correction of a mean field theory, in contrast to the naïve and simplified Curie–Weiss model.

Glauber + Kawasaki processes A spin model related to the above equation can be constructed through the superposition of a spin-flip type dynamics and a fast stirring process, having $\mathcal{X} = \{-1, +1\}^{\mathbb{Z}^d}$ as state space. This model was introduced by De Masi, Ferrari and Lebowitz in [80] and investigated by several authors (see [78] and references therein). It is generally non-reversible, which brings extra difficulties and enriches the study of relations between macroscopic and microscopic descriptions, providing an interesting 'laboratory' to test several ideas. Under very fast stirring and in some continuum limit (in a proper spatial scale) this is described by an equation very similar to (5.82), with a different noise. More precisely, if we take a continuous time Markov process with values in $\mathcal{X}$, with generator

$$L_\varepsilon = \varepsilon^{-2} L_0 + L_G,$$

where

$$L_0 f(\sigma) = \frac{1}{2} \sum_{|x-y|=1} (f(\sigma^{x,y}) - f(\sigma)),$$

$$L_G f(\sigma) = \sum_x c(x,\sigma)(f(\sigma^x) - f(\sigma)),$$

$\sigma^{x,y}$ denotes the configuration obtained from σ by exchanging the values of the spins at x and y, and σ^x represents the configuration obtained from σ by flipping the spin at x. With proper assumptions on the rate function c (including finite range and translation invariance, as in the next chapter) one can prove propagation of chaos on the spatial scale $r = \varepsilon x$ in the limit as $\varepsilon \to 0$. The continuum limit is described in terms of a reaction diffusion equation

$$\frac{\partial m}{\partial t}(r,t) = \frac{1}{2}\Delta m(r,t) + F(m(r,t)),$$

$$m(r,0) = m(r),$$

with the reactive term $F(m) = -2\nu_m(\sigma(0)c(0,\sigma))$, where ν_m denotes the product Bernoulli measure on $\mathcal{X}$ with average spin m (see [78] for a precise statement). We can easily give many examples where F is any given polynomial. A particular interesting case is when $F = -V'$, analogously to (5.82). After deriving the continuum limit (law of large numbers) it is natural to ask about large deviations. For a finite volume version, a l.d.p. was described in [171], extending ideas in [179]. Though the rate function is in general quite complicated, it allows some simple evaluations of the quasi-potential $V(m_-, m)$, where m_- is a stable equilibrium point and m is in its domain of attraction. This furnishes a class of examples where, in spite of non-reversibility, the optimal path connecting an attractor m_- to an element m of its basin, is simply the time reversed deterministic trajectory, as in Section 2.6 (see [126]).

Some further problems on the long time behaviour of Glauber+Kawasaki dynamics have been treated with the analysis of the truncated correlation functions, also called v-functions ([42, 81, 137]). Other estimates related to metastability appear in [188], based on methods developed initially by [298] for Ginzburg–Landau models. The ergodic behaviour of a large class of such dynamics has been treated in [103]. In [35] and [36], the ergodicity of the process was studied in detail for a particular family, where attractiveness was used in a crucial manner. The estimates obtained there were applied in [154] to verify the asymptotic exponentiality of certain escape times. A full description of metastability is still lacking, as in the analogous situation of the Glauber dynamics for the Kac model mentioned in Section 4.5.

6

Reversible Markov chains in the Freidlin–Wentzell regime

We start by defining the Markov chains that we want to study in the present chapter. After general Freidlin–Wentzell Markov chains, we move on to the simple and interesting particular case of reversible Freidlin–Wentzell Markov chains; finally we introduce the more particular case of Metropolis Markov chains which constitute the main object of our study in this chapter. We also introduce particular Markov chains satisfying a non-degeneracy condition which further simplifies the theory. Sometimes we adopt a different notation with respect to the previous chapters.

Condition FW (Freidlin–Wentzell) *We say that an ergodic aperiodic Markov chain on a finite state space S satisfies condition FW if its transition probabilities $P(x, y)$ decay exponentially fast in a large parameter β, satisfying the following estimates*:

$$\exp[-\beta(\Delta(x, y) + \gamma(\beta))] \leq P(x, y) \leq \exp[-\beta(\Delta(x, y) - \gamma(\beta))]$$

where Δ is a non-negative matrix

$$\Delta : S \times S \to \mathbb{R}^{+} \cup \{+\infty\}$$

and $\gamma(\beta) \to 0$ as $\beta \to \infty$.

The elements of S will be called indifferently points or states.

FW Markov chains can be regarded as small random perturbations of $\beta = \infty$ dynamics, where only transitions with $\Delta(x, y) = 0$ are allowed.

Condition R (reversibility) *We say that a Markov chain on a finite state space S satisfies condition R if, considering the set $\mathcal{V}_0 := \{(x, y) : x \in S, y \in S, x \neq y\}$ of pairs of distinct states,*

(*i*) *there exists a connectivity function* $q : \mathcal{V}_0 \to [0, 1]$ *satisfying*
- *for any* $x \in S$, $\sum_{y \neq x} q(x, y) \leq 1$ *(normalizability)*
- *for any* $x, y \in S$, $q(x, y) = q(y, x)$ *(symmetry)*
- *for any* $x, y \in S$ *there exists* $n \in \mathbb{N}$ *and* $x_0, \ldots, x_n \in S$ *with* $x_0 = x$, $x_n = y$ *and* $q(x_i, x_{i+1}) > 0$, $i = 0, \ldots, n-1$ *(irreducibility),*

(*ii*) *there exists an energy function* $H : S \to \mathbb{R}$,

(*iii*) *let* $\mathcal{V}$ *be the set of pairs of communicating states* $\mathcal{V} := \{(x, y) \in \mathcal{V}_0 : q(x, y) > 0\}$, *then there exists a function* $\tilde{H} : \mathcal{V} \to \mathbb{R}$ *such that* $\tilde{H}(x, y) = \tilde{H}(y, x)$ *and* $\tilde{H}(x, y) \geq H(x) \vee H(y)$, *so that the transition probabilities* $P(x, y)$ *of the Markov chain are given by*

$$P(x, y) = \begin{cases} q(x, y) \exp(-\beta[\tilde{H}(x, y) - H(x)]) & \text{if } x \neq y \\ 1 - \sum_{z \neq x} P(x, z) & \text{if } x = y. \end{cases} \tag{6.1}$$

Condition R is a particular case of FW with

$$\Delta(x, y) := \tilde{H}(x, y) - H(x). \tag{6.2}$$

The above choice corresponds to a *reversible Markov chain*:

$$\forall\, x, x' \in S, \quad \mu(x) P(x, x') = \mu(x') P(x', x) \tag{6.3}$$

with

$$\mu(x) = \frac{\exp(-\beta H(x))}{\sum_{z \in S} \exp(-\beta H(z))}$$

(see Chapter 4, equation (4.52), Section 5.2). It is natural to interpret β as the inverse temperature and μ as the Gibbs measure. From (6.3) one immediately deduces that μ is the unique invariant probability measure of the chain.

Condition RND (reversibility, non-degeneracy) *We say that a Markov chain satisfies condition RND if it satisfies condition R and*

(*i*) H *is invertible,* $H(x) \neq H(y)$ *if* $x \neq y$,

(*ii*) *for any* $x \in S$, *there exists a unique state* y *such that* $\min_{z \neq x} \tilde{H}(x, z) = \tilde{H}(x, y)$.

Condition RND highly simplifies the theory and it is useful for didactical purposes. It is in general very restrictive; however any reversible Markov chain can be transformed into a non-degenerate reversible Markov chain by an arbitrarily small change of the functions H, $\tilde{H}$ (recall that $|S| < \infty$).

Condition M (Metropolis) *We say that a Markov chain on a finite state space* S *with transition probabilities* $P(x, y)$ *satisfies condition M if there exists a connectivity function* q *satisfying (i) in condition R and an energy function* $H : S \to \mathbb{R}$

so that:

$$P(x, y) = \begin{cases} q(x, y)\exp(-\beta[H(y) - H(x)]^+) & \text{if} \quad x \neq y \\ 1 - \sum_{z \neq x} P(x, z) & \text{if} \quad x = y. \end{cases} \tag{6.4}$$

The above defined Metropolis Markov chains are a particular case of reversible Markov chains: they correspond to the choice $\tilde{H}(x, y) = H(x) \vee H(y)$. In this case the $\beta = \infty$ dynamics only allows transitions that lower the energy.

Condition MND (Metropolis, non-degeneracy) *We say that a Markov chain satisfies condition MND if it satisfies condition M and*

(i) *H is invertible, $H(x) \neq H(y)$ if $x \neq y$,*
(ii) *for any $x \in S$, there exists at most one state $y \in S$ with $q(x, y) > 0$, $H(x) > H(y)$.*

Notice that for MND Markov chains the $\beta = \infty$ dynamics is deterministic.

A particular example satisfying condition M (but not satisfying MND) is given by the standard Ising model under Glauber–Metropolis dynamics (see Chapter 7, equation (7.2)).

In what follows we always assume that our Markov chain satisfies condition M, unless stated explicitly otherwise.

Notice that we can treat other reversible dynamics practically in the same way, for instance the so-called heat bath dynamics (see, for instance, [211–213]) whose transition probabilities are 'logarithmically equivalent' as $\beta \to \infty$ (see Chapter 1, equation (1.1)) to those of Metropolis Markov chains.

Introduction

We now discuss the first exit problem from a general set G for a general Metropolis Markov chain with finite state space, in the limit of large β. It turns out that the most relevant case corresponds to G given by a *cycle* (see Definition 6.5 below), namely a set whose internal points, for large β, are typically visited many times by our process before exiting. As we shall see, some specific difficulties arise when G contains many stable equilibria, namely many local minima for the energy, which can be considered as attractors with respect to the $\beta = \infty$ dynamics (where only moves decreasing the energy are allowed).

In the case of a set G 'completely attracted' by a unique stable point, it turns out that the exit from G, when it occurs, follows a quite fast path, taking place in an interval of time T independent of β, without 'hesitations'. The exponentially long (in β) time needed by the process to exit from G is a consequence of the exponentially small probability that an excursion outside G takes place in a finite time independent of β. So the typical time needed to see the first excursion outside G is very long as a consequence of a very long series of unlikely attempts before the successful one, but the time spent during this first excursion is relatively short.

On the contrary, in the case of a set G containing many minima for the energy, the time needed to see the first excursion is still exponentially large in β but, in general, also the time spent during the first excursion from G will typically be exponentially large in β, even though with a smaller rate. Indeed the first excursion will involve random fluctuations inside suitable 'permanence sets' during exponentially long random intervals of time ('resistance times'). As will appear clear, a mechanism of exit in a finite interval of time and without hesitation is, in general, very low in probability.

We shall see that when G contains many minima for H, even the formulation of the problem of determination of typical paths of exit from G changes drastically with respect to the case of completely attracted domains. Whereas in the completely attracted case we have to determine single optimal paths of exit, in the general case we have to introduce generalized paths given by sequences of permanence sets with relative permanence times. The single trajectory inside a permanence set cannot be specified. The specification of a typical tube of exit from a general domain G is intrinsically stochastic: the exit from a permanence set is exponentially long in β and tends to be unpredictable for large β; moreover the behaviour inside a permanence set is well described by a restricted equilibrium (conditional Gibbs) measure.

Let us start with a list of basic definitions. As we said before, for the sake of simplicity of the exposition, we shall consider Metropolis Markov chains. Our definitions and results can be easily extended to the case of general reversible Markov chains; here we only give explicitly, in this case, the crucial definition of cycle (see Definition 6.6) and an estimate of the first exit time (see Lemma 6.22). The case of general non-reversible Markov chains is much more difficult. We refer to [55, 232] for a general treatment of this case.

6.1 Definitions and notation

Consider a Markov chain X_t satisfying condition M. As usual we denote by P_x the law of the process starting at $x \in S$.

A *path* ω is a sequence $\omega = (\omega_0, \ldots, \omega_N)$, $N \in \mathbb{N}$, with ω_j, ω_{j+1}, $j = 0, \ldots, N-1$, communicating states (i.e. $P(\omega_j, \omega_{j+1}) > 0$). We often write $\omega : x \to y$ to denote a path joining x to y. Given $G_1 \subseteq S$, $G_2 \subseteq S$, we write $\omega : G_1 \to G_2$ to denote a path joining x to y for some $x \in G_1$, $y \in G_2$. If ω is a path and $\omega_i = x$ for some i, then we say $x \in \omega$.

We denote by $\Omega \subseteq S^{\mathbb{N}}$ the set of all possible paths.

A path $\omega = (\omega_0, \ldots, \omega_k)$ is called *downhill* (*strictly downhill*) if $H(\omega_{i+1}) \leq H(\omega_i)$ $(H(\omega_{i+1}) < H(\omega_i))$ for $i = 0, \ldots, k-1$.

We say that a state x is *downhill connected* (*strictly downhill connected*) to a state y if there exists a downhill (strictly downhill) path $\omega : x \to y$.

A path $\omega = (\omega_0, \dots, \omega_k)$ with $H(\omega_i) \leq H(\omega_0)$ for $i = 0, \dots, k$ and $H(\omega_k) < H(\omega_0)$ is called *declining*. Notice that a declining path is not necessarily downhill.

A set $G \subseteq S$ is *connected* if for any $x, x' \in G$ there exists a path $\omega : x \to x'$ totally contained in G.

Definition 6.1 Given $G \subseteq S$, we define

(i) its exterior boundary

$$\partial G \equiv \partial^+ G := \{y \notin G : P(x, y) > 0 \quad \text{for some } x \in G\},$$

(ii) its interior boundary

$$\partial^- G = \{y \in G : P(x, y) > 0 \quad \text{for some } x \notin G\}. \tag{6.5}$$

Definition 6.2 Given $G \subseteq S$ we denote by $F = F(G)$ the set of all absolute minima of the energy on G:

$$F(G) := \left\{ y \in G : \min_{x \in G} H(x) = H(y) \right\}. \tag{6.6}$$

The set $F(G)$ is called the ground of G.

Definition 6.3 A maximal connected set of equal energy states is called a *plateau*.

Given $z \in S$ we denote by $D(z)$ the plateau containing z.

We observe that in the context of our treatment, involving asymptotic estimates of probabilities and times that are exponential in β, a plateau could be identified with a unique point since the motion on a plateau takes place without exponential costs, in times of order one. With the non-degeneracy hypothesis MND, a plateau coincides with a single site.

Definition 6.4 We say that the real positive function $\beta \to f(\beta)$ is superexponentially small and we use the notation $f(\beta) = \text{SES}$, if $\lim_{\beta \to \infty} \beta^{-1} \log f(\beta) = -\infty$.

Notation Given a set G on which H is constant, $H(x) = H_0$, for any $x \in G$, we write by abuse of notation $H(G)$ for the common value.

A plateau D such that $H(F(\partial D)) > H(D)$ is said to be *stable*; otherwise it is called *transient*. In particular the set of local minima for the energy

$$M = \{x : H(F(\partial\{x\})) > H(x)\} \tag{6.7}$$

is made of stable points.

We denote by $\bar{M}$ the collection of all stable plateaux in S.

Definition 6.5 A connected set $A \subseteq S$ which satisfies

$$\max_{x \in A} H(x) < \min_{z \in \partial A} H(z) = H(F(\partial A)) \tag{6.8}$$

is called a *non-trivial cycle*.

A cycle is either a singleton $\{x\}$ or a non-trivial cycle.

Notice that in our definition a singleton $\{x\}$ is a non-trivial cycle if and only if it is a local minimum $x \in M$.

Consider a Markov chain satisfying condition R. Notions of connectedness of a set of states G and boundary ∂G are immediately extended in the natural way. In this case we want now to give the following.

Definition 6.6 Given a Markov chain satisfying condition R, a connected set $A \subseteq S$ which satisfies

$$\max_{(x,y) \in \mathcal{V}: x, y \in A} \tilde{H}(x, y) < \min_{(u,z) \in \mathcal{V}: u \in A, z \in \partial A} \tilde{H}(u, z) \tag{6.9}$$

is called a *non-trivial cycle*.

A cycle is either a singleton $\{x\}$ or a non-trivial cycle.

It is immediate to see that the above definition of cycle, for RND Markov chains, coincides with that given, in terms of graphs, by Freidlin and Wentzell in the context of a (not necessarily reversible) Markov chain satisfying condition FW when

(i) every state is stable in the sense that, for any $x, y, \in S$, with $x \neq y$ we have $\Delta(x, y) > 0$ and

(ii) a non-degeneracy hypothesis holds which, in the reversible case, amounts to assuming our condition RND and in the Metropolis case, MND (see Chapter 5 after Assumptions 5.24), [122] p. 198).

Indeed Freidlin and Wentzell, for the sake of simplicity only gave their definition of cycle in the non-degenerate case. It is possible to extend the FW graphical definition to the fully general non-reversible and possibly degenerate case (see, for instance, [232]). It is also possible to show that this general graphical definition is equivalent, in the Metropolis (but not necessarily MND) case, to our Definition 6.5.

Let us now go back to our treatment of the Markov chains just satisfying condition M.

Notation Given a non-trivial cycle A we write

$$U(A) := F(\partial A). \tag{6.10)}$$

For a trivial cycle $\{x\}$, we set $U(\{x\}) \equiv x$.

The *height* of a non-trivial cycle A is given by

$$\Phi(A) := H(U(A)). \tag{6.11}$$

The *depth* of a non-trivial cycle A is given by

$$\Gamma(A) := \Phi(A) - H(F(A)) = H(U(A)) - H(F(A)). \tag{6.12}$$

For a trivial cycle $\{x\}$ we set $\Phi(\{x\}) = H(x)$ and $\Gamma(\{x\}) = 0$.

We now give some propositions (in addition to other definitions). Most of the propositions refer to simple properties of the cycles. They are relevant in view of the basic results given in Theorem 6.23. For some of them the proof is immediate and we omit it.

Proposition 6.7 *Given a state $\bar{x} \in S$ and a real number c, the set of all points x connected to $\bar{x}$ by paths whose points have an energy strictly less than c, if non-empty, either coincides with S or is a non-trivial cycle A containing $\bar{x}$ with $\Phi(A) \geq c$.*

Proof Immediate. □

Proposition 6.8 *Consider two cycles A_1, A_2 with $A_1 \cap A_2 \neq \emptyset$. Then either $A_1 \subseteq A_2$ or $A_2 \subseteq A_1$.*

Proof It suffices to check that $\partial A_1 \cap A_2 \neq \emptyset$ simultaneously with $\partial A_2 \cap A_1 \neq \emptyset$ is impossible. For this, just notice that if $x_1 \in \partial A_1 \cap A_2$ and $x_2 \in \partial A_2 \cap A_1$ we should have, at the same time, $H(x_1) < H(x_2)$ as well as $H(x_2) < H(x_1)$. □

The above Proposition 6.8 establishes a partial ordering among cycles. Notice that the cycles containing a given point x are totally ordered by inclusion.

Definition 6.9 For each pair of states $x, y \in S$ we define their communication height $\Phi(x, y)$ as follows:

$$\Phi(x, y) := \min_{\omega : x \to y} \max_{z \in \omega} H(z).$$

Given a subset Q of S containing both x and y, we write

$$\Phi_Q(x, y) := \min_{\substack{\omega : x \to y \\ \omega \subseteq Q}} \max_{z \in \omega} H(z) \tag{6.13}$$

to denote the communication height between x and y in Q.

Definition 6.10 The set of minimal saddles $\mathcal{S}(x, y)$ from x to y in S is given by:

$$\begin{aligned} \mathcal{S}(x, y) &= \{z : \exists\ \omega : x \to y, \quad \text{such that } z \in \omega \text{ and } H(z) \\ &= \max_{u \in \omega} H(u) = \Phi(x, y)\}. \end{aligned} \tag{6.14}$$

The set of optimal paths between x and y, that we denote by $(x \to y)_{opt}$, is given by

$$(x \to y)_{opt} := \left\{ \omega : x \to y, \text{ such that } \max_{z \in \omega} H(z) = \Phi(x, y) \right\}. \tag{6.15}$$

A set $W \subseteq \mathcal{S}$ is called a *gate* for the transition $x \to y$ if for any $\omega \in (x \to y)_{opt}$, one has $\omega \cap W \neq \emptyset$. A gate W is called a *minimal gate* if for any $W' \subsetneq W$ there exists $\omega^* \in (x \to y)_{opt}$ such that $\omega^* \cap W' = \emptyset$.

In other words, a minimal gate is a minimal (by inclusion) subset of $\mathcal{S}(x, y)$ that is crossed by all optimal paths.

For a given transition $x \to y$, there can be many different minimal gates. Let $\mathcal{W}$ be the set of minimal gates. We denote by $\mathcal{G}(x, y)$ the union of minimal gates:

$$\mathcal{S}(x, y) \supseteq \mathcal{G}(x, y) := \cup_{W \in \mathcal{W}} W. \tag{6.16}$$

The configurations in $\mathcal{S}(x, y) \backslash \mathcal{G}(x, y)$ (if any) are called *dead-ends*. As will be made clear by Proposition 6.11, given any path $\omega \in (x \to y)_{opt}$ passing through a dead-end $z \in \mathcal{S}(x, y)$, there exists another path ω' playing the role of a short-cut for going to y by skipping z.

The set $\mathcal{S}(x, y)$ is a trivial gate for the transition $x \to y$, but in general it is not a minimal gate. Actually, it can contain configurations that are unessential in the following sense.

Given $z \in \mathcal{S}(x, y)$, we say that it is *unessential* if for any $\omega \in (x \to y)_{opt}$ such that $\omega \cap z \neq \emptyset$, defining $\mathcal{B}(\omega) = [\omega \cap \mathcal{S}(x, y)] \setminus z$, we have:

(i) $\mathcal{B}(\omega) \neq \emptyset$ and
(ii) there exists a path $\omega' \in (x \to y)_{opt}$ with $\omega' \cap \mathcal{S}(x, y) \subseteq \mathcal{B}(\omega)$.

We say that $z \in \mathcal{S}(x, y)$ is *essential* if it is not unessential, namely if either

(i) there exists a path $\omega \in (x \to y)_{opt}$ such that $\omega \cap \mathcal{S}(x, y) \equiv \{z\}$, i.e. $\mathcal{B}(\omega) = [\omega \cap \mathcal{S}(x, y)] \setminus \{z\} = \emptyset$, or
(ii) there exists a path $\omega \in (x \to y)_{opt}$ such that $\mathcal{B}(\omega) \neq \emptyset$ and $\omega' \cap \mathcal{S}(x, y) \nsubseteq \mathcal{B}(\omega)$ for all $\omega' \in (x \to y)_{opt}$.

Unessential configurations are indeed dead-ends for the transition $x \to y$ and, in a certain sense, can be skipped. This is the result of the following.

Proposition 6.11 *$z \in \mathcal{S}(x, y)$ is essential if and only if $z \in \mathcal{G}(x, y)$.*

Proof Abbreviate $\mathcal{S} = \mathcal{S}(x, y)$, $\mathcal{G} = \mathcal{G}(x, y)$, $\Omega = (x \to y)_{opt}$. We start by proving that $z \in \mathcal{S}$ unessential implies $z \in \mathcal{S} \setminus \mathcal{G}$. We show that assuming the existence of a minimal gate W containing z, we get a contradiction. Indeed, let us first suppose $W \equiv z$. Take $\omega \in \Omega$, $\omega \ni z$. Since z is unessential we know that $\omega \cap \mathcal{S} = z \cup \mathcal{B}(\omega)$, $\mathcal{B}(\omega) \not\ni z$, $\mathcal{B}(\omega) \neq \emptyset$ and there exists $\omega' \in \Omega$ such that $\omega' \cap \mathcal{S} \subseteq \mathcal{B}(\omega)$ so that $\omega' \cap z = \emptyset$ contradicting that z is a gate. Suppose now the minimal gate W is given as $W = W' \cup z$ with $W' \neq \emptyset$; necessarily, by minimality, W' is not a gate, there exists ω^* such that $\omega^* \cap W' = \emptyset$. But, since W is a gate, we must have $\omega^* \cap z \neq \emptyset$. Recalling that z is unessential, we have $\omega^* \cap \mathcal{S} = z \cup \mathcal{B}(\omega^*)$ with $\mathcal{B}(\omega^*) \not\ni z$, $\mathcal{B}(\omega^*) \neq \emptyset$ and we know that there exists ω' such that $\omega' \cap \mathcal{S} \subseteq \mathcal{B}(\omega^*)$; thus $\omega' \cap z = \emptyset$ and, since $W' \cap \mathcal{B}(\omega^*) = \emptyset$, we have $\omega' \cap W' = \emptyset$; then $\omega' \cap W = \emptyset$ contradicting that W is a gate and then proving that $z \in \mathcal{S} \setminus \mathcal{G}$.

We next prove that $z \in \mathcal{S}$ essential implies $z \in \mathcal{G}$, namely, there exists a minimal gate W containing z. If there exists $\omega \in \Omega$ such that $\omega \cap \mathcal{S} = z$ it follows that z belongs to *all* minimal gates; thus we can suppose that for any $\omega \in \Omega$, $\omega \ni z$, we have $\omega \cap \mathcal{S} = z \cup \mathcal{B}(\omega)$ with $\mathcal{B}(\omega) \neq \emptyset$, $\mathcal{B}(\omega) \not\ni z$. Since z is essential, we know that there exists $\bar{\omega} \in \Omega$, $\bar{\omega} \ni z$, such that for any $\omega' \in \Omega$, $\omega' \cap \mathcal{S} \not\subseteq \mathcal{B}(\bar{\omega})$. In what follows we suppose such a $\bar{\omega}$. Let

$$\Omega = \Omega_z \cup (\Omega_z)^c \tag{6.17}$$

with $\Omega_z := \{\omega \in \Omega : \omega \ni z\}$, $(\Omega_z)^c = \Omega \setminus \Omega_z$. We can suppose that $(\Omega_z)^c$ is non-empty, otherwise z is obviously a minimal gate and we have concluded. We know that for any $\omega' \in (\Omega_z)^c$, $\omega' \cap \mathcal{S}' \neq \emptyset$, where $\mathcal{S}' := \mathcal{S} \setminus (z \cup \mathcal{B}(\bar{\omega})) \neq \emptyset$.

Let W' be a minimal partial gate for $(\Omega_z)^c$ in $\mathcal{S}'$:

$$\begin{aligned} &W' \subseteq \mathcal{S}' : \forall \omega' \in (\Omega_z)^c : \omega' \cap W' \neq \emptyset;\ \forall W'' \subseteq W', \\ &\exists \omega'' \in (\Omega_z)^c : \omega'' \cap W'' = \emptyset. \end{aligned} \tag{6.18}$$

This W' certainly exists since $\mathcal{S}'$ is a partial gate for $(\Omega_z)^c$ and we can always deduce from it a minimal partial gate. Now we claim that $W := W' \cup z$ is a minimal gate. Indeed every path in Ω clearly intersects W; moreover it is minimal because, by partial minimality we cannot exclude any point from W' preserving the gate property; nor can we exclude z since the path $\bar{\omega}$ does not intersect $W' \subseteq \mathcal{S}'$. □

From the reversibility of our chain it immediately follows that $\Phi(x, y) = \Phi(y, x)$, $\mathcal{S}(x, y) = \mathcal{S}(y, x)$ and $\mathcal{G}(x, y) = \mathcal{G}(y, x)$, for any $x, y \in S$.

For every $G_1, G_2 \subseteq S$, $G_1 \cap G_2 = \emptyset$ we write:

$$\begin{aligned} &\Phi(G_1, G_2) := \min_{x \in G_1, y \in G_2} \Phi(x, y), \\ &\mathcal{S}(G_1, G_2) = \left\{ z \in \cup_{\substack{x \in G_1 \\ y \in G_2}} \mathcal{S}(x, y) \ : \ \min_{\substack{u \in G_1, \\ v \in G_2}} \Phi(u, v) = H(z) \right\}, \\ &(G_1 \to G_2)_{opt} := \left\{ \omega : G_1 \to G_2 \ \text{ with } \max_{z \in \omega} H(z) = \Phi(G_1, G_2) \right\}. \end{aligned} \tag{6.19}$$

A set $W \subseteq \mathcal{S}(G_1, G_2)$ is called a gate for the transition $G_1 \to G_2$ if $\omega \cap W \neq \emptyset$ for all $\omega \in (G_1 \to G_2)_{opt}$. Similarly to what we did for $G_1 = \{x\}$, $G_2 = \{y\}$ we define the minimal gates and the set $\mathcal{G}(G_1, G_2)$ given as the union of minimal gates.

We can have the absence of any energy barrier between a pair x and y. In this case the minimax problem has a trivial solution and $x \in \mathcal{S}(x, y)$ or $y \in \mathcal{S}(x, y)$.

This cannot happen if x and y belong to different stable plateaux.

Proposition 6.12 *If A is a non-trivial cycle then:*

(i) *for any $x, y, z \in A$ and $w \notin A$,*

$$\Phi(x, y) < \Phi(z, w);$$

(ii) *for any $x \in A$ the set $\mathcal{S}(x, A^c)$ does not depend on x and contains $U(A)$.*

Proof By definition of a non-trivial cycle, for any $x, y \in A$ there exists a path $\omega : x \to y$ contained in A such that for any $\omega_i \in \omega$ one has $H(\omega_i) < \Phi(A)$; this proves (i). On the other hand any path joining $z \in A$ to $w \in \partial A$ maximizes H in ∂A.

Thus we have, for any $x \in A$

$$\min_{w \in \partial A} \Phi(x, w) = \Phi(A)$$

implying (ii). □

Definition 6.13 A cycle A such that for each $z \in U(A)$ the states x to which the plateau $D(z)$ is strictly downhill connected belong to A, is called a *stable* or *attractive* cycle. A cycle A which is not stable is called *transient*.

Suppose that the cycle A is transient, then there exists $y^* \in U(A)$ with $D(y^*)$ strictly downhill connected to some point x in A^c; given a transient cycle A the points y^* downhill connected to A^c are called *minimal saddles of* A. The set of all minimal saddles of a transient cycle A is denoted by $\mathcal{S}(A)$.

Notice that besides $\mathcal{S}(A) \subseteq U(A)$, we also have $\mathcal{S}(A) \subseteq \mathcal{S}(A, A^c)$; for a trivial transient cycle $\{x\}$, we set $\mathcal{S}(\{x\}) = U(\{x\}) = \{x\}$.

Definition 6.14 A transient cycle A such that there exist $\bar{x} \notin A$ with $H(\bar{x}) \le H(F(A))$ $(H(\bar{x}) < H(F(A)))$, $y^* \in \mathcal{S}(A)$ and a declining path $\omega : y^* \to \bar{x}$, is called *metastable* (*strictly metastable*).

Definition 6.15 Given a stable plateau D (see Definition 6.3 and before (6.7)), we define the following *basins* for D:

(i) the *wide basin of attraction of* D

$$\hat{B}(D) = \{z : \exists \text{ downhill path } \omega : z \to D\}, \tag{6.20}$$

(ii) the *basin of attraction of* D given by

$$\bar{B}(D) = \{z : \text{every downhill path starting from } z \text{ and ending in } \bar{M}, \text{ ends in } D\}, \tag{6.21}$$

(iii) the *strict basin of attraction of* D, $B(D)$, given by the whole set of states S if $\bar{B}(D)$ coincides with S, otherwise

$$B(D) = \{z \in \bar{B}(D) \; : \; H(z) < H(F(\partial \bar{B}(D)))\}. \tag{6.22}$$

Remarks

(1) $\hat{B}(D)$ is necessarily non-empty whereas $\bar{B}(D)$ could be empty. $\bar{B}(D)$ can be seen as the usual basin of attraction of D with respect to the $\beta = \infty$ dynamics.

(2) For any a such that $a \leq H(F(\partial \bar{B}(D)))$, the set $B_a(D) := \{z \in \bar{B}(D) : H(z) < a\}$, if non-empty, is a non-trivial cycle contained in $B(D)$.

(3) The above definitions of different basins of attraction can be extended immediately to the case in which, instead of a stable plateau D, there is a generic cycle A. In this way one gets $\hat{B}(A)$, $\bar{B}(A)$, $B(A)$. Of course one can even consider the basins of attraction of more general sets (not necessarily cycles).

(4) The case of a Markov chain for which the state space S just coincides with the basin of attraction of a unique stable plateau D, constituting the set of absolute minima for the energy H, is in a sense almost trivial; we refer to this case as the *one well* case.

Definition 6.16 Given a non-trivial cycle A, we call a non-trivial saddle internal to A any saddle between a pair of stable plateaux contained in A.

Notation Given a cycle A we denote by $C(A)$ the (possibly empty) set of its internal non-trivial saddles, namely

$$\begin{aligned} C(A) = \{ & y \in \mathcal{S}(x, x') \text{ for some } x, x' \\ & \text{belonging to different stable plateaux in } A\} \subseteq A. \end{aligned} \tag{6.23}$$

Notice that $C(A)$ is empty if and only if $F(A)$ consists of a unique plateau.

We denote by $\bar{C}(A)$ the set of its internal non-trivial saddles with maximal energy, namely

$$\max_{z \in C(A)} H(z) = H_0, \qquad \bar{C}(A) = \{y \in C(A) : H(y) = H_0\}. \tag{6.24}$$

Given a cycle A with $C(A) \neq \emptyset$, we use $\bar{\Gamma}(A)$ to denote the height in energy of the maximal internal non-trivial saddle with respect to the ground $F(A)$

$$\bar{\Gamma}(A) := H(\bar{C}(A)) - H(F(A)). \tag{6.25}$$

We set $\bar{\Gamma}(A) = 0$ in the particular case when A is completely attracted by a unique stable plateau.

The following proposition provides a constructive criterion to find the saddles between different points.

Proposition 6.17 *Suppose that there exist a subset G of S and two points x, x' with $x \in G$, $x' \notin G$, with the following properties:*

(i) *G is connected,*

(ii) *for some $y^* \in F(\partial G)$ with $H(x') < H(y^*)$ there exists a declining path $\omega : y^* \to x$, $\omega \setminus y^* \subseteq G$,*

(iii) *there exists a declining path $\omega' : y^* \to x'$, $\omega' \cap G = \emptyset$.*

Then, if $A = \{z \ : \ \exists\ \tilde{\omega} : z \to x : \ \forall\, y \in \tilde{\omega}, \quad H(y) < H(y^*)\}$, $A' = \{z' \ : \ \exists\ \tilde{\omega} : z' \to x' : \ \forall\, y' \in \tilde{\omega}, \quad H(y') < H(y^*)\}$ so that A (A') $\equiv$ maximal connected set containing x (x') with energy less than $H(y^*)$. Then:

(1) *A, A' are non-trivial cycles and $A \subseteq G$, $A' \subseteq S \setminus G$,*

(2) *$\Phi(x, x') = \Phi(G, S \setminus G) = H(y^*)$, moreover $y^* \in \mathcal{S}(x, x')$,*

(3) *every path in $(x \to x')_{opt}$ has to cross ∂G in $F(\partial G)$ so that $F(\partial G)$ contains a minimal gate for the transition $x \to x'$.*

If, moreover, we have the following strict inequality,

$$H(z) < H(y^*) \text{ for any } z \in (\omega \cup \omega') \setminus y^*,$$

then

(4) *$y^* \in U(A) \cap U(A')$.*

Proof Immediate. □

Given a cycle A containing several stable plateaux, we consider some smaller cycles A_i contained in A, by looking at the internal non-trivial saddles between these plateaux in A. Let us start with an immediate corollary of Proposition 6.12.

Proposition 6.18 *If a cycle A contains an internal non-trivial saddle y, then it contains all the cycles A_j such that $y \in \cup_{z \in U(A_j)} D(z)$.*

Proof Immediate. □

Proposition 6.19 *Let A be a non-trivial cycle and suppose that it is not contained in the strict basin of attraction of any stable plateau (i.e. $C(A)$ is non-empty).*

Then the following are true.

(i) *The cycle A can be decomposed as the union of the highest internal transient cycles and the highest internal saddles $\bar{C}(A)$ plus, possibly, a part attracted by this union, namely*

$$A = \tilde{A} \cup \bar{C}(A) \cup V, \tag{6.26}$$

$$\tilde{A} = A_1 \cup \cdots \cup A_k, \tag{6.27}$$

where $A_1, \ldots, A_k$ are transient cycles with $\Phi(A_i) = H_0 := H(\bar{C}(A))$, $i = 1, \ldots, k$ and for any $x_i \in A_i, x_j \in A_j, \mathcal{S}(x_i, x_j) \subseteq \bar{C}(A)$.

(ii) *$F(A) \subseteq \{\cup_j A_j\}$ so that at least one of the A_j's must contain points in $F(A)$.*

(iii) *The set V, if it is non-empty, is made of points with energy greater than H_0, does not contain non-trivial saddles and is completely attracted by the set*

$\tilde{A} \cup \bar{C}(A)$, *namely*

$$V \cap \bar{M} = \emptyset, \qquad V \subseteq \bar{B}(\tilde{A} \cup \bar{C}(A)).$$

(iv) *For all* $y \in \bar{C}(A)$ *there exists a declining path* $\omega : y \to F(A)$, $\omega \subseteq \tilde{A} \cup \bar{C}(A)$.

Proof Since A is a cycle then $\Phi(A) > H_0$. By Proposition 6.7 and Proposition 6.18 the maximal connected components $A_1, \ldots, A_k \subseteq A$ of states in A with energy strictly less than H_0 are cycles contained in A with $\Phi(A_i) = H_0$.

Given A_i, A_j, for any $x_i \in A_i$, $x_j \in A_j$, by Proposition 6.12 $\mathcal{S}(x_i, x_j)$ takes the same value for every $x_i \in A_i$ and $x_j \in A_j$; moreover, for any $x_i \in A_i$, $x_j \in A_j$ $\mathcal{S}(x_i, x_j) = \mathcal{S}(F(A_i), F(A_j)) \subseteq \bar{C}(A)$.

From this, points (i), (ii), (iii) immediately follow.

Point (iv) easily follows from the fact that for any $x \in A \cap \bar{M}$, we have $\Phi(x, F(A)) \leq H_0$ since $\bar{C}(A)$ contains the internal non-trivial saddles with maximal energy. □

Remarks

(1) If we consider a cycle A being a subset of the basin of attraction $\bar{B}(x)$, for some $x \in M$, then we have $C(A) = \emptyset$, $\tilde{A} = \{x\}$ and $V = A \setminus \{x\}$.

(2) The simplest non-trivial example is the one in which $k = 2$ and the number of internal saddles of maximal energy is one ($\bar{C}(A) \equiv \{y_1\}$):

$$A = A_1 \cup A_2 \cup V \cup \{y_1\},$$

and

$$F(A) \equiv \{x\} \ \text{for} \ \ \text{some } x.$$

Notation Given a cycle A satisfying the hypotheses of Proposition 6.19 and the set $\tilde{A}$ given by (6.27) (see (6.26)), the integer $h(A) \geq 1$ denotes the number of cycles in $\tilde{A}$ with non-empty intersection with $F(A)$; $j(A) \geq 0$ is the number of cycles in $\tilde{A}$ with empty intersection with $F(A)$. We can write

$$A = A_1 \cup A_2 \cup \cdots \cup A_h \cup A_{h+1} \cup \cdots \cup A_{h+j} \cup \bar{C}(A) \cup V \tag{6.28}$$

where $A_i \cap F(A) \neq \emptyset$, $i = 1, \cdots, h$; $A_i \cap F(A) = \emptyset$, $i = h+1, \cdots, h+j$. With our convention A_1 always intersects $F(A)$.

Definition 6.20 Given a non-trivial cycle A we speak of its *maximal internal barrier* $\Theta(A)$ given by the maximal depth of the cycles A' contained in A which do not contain entirely the ground $F(A)$.

Remark Of course $\Theta(A) < \Gamma(A)$. If $h(A) > 1$ (there are at least two cycles in $\tilde{A}$ with non-empty intersection with $F(A)$) we have $\Theta(A) = \Gamma(A_1) = \bar{\Gamma}(A)$. If $h = 1$ we have $\Theta(A) = \max\{\Gamma(A_2), \ldots, \Gamma(A_{j+1}), \Theta(A_1)\}$; in any case $\Theta(A) \leq \bar{\Gamma}(A) = \Gamma(A_1)$.

6.2 Main results

We recall now a simple but useful result based on the stationarity of μ providing a lower bound in probability to the first hitting time to a particular state.

Lemma 6.21 *For any pair of states y, x such that $H(y) > H(x)$ and any positive $\varepsilon < H(y) - H(x)$, we have:*

$$P_x(\tau_y \le \exp[\beta(H(y) - H(x) - \varepsilon)]) \le \exp(-\beta\varepsilon)\{1 + \exp(-\beta[H(y) - H(x) - \varepsilon])\}.$$

Proof See the proof of Proposition 4.7 in Chapter 4. □

We now give a simple lower bound in probability to the first exit time from a cycle, for a reversible, not necessarily Metropolis, Markov chain. It is useful to extend to general reversible Markov chains the results we are giving on the first exit problem for Metropolis Markov chains.

Lemma 6.22 *Consider a Markov chain satisfying condition R. Given a non-trivial cycle A let $F(A)$ be its ground defined exactly as in (6.6) and let $\Phi(A) = \min_{x\in\partial^- A, y\in\partial A} \tilde{H}(x, y)$, then for any $x \in F(A)$, ε such that $0 < \varepsilon < \Phi(A) - H(F(A))$,*

$$\lim_{\beta\to\infty} P_x\big(\tau_{\partial A} \le e^{\beta[\Phi(A) - H(F(A)) - \varepsilon]}\big) = 0. \tag{6.29}$$

Proof Given $T \in \mathbb{N}\setminus\{1\}$ we have, using reversibility:

$$\begin{aligned} &P_x(\tau_{\partial A} \le T) \\ &= \sum_{y\in\partial A}\left[P(x, y) + \sum_{n=1}^{T-1} \sum_{\substack{x_1,\dots,x_{n-1}\in A \\ x_n\in\partial^- A}} P(x, x_1)\cdots P(x_{n-1}, x_n)P(x_n, y)\right] \\ &= \sum_{y\in\partial A}\left[P(x, y) + \sum_{n=1}^{T-1} \sum_{\substack{x_1,\dots,x_{n-1}\in A \\ x_n\in\partial^- A}} e^{-\beta[H(x_n) - H(x)]}\right. \\ &\qquad\left.\times P(x_n, x_{n-1})\cdots P(x_1, x)P(x_n, y)\right] \\ &\le \max_{z\in\partial^- A, y\in\partial A} e^{-\beta[\tilde{H}(z,y) - H(x)]}|\partial^- A|\,|\partial A| T \end{aligned} \tag{6.30}$$

since for $z \in \partial^- A$, $y \in \partial A$ we have $e^{-\beta[H(z) - H(x)]}P(z, y) \le e^{-\beta[\tilde{H}(z,y) - H(x)]}$ and, for any $n \in \mathbb{N}$, $\sum_{x_1,\dots,x_{n-1}\in A} P(z, x_{n-1})\cdots P(x_1, x) \le 1$.

From (6.30), choosing $T = \inf\{n \in \mathbb{N} : n > \exp[\beta(\Phi(A) - H(F(A)) - \varepsilon)\}$ we conclude the proof. □

Let us go back to Markov chains satisfying condition M. The following theorem contains the basic results on the first exit from a cycle. These results are less immediate than the previous ones. As a matter of fact they are known in a more general situation, such as the one considered by Freidlin and Wentzell and discussed in Section 5.3 with the help of suitable graphs. We provide here, in the reversible case, a proof based on the analysis of typical paths. We use a simple intuitive argument involving the construction of appropriate events taking place on suitable, generally exponentially large in β, intervals of time. We see here the appearance of what can be called the *resistance times* and why they play an important role in providing an efficient mechanism of escape.

Theorem 6.23 *Consider a Markov chain satisfying condition M. Given a non-trivial cycle A:*

(i) *for all $\varepsilon > 0$, there exist $\beta_0 > 0$ and $k > 0$ such that for any $\beta > \beta_0, x \in A$*

$$P_x(\tau_{\partial A} > \exp(\beta [\Gamma(A) - \varepsilon])) \geq 1 - e^{-k\beta};$$

(ii) *for all $\varepsilon > 0$ there exist $\beta_0 > 0$ and $k' > 0$ such that for any $\beta > \beta_0$ and for any $x \in A$*

$$P_x(\tau_{\partial A} < \exp(\beta [\Gamma(A) + \varepsilon])) \geq 1 - e^{-e^{k'\beta}};$$

(iii) *there exist $\kappa = \kappa(A) > 0$, $\beta_0 > 0$ and $k'' > 0$ such that for all $\beta > \beta_0$ and for any $x, x' \in A$*

$$P_x(\tau_{x'} < \tau_{\partial A} \; ; \; \tau_{x'} < \exp(\beta[\Gamma(A) - \kappa]) \geq 1 - e^{-k''\beta};$$

(iv) *there exists $c > 0$ such that for any $x \in A$, $\hat{y} \in \partial A$ and β sufficiently large*

$$P_x(X_{\tau_{\partial A}} = \hat{y}) \geq c \exp(-\beta[H(\hat{y}) - \Phi(A))]).$$

Proof Recall that $\Gamma(A) = \Phi(A) - H(F(A))$ (see (6.11)).

The proof uses induction on the total number of internal non-trivial saddles $|C(A)|$.

We first assume that properties (i), (ii), (iii), (iv) are verified for all non-trivial cycles A with $|C(A)| \leq n$, for a given integer $n \geq 0$, and we prove them for all non-trivial cycles A with $|C(A)| = n + 1$; then we verify (i), (ii), (iii), (iv) for the case $n = 0$, the basis of the induction. This case corresponds to a cycle A being (part of) the strict basin of attraction of a stable plateau $F(A)$ (in particular $F(A)$ could be a single local minimum x).

Consider a generic non-trivial cycle A with $|C(A)| = n + 1$. We can use the decomposition given by Proposition 6.19, namely:

$$A = \tilde{A} \cup \bar{C}(A) \cup V$$

where $\tilde{A}$, defined in (6.26), (6.27) is a union of the highest internal transient cycles A_j which, beyond satisfying the properties specified in Proposition 6.19, have, for any j, a number of internal saddles $|C(A_j)|$ less than or equal to n so that they satisfy the recursive hypotheses (i), (ii), (iii), (iv).

The scheme of the proof is the following.

Step (a). Prove that (ii), (iv) for $|C(A)| \leq n$ imply (ii) for $|C(A)| = n + 1$.

Step (b). Prove that (i), (ii), (iii), (iv) for $|C(A)| \leq n$ imply (i) for $|C(A)| = n + 1$.

Step (c). Prove that (ii), (iii), (iv) for $|C(A)| \leq n$ and (i) for $|C(A)| = n + 1$. imply (iii) for $|C(A)| = n + 1$.

Step (d). Prove that (ii), (iii) for any non-trivial cycle A imply (iv) for the same cycle.

Step (e). Prove that (i), (ii), (iii) hold true for $n = 0$, the basis of the induction.

Step (a) We want to prove (ii) for our cycle A with $|C(A)| = n + 1$. Given $\varepsilon > 0$ let

$$T_1 = T_1(\varepsilon) := [\exp(\beta(\bar{\Gamma}(A) + \varepsilon/2))] \tag{6.31}$$

and

$$T_2 = T_2(\varepsilon) := [\exp(\beta(\Gamma(A) + \varepsilon))]. \tag{6.32}$$

Recall $\bar{\Gamma}(A) = H(\bar{C}(A)) - H(F(A))$, see (6.25).

Then the argument goes as follows. We construct, for every state $x \in A$, an event $\mathcal{E}_{x,T_1}$ containing trajectories starting from x at time $t = 0$, defined in the interval of time $[0, T_1]$ and satisfying the following conditions:

(1) if $\mathcal{E}_{x,T_1}$ takes place, our Markov chain X_t hits ∂A before T_1 and

(2) $\inf_{x \in A} P(\mathcal{E}_{x,T_1}) \geq \alpha_{T_1} > 0$ with $\lim_{\beta\to\infty}(1 - \alpha_{T_1})^{T_2/T_1} = 0.$ (6.33)

In particular we take

$$\alpha_{T_1} = \exp(-\beta\{\Phi(A) - H(\bar{C}(A)) + \varepsilon/4\}). \tag{6.34}$$

Suppose first that T_2/T_1 is not integer; let us divide the interval $[0, T_2]$ into $q = [T_2/T_1] + 1$ intervals such that the first $q - 1$ are of length T_1 and the last one is of length less than T_1; if T_2/T_1 is integer we just divide T_2 into $q = T_2/T_1$ intervals of length T_1. By properties (1) and (2) above of $\mathcal{E}_{x,T_1}$, we easily get the proof of our theorem. For, if $\tau_{\partial A} > T_2$, necessarily, by property (1), in none of the q intervals of length T_1 can the (translated of the) event $\mathcal{E}_{x,T_1}$ have taken place; by (2) and the Markov property, part (ii) of our proposition follows directly.

More explicitly, let $t_i = iT_1,\ i = 0, \ldots, q-1,\ t_q = T_2$; we have $t_{i+1} - t_i = T_1, i = 0, \ldots, q-2;\ t_q - t_{q-1} \le T_1$. Let $x_0 = x$. For $i = 0, \ldots, q-1$, given $x_i, x_{i+1} \in A$, we set, for $i = 0, \ldots, q-1$:

$$\mathcal{F}^{(i)}_{x_i,x_{i+1}} := \{\omega \in \Omega : \omega_{t_i} = x_i, \omega_{t_{i+1}} = x_{i+1}, \omega_s \in A\ \forall\, s \in [t_i, t_{i+1}]\}. \quad (6.35)$$

We have, for $x \in A$:

$$\begin{aligned} P_x(\tau_{\partial A} > T_2) &= \sum_{x_1,x_2,\ldots,x_q \in A} P(\mathcal{F}^{(0)}_{x,x_1} \cap \mathcal{F}^{(1)}_{x_1,x_2} \cap \cdots \cap \mathcal{F}^{(q-2)}_{x_{q-2},x_{q-1}} \cap \mathcal{F}^{(q-1)}_{x_{q-1},x_q}) \\ &= \sum_{x_1,x_2,\ldots,x_q \in A} P(\mathcal{F}^{(q-1)}_{x_{q-1},x_q} | \mathcal{F}^{(0)}_{x,x_1} \cap \mathcal{F}^{(1)}_{x_1,x_2} \cap \cdots \cap \mathcal{F}^{(q-2)}_{x_{q-2},x_{q-1}}) \\ &\quad \times P(\mathcal{F}^{(q-2)}_{x_{q-2},x_{q-1}} | \mathcal{F}^{(0)}_{x,x_1} \cap \mathcal{F}^{(1)}_{x_1,x_2} \cap \cdots \cap \mathcal{F}^{(q-3)}_{x_{q-3},x_{q-2}}) \\ &\quad \cdots P(\mathcal{F}^{(1)}_{x_1,x_2} | \mathcal{F}^{(0)}_{x,x_1}) P(\mathcal{F}^{(0)}_{x,x_1}). \end{aligned} \quad (6.36)$$

By the Markov property

$$P(\mathcal{F}^{(i)}_{x_i,x_{i+1}} | \mathcal{F}^{(0)}_{x,x_1} \cap \cdots \cap \mathcal{F}^{(i-1)}_{x_{i-1},x_i}) = P(\mathcal{F}^{(i)}_{x_i,x_{i+1}} | X_{t_i} = x_i). \quad (6.37)$$

Let

$$\mathcal{F}_x := \{\omega \in \Omega : \omega_0 = x;\ \ \omega_t \in A\ \forall\, t \in \{0, \ldots, t_1\}\} \quad (6.38)$$

by the homogeneity of our chain we have:

$$\sum_{x_{i+1} \in A} P(\mathcal{F}^{(i)}_{x_i,x_{i+1}} | X_{t_i} = x_i) = P(\mathcal{F}_{x_i}). \quad (6.39)$$

Suppose we know that for a suitable positive α:

$$\sup_{x \in A} P(\mathcal{F}_x) < 1 - \alpha, \quad (6.40)$$

from (6.36), (6.37), (6.39) (6.40) it follows immediately that

$$P_x(\tau_{\partial A} > T_2) \le (1-\alpha)^{q-1}. \quad (6.41)$$

Now, if there exists $\kappa > 0$ such that

$$\alpha(\beta) \frac{T_2(\beta)}{T_1(\beta)} \ge e^{\kappa\beta}, \quad (6.42)$$

we conclude the proof of point (ii) of the theorem. To get (6.40) it is sufficient to construct, for any $x \in A$, an event $\mathcal{E} = \mathcal{E}_{x,T_1}$ such that

$$(\mathcal{F}_x)^c \supseteq \mathcal{E}_{x,T_1}, \qquad \inf_{x \in A} P(\mathcal{E}_{x,T_1}) > \alpha.$$

Then the proof of part (ii) of the theorem is reduced to the construction, for any $x \in A$, of such an event $\mathcal{E}_{x,T_1}$, namely a set of trajectories $\mathcal{E}_{x,T_1}$ starting at x and exiting, in a particular way, from A within T_1.

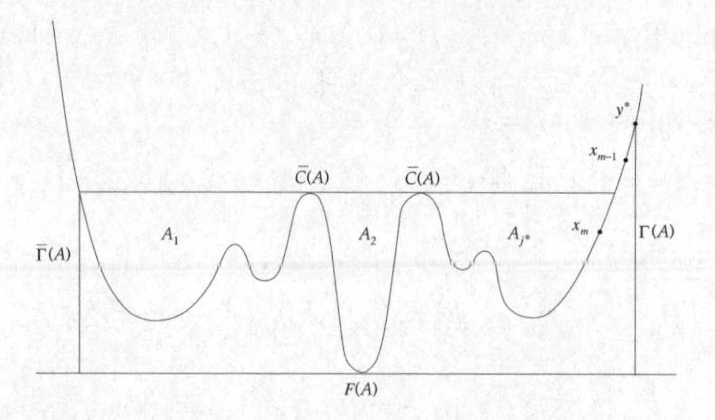

Figure 6.1

Let us give a description of $\mathcal{E}_{x,T_1}$.

Let y^* be a state in $U(A) \equiv F(\partial A)$; it will be our target in the sense that the trajectories in $\mathcal{E}_{x,T_1}$ will exit from A through y^*. By definition, there exists a downhill path from y^* to the set $\tilde{A}$:

$$\bar{x}_0 = y^*, \bar{x}_1, \bar{x}_2, \ldots, \bar{x}_m : \quad \bar{x}_1, \ldots, \bar{x}_{m-1} \in V \cup \bar{C}(A); \quad \bar{x}_{m-1} \in \partial \tilde{A}, \ \bar{x}_m \in \tilde{A}$$

(with $H(\bar{x}_{i+1}) \leq H(\bar{x}_i)$). We set $m = 1$ if y^* is communicating with $\tilde{A}$, i.e. $y^* \in \partial \tilde{A}$. Let A_{j^*} be the particular component (one of the highest internal transient cycles) of $\tilde{A}$ hit by this path (i.e. $\bar{x}_m \in A_{j^*}$), see Figure 6.1. The event $\mathcal{E}_{x,T_1}$ is then defined by requiring that the process, starting from x, hits the set A_{j^*} in a time shorter than $T_1/2$, and then, after reaching the boundary ∂A_{j^*} of A_{j^*} at $\bar{x}_{m-1}$ within a time $T_1/4$, it follows the path obtained from $\bar{x}_0, \ldots, \bar{x}_{m-1}$ by time reversal.

More precisely let

$$\tau_{\partial A_{j^*}}^{(>\tau_{A_{j^*}})} := \min\{t > \tau_{A_{j^*}};\ X_t \in \partial A_{j^*}\}$$

where, as before, $\tau_{A_{j^*}}$ is the first hitting time to the set A_{j^*}. Then:

$$\mathcal{E}_{x,T_1} := \left\{\tau_{A_{j^*}} < \frac{T_1}{2}\right\} \cap \left\{\tau_{\partial A_{j^*}}^{(>\tau_{A_{j^*}})} - \tau_{A_{j^*}} < \frac{T_1}{4}\right\} \cap \left\{X_{\tau_{\partial A_{j^*}}^{(>\tau_{A_{j^*}})}} = y^*\right\},$$

for $m = 1$;

$$\mathcal{E}_{x,T_1} := \left\{\tau_{A_{j^*}} < \frac{T_1}{2}\right\} \cap \left\{\tau_{\partial A_{j^*}}^{(>\tau_{A_{j^*}})} - \tau_{A_{j^*}} < \frac{T_1}{4}\right\} \cap \left\{X_{\tau_{\partial A_{j^*}}^{(>\tau_{A_{j^*}})}} = \bar{x}_{m-1}\right\}$$

$$\cap \left\{X_{\tau_{\partial A_{j^*}}^{(>\tau_{A_{j^*}})}+s} = \bar{x}_{m-1-s}, \quad \forall s = 1, \ldots, m-1\right\}$$

for $m > 1$.

Using the Markov property we have, for $m > 1$,

$$P(\mathcal{E}_{x,T_1}) = E_x\left(\mathbf{1}_{\tau_{A_{j^*}} < \frac{T_1}{2}} P_{X_{\tau_{A_{j^*}}}}\left(\left\{\tau_{\partial A_{j^*}} < \frac{T_1}{4}\right\} \cap \{X_{\tau_{\partial A_{j^*}}} = \bar{x}_{m-1}\}\right)\right)$$
$$\times\ P(\bar{x}_{m-1}, \bar{x}_{m-2}) P(\bar{x}_{m-2}, \bar{x}_{m-3}) \cdots P(\bar{x}_1, y^*) \tag{6.43}$$

where E_x denotes expectation over the process starting from x and $\mathbf{1}_{\mathcal{E}}$ denotes the characteristic function of the event $\mathcal{E}$; similarly for $m = 1$.

Let $T_0 = |S|$; notice that, for β sufficiently large, $T_0 < T_1/4$. If $x \notin \tilde{A}$ certainly there exists a downhill path ω entering $\tilde{A}$ in a time shorter than T_0. Suppose that this path ends in $A_{\hat{j}}$ or that $x \in A_{\hat{j}}$. From Proposition 6.19 we know that there is a sequence of contiguous highest internal transient cycles joining $A_{\hat{j}}$ to A_{j^*}; namely there exist $l \in \mathbb{N}$ and cycles $A_{j_1}, \dots, A_{j_l}$ among the highest internal transient cycles in $\tilde{A}$, with the properties:

(i) $A_{j_1} = A_{\hat{j}}$, $A_{j_l} = A_{j^*}$, moreover, either $l = 1$ and $\hat{j} = j^*$ or

(ii) the A_{j_k} are contiguous in the sense that there exists a sequence of plateaux $D_1, \dots, D_{l-1}$ with $D_i \subseteq \bar{C}(A)$, $i = 1, \dots, l-1$ (D_i are made of maximal saddles) and

$$\mathcal{S}(A_{j_k}) \cap D_k \neq \emptyset, \quad \mathcal{S}(A_{j_{k+1}}) \cap D_k \neq \emptyset, \quad k = 1, \dots, l-1.$$

Consider the set of paths

$$\bar{\Omega}_{x,A_{j_1},\dots A_{j_l}} := \Bigg\{\omega \in \Omega : \exists s_i, t_{i+1} \in \mathbb{N}, i = 0, \dots, l-1, \text{ with}$$
$$s_0 = 0,\ t_{i+1} \in [s_i, s_i + T_0], s_{i+1} \in \left[t_{i+1}, t_{i+1} + \frac{T_1}{8l}\right] \text{ such that} \tag{6.44}$$
$$\omega_0 = x;\ \omega_{t_i} \in A_{j_i}, i = 1, \dots, l;\ \omega_{s_i} \in D_i, i = 1, \dots, l-1\Bigg\}.$$

In words: $\bar{\Omega}_{x,A_{j_1},\dots A_{j_l}}$ contains the set of paths ω starting from x such that they enter A_{j_1} (if x were not already there) within T_0. If $\hat{j} = j^*$ the paths are concluded; otherwise they get out of A_{j_1} through a minimal saddle belonging to the plateau D_1 within a time $T_1/8l$. Then they stay in D_1 and, within an interval of time T_0, they enter A_{j_2} ($T_0 < T_1/8l$ for β sufficiently large.) They continue in this way up to the entrance into $A_{j_l} = A_{j^*}$. Fix $\varepsilon > 0$ in (6.31). We have, for β sufficiently large,

$$P(\mathcal{E}_{x,T_1}) \geq P(\bar{\Omega}_{x,A_{j_1},\dots,A_{j_l}}) \inf_{y \in A_{j^*}} P_y\left(\left\{\tau_{\partial A_{j^*}} < \frac{T_1}{4}\right\} \cap \{X_{\tau_{\partial A_{j^*}}} = \bar{x}_{m-1}\}\right)$$
$$\times \exp\left[-\beta(H(y^*) - H(\bar{x}_{m-1}) + \frac{\varepsilon}{12})\right] \tag{6.45}$$

where we have used reversibility and the fact that the path $y^*, \bar{x}_1, \dots, \bar{x}_{m-1}$ is downhill.

To estimate $P(\bar{\Omega}_{x,A_{j_1},\dots,A_{j_l}})$ we observe

(1) that there are downhill paths joining x with $\tilde{A}$ in a time shorter than T_0, whenever $x \in V \cup \bar{C}(A)$,

(2) the exit from A_j through a suitable minimal saddle leads to D_j, so that we can apply iterative hypothesis (iv) valid, together with (ii), for any A_j, since $|C(A_j)| \le n$,

(3) in $\bar{\Omega}_{x,A_{j_1},\dots A_{j_l}}$ the exit times from the A_j are supposed to be smaller than $T_1/8l$ which, in turn, for large enough β, is exponentially larger than the maximal typical exit time ($\sim \exp \beta[H(\bar{C}(A_j)) - H(F(A_j))]$) from A_j so that we can apply iterative hypotheses (ii).

From these considerations and from the definition of plateau it follows immediately that for all sufficiently large β,

$$P(\bar{\Omega}_{x,A_{j_1},\dots,A_{j_l}}) > \exp\left(-\frac{\varepsilon}{12}\beta\right). \tag{6.46}$$

Again, by the iterative hypotheses (ii), (iv), the second term on the right-hand side of (6.45) is estimated as follows for β sufficiently large:

$$\inf_{x \in A_{j^*}} P_x\left(\left\{\tau_{\partial A_{j^*}} < \frac{T_1}{4}\right\} \cap \{X_{\tau_{\partial A_{j^*}}} = \bar{x}_{m-1}\}\right) \ge e^{-\frac{\varepsilon}{12}\beta} e^{-\beta[H(\bar{x}_{m-1}) - \Phi(A_{j^*})]}. \tag{6.47}$$

In conclusion, from (6.45), (6.46), (6.47), we obtain the following estimate, valid for β sufficiently large:

$$P(\mathcal{E}_{x,T_1}) \ge e^{-\frac{\varepsilon}{4}\beta} e^{-\beta[H(y^*) - \Phi(A_{j^*})]} = e^{-\frac{\varepsilon}{4}\beta} e^{-\beta[\Phi(A) - H(\bar{C}(A))]} =: \alpha. \tag{6.48}$$

This shows the validity of (6.42) and concludes the proof of part (ii) of the theorem with, say, $k' = \varepsilon/5$.

Step (b) We want now to prove (i) for $|C(A)| = n+1$, by supposing the validity of (i), (ii), (iii), (iv) for $|C(A)| \le n$, namely we want to prove that, for any $\varepsilon > 0$, there exists $k > 0$ such that for any $x \in A$ and β sufficiently large:

$$P_x(\tau_{\partial A} < e^{\beta(\Gamma(A) - \varepsilon)}) \le e^{-k\beta}. \tag{6.49}$$

For $x \in F(A)$ (6.49) is an immediate consequence of Lemma 6.21; then, using the Markov property, it will be proved in general once we have that there exists $k_1 > 0$, such that for any $x \in A$:

$$P_x(\tau_{F(A)} < \tau_{\partial A} \ ; \ \tau_{F(A)} < e^{\beta \bar{\Gamma}(A)}) \ge 1 - e^{-k_1 \beta}. \tag{6.50}$$

Indeed from (6.50) we get in particular for any $\varepsilon > 0$

$$P_x(\tau_{\partial A} < e^{\beta(\Gamma(A)-\varepsilon)}) \le \sup_{z\in F(A)} P_z(\tau_{\partial A} < e^{\beta(\Gamma(A)-\varepsilon)}) + e^{-k_1\beta} < e^{-k\beta}, \tag{6.51}$$

for a suitable positive k and any sufficiently large β, by Lemma 6.21.

Now, to prove (6.50) we first note that for any $\varepsilon > 0$ there exists $k_2 > 0$ such that for any $x \in V \cup \bar{C}(A)$:

$$P_x(\tau_{\tilde{A}} < \tau_{\partial A}; \tau_{\tilde{A}} < e^{\beta\varepsilon}) \ge 1 - e^{-k_2\beta}. \tag{6.52}$$

Indeed from Lemma 6.21 we know that there exist constants $\tilde{k}_1, \tilde{k}_2 > 0$ such that for all $x \in V \cup \bar{C}(A)$

$$P_x(\tau_{\partial A} < e^{\tilde{k}_1\beta}) < e^{-\tilde{k}_2\beta}; \tag{6.53}$$

moreover, using the Markov property as in (6.41), since there exist c such that $0 < c < 1$ and $t_0 \in \mathbb{N}$ such that $\inf_{x\in V\cup\bar{C}(A)} P_x(\tau_{\tilde{A}} < t_0) > c$, we get for any ε such that $0 < \varepsilon < \tilde{k}_1$ and β sufficiently large,

$$P_x(\tau_{\tilde{A}} > e^{\varepsilon\beta}; \tau_{\partial A} > e^{\tilde{k}_1\beta}) < (1-c)^{e^{\frac{\varepsilon}{2}\beta}}, \tag{6.54}$$

and from (6.53), (6.54) we get

$$P_x(\tau_{\tilde{A}} > e^{\varepsilon\beta}) \le P_x(\tau_{\tilde{A}} > e^{\varepsilon\beta}; \tau_{\partial A} \ge e^{\tilde{k}_1\beta}) + e^{-\tilde{k}_2\beta} \le (1-c)^{e^{\frac{\varepsilon}{2}\beta}} + e^{-\tilde{k}_2\beta}. \tag{6.55}$$

From (6.53), (6.55), we easily get (6.52). Thus we can suppose our starting point x to belong to $\tilde{A}$.

Suppose x belongs to the transient cycle A_j (among the highest internal cycles). If $A_j \cap F(A) \neq \emptyset$, using the iterative hypothesis (iii) valid for A_j since $|C(A_j)| \le n$, we conclude the proof of (6.50). Let us then assume that $A_j \cap F(A) = \emptyset$. From the recursive hypothesis (ii), since $\Gamma(A_j) < \bar{\Gamma}(A)$, as a consequence of $A_j \cap F(A) = \emptyset$, we have that there exist $k_3 > 0$, $\delta > 0$ such that for any $x \in A_j$,

$$P_x(\tau_{\partial A_j} > e^{\beta(\bar{\Gamma}(A)-\delta)}) \le e^{-k_3\beta}. \tag{6.56}$$

Moreover we have that there exists $k_4 > 0$ such that for any $x \in A_j$,

$$P_x(X_{\tau_{\partial A_j}} \notin U(A_j)) \le e^{-k_4\beta}. \tag{6.57}$$

Indeed from the recursive hypothesis (ii) we know that for any $\varepsilon > 0$ there exist positive constants κ_1, κ_2 such that for all sufficiently large β,

$$\begin{aligned} P_x\big(X_{\tau_{\partial A_j}} \notin U(A_j)\big) &\le P_x\big(X_{\tau_{\partial A_j}} \notin U(A_j), \tau_{\partial A_j} < e^{\beta(\Gamma(A_j)+\varepsilon)}\big) + e^{-e^{\kappa_1\beta}} \\ &\le P_x\big(\tau_{(\partial A_j\setminus U(A_j))} \le e^{\beta(\Gamma(A_j)+\varepsilon)}\big) + e^{-e^{\kappa_1\beta}}. \end{aligned} \tag{6.58}$$

Let us pick $y \in F(A_j)$; by recursive hypothesis (iii), we have, for a suitable $\kappa_2 > 0$ and β sufficiently large,

$$P_x(\tau_{\partial A_j \setminus U(A_j)} \leq e^{\Gamma(A_j)+\varepsilon}) \leq P_x(\tau_{\partial A_j \setminus U(A_j)} \leq e^{\Gamma(A_j)+\varepsilon}, \tau_y \leq \tau_{\partial A_j}) + e^{-2\kappa_2\beta}$$
$$\leq P_y(\tau_{\partial A_j \setminus U(A_j)} \leq e^{\Gamma(A_j)+\varepsilon}) + e^{-\kappa_2\beta}. \tag{6.59}$$

Choosing

$$\varepsilon = \varepsilon_0 = \frac{1}{2} \min_{x \in \partial A_j \setminus U(A_j)} (H(x) - \Phi(A_j))$$

we have, from Lemma 6.21,

$$P_y\big(\tau_{(\partial A_j \setminus U(A_j))} \leq e^{\beta(\Gamma(A_j)+\varepsilon_0)}\big) \leq e^{-\varepsilon_0\beta}. \tag{6.60}$$

From (6.58), (6.59), (6.60), we get (6.57).

Now, given $A_l \in \tilde{A}$ contiguous to A_j so that there is a plateau $D_{j,l} \subseteq \bar{C}(A)$ connecting A_j to A_l, we have, using the recursive hypothesis (iv), for any $x \in A_j$, $y \in D_{j,l} \cap \partial A_j$, for all $\varepsilon > 0$ and β sufficiently large:

$$P_x(X_{\tau_{\partial A_j}} = y) \geq e^{-\varepsilon\beta}. \tag{6.61}$$

Moreover, using $U(A_j) \subseteq \bar{C}(A)$, by (6.52) and the definition of plateau, we have, for any $y \in D_{j,l}$, $\varepsilon_1, \varepsilon_2 > 0$ and β sufficiently large,

$$P_y(\tau_{\tilde{A}} < \tau_{\partial A} \ ; \ X_{\tau_{\tilde{A}}} \in A_l \ ; \ \tau_{\tilde{A}} < e^{\beta\varepsilon_1}) \geq e^{-\varepsilon_2\beta}. \tag{6.62}$$

Let us define the following times

$$\begin{aligned} \tau_0 &:= \inf\{t \geq 0 : X_t \notin \tilde{A}\} \\ \sigma_1 &:= \inf\{t > \tau_0 : X_t \in \tilde{A}\}, \end{aligned} \tag{6.63}$$

and for $j = 1, 2, \ldots$

$$\begin{aligned} \tau_j &:= \inf\{t > \sigma_j : X_t \notin \tilde{A}\} \\ \sigma_{j+1} &:= \inf\{t > \tau_j : X_t \in \tilde{A}\}. \end{aligned} \tag{6.64}$$

Let A_1 be one of the highest internal transient cycles such that $A_1 \cap F(A) \neq \emptyset$; let

$$\bar{\tau} := \tau_{A_1 \cup \partial A} \tag{6.65}$$

and

$$\nu := \inf\{i, \text{ such that } \ \tau_i \leq \bar{\tau} \leq \sigma_{i+1}\}. \tag{6.66}$$

For any pair $A_i, A_{i'}$ among the highest internal transient cycles, there exists at least a sequence of contiguous cycles joining them. Without loss of generality we can suppose A_1 such that there exists a sequence of contiguous cycles $A_{j_1}, \ldots, A_{j_k}$ (for some integer k) with $A_{j_1} = A_j$ (our initial cycle), $A_{j_k} = A_1$, $A_{j_l} \cap F(A) = \emptyset, l = 1, \ldots, k-1$.

From (6.61), (6.62), given $x \in A_j$ we know, in complete analogy with $\bar{\Omega}_{x,A_{j_1},\dots,A_{j_l}}$ defined in (6.44), that there exists a set $\tilde{\Omega}_{x,A_{j_1},\dots,A_{j_k}}$ of paths ω starting from x and joining A_j to A_1 through the sequence $A_{j_1},\dots,A_{j_k}$ within a time $e^{\beta(\bar{\Gamma}-\frac{\delta}{2})}$ for which we have, for any $\varepsilon > 0$ and β sufficiently large,

$$P(\tilde{\Omega}_{x,A_{j_1},\dots,A_{j_k}}) > e^{-\varepsilon\beta}. \tag{6.67}$$

From (6.52) and (6.67) we easily get, using the Markov property, that for all $\varepsilon > 0$, there exists $c > 0$ such that for any $x \in \tilde{A}$:

$$P_x(\nu > e^{\varepsilon\beta}) \le e^{-e^{c\beta}}. \tag{6.68}$$

We say that an interval $[\sigma_l, \sigma_{l+1}]$ is good if

(1) $\tau_l - \sigma_l < e^{\beta(\bar{\Gamma}(A)-\delta)}$,
(2) $X_{\tau_l} \in U(A_{j_l})$ if $X_{\sigma_l} \in A_{j_l}$,
(3) $\tau_{\partial A} \notin [\tau_l, \sigma_{l+1}]$,
(4) $\sigma_{l+1} - \tau_l < e^{\beta\frac{\delta}{8}}$.

From (6.56), (6.57) and (6.52), we easily get that there exists a positive constant k_5 such that for any $x \in \tilde{A}$,

$$P_x([\sigma_l, \sigma_{l+1}] \text{ is not good}) \le e^{-k_5\beta}, \tag{6.69}$$

implying

$$P_x([\sigma_l, \sigma_{l+1}] \text{ are all good for } l = 1, \dots, e^{\varepsilon\beta}) \ge 1 - e^{-(k_5-\varepsilon)\beta}. \tag{6.70}$$

We have, for all $x \in \tilde{A}$, ε such that $0 < \varepsilon < \delta/4$, using (6.68):

$$\begin{aligned} P_x(X_{\bar{\tau}} \in A_1; \bar{\tau} < e^{\beta(\bar{\Gamma}(A)-\frac{\delta}{2})}) &\ge P_x(X_{\bar{\tau}} \in A_1; \bar{\tau} < e^{\beta(\bar{\Gamma}(A)-\frac{\delta}{2})}; \nu \le e^{\varepsilon\beta}) \\ &\ge P([\sigma_l, \sigma_{l+1}] \text{ are all good for } l = 1, \dots, e^{\varepsilon\beta}, \nu \le e^{\varepsilon\beta}) \\ &\ge P([\sigma_l, \sigma_{l+1}] \text{ are all good for } l = 1, \dots, e^{\varepsilon\beta}) - \text{ SES} \\ &\ge 1 - \text{ SES } - e^{-(k_5-\varepsilon)\beta} \end{aligned} \tag{6.71}$$

where SES stands for any superexponentially small quantity as β goes to infinity (see Definition 6.4).

From (6.71) and from the recursive hypothesis (iii) applied to A_1 we conclude the proof of (6.50).

Step (c) We proceed similarly to the proof of (ii) using recurrence in $\tilde{A}$ and the Markov property. We first prove that there exist $\kappa > 0, k_6 > 0$ such that for any $x, x' \in A$ and β sufficiently large:

$$P_x\big(\tau_{x'} < \exp(\beta[\Gamma(A) - \kappa])\big) \ge 1 - e^{-k_6\beta}. \tag{6.72}$$

Let us first consider the case $x' \in \tilde{A}$; we construct an event $\mathcal{E}'_{x,x',T_1}$ similar to $\mathcal{E}_{x,T_1}$; $\mathcal{E}'_{x,x,'T_1}$ contains paths of our process starting from x, ending at x' during a time interval shorter than $T_1(\varepsilon_0) = \exp[\beta(\bar{\Gamma}(A) + \varepsilon_0)]$, with

$$\varepsilon_0 = \frac{\Phi(A) - H(\bar{C}(A))}{4} = \frac{\Gamma(A) - \bar{\Gamma}(A)}{4}$$

and following a suitable sequence of cycles A_j and plateaux contained in $\bar{C}(A)$.

We describe $\mathcal{E}'_{x,x',T_1}$ in words leaving to the reader the easy task of a precise definition, along the same lines seen previously for $\mathcal{E}_{x,T_1}$.

$\mathcal{E}'_{x,x',T_1}$ contains the set of trajectories which, starting from x, first descend, in a time much shorter than $T_1(\varepsilon_0)/3$, to $\tilde{A}$ (if they were not still there) and this happens with a probability almost one for β large; then, within a time of order $T_1(\varepsilon_0)/3$, following a suitable sequence of cycles A_i and plateaux contained in $\bar{C}(A)$ the paths in $\mathcal{E}'_{x,x',T_1}$ enter the cycle A_j, say A', containing x'; by the recursive hypothesis (ii), as before, this happens with probability logarithmically equivalent to one. Now, it follows from the recursive hypothesis (iii), valid for A', that, with probability almost one for large β, before leaving A', and in a time typically much shorter that $T_1(\varepsilon_0)/3$, our process touches x'.

In conclusion we get that for all $\varepsilon > 0$ and β sufficiently large

$$P(\mathcal{E}'_{x,x',T_1(\varepsilon_0)}) \geq e^{-\varepsilon\beta}. \tag{6.73}$$

From (6.49) we have, for some $k > 0$ and for all sufficiently large β:

$$P_x(\tau_{x'} > T_1(2\varepsilon_0)) \leq P_x(\tau_{x'} > T_1(2\varepsilon_0)\ ;\ \tau_{\partial A} > T_1(2\varepsilon_0)) + e^{-k\beta}. \tag{6.74}$$

Using recurrence as in (6.36), from (6.73) and the Markov property we get that for β sufficiently large

$$P_x(\tau_{x'} > T_1(2\varepsilon_0); \tau_{\partial A} > T_1(2\varepsilon_0)) \leq (1 - e^{-\beta\frac{\varepsilon_0}{2}})^{\frac{T_1(2\varepsilon_0)}{T_1(\varepsilon_0)}} \tag{6.75}$$

vanishing superexponentially fast for large β. This, together with (6.74), concludes the proof of (6.72) for $x' \in \tilde{A}$ with $\kappa = \varepsilon_0$.

When $x' \in V \cup \bar{C}(A)$, we proceed again similarly to the proof of point (ii). We know that there exists a downhill path from x' to the set $\tilde{A}$:

$$\bar{x}_0 = x', \bar{x}_1, \bar{x}_2, \ldots, \bar{x}_m :\quad \bar{x}_1, \ldots, \bar{x}_{m-1} \in V \cup \bar{C}(A);\quad \bar{x}_{m-1} \in \partial\tilde{A},\ \bar{x}_m \in \tilde{A}.$$

We set $m = 1$ if x' is communicating with $\tilde{A}$ ($x' \in \partial\tilde{A}$). Let A_{j^*} be the particular component of $\tilde{A}$ hit by this path (i.e. $\bar{x}_m \in A_{j^*}$). We define now a new event $\mathcal{E}''_{x,x',T_1}$ containing paths starting from x and ending in x' within the time $T_1 = T_1(\varepsilon_0)$. The first part of $\mathcal{E}''_{x,x',T_1}$ is defined similarly to $\mathcal{E}'$ above with $A' \equiv A_{j^*}$ but now we add an exit event from A_{j^*} through $\bar{x}_{m-1}$ in a time of order $\exp(\beta\Gamma(A_{j^*}))$ followed by the single uphill path $\bar{x}_{m-2}, \ldots, \bar{x}_1 = x'$. We leave the details to the reader. We have, using the recursive hypothesis, that for all $\varepsilon > 0$ and β sufficiently large

$$P(\mathcal{E}''_{x,x',T_1}) \geq \exp(-\beta[H(x') - H(\bar{C}(A)) + \varepsilon]) \tag{6.76}$$

which, again by recurrence like in (6.36) and the Markov property proves (6.72) with $\kappa = (\Phi(A) - H(x'))/2$.

Now, by using point (i) of the theorem, we get $P_x(\tau_{x'} < \tau_{\partial A}) > 1 - e^{-k_7 \beta}$ for a suitable $k_7 > 0$ and then the whole point (iii) since, by choosing ε sufficiently small, we obviously have

$$\exp(\beta[\Gamma(A) - \kappa]) < \exp(\beta[\Gamma(A) - \varepsilon]).$$

Step (d) Let us now prove point (iv); it follows directly from the previous points (ii),(iii).

Given $x \in A$, we first introduce an auxiliary Markov chain ξ_t with state space $\{x\} \cup \partial A$ defined as follows.

Let

$$\begin{aligned} &\tau_0 = 0, \quad \sigma_k = \min\{t > \tau_{k-1} \ : \ X_t \neq x\}, \\ &\tau_k = \min\{t \geq \sigma_k \ : \ X_t \ \in \{x\} \cup \partial A\}, \quad k = 1, \ldots \end{aligned} \tag{6.77}$$

then

$$\xi_k := X_{\tau_k}, \qquad k = 1, 2, \ldots \tag{6.78}$$

Given $\hat{y} \in \partial A$, we have

$$\begin{aligned} P_x(X_{\tau_{\partial A}} = \hat{y}) &= \sum_{m=0}^{\infty} P_x(\xi_1 = x)^m P_x(\xi_1 = \hat{y}) = \frac{P_x(\xi_1 = \hat{y})}{1 - P_x(\xi_1 = x)} \\ &= \frac{P_x(\xi_1 = \hat{y})}{P_x(\xi_1 \in \partial A)} = P_x(\xi_1 = \hat{y} | \xi_1 \in \partial A). \end{aligned} \tag{6.79}$$

Writing

$$\begin{aligned} &P_x(\xi_1 = \hat{y}) \\ &= \sum_{n=0}^{\infty} P(x,x)^n \left[P(x,\hat{y}) + \sum_{t=1}^{\infty} \sum_{\bar{x}_1, \bar{x}_2, \ldots, \bar{x}_t \in A \setminus x} P(x, \bar{x}_1, \bar{x}_2, \ldots, \bar{x}_t, \hat{y}) \right], \end{aligned} \tag{6.80}$$

and using reversibility we get:

$$\begin{aligned} &[1 - P(x,x)] P_x(\xi_1 = \hat{y}) = P(x,\hat{y}) + \sum_{t=1}^{\infty} \sum_{\bar{x}_1, \bar{x}_2, \ldots, \bar{x}_t \in A \setminus x} P(x, \bar{x}_1, \bar{x}_2, \ldots, \bar{x}_t, \hat{y}) \\ &= \left[P(\hat{y}, x) + \sum_{t=1}^{\infty} \sum_{\bar{x}_1, \bar{x}_2, \ldots, \bar{x}_t \in A \setminus x} P(\hat{y}, \bar{x}_t, \ldots, \bar{x}_2, \bar{x}_1, x) \right] \\ &\quad \times \exp[-\beta(H(\hat{y}) - H(x))]. \end{aligned} \tag{6.81}$$

We have:

$$\sum_{t=1}^{\infty} \sum_{\bar{x}_1,\bar{x}_2,\dots,\bar{x}_t \in A\setminus x} P(\hat{y}, \bar{x}_t, \dots, \bar{x}_2, \bar{x}_1, x) = \sum_{y\in A\setminus x} P(\hat{y}, y) P_y(\tau_x < \tau_{\partial A}). \tag{6.82}$$

By the already proved part (iii) of the theorem we get that there exists $c > 0$ and β sufficiently large:

$$P(\hat{y}, x) + \sum_{y\in A\setminus x} P(\hat{y}, y) P_y(\tau_x < \tau_{\partial A}) > c. \tag{6.83}$$

On the other hand we have:

$$\begin{aligned}
&[1 - P(x, x)] P_x(\xi_1 \in \partial A) \\
&= \sum_{z\in\partial A} P(x, z) + \sum_{t=1}^{\infty} \sum_{\bar{x}_1,\dots,\bar{x}_t \in A\setminus x; \bar{x}_{t+1}\in\partial A} P(x, \bar{x}_1, \dots, \bar{x}_t, \bar{x}_{t+1}) \\
&= \sum_{z\in\partial A} P(z, x) \exp[-\beta(H(z) - H(x))] \\
&\quad + \sum_{t=1}^{\infty} \sum_{\bar{x}_1,\dots,\bar{x}_t \in A\setminus x; \bar{x}_{t+1}\in\partial A} P(\bar{x}_{t+1}, \bar{x}_t, \dots, \bar{x}_2, \bar{x}_1, x) \\
&\quad \times \exp[-\beta(H(\bar{x}_{t+1}) - H(x))] \le \exp[-\beta(\Phi(A) - H(x))] \\
&\quad \times \sum_{z\in\partial A} \Big[P(z, x) + \sum_{y\in A\setminus x} P(z, y) P_y(\tau_x < \tau_{\partial A})\Big] \\
&\le |\partial A| \exp[-\beta(\Phi(A) - H(x))]
\end{aligned} \tag{6.84}$$

From (6.83), (6.84) we conclude the proof of (iv)

Step (e) Since, as we have seen in step (d), point (iv) follows directly, in general, from (ii), (iii), to conclude the proof of our theorem we have to show that properties (i), (ii), (iii) are true for A such that the number $|C(A)|$ of internal non-trivial saddles is zero, namely when the cycle A is part of (or coincides with) the strict basin of attraction of $F(A)$, $F(A)$ being a single plateau.

It will suffice to prove (ii), (iii) since then (i) follows easily by the following argument. When $x \in F(A)$ point (i) is an immediate consequence of Lemma 6.21. Otherwise it follows from Lemma 6.21 and from point (iii).

Property (ii) follows easily by the same argument used before: we construct, for any $x \in A$, $\varepsilon > 0$ and β sufficiently large, an event $\mathcal{E}_{x,T}$ with $T = T(\varepsilon) = \exp(\beta\varepsilon/2)$ which consists in descending from x to $F(A)$ in a time shorter than $T/2$ following a downhill path ω from x to $F(A)$; then in following an uphill path ω' from $F(A)$ up to $U(A)$ in a time shorter than $T/2$. This path ω' is the time-reversed of a path going downhill from $U(A)$ to $F(A)$. The paths ω

and ω' certainly exist as the cycle A is part of the strict basin of attraction of $F(A)$.

With $T_2 = T_2(\varepsilon)$ given by (6.32) we verify easily, in the present case, (6.33) and (6.34) with $T_1 := T(\varepsilon)$ since (1) for every $\varepsilon > 0$ the descent to $F(A)$, along a downhill path, takes place in a suitable finite time, much smaller than $T(\varepsilon)$, with a probability approaching one as $\beta \to \infty$ and (2) the ascent from $F(A)$ to $U(A)$, along an uphill path, in a suitable finite time, much smaller than $T(\varepsilon)$, for all $\varepsilon > 0$ and β sufficiently large, takes place with a probability larger than $\exp(-\beta[\Gamma(A) + \varepsilon/4])$. Then property (ii) follows easily.

The proof of property (iii) is completely analogous.

The proof of the theorem is concluded. □

We now give two useful corollaries.

Corollary 6.24 *Given a non-trivial cycle A, let $\Theta(A)$ be its maximal internal barrier (see Definition 6.20). Then for all $\varepsilon > 0$, there exists $\hat{k} > 0$ such that for any $x \in A$, $z \in F(A)$ and β sufficiently large,*

$$P_x(\tau_z > e^{\beta(\Theta(A)+\varepsilon)}) < e^{-\beta\hat{k}}. \tag{6.85}$$

Proof The proof can be obtained easily by readapting the proof of point (iii) in Theorem 6.23. □

Corollary 6.25 *Given a non-trivial cycle A there exists $\bar{\kappa} > 0$ such that*

$$\sup_{x\in A} P_x\big(X_{\tau_{\partial A}} \notin U(A)\big) < e^{-\bar{\kappa}\beta}. \tag{6.86}$$

Proof The proof is totally analogous to that of (6.57): it suffices to use Theorem 6.23 instead of the iterative hypothesis. □

We want now to give a result concerning a lower bound on the probability of the rare event consisting in exiting from a non-trivial cycle A too early, passing through a given point $y \in \partial A$.

This is contained in the following proposition which provides a counterpart to Theorem 6.23 (i) and extends Theorem 6.23 (iv).

Proposition 6.26 *Given a non-trivial cycle A and a positive number k,*

$$\Theta(A) < k \le \Gamma(A)$$

we have for any $x \in A$, $y \in \partial A$, $\varepsilon > 0$ and β sufficiently large:

$$P_x\left(\tau_{\partial A} < e^{\beta k}, X_{\tau_{\partial A}} = y\right) \ge e^{-\beta(H(y)-H(F(A))-k+\varepsilon)}. \tag{6.87}$$

Proof The proof is very similar to that of Theorem 6.23 (ii); again, we use induction on the number of internal saddles.

We suppose $|C(A)| = n + 1$ and assume (6.87) to be valid for every non-trivial cycle A' with $|C(A')| \le n$.

We recall the partition $A = \tilde{A} \cup \bar{C}(A) \cup V$, with $\tilde{A} = A_1 \cup \cdots \cup A_h \cup A_{h+1} \cup \cdots \cup A_{h+j}$, $h \ge 1$, $j \ge 0$, $A_i \cap F(A) \ne \emptyset$ for $i = 1, \ldots, h$, whereas $A_i \cap F(A) = \emptyset$ for $i = h+1, \ldots, h+j$ (see (6.28)); moreover $U(A_i) \cap \bar{C}(A) \ne \emptyset$ for $i = 1, \ldots, h+j$ where $\bar{C}(A)$ denotes the set of maximal (highest in energy) internal non-trivial saddles and the set V is completely attracted by $\tilde{A} \cup \bar{C}(A)$.

We can use the inductive hypothesis on the cycles in $\tilde{A}$ which is certainly valid since, of course, $|C(A_i)| \le n, i = 1, \ldots, h+j$.

We distinguish two cases (see Figure 6.2):

(a) $k > H(\bar{C}(A)) - H(F(A)) = \bar{\Gamma}(A)$,

(b) $k \le \bar{\Gamma}(A)$.

Notice that when $h(A) \ge 2$ (i.e. when in $\tilde{A}$ there are at least two distinct cycles with non-empty intersection with $F(A)$) we necessarily have $\Theta(A) = \Gamma(A_1) = \bar{\Gamma}(A)$ and, since we assume $k > \Theta(A)$, only case (a) is possible. It will appear clear from the proof concerning case (b) that if we considered the case $k \le \Theta(A)$ we would not be able to treat the problem with our construction. This is not a limit of our proof but a crucial feature of the exit problem at very small times. When

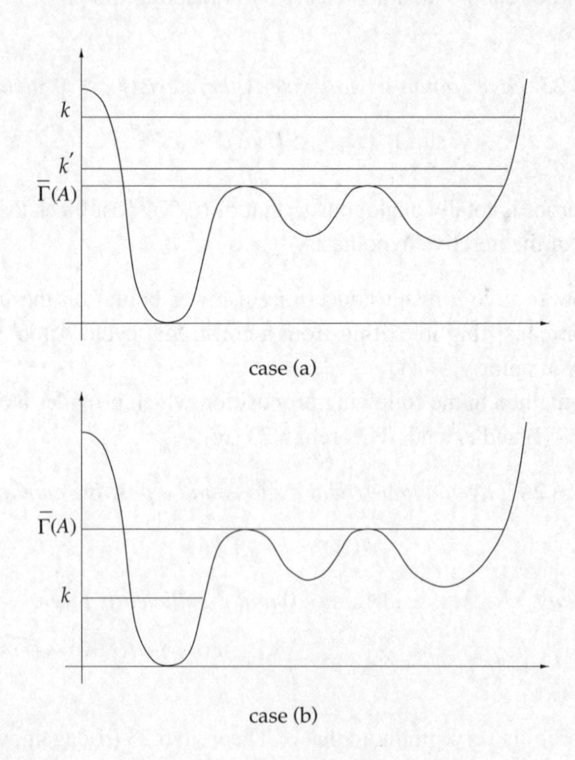

Figure 6.2

considering such small times, the inner structure of the cycles becomes relevant and counter-examples where (6.87) is not valid are easy to find.

Case (a) Let us first prove

$$P_x\left(\tau_{\partial A} < e^{\beta k}\right) \geq e^{-\beta(\Gamma(A)-k+\varepsilon)} \tag{6.88}$$

for all $\varepsilon > 0$, β sufficiently large.

Given $k' \in (\bar{\Gamma}(A), k)$, let us divide, like in the proof of Theorem 6.23(ii), the time interval $[e^{\beta k}]$ into subintervals of equal length $[e^{\beta k'}]$ (except, possibly, for the last one whose length can be smaller); see after (6.34) for more details.

We take $k' = \bar{\Gamma}(A) + \varepsilon'$ with $0 < 2\varepsilon' < k - \bar{\Gamma}(A)$.

We have, using the Markov property as in (6.36), (6.41),

$$P_x\left(\tau_{\partial A} > e^{\beta k}\right) \leq \left[\sup_{z\in A} P_z\left(\tau_{\partial A} > e^{\beta k'}\right)\right]^{\frac{1}{2}e^{\beta(k-k')}}. \tag{6.89}$$

Now we use the event $\mathcal{E}_{x,T_1}$ that we introduced in the proof of Theorem 6.23 with $x \in A$, $T_1 = [e^{\beta k'}]$. We recall the definition briefly: $\mathcal{E}_{x,T_1}$ contains a set of trajectories starting from x and exiting A within the time T_1 in a particular maner.

Given y^* in $U(A)$, there is a downhill path $\bar{x}_0, \bar{x}_1, \ldots, \bar{x}_{m-1}, \bar{x}_m$ with $\bar{x}_0 = y^*$, $\bar{x}_1, \ldots, \bar{x}_{m-2} \in V \cup \bar{C}(A)$, $\bar{x}_{m-1} \in \partial\tilde{A}$, $\bar{x}_m \in \tilde{A}$. It may happen that $y^* \in \partial\tilde{A}$: in that case, we set $m = 1$ and the path is just one step.

We call A_{j^*} the cycle in $\tilde{A}$ to which $\bar{x}_m$ belongs. The trajectories in $\mathcal{E}_{x,T_1}$ first enter $\tilde{A}$ within time $T_1/4$ then, by following a suitable finite sequence of contiguous cycles in $\tilde{A}$ they hit A_{j^*} in a time shorter than $T_1/4$; subsequently, they exit from A_{j^*} within another interval $T_1/4$, through $\bar{x}_{m-1}$. Finally, they follow, without hesitation, the uphill path $\bar{x}_{m-1}, \bar{x}_{m-2}, \ldots, \bar{x}_1, y^*$; obviously, for β large enough this last segment takes place within $T_1/4$. We refer to the proof of Theorem 6.23 (ii) for more details.

We have, like in (6.48),

$$\inf_{x\in A} P\left(\mathcal{E}_{x,T_1}\right) \geq e^{-\beta(\Phi(A)-H(\bar{C}(A))+\frac{\varepsilon'}{2})} \tag{6.90}$$

for β sufficiently large.

Now, since

$$\{\tau_{\partial A} < e^{\beta k'}, X_0 = x\} \supseteq \mathcal{E}_{x,T_1}, \tag{6.91}$$

from (6.89) and (6.90), we have for β sufficiently large

$$P_x\left(\tau_{\partial A} > e^{\beta k}\right) \leq \left[1 - e^{-\beta(\Phi(A)-H(\bar{C}(A))+\frac{\varepsilon'}{2})}\right]^{\frac{1}{2}e^{\beta(k-\bar{\Gamma}(A)-\varepsilon')}}$$
$$\leq 1 - e^{-\beta[\Gamma(A)-k+2\varepsilon']},$$

and (6.88) follows with $\varepsilon = 2\varepsilon'$.

To conclude the proof we note that it is sufficient to prove (6.87) for $x \in F(A)$. Indeed, given $z \in F(A)$, just using the Markov property we can write:

$$\begin{aligned} P_x\left(\tau_{\partial A} < e^{\beta k}, X_{\tau_{\partial A}} = y\right) &\geq P_x\left(\tau_{\partial A} < e^{\beta k}, X_{\tau_{\partial A}} = y,\ \tau_z < \tau_{\partial A} \wedge \frac{e^{\beta k}}{2}\right) \\ &\geq P_x\left(\tau_z < \frac{e^{\beta k}}{2}, \tau_{\partial A} > \tau_z\right) P_z\left(\tau_{\partial A} < \frac{e^{\beta k}}{2}, X_{\tau_{\partial A}} = y\right). \end{aligned} \tag{6.92}$$

By Theorem 6.23(ii) and Corollary 6.24, we get, for β sufficiently large,

$$P_x\left(\tau_z < \frac{e^{\beta k}}{2}, \tau_{\partial A} > \tau_z\right) \geq \frac{1}{2}. \tag{6.93}$$

Then, we can assume that $x \in F(A)$.

We need, now, some more definitions. Given $x \in A$ we set

$$\begin{aligned} \theta_x &= \sup\{t : 0 \leq t < \tau_{\partial A}; X_t = x\} \\ \tau'_x &= \tau_{\partial A} - \theta_x \end{aligned} \tag{6.94}$$

(we set $\theta_x = \tau_{\partial A}$; $\tau'_x = 0$ when $\tau_x \geq \tau_{\partial A}$).

It follows from Corollary 6.24 and the Markov property, that for all $\varepsilon > 0$ there exists $c > 0$:

$$\sup_{z \in A}\ \sup_{x \in F(A)}\ P_z\left(\tau_x > e^{\beta(\Theta(A)+\varepsilon)}, \tau_{\partial A} > e^{\beta(\Theta(A)+\varepsilon)}\right) \leq e^{-e^{c\beta}} \tag{6.95}$$

for β sufficiently large.

It follows from (6.95) that for any $\varepsilon > 0$, there exists $c > 0$ such that, for any $z \in A,\ x \in F(A)$,

$$P_z\left(\tau'_x > e^{\beta(\Theta(A)+\varepsilon)}\right) \leq e^{-e^{c\beta}}.$$

From (6.94), (6.95), taking into account that $k > \Theta(A)$, we have, for any $x \in F(A)$,

$$\begin{aligned} P_x\left(\tau_{\partial A} \leq e^{\beta k}, X_{\tau_{\partial A}} = y\right) &\geq P_x\left(\theta_x \leq \frac{e^{\beta k}}{2}, \tau'_x \leq \frac{e^{\beta k}}{2}, X_{\tau_{\partial A}} = y\right) \\ &= P_x\left(\theta_x \leq \frac{e^{\beta k}}{2}, X_{\tau_{\partial A}} = y\right) - P_x\left(\theta_x \leq \frac{e^{\beta k}}{2}, \tau'_x > \frac{e^{\beta k}}{2}, X_{\tau_{\partial A}} = y\right) \\ &\geq P_x\left(\theta_x \leq \frac{e^{\beta k}}{2}, X_{\tau_{\partial A}} = y\right) - \text{SES} \end{aligned} \tag{6.96}$$

where we use SES to denote a superexponentially (in β) small quantity (see Definition 6.4).

Now, it is easy to see that for any $T \in \mathbb{N}, x \in A$

$$P_x\left(\theta_x \le T, X_{\tau_{\partial A}} = y\right) = P_x\left(\theta_x \le T\right) P_x\left(X_{\tau_{\partial A}} = y\right). \tag{6.97}$$

Indeed,

$$P_x\left(\theta_x \le T, X_{\tau_{\partial A}} = y\right) = \left[1 + P(x,x) + \sum_{s=1}^{T-1} \sum_{x_1,\dots,x_s \in A} P(x, x_1, \dots, x_s, x)\right] \times \left[P(x,y) + \sum_{l=1}^{\infty} \sum_{y_1,\dots,y_l \in A\setminus\{x\}} P(x, y_1, \dots, y_l, y)\right]. \tag{6.98}$$

On the other hand we have

$$P_x\left(\theta_x \le T\right) = \left[1 + P(x,x) + \sum_{s=1}^{T-1} \sum_{x_1,\dots,x_s \in A} P(x, x_1, \dots, x_s, x)\right] \times \left[\sum_{z \in \partial A}\left(P(x,z) + \sum_{l=1}^{\infty} \sum_{y_1,\dots,y_l \in A\setminus\{x\}} P(x, y_1, \dots, y_l, z)\right)\right]. \tag{6.99}$$

One has:

$$\lim_{T\to\infty} P_x\left(\theta_x \le T\right) = P_x\left(\theta_x < \infty\right) = 1, \tag{6.100}$$

as follows from an immediate estimate based on ergodicity of our chain with finite state space and the Borel–Cantelli lemma.

From (6.98), (6.99), passing to the limit $T \to \infty$, we get:

$$\frac{P(x,y) + \sum_{l=1}^{\infty} \sum_{y_1,\dots,y_l \in A\setminus\{x\}} P(x, y_1, \dots, y_l, y)}{\sum_{z\in\partial A}(P(x,z) + \sum_{l=1}^{\infty} \sum_{y_1,\dots,y_l \in A\setminus\{x\}} P(x, y_1, \dots, y_l, z))} = P_x\left(X_{\tau_{\partial A}} = y\right). \tag{6.101}$$

Recall that we already obtained expression (6.101) for $P_x\left(X_{\tau_{\partial A}} = y\right)$ by a very similar argument, using the auxiliary Markov chain ξ_k, see (6.79)

From (6.101), (6.98), (6.99), we conclude the proof of (6.97).

From (6.96), (6.97), we get,

$$\begin{aligned} P_x\left(\tau_{\partial A} \le e^{\beta k}, X_{\tau_{\partial A}} = y\right) &\ge P_x\left(\theta_x \le \frac{e^{\beta k}}{2}\right) P_x\left(X_{\tau_{\partial A}} = y\right) - \text{ SES} \\ &\ge P_x\left(\tau_{\partial A} \le \frac{e^{\beta k}}{2}\right) P_x\left(X_{\tau_{\partial A}} = y\right) - \text{ SES}. \end{aligned} \tag{6.102}$$

From (6.102), (6.88) and Theorem 6.23(iv) we conclude the proof in case (a).

Case (b) Here we cannot apply Theorem 6.23 and we can only use the iterative hypothesis. Again we first prove (6.88). We know that $h = 1$, namely, in the set $\tilde{A}$

of highest internal transient cycles there is a unique cycle A_1 containing the whole $F(A)$; we have $\Theta(A_1) < k$ (since $k > \Theta(A)$ and $\Theta(A) \geq \Theta(A_1)$) together with $k < \Gamma(A_1)$, since $\bar{\Gamma}(A) = \Gamma(A_1)$ and we are in case (b). Then the hypotheses on k are satisfied for A_1 and $|C(A_1)| \leq n$. On the other hand, $k > \Gamma(A_j)$ for any $j \geq 2$ since $\Theta(A) \geq \Gamma(A_j)$ for any $j \geq 2$. So, to treat the exit from A_1, we use the inductive hypothesis, whereas to treat the exit from $A_2, A_3, \ldots$, we use Theorem 6.23

For any $x \in A$ we consider again an event $\mathcal{E}_{x,T_1}$ with $T_1 = e^{\beta k}$.

First we go to $\tilde{A}$ within a time $T_1/6$; then we try to reach our target $y^* \in A_{j^*}$.

If there exists a path of cycles in $\tilde{A}$ joining the first cycle that we touch entering $\tilde{A}$ with A_{j^*}, without touching A_1, it is easily seen, using $k > \Gamma(A_j)$, $j \neq 1$, that we can proceed as in case (a) to get, for all $\varepsilon > 0$ and β sufficiently large,

$$P\left(\mathcal{E}_{x,T_1}\right) \geq e^{-\beta(\Phi(A)-H(\bar{C}(A))+\varepsilon)} = e^{-\beta(\Gamma(A)-\bar{\Gamma}(A)+\varepsilon)} \geq e^{-\beta(H(\Gamma(A))-k+\varepsilon)}. \tag{6.103}$$

If, on the contrary, all cycle paths in $\tilde{A}$ reaching A_{j^*} have to pass through A_1, we have to modify our strategy. After entering $\tilde{A}$ we follow, within another $T_1/6$, a suitable finite sequence of contiguous distinct cycles in $\tilde{A}$ (by always exiting through $\bar{C}(A)$) up to A_1. Then, in a time not exceeding $T_1/6$, we exit from A_1 through $U(A_1) \cap \bar{C}(A)$ and, during another interval of time $T_1/6$ we follow another suitable finite sequence of contiguous distinct cycles in $\tilde{A}$ (by always exiting through $\bar{C}(A)$) up to A_{j^*}. Then we exit through a suitable $\bar{x}_{m-1}$ within another $T_1/6$ and finally we go uphill up to $y^* \in U(A)$, as before, without exceeding $T_1/6$. Notice that the trajectories in $\mathcal{E}_{x,T_1}$ visit at most once each A_i and, in particular, A_1.

The first and second parts of our event (arrival at $\tilde{A}$ and at A_1) as well as the fourth part (going from $U(A_1) \cap \bar{C}(A)$ to A_{j^*}) have a probability estimated from below by $e^{-\beta\varepsilon}$ (for all $\varepsilon > 0$, β sufficiently large)

For the third part (exit from A_1), since $|C(A_1)| \leq n$ and $\Theta(A_1) < k < \Gamma(A_1)$ we can use the recursive hypothesis to estimate from below the probability to exit within $T_1/6 = e^{\beta k}/6$ from A_1 through $U(A_1)$ by $\exp(-\beta[\Phi(A_1) - H(F(A)) - k + \varepsilon])$, for all $\varepsilon > 0$ and β sufficiently large.

To estimate the fifth and the sixth parts (exit from A_{j^*} through x_{m-1} and fast ascent to y^*) we proceed as in the proof of point (ii) of Theorem 6.23 and in the proof concerning case (a) of the present proposition; we get a lower bound to the probability of these fifth and sixth parts of the form: $e^{-\beta(\Phi(A)-H(\bar{C}(A))+\varepsilon)}$ (for all $\varepsilon > 0$, β sufficiently large).

So we get

$$P\left(\mathcal{E}_{x,T_1}\right) \geq e^{-\beta(\Phi(A_1)-H(F(A))-k+\varepsilon)} e^{-\beta(\Phi(A)-H(\bar{C}(A))+\varepsilon)}, \tag{6.104}$$

From (6.104), using $\Phi(A_1) = H(\bar{C}(A))$, we conclude the proof of

$$P_x\left(\tau_{\partial A} < e^{\beta k}\right) \geq e^{-\beta(\Gamma(A)-k+\varepsilon)}$$

and then, with the previous argument, the proof of (6.87). Notice that in the present case (b) we do not use recurrence and the Markov property as in (6.89); also notice that we could have obtained the result directly by simply substituting, in our construction, for y^* a $y \in \partial A$ not necessarily belonging to $U(A)$, avoiding in this way the use of (6.97).

To conclude the proof of the proposition we only need to prove the basis of the induction, namely the case of A completely attracted by a unique stable plateau D.

In this case $0 < k \leq \Gamma(A)$, $\Theta(A) = \bar{\Gamma}(A) = 0$ we are always in case (a); again we have only to prove (6.88). To this end we construct an event taking place in a finite (independent of β), sufficiently large interval T_0 consisting in descending from a generic $z \in A$ to D and then ascending to $U(A)$. □

Remark It is very easy to give an upper bound to $P_x\left(\tau_{\partial A} < e^{\beta k}, X_{\tau_{\partial A}} = y\right)$ when $x \in F(A)$.

Indeed

$$P_x\left(\tau_{\partial A} < e^{\beta k}, X_{\tau_{\partial A}} = y\right) \leq P_x\left(\tau_y < e^{\beta k}\right) \leq e^{-\beta(H(y)-H(F(A))-k)} \quad (6.105)$$

as follows immediately from Lemma 6.21. Notice that, in general, when we start from $x \notin F(A)$, the above inequality (6.105) is false.

6.3 Restricted dynamics

In what follows we use an auxiliary dynamics obtained by restricting our original Metropolis Markov chain with state space S to a given connected set $\overline{G} \subseteq S, \overline{G} = G \cup \partial G$, via the introduction of a 'reflecting barrier' on $\partial G \equiv \partial^- \overline{G}$ (see Definition 6.1). This auxiliary dynamics is defined in the following way. The state space is $\overline{G}$; the transition probabilities $P^G(x, y)$ are given by:

$$P^G(x, y) = \begin{cases} P(x, y) & \text{if } x, y \in \overline{G}, x \neq y \\ P(x, x) + \sum_{z \in S \setminus \overline{G}} P(x, z) & \text{if } x = y \in \overline{G} \end{cases} \quad (6.106)$$

i.e. we have set equal to zero the probability of exiting from $\overline{G}$ by properly renormalizing the probability of staying in the same state of ∂G.

It is easy to see that the above defined Markov chain is reversible with respect to the restricted Gibbs measure on $\overline{G}$, that we denote by $\mu^{\overline{G}}$, given by:

$$\mu^{\overline{G}}(x) = \frac{\mu(x)\mathbf{1}_{\overline{G}}(x)}{\mu(\overline{G})} \qquad \text{where } \mu(x) := \frac{\exp(-\beta H(x))}{\sum_{z \in S} \exp(-\beta H(z))}. \quad (6.107)$$

Notice that, to be consistent with the notation used in Chapter 4, we should denote by $\mu_{\overline{G}}$ the restricted Gibbs measure that, in analogy with (6.106), we are now denoting by $\mu^{\overline{G}}$.

Given two ergodic aperiodic Markov chains with the same state spaces S, transition probabilities $P(\cdot, \cdot)$ but different initial conditions $x, y \in S$, we want to couple them, namely we want to introduce a particular probability measure $P_{x,y}$ on the Cartesian product $\Omega_x \times \Omega_y$ of the spaces of trajectories Ω_x, Ω_y emerging from x, y, respectively, such that the first (second) marginal corresponds to the law of the Markov chain starting from x (y), respectively. Our coupled process, that we denote by (X_t^x, X_t^y) and call 'basic coupling', is defined in the following way. Let

$$\tau^{join} := \inf\{t : X_t^x = X_t^y\} \tag{6.108}$$

(we set $\tau^{join} = \infty$ if such a time does not exist). We take the two processes to be independent up to τ^{join} (product measure) and we make them proceed together after τ^{join}. We leave the formal definition to the reader. See Chapters 4 and 5 (and references therein) for a discussion of the notion of coupling (see also [116]). Given the subset $B \subseteq S$, sometimes we write:

$$\tau_B^x := \inf\{t : X_t^x \in B\} \tag{6.109}$$

similarly for τ_B^y.

In the case of Metropolis Markov chains, we are interested, in particular, in the basic coupling $P_{x,y}^A$ between two processes restricted to a non-trivial cycle A starting from the initial states x, y, respectively. We shall prove, in this case, that our two processes will join, with very high probability for large β, before a time $e^{v\beta}$, whenever $v > \Theta(A)$; $\Theta(A)$ was defined in Definition 6.20 and represents the maximal internal barrier in A towards $F(A)$. The result is the content of the following.

Proposition 6.27 *For any $\delta > 0$, there exists $c > 0$ such that for β large enough*

$$\sup_{x,y \in A} P_{x,y}^A\left(\tau^{join} > e^{\beta(\delta+\Theta(A))}\right) < e^{-e^{c\beta}}. \tag{6.110}$$

Proof We use again induction on the number $|C(A)|$ of internal non-trivial saddles. We suppose (6.110) true for every A' with $|C(A')| \leq n$ and we want to prove it for any A such that $|C(A)| = n+1$. The case $n = |C(A)| = 0$ can be treated along the same lines. Consider indeed such a cycle A and fix $\delta > 0$. Starting from $x, y \in A$ we want to construct an event $\mathcal{E}_{(x,y),T}^{join}$ containing coupled trajectories starting from (x, y) and joining before the suitably chosen time T, in a particular manner.

Like in (6.28) we write $A = A_1 \cup A_2 \cup \cdots \cup A_h \cup A_{h+1} \cup \cdots \cup A_{h+j} \cup \bar{C}(A) \cup V$ where $A_i \cap F(A) \neq \emptyset$, $i = 1, \ldots, h$; $A_i \cap F(A) = \emptyset$, $i = h+1, \ldots, h+j$. We distinguish two cases:

(a) $h = 1$ wherc $\Theta(A) = \max\{\Gamma(A_2), \dots, \Gamma(A_{j+1}), \Theta(A_1)\}$,
(b) $h > 1$ where $\Theta(A) = \Gamma(A_1)$.

Fix $\delta > 0$. In case (a) we take $T = e^{k\beta}$ with $k = \Theta(A) + \varepsilon$ and

$$\varepsilon \leq \min\left\{ \frac{\delta}{2}, \frac{\Gamma(A_1) - \Theta(A)}{2} \right\} \tag{6.111}$$

and for the lower estimates of the probability $P(\mathcal{E}^{join}_{(x,y),T})$ we use Theorem 6.23; in case (b) we have to take $T = e^{k'\beta} = e^{(\Theta(A)-\varepsilon')\beta} = e^{(\Gamma(A_1)-\varepsilon')\beta}$ with a suitable choice (in terms of δ) of $\varepsilon' > 0$, and for its estimate we use Proposition 6.26. We give the definition of $\mathcal{E}^{join}_{(x,y),T}$ in words, leaving to the reader the easy task of the formal definition. $\mathcal{E}^{join}_{(x,y),T}$ contains a set of pairs of trajectories emerging from (x, y); in a time $T/3$ boths trajectories enter $\tilde{A} = A_1 \cup A_2 \cup \dots \cup A_h \cup A_{h+1} \cup \dots \cup A_{h+j}$ via a downhill path if they are in $\bar{C}(A) \cup V$ at $t = 0$. This happens with a probability of order one; otherwise, if both $x, y \in \tilde{A}$, we simply disregard this first interval. Let A_x, A_y be, respectively, the cycles in $\tilde{A}$ reached in this way. Then within a time $T/3$ both trajectories reach A_1: after the first one has reached A_1 it stays there up to the arrival of the second one. This happens with a probability of order one in case (a) as a consequence of Theorem 6.23 and with a probability larger than $e^{-\beta(\Theta(A)-k)4|\bar{C}(A)|}$ in case (b) as a consequence of Proposition 6.26.

Indeed in case (a) if the single trajectories are not already both in A_1 let us consider one of them, say X_t^x. Since $k > \Theta(A)$, by Corollary 6.24, X_t^x can follow a suitable finite sequence of contiguous distinct cycles in $\tilde{A}$ (by always exiting through $\bar{C}(A)$) up to A_1 within a time $T/3$ with a probability tending to one as $\beta \to \infty$. Moreover, since $k < \Gamma(A_1)$ (T is much smaller than the typical exit time from A_1), by Theorem 6.23, the trajectory X_t^x, once it reaches A_1, stays there with a probability of order one at least for a time $2T/3$. Simultaneously the other trajectory X_t^y with a similar mechanism reaches A_1 within $T/3$ and afterwards it stays there a time $2T/3$ with a probability of order one as well. So for a time at least $T/3$ both belong to A_1 with a probability tending to one as $\beta \to \infty$ and we can substitute P^A with P^{A_1} since, before $\tau_{\partial A_1}$, they coincide. Now, since $\Theta(A_1) < k$ and $|C(A_1)| \leq n$ we can use the inductive hypothesis so that, with a probability tending to one as $\beta \to \infty$, the two processes join before $T/3$.

In case (b) we proceed in a similar way. The only difference is that now, since $T = e^{(\Theta(A)-\varepsilon')\beta}$, the arrival, within this time, from a generic cycle $A' \subseteq \tilde{A}$ to A_1 following a suitable finite sequence of distinct cycles in $\tilde{A}$ (exiting through optimal saddles in $\bar{C}(A)$) costs, by Proposition 6.26, a probability bounded from below by $e^{-2\beta\varepsilon'|\bar{C}(A)|}$ for each one of the two processes. On the other hand the choice $T = e^{(\Theta(A)-\varepsilon')\beta} = e^{(\Gamma(A_1)-\varepsilon')\beta}$ ensures that T is much smaller than the typical exit time from A_1 so, again, by Theorem 6.23 both trajectories X_t^x, X_t^y, once they have reached A_1, stay there with a probability of order one at least for a time $2T/3$ and we can substitute P^A with P^{A_1}. If we choose

$$\varepsilon' < \frac{\Gamma(A_1) - \Theta(A_1)}{2} \tag{6.112}$$

we get $k' > \Theta(A_1)$ and so we can use the inductive hypothesis on A_1. Then we get in case (a)

$$P(\mathcal{E}^{join}_{(x,y),T}) \geq \frac{1}{2}, \tag{6.113}$$

and in case (b)

$$P(\mathcal{E}^{join}_{(x,y),T}) \geq e^{-\beta\varepsilon' 4|\bar{C}(A)|}, \tag{6.114}$$

for β sufficiently large. Now we divide the time interval $[0, e^{(\Theta(A)+\delta)\beta}]$ into subintervals of amplitude T like in the proof of Theorem 6.23 and, using the Markov property, we get in case (a):

$$P_{x,y}(\tau^{join} > e^{(\Theta(A)+\delta)\beta}) < \left(\frac{1}{2}\right)^{e^{(\delta-\varepsilon)\beta}}, \tag{6.115}$$

implying (6.110) for, say, $c = \delta/4$.

In case (b):

$$P_{x,y}(\tau^{join} > e^{(\Theta(A)+\delta)\beta}) < (1 - e^{-4\beta\varepsilon'|\bar{C}(A)|})^{e^{\delta\beta}}. \tag{6.116}$$

We choose ε' so that $4\varepsilon'|\bar{C}(A)| < \delta/2$. A choice also satisfying (6.112) is

$$\varepsilon' = \min\left\{\frac{\Gamma(A_1) - \Theta(A_1)}{2}, \frac{\delta}{8|\bar{C}(A)|}\right\} \tag{6.117}$$

by 4.17, (6.117) we conclude the proof of (6.110) with $c = \delta/8$. □

6.4 Conditional ergodic properties

We want now to analyse the 'conditional ergodic properties' of our Metropolis Markov chain. We state and prove a result saying that before the exit from a cycle our process will reach an apparent equilibrium. After a sufficiently large time, but before the exit, our system will be well described in terms of the Gibbs conditional distribution. In this way we see the importance, also in the framework of the pathwise approach, of the notion of restricted ensemble introduced by Penrose and Lebowitz (see Chapter 4).

Consider a Metropolis Markov chain with state space S and energy function H; given a non-trivial cycle $A \subseteq S$ the following result holds true.

Theorem 6.28 *For any* $B \subseteq A$, *for any* $x \in A, t = e^{k\beta}, \Theta(A) < k < \Gamma(A)$

$$\left|\frac{P_x(X_t \in B | \tau_{\partial A} > t)}{\mu^A(B)} - 1\right| < o(1) \tag{6.118}$$

where $o(1)$ is an infinitesimal quantity for large β and the conditional Gibbs measure μ^A has been defined in (6.107).

Proof Since by Theorem 6.23(i) we know that

$$P_x(\tau_{\partial A} > t) > 1 - o(1) \tag{6.119}$$

it will be sufficient to prove that

$$\left| \frac{P_x(X_t \in B, \tau_{\partial A} > t)}{\mu^A(B)} - 1 \right| < o(1). \tag{6.120}$$

We have:

$$P_x(X_t \in B, \tau_{\partial A} > t) = P_x^A(X_t \in B, \tau_{\partial A} > t) \tag{6.121}$$

since before $\tau_{\partial A}$, P^A and P coincide. Now we make use of the basic coupling to compare two processes starting from two different initial states. Using the notation introduced in (6.109) we write, for $x, y \in A$:

$$\begin{aligned} &|P^A(X_t^x \in B, \tau_{\partial A}^x > t) - P^A(X_t^y \in B, \tau_{\partial A}^y > t)| \\ &\leq E(|\mathbf{1}_{X_t^x \in B}\mathbf{1}_{\tau_{\partial A}^x > t} - \mathbf{1}_{X_t^y \in B}\mathbf{1}_{\tau_{\partial A}^y > t}|). \end{aligned} \tag{6.122}$$

We have:

$$\begin{aligned} &|\mathbf{1}_{X_t^x \in B}\mathbf{1}_{\tau_{\partial A}^x > t} - \mathbf{1}_{X_t^y \in B}\mathbf{1}_{\tau_{\partial A}^y > t}| \leq |\mathbf{1}_{X_t^x \in B} - \mathbf{1}_{X_t^y \in B}| \\ &\quad + \mathbf{1}_{X_t^y \in B}\mathbf{1}_{\tau_{\partial A}^x < \tau^{join} < \bar{t} < t < \tau_{\partial A}^y} + \mathbf{1}_{X_t^x \in B}\mathbf{1}_{\tau_{\partial A}^y < \tau^{join} < \bar{t} < t < \tau_{\partial A}^x} + \mathbf{1}_{\tau^{join} \geq \bar{t}} \end{aligned} \tag{6.123}$$

where we chose $\bar{t} := e^{\bar{k}\beta}$ with $\Theta(A) < \bar{k} < k$. By (6.123) and Proposition 6.27, using $|\mathbf{1}_{X_t^x \in B} - \mathbf{1}_{X_t^y \in B}| \leq \mathbf{1}_{\tau^{join} > t}$, we get

$$\begin{aligned} &|P^A(X_t^x \in B, \tau_{\partial A}^x > t) - P^A(X_t^y \in B, \tau_{\partial A}^y > t)| \\ &\leq \text{SES} + 2 \sup_{x,y \in A} P^A(\tau_{\partial A}^x < \tau^{join} < \bar{t} < t < \tau_{\partial A}^y; X_t^y \in B) \\ &\leq \text{SES} + 2 \sup_{x \in A} P^A(\tau_{\partial A}^x < \bar{t}) \sup_{z \in A} P^A(X_{t-\bar{t}}^z \in B). \end{aligned} \tag{6.124}$$

Now, using Proposition 6.27 again we have for $s = e^{(\Theta(A)+\delta)\beta}$, $\delta > 0$

$$|P^A(X_s^x \in B) - P^A(X_s^y \in B)| \leq \sup_{x,y \in A} P_{x,y}^A(\tau^{join} > s) < \text{SES}; \tag{6.125}$$

using the stationarity of μ^A:

$$P_{\mu^A}^A(X_s \in B) := \sum_{z \in A} \mu^A(z) P_z^A(X_s \in B) = \mu^A(B) \tag{6.126}$$

and from (6.125), (6.126), we get

$$|P_x^A(X_s \in B) - \mu^A(B)| < \text{SES}. \tag{6.127}$$

In conclusion, using that, by Theorem 6.23, $P^A(\tau^x_{\partial A} < \bar{t}) = o(1)$, since for large enough β, $t - \bar{t} > t/2$, we get

$$|P^A(X^x_t \in B, \tau^x_{\partial A} > t) - P^A(X^y_t \in B, \tau^y_{\partial A} > t)| < \text{SES} + o(1)\mu^A(B). \tag{6.128}$$

By (6.128) we get, for β large

$$|P^A_x(X_t \in B, \tau_{\partial A} > t) - P^A_{\mu^A}(X_t \in B, \tau_{\partial A} > t)| < o(1)\mu^A(B). \tag{6.129}$$

On the other hand if we prove that

$$P^A_{\mu^A}(X_t \in B, \tau_{\partial A} \le t) < o(1)\mu^A(B) \tag{6.130}$$

we get

$$|P^A_x(X_t \in B, \tau_{\partial A} > t) - P^A_{\mu^A}(X_t \in B)| < o(1)\mu^A(B) \tag{6.131}$$

and by the stationarity of μ^A we conclude the proof.

To get (6.130) we just use reversibility (the event $\{\tau_{\partial A} \le t\}$ being invariant under time-reversal):

$$\begin{aligned} P^A_{\mu^A}(X_t \in B, \tau_{\partial A} \le t) &= P^A_{\mu^A}(X_0 \in B, \tau_{\partial A} \le t) \\ &= \sum_{z \in B} \mu^A(z) P^A_z(\tau_{\partial A} \le t) \le o(1)\mu^A(B). \end{aligned} \tag{6.132}$$

□

6.5 Escape from the boundary of a transient cycle

We want now to state and prove a proposition referring to the first escape from a transient cycle A and saying, roughly, that, under general hypotheses, with high probability, after many attempts, sooner or later our process will *really* escape from A entering into another different cycle by passing through one of the minimal saddles of the boundary of A.

The time for this transition has the same asymptotics, in the sense of logarithmic equivalence, as the first hitting time to the boundary of A.

The proposition is, in fact, a simple consequence of the Markov property but we think that it is useful to provide an explicit proof.

Given a transient cycle A, consider the set $U(A) = F(\partial A)$ and the set $\mathcal{S}(A) \subseteq U(A)$ of the minimal saddles of A.

Let

$$\bar{\partial} A := \cup_{z \in \partial A} D(z). \tag{6.133}$$

Let $U^- = U^-(A)$ be the subset of A to which some point in $U(A)$ is downhill connected:

$$U^-(A) := \{x \in A \ \text{ such that } \exists z \in U(A) \ \text{ with } \ P(z, x) > 0\}. \tag{6.134}$$

Let $\mathcal{S}^+ = \mathcal{S}^+(A)$ be the analogue of U^- outside A:

$$\begin{aligned}\mathcal{S}^+(A) := \{x \notin A : \exists z \in \mathcal{S}(A), y \in D^-(z) \\ \text{such that } H(x) < H(y),\ P(x, y) > 0\}.\end{aligned} \tag{6.135}$$

In words, $\mathcal{S}^+(A)$ is the set of points x outside $A \cup \bar{\partial} A$ such that there exists a downhill path starting from some point $z \in \mathcal{S}(A)$ and staying inside the plateau $D(z)$ up to the exit from $\bar{\partial} A$ through x. We set:

$$\bar{\mathcal{S}} = \bar{\mathcal{S}}(A) := U^-(A) \cup \mathcal{S}^+(A). \tag{6.136}$$

Proposition 6.29 *Consider a transient cycle A. Given $\varepsilon > 0$ let*

$$T(\varepsilon) := \exp \beta[H(\mathcal{S}(A)) - H(F(A)) + \varepsilon]. \tag{6.137}$$

Then, for every $\varepsilon > 0,\ x \in A$,

$$\lim_{\beta \to \infty} P_x(\tau_{(A \cup \bar{\partial} A)^c} > T(\varepsilon)) = 0, \tag{6.138}$$

and

$$\lim_{\beta \to \infty} P_x(X_{\tau_{(A \cup \bar{\partial} A)^c}} \in \mathcal{S}^+(A)) = 1. \tag{6.139}$$

Proof From the definition of $\mathcal{S}^+(A)$ we know that there exists a positive constant κ, independent of β, such that

$$\inf_{x \in \mathcal{S}(A)} P_x(X_{\tau_{(D(x))^c}} \in \mathcal{S}^+(A)) > \kappa, \tag{6.140}$$

moreover, by the Markov property we know that for any $\varepsilon > 0$, there exists $c > 0$ such that, for sufficiently large β,

$$\sup_{x \in U(A)} P_x(\tau_{(D(x))^c} > e^{\varepsilon\beta}) < e^{-e^{c\beta}}, \tag{6.141}$$

so that, using Lemma 6.21, we know that there exists $\kappa' > 0$ such that

$$\sup_{x \in U(A)} P_x(X_{\tau_{(D(x))^c}} \notin \bar{\mathcal{S}}(A)) \le e^{-\kappa'\beta}. \tag{6.142}$$

By Theorem 6.23(iv) we know that

$$\inf_{x \in A} P_x(X_{\tau_{\partial A}} \in \mathcal{S}(A)) > e^{-\delta\beta}, \qquad \forall \delta > 0 \text{ and } \beta \text{ sufficiently large.} \tag{6.143}$$

We define, now, the sequence τ_i of stopping times corresponding to subsequent passages of our chain X_t in $\bar{\partial} A$:

$$\begin{aligned}\tau_0 &:= \inf\{t \ge 0 : X_t \in \bar{\partial} A\} \\ \sigma_1 &:= \inf\{t > \tau_0 : X_t \notin \bar{\partial} A\},\end{aligned} \tag{6.144}$$

and for $j = 1, 2, \ldots,$

$$\begin{aligned} \tau_j &:= \inf\{t > \sigma_j : X_t \in \bar{\partial} A\} \\ \sigma_{j+1} &:= \inf\{t > \tau_j : X_t \notin \bar{\partial} A\}. \end{aligned} \tag{6.145}$$

We set, for $j = 1, 2, \ldots,$

$$\mathcal{I}_j = [\tau_{j-1} + 1, \tau_j], \qquad T_j := |\mathcal{I}_j| = \tau_j - \tau_{j-1} - 1. \tag{6.146}$$

Let

$$\bar{\sigma}_j := \min\{t > \tau_j : X_t \notin D(X_{\tau_j})\}; \tag{6.147}$$

of course $\bar{\sigma}_j \le \sigma_{j+1}$. Fix ε in (6.137); we say that the interval $\mathcal{I}_j$ is *good* if the following conditions are satisfied:

$$\begin{aligned} &T_j < T(\varepsilon/2) \\ &X_{\tau_j} \in U(A) \\ &X_{\bar{\sigma}_j} \in \bar{\mathcal{S}}(A). \end{aligned}$$

Let

$$j^* := \min\{j : T_j \text{ is not good}\}.$$

Given the integer N we write, for every $x \in A$:

$$P_x(\tau_{(A \cup \bar{\partial} A)^c} > NT(\varepsilon/2)) \le P_x(\tau_{\mathcal{S}^+(A)} > NT(\varepsilon/2) ; j^* > N) + P_x(j^* \le N). \tag{6.148}$$

Let us consider the event $\{\tau_{\mathcal{S}^+(A)} > NT(\varepsilon/2); \; j^* > N\}$.

Using (6.140), (6.143), we have, by the Markov property,

$$\begin{aligned} &P_x(\tau_{\mathcal{S}^+(A)} > NT(\varepsilon/2) ; j^* > N) \\ &\quad \le P_x(X_{\tau_1} \in U(A); X_{\bar{\sigma}_1} \in U^-(A), \ldots, X_{\tau_N} \in U(A); X_{\bar{\sigma}_N} \in U^-(A)) \\ &\quad \le (1 - e^{-\delta\beta})^N, \end{aligned} \tag{6.149}$$

for all $\delta > 0$ and β sufficiently large

Now, from Theorem 6.23(ii), Corollary 6.25 and (6.142) we have that there exists $\kappa'' > 0$ such that for any $x \in A$,

$$P_x(\mathcal{I}_j \text{ is not good}) \le e^{-\kappa''\beta}. \tag{6.150}$$

From (6.150) we get:

$$P_x(j^* \le N) \le e^{-\kappa''\beta} N. \tag{6.151}$$

To conclude the proof of (6.138) it suffices to choose

$$N = N(\beta) = e^{\frac{\varepsilon\beta}{2}} \wedge e^{\frac{\beta\kappa''}{2}}.$$

On the other hand let us consider the reduced Markov chain ξ_j with state space $S \setminus \bar{\partial} A$, given by $\xi_j := X_{\sigma_j}$.

Let $R(A) := (A \cup \bar{\partial} A)^c \setminus \mathcal{S}^+(A)$. We have, similarly to (6.79):

$$\sup_{x\in A} P_x(X_{\tau_{(A\cup\bar{\partial}A)^c}} \notin \mathcal{S}^+(A)) \le \frac{\sup_{x\in A} P_x(\xi_1 \in R(A))}{\inf_{x\in A} P_x(\xi_1 \notin A)} \le \frac{\sup_{x\in A} P_x(\xi_1 \in R(A))}{1 - \sup_{x\in A} P_x(\xi_1 \in R(A))}. \tag{6.152}$$

But

$$\sup_{x\in A} P_x(\xi_1 \in R(A)) \le \sup_{x\in A} P_x(X_{\tau_0} \notin U(A)) + \sup_{x\in \mathcal{S}(A)} P_x(X_{\tau_{(D(x))^c}} \notin \bar{\mathcal{S}}(A)). \tag{6.153}$$

From Corollary 6.25, (6.152), (6.153), (6.142) we conclude the proof of (6.139) and that of the proposition. □

6.6 Asymptotic exponentiality of the exit time

We now analyse, in more detail, the first exit from a non-trivial cycle A. In particular, following the ideas developed in the framework of the pathwise approach to metastability (see Chapter 4), we prove the asymptotic exponentiality of the properly renormalized first exit time from any non-trivial cycle, in the limit $\beta \to \infty$. Then we deduce the asymptotic behaviour of the expectation of this exit time; notice that the methods developed in the previous part of this chapter led naturally only to estimates in probability of the exit times but, as we shall see, we can even get good control on the tails of the distribution of these random variables and this will allow us to get the asymptotics of the averages.

Let A be a given non-trivial cycle.

Given a point $x \in A$, let the time $T_\beta = T_\beta(x)$ be defined by

$$T_\beta(x) := \min\{T \in \mathbb{N} : P(\tau^x_{\partial A} \le T) \ge 1 - e^{-1}\}, \tag{6.154}$$

where, given $G \subseteq S$, τ^x_G represents the first hitting time to G under P_x (see (6.109)); so that

$$P_x(\tau_{\partial A} < T_\beta(x)) < 1 - e^{-1} \le P_x(\tau_{\partial A} \le T_\beta(x)). \tag{6.155}$$

The above definition is interesting since T_β does not depend on $x \in A$, in the sense of logarithmic equivalence; namely, as a consequence of Theorem 6.23(i), (ii), we have for any $x, y \in A$:

$$\lim_{\beta\to\infty} \frac{1}{\beta} \log[T_\beta(x)] = \Phi(A) - H(F(A)); \qquad \lim_{\beta\to\infty} \frac{1}{\beta} \log\left[\frac{T_\beta(x)}{T_\beta(y)}\right] = 0. \tag{6.156}$$

Moreover, for any $x, y \in A$, $\varepsilon > 0$,

$$\lim_{\beta\to\infty} P_x(T_\beta(y)e^{-\varepsilon\beta} < \tau_{\partial A} < T_\beta(y)e^{\varepsilon\beta}) = 1. \tag{6.157}$$

The following theorem states the asymptotic exponentiality of the properly renormalized exit time from a non-trivial cycle A.

Theorem 6.30 *Let $T^*_\beta = T_\beta(x^*)$ where x^* is a particular point in A chosen once for all. Then for any $x \in A$, $\tau^x_{\partial A}/T^*_\beta$ converges in law to a unit mean exponential random time; in particular, for any $s \in \mathbb{R}^+$*

$$\lim_{\beta\to\infty} P_x\left(\frac{\tau_{\partial A}}{T^*_\beta} > s\right) = e^{-s}. \tag{6.158}$$

Moreover, for any $x \in A$,

$$\lim_{\beta\to\infty} \frac{E_x\tau_{\partial A}}{T^*_\beta} = 1. \tag{6.159}$$

Proof The proof is very similar to that of Theorem 4.6 in Chapter 4 to which we refer for more details.

Given $s,\ t \in \mathbb{R}^+$ we write:

$$\begin{aligned} P_x(\tau_{\partial A} > (t+s)\, T_\beta(x)) &= \sum_{y\in A} P_x(\tau_{\partial A} > (t+s)\, T_\beta(x)\,;\ X_{T_\beta(x)t} = y) \\ &= \sum_{y\in A} P_x(\tau_{\partial A} > t\, T_\beta(x)\,;\ X_{T_\beta(x)t} = y)\ \ P_y(\tau_{\partial A} > s\, T_\beta(x)). \end{aligned} \tag{6.160}$$

We can write, for any $T \in \mathbb{N}$:

$$\begin{aligned} P_y(\tau_{\partial A}\ >\ s\, T_\beta(x)\,) = P_y(\tau_{\partial A}\ >\ s\, T_\beta(x)\,;\, \tau_x < T) \\ + P_y(\tau_{\partial A} >\ s\, T_\beta(x)\,;\, \tau_x \geq T). \end{aligned} \tag{6.161}$$

We have, for any $T \in \mathbb{N}$ such that $T \leq sT_\beta(x)$

$$\begin{aligned} &P_y(\tau_{\partial A} > sT_\beta(x)\,;\, \tau_x < T) \\ &= \sum_{t=0}^{T-1} P_x(\tau_{\partial A} > sT_\beta(x)\ - t) P_y(\tau_{\partial A\cup\{x\}} = t,\ X_{\tau_{\partial A\cup\{x\}}} = x). \end{aligned} \tag{6.162}$$

By Theorem 6.23(iii) we know that there exists $\kappa > 0$ such that if the time T_1 is defined as

$$T_1 = \exp(\beta[\Gamma(A) - \kappa]), \tag{6.163}$$

we have

$$P_y(\tau_{\partial A} > s\, T_\beta(x)\,;\, \tau_x \geq T_1) \leq P_y(\tau_x \geq T_1) = o(1). \tag{6.164}$$

On the other hand, using (6.162) and taking also into account that for β sufficiently large $T_1 < sT_\beta(x)$, we have

$$P_y(\tau_{\partial A} > s\, T_\beta(x)\, ;\, \tau_x < T_1)\ \le P_x(\tau_{\partial A} > s\,(T_\beta(x) - T_1/s)). \tag{6.165}$$

Similarly we get from (6.162) and from Theorem 6.23(iii):

$$P_x(\tau_{\partial A} > s\, T_\beta(x)\,)[1 - \mathrm{o}(1)] \le P_y(\tau_{\partial A} > s\, T_\beta(x)\, ;\, \tau_x < T_1). \tag{6.166}$$

From (6.160), (6.164), (6.165), (6.166), we get

$$\begin{aligned} P_x(\tau_{\partial A} > t\, T_\beta(x)\,)P_x(\tau_{\partial A} > s\, T_\beta(x)\,)[1 - \mathrm{o}(1)] &\le P_x(\tau_{\partial A} > (t+s)\, T_\beta(x)) \\ &\le P_x(\tau_{\partial A} > t\, T_\beta(x)\,)\big[P_x(\tau_{\partial A} > s\,(T_\beta(x) - T_1/s)) + o(1)\big]. \end{aligned} \tag{6.167}$$

Which is totally analogous to (4.88).

For any integer $k \ge 2$ and β suffficiently large, we write

$$P_x(\tau_{\partial A} > (k+2)T_\beta(x)) \le P_x(\tau_{\partial A} > kT_\beta)(o(1) + P_x(\tau_{\partial A} > T_{\beta(x)})) \tag{6.168}$$

and taking $r := 2e^{-1}$ we may assume that $o(1) + P(\tau_{\partial A} > T_\beta(x)) \le o(1) + e^{-1} \le r < 1$, for each β sufficiently large. Thus, for each $k \ge 3$ we get

$$P_x\big(\tau_{\partial A} > kT_\beta(x)\big) \le r^{[k/2]} \tag{6.169}$$

for any sufficiently large β which immediately implies the tightness of the family $\{\tau_{\partial A}/T_\beta(x)\}$, on $[0, +\infty)$.

Now, proceeding exactly like in the proof of Theorem 4.6 we get that $\tau^x_{\partial A}/T_\beta(x)$ converges in law to a unit mean exponential random time and $\lim_{\beta\to\infty} E_x\tau_{\partial A}/T_\beta(x) = 1$.

On the other hand, by (6.164), (6.165), (6.166), exploiting continuity of the limiting distribution, we get, for any $x, y\ \in A,\ s\ \in \mathbb{R}^+$:

$$\lim_{\beta\to\infty}\left[P_x\left(\frac{\tau_{\partial A}}{T_\beta(x)} > s\right) - P_y\left(\frac{\tau_{\partial A}}{T_\beta(x)} > s\right)\right] = 0. \tag{6.170}$$

By (6.170), again using (6.169), we conclude the proof of the theorem. □

Remarks In particular, it follows from the above theorem that for every $x \in A$ and $\varepsilon > 0$, we have for β large enough:

$$\exp(\beta\,[\Gamma(A) - \varepsilon]) < E_x(\tau_{\partial A}) < \exp(\beta[\Gamma(A) + \varepsilon]). \tag{6.171}$$

To get the above estimate on the expectation of the exit time it was necessary to have the control (6.169) on the tail of the distribution.

6.7 The exit tube

This section is devoted to the study of the typical trajectories of the first excursion outside a non-trivial cycle A (see [231, 233]).

We start by considering the first descent from any point y_0 in A to $F(A)$. Then we analyse the problem of the typical tube of first exit trajectories. It will turn out, using reversibility, that this tube is simply related, via a time reversal transformation, to the typical tube followed by the process during the first descent to the bottom $F(A)$ of A starting from suitable points in $\partial^- A$.

In order to define the tube of typical trajectories of this first descent, the basic objects will be what we call *standard cascades* emerging from y_0. They specify the geometric characteristics of the tube; roughly speaking these cascades consist in sequences of minimaxes $y_1, \dots, y_n$ towards $F(A)$, decreasing in energy, intercalated by sequences of downhill paths $\omega^{(1)}, \dots, \omega^{(n)}$ and *permanence sets* $Q_1, \dots, Q_n$ which are a sort of generalized cycle.

More generally $\omega^{(i)}$ has to be seen as a downhill sequence of transient plateaux. We call *p-path* a sequence $\omega = D_1, \dots, D_k$ with D_i communicating with D_{i+1}; it can also be identified with the set of usual paths following the sequence of plateaux $D_1, \dots, D_k$. A p-path is said to be downhill if $H(D_{i+1}) \leq H(D_i)$, $i = 1, \dots, k$. Indeed a downhill p-path, due to the maximality in the definition of plateau, is strictly downhill and the corresponding paths are declining.

We prove that during its first descent to $F(A)$, with high probability, our system will follow one of the possible standard cascades; moreover we also give information about the temporal law of the descent by specifying the typical values of the random times spent inside each one of the sets Q_i.

Given any point y_0 in A and a downhill p-path $\omega^{(1)}$ starting from y_0, we define a set $Q_1 = Q_1(y_0, \omega^{(1)})$. This set Q_1 is a union of cycles with pairwise common minimal saddles of the same height or, more generally, with minimal saddles sharing, pairwise, the same plateau. Q_1 represents the first set where our process, during its first excursion to $F(A)$, is captured if it follows the path $\omega^{(1)}$; after entering into Q_1 it will spend some time inside it before leaving through a minimal saddle, to enter, after another downhill p-path $\omega^{(2)}$, into another similar set of lower height Q_2 and so on, untill it enters a cycle containing part of $F(A)$.

Let $y_0 \in A$; suppose first that y_0 does not belong to the union $\bar{M}$ of stable plateaux. Consider a downhill p-path $\omega^{(1)}$ starting from y_0 (such a path certainly exists since $y_0 \notin \bar{M}$). We stress that this path is not, in general, unique. This means that the whole construction we are making must be repeated for each path.

Let $\hat{D}_1$ be the first plateau in $\omega^{(1)}$ with $H(\hat{D}_1) < H(y_0)$ and belonging to $\bar{M}$ (see Figure 6.3 as an example). If, instead, $y_0 \in \bar{M}$, we take $\hat{D}_1 = D(y_0)$. If $\hat{D}_1$

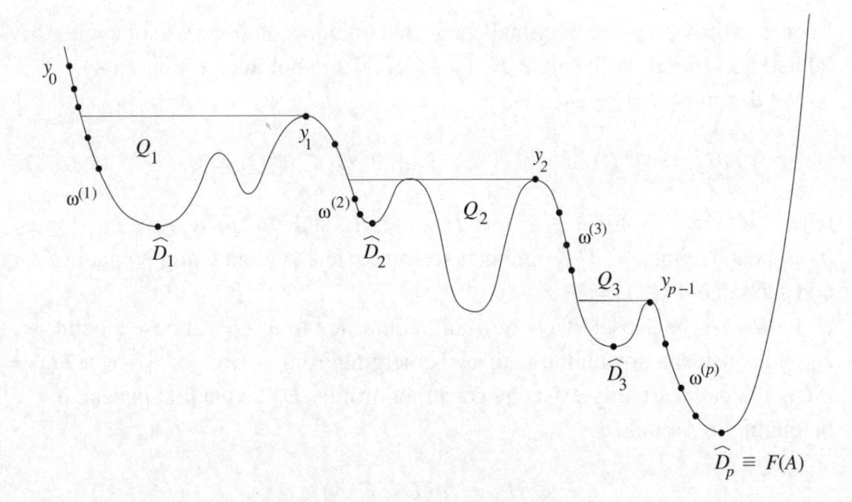

Figure 6.3

is contained in $F(A)$, then y_0 belongs to the wide basin of attraction $\hat{B}(\hat{D}_1)$. We set $\hat{D}_1 = D^*$; this connected component of $F(A)$ depends on A, y_0, $\omega^{(1)}$.

In this degenerate case we set $Q_1 := D^*$ and the *cascade of saddles* y_0, $y_1, \dots, y_n$ reduces to y_0.

Let us now suppose that $\hat{D}_1 \nsubseteq F(A)$. Let H_1 be the communication height between $\hat{D}_1$ and $F(A)$:

$$\Phi(\hat{D}_1, F(A)) = H_1.$$

We call Q_1 the maximal connected set, containing $\hat{D}_1$, such that $\Phi(Q_1, F(A)) = H_1$. We have:

$$Q_1 = \{y \text{ such that } \exists \omega : y \to \hat{D}_1 \text{ with } \max_{z \in \omega} H(z) \le H_1 \text{ and } \Phi(y, F(A)) = H_1\}. \tag{6.172}$$

It is seen immediately that Q_1 is of the form

$$Q_1 = (\cup_j A_j^{(1)}) \cup (\cup_i D_i) \tag{6.173}$$

where

- $A_j^{(1)}$ are disjoint cycles with $\Phi(A_j^{(1)}) = H_1$,
- D_i are plateaux such that $H(D_i) = H_1$, $D_i \cap (\cup_j U(A_j^{(1)})) \neq \emptyset$,
- one of the $A_j^{(1)}$, say $A_1^{(1)}$, contains $\hat{D}_1$,
- $A_j^{(1)} \cap F(A) = \emptyset$.

We can also write:

$$Q_1 = \cup_j \big[(A_j^{(1)}) \cup (\cup_{z \in U(A_j^{(1)})} D(z)) \big]. \tag{6.174}$$

In other words Q_1 is the maximal connected union, containing $\hat{D}_1$, of cycles (trivial and non-trivial) with height H_1 (see (6.11)) and not intersecting $F(A)$.

We decompose ∂Q_1 as

$$\partial Q_1 = \partial^u Q_1 \cup \partial^d Q_1, \qquad \text{with } \partial^u Q_1 \cap \partial^d Q_1 = \emptyset, \tag{6.175}$$

where $H(x) > H_1$ for $x \in \partial^u Q_1$; $H(x) < H_1$ and $\Phi(x, F(A)) < H_1$, for $x \in \partial^d Q_1$; each point $y \in \partial^d Q_1$ belongs to some cycle A' containing points in $F(A)$ and such that $\Phi(A') = H_1$.

Let $\mathcal{S}_1$ be the subset of Q_1 downhill connected to $\partial^d Q_1$. Choose a point $y_1 \in \mathcal{S}_1$ and consider a downhill p-path $\omega^{(2)}$ emerging from y_1 with $(\omega^{(2)} \setminus y_1) \cap Q_1 = \emptyset$ (such a path certainly exists by construction); let $\hat{D}_2$ be the first plateau in $\omega^{(2)}$ belonging to $\bar{M}$ and let

$$H_2 := \Phi(\hat{D}_2, F(A)).$$

We can now repeat exactly the previous construction and get the sets $Q_2, \mathcal{S}_2$, depending on the choice of $\omega^{(2)}$; we decompose Q_2 as $Q_2 = \cup_j \big[(A_j^{(2)}) \cup (\cup_{z \in U(A_j^{(2)})} D(z)) \big]$ where the disjoint cycles $A_j^{(2)}$ have properties analogous to those of the $A_j^{(1)}$. We continue in this way and recursively construct a sequence of the form $\omega^{(3)}, Q_3, \omega^{(4)}, \dots,$ where $\omega^{(i)} : y_{i-1} \to \hat{D}_i$ with $y_i \in \mathcal{S}_i$, $\hat{D}_i \in \bar{M}$, $H_i := \Phi(\hat{D}_i, F(A)) = H(y_i) > \Phi(\hat{D}_{i+1}, F(A)) = H(y_{i+1})$. Certainly this procedure ends up with some $y_{p-1}, \omega^{(p)}$, for some finite p, with $\hat{D}_p$ being a connected component D^* of $F(A)$ (see Figure 6.3).

Suppose that a particular choice is made of $y_0, \omega^{(1)}, \omega^{(2)}, \dots, \omega^{(p)}$, compatible with the above construction.

Then, automatically, $\hat{D}_2, \dots, \hat{D}_{p-1}, \hat{D}_p, Q_1, \dots, Q_{p-1}, Q_p$ are given.

Let

$$\mathcal{T}(y_0, \omega^{(1)}, \omega^{(2)}, \dots, \omega^{(p)}) = y_0 \cup \omega^{(1)} \cup Q_1 \cup \omega^{(2)} \cdots Q_{p-1} \cup \omega^{(p)} \cup Q_p. \tag{6.176}$$

Any sequence like $\mathcal{T}(y_0, \omega^{(1)}, \omega^{(2)}, \dots, \omega^{(p)})$, obtained via the above construction, will be called a *standard cascade*; it can be visualized as a sequence of falls and communicating lakes.

Notice that, by construction, $F(Q_1), \dots, F(Q_{p-1})$ are strictly higher, in energy, than $F(A)$.

A particularly simple case is when each Q_j is just given by a single cycle A_j.

Notice that when A is completely attracted (in the $\beta = \infty$ dynamics) by a unique plateau $D^* \equiv F(A)$, every standard cascade is just given by a downhill p-path and we always have $p = 1$, $Q_1 = D^*$.

Now we can state our main result.

Theorem 6.31 *Given a non-trivial cycle A, for every $y_0 \in A$, $y_0 \notin F(A)$.*

(i) *With a probability tending to one as β goes to infinity, the following happens: there exists a sequence $y_0, \omega^{(1)}, \omega^{(2)}, \ldots, \omega^{(p)}$ such that our process, starting at $t = 0$ from y_0, between $t = 0$ and $t = \tau_{F(A)}$, belongs to $\mathcal{T}(y_0, \omega^{(1)}, \omega^{(2)}, \ldots, \omega^{(p)})$; moreover after having followed the initial downhill p-path $\omega^{(1)}$, it visits sequentially the sets $Q_1, Q_2, \ldots, Q_{p-1}$ exiting from Q_j through $y_j \in S_j$ and then following the p-path $\omega^{(j+1)}$ before entering Q_{j+1}.*

(ii) *For every $\varepsilon > 0$ with a probability tending to one as $\beta \to \infty$, the process spends inside each plateau of the p-paths $\omega^{(1)}, \ldots, \omega^{(p)}$ a time shorter than $e^{\varepsilon\beta}$ and inside each Q_j a time lying in the interval*

$$[\min_k \exp(\beta\{H(y_j) - H(F(A_k^{(j)})) - \varepsilon\}),$$
$$\max_k \exp(\beta\{H(y_j) - H(F(A_k^{(j)})) + \varepsilon\})];$$

before exiting from Q_j it can perform an arbitrary sequence of passages through the cycles $A_k^{(j)}$ belonging to Q_j. Each transition between different cycles $A_k^{(j)}$ is made through a minimal saddle z_j in the boundary of $A_k^{(j)}$ ($H(z_j) = H_j$); once the process enters into a particular $A_k^{(j)}$, it spends there a time T such that

$$\exp(\beta[H_j - H(F(A_k^{(j)})) - \varepsilon]) < T < \exp(\beta[H_j - H(F(A_k^{(j)})) + \varepsilon]),$$

whereas it spends at most a time $e^{\varepsilon\beta}$ inside the plateaux of Q_j, at height H_j.

Proof The proof follows immediately by Theorem 6.23, Corollary 6.25 and Proposition 6.29; it suffices to make the following remarks.

- Once entered a plateau D strictly downhill connected with some point lower in energy, then our process, with a probability tending to one as $\beta \to \infty$, will exit from D downhill within a time $e^{\varepsilon\beta}$, for any $\varepsilon > 0$ and β sufficiently large; this is true, in particular, for plateaux consisting of single points.
- By Proposition 6.29 we know that, once entered a transient cycle $A_k^{(j)}$, in a time shorter than $\exp(\beta[\Gamma(A_k^{(j)}) + \varepsilon])$, our process, with a probability tending to one as $\beta \to \infty$, will enter another cycle, contiguous to $A_k^{(j)}$.
- For any $\varepsilon > 0$, with a probability tending to one as $\beta \to \infty$, our process will not perform more than $\exp(\varepsilon\beta)$ transitions between the cycles $A_k^{(j)}$ (through minimal saddles at a height H_j, by Corollary 6.25) before touching some point in $\partial^d Q_j$.
- The bounds of point (ii) of the theorem on permanence times inside the Q_j and $A_k^{(j)}$ follow immediately from Theorem 6.23.

We leave the details to the reader. □

Now, given any non-trivial cycle A, we want to describe, in the maximal possible detail, the first excursion from $F(A)$ to ∂A.

Following Schonmann [267], we first give some simple general definitions.

Given $x, y \in S$, $x \neq y$, we define the following functions on the space of trajectories Ω:

$$\tau_y(\omega) = \inf\{t \geq 0 : \omega_t = y\}, \qquad \tau_y(\omega) = \infty \qquad \text{if } \omega_t \neq y, \qquad \forall t \geq 0, \tag{6.177}$$

$$\theta_{x,y}(\omega) = \sup\{t < \tau_y(\omega) : \omega_t = x\}. \tag{6.178}$$

Let

$$\begin{aligned} \tilde{\Omega}(x, y) :&= \{\omega : \exists t \in \mathbb{N};\ \omega = \omega_0, \omega_1, \ldots, \omega_t;\ \omega_0 = x, \\ &\omega_t = y;\ \omega_s \neq x, y, \text{ for } s \in [1, t-1]\} \end{aligned} \tag{6.179}$$

namely $\tilde{\Omega}(x, y)$ is the set of finite paths with $\theta_{x,y} = 0$ and ending at $\tau_y < \infty$.

We say that an infinite path $\{\omega_t\}_{t=0,1,\ldots}$ *contains a $\theta\tau$-segment* if

$$\begin{aligned} &\exists t_1 < t_2 : \{\omega_{t_1}, \ldots, \omega_{t_2}\} \in \tilde{\Omega}(x, y);\ \forall s_1, s_2 :\ 1 \leq s_1 < s_2 \leq t_1, \\ &\quad \{\omega_{s_1}, \ldots, \omega_{s_2}\} \notin \tilde{\Omega}(x, y); \end{aligned}$$

this $\theta\tau$-segment contains the first excursion between x and y namely a path between the last visit to x before the first visit to y.

We set

$$\bar{\Omega}(x, y) := \{\omega \in \Omega \ \text{ containing a } \theta\tau\text{-segment}\}. \tag{6.180}$$

It is easily seen that, under our hypotheses on the Markov chain, we have

$$P(\bar{\Omega}(x, y)) = 1.$$

Let

$$\Gamma : \bar{\Omega}(x, y) \to \tilde{\Omega}(x, y)$$

be the map associating to $\omega \in \bar{\Omega}(x, y)$ its $(\theta_{x,y}, \tau_y)$-segment. We use the following notation: given $\omega \in \bar{\Omega}(x, y)$ we write $\tilde{\omega}$ to denote

$$\tilde{\omega} = \Gamma\omega, \qquad \tilde{\omega} \in \tilde{\Omega}(x, y).$$

We consider the set of infinite paths $\Omega^*(x, y)$ starting with an element of $\tilde{\Omega}(x, y)$:

$$\Omega^*(x, y) := \{\omega \in \bar{\Omega}(x, y) : \theta_{x,y} = 0\}. \tag{6.181}$$

Given $\omega \in \Omega^*(x, y)$, let $\tilde{\omega} = \Gamma\omega$; we say that ω starts as $\tilde{\omega}$.

We denote by R the time reversal operator defined on finite paths:

$$\forall\, \omega := (\omega_0, \ldots, \omega_t),\ R\omega := \bar{\omega} := (\omega_t, \ldots, \omega_0). \tag{6.182}$$

We naturally define, for every set of finite paths Δ,

$$R\Delta = \{\bar{\omega} = R\omega;\ \omega \in \Delta\}.$$

For any $x, y \in S$ we define a measure ρ on Ω as follows: if $\omega \notin \Omega^*(x, y)$ then $\rho(\omega) = 0$; if $\omega \in \Omega^*(x, y)$ let $\tilde{\omega} = \Gamma\omega$. Then

$$\rho(\omega) = P_x(\ \{X_t\}_{t\geq 0} \text{ starts as } \tilde{\omega}\ |\omega \in \Omega^*(x, y))$$

and by $\bar{\rho}(\omega)$ the measure on Ω given by:

$$\bar{\rho}(\omega) = P_y(\ \{X_t\}_{t\geq 0} \text{ starts as } \tilde{\omega}\ |\omega \in\ \Omega^*(y, x)\ ), \qquad \text{if } \omega \in \Omega^*(y, x),$$

$$\bar{\rho}(\omega) = 0 \quad \text{otherwise.}$$

Proposition 6.32 [267] *For every $x, y \in S$, every $\Xi \subseteq \tilde{\Omega}(x, y)$,*

$$\begin{aligned} P_x(X_t \in \Xi,\ \forall t \in [\theta_{x,y}, \tau_y]) &= \rho(\Xi) = \bar{\rho}(R\Xi) \\ &= P_y(X_t \in R\Xi,\ \forall t \in [\theta_{y,x}, \tau_x]). \end{aligned} \tag{6.183}$$

Proof Given t we define, on $\bar{\Omega}(x, y)$, the following event:

$$\mathcal{B}_t = \{\omega : \{\omega_0, \omega_1, \ldots, \omega_t\} \text{ does not contain a } (\theta_{x,y}, \tau_y)\text{-segment}\}.$$

Given $\tilde{\omega} \in \tilde{\Omega}$, we have

$$P(\Gamma(\{X_t\}_{t=1,2,\ldots}) = \tilde{\omega}) = \sum_{s=1}^{\infty} P(\mathcal{B}_s) P(\{X_t\}_{t=0,1,\ldots} \text{ starts as } \tilde{\omega}). \tag{6.184}$$

Summing over $\tilde{\omega}$ we get:

$$\sum_{s=1}^{\infty} P(\mathcal{B}_s) = \frac{1}{P(\tilde{\Omega}(x, y))} \tag{6.185}$$

and from (6.184), (6.185), we get

$$P(\Gamma(\{X_t\}_{t=1,2,\ldots}) = \tilde{\omega}) = \frac{P(\tilde{\omega})}{P(\tilde{\Omega}(x, y))} = \rho(\omega). \tag{6.186}$$

Using reversibility we get, for $\tilde{\omega} \in \tilde{\Omega}(x, y)$,

$$\rho(\omega) = \frac{\exp[-\beta(H(y) - H(x))] P(R\tilde{\omega})}{\exp[-\beta(H(y) - H(x)) P(\tilde{\Omega}(y, x))} = \bar{\rho}(R\omega). \tag{6.187}$$

□

We recall that we already know, by Corollary 6.25 that, with probability tending to one as $\beta \to \infty$, we have $X_{\tau_{\partial A}} \in U(A)$.

Let us now denote by $\partial^- F(A)$ the set of all $\hat{x} \in F(A)$, (uphill) communicating with $A \setminus F(A)$.

We call $\partial^* F(A)$ the set of points in $\partial^- F(A)$ for which the set $U^-_{(\searrow \hat{x})}$ of all points $x \in U^-(A)$ (see (6.134)), such that there exists a standard cascade

$\mathcal{T}(y_0 = x, \omega^{(1)}, \omega^{(2)}, \dots, \omega^{(p)})$ starting from x and ending in $\hat{x}$ is not empty; $\hat{x}$ will belong to some component D^* of $F(A)$; $D^* := Q_p$ and $\omega^{(p)}$ will end entering into Q_p at $\hat{x}$.

It is easy to verify that $y \in U^-_{(\searrow \hat{x})}$ if and only if there exists a path $\omega : y \to \hat{x}$, $\omega \subseteq A$, such that $\omega_{\tau_{F(A)}} = \hat{x}$ where, for a set $G \subseteq A$, we write $\tau_G(\omega) := \inf\{i : \omega_i \in G\}$.

Now we are able to state our main result about the typical trajectories realizing the escape from a non-trivial cycle A.

Theorem 6.33 *Let*

$$\theta_{F(A)} = \max\{t < \tau_{\partial A} : X_t \in F(A)\}.$$

Given $\hat{x} \in \partial^* F(A)$ and any $z \in A$, we have:

$$\begin{aligned} &P_z(\exists\, \hat{y} \in U(A),\ y_0 \in U^-_{(\searrow \hat{x})} \ \textit{with}\ P(\hat{y}, y_0) > 0,\ \textit{and}\ H(\hat{y}) > H(y_0);\\ &\quad \exists\, \omega^{(1)}, \omega^{(2)}, \dots, \omega^{(p)} : X_t \in R\mathcal{T}(y_0, \omega^{(1)}, \omega^{(2)}, \dots, \omega^{(p)})\\ &\quad\ \forall\, t \in [\theta_{\hat{x}, y_0}, \tau_{\partial A} - 1], X_{\tau_{\partial A}} = \hat{y} | X_{\theta_{F(A)}} = \hat{x}) \to 1 \ \textit{as}\ \beta \to \infty. \end{aligned} \tag{6.188}$$

Moreover, with a probability tending to one as $\beta \to \infty$, the following happens: during the first excursion from $F(A)$ to $S \setminus A$, conditioning to $X_{\theta_{F(A)}} = \hat{x}$ (for some $\hat{x} \in \partial^* F(A)$), to $X_{\tau_{\partial A} - 1} = y_0$ (for some $y_0 \in U^-_{(\searrow \hat{x})}$) and to follow a particular 'anticascade' $R\mathcal{T}(y_0, \omega^{(1)}, \omega^{(2)}, \dots, \omega^{(p)})$ between $\theta_{\hat{x}, y_0}$ and $\tau_{\partial A} - 1$, all the 'time reversed' of the properties specified in Theorem 6.31 *hold true*; *namely for all $\varepsilon > 0$, with a probability tending to one as $\beta \to \infty$ our process, during the above mentioned first excursion, visits, sequentially $Q_{p-1}, \dots, Q_2, Q_1$, following, when exiting from Q_j, the uphill p-path $R\omega^{(j)}$ before entering through y_{j-1} into Q_{j-1}. Finally inside each plateau in the $R\omega^{(i)}$ and inside the Q_j our process has exactly the same behaviour as specified in Theorem* 6.31 (ii).

Proof It follows immediately from Corollary 6.25, Theorem 6.31 and Proposition 6.32. □

Remark In the particular case (relevant for the applications to stochastic Ising models) where the sets Q_i always coincide with a single cycle A_i, it follows immediately from Theorem 6.33 that the typical tube of trajectories, during the first excursion from $F(A)$ to ∂A, is an anticascade starting from some $\hat{x} \in \partial^* F(A)$ and ending in some $y^* \in U(A)$ given by a sequence $\bar{A}_1, \bar{\omega}^{(1)}, \bar{y}_1, \bar{A}_2, \bar{\omega}^{(2)}, \bar{y}_2, \dots, \bar{A}_p, \bar{\omega}^{(p)}, y^*$ with the properties:

(i) $H(\bar{y}_i) < H(\bar{y}_{i+1}),\ 1 = 1, \dots, p-1$,

(ii) $\bar{y}_i \in \mathcal{S}(\bar{A}_{i+1})$.

Some $\bar{\omega}^{(i)}$ can be empty; in that case $\bar{y}_i$ is also a saddle point in $\partial \bar{A}_i$.

6.8 Decomposition into maximal cycles

Given a non-trivial cycle A, we want now to discuss further the typical first excursion from $F(A)$ to ∂A as well as the first excursion from $U(A)$ to $F(A)$.

Using reversibility, it suffices to study this second case, as we have already seen.

Consider the decomposition of $A \setminus F(A)$ into maximal (with respect to inclusion) cycles. This makes sense due to the property of ordering by inclusion satisfied by cycles (see Proposition 6.8). We write

$$A \setminus F(A) = A_1 \cup A_2 \cup \cdots \cup A_m \tag{6.189}$$

where A_i can be either a singleton $\{x\} \not\subseteq M$ or a non-trivial cycle and $A_i \cap A_j = \emptyset, i, j = 1, \ldots, m$.

Proposition 6.34 *If A_i is a non-trivial cycle in the decomposition* (6.189), *then*

$$\mathcal{S}(A_i) \cap \mathcal{S}(F(A_i), F(A)) \neq \emptyset \tag{6.190}$$

so that A_i is a strictly metastable cycle (see Definition 6.14) *and there is a declining path joining some point in $\mathcal{S}(A_i)$ with $F(A)$. If $A_i \equiv \{x\} \not\subseteq M$ is a singleton appearing in the decomposition* (6.189) *then there exists a declining path $\omega : x \to F(A)$.*

Proof Let us prove (6.190). Suppose first that A_i is a non-trivial cycle. Necessarily, $\Phi(A_i) \leq \Phi(F(A_i), F(A))$, otherwise A_i would have a non-empty intersection with $F(A)$. If we had the strict inequality $\Phi(A_i) < \Phi(F(A_i), F(A)) =: \hat{H}$, but still, by the cycle property of A, $\hat{H} < \Phi(A)$, then the maximal connected component containig $F(A_i)$ with energy less than $\hat{H}$ would be a cycle contained in A with empty intersection with $F(A)$ and strictly containing A_i, thus contradicting the maximality of A_i. For $A_i \equiv \{x\} \not\subseteq \bar{M}$, a similar argument shows the existence of a declining path $\omega : x \to F(A)$. Thus $\Phi(A_i) = \hat{H}$ and (6.190) follows. The rest is immediate. □

Two maximal cycles $A', A'' \subseteq A \setminus F(A)$ are *equi-elevated* if either they are trivial cycles belonging to the same plateau, or there exists a sequence of maximal cycles $A_{i_1}, \ldots, A_{i_k} \subseteq A \setminus F(A)$ with $A_{i_1} = A', A_{i_k} = A''$ and plateaux $D_1, \ldots, D_{k-1}$ (whose elements are singletons in the partition (6.189)), all with the same height $\Phi(A_{j_l}) = H(D_l) = \hat{H}$ and such that $D_l \cap U(A_{i_l}) \neq \emptyset \neq D_l \cap U(A_{i_{l+1}}), l = 0, \ldots, k-1$. It is seen immediately that equi-elevation is an equivalence relation and then we can partition $A \setminus F(A)$ as:

$$A \setminus F(A) = Y_1 \cup Y_2 \cup \cdots \cup Y_l \quad \text{for some integer } l \tag{6.191}$$

where the set Y_i are maximal *extended cycles* namely maximal collections of equi-elevated cycles appearing in the partition (6.189). A maximal extended cycle Y

appearing in the partition (6.191) is a set of the form

$$Y = \cup_s [A_s \cup (\cup_{z \in U(A_s)} D(z))] \tag{6.192}$$

with $\Phi(A_s) = \hat{H}, \ \forall s$; Y is a maximal connected set in $A \setminus F(A)$ whose elements z have $\Phi(\mathcal{S}(z, F(A))) = \hat{H}$. It is easy to see that Y can be either a plateau $\{D\}$ (trivial extended cycle) with $H(D) = \hat{H}$ or coincide with one of the permanence sets Q introduced in (6.174) (non-trivial extended cycle).

A *generalized path* is a sequence $Y_1, \ldots, Y_l$ such that Y_i is communicating (i.e. connected in one step) with Y_{i+1} for $i = 1, \ldots, l-1$. A standard generalized path starting from $x \in U(A)$ and ending in $y \in F(A)$ is a sequence $Y_0 = x, Y_1, \ldots, Y_k = y$ such that Y_i is strictly downhill communicating with Y_{i+1} for $i = 1, \ldots, k-1$.

It is seen immediately that a standard generalized path coincides with a standard cascade (see (6.176)).

We want to stress that the typical first descent from $x \in U(A)$ to $y \in F(A)$ can be specified, in general, from the geometric point of view, only by giving a specific generalized path. The rule, in this language, is very simple: as a consequence of Theorem 6.31 we have that, with probability tending to one in the limit of large β, this first descent will take place following a standard generalized path. It may happen that there is a trivial standard path from $x \in U(A)$ to $y \in F(A)$ all made by plateaux; indeed this is always the case when the cycle A is completely attracted by a unique stable plateau; but in general, to join x to y via a 'downhill' sequence of extended cycles (namely a standard one) we *have* to choose non-trivial extended cycles.

Thus, in general, to a given generalized path will correspond a wide set of single trajectories; but in general we cannot give significant further specifications about the history of our process when it enters some extended cycle (except for bounds on the permanence time and the exit from a minimal saddle), without renouncing statements which are true with probability of order one for large β.

Let us now analyse the special interesting case when a permanence set Q is a single non-trivial cycle A'.

The typical trajectories, after reaching A', in a way described by Theorem 6.31, will first descend to $F(A')$ and then they will start randomly oscillating around $F(A')$ up to the almost unpredictable exit time $\tau_{\partial A'}$ as specified by Theorem 6.30.

Thus, the permanence inside $A' \subseteq A$, during the first descent from $x \in U(A)$ to $y \in F(A)$, is an intrinsically random phenomenon: more deterministic behaviour between $\tau_{A'}$ and $\tau_{\partial A'}$ would be very low in probability.

Actually much before the exit from A' a sort of apparent equilibrium is established as is stated by Theorem 6.28.

In order to realize the exit, instead of following a single optimal path, it is more convenient to use a mechanism involving a large set of single paths. An energy–entropy balance argument enters into the game; a very large set of singularly very unlikely individual trajectories (corresponding to exponentially long hesitation) is more convenient with respect to a fast exit realized by more likely but not enough numerous individual trajectories.

Theorem 6.28 shows that in the description of the behaviour of the process between $\tau_{A'}$ and $\tau_{\partial A'}$, the notion of entropy enters in a more specific sense (see [230]). Indeed during the permanence in A' our system reaches a quasi-equilibrium state corresponding to the principle of maximal entropy subject to some constraints. First of all the thermal equilibrium under our conditions (fixed temperature) corresponds to minimizing the free energy rather than maximizing the entropy as would be the case for a system with fixed energy. Moreover there is the constraint of belonging to A' which leads to the restricted canonical ensemble. The general rule for the asymptotic behaviour of a Metropolis Markov chain, during the first descent to the ground $F(A)$, is that once the process enters into a non-trivial cycle A', its behaviour is metastable as prescribed by the pathwise approach and in particular:

(1) its properly renormalized exit time tends to be exponential;
(2) the empirical measure is described, for large times prior to the exit, by the conditional Gibbs measure;
(3) the exit from A' takes place through a minimal saddle.

Summarizing, when the energy function H is invertible, in the case of a cycle $A = B(x) \equiv$ the strict basin of attraction of a stable state $x \in M$, the natural problem of the first descent from $z \in A$ to x consists in finding an optimal path $\omega : z \to x$ and the solution is given by any downhill path $\bar{\omega} : z \to x$ (no uphill steps, no hesitations). This exhausts the possible information that we can give with probability of order one.

For a general energy function and a general cycle A with many internal saddles, the problem of the first descent from $z \in A$ to $F(A)$ can be posed naturally in the following way. Given the partition $A \setminus F(A) = Y_1 \cup Y_2 \cup \ldots \cup Y_l$ into maximal extended cycles,

(1) find the optimal generalized paths $Y_{i_1}, \ldots, Y_{i_k}$ with $z \in Y_{i_1}$, $Y_{i_k} \cap F(A) \neq \emptyset$,
(2) determine the typical exit point from Y_{i_l},
(3) determine the typical exit time from Y_{i_l},
(4) describe the behaviour of the process in the interior of Y_{i_l}.

The answer to (1) is that optimal generalized paths are standard generalized paths; the answer to (2) is given by Corollary 6.25; the answer to (3) contains, in general, a large indetermination and is given in points (i), (ii), of Theorem 6.23; when Y_{i_l} is a single cycle, the typical exit time can be determined but only in the sense of logarithmic equivalence. The answer to (4) is given by Theorem 6.31, Theorem 6.30 and Theorem 6.28.

On the other hand the problem of first exit from a cycle A consists first in determining $U(A)$ then, using reversibility, in determining the possible standard generalized paths emerging from the points in $U^-(A)$ (see Theorem 6.33).

Conclusions To simplify the discussion let H be invertible (every plateau is a singleton). The typical descent to the bottom of a cycle containing many attractors takes place in a way that can be considered as the natural generalization of the typical descent to the bottom x of a domain $B(x)$ completely attracted by the unique stable equilibrium point x. The main difference is the following: in the completely attracted case the system does not 'hesitate' and it always follows the drift up to the arrival at x in a finite time, uniformly bounded in β, whereas in the general not completely attracted case the process tries to follow the drift in finite times as far as possible but sometimes it has to enter into suitable permanence sets Q_i, waiting suitable random times T_i and then getting out from Q_i through suitable optimal points. The fact that the permanence times T_i are close to the typical escape times from Q_i and that this escape takes place in the optimal way is the counterpart of the fact that, in the completely attracted case, the only permanence sets are trivial in the sense that they reduce to single points and the method of exit from these single unstable points (after a permanence time of order one) is optimal in the sense that it is along the drift.

As has been shown in [232] a similar picture is still valid in the non-reversible case (see also [55, 289]). The example treated in [237] is also non-reversible.

The situation when analysing the typical 'ascent' against the drift is more complicated and, of course, in the non-reversible case typical ascents outside a generalized basin Q and typical descents to the bottom of Q are not related by time reversal.

6.9 Renormalization

We want now to describe some results concerning the so-called renormalization procedure for Freidlin–Wentzell Markov chains that were introduced by Scoppola in [272] (see also [273, 274]).

We consider a Markov chain satisfying an approximate reversibility condition. The general case of FW, not necessarily reversible, Markov chains can be considered as well.

The aim of the renormalization approach is to control, by means of an iterative argument, with estimates from above and from below, the following quantities characterizing the long time behaviour of the chain X_t:

$$\mu(G), \qquad E\tau_G^x, \qquad P(\tau_G^x > t), \qquad P(X_{\tau_G}^x = y), \tag{6.193}$$

for all $G \subseteq S$, $x \in S \setminus G$, $y \in G$, $t \in \mathbb{N}$, where μ is the invariant measure and τ_G^x is defined in (6.109).

Let us start from our Markov chain X_t on the state space S. We introduce a classification of the states in S in terms of their increasing stability $S \supseteq S^{(1)} \supseteq S^{(2)} \supseteq \ldots \supseteq S^{(n)}$. This classification allows us to introduce a sequence of Markov chains $X_t^{(k)}$ defined over the sequence of state spaces $S^{(k)}$ and corresponding to our original chain X_t viewed on a sequence of increasing time scales $T_1, T_2, T_3, \ldots$ This means that the chain $X_t^{(k)}$ is a coarse graining version of the chain X_t in the sense that, passing from the chain X_t to the chain $X_t^{(k)}$, we give a less detailed description of the process by only analysing events which occur in a typical time of order T_k.

At each step of the iteration the quantities (6.193) are estimated in terms of the same quantities relative to the chain of the next step. Since in the construction $S^{(k)} \subseteq S^{(k-1)}$, the basic idea of the method is to iterate the argument up to a sufficiently large n, such that the space $S^{(n)}$ is sufficiently small and the quantities (6.193) are easy to evaluate.

We want now to introduce a class of Markov chains satisfying only approximately condition RND.

Condition ARND (approximately reversible non-degenerate) *We say that an ergodic, aperiodic Markov chain with finite state space S and set of pairs of communicating states $\mathcal{G}$, satisfies condition ARND if there exists $\gamma = \gamma(\beta) \to 0$ as $\beta \to \infty$ such that the transition probabilities satisfy*:

$$\exp[-\beta(\Delta(x, y) + \gamma(\beta))] \leq P(x, y) \leq \exp[-\beta(\Delta(x, y) - \gamma(\beta))] \quad (6.194)$$

where

$$\Delta(x, y) = +\infty, \qquad if\ (x, y) \notin \mathcal{G},$$

and

$$\Delta(x, y) = \tilde{H}(x, y) - H(x), \qquad for\ (x, y) \in \mathcal{G}, \quad (6.195)$$

with $H : S \to \mathbb{R}$, $\tilde{H} : \mathcal{G} \to \mathbb{R}$, enjoying the properties:

(i) $H(x) \neq H(y)$ *if* $x \neq y$,

(ii) $\tilde{H}(x, y) = \tilde{H}(y, x)$ *and* $\tilde{H}(x, y) \geq H(x) \vee H(y)$.

We can extend $\tilde{H}$ to the whole space $S \times S$ by setting $\tilde{H}(x, y) = +\infty$ whenever $\Delta(x, y) = +\infty$.

Notice that we are not assuming point (ii) of the RND condition since it is not important for the results on renormalization that we are reporting here.

Let us state the main results by assuming condition ARND for the sake of simplicity. General FW Markov chains can be considered as well (see [272]).

Theorem 6.35 [274] *Let X_t be a Markov chain satisfying condition ARND. Then it is possible to define a sequence of time scales T_k exponentially long in β*

$$T_k \equiv e^{(\Phi_1+\Phi_2+\cdots+\Phi_k)\beta}, \qquad k = 1, 2, \ldots \quad (6.196)$$

with Φ_i *positive constants, and a finite sequence of Markov chains* $X_t^{(1)}, X_t^{(2)}, \ldots, X_t^{(n)}$ *on state spaces* $S \supseteq S^{(1)} \supseteq S^{(2)} \supseteq \ldots \supseteq S^{(n)}$, *with* $S^{(n)} = \{x_m\}$, *where* x_m *is the state of absolute minimum for* H, *such that each Markov chain* $X_t^{(k)}$ *satisfies condition ARND with* $H^{(k)}(x) = H(x)$ *and new functions* $\tilde{H}^{(k)}(x, y)$. *We set* $S = S^{(0)}$, $X_t = X_t^{(0)}$. *The processes* $X_t^{(k)}$ *correspond to the chain* X_t *on the time scale* T_k *in the following sense. Given* $G \subseteq S^{(k)}$ *let* $\tau_G^{(k)} = \inf\{t : X_t^{(k)} \in G\}$. *Let* $\mu^{(k)}$ *denote the unique invariant measure for* $X_t^{(k)}$. *Then we have, for* $k = 1, \ldots, n$, $G \subseteq S^{(k)}$, $x \in S^{(k)} \backslash G$, $y \in G$ *and* β *sufficiently large*:

(*i*) $P_x(X_{\tau_G} = y) = P_x(X^{(k)}_{\tau_G^{(k)}} = y)$; (6.197)

(*ii*) *for any* $\eta > 0$ *and* β *sufficiently large*

$$e^{-\eta\beta} T_k E_x \tau_G^{(k)} \leq E_x \tau_G \leq e^{\eta\beta} T_k E_x \tau_G^{(k)}; \tag{6.198}$$

(*iii*) *there exists* $\gamma' = \gamma'(\beta)$, *with* $\gamma'(\beta) \to 0$ *as* $\beta \to \infty$, *such that for any* k *and* $G, G' \subseteq S^{(k)}$ *with* $\mu^{(k)}(G') \neq 0$,

$$e^{-\gamma'\beta} \frac{\mu^{(k)}(G)}{\mu^{(k)}(G')} \leq \frac{\mu(G)}{\mu(G')} \leq e^{\gamma'\beta} \frac{\mu^{(k)}(G)}{\mu^{(k)}(G')}; \tag{6.199}$$

(*iv*) *for any* $t > T_k e^{\delta\beta} 2^k$, *for any* $G \subseteq S^{(k)}$ *and for any* $x \in S^{(k)}$

$$P_x(\tau_G > t) \leq P_x(\tau_G^{(k)} > \frac{t}{T_k 2^k}) + SES. \tag{6.200}$$

The quantities $\tilde{H}^{(k)}(x, y)$, Φ_k *and the state spaces* $S^{(k)}$ *are defined explicitly in terms of the quantities* $\{\Delta(u, v)\}_{u,v \in S}$:

$$\Delta(u, v) = \tilde{H}(u, v) - H(v). \tag{6.201}$$

For the proof we refer to [274] (see also [272, 273]). Here we just give the definitions concerning the chains $X_t^{(k)}$.

Let us first construct the first chain $X_t^{(1)}$. Its state space will be $S^{(1)} = M \equiv$ the set of stable states

$$M = \{x \in S : \forall\, y \in S,\ y \neq x, \quad H(x) < \tilde{H}(x, y)\}, \tag{6.202}$$

$$\Phi_1 \equiv \min_{x \in M,\, y \in S,\, x \neq y} \tilde{H}(x, y) - H(x), \tag{6.203}$$

$$t_1 \equiv e^{\Phi_1 \beta}. \tag{6.204}$$

We introduce now a sequence of stopping times:

$$\begin{aligned} \sigma_1 &\equiv \min\{t > 0;\ X_t \neq X_0\}, \\ \tau_1 &\equiv \min\{t \geq \sigma_1;\ X_t \in M\}, \\ \zeta_1 &= \begin{cases} t_1 & \text{if } \sigma_1 > t_1 \\ \tau_1 & \text{if } \sigma_1 \leq t_1, \end{cases} \end{aligned} \tag{6.205}$$

and for each $n > 1$

$$\begin{aligned} \sigma_n &\equiv \min\{t > \zeta_{n-1};\ X_t \neq X_{\zeta_{n-1}}\}, \\ \tau_n &\equiv \min\{t \geq \sigma_n;\ X_t \in M\}, \\ \zeta_n &= \begin{cases} \zeta_{n-1} + t_1 & \text{if } \sigma_n - \zeta_{n-1} > t_1 \\ \tau_n & \text{if } \sigma_n - \zeta_{n-1} \leq t_1. \end{cases} \end{aligned} \tag{6.206}$$

It is easily seen (see [272]) that $X_n^{(1)} := X_{\zeta_n}$ is a homogeneous Markov chain. For any $x \in M$, we can then consider the new Markov chain $X_t^{(1),x}$ with $X_0^{(1),x} = x$ on the state space $M =: S^{(1)}$.

Notice that this new Markov chain is strictly related to the time scale t_1 (see (6.204)) in the following sense:

$$P(\zeta_{n+1} - \zeta_n \in [t_1 e^{-\gamma\beta}, t_1 e^{\gamma\beta}]) \sim 1$$

where γ is the infinitesimal quantity appearing in (6.194).

For any pair of states $x, y \in M$ we denote by $P^{(1)}(x, y)$ the transition probability of the chain $X_n^{(1)}$ that is

$$P^{(1)}(x, y) = P(X_{\zeta_n} = y |\, X_{\zeta_{n-1}} = x). \tag{6.207}$$

Let us define for any trajectory $\omega : \mathbb{N} \to S$:

$$I_{[0,t]}(\omega) \equiv \sum_{i=0}^{t-1} \Delta(\omega_i, \omega_{i+1}). \tag{6.208}$$

Consider the quantity:

$$\bar{\Delta}(x, y) = \inf_{t, \omega : \omega_0 = x,\, \omega_t = y,\, \omega_s \notin M,\, \forall s \in (0,t)} I_{[0,t]}(\omega). \tag{6.209}$$

It is possible to prove (see [272, 273]) that:

$$t_1\, e^{-\bar{\Delta}(x,y)\beta - \bar{\gamma}\beta} \leq P^{(1)}(x, y) \leq t_1\, e^{-\bar{\Delta}(x,y)\beta + \bar{\gamma}\beta} \tag{6.210}$$

with $\bar{\gamma} \to 0$, as $\beta \to \infty$.

The quantity $\bar{\Delta}(x, y)$ can assume the values $\bar{\Delta}_1 = \Phi_1 < \bar{\Delta}_2 < \ldots < \bar{\Delta}_{\bar{m}}$ with $\bar{\Delta}_{\bar{m}} < |S| \Delta_m$.

We define

$$H^{(1)}(x) = H(x), \quad \forall x \in M \equiv S^{(1)}. \tag{6.211}$$

For any $x, y \in S^{(1)}$ with $x \neq y$, let

$$\Delta^{(1)}(x, y) \equiv \bar{\Delta}(x, y) - \Phi_1. \tag{6.212}$$

Let $\mathcal{G}^{(1)} \subseteq (S^{(1)})^2$ be the space of pairs of states (x, y) such that $P^{(1)}(x, y) > 0$. Then there exists a function $\tilde{H}$ on $\mathcal{G}^{(1)}$ with values in $\mathbb{R}$ such that:

$$\tilde{H}^{(1)}(x, y) = \tilde{H}^{(1)}(y, x), \tag{6.213}$$

$$\tilde{H}^{(1)}(x, y) \geq H^{(1)}(x) \vee H^{(1)}(y), \tag{6.214}$$

$$\Delta^{(1)}(x, y) = \tilde{H}^{(1)}(x, y) - H^{(1)}(x). \tag{6.215}$$

The iteration scheme now is the following: for any $k \geq 1$ we define the following quantities

$$H^{(k+1)}(x) \equiv H(x), \quad \forall x \in S^{(k)}. \tag{6.216}$$

For any $\omega : \mathbb{N} \to S^{(k)}$

$$I^{(k)}_{[0,t]}(\omega) = \sum_{i=0}^{t-1} \Delta^{(k)}(\omega_i, \omega_{i+1}), \tag{6.217}$$

$$M^{(k)} = \{x \in S^{(k)}; \forall y \in S^{(k)},\ y \neq x,\ H^{(k)}(x) < \tilde{H}^{(k)}(x, y)\}, \tag{6.218}$$

$$\bar{\Delta}^{(k)}(x, y) = \inf_{t,\omega:\omega_0=x,\omega_t=y,\ \omega_s \notin M^{(k)},\ \forall s \in [0,t]} I^{(k)}_{[0,t]}(\omega), \qquad \forall x, y \in M^{(k)}, \tag{6.219}$$

$$S^{(k+1)} = M^{(k)}, \tag{6.220}$$

$$\Phi_{k+1} = \inf_{x \in M^{(k)}, y \in S^{(k)},\ x \neq y} \tilde{H}^{(k)}(x, y) - H^{(k)}(x), \tag{6.221}$$

$$t_{k+1} = e^{\Phi_{k+1}\beta}, \tag{6.222}$$

$$T_1 = t_1,$$

$$T_{k+1} = t_1 t_2 \cdots t_k t_{k+1}, \tag{6.223}$$

$$\Delta^{(k+1)}(x, y) = \bar{\Delta}^{(k)}(x, y) - \Phi_{k+1}, \qquad \forall x, y \in S^{(k+1)}, \tag{6.224}$$

$$\tilde{H}^{(k+1)}(x, y) \equiv \Delta^{(k+1)}(x, y) + H(x). \tag{6.225}$$

It is natural to try to analyse the set of trajectories of the original chain X_t corresponding to a given trajectory of the chain $X_t^{(k)}$. This happens to be related to the characterization of the typical tube of trajectories followed by the original chain X_t during the first exit from a generic set G. Indeed, let $N = N(G)$ be the level of renormalization such that the $(N+1)$st renormalized Markov chain does not contain states inside G: $S^{(N+1)} \cap G = \emptyset$. This means that the first excursion outside G for the chain $X_t^{(N)}$ is a sort of 'descent' along the drift. This provides a first rough approximation of the typical tube of escape from G: it is given by the set of typical trajectories followed during the first excursion outside G by the chain $X_t^{(N)}$.

The question remains of 'reading' the result in terms of the paths followed on the original time scale by our original chain X_t.

In [232] it was shown, even in the more complicated non-reversible and degenerate case, that it is possible to associate to any state $x^{(N)}$ of $S^{(N)}$ a suitable extended cycle $Q_{x^{(N)}} \subseteq S$ representing the set where the original process X_t typically remains in the interval of time corresponding to a jump of the chain $X_t^{(N)}$.

In the particular reversible, non-degenerate case, following these ideas when studying the first exit problem from a non-trivial cycle G provides an alternative way to deduce the results of Theorems 6.31 and 6.33.

It is immediate to see that the quantities

$$\Psi_k := \Phi_1 + \cdots + \Phi_k \tag{6.226}$$

represent the possible growing barriers of the minima in S.

Let us introduce these quantities.

Given a point $x \in S$ let

$$W_x := \{z \in S : H(z) < H(x)\}. \tag{6.227}$$

The quantity

$$\gamma(x) := \Phi(x, W_x) - H(x), \tag{6.228}$$

is called the stability level of the state x.

It is easily seen that

$$\gamma(x) = \Gamma(A^*(x)) \tag{6.229}$$

where $A^*(x)$ is the largest cycle containing x in its ground $F(A^*)$. $A^*(x)$ coincides with the minimal strictly metastable cycle containing x. If x does not belong to a stable plateau, we simply have $A^*(x) \equiv x$.

We have that the possible values of the barrier $\gamma(x)$ are the Ψ_k; this comes from the very definition of renormalized chains.

We use these notions in Theorem 6.36 and Proposition 6.37, below.

It is natural to ask oneself about the relationship between different approaches to the study of FW Markov chains and in particular the first exit problem from a set G: the graphical approach introduced by Freidlin and Wentzell, the renormalization procedure and the methods developed above for the simpler reversible case.

In particular three partitions emerge: the partition introduced by Freidlin and Wentzell in terms of cycles of increasing rank, the partition introduced in [272–274] in terms of state spaces of increasing stability, and the partition into maximal cycles of $A \setminus F(A)$ that we introduced in the reversible case.

We remark that the rank of the cycles is necessary to the construction, but it does not have an intrinsic meaning. The iteration of this cycle construction is completely different from the renormalization procedure, where the iterative parameter has an immediate interpretation in terms of time rescaling. On the other hand the permanence set that we found in the reversible case is exactly the same as the set $Q_{x^{(N)}}$ associated with single points of states of increasing stability.

We want to quote, now, an interesting application of the renormalization procedure, namely a simple formula providing an asymptotic expression, at large β, for the eigenvalues of the generator of our dynamics.

Theorem 6.36 [274] *Let $\mathbf{L} = \mathbf{P} - \mathbb{I}$ be the generator associated with the Markov chain X_t with transition matrix $\mathbf{P}$ ($\mathbb{I}$ is the identity matrix). We denote by $\lambda_1 = 0, \lambda_2, \dots, \lambda_i, \dots$ the eigenvalues of the matrix $-\mathbf{L}$ arranged in increasing order. Suppose that our Markov chain satisfies condition R. Then we have:*

$$\lim_{\beta\to\infty} -\frac{1}{\beta}\log(\lambda_j) = \lim_{\beta\to\infty} \frac{1}{\beta}\log T_{N_j} = \Phi_1 + \Phi_2 + \cdots + \Phi_{N_j} = \Psi_{N_j} \quad (6.230)$$

where $N_j \equiv \min\{n;\ |S^{(n+1)}| < j\}$.

Equation (6.230) says that the eigenvalues, for large β, are determined by the values assumed by the depths of metastable cycles. In particular, the minimal non-vanishing eigenvalue λ_{min} ($\equiv$ gap in the spectrum) corresponding to the maximum value of minus the logarithm of the eigenvalues, is determined, for large β, by the depth of the deepest metastable cycle in S:

$$\lim_{\beta\to\infty} -\frac{1}{\beta}\log(\lambda_{min}) = \Psi_{k_{max}}. \quad (6.231)$$

Suppose we apply Theorem 6.36 to the chain X_t^A, restricted to a non-trivial cycle A (see (6.106)). We have in this case

$$\lim_{\beta\to\infty} -\frac{1}{\beta}\log(\lambda_k) = \Theta(A) \quad (6.232)$$

saying that the 'relaxation time' in the interior of a cycle A is determined by the maximal internal resistance which, in turn, is related to the maximal time needed to reach the ground $F(A)$. In the totally non-degenerate case, when there is always only one metastable cycle with a given depth, the theory simplifies further. In particular we have that the cardinality $|S^{(k)}|$ of $S^{(k)}$ is given as $|S^{(k)}| = |S^{(k-1)}| - 1$, namely, at each step of renormalization we just lose one point.

In this case the eigenvalues are not degenerate and the statement of Theorem 6.36 becomes

$$\lim_{\beta\to\infty} -\frac{1}{\beta}\log(\lambda_k) = \lim_{\beta\to\infty}\frac{1}{\beta}\log T_k = \Phi_1 + \Phi_2 + \cdots + \Phi_k = \Psi_k. \tag{6.233}$$

6.10 Reduction and recurrence

We want now to analyse, from a slightly different point of view, the sets of states in S with increasing stability. In particular we prove that, with a probability extremely close to one for β large, our process will recur on these states on suitable increasing time scales.

Given $v \geq 0$ we define the set S_v of *v-irreducible* points as the set of points with stability level larger than v:

$$S_v := \{x \in S : \gamma(x) > v\} \tag{6.234}$$

(see (6.228)). The set $S \setminus S_v$ is the set of *v-reducible* points.

The elements of the set of maximal stability level in $S \setminus F(S)$:

$$S^m := \left\{ x \in S : \gamma(x) = \max_{y \in S\setminus F(S)} \gamma(y) \right\} \tag{6.235}$$

are called *metastable states*.

A crucial step in our construction is the partition of $S \setminus S_v$ into maximal (by inclusion) cycles

$$S \setminus S_v := A_1 \cup A_2 \cup \cdots \cup A_n. \tag{6.236}$$

Notice that the partition considered in (6.191) is a particular case of (6.236), corresponding to S equal to a non-trivial cycle A, and v the maximal possible value v_{max} of the barrier $\gamma(x)$, for which $S_v \equiv F(A)$.

It is seen immediately that

$$\Gamma(A_i) \leq v. \tag{6.237}$$

Indeed if $\Gamma(A_i) > v$, we would have $\gamma(x) > v$, for any $x \in F(A_i)$, contradicting $x \in S \setminus S_v$.

A maximal-cycle-path (mc-path) is a sequence $A'_1, \ldots, A'_k$, $A'_j \in \{A_1, \ldots, A_n\}$, with A'_j connected (in the usual sense) to A'_{j+1}.

Notice that, in our Metropolis dynamics, if a non-trivial cycle A' appears in an mc-path, the subsequent and previous elements must be trivial cycles in its boundary.

A downhill mc-path $A'_1, \dots, A'_k$ is characterized by $\Phi(A'_{j+1}) \le \Phi(A'_j)$, $j = 1, \dots, k-1$.

A set of cycles $\mathbf{A} = \{A'_1, \dots, A'_m\}$ is mc-connected if for any pair of its cycles there is an mc-path in $\mathbf{A}$ joining them.

Given $A'' \in \{A_1, \dots, A_n\}$, A_i on the righ-hand side of (6.236), let us denote by $Y(A'')$ the maximal mc-connected set of cycles $\{A'_1, \dots, A'_k\}$ containing A'' ($A'_j \in \{A_1, \dots, A_n\}$) such that $\Phi(A'_j) = \hat{H} := \Phi(A'')$. The sets Y are extended cycles like those appearing in the partition (6.191) of $A \setminus F(A)$ for a non-trivial cycle A. The reader should think of Y as a set of communicating lakes.

We are now ready to formulate our main recurrence result. Let $T_v = e^{v\beta}$, and let

$$\tau_{S_v} = \min\{t \in \mathbb{N} \colon \ X_t \in S_v\}. \tag{6.238}$$

Proposition 6.37 *Let $T_v = e^{v\beta}$. For every $\delta > 0$,*

$$\max_{x \in S \setminus S_v} P_x(\tau_{S_v} > T_v e^{\delta\beta}) = \text{SES} \tag{6.239}$$

(*see Definition* 6.4).

Proof Using the Markov property it is sufficient to define, for all $x \in S \setminus S_v$, an event $\mathcal{E}_x^T \subseteq \Omega$ with $T = T_v e^{\frac{\delta}{2}\beta}$ such that

$$\mathcal{E}_x^T \subseteq \{x_0 = x, \exists t \in [0, T] \cap \mathbb{N} \colon \ x_t \in S_v\}, \qquad \min_{x \in S} P(\mathcal{E}_x^T) \ge p > 0, \tag{6.240}$$

with p not exponentially small in β in the sense that for all $\varepsilon > 0$, $p > e^{-\varepsilon\beta}$ for large enough β.

Let us denote by $A^{init} = A^{init}(x)$ the cycle in the partition (6.236) containing the initial point x and set $\Phi(A^{init}) =: \hat{H}$, $Y = Y(A^{init})$.

We prove the following by contradiction.

Claim. Y is (necessarily downhill) connected either to S_v or to some $\hat{A}$ in the partition (6.236), *with $\Phi(\hat{A}) < \hat{H}$.*

Indeed let us suppose that ∂Y neither intersects S_v nor any $\hat{A}$ in the partition (6.236), with $\Phi(\hat{A}) < \hat{H}$.

We have that in ∂Y we cannot find points x with $H(x) = \hat{H}$ because of the maximality of Y. Nor can we find any point x with $H(x) < \hat{H}$ because, since such a point x, by hypothesis, can belong neither to S_v nor to some $\hat{A}$ in the partition (6.236), with $\Phi(\hat{A}) < \hat{H}$, certainly it has to belong to a cycle in the partition (6.236), with $\Phi(\hat{A}) = \hat{H}$, violating again the maximality of Y. Moreover, certainly ∂Y cannot be empty otherwise we would violate the ergodicity of our Markov chain. Thus, necessarily, ∂Y is non-empty and made of points x with $H(x) > \hat{H}$. We conclude that Y is a cycle contained in $S \setminus S_v$; but in this way we violate the maximality of the cycle A^{init}; so necessarily we have that either Y is connected to S_v or to some $\hat{A}$ in the partition (6.236), with $\Phi(\hat{A}) < \hat{H}$.

We are now able to construct, for any A^{init} a downhill mc-path $A'_1, \ldots, A'_k$ with $A'_1 = A^{init}$ ending in S_v in the sense that $\partial A'_k \cap S_v \neq \emptyset$.

Indeed by the above claim either by first following a suitable path $A'_1, \ldots, A'_k$ in Y, with constant height $\Phi(A'_i) = \hat{H}$ and then downhill, we end directly in S_v ($S_v \cap \partial Y \neq \emptyset$); or we have that there exists A'_{k+1} on the right-hand side of (6.236) with $\partial A'_k \cap A'_{k+1} \neq \emptyset$, $\Phi(A'_{k+1}) < \hat{H}$; then, by exploiting the arbitrariness of the initial cycle A^{init}, we can iterate the procedure so that, simply using that the energy is bounded from below, we end up with a downhill mc-path ending in S_v.

Consider the completed downhill mc-path $\phi(A^{init}) = A'_1, \ldots, A'_k, \xi$ with $A'_1 = A^{init}, \xi \in S_v \cap \partial A'_k$.

We say that a trajectory $\{x_s\}_{s=0,1,\ldots}$ follows ε-regularly the path $\phi(A^{init})$ if it belongs to the sequence of cycles in ϕ, stays in each cycle A'_i for a time in the range $\left[\exp(\beta[\Gamma(A'_i) - \varepsilon]), \exp(\beta[\Gamma(A'_i) + \varepsilon])\right]$, and ends in ξ.

The following is true: for every $\varepsilon, \varepsilon' > 0$ there exist $\beta_0 > 0$ such that, for all $\beta > \beta_0$,

$$\min_{x \in S \setminus S_v} P\big(\{x_s\}_{s=0,1,\ldots} \text{ follows } \varepsilon\text{-regularly } \phi(A^{init}(x))\big) > e^{-\varepsilon'\beta}. \tag{6.241}$$

Indeed (6.241) is an immediate consequence of Theorem 6.23.

To conclude the proof of Proposition 6.37, we now pick $x \in S \setminus S_v$ and take for $\mathcal{E}_T^v$ the event where $(x_t)_{t \in \mathbb{N}}$ follows, $\frac{\delta}{4}$-regularly, the mc-path from $A^{init}(x)$ to S_v within time $T = T_v e^{\delta\beta/2}$. By Theorem 6.23, we have, for β sufficiently large,

$$\min_{x \in S \setminus S_v} P_x(\mathcal{E}_T^v) \geq e^{-\delta'\beta}, \qquad \delta' < \delta/4. \tag{6.242}$$

Hence by the Markov property,

$$\max_{x \in S \setminus S_v} P_x(\tau_{S_v} > T_v e^{\delta\beta}) \leq (1 - e^{-\delta'\beta})^{e^{\frac{\delta}{2}\beta}} = \text{SES}. \tag{6.243}$$

□

6.11 Asymptotics in probability of tunnelling times

We want now to give an application to the evaluation of the tunnelling time between an element x_m of a stable plateau and the set $F(S)$ of the absolute minima (in the full state space S) of H. The following theorem states that the typical value of the first hitting time to $F(S)$ starting from x_m will be $\sim e^{\beta\Gamma}$ provided $\{x_m\} \cup F(S) \equiv S_{\Gamma_0}$ for some $\Gamma_0 \leq \Gamma$ and that $\Phi(x_m, F(S)) = \Gamma$.

Theorem 6.38 *Suppose that* (i) *the communication height between x_m and $F(S)$ is $\Phi(x_m, F(S)) = H(x_m) + \Gamma$ and* (ii) *there exists $\Gamma_0 \leq \Gamma$ such that every state $x \in S \setminus (\{x_m\} \cup F(S))$ is Γ_0-reducible in the sense that there exists $x' \in S$ such* *that* $H(x') < H(x)$, $\Phi(x, x') \le H(x) + \Gamma_0$. *Then, for any* $\delta > 0$:

$$\lim_{\beta \to \infty} P_{x_m}\left(e^{\beta(\Gamma-\delta)} \le \tau_{F(S)} \le e^{\beta(\Gamma+\delta)}\right) = 1. \tag{6.244}$$

Proof Let A_{x_m} be the cycle given as the maximal connected set containing x_m with energy smaller than $H(x_m) + \Gamma$. We have

$$\tau_{\partial A_{x_m}} < \tau_{F(S)}. \tag{6.245}$$

The lower bound on $\tau_{F(S)}$ in (6.244) follows immediately from (6.245) and from Theorem 6.23(i).

The upper bound on $\tau_{F(S)}$ follows immediately from Proposition 6.37 by taking $v = \Gamma$ and noticing that in this case $S_v \equiv F(S)$. Indeed in this way we prove that for any $x \in S$, $\delta > 0$:

$$P_x\left(\tau_{F(S)} \le e^{\beta(\Gamma+\delta)}\right) = 1 - \text{SES}. \tag{6.246}$$

□

From the same hypotheses of the above theorem we have, as a corollary, that x_m is a metastable state: $x_m \in S^m$ (see (6.235)) and Γ is the maximal stability level in $S \setminus F(S)$.

In the case of several ground states ($|F(S)| > 1$), we can obtain a similar result for the tunnelling time $\tau_{x_1}^{x_0}$ between two ground states x_0, x_1, by substituting Γ with the maximal internal barrier $\Theta(S)$ defined in analogy with Definition 6.20.

7

Metastable behaviour for lattice spin models at low temperature

Introduction

In this chapter we discuss metastability and nucleation for several short range lattice spin systems at low temperature. We already analysed the case of the Curie–Weiss model in Chapter 4. Due to the fact that the intermolecular interaction does not decay at infinity, the Curie–Weiss model exhibits a mean field behaviour without any spatial structure; the configurations are well described, especially for large volumes, by the values of a unique macroscopic order parameter, i.e. the magnetization, so that the configuration space becomes one-dimensional. In contrast, for short range stochastic Ising models the geometrical aspects are particularly relevant. In this last case the configuration space can be viewed as a space of families of contours and it tends to be infinite dimensional in the thermodynamic limit. Indeed it is well known that the description of pure coexisting equilibrium phases at low temperature is naturally given in terms of a gas of contours (see Chapter 3, proof of Proposition 3.30, and [280]). It is clear that in the analysis of dynamical phenomena taking place in large finite systems, the geometrical description in terms of contours will also play a relevant role.

The central question in the description of the decay from a metastable to a stable phase for short range stochastic Ising models is, in addition to determination of the typical escape time, the characterization of the typical 'nucleation pattern' namely the typical sequences, in shape and size, of droplets along which nucleation of the stable phase takes place.

We shall see that the Curie–Weiss model and short range stochastic Ising models share many characteristic features of metastable behaviour but, at the same time, they show some very relevant differences.

To be concrete let us consider the prototype of the class of short range models that we want to analyse in this chapter: the two-dimensional Metropolis standard stochastic Ising model with small but fixed, say positive, magnetic field h in a large but fixed squared domain Λ with periodic boundary conditions, in the limit of large inverse temperature β. A continuous time version was already introduced in Chapter 4 (see (4.54)). A precise definition of the discrete time version will be given later on.

Let us denote by $-\underline{1}$, $+\underline{1}$ the configurations with all spins $-1, +1$ in Λ, respectively. By choosing a positive and sufficiently small h we see immediately that $-\underline{1}$ is a local minimum whereas $+\underline{1}$ is the absolute minimum for the energy, so they are naturally associated with the metastable and stable equilibrium, respectively.

As will emerge from our analysis, in the case of low temperature short range stochastic Ising models, it is natural to introduce a sort of 'basin of metastability', namely a set of configurations where our system will be confined before transition to the stable situation. The configurations in the basin of metastability will be close to $-\underline{1}$. The meaning of 'closeness' in this case is not, a priori, clear at all; intuitively we expect that 'close to $-\underline{1}$' means 'without too large droplets of pluses'.

The equivalent notion of metastable basin for the Curie–Weiss model simply refers to configurations whose global magnetization is close, in the usual metric on $\mathbb{R}$, to the local minimum $m_-(\beta, h)$ of the canonical free energy $f_{\beta,h}(m)$ (see Chapter 4). In both Curie–Weiss and short range Ising models the system spends a very long time performing random fluctuations in the metastable basin before an almost unpredictable jump leads eventually to the stable situation. In both Curie–Weiss and short range Ising models there is a notion of 'coercive field' $\hat{h} = \hat{h}(\beta)$ representing the threshold to instability. For $h > \hat{h}$ no metastability is possible: the state that was metastable for $h < \hat{h}$ now becomes unstable. In the Curie–Weiss model $\hat{h}$ is the maximal value of h for which the canonical free energy $f_{\beta,h}(m)$ has double well behaviour; for $h > \hat{h}$, $f_{\beta,h}(m)$ has only one well and for $h = \hat{h}$ it has a horizontal inflection point. For the short range Ising model the coercive field can be identified as the minimal value of h for which even the smallest possible droplets have a tendency to grow. It will turn out that for both models the 'lifetime' increases as h decreases.

Let us discuss the differences. Whereas for the Curie–Weiss model the free energy barrier to be overcome is already implicit in the dynamics as it is given by the (unphysical) double well structure of the canonical free energy, in the short range Ising model this barrier is not directly related to the microscopic single spin-flip elementary process of the dynamics but, rather, comes from the collective behaviour of the system. The height and location in the configuration space of this barrier are not given at all a priori, and they have to be determined by looking at the whole, geometrically complicated, configuration space. This is somehow reminiscent of the equilibrium description of phase transitions for short range systems where, in general, the equilibrium coexistence line is unknown as its location is a result of the collective behaviour of the system. The description of how a short range Ising model spends its time in the metastable basin before the jump to stable equilibrium is much more complicated than the corresponding description for the Curie–Weiss model. In both cases the fraction of time spent in the different subsets of the metastable basin before the jump tends to be proportional to an almost stationary measure: the restricted Gibbs ensemble (see Chapter 4 and Chapter 6, Theorem 6.28). In the case of the short range Ising model this is a measure on a complicated configuration space whereas in the case of the Curie–Weiss model we have a simple measure on the line. Moreover the exit path from the metastable basin in the Curie–Weiss model is determined almost immediately, using reversibility, whereas in the short range Ising model it involves complicated geometrical considerations.

The crucial point is that the basin of metastability for short range Ising systems will contain many stable equilibrium configurations, namely many local minima of the energy H. These minima will correspond to particular sets of stable isolated 'droplets'; for the standard Ising model they will be rectangles of pluses in a sea of minuses.

Before the decay to the stable situation, characterized by a sea of pluses with small islands of minuses, the system will visit many times configurations with a sea of minuses with small droplets of pluses.

Since β is very large, only moves that decrease the energy will typically take place; our system will spend the largest part of its time in the local minima for the energy. Sometimes, moves against the drift, with an increase of energy, will take place and, when entering into the basin of attraction of some local minimum, the system will typically take a suitable exponentially long (in β) time to get out of it. On a smaller time scale, a situation similar to the global escape from metastability will occur, at a local level, in the escape from the basin of attraction of a local minimum. It will also occur that, during the first excursion from $-\underline{1}$ to $+\underline{1}$, namely during the decay from the metastable to stable situation, the system will typically visit many basins of local minima, remaining exponentially long intervals of time there. This is the main new difficulty with respect to the description of the first escape from a completely attracted domain. To treat this phenomenon we make use of conceptual categories and results developed in Chapter 6 for Freidlin–Wentzell reversible Markov chains.

7.1 The standard stochastic Ising model in two dimensions

Let us now define the standard stochastic Ising model.

It is a discrete time stochastic dynamics given by a Metropolis Markov chain, reversible with respect to the Gibbs measure for a standard Ising model, when the elementary process is a single spin-flip.

We shall take our Ising spin system enclosed in a two-dimensional torus Λ of edge L namely, an $L \times L$ square with periodic boundary conditions. Other boundary conditions can be considered as well: this point will be discussed later on.

With any lattice site $x \in \Lambda$ we associate a spin variable $\sigma(x)$ taking values $+1, -1$: the configuration space is $\mathcal{X} := \{-1, +1\}^{\Lambda}$. The energy $H(\sigma)$, associated with the configuration $\sigma \in \mathcal{X}$, is given by:

$$H(\sigma) = -\frac{J}{2} \sum_{\substack{\{x,y\} \subseteq \Lambda: \\ |x-y|=1}} \sigma(x)\sigma(y) - \frac{h}{2} \sum_{x \in \Lambda} \sigma(x). \tag{7.1}$$

Notice that in this chapter we use the notation $J/2$ for the coupling constant and $h/2$ for the external magnetic field. We take $J > 0$ (ferromagnetic case). The external magnetic field $h/2$ is taken positive. Notice that at $h = 0$, with our parametrization, the energy necessary to break a bond, namely to pass from a parallel pair of nearest neighbour spins to an opposite pair, is J.

The transition probabilities of our Ising–Metropolis Markov chain are given, for $\sigma \neq \sigma'$, by:

$$P(\sigma, \sigma') = \begin{cases} 0 & \text{if} \quad \sigma' \neq \sigma^{(x)} \text{ for all } x \in \Lambda \\ \frac{1}{|\Lambda|} \exp\{-\beta(\Delta_x H(\sigma) \vee 0)\} & \text{if} \quad \sigma' = \sigma^{(x)} \text{ for some } x, \end{cases} \tag{7.2}$$

where

$$\Delta_x H(\sigma) = H(\sigma^{(x)}) - H(\sigma) \tag{7.3}$$

and

$$\sigma^{(x)}(y) = \begin{cases} \sigma(y) & \text{if } y \neq x \\ -\sigma(x) & \text{if } y = x. \end{cases} \tag{7.4}$$

For $\sigma' = \sigma$ we set:

$$P(\sigma, \sigma) = 1 - \sum_{\sigma' \in \mathcal{X}, \sigma' \neq \sigma} P(\sigma, \sigma'). \tag{7.5}$$

This is a particular case of Markov chains in the Freidlin–Wentzell regime which are characterized by finite state space and transition probabilities exponentially decreasing in a large parameter β. More specifically it is a particular case of a dynamics satisfying condition M of Chapter 6; indeed it is commonly called the Metropolis algorithm. It corresponds to the following updating rule: given a configuration σ at time t we choose at random a site $x \in \Lambda$ and compute the increment of energy $\Delta_x H(\sigma)$; if this quantity is non-positive we flip the spin at x; otherwise, if it is strictly positive, we flip it with a probability $\exp[-\beta \Delta_x H(\sigma)]$. We denote by σ_t our stochastic trajectory.

We consider a situation in which h, J, L, are fixed in such a way that $0 < h < 2J < h(L-3)$ and we take the limit of large β. The condition $h < 2J$ means critical length (see (4.57)) larger than two (twice the spacing of the lattice) whereas $2J < h(L-3)$ means that the side of the box Λ exceeds the critical length by at least three units. Notice that with our choice of parameters J, h we always have $\Delta_x H(\sigma) \neq 0$. From a physical point of view the above described asymptotic regime corresponds to analysing local aspects of nucleation at very low temperature.

Other asymptotic regimes will be quoted at the end of this chapter.

We want to note that in our finite volume system, at equilibrium, for large β the energy dominates with respect to the entropy so that the Gibbs measure is a small perturbation of a δ-mass concentrated on the unique ground state.

As we have already said in the introduction, it is easily seen that this unique ground state is given by the configuration $+\underline{1}$ where all spins are plus, whereas the natural candidate to describe the metastable situation is the configuration $-\underline{1}$ where all spins are minus.

Given a set of configurations $G \subseteq \mathcal{X}$ we denote, as usual, by τ_G the first hitting time to G:

$$\tau_G := \min\{t > 0 : \sigma_t \in G\}. \tag{7.6}$$

We are interested in the asymptotic behaviour, for large β, of the first hitting time $\tau_{+\underline{1}}$ to the configuration $+\underline{1}$, starting from $-\underline{1}$.

A particularly interesting set of configurations that we call 'critical configurations' and denote by $\mathcal{P}$, is the set of configurations in which the plus spins are precisely the spins contained in a polygon given by a rectangle with sides $l^*, l^* - 1$ plus a unit square protuberance attached to one of the longest sides (see Figure 7.1). Here

$$l^* := \left\lceil \frac{2J}{h} \right\rceil + 1, \tag{7.7}$$

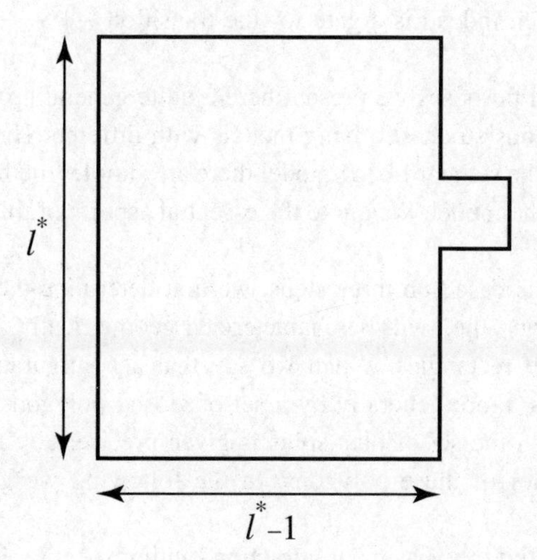

Figure 7.1

where $[\cdot]$ denotes the integer part. We always suppose that $2J/h$ is not an integer. Recall that in the simple static analysis developed in Chapter 4 this value came out as the critical length.

We denote by Γ the formation energy of a critical 'droplet':

$$\Gamma := H(\mathcal{P}) - H(-\underline{1}) = 4Jl^* - h(l^*)^2 + h(l^* - 1). \tag{7.8}$$

$\theta_{-\underline{1},+\underline{1}}$ is the last instant in which $\sigma_t = -\underline{1}$ before $\tau_{+\underline{1}}$:

$$\theta_{-\underline{1},+\underline{1}} := \max\{t < \tau_{+\underline{1}} : \sigma_t = -\underline{1}\}. \tag{7.9}$$

$\theta_{-\underline{1},\mathcal{P},+\underline{1}}$ is the first instant after $\theta_{-\underline{1},+\underline{1}}$ in which the process visits $\mathcal{P}$:

$$\theta_{-\underline{1},\mathcal{P},+\underline{1}} := \min\{t > \theta_{-\underline{1},+\underline{1}} : \sigma_t \in \mathcal{P}\}. \tag{7.10}$$

The main result is contained in the following theorem due to Neves and Schonmann who also derived several other results (see [225, 226]).

Theorem 7.1 [225] *Let* $h < 2J$, $2J/h \notin \mathbb{N}$, $L \geq l^* + 3$; *then for every* $\delta > 0$:

(*i*) $\lim_{\beta\to\infty} P_{-\underline{1}}(\exp[\beta(\Gamma - \delta)] < \tau_{+\underline{1}} < \exp[\beta(\Gamma + \delta)]) = 1;$

(*ii*) $\lim_{\beta\to\infty} P_{-\underline{1}}(\theta_{-\underline{1},\mathcal{P},+\underline{1}} < \tau_{+\underline{1}}) = 1.$

Proof We present a proof based on Proposition 6.17 and Theorem 6.23 in Chapter 6. The main point is the determination of the communication height between $-\underline{1}$ and $+\underline{1}$ (see Definition 6.9 in Chapter 6). □

We make a geometrical construction from which we deduce that $\mathcal{S}(-\underline{1}, +\underline{1})$ contains $\mathcal{P}$, which indeed is a gate for the transition $-\underline{1} \to +\underline{1}$ (see Definition 6.10).

For didactical purposes we present here a quite general approach that can be extended to various stochastic Ising models with different Hamiltonians. In the specific case of the standard Ising model there are simpler methods exploiting the peculiarities of the model. We quote the essential aspects of this simple approach later on.

Our strategy is based on three steps: we first determine the set of local minima for the energy; they will be characterized geometrically as sets of isolated, sufficiently large rectangles. When we say that a configuration is given by a set of rectangles, more generally by a set of closed polygons, we mean that in that configuration the set of plus spins is given precisely by the set of sites lying in the interior of these polygons. In the following we give more detailed definitions.

Second, we study the basins of attraction (with respect to the $\beta = \infty$ dynamics) of these minima, in particular we determine the saddles between them. This

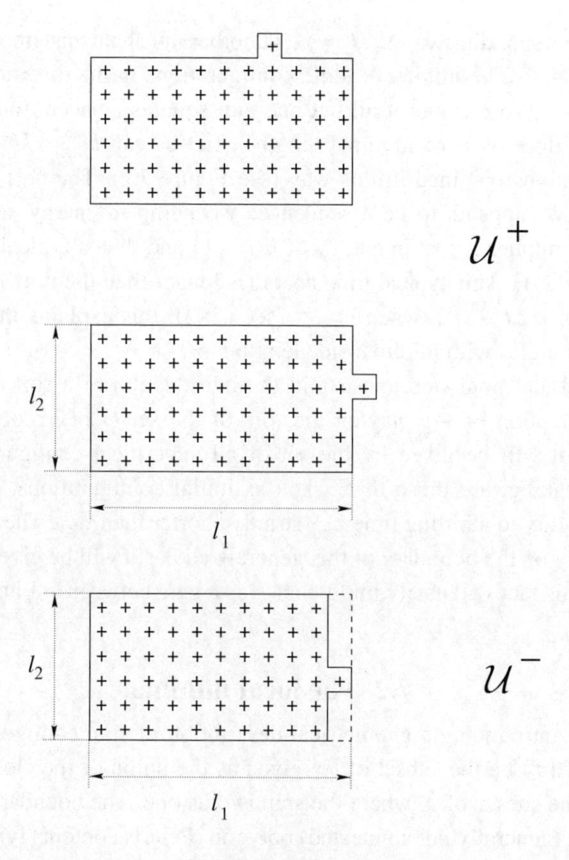

Figure 7.2

will allow us to characterize the rectangles that have a tendency to shrink and those that have a tendency to grow; the value l^* will emerge from this analysis as the 'critical length' that discriminates between growing, i.e. supercritical, and shrinking, i.e. subcritical, rectangles. The heuristics behind this is related to a comparison between the speed of growth and the speed of contraction of a given rectangle R. The mechanism of growth implies the formation of a unit square protuberance from the exterior, adjacent to one of the edges of R (see Figure 7.2). Let us call $\mathcal{U}^+ = \mathcal{U}^+(R)$ the class of configurations obtained in this way. The corresponding positive increment in energy is $2J - h$ so that the typical time needed to reach $\mathcal{U}^+$ is of the order $\exp[\beta(2J - h)]$. $\mathcal{U}^+$ is a saddle configuration. Indeed, starting from $\sigma \in \mathcal{U}^+$, two transitions with negative $\Delta_x H$ can take place: a spin-flip of a minus adjacent both to the unit square protuberance and to R, leading to the formation of a stable 1×2 protuberance attached to R, which would lead into the basin of attraction of a larger rectangle; or the flip of the plus spin inside the unit square protuberance, leading back to R. On the other hand, it is not difficult to convince oneself that the best mechanism of contraction is the 'corner erosion' corresponding to $\Delta_x H = h$, all other mechanisms implying at least $\Delta_x H = 2J + h$. The minimal saddle configurations in the direction of contraction correspond to erosion of all but one unit squares adjacent to the interior to one of the edges of R of minimal length l. Let us call $\mathcal{U}^- = \mathcal{U}^-(R)$ the class of configurations obtained in this way (see Figure 7.2). The best way of going from R to $\mathcal{U}^-$ appears to be a sequence, ascending in energy, of 'corner erosions'. This implies a cost in energy of $h(l-1)$ and then a typical time of order $\exp[\beta(h(l-1))]$. This typical time becomes longer than the time needed for the growth, $\exp[\beta(2J-h)]$, when $l \geq l^*$ (see (7.7)); this explains the tendency to grow of rectangles with minimal side length $l \geq l^*$.

The third and final step consists in the introduction of a sort of generalized basin of attraction of $-\underline{1}$ playing the role of the set G of Proposition 6.17 in Chapter 6. It will be given by the set of all subcritical configurations which, roughly speaking, are those that, taken as initial configurations, give rise with high probability to a hitting time $\tau_{-\underline{1}}$ strictly shorter than $\tau_{+\underline{1}}$. The set of minima of the energy on the boundary of this generalized basin will be given precisely by the set $\mathcal{P}$; this fact will easily imply that $\mathcal{P}$ is a gate between $-\underline{1}$ and $+\underline{1}$.

7.2 The local minima

We start by introducing a geometrical description of spin configurations. Given $\sigma \in \mathcal{X}$ let $C(\sigma)$ be the subset of $\mathbb{R}^2$ given as the union of the closed unit cubes centred at the sites x of Λ where the spin is plus one. The boundary of $C(\sigma)$ can be seen as a (generally not connected) polygon (Peierls contour) lying on the dual lattice $\mathbb{Z}^2 + (1/2, 1/2)$. Given $C(\sigma)$, let us decompose it into maximal connected components $C(\sigma) = C_1 \cup C_2 \cup \cdots \cup C_k$. The C_i are called 'clusters of σ'. The sites of the original lattice $\mathbb{Z}^2$ lying inside a C_i form a *-cluster in the sense of site percolation; namely they are connected via nearest and next to nearest neighbours. Of course there is a one-to-one correspondence between the spin configurations and the set of collections of non-overlapping clusters.

We often identify a configuration σ with the set of unit cubes centred at the sites containing plus spins. Actually, with an abuse of notation, we denote by the same symbol $C_1, C_2, \cdots, C_k$ a set of non-overlapping clusters and the configuration where the plus spins are precisely those sitting on the sites internal to $C_1 \cup C_2 \cup \cdots \cup C_k$.

The total boundary $\partial C(\sigma)$ of a configuration σ is given as the union of the boundaries of the clusters of σ; it consists of a collection of unit segments with extrema in the dual lattice $\mathbb{Z}^2 + (1/2, 1/2)$ with the property that at each site of $\mathbb{Z}^2 + (1/2, 1/2)$ an even number of segments (0, 2 or 4) converge. We write

$$p(\sigma) := |\partial C(\sigma)| = \sum_{i=1}^{k} |\partial C_i|; \tag{7.11}$$

we call $p(\sigma)$ the *perimeter* of σ; it equals the total number of pairs of n.n. sites with opposite spins in σ. The energy $H(\sigma)$ (see (7.1)) associated with σ can be written as

$$H(\sigma) = H(-\underline{1}) + Jp(\sigma) - h\sum_{i=1}^{k} |C_i|. \tag{7.12}$$

We also consider the partition of the boundary $\partial C(\sigma)$ into connected components:

$$\partial C(\sigma) = \bigcup_{i=1}^{m} \gamma_i \tag{7.13}$$

where the γ_i are called *contours* of σ. Of course we have

$$\bigcup_{i=1}^{m} \gamma_i = \bigcup_{i=1}^{k} \partial C_i = \partial C \quad \text{and} \quad p(\sigma) = \sum_{i=1}^{m} |\gamma_i|. \tag{7.14}$$

Definition 7.2 A configuration $\sigma \in \mathcal{X}$ is called 'non-winding' if all its clusters are not encircling the torus i.e. connecting two opposite sides of Λ. We denote by $\mathcal{X}_{nw}$ the set of all non-winding configurations.

For any non-winding configuration we have a well defined 'sea of minuses' given by the unique component of minus sites encircling the torus; inside this 'sea' the C_i can be seen as islands possibly having in their interior lakes with possibly other islands in their interiors and so on.

Given a non-winding configuration $\sigma \in \mathcal{X}_{nw}$, the maximal connected components of its boundary $\partial C(\sigma)$ that are all made by unit segments touching the sea of minuses are called *outer contours*.

Definition 7.3 We denote by $\hat{\mathcal{R}}(l_1, l_2)$ the set of configurations whose plus spins are precisely those sitting on the sites internal to some rectangle $R(l_1, l_2)$ with edges parallel to the lattice axes (and vertices on the dual lattice $\mathbb{Z}^2 + (1/2, 1/2)$) of horizontal and vertical side lengths l_1, l_2, respectively.

We want to stress that by $R(l_1, l_2)$ we mean a particular rectangle with side lengths l_1, l_2 and a particular location, for example centred at the origin, but we do not make the location explicit in the notation. Using the identification of $\sigma \in \mathcal{X}$ with $C(\sigma)$, $\hat{\mathcal{R}}(l_1, l_2)$ can be viewed as an equivalence class of a given rectangle $R(l_1, l_2)$, modulo translations.

Let

$$l := l_1 \wedge l_2, \qquad m := l_1 \vee l_2 \tag{7.15}$$

be, respectively, the minimal and maximal sides of a rectangle $R(l_1, l_2)$.

Definition 7.4 A rectangle $R(l_1, l_2)$ with $|l_1 - l_2| \le 1$ is called a quasi-square.

Notation We write

$$\mathcal{R}(l_1, l_2) = \hat{\mathcal{R}}(l_1, l_2) \cup \hat{\mathcal{R}}(l_2, l_1) \tag{7.16}$$

to denote the set of configurations whose plus spins are precisely those lying inside some rectangle with side lengths l_1 and l_2 (independently of the orientation).

We now give some definitions concerning rectangles seen as geometrical objects; from now on we only consider rectangles (seen as closed subsets of $\mathbb{R}^2$) with sides parallel to the lattice axes and vertices in the dual lattice.

Definition 7.5 Two disjoint rectangles R, R' are mutually 'isolated' if there does not exist any site $x \in \mathbb{Z}^2 \setminus (R \cup R')$ having two distinct nearest neighbour sites y, y' lying inside R, R' respectively.

In other words R, R' are mutually isolated if, given $x \in \mathbb{Z}^2 \cap R$, $x' \in \mathbb{Z}^2 \cap R'$, the Euclidean distance $|x - x'|$ is $|x - x'| \geq \sqrt{5}$; thus, given two isolated rectangles R, R' either they lie at a distance greater than one or the minimal distance is one but it is realized only on two of their vertices. $R_1, \ldots, R_n$ form a set of isolated rectangles if they are pairwise mutually isolated.

Definition 7.6 We call stable the rectangles with $m \leq L - 2$ and $2 \leq l \leq L - 2$ or with $m = L$ (encircling the torus) and any l such that $1 \leq l \leq L - 2$.

A rectangle with $l = 1$ and $m \leq L - 1$ (segment) is said to be 'ephemere'.

Notice that a configuration $\bar{\sigma}$ containing a unique segment with $l = 1$ and $m = L - 1$ can be connected by a downhill path to two distinct local minima: the circle with $l = 1$ and $m = L$ or the configuration $-\underline{1}$. Thus $\bar{\sigma}$ is a saddle configuration.

Definition 7.7 We say that a single cluster is monotonous if it is simply connected (does not have holes) and it has a perimeter equal to that of the circumscribed rectangle (see Figure 7.3).

Definition 7.8 Two rectangles R and R' are said to be interacting if one of the following two circumstances occurs:

(i) the rectangles R and R' intersect, or

(ii) R and R' are disjoint but not mutually isolated so that there exists a site $x \in \Lambda \setminus (R \cup R')$ having two distinct nearest neighbour sites y, y' lying inside R, R' respectively.

Lemma 7.9 *The set of local minima for the energy is given precisely by the set of collections of isolated stable rectangles including the degenerate cases of* $-\underline{1}$ *(absence of rectangles) and* $+\underline{1}$ *(unique rectangle* $\equiv \Lambda$*).*

Proof σ is stable if for all $x \in \Lambda$, $\Delta_x H(\sigma)$ is positive (recall that, with our choice of parameters, the quantity $\Delta_x H(\sigma)$ is never zero). Let us look at the catalogue

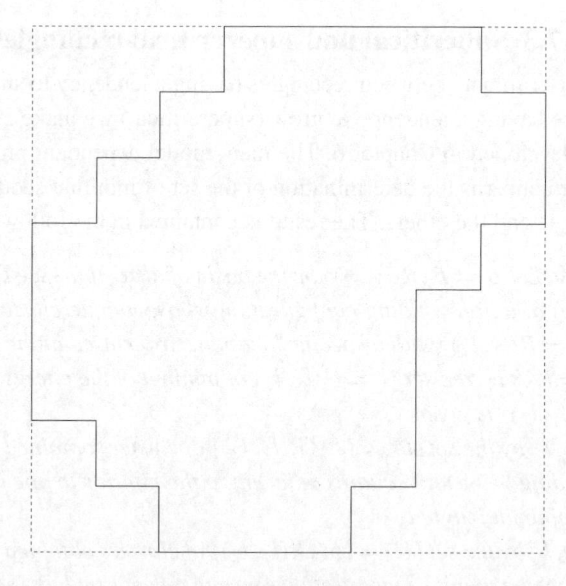

Figure 7.3

of all possible positive increments

$$\Delta_x H(\sigma) = 2\sigma(x) \left[\frac{J}{2} \sum_{y \in \Lambda : |x-y|=1} \sigma(y) + \frac{h}{2} \right],$$

in increasing value; we get $h, 2J - h, 2J + h, 4J - h, 4J + h$. We see that the only possibility for σ to be stable is that all minus spins have at least three positive nearest neighbours, whereas all plus spins have at least two positive nearest neighbours. From this we have the following

(1) Every single cluster C of a local minimum configuration σ must be a stable rectangle (possibly encircling the torus). Indeed, if C does not coincide with its circumscribed rectangle, certainly there is a minus spin in σ with less than three negative neighbouring spins, that can flip into plus by strictly decreasing the energy. C cannot be ephemere, otherwise the two extreme plus spins can flip into minus by decreasing the energy. Finally C cannot be a rectangle with $m = L - 1$, namely almost encircling Λ and then self-interacting, otherwise there would be sites containing -1 spins and adjacent to the exterior to two opposite sides of C, that could flip by decreasing the energy (self-coalescence of C).

(2) Every rectangular cluster of σ must be isolated from the others, otherwise there would exist, between two rectangles, a minus spin with less than three negative neighbouring spins that could flip by strictly decreasing the energy; in other words an energetically favourable coalescence between two rectangles could take place.

This proves the assertion of the lemma. □

7.3 Subcritical and supercritical rectangles

In order to discriminate between rectangles having a tendency to shrink (subcritical) and those having a tendency to grow (supercritical) we make use of methods and results developed in Chapter 6. The main model dependent problem that we have to solve concerns the determination of the set of minimal saddles between a given rectangle and the others. The result is contained in the following.

Lemma 7.10 *Let $\bar{B} = \bar{B}(R(l_1, l_2))$ be the basin of attraction* (*see Definition* 6.15 *in Chapter* 6) *of a non-winding configuration whose unique cluster is the stable rectangle $R = R(l_1, l_2)$* (*with a specific location, say centred at the origin*). *Then, recalling $l = l_1 \wedge l_2$, the set $U = U(l_1, l_2)$ of minima of the energy in the boundary of $\bar{B}(R(l_1, l_2))$ is given*

(*a*) *for $l < l^*$ by the set $\mathcal{U}^- = \mathcal{U}^-(R(l_1, l_2))$ of clusters obtained from $R(l_1, l_2)$ by eroding $l - 1$ unit squares adjacent to the interior to one of the edges of R of minimal length l,*

(*b*) *for $l \geq l^*$ by the set $\mathcal{U}^+ = \mathcal{U}^+(R(l_1, l_2))$ of clusters obtained from $R(l_1, l_2)$ by attaching a unit square protuberance to one of its edges* (*see Figure* 7.2).

Proof To get the proof it is sufficient to prove that

$$U(l_1, l_2) \subseteq (\mathcal{U}^+ \cup \mathcal{U}^-) \tag{7.17}$$

since then the lemma follows by direct comparison of

$$E^+ := H(\mathcal{U}^+) - H(R) = 2J - h, \quad \text{with} \quad E^- := H(\mathcal{U}^-) - H(R) = h(l - 1). \tag{7.18}$$

We first observe that $\mathcal{U}^+$, $\mathcal{U}^-$ are 'saddle configurations' between R and the exterior of $\bar{B}$ in the sense that they belong to $\partial\bar{B}$; thus they are candidates for belonging to $U(l_1, l_2)$. Indeed, as one can easily verify and as we have explained in our heuristic discussion, from any $\sigma \in \mathcal{U}^+ \cup \mathcal{U}^-$, with a suitable single spin-flip we enter into $\bar{B}$; on the other hand, from any $\sigma \in \mathcal{U}^+ \cup \mathcal{U}^-$ there exists a downhill path leading to $R' \neq R$.

Then we can certainly use the criterion of comparison with $\mathcal{U}^+$, $\mathcal{U}^-$ to exclude the possibility that other configurations in $\partial\bar{B}$, with energy strictly higher than $H(\mathcal{U}^+) \wedge H(\mathcal{U}^-)$, could be elements of U. We know, by the very definition of basin of attraction, that certainly, for any configuration η in $\partial\bar{B}$, there exists an uphill path $\omega : R \to \eta$; namely $\exists\, T, \omega_1, \ldots, \omega_T$, $\omega_i \in \mathcal{X}$, $P(\omega_i, \omega_{i+1}) > 0$, $H(\omega_{i+1}) > H(\omega_i)$, $i = 1, \ldots, T - 1$, $\omega_1 = R$, $\omega_T = \eta$.

Indeed, given $\eta \in \partial\bar{B}$, certainly for all $\sigma \in \bar{B}$ with $P(\sigma, \eta) > 0$ we must have $H(\eta) > H(\sigma)$, otherwise we would have a downhill path emerging from σ through η and then not ending in R, which would contradict $\sigma \in \bar{B}(R)$.

So we have

$$\hat{B}(R) \supseteq \partial\bar{B} \tag{7.19}$$

(see Definition 6.15 in Chapter 6).

We notice that no uphill path, starting from R and ascending to some $\eta \in U$, can contain a flip of a spin surrounded by four opposite spins, which would involve an increment of the energy of an amount $4J - h$ or $4J + h$. Moreover we notice that no such uphill path, ascending to some $\eta \in U$, can contain even only one 'edge erosion', namely a flip of a $+1$ spin adjacent to the interior to a flat portion of the boundary of a cluster and then having three neighbouring positive spins. Indeed this involves an increment $\Delta_x H(\sigma) = 2J + h > H(\mathcal{U}^+) - H(R) = 2J - h$. Thus we are left to consider only two possible moves which increase the energy: a flip of a minus with three neighbouring minus (formation of a plus unit square protuberance) or a flip of a plus with two neighbouring minus (a corner erosion or the break of a segment).

Notice that an uphill path that joins R to U can contain a step with $\Delta H = 2J - h$, corresponding to the formation of a unit square protuberance, only once; in fact such a path can only consist of a single step starting from R itself and leading to $\mathcal{U}^+$; otherwise, since $F(\bar{B}) = R$, we would overpass the energy of $\mathcal{U}^+$, so $\mathcal{U}^+$ is the only candidate to be an element $\eta \in U$ with plus spins outside R.

Then the allowed single step, in an uphill path ascending to $U \setminus \mathcal{U}^+$, can only involve an increment $\Delta_x H(\sigma) = h$ which corresponds to shrinking the region occupied by plus spins. Thus, in the search for a configuration $\eta \in U \setminus \mathcal{U}^+$, we can restrict ourselves to the case with a circumscribing rectangle $R' \subseteq R$. In fact, it is easy to convince oneself that it must be $R' = R$ since any configuration with circumscribed rectangle R', strictly contained in R, cannot be connected to $\bar{B}$ by a single spin-flip.

Starting from R the first steps with $\Delta_x H(\sigma) = h$ certainly correspond to corner erosions, i.e., flips of plus spins sitting inside an external corner. The other possible step with $\Delta_x H(\sigma) = h$ corresponds to the flip of a plus spin sitting inside a segment (a portion of cluster with thickness one) which induces fragmentation into two clusters. It is easy to see that if $\eta \in U$ is obtained by R via an uphill sequence of corner erosions, it must contain a unique connected cluster inscribed in R and this cluster must be monotonous. Indeed a sequence of angle erosions preserving connection also preserves monotonicity. On the other hand, if by a sequence of corner erosions we go to a non-connected configuration, we certainly have to cross a configuration $\bar{\eta}$ with a unique connected monotonous cluster such that a suitable single spin-flip is able to disconnect it (last passage to the set of connected configurations). It is easy to convince oneself that since the rectangle circumscribing $\bar{\eta}$ is the original one, namely $R(l_1, l_2)$, necessarily the number of minus spins in η internal to $R(l_1, l_2)$ is at least equal to $l_1 + l_2 - 1$ which is strictly larger than $l - 1$, so that $H(\eta) - H(R(l_1, l_2)) > H(\mathcal{U}^-) - H(R) = h(l - 1)$. Thus to find $U \setminus \mathcal{U}^+$ actually we can restrict ourselves to a connected monotonous cluster inscribed in R. We notice that, in order to be in $\partial \bar{B}$, a single monotonous cluster η must have an intersection with at least one of the four sides of R given by a unit segment. Otherwise, if all intersections of the boundary of the unique cluster of η with the four edges of R were of length ≥ 2, then, necessarily, we would have $\eta \in \bar{B}$.

In other words the intersection of (the region occupied by plus spins in) η with at least one row or column of thickness one adjacent to the interior to the boundary of R must be given by a unit square q. Then η is the union of q with η', given by a unique monotonous cluster, inscribed in $R(l_1', l_2')$ with $l_1', l_2' = l_1 - 1, l_2$ or $l_1', l_2' = l_1, l_2 - 1$. Among these configurations, those of minimal energy are given by $\eta' \equiv R(l_1', l_2')$ where $R(l_1', l_2')$ is obtained from $R(l_1, l_2)$ by cutting one of the shorter sides so that $\eta \in \mathcal{U}^-$.

The lemma is proved. □

The above lemma justifies the following

Definition 7.11 We call a rectangle $R(l_1, l_2)$ *supercritical* if either $l = l_1 \wedge l_2 \geq l^*$ or $m = l_1 \vee l_2 \geq L - 1$ ($R(l_1, l_2)$ is encircling or almost encircling the torus). A rectangle that is not supercritical is called subcritical.

Notice that the case of an ephemere almost encircling rectangle is considered here to be supercritical even though, as we have already said, it is easily seen to be a 'saddle configuration' between $-\underline{1}$ and $+\underline{1}$.

Remark We notice that given any rectangle encircling the torus (winding configuration) $R(l_1, l_2)$, with $m = L$ and $1 \leq l \leq L - 2$, we have that $U(l_1, l_2)$ is given by the set $\mathcal{U}^+$ obtained from $R(l_1, l_2)$ by attaching a unit square protuberance to one of its edges of length L.

7.4 Subcritical configurations and global saddles

We call *global saddles* the elements of $\mathcal{S}(-\underline{1}, +\underline{1})$.

Claim 7.12 *A set G exists satisfying the stronger version of the properties introduced in Proposition* 6.17 i.e., recalling $\mathcal{P}$ has been defined before (7.7),

(*a*) $G \ni -\underline{1}$, $G \not\ni +\underline{1}$,

(*b*) *G is connected,*

(*c*) $\mathcal{P} \subseteq \partial G$, $\forall\, \eta \in \mathcal{P}$ *there exists a path* $\omega : -\underline{1} \to \eta$, $\omega \setminus \eta \subseteq G$, *with*

$$H(\zeta) < H(\mathcal{P}), \qquad \forall\, \zeta \in \omega, \zeta \neq \eta,$$

moreover there exists a path $\omega' : \eta \to +\underline{1}$, $\omega' \cap G = \emptyset$ with

$$H(\zeta) < H(\mathcal{P}), \quad \forall \zeta \in \omega, \zeta \neq \eta,$$

(*d*) $\mathcal{P} \equiv F(\partial G)$.

Then, taking this claim for granted, it easily follows by Proposition 6.17 that if $A_{-\underline{1}} = \{\zeta \, : \, \exists \; \omega : \zeta \to -\underline{1}; \; \forall \eta \in \omega, \quad H(\eta) < H(\mathcal{P})\} \equiv$ maximal connected set containing $-\underline{1}$ with energy less than $H(\mathcal{P})$, $A_{+\underline{1}} = \{\zeta \, : \, \exists \; \omega : \zeta \to +\underline{1}; \; \forall \eta \in \omega, \quad H(\eta) < H(\mathcal{P})\} \equiv$ maximal connected set containing $+\underline{1}$ with energy less than $H(\mathcal{P})$, then:

(1) $A_{-\underline{1}} \subseteq G$, $A_{+\underline{1}} \subseteq \mathcal{X} \setminus G$,

(2) $A_{-\underline{1}}, A_{+\underline{1}}$ are cycles with $U(A_{-\underline{1}}), U(A_{+\underline{1}}) \supseteq \mathcal{P}$,

(3) $H(\mathcal{P}) = \Phi(-\underline{1}, +\underline{1})$. $\mathcal{P} \subseteq \mathcal{S}(-\underline{1}, +\underline{1})$.

Moreover, from the strong property $F(\partial G) \equiv \mathcal{P}$, we deduce that $\mathcal{P}$ is a gate for the transition $-\underline{1} \to +\underline{1}$.

The idea in constructing such a G is to find a criterion to decide whether or not a given configuration η is subcritical, namely whether $\Phi(\eta, -\underline{1}) < \Phi(\eta, +\underline{1})$. It is clear that a natural candidate for G is the set of all configurations that are subcritical according to that criterion. Using Lemma 7.10 and the remark following Definition 7.11, we are already able to decide whether or not a configuration with a *single* rectangle is subcritical or supercritical, similarly for a set of non-interacting rectangles. It is clear that the set $\mathcal{X}$ of all configurations is just a finite set that, however, tends to be geometrically very complicated for large volumes Λ: indeed for large Λ it tends to be an infinite dimensional space whose elements are sets of polygons (the boundaries of the clusters). Thus it is convenient to introduce a rough description of a generic configuration in terms of isolated rectangles. We stress that the set G will be just an *estimate* of the 'true' set of subcritical configurations.

We introduce a map $\mathcal{F}$ on the set $\mathcal{X}_{nw}$ of non-winding configurations (see Definition 7.2) with values in the set of configurations whose set of clusters are collections of isolated rectangles and for $\sigma \in \mathcal{X}_{nw}$ we write:

$$\hat{\sigma} = \mathcal{F}\sigma. \tag{7.20}$$

$\mathcal{F}$ will satisfy the properties:

(i) $\mathcal{F}$ decreases the energy

$$H(\hat{\sigma}) \leq H(\sigma), \tag{7.21}$$

(ii) $\mathcal{F}$ increases the set of pluses

$$\sigma \prec \hat{\sigma} \tag{7.22}$$

where $\sigma \prec \eta$ means $\sigma(x) \leq \eta(x), \quad \forall x \in \Lambda$,

(iii) $\mathcal{F}$ is a monotonous increasing map

$$\hat{\sigma} \prec \hat{\eta} \quad \text{for } \sigma \prec \eta. \tag{7.23}$$

For a set Q given as the union of disjoint clusters $Q = C_1 \cup \ldots \cup C_k$, we denote by $R(Q)$ its rectangular envelope given by the smallest rectangle containing Q. Notice that every edge of $R(Q)$ contains at least one unit segment belonging to the boundary of Q.

Now, for any non-winding σ, we construct the new (not necessarily non-winding) configuration

$$\hat{\sigma} = \mathcal{F}\sigma$$

by 'filling up' and 'gluing' together some of its rectangular envelopes. To this end we first introduce the notion of chains of rectangles.

A set of rectangles $R_1, \ldots, R_m$ is said to form a *chain* Ξ if every pair (R_i, R_j) of them can be linked by a sequence $\{R_{i_1}, \ldots, R_{i_n}\}$ of pairwise interacting rectangles from Ξ; $R_{i_1} = R_i$, $R_{i_n} = R_j$, and R_{i_s}, $R_{i_{s+1}}$ are interacting pairs for all $s = 1, \ldots, n-1$.

Given a collection of chains $\Xi_1, \ldots, \Xi_n$ we start the following iterative procedure:

(1) the chains $\Xi_j^{(1)}$ of the 'first generation' are identical to Ξ_j; $j = 1, \ldots, n$;

(2) having defined $\Xi_j^{(r)}$, $j = 1, \ldots, n_r$, we construct rectangular envelopes $R_j^{(r)}$ of the sets

$$\bigcup_{R \in \Xi_j^{(r)}} R$$

and the maximal chains $\Xi_j^{(r+1)}$ of them $(j = 1, \ldots, n_{r+1})$.

The procedure ends once we reach a set of chains, each consisting of a single rectangle. Notice that every pair from the resulting set is non-interacting.

Starting now from any non-winding configuration σ, we apply the above construction on chains of rectangular envelopes of its outer contours and define $\tilde{\sigma}$ as the configuration obtained by placing the spin $+1$ at all sites inside the resulting rectangles $\bar{R}_1, \ldots, \bar{R}_s$ (filling up the rectangles).

Notice that it may happen that a configuration σ is non-winding but, by the above construction, we get some rectangles $R(l_1, l_2)$ encircling or almost encircling the torus; we recall that by 'almost encircling the torus' we mean that the maximal side length $m := l_1 \vee l_2$ equals $L - 1$ and thus, so to speak, $R(l_1, l_2)$ interacts with itself. If we do not have any such rectangles we define $\hat{\sigma} = \tilde{\sigma}$, otherwise we complete our construction in the following way. First change into pluses the residual minuses sitting on sites x having two of its neighbours in one of the above rectangles. If the new configuration $\tilde{\sigma}'$ obtained in this way contains only non-interacting rectangles we stop our construction and set $\hat{\sigma} := \tilde{\sigma}'$. However it may happen that $\tilde{\sigma}'$ contains pairs of interacting rectangles made of an efhemere and a winding rectangle. In this case we continue our construction in terms of chains like before and in this way we finally obtain the configuration $\hat{\sigma}$ containing neither interacting nor self-interacting rectangles. Notice that $\hat{\sigma}$ does not necessarily correspond to a minimum for H; indeed it may contain ephemere rectangles. Moreover, we repeat, in $\hat{\sigma}$ we may have rectangles encircling the torus but we cannot have rectangles almost encircling the torus.

The above construction can be easily described in terms of 'bootstrap percolation' (see [2]). Consider the following updating rule that can be seen as a deterministic cellular automaton. Given a configuration $\sigma \in \mathcal{X}$ at time t, the configuration σ' at time $t+1$ is obtained as follows:

(i) occupied sites (i.e. sites $x \in \Lambda$ such that $\sigma(x) = +1$) remain occupied,
(ii) if an empty site has two or more occupied nearest neighbours it becomes occupied. Otherwise it stays empty.

It is easy to see that for every $\sigma \in \mathcal{X}$ there exists $T = T(\sigma)$ such that if we repeat T times the above updating we end up with exactly the same invariant configuration $\hat{\sigma}$ that we have defined before, containing a finite set of non-interacting and non-self-interacting rectangles, some of which can encircle the torus. Notice that $+\underline{1}$ is a particular case of $\hat{\sigma}$. We have given the definition of $\hat{\sigma}$ also in terms of a hierarchy of chains because this approach can be extended to different models whereas bootstrap is a notion specifically relevant only for the standard Ising model.

The property $H(\sigma) \geq H(\hat{\sigma})$ is verified immediately.

Indeed, given any configuration giving rise to a single cluster C not coinciding with its rectangular envelope, certainly there exists a sequence of spin-flips decreasing the energy and increasing the plus spins, that leads to $\hat{\sigma}$. Moreover whenever a configuration σ has clusters C', C'' with interacting rectangular envelopes $R' = R(C')$, $R'' = R(C'')$, we decrease the energy by filling the rectangular envelope of the union of R' and R''. This is evident in both cases (i) and (ii) of the definition of interacting rectangles (see Definition 7.8), since the number of broken bonds (i.e. pairs of nearest neighbour opposite spins) is non-increasing, whereas the volume occupied by plus spins is increasing.

Using this observation in an iterative manner, we can construct a sequence of configurations of decreasing energy starting with σ and ending with $\hat{\sigma}$.

The monotonicity of the map $\mathcal{F}$ is also evident.

Now we are ready to define the set G, namely, we introduce G as the set of all non-winding configurations σ such that every resulting rectangle $\bar{R}(l_1, l_2)$ from the configuration $\hat{\sigma}$ is subcritical ($l = l_1 \wedge l_2 < l^*$ and $m = l_1 \vee l_2 < L - 1$) and not interacting with any other rectangle.

Notice that whenever $\hat{\sigma} \neq \tilde{\sigma}$, i.e. $\tilde{\sigma}$ contains almost winding rectangles, then certainly σ is out of G. Thus we could have avoided introduction of the completion $\hat{\sigma}$ of $\tilde{\sigma}$; we did it just to be completely coherent with the boostrap transformation of Aizenman and Lebowitz.

The required property (a) of Claim 7.12 of G is obvious; property (b) saying that G is connected is easily verified since, given $\sigma, \eta \in G$ it is easy to construct a path $\hat{\omega} : \sigma \to \eta, \hat{\omega} \subseteq G$; indeed it suffices to note that there exist two paths $\bar{\omega} : \sigma \to -\underline{1}, \tilde{\omega} : \eta \to -\underline{1}, \bar{\omega}, \tilde{\omega} \in G$; such paths $\bar{\omega}, \tilde{\omega}$ are obtained by successively flipping all plus spins of σ, η, respectively, following a sequence of corner erosions.

It is also immediate to verify that $\mathcal{P} \subseteq \partial G$; moreover the path ω whose existence is required in property (c) is, for example, a step by step contraction of $\mathcal{P}$

obtained by successively eliminating, using only the corner erosion mechanism, all its plus spins whereas ω' is a growth obtained by successively creating plus spins, around a unique cluster, up to total invasion of the whole Λ.

More explicitly, given a configuration in $\mathcal{P}$, the first step in ω consists in removing the unit square protuberance, by gaining an amount $2J - h$ of energy. Subsequently, we erode, by corners, all but one of the plus spins adjacent to the interior to a side of the rectangle of length $l = l^* - 1$ (notice that the length of the side to which the unit square protuberance was adjacent is l^*). In this way, we increase the energy an amount $h(l^* - 2) < 2J - h$ so we stay strictly below $\mathcal{P}$ in energy. Continuing in this way we reach $-\underline{1}$ always remaining below $H(\mathcal{P})$.

Given $\sigma \in \mathcal{P}$, the first step in ω' consists in flipping into plus one of the minus spins adjacent to the rectangle with $l = l^* - 1$, $m = l^*$ and to the unit square protuberance, by getting, in this way, a stable 2×1 rectangular protuberance, attached to the rectangle with $l = l^* - 1$, $m = l^*$. Then we continue by successively flipping the minus spins adjacent to the growing protuberance and to the rectangle, until we get to a configuration consisting of the square $l^* \times l^*$. In this way we decrease the energy an amount $h(l^* - 1)$. Then we create a unit square protuberance adjacent to the exterior of one of the four sides, by paying an amount $2J - h < h(l^* - 1)$ of energy; thus we stay below $\mathcal{P}$. Then we continue in the same way up to $+\underline{1}$.

Note that joining the time reversal of ω with ω' we get a *reference path* according to the definition in equation (7.46) below.

Our next task is to analyse the boundary ∂G and to prove property (d).

Recall that L is only supposed to be strictly larger than $l^* + 2$; it will appear clear from what follows that choosing L much larger than l^* would simplify the treatment of those configurations in ∂G that correspond to supercriticality given by the case $m = l_1 \vee l_2 \in \{L - 1, L\}$.

Let $\eta \in \partial G$; then there exists $\sigma \in G$ and $x \in \Lambda$ such that $\eta = \sigma^{(x)} \notin G$. As a consequence of the monotonicity of the map $\mathcal{F}$, $\sigma(x)$ is necessarily -1; otherwise $\sigma \in G$ would imply also $\eta \in G$. Moreover, for the same reason, the site x lies outside of all rectangles $\bar{R}_1, \ldots, \bar{R}_s$ corresponding to $\hat{\sigma}$. Among the rectangles corresponding to $\hat{\eta}$ there exists a rectangle $\bar{R}(l_1, l_2)$ with the following properties:

(i) $\bar{R}$ is supercritical,

(ii) it contains the site x and several rectangles $\bar{R}_i$, say, $\bar{R}_1, \ldots, \bar{R}_k$, corresponding to $\hat{\sigma}$; the remaining rectangles $\bar{R}_{k+1}, \ldots, \bar{R}_s$ are also rectangles of $\hat{\eta}$.

Our aim now is to prove that

$$H(\eta) - H(-\underline{1}) \geq \Gamma = 4Jl^* - h\left[(l^*)^2 - l^* + 1\right]. \tag{7.24}$$

Consider first the configuration $\tilde{\eta}$, $H(\eta) \geq H(\tilde{\eta})$, whose set of plus spins consists of the site x and the rectangles $\bar{R}_1, \ldots, \bar{R}_k$ (the energy decreases when skipping the subcritical rectangles $\bar{R}_{k+1}, \ldots, \bar{R}_s$). Further, consider the set $C^{(0)}$

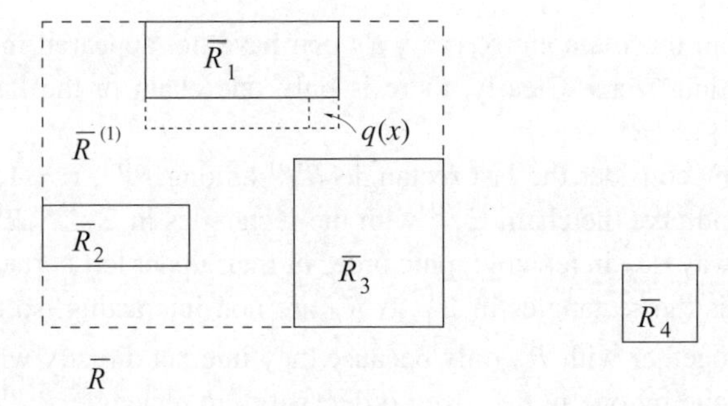

Figure 7.4

consisting of the union of the unit square $q(x)$ centred at the site x and the rectangle (if any) among $\bar{R}_1, \ldots, \bar{R}_k$ that intersects $q(x)$ along one of its edges; given the fact that the rectangles $\bar{R}_1, \ldots, \bar{R}_k$ are mutually non-interacting, there is at most one among these rectangles, say $\bar{R}_1$, touching an edge of $q(x)$.

Let us now take the rectangular envelope $\bar{R}^{(1)}$ of $C^{(0)}$ and distinguish two cases, either the rectangle $\bar{R}^{(1)}$ is supercritical or not. (Notice that both $C^{(0)}$ and $\bar{R}^{(1)}$ may actually coincide with $q(x)$.) See Figure 7.4. If $\bar{R}^{(1)}$ is supercritical, we decrease the energy of $\tilde{\eta}$ further by erasing all rectangles among $\bar{R}_1, \ldots, \bar{R}_k$ that were not contributing to the set $C^{(0)}$ and consider the configuration yielded by the set $C^{(0)}$.

Suppose that $\bar{R}^{(1)}$ is supercritical because its larger side is $m = L - 1$ (m cannot be equal to L otherwise $\hat{\sigma}$ would have almost encircled Λ); since it is energetically favourable to add edges of length $L - 1 > l^*$, we see that the case with minimal energy corresponds to $l = l^* - 1$ so that

$$H(C^{(0)}) > H(\mathcal{P}), \qquad \forall L \geq l^* + 3. \tag{7.25}$$

Notice that if we had chosen $L = l^* + 2$ we would have had the possibility $H(C^{(0)}) = H(\mathcal{P})$.

On the other hand, it is easy to convince oneself that among the other cases, with $\bar{R}^{(1)}$ supercritical because it has minimal edge $l \geq l^*$, the one with minimal energy corresponds to $\bar{R}_1$ with edges $l^* - 1, l^*$ and $q(x)$ touching one of the sides of length l^* so that $C^{(0)} \in \mathcal{P}$. Indeed $\bar{R}_1$ must be of the form $R(l^* - 1, m)$ (or $R(m, l^* - 1)$) with $m \geq l^*$ and we know that contracting orthogonally to the edge $l^* - 1$, i.e. cutting a row or column of length $l^* - 1$, decreases the energy. This yields the bound (7.24) with the equality holding if and only if $C^{(0)} \in \mathcal{P}$.

Next, consider the case when $\bar{R}^{(1)}$ is subcritical. Take $\bar{R}^{(1)}$ and all rectangles $\bar{R}_2, \ldots, \bar{R}_k$ (that were not used for $C^{(0)}$) and construct from them the set of chains $\{\Xi_j^{(1)}\}_{j=1,\ldots,n_1}$ of the first generation. A sequence $\{\Xi_j^{(r)}\}_{j=1,\ldots,n_r}, r = 1, \ldots, s$ of sets of chains of following generations is obtained from it by iteration. Since the rectangles $\bar{R}_1, \ldots, \bar{R}_k$ are mutually non-interacting, for every generation r we get a chain, say $\Xi_1^{(r)}$, consisting of a rectangle $\bar{R}^{(r)}$ containing the site x and certain subset of $\bar{R}_1, \ldots, \bar{R}_k$. The remaining chains $\Xi_j^{(r)}$, $j = 2, \ldots$ contain each just one rectangle from those among $\bar{R}_1, \ldots, \bar{R}_k$ that have not appeared in $\Xi_1^{(p)}$, $p \le r$, in the preceding steps. Clearly, there is only one chain in the last generation, $\Xi_1^{(s)} = \{\bar{R}^{(s)}\} \equiv \{\bar{R}\}$.

Let us now consider the last rectangle $\bar{R}^{(p)}$ among $\bar{R}^{(r)}$, $r = 1, \ldots, s$ that is subcritical and take the chain $\Xi_1^{(p)}$ with the rectangles in $\Xi_1^{(p)} \setminus \bar{R}^{(p)}$ ordered in a particular way, say in lexicographic order of their upper left corner. Notice that, by hypothesis the rectangles in $\Xi_1^{(p)} \setminus \bar{R}_p$ are non-interacting, so they form the chain $\Xi_1^{(p)}$ together with $\bar{R}_p$ only because they interact directly with $\bar{R}_p$. Let us unite them, one by one in the given order, with the rectangle $\bar{R}^{(p)}$ until the circumscribed rectangle is supercritical. Cutting off the remaining rectangles from the chain $\Xi_1^{(p)}$ we get the chain $\tilde{\Xi}_1^{(p)} \subseteq \Xi_1^{(p)}$. Let us use $\tilde{R}'$ to denote the last rectangle that was attached to form the chain $\tilde{\Xi}_1^{(p)}$, and $\tilde{R}$ the circumscribed rectangle to the union of rectangles from $\tilde{\Xi}_1^{(p)} \setminus \{\tilde{R}'\}$. Clearly, $\tilde{R}$ and $\tilde{R}'$ are subcritical interacting rectangles with a supercritical envelope R^* of their union. Then we have two possible situations.

Case (A). There exists a rectangle $R(l_1, l_2) =: Q^*$ with $l_1 = l_2 = l^*$ (first supercritical square), contained in R^* such that it intersects both rectangles $\tilde{R}$ and $\tilde{R}'$ in non-degenerated rectangles $\hat{R}$ and $\hat{R}'$, $\hat{R} = \tilde{R} \cap Q^*$ and $\hat{R}' = \tilde{R}' \cap Q^*$, and has non-empty intersection also with the intersection $\tilde{R} \cap \tilde{R}'$ (if it is non-empty).

Case (B). R^* has its larger side of length $l_1 \vee l_2 \in \{L - 1, L\}$, i.e. R^* is encircling or almost encircling the torus; moreover the shorter side is $l_1 \wedge l_2 = l \in [1, l^* - 1]$.

Now we observe that, given a rectangle $R(l_1, l_2)$ we decrease the energy if we contract it along a direction orthogonal to a side $l \le l^* - 1$; moreover we decrease the energy if we expand it along a direction orthogonal to a side $l \ge l^*$. In other words if we erase from a rectangle a column or row adjacent to the interior from a side $l \le l^* - 1$ or we add a column or row adjacent to the exterior to a side $l \ge l^*$, we decrease the energy.

Consider first case (A). Taking into account the above remark we further decrease the energy with the following transformation $\mathcal{M}$ that also further shrinks the corresponding configuration.

Let $\tilde{l}_1, \tilde{l}_2$ be the edges of $\tilde{R}$ and $\hat{l}_1, \hat{l}_2$ be the edges of $\hat{R} \equiv \tilde{R} \cap Q^*$.

We distinguish two cases.

(1) If $m = \hat{l}_1 \vee \hat{l}_2 < l^*$ we just substitute $\tilde{R}$ with $\hat{R}$. We set $R := \mathcal{M}\tilde{R} \equiv \hat{R}$.

(2) if $m \ge l^*$ we substitute $\tilde{R}$ with the rectangle $R := \mathcal{M}\tilde{R} = R(l, l^*)$ obtained as intersection of $\tilde{R}$ with the strip of thickness l^* containing Q^* orthogonal to the direction of the side of length $\ge l^*$ of $\tilde{R}$ (see Figure 7.5).

We act in a totally analogous way with $\tilde{R}'$ to get $R' := \mathcal{M}\tilde{R}'$.

One has

$$H(\tilde{\eta}) \ge H(\tilde{R}) + H(\tilde{R}') \ge H(R) + H(R'), \tag{7.26}$$

where $H(R)$ denotes the energy of the configuration with plus in R and so on.

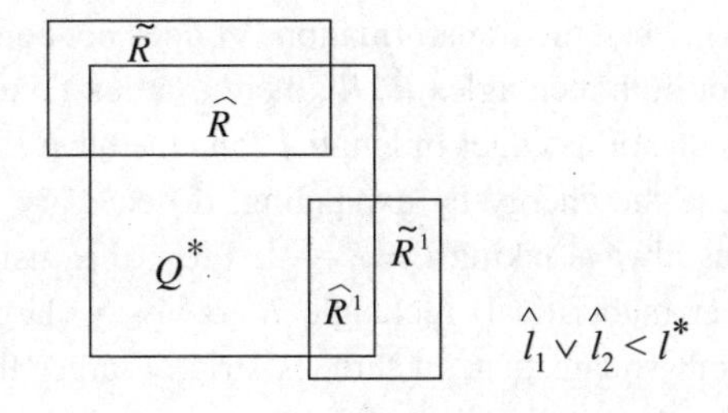

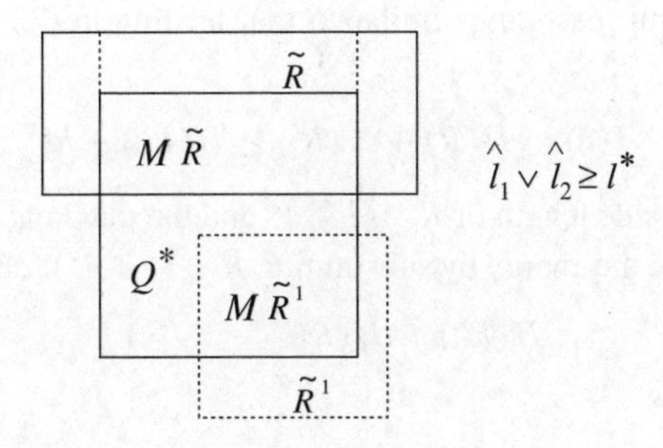

Figure 7.5

To see this, it is enough to realize that the energy of $\tilde{\eta}$ is certainly higher than the energy of all among the rectangles $\bar{R}^{(1)}, \bar{R}_2, \ldots, \bar{R}_k$ that were subsequently used in the construction of $\tilde{\Xi}_1^{(p)} \setminus \{\tilde{R}'\}$ plus the energy associated with the rectangle $\tilde{R}'$ (notice that $\tilde{R}'$ does not intersect the remaining rectangles from the original set $\bar{R}_1, \ldots, \bar{R}_k$ used to construct $\tilde{\Xi}_1^{(p)}$ and it can only touch by its corners the set $C^{(0)}$). The first term can be subsequently bounded from below by $H(\tilde{R})$ and we get the last inequality in (7.26) by observing that R, R' are obtained from $\tilde{R}$ and $\tilde{R}'$ by applying the transformation $\mathcal{M}$; $\tilde{R}$ and $\tilde{R}'$ are subcritical and thus one gains energy by reducing them to R and R' via successive cuttings of rows or columns shorter than l^*.

Now, consider the case when both $R = \hat{R}$ and $R' = \hat{R}'$. We have

$$H(R) + H(R') > H(Q^*) + h(l^* - 1) \tag{7.27}$$

which implies (7.24). Indeed, if the rectangles R and R' intersect in more than one point, there is a surplus of at least two bonds in the sum of their boundaries yielding at least $2J > (l^* - 1)h$. If R and R' just touch in the corner, the boundary has the same number of bonds as in Q^* and there are at least l^* minus sites inside Q^*. If R and R' are interacting according to case (ii) from Definition 7.8 of interacting rectangles, the number of unit edges is at least that in the configuration Q^* but, again, there are at least l^* minus inside the square Q^*; thus in all analysed cases we get (7.27).

Still in case (A), when the transformation $\mathcal{M}$ does not coincide with the intersection with Q^* for both rectangles $\tilde{R}$, $\tilde{R}'$, there is at least one rectangle, between R, R', say R, with one of its edges of length l^* and the other $l = l_1 \wedge l_2 \leq l^* - 1$. We further decrease the energy by expanding, if necessary, orthogonally to the large side and thus always taking $l = l^* - 1$. From this, using (7.26), since the energy of the other (subcritical) rectangle R' is always larger than or equal to the energy of a unit square that, in turn, is strictly larger than $2J - h$, we get (7.24); indeed $2J - h$ represents the difference between Γ and the energy of an $l^* \times (l^* - 1)$ rectangle.

Let us now analyse case (B).

In this case, with reasoning similar to that leading to (7.27) in case (A), we easily get:

$$H(\tilde{\eta}) \geq H(\tilde{R}) + H(\tilde{R}') \geq H(R^*) + hl^*. \tag{7.28}$$

Since the minimal side length of R^* is $l < l^*$ and the maximal one is $m \geq l^* + 2$, we strictly decrease the energy by substituting $R(l^* - 1, l^*)$ for R^*, so that we get:

$$H(R^*) > H(R(l^* - 1, l^*)). \tag{7.29}$$

We conclude that

$$H(\tilde{\eta}) > H(\mathcal{P}). \tag{7.30}$$

From (7.30) and (7.25) we deduce that when $L > l^* + 2$ the minimal saddles on ∂G and then the minimal saddles between $-\underline{1}$ and $+\underline{1}$ never correspond to the formation of a cluster encircling the torus.

Thus, we are left with the task of finding configurations in ∂G for which the energy equals the right-hand side in (7.24). Let us suppose that η satisfies the equality in (7.24). Then, necessarily, $\tilde{R}^{(1)}$ in the construction above is supercritical. If it were not and other steps of the construction and filling of final chains followed, we would run into a contradiction since in every one of those steps the energy strictly decreases with, as the proof shows, the inequality (7.24) maintained. This excludes the possibility of having an equality for the starting configuration η. Similar reasoning also shows that $\eta = \hat{\eta}$. Hence, $C^{(0)}$ consists of $q(x)$ attached to a single rectangle and to get the equality in (7.24) it must be a $(l^* - 1) \times l^*$ rectangle (see Figure 7.4). Thus, the only possibility of achieving an equality in (7.24) is to take $\eta \in \mathcal{P}$ and we can conclude that

$$\min_{\sigma \in \partial G} H(\sigma) = H(\mathcal{P})$$

and

$$\min_{\sigma \in \partial G \setminus \mathcal{P}} H(\sigma) = H(\mathcal{P}) + h,$$

which implies property (d) of G.

This concludes the proof of the claim.

Now it is easy to conclude the proof of Theorem 7.1. Indeed let M^{sub} denote the set of subcritical minima $\zeta \in \mathcal{X}_{nw}$, i.e. no rectangles in ζ (if any) encircle (or almost encircle) the torus and all the rectangles in ζ have minimal side length smaller than the critical value l^*. Suppose, for simplicity, that $\zeta \in M^{sub}$ contains a unique subcritical rectangle. Let $B(\zeta)$ be the strict basin of attraction of ζ (see Definition 6.15 in Chapter 6); we know, by Lemma 7.10, that the set of minimal saddles in its boundary (see Chapter 6 after Definition 6.13) is given by $\mathcal{U}^-(\zeta)$. It follows from Theorem 6.23 and Proposition 6.29 applied to the cycle $B(\zeta)$ that for all $\epsilon > 0$, within a time $e^{\beta(h(l-1)+\epsilon)}$ our process, with a probability tending to one as $\beta \to \infty$, will escape from $B(\zeta) \cup \partial B(\zeta)$ through $\mathcal{U}^-$. Moreover, looking at $\mathcal{U}^-$, we deduce that it will enter into $B(\zeta')$ where ζ' is a rectangular configuration obtained from ζ by cutting one row or column of minimal length. Indeed all downhill paths emerging from $\mathcal{U}^- \equiv F(\partial B(\zeta))$ and going outside $B(\zeta)$), enter $B(\zeta')$) for one of the above specified ζ'. The general case of ζ containing several rectangles is analogous. In this way, by iterating the argument we conclude that for all $\zeta \in G$ and $\epsilon > 0$,

$$P_\zeta\big(\tau_{-\underline{1}} < e^{\beta[h(l^*-1)+\epsilon]}\,;\ \tau_{-\underline{1}} < \tau_{\partial G}\big) \to 1 \text{ as } \beta \to \infty. \tag{7.31}$$

Next, by applying Theorem 6.23(ii), (iv) to the cycle $A_{-\underline{1}} \subseteq G$ we know that for all $\epsilon, \epsilon' > 0$ and β sufficiently large:

$$P_{-\underline{1}}\big(\tau_{\mathcal{P}} < e^{\beta(\Gamma+\epsilon)}\big) > e^{-\epsilon'\beta}. \tag{7.32}$$

So, by (7.31), (7.32), using recurrence as in (6.41) in Chapter 6, we deduce that for all $\epsilon > 0$, $\zeta \in G$,

$$P_\zeta(\tau_{\partial G} < e^{\beta(\Gamma+\epsilon)}) \to 1 \text{ as } \beta \to \infty. \tag{7.33}$$

On the other hand, using (d) of Claim 7.12 we know that

$$\Gamma' := \min_{\eta \in \partial G \setminus \mathcal{P}} H(\eta) - H(-\underline{1}) > \Gamma. \tag{7.34}$$

Combining (7.33), (7.31), (7.34), with the simple consequence of stationarity of the Gibbs measure given by Lemma 6.21, we get that, analogous to Corollary 6.25, for all $\zeta \in G$,

$$P_\zeta(\sigma_{\tau_{\partial G}} \notin \mathcal{P}) \to 0 \text{ as } \beta \to \infty. \tag{7.35}$$

Using an argument like the one leading to the proof of Proposition 6.29 we conclude that

$$P_\zeta(\tau_{\bar{\mathcal{P}}} < e^{\beta(\Gamma+\epsilon)}) \to 1 \text{ as } \beta \to \infty, \tag{7.36}$$

where $\bar{\mathcal{P}}$ is the set of configurations obtained from $\mathcal{P}$ by flipping a minus spin adjacent both to the quasi-square and to the unit square protuberance; in other words the configurations in $\bar{\mathcal{P}}$ are obtained from those in $\mathcal{R}(l^*-1, l^*)$ by attaching to one of the longer sides of the the quasi-square a stable 1×2 rectangular protuberance.

Starting from $\bar{\mathcal{P}}$, which certainly belongs to the cycle $A_{+\underline{1}}$, by Corollary 6.24 we deduce that for all $\epsilon > 0$

$$P_{\bar{\mathcal{P}}}(\tau_{+\underline{1}} < e^{\beta(2J-h-\epsilon)}) \to 1 \text{ as } \beta \to \infty, \tag{7.37}$$

since the maximal internal barrier $\Theta(A_{+\underline{1}})$ is seen immediately to be $2J - h$.

Again by Lemma 6.21, since $\Gamma = \min_{\sigma \in \partial G} H(\sigma) - H(-1)$, we have for all $\epsilon > 0$,

$$P_{-\underline{1}}(\tau_{\partial G} > e^{\beta(\Gamma-\epsilon)}) \to 1 \text{ as } \beta \to \infty. \tag{7.38}$$

By (7.36) and (7.37) on the one hand, and (7.38) on the other hand, using that G is connected, contains $-\underline{1}$ but not $+\underline{1}$, so that any path $\omega : -\underline{1} \to +\underline{1}$ has to cross its boundary, we conclude the proof of point (i) of Theorem 7.1.

To prove point (ii) we note that the fact that $\mathcal{P}$ is a gate for the transition $-\underline{1} \to +\underline{1}$ (consequence of Claim 7.12) implies that with a probability tending to one as β tends to infinity, our process will visit $\mathcal{P}$ during the first excursion between $-\underline{1}$ and $+\underline{1}$. Indeed a path not visiting $\mathcal{P}$ between $\theta_{-\underline{1},+\underline{1}}$ and $\tau_{+\underline{1}}$ has to cross ∂G at an energy strictly larger than $H(\mathcal{P})$:

$$P_{-\underline{1}}(\theta_{-\underline{1},\mathcal{P},+\underline{1}} > \tau_{+\underline{1}}) \leq P_{-\underline{1}}(\tau_{\partial G \setminus \mathcal{P}} < \tau_{+\underline{1}}). \tag{7.39}$$

Given any $\delta > 0$, we write:

$$P_{-\underline{1}}(\tau_{\partial G\setminus\mathcal{P}} < \tau_{+\underline{1}}) \leq P_{-\underline{1}}(\tau_{\partial G\setminus\mathcal{P}} < e^{\beta(\Gamma+\delta)}) + P_{-\underline{1}}(\tau_{+\underline{1}} > e^{\beta(\Gamma+\delta)}). \tag{7.40}$$

By point (i) of Theorem 7.1 and Lemma 6.21, choosing $\delta < \min_{\sigma\in\partial G\setminus\mathcal{P}} H(\sigma) - H(-\underline{1}) - \Gamma$ we get point (ii). The proof of Theorem 7.1 is concluded. □

Remark We have already seen that $\mathcal{P}$, as a consequence of Claim 7.12, is a gate for the transition $-\underline{1} \to +\underline{1}$. Moreover it also follows from point (c) of Claim 7.12 that $\mathcal{P}$ is a minimal gate; indeed for all $\eta \in \mathcal{P}$ there is a path in $(-\underline{1} \to +\underline{1})_{opt}$ intersecting $\mathcal{S}(-\underline{1}, +\underline{1})$ only in η. This also shows that $\mathcal{P} \equiv \mathcal{G}(-\underline{1}, +\underline{1})$ (see Definition 6.10). Indeed take any $\xi \in \mathcal{S}(-\underline{1}, +\underline{1}) \setminus \mathcal{P}$; given any $\omega \in (-\underline{1} \to +\underline{1})_{opt}$ such that $\omega \cap \mathcal{S}(-\underline{1}, +\underline{1}) \ni \xi$, we know that it has to visit $\mathcal{P}$ but since, as already noticed, for all $\eta \in \mathcal{P}$ there exists $\omega' \in (-\underline{1} \to +\underline{1})_{opt}$ such that $\omega' \cap \mathcal{S}(-\underline{1}, +\underline{1}) = \eta$, we deduce that ξ is unessential (see after Definition 6.10) and, by Proposition 6.11, we conclude that $\xi \in \mathcal{S}(-\underline{1}, +\underline{1}) \setminus \mathcal{G}(-\underline{1}, +\underline{1})$ so that $\mathcal{G}(-\underline{1}, +\underline{1}) \equiv \mathcal{P}$.

We notice that the argument of the proof of Theorem 7.1 and in particular Claim 7.12 justifies the choice of $A_{-\underline{1}}$ as the basin of metastability. Indeed our process, with high probability for large β, will stay inside $A_{-\underline{1}}$ before performing a transition to $+\underline{1}$ through $\mathcal{P}$.

Finally it is immediate to show that with high probability for large β, our process visits $\mathcal{P}$ between $\theta_{-\underline{1},+\underline{1}}$ and $\tau_{+\underline{1}}$ only once.

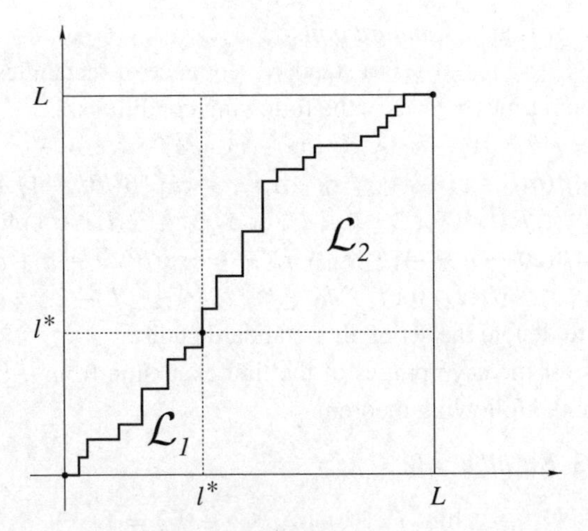

Figure 7.6

Let us now state a more detailed result concerning the tube of typical trajectories during the first excursion from $-\underline{1}$ to $+\underline{1}$.

We first introduce a *standard tube* (of rectangles) as a subset $\mathcal{T}$ of $(\mathbb{Z}_+)^2$ consisting of points corresponding either to 'small quasi-squares' (with $l = l_1 \wedge l_2 < l^*$) or 'large rectangles' (with $l = l_1 \wedge l_2 \geq l^*$):

$$\mathcal{T} = \mathcal{L}_1 \cup \mathcal{L}_2. \tag{7.41}$$

Here (see Figure 7.6)

$$\begin{aligned} \mathcal{L}_1 &= \{\underline{l} \equiv (l_1, l_2) \in (\mathbb{Z}_+)^2 : 1 \leq l_1 \wedge l_2 \leq l^* - 1,\ |l_1 - l_2| \leq 1\}, \\ \mathcal{L}_2 &= \{(l_1, l_2) \in (\mathbb{Z}_+)^2 : l^* \leq l_1 \wedge l_2 \leq L\}. \end{aligned} \tag{7.42}$$

We call a *standard sequence of rectangles* any sequence $\underline{l}^{(1)}, \dots, \underline{l}^{(2L-1)}, \underline{l}^{(i)} \in (\mathbb{Z}_+)^2$ such that

$$\{\underline{l}^{(i)}\}_{i=1,\dots,2L-1} \in \mathcal{T},$$

$\underline{l}^{(1)} = (1, 1)$ and the sequence $\{\underline{l}^{(i)}\}_{i=1,\dots,2L-1}$ is monotonous and consists of nearest neighbours in the sense

$$\underline{l}^{(i+1)} \equiv (l_1^{(i+1)}, l_2^{(i+1)}) = (l_1^{(i)}, l_2^{(i)}) + \mathrm{e},$$

where e is either $\mathrm{e}_1 = (1, 0)$ or $\mathrm{e}_2 = (0, 1)$.

Let $\tau_0, \tau_1, \dots, \tau_n, \dots$ be random times after $\theta_{-\underline{1},+\underline{1}}$ in which σ_t visits the set $\mathcal{R}$ of rectangular configurations (after a change):

$$\tau_0 = \theta_{-\underline{1},+\underline{1}}$$

(see (7.9))

$$\tau_{n+1} = \min\{t > \tau_n : \sigma_t \in \mathcal{R} \setminus \{\sigma_{\tau_n}\}\}, \qquad n = 0, 1, 2, \dots \tag{7.43}$$

We say that σ_t is an ε-*standard path* if

(1) $\sigma_{\tau_0} = -\underline{1}$, $\{\sigma_{\tau_n}\}_{n=0,1,\dots}$ is a standard sequence of rectangles,

(2) the random times τ_n satisfy the following conditions:

(a) $\tau_1 < e^{\varepsilon\beta}, \quad \tau_2 - \tau_1 < e^{\varepsilon\beta}, \tau_3 - \tau_2 < e^{\varepsilon\beta}$,

(b) $\exp\{\beta(h(l-1)-\varepsilon)\} \le \tau_n - \tau_{n-1} \le \exp\{\beta(h(l-1)+\varepsilon)\}$, whenever $\sigma_{\tau_{n-1}} \in \mathcal{R}(l,l)$ for $2 \le l < l^*$, or $\sigma_{\tau_{n-1}} \in \mathcal{R}(l,l+1)$ for $2 \le l < l^*$,

(c) $\exp\{\beta(2J-h-\varepsilon)\} \le \tau_n - \tau_{n-1} \le \exp\{\beta(2J-h+\varepsilon)\}$, whenever $\sigma_{\tau_{n-1}} \in \mathcal{R}(l_1,l_2)$ for $l_1 \wedge l_2 \ge l^*$, $l_1 \vee l_2 \le L-2$.

We use $\mathcal{T}_\varepsilon$ to denote the set of all ε-standard paths.

The results for the asymptotics of the first excursion from $-\underline{1}$ to $+\underline{1}$ are then summarized in the following theorem.

Theorem 7.13 *For all* $\varepsilon > 0$:

$$\lim_{\beta\to\infty} P_{-\underline{1}}(\{\sigma_t\}_{t=1,2,\dots} \in \mathcal{T}_\varepsilon) = 1. \tag{7.44}$$

Proof Consider the cycles $A_{-\underline{1}}$, $A_{+\underline{1}}$ given as the maximal sets of configurations with energy smaller than $H(\mathcal{P})$ containing $-\underline{1}$, $+\underline{1}$, respectively.

Recall that $\mathcal{P}$ is a minimal gate coinciding with $\mathcal{G}(-\underline{1},+\underline{1})$. Suppose we condition to cross between $\theta_{-\underline{1},+\underline{1}}$ and $\tau_{+\underline{1}}$ the minimal gate $\mathcal{P}$ in η; it is seen immediately that there are two classes of configurations, say $\eta^{(-)}$ and $\eta^{(+)}$, with $\eta^{(-)}$ a singleton in $A_{-\underline{1}}$ and $\eta^{(+)} \subseteq A_{+\underline{1}}$, such that between $\theta_{-\underline{1},+\underline{1}}$ and $\tau_{+\underline{1}}$ our process performs the path $\eta^{(-)}, \eta, \eta^{(+)}$. $\eta^{(-)}$ is the quasi-square obtained from η by removing the unit square protuberance, whereas $\eta^{(+)}$ is the set of configurations obtained from η by flipping a minus spin adjacent both to the quasi-square and to the unit square protuberance producing a stable 1×2 protuberance. Then, to find the tube of typical trajectories between $\theta_{-\underline{1},+\underline{1}}$ and $\tau_{+\underline{1}}$, we have to find the typical tube of first exit from $A_{-\underline{1}}$ conditioning to $\sigma_{\tau_{\partial A_{-\underline{1}}}} = \eta$ and the tube of first descent to the bottom $+\underline{1}$ of the cycle $A_{+\underline{1}}$ starting from $\eta^{(+)}$; to this end, by applying Theorems 6.31 and 6.33, we have only to find the standard cascades (see (6.176)) from $\eta^{(-)}$ in $A_{-\underline{1}}$ and from $\eta^{(+)}$ in $A_{+\underline{1}}$. The sequence of minimaxes necessary to determine the standard cascades is easily deduced by Lemma 7.10. This concludes the proof of the theorem. □

Notice that the case $l_1 \vee l_2 > L-2$ has to be treated separately, in the obvious way.

7.5 Alternative method to determine $\Phi(-\underline{1},+\underline{1})$

Let us now present, always for $L \ge l^* + 3$, another method to solve the main model-dependent variational problem, i.e. to determine the communication height $\Phi(-\underline{1},+\underline{1})$ between $-\underline{1}$ and $+\underline{1}$. It is much simpler but it is specific to the standard stochastic Ising model since it exploits its peculiar properties. For n

integer, $0 \le n \le |\Lambda|$, we introduce the set

$$\mathcal{V}_n := \left\{ \sigma \in \mathcal{X} : \sum_{x \in \Lambda} \frac{1 + \sigma(x)}{2} = n \right\}, \tag{7.45}$$

namely $\mathcal{V}_n$ is the set of configurations with a number of plus spins fixed at the value n. It is clear that, by continuity of the dynamics, every path going from $-\underline{1}$ to $+\underline{1}$ has to cross each one of the sets $\mathcal{V}$.

The reference path ω^R Let us now introduce a particular path $\omega^R : -\underline{1} \to +\underline{1}$ that we call a *reference path*. We set $\omega^R = \{\omega_n^R\}$ with $n = 0, \dots, |\Lambda|$, namely,

$$\omega_0^R = -\underline{1}, \omega_1^R = \{x_0\}, \dots, \omega_{|\Lambda|}^R = +\underline{1} \tag{7.46}$$

where x_0 is a fixed site in Λ, say the origin, and ω^R consists of a growing sequence of single clusters given by a quasi-square ($R(l, l)$ or $R(l-1, l)$) with a bar attached to one of its largest sides (here by 'bar' we mean a rectangle $R(1, l')$ with $l' \le l$). These clusters are called *standard polyominoes* (for the two-dimensional case) in [3].

Thus ω^R hits every $\mathcal{V}_n$ with $n = l^2$, $n \in \{1, \dots, |\Lambda| - 1\}$, in a square $R(l, l)$ and every $\mathcal{V}_n$ with $n = l(l+1)$, $n \in \{1, \dots, |\Lambda| - 1\}$ in a quasi-square $R(l, l+1)$. The part of ω^R between $R(l, l+1)$ and $R(l+1, l+1)$ is as follows: first we add a unit square protuberance to one of the longest sides of $R(l, l+1)$; then we flip, successively, the minus spins adjacent to the exterior to $R(l, l+1)$ and to the growing bar, until we get $R(l+1, l+1)$. Similarly we go from $R(l, l)$ to $R(l, l+1)$. The order of the directions of growth of the standard polyominoes, as well as the sequence of growing bars attached to the quasi-square, may be picked arbitrarily but are fixed.

An easy computation shows that $\max_{\sigma \in \omega^R} H(\sigma) = H(\mathcal{P})$; we immediately get

$$\Phi(-\underline{1}, +\underline{1}) = H\big(\mathcal{S}(-\underline{1}, +\underline{1})\big) \le H(\mathcal{P}). \tag{7.47}$$

We observe that we can use ω^R as reference path, for a comparison, for both the transitions from $\mathcal{V}_{l^2}$ to $\mathcal{V}_{l(l+1)}$ and from $\mathcal{V}_{l(l+1)}$ to $\mathcal{V}_{(l+1)^2}$.

It is easy to solve the isoperimetric problem for $n = l^2, n = l(l+1)$, i.e. to find the configuration with minimal perimeter for a given number of plus spins. We get

$$F(\mathcal{V}_{l^2}) = \mathcal{R}(l, l), \qquad F(\mathcal{V}_{l(l+1)}) = \mathcal{R}(l, l+1); \tag{7.48}$$

recall that by $F(Q)$ we denote the set of ground states in $Q \subseteq \mathcal{X}$ (see Definition 6.2 in Chapter 6).

Indeed, let us first note that for any $n \in \mathbb{N}$, given a number of up spins n, to minimize the energy, i.e. the total perimeter p, it is always convenient to collect the up spins into a single cluster without 'holes', in the following sense. Consider a configuration $\eta \in \{-1, +1\}^\Lambda$ containing at least two disjoint clusters C_1

and C_2; with a suitable translation of C_1 towards the other clusters we can join C_1 to the rest in such a way that the translated C_1 will have at least one unit edge of its boundary in common with the boundary of another cluster; then the total perimeter strictly decreases. In this way, by uniting the clusters of η we can always transform it into another configuration η_1 containing a single cluster and having strictly smaller energy. Moreover if in η_1 there are holes, namely minus spins surrounded by a circuit of pluses, by moving plus spins lying in some 'external corners' where the number of n.n. plus spins is ≤ 2 to some internal hole whith number of n.n. plus spins ≥ 2, it is possible to construct a configuration η_2 without holes eventually by strictly decreasing the energy. Similarly, if η_2 is not monotonous, i.e. its perimeter is strictly larger than that of the circumscribed rectangle, it is always possible to transform η_2 into a configuration η_3 containing a single monotonous cluster by strictly decreasing the energy.

The energy of this configuration η_3, when l_1, l_2 are the side lengths of the circumscribed rectangle, is $2J(l_1 + l_2) - hn + H(\underline{-1})$. Thus, to minimize the energy in $\mathcal{V}_n$ we are left to find $l_1, l_2 \in \mathbb{N}$ minimizing $(l_1 + l_2)$ in the set $l_1 l_2 \geq n$, $1 \leq l_1, l_2 \leq n$. It is an easy exercise to prove that if n is of the form $n = l^2$, $n = l(l+1)$, $l \in \mathbb{N}$, this minimum is uniquely achieved for $(l_1, l_2) = (l, l)$, $(l_1, l_2) \in \{(l, l+1), (l+1, l)\}$, respectively, thus proving (7.48).

We have

$$\Phi(\mathcal{V}_{l^2}, \mathcal{V}_{(l^2+1)}) = \Phi(\mathcal{R}(l,l), \mathcal{V}_{(l^2+1)}) = H(\mathcal{Q}_l)$$
$$\Phi(\mathcal{V}_{l(l+1)}, \mathcal{V}_{[(l+1)l+1]}) = \Phi(\mathcal{R}(l, l+1), \mathcal{V}_{[(l+1)l+1]}) = H(\mathcal{P}_l) \qquad (7.49)$$

where $\mathcal{Q}_l$ is the set of clusters obtained from the squares in $\mathcal{R}(l,l)$ by adding a unit square protuberance to one of its sides; $\mathcal{P}_l$ is the set of clusters obtained from the quasi-squares in $\mathcal{R}(l, l+1)$ by adding a unit square protuberance to one of its longer sides. To prove (7.49) we first notice that in order to determine $\Phi(\mathcal{V}_{l^2}, \mathcal{V}_{(l^2+1)})$ we can restrict ourselves, in $\mathcal{V}_{l^2}$, to $\mathcal{R}(l,l)$, otherwise we would overpass the energy of the reference saddle given by ω^R. Indeed we have

$$\min_{\sigma \in \mathcal{V}_{l^2} \setminus \mathcal{R}(l,l)} H(\sigma) - H(\mathcal{R}(l,l)) \geq 2J. \qquad (7.50)$$

It is immediate to show, using comparison with ω^R, that $\Phi(\mathcal{R}(l,l), \mathcal{V}_{(l^2+1)}) \leq H(\mathcal{Q}_l) = H(\mathcal{R}(l,l)) + 2J - h$; thus we have

$$\Phi\big(\mathcal{V}_{l^2} \setminus \mathcal{R}(l,l), \mathcal{V}_{(l^2+1)}\big) \geq \Phi(\mathcal{R}(l,l), \mathcal{V}_{(l^2+1)}) + h. \qquad (7.51)$$

But, evidently, $\Phi(\mathcal{R}(l,l), \mathcal{V}_{(l^2+1)}) \geq H(\mathcal{Q}_l)$.

The case of $\Phi(\mathcal{V}_{l(l+1)}, \mathcal{V}_{[(l+1)l+1]})$ is analogous; the proof of (7.49) is concluded.

We have, in particular:

$$\Phi(\mathcal{V}_{(l^*-1)l^*}, \mathcal{V}_{[(l^*-1)l^*+1]}) = H(\mathcal{P}_{l^*-1}) \equiv H(\mathcal{P}). \qquad (7.52)$$

Since any path $\omega \in (-\underline{1} \to +\underline{1})_{opt}$ has to make the one-step transition between $\mathcal{V}_{(l^*-1)l^*}$ and $\mathcal{V}_{[(l^*-1)l^*+1]}$ we get from (7.52) that $\Phi(-\underline{1}, +\underline{1}) \geq H(\mathcal{P})$ which, together with (7.47), implies $\Phi(-\underline{1}, +\underline{1}) = H(\mathcal{P})$. This concludes the alternative computation of the communication height between $-\underline{1}$ and $+\underline{1}$ and, looking at ω^R, also concludes the proof that $\mathcal{S}(-\underline{1}, +\underline{1}) \supseteq \mathcal{P}$. (Notice that the above method does not prove that $\mathcal{P}$ is a gate.)

We remark that for $L = l^* + 1$ the global saddle is just the set of rectangles $\mathcal{R}(l^* - 1, l^*)$ whereas for $L = l^* + 2$ the global saddle (being also a minimal gate) is given by the set (containing $\mathcal{P}$) of clusters obtained by attaching a unit square protuberance to *any* of the edges of a rectangle in $\mathcal{R}(l^* - 1, l^*)$

7.6 Extensions and generalizations

We want now to describe metastability and nucleation for other stochastic dynamics extending and generalizing, in various ways, the previously described two-dimensional standard stochastic Ising model. We mainly consider the same asymptotic regime as that of the previous chapter: sufficiently large but fixed volume, thermodynamical parameters such as the magnetic field fixed at values sufficiently close to the coexistence region, in the limit of large inverse temperature β. As previously noticed, from a physical point of view this means that we analyse the local aspects of nucleation at very low temperature; mathematically we remain in the Freidlin–Wentzell scenario so that we are still able to use the general results of Chapter 6.

We always have, in the class of models we consider in this chapter, two configurations, say η^m and η^s, corresponding to the initial metastable configuration and to the stable configuration. Three groups of problems naturally arise.

(a) The asymptotics, namely to evaluate the asymptotic behaviour, for large β, of the first hitting time to η^s starting from η^m; in other words the analogue of Theorem 7.1(i).

(b) The gate, namely to characterize, in the sharpest possible way, the set of configurations that constitute the gate to the stable configuration η^s; that is to say the analogue of Theorem 7.1(ii).

(c) The tube, namely to characterize, in the sharpest possible way, the tube of typical trajectories realizing the transition from η^m to η^s, in practice a suitable subset of $(\eta^m \to \eta^s)_{opt}$. This can be seen as strengthening item (b) up to determination of the 'nucleation pattern'. It is the analogue of Theorem 7.13.

As we have seen in the previous part of this chapter, to solve the above problems, we can use the general results of Chapter 6 and in particular Theorem 6.23, provided we are able to solve some suitable model dependent variational problems. For this purpose we can follow, in some cases, a general approach generalizing that used for the two-dimensional standard stochastic Ising model. It consists of three steps.

(1) Give a geometrical description of those local minimum configurations for the energy that are 'close' to η^m. In the two-dimensional standard stochastic Ising model this corresponds to saying that among the non-winding configurations the local minima are given by sets of isolated rectangles with minimal side larger than or equal to 2.
(2) Determine the tendency to grow or shrink for the droplets representing local minima. For the two-dimensional standard stochastic Ising model this corresponds to the fact that a non-winding rectangle has a tendency to grow if and only if its minimal side length is larger than or equal to the critical length l^*.
(3) Define a set G with the properties given in Proposition 6.17 with $x = \eta^m$, $x' = \eta^s$. This, in particular, implies that we are able to determine the communication height between η^m, η^s. The crucial point, and by far the most difficult one, is to prove that the minimum of the energy on ∂G is achieved in a suitable set $\mathcal{P}$, which turns out to be contained in the set $\mathcal{S}(\eta^m, \eta^s)$ of minimal saddles between η^m, η^s.

Intuitively G represents an estimate of the set of subcritical configurations. We recall that for the two-dimensional standard stochastic Ising model, G was the set of configurations σ whose image $\hat{\sigma}$, via the bootstrap percolation map, contains only subcritical rectangles.

If we succeed in determining a set G with the required properties, then the maximal connected component, containing η^m, with energy less than $H(\mathcal{P}) \equiv \Phi(\eta^m, \eta^s)$, is a cycle representing the basin of metastability. It turns out that it is not particularly important that we include in G some supercritical configurations or that we exclude from G some subcritical configuration, provided these configurations have an energy strictly larger than $H(\mathcal{P})$.

It is also possible to follow another strategy that requires less complete control of the energy landscape, but is only able in general to give the asymptotics, i.e. the solution of problem (a). This approach is based on Theorem 6.38; it requires the following items to be solved:

(1) evaluation of the communication height between η^m and η^s, i.e. the determination of $\Gamma := \Phi(\eta^m, \eta^s)$,
(2) proof that every $\sigma \notin \{\eta^m, \eta^s\}$ is Γ_0-reducible with $\Gamma_0 \leq \Gamma$, namely for all $\sigma \notin \{\eta^m, \eta^s\}$ there exists a path $\omega : \sigma \to \{\sigma' : H(\sigma') < H(\sigma)\}$ with $\max_i H(\omega_i) \leq \Gamma_0 + H(\sigma)$; this corresponds to a recurrence property of the dynamics to the set $\{\eta^m, \eta^s\}$, in a time of order at most $e^{\Gamma_0 \beta}$.

We illustrate both methods in the case of the anisotropic Ising model.

7.7 The anisotropic ferromagnetic Ising model in two dimensions

We consider a discrete time single spin-flip Metropolis dynamics for the two-dimensional ferromagnetic Ising model with n.n. interaction and different couplings in horizontal and vertical directions. We call this the $J_1 J_2$ model. The configuration space is $\mathcal{X} := \{-1, +1\}^{\Lambda}$ with Λ given by an $L \times L$ torus (square with periodic boundary conditions). The energy associated with the configuration σ is:

$$H(\sigma) = -\frac{J_1}{2} \sum_{\{x,y\} \in \mathcal{H}(\Lambda)} \sigma(x)\sigma(y) - \frac{J_2}{2} \sum_{\{x,y\} \in \mathcal{V}(\Lambda)} \sigma(x)\sigma(y) - \frac{h}{2} \sum_{x \in \Lambda} \sigma(x), \tag{7.53}$$

where $\mathcal{H}(\Lambda)$ is the set of (unordered) horizontal nearest neighbour pairs in Λ and $\mathcal{V}(\Lambda)$ is the set of (unordered) vertical nearest neighbour pairs in Λ. We suppose that

$$J_1 \geq J_2 \gg h > 0 \quad \text{and} \quad L \gg \left(\frac{2J_1}{h}\right). \tag{7.54}$$

Here $\gg$ stands for 'sufficiently larger than'. The interest in the study of metastability for the $J_1 J_2$ model is essentially related to the fact that its zero temperature Wulff shape is not a square. Indeed it is natural to define the zero temperature Wulff shape as the shape minimizing the energy at fixed volume; it is easy to see that in the case of the $J_1 J_2$ model it is given (modulo diophantine problems) by a rectangle with side lengths l_1, l_2 proportional, respectively, to J_1, J_2.

The main question that we address is whether or not the Wulff shape is relevant for the nucleation pattern in the limit of large β; naïvely, since Metropolis dynamics tends to minimize energy, one could conjecture a nucleation through Wulff shapes. The relatively unexpected result that was obtained in [184] is that the shape of the critical nucleus is square (then not Wulff) and, moreover, the typical nucleation pattern is given initially by a sequence of growing quasi-squares.

There are two relevant lengths given by $l_1^* = [2J_1/h] + 1$, $l_2^* = [2J_2/h] + 1$ ($l_2^* \leq l_1^*$). We suppose both $2J_1/h$ and $2J_2/h$ non-integer. The critical droplet turns out to be a square with side length l_2^*.

Indeed, as in the isotropic case, the set of global saddles typically visited during the first excursion between $-\underline{1}$ and $+\underline{1}$ coincides with the set $\mathcal{P}$ of protocritical configurations, given by the set of configurations in which the plus spins are precisely those contained in a polygon given by a rectangle with sides $l_2^*, l_2^* - 1$ plus a unit square protuberance attached to one of the longest sides that can be, independently of the fact that $J_1 \geq J_2$, either horizontal or vertical.

The statements of Theorem 7.1 hold unaltered in the case of the $J_1 J_2$ model, with l_2^* in place of l^* and $\Gamma := H(\mathcal{P}) - H(-\underline{1})$ with H given by (7.53).

Indeed with the same definitions and notation as used in Theorem 7.1 we have the following.

Theorem 7.14 [184] *Let* $h < 2J_2 \wedge 2(J_1 - J_2)$; *there exists* L_0 *such that for all* $L \geq L_0, \delta > 0$

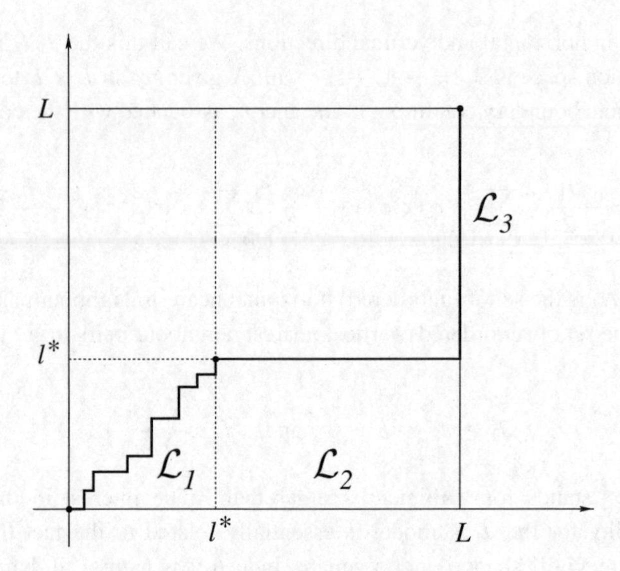

Figure 7.7

(*i*) $\lim_{\beta\to\infty} P_{-\underline{1}}(\exp[\beta(\Gamma-\delta)] < \tau_{+\underline{1}} < \exp[\beta(\Gamma+\delta)]) = 1,$

(*ii*) $\lim_{\beta\to\infty} P_{-\underline{1}}(\theta_{-\underline{1},\mathcal{P},+\underline{1}} < \tau_{+\underline{1}}) = 1,$
with $\theta_{-\underline{1},\mathcal{P},+\underline{1}}$ *as in* (7.10).

Also the tube of typical trajectories during the first excursion from $-\underline{1}$ to $\mathcal{P}$ is unchanged; the behaviour during supercritical growth is very different. During this epoch, the system behaves in a very anisotropic way.

To state this result, we first introduce a *standard tube* (of rectangles) as a subset $\mathcal{T}$ of $(\mathbb{Z}_+)^2$ consisting of points corresponding either to 'small quasi-squares' (with $l = l_1 \wedge l_2 < l_2^*$) or to 'large rectangles' (with either $l_2 = l_2^*$ or $l_1 = L$):

$$\mathcal{T} = \mathcal{L}_1 \cup \mathcal{L}_2 \cup \mathcal{L}_3. \tag{7.55}$$

Here (see Figure 7.7)

$$\begin{aligned} \mathcal{L}_1 &= \{\underline{l} \equiv (l_1,l_2) \in (\mathbb{Z}_+)^2 : 1 \le l_1 \wedge l_2 \le l_2^* - 1,\ |l_1 - l_2| \le 1\}, \\ \mathcal{L}_2 &= \{(l_1,l_2) \in (\mathbb{Z}_+)^2 : l_2^* \le l_1 \le L, l_2 = l_2^*\}, \\ \mathcal{L}_3 &= \{(l_1,l_2) \in (\mathbb{Z}_+)^2 : l_1 = L, l_2^* \le l_2 \le L\}. \end{aligned} \tag{7.56}$$

We call a *standard sequence of rectangles* any sequence $\underline{l}^{(1)}, \ldots, \underline{l}^{(2L-1)}, \underline{l}^{(i)} \in (\mathbb{Z}_+)^2$ such that

$$\{\underline{l}^{(i)}\}_{i=1,\ldots,2L-1} \in \mathcal{T},$$

$\underline{l}^{(1)} = (1,1)$ and the sequence $\{\underline{l}^{(i)}\}_{i=1,\dots,2L-1}$ is monotonous and consists of nearest neighbours in the sense

$$\underline{l}^{(i+1)} \equiv (l_1^{(i+1)}, l_2^{(i+1)}) = (l_1^{(i)}, l_2^{(i)}) + \mathrm{e},$$

where e is either $\mathrm{e}_1 = (1,0)$ or $\mathrm{e}_2 = (0,1)$. Let $\theta_{-\underline{1},+\underline{1}}$ be the last instant in which $\sigma_t = -\underline{1}$ before $\tau_{+\underline{1}}$:

$$\theta_{-\underline{1},+\underline{1}} = \max\{t < \tau_{+\underline{1}} : \sigma_t = -\underline{1}\}. \tag{7.57}$$

Let $\tau_0, \tau_1, \dots, \tau_n, \dots$ be random times after $\theta_{-\underline{1},+\underline{1}}$ in which σ_t visits the set $\mathcal{R}$ of rectangular configurations (after a change):

$$\begin{aligned} \tau_0 &= \theta_{-\underline{1},+\underline{1}} \\ \tau_{n+1} &= \min\{t > \tau_n : \sigma_t \in \mathcal{R} \setminus \{\sigma_{\tau_n}\}\}, n = 0, 1, 2, \dots \end{aligned} \tag{7.58}$$

We say that σ_t is an *ε-standard path* if

(1) $\sigma_{\tau_0} = -\underline{1}$, $\{\sigma_{\tau_n}\}_{n=0,1,\dots}$ is a standard sequence of rectangles,
(2) the random times τ_n satisfy the following conditions:
  (a) $\tau_1 < e^{\varepsilon\beta}$, $\tau_2 - \tau_1 < e^{\varepsilon\beta}$, $\tau_3 - \tau_2 < e^{\varepsilon\beta}$,
  (b) $\exp\{\beta(h(l-1) - \varepsilon)\} \le \tau_n - \tau_{n-1} \le \exp\{\beta(h(l-1) + \varepsilon)\}$, whenever $\sigma_{\tau_{n-1}} \in \mathcal{R}(l,l)$ for $2 \le l < l_2^*$, or $\sigma_{\tau_{n-1}} \in \mathcal{R}(l, l+1)$ for $2 \le l < l_2^*$,
  (c) $\exp\{\beta(2J_2 - h - \varepsilon)\} \le \tau_n - \tau_{n-1} \le \exp\{\beta(2J_2 - h + \varepsilon)\}$, whenever $\sigma_{\tau_{n-1}} \in \hat{\mathcal{R}}(l, l_2^*)$ for $l_2^* \le l \le L-2$,
  (d) $\exp\{\beta(2J_1 - h - \varepsilon)\} \le \tau_n - \tau_{n-1} \le \exp\{\beta(2J_1 - h + \varepsilon)\}$, whenever $\sigma_{\tau_{n-1}} \in \hat{\mathcal{R}}(L, p)$ for $l_2^* \le p \le L-2$.

We use $\mathcal{T}_\varepsilon$ to denote the set of all ε-standard paths.

The results for the asymptotics of the first excursion from $-\underline{1}$ to $+\underline{1}$ are then summarized in the following theorem.

Theorem 7.15 [184] *For any $\varepsilon > 0$*

$$\lim_{\beta\to\infty} P_{-\underline{1}}(\{\sigma_t\}_{t=1,2,\dots} \in \mathcal{T}_\varepsilon) = 1. \tag{7.59}$$

Sketch of the proof As in the isotropic Ising model, a configuration containing a single non-winding cluster is stable (i.e. a local minimum for H) if and only if this cluster is a rectangle with minimal side length $l \ge 2$. Then, to understand the result, the basic point is the following: the height of the energy barrier between a rectangular configuration in $\mathcal{R}(l_1, l_2)$ (see Definition 7.3) and the set of all smaller rectangles, is given by $h(l-1)$ with $l = l_1 \wedge l_2$. Then the speed of contraction of a rectangle does not depend on J_1, J_2 but only on the smaller side length l. This implies that a rectangle is subcritical if and only if $h(l-1) < 2J_2 - h$ as $2J_2 - h$ is the minimal energy barrier between $\mathcal{R}(l_1, l_2)$ and all larger rectangles. In other words, as far as the contraction of a subcritical rectangle is concerned, the behaviour of the $J_1 J_2$ model is exactly the same as that of the isotropic Ising model. Thus, since by reversibility the law of the first growth from $-\underline{1}$ to the protocritical droplet $\mathcal{P}$ is the time reversal of the law of the contraction starting from $\mathcal{P}$, the epoch of nucleation before reaching the critical droplet is exactly as in the isotropic case; on the contrary, supercritical growth is totally anisotropic since the typical time needed to grow in the 1-direction is $\exp[\beta(2J_2 - h)]$ whereas in the 2-direction it is $\exp[\beta(2J_1 - h)]$ so that, before seeing the first vertical protuberance, the growing supercritical droplet has already invaded the whole volume Λ in the 1-direction.

The strategy of proof for Theorems 7.14 and 7.15 is that described previously in the isotropic case. There is only one main new difficulty related to the notion of 'interacting rectangles', namely to the possible mechanisms of coalescence. Now, since the horizontal coupling constant is J_1, which is larger than the vertical coupling constant J_2, if we start from a configuration with two rectangles at a distance one in the vertical direction then, typically, they will not coalesce in a time of order one, as in the isotropic case; indeed coalescence, in this case, will imply a strictly positive increment of energy equal to $2(J_1 - J_2) - h$. Then this mechanism can be active only if both rectangles have a minimal edge greater than or equal to $[2(J_1 - J_2)/h] + 1 =: l^*$; otherwise one of the two rectangles will typically disappear before coalescing. This observation makes it natural to call the rectangles with minimal side-length smaller than l^* 'ephemere' and not to call two rectangles 'interacting', if one of them is ephemere, even though they are at a distance one in the vertical direction.

More precisely we give the following definition.

Two rectangles $R = R(l_1, l_2)$ and $R' = R(l'_1, l'_2)$ are said to be *interacting* if one of the following three possibilities occurs:

(i) the rectangles R and R' intersect, or

(ii) there exists a unit square centred at some lattice site such that one of its vertical edges is contained in R and the other in R', or

(iii) there exists a unit square centred at some lattice site such that one of its horizontal edges is contained in R and the other in R' and, at the same time, $\min(l_1, l_2, l'_1, l'_2) \geq l^*$, where

$$l^* = \left[\frac{2(J_1 - J_2)}{h}\right] + 1.$$

(Neither R nor R' is *ephemere*.)

Also in this $J_1 J_2$ case we introduce a set G defined in analogy with the corresponding set in the isotropic case but using the new notion of interaction. In the proof that the set of minima for the energy in ∂G coincides with $\mathcal{P}$ a crucial role is played by the fact that two rectangles, that could be responsible for coalescence in the vertical direction, must both be non-ephemere.

7.8 Alternative approach to study of the $J_1 J_2$ model

Let us now show another, simpler method to treat metastability for the $J_1 J_2$ model. This approach is aimed at obtaining the basic results given in Theorem 7.14 concerning the asymptotics, for large β, of the tunnelling time and at 'estimating' the 'gate' to stability, but it is not able, in general, to provide the detailed results of Theorem 7.15 concerning the tube of typical trajectories during tunnelling. This strategy is based on the application of Theorem 6.38 in Chapter 6. To determine the communication height between $-\underline{1}$ and $+\underline{1}$ and to find some gate between them we introduce a suitable foliation of the configuration space.

Let us give a few definitions.

Given $\sigma \in \mathcal{X}$ we denote by $l_1(\sigma), l_2(\sigma)$, respectively, the horizontal and vertical 'shadows' of σ, namely the lengths of the horizontal and vertical projections of $C(\sigma)$ ($\equiv$ union of closed unit cubes centred at plus sites in σ). We say that $l_1(\sigma)$ is the shadow in the 2-direction imagining to lighten $C(\sigma)$ in the 2-direction; similarly $l_2(\sigma)$ is the shadow in the 1-direction. We denote by $l(\sigma)$ the minimal shadow:

$$l(\sigma) := l_1(\sigma) \wedge l_2(\sigma). \tag{7.60}$$

We introduce the manifolds with fixed minimal shadow

$$\mathcal{X}_l = \{\sigma \in \mathcal{X} \; : \; l(\sigma) = l\}. \tag{7.61}$$

They induce a partition of $\mathcal{X}$:

$$\mathcal{X} = \bigcup_{l=0}^{L} \mathcal{X}_l. \tag{7.62}$$

We set, for $l \in \{0, \dots, L-1\}$

$$\begin{aligned} &\Phi(l, l+1) := \Phi(\mathcal{X}_l, \mathcal{X}_{l+1}), \qquad \mathcal{S}(l, l+1) := \mathcal{S}(\mathcal{X}_l, \mathcal{X}_{l+1}), \\ &\mathcal{G}(l, l+1) := \mathcal{G}(\mathcal{X}_l, \mathcal{X}_{l+1}), \end{aligned} \tag{7.63}$$

i.e. $\Phi(l, l+1)$ is the communication height between $\mathcal{X}_l$ and $\mathcal{X}_{l+1}$ whereas $\mathcal{S}(l, l+1)$ is the set of configurations where this communication height is achieved; $\mathcal{G}(l, l+1)$ is the set of essential saddles between $\mathcal{X}_l$ and $\mathcal{X}_{l+1}$ (see Proposition 6.11).

We have:

$$\Phi(l, l+1) = \min_{\substack{\sigma \in \hat{\mathcal{X}}_{l+1} \\ x : \sigma^{(x)} \in \mathcal{X}_l}} H(\sigma^{(x)}) \vee H(\sigma) \tag{7.64}$$

where

$$\hat{\mathcal{X}}_{l+1} = \{\sigma \in \mathcal{X}_{l+1} \; : \; \exists x \in \Lambda \text{ with } \sigma^{(x)} \in \mathcal{X}_l\}. \tag{7.65}$$

The case $l = 0$ is trivial: we have $\mathcal{X}_0 = -\underline{1}$ and $\mathcal{S}(0,1) = \mathcal{P}_0$ equal to the set of configurations containing a unique plus spin. For $l = 1, \ldots, L-1$, recall that we denote by $\mathcal{P}_l$ the set of configurations containing a unique cluster made of a rectangle $l \times (l+1)$ or $(l+1) \times l$ with a unit square protuberance attached to one of the longer sides. We stress that in the definition of $\mathcal{P}_l$ the orientation of the cluster is not important and the short side of the rectangle can be either horizontal or vertical. For $l = 1, \ldots, L-1$ we denote by $\mathcal{B}_l$ the set of configurations where the plus spins are precisely those contained in a horizontal strip encircling the torus, of width l, with a unit square attached to one of the sides of the strip; notice that for $l = L-1$ the unit square is attached to both sides of the strip (that have distance one).

Proposition 7.16 *Let $0 \le l \le l_2^* - 1$. We have*

$$\mathcal{S}(l, l+1) \supseteq \mathcal{P}_l,$$

$$\Phi(l, l+1) = \min_{\sigma \in \hat{\mathcal{X}}_{l+1}} H(\sigma) = (2J_1 + 2J_2)(l+1) - h(l^2 - l + 1) + H(-\underline{1}) \tag{7.66}$$

and $\mathcal{P}_l$ is a gate for the transition $\mathcal{X}_l \to \mathcal{X}_{l+1}$.

Let $l_2^* \le l \le L$. We have

$$\mathcal{S}(l, l+1) \supseteq \mathcal{B}_l,$$

$$\Phi(l, l+1) = \min_{\sigma \in \hat{\mathcal{X}}_{l+1}} H(\sigma) = -hlL + 2J_2 L + 2J_1 - h + H(-\underline{1}) \tag{7.67}$$

and $\mathcal{B}_l$ is a gate for the transition $\mathcal{X}_l \to \mathcal{X}_{l+1}$.

Proof For $\sigma \in \hat{\mathcal{X}}_{l+1}$, let $x \in \Lambda$ be such that $\sigma^{(x)} =: \eta \in \mathcal{X}_l$. Certainly $\eta(x) = -1$ and there exists a line $\mathcal{T} \subsetneq \Lambda$, namely a rectangle $L \times 1$, or $1 \times L$, containing x, encircling the torus Λ and with $\eta(y) = -1 \, \forall \, y \in \mathcal{T}$; the minimal shadow of η is in the direction $\mathcal{T}$ and equals l.

Consider the sites x', x'' adjacent to x from the exterior of $\mathcal{T}$, so that $\{x', x, x''\}$ define a line orthogonal to $\mathcal{T}$. There are three possibilities:

(a) $\eta(x') = \eta(x'') = +1$,

(b) the number of plus spins in $\{x', x''\}$ is one; $\eta(x') = -1, \eta(x'') = +1$ or $\eta(x') = +1, \eta(x'') = -1$,

(c) $\eta(x') = \eta(x'') = -1$.

It is immediate to see that in cases (b),(c), since we are assuming $h < 2J_2$, we have $H(\eta) < H(\sigma)$.

Let us analyse case (b). For a given l_1, l_2, let $\bar{\mathcal{X}}_{l_1,l_2}$ be the set of configurations such that $l_1(\eta) = l_1, l_2(\eta) = l_2$.

In case (b) we have

$$\Phi(l, l+1) = \left(\min_{l' \geq l+1,\, \eta \in \bar{\mathcal{X}}_{l,l'}} H(\eta) + 2J_2 - h \right) \wedge \left(\min_{l' \geq l+1,\, \eta \in \bar{\mathcal{X}}_{l',l}} H(\eta) + 2J_1 - h \right). \tag{7.68}$$

Thus, to find $\Phi(l, l+1)$, we have first to find the best $\eta \in \bar{\mathcal{X}}_{l_1,l_2}$, i.e. we have to optimize the choice of l_1, l_2 according to (7.68).

It is immediate to see that the configurations of minimal energy in $\bar{\mathcal{X}}_{l_1,l_2}$ are those with a unique rectangle with side lengths l_1, l_2. Indeed the energy of a configuration η with shadows l_1, l_2 has both surface and volume energy larger than or equal to those of a rectangle with sides l_1, l_2, being equal only when η coincides with the rectangle. It follows by direct computation that if $l < l_2^*$ (subcritical case) it is energetically convenient to take, for l', the smallest possible value, namely $l' = l + 1$, and we have a degeneracy in the sense that both when the shorter side of η is horizontal and when it is vertical, $H(\sigma)$ takes the same value $(2J_1 + 2J_2)(l + 1) - h(l^2 - l + 1)$ as is verified immediately; this corresponds to $\mathcal{P}_l$. On the other hand, for $l \geq l_2^*$ (supercritical case) we have to take the largest possible value $l' = L$ and, by direct comparison, we see that we have to choose the case with horizontal long side; this corresponds to $\mathcal{B}_l$. This concludes the proof in case (b).

In the general case ((a),(b),(c)), using (7.64), we want to determine $\min_{\sigma \in \hat{\mathcal{X}}_{l+1}, x: \sigma^{(x)} \in \mathcal{X}_l} H(\sigma^{(x)}) \vee H(\sigma)$. To this end we can, of course, disregard the configurations $\sigma \in \hat{\mathcal{X}}_{l+1}$ for which we can exhibit another $\tilde{\sigma} \in \hat{\mathcal{X}}_{l+1}$ and $y \in \Lambda$ such that $H(\tilde{\sigma}^{(y)}) \vee H(\tilde{\sigma}) < H(\sigma^{(y)}) \vee H(\sigma)$.

In particular, using this comparison criterion, we see that we can disregard case (c) ending up with either case (a) or case (b). Indeed, given $\eta \in \mathcal{X}_l$, in case (c), in order to increase the minimal shadow from l to $l + 1$ it is energetically convenient to flip an always existing minus spin $\eta(y)$ with y adjacent to some plus spin in η. So the communication height $\Phi(l, l+1)$ cannot be realized in case (c). Actually we shall see that we can restrict ourselves to consider only the already analysed case (b) unless $l = L - 1$.

Indeed suppose $\eta \in \mathcal{X}_l$ falls into case (a). If $l = L - 1$, by direct inspection we see that for L sufficiently large, $\mathcal{S}(L-1, L) = \mathcal{B}_{L-1}$. Suppose $l < L - 1$; then there exists another line $\mathcal{T}'$, parallel to $\mathcal{T}$ with $\eta(x) = -1$, $\forall x \in \mathcal{T}'$. As usual we identify η with the set of its plus spins. Let us consider case (a) with $l < L - 1$ and $\mathcal{T}$ vertical. We have that $\eta := \sigma^{(x)}$ must have a horizontal shadow l and a vertical shadow $\geq l + 1$.

In this case (a) with vertical $\mathcal{T}$, we have that, contrary to cases (b), (c), $H(\sigma) < H(\eta)$; indeed we have

$$H(\sigma) = H(\eta) - 2(J_1 - J_2) - h < H(\eta).$$

It is energetically convenient to use the transition $\tilde{\eta} \to \tilde{\sigma} = \tilde{\eta}^{(y)}$, where $\tilde{\eta} \in \mathcal{X}_l$ is obtained from η by translating one unit horizontally towards $\mathcal{T}$, the part of η lying between $\mathcal{T}$ and $\mathcal{T}'$ in such a way that in $\tilde{\eta}$ the empty (of plus) line $\mathcal{T}$ does not exist anymore; the two plus spins of η sitting in x, x' are now (in $\tilde{\eta}$) at distance one and there exists, in $\tilde{\eta}$, an empty vertical strip of width at least two that contains $\mathcal{T}'$.

Then, to get $\sigma \in \mathcal{X}_{l+1}$ we flip into plus an always existing minus spin at a suitable site y adjacent to the exterior of $\tilde{\eta}$ and nearest neighbour to only one plus spin in $\tilde{\eta}$. We have

$$H(\tilde{\eta}) \leq H(\eta) - 2J_1, \qquad H(\tilde{\sigma}) = H(\tilde{\eta}) + 2J_2 - h \leq H(\eta) - 2J_1 + 2J_2 - h \tag{7.69}$$

then

$$H(\tilde{\sigma}) = H(\tilde{\sigma}) \vee H(\tilde{\eta}) < H(\sigma) \vee H(\eta) = H(\eta) \tag{7.70}$$

so that the transition $\tilde{\eta} \to \tilde{\sigma}$, that falls into case (b), is always convenient with respect to the transition $\eta \to \sigma$. We conclude that when $\mathcal{T}$ is vertical it is sufficient to analyse case (b).

Consider now case (a) with $l < L - 1$ and $\mathcal{T}$ horizontal; we have that

$$H(\sigma) = H(\eta) + 2(J_1 - J_2) - h > H(\eta). \tag{7.71}$$

Consider the configuration $\tilde{\eta}$ obtained from η via a vertical translation in complete analogy to what we did in the case of vertical $\mathcal{T}$. We have

$$H(\tilde{\eta}) \leq H(\eta) - 2J_2. \tag{7.72}$$

Then, as before, to get $\sigma \in \hat{\mathcal{X}}_{l+1}$, we flip from minus into plus the spin at y, a suitable site adjacent to the exterior of $\tilde{\eta}$ and nearest neighbour to only one plus spin in $\tilde{\eta}$. We have

$$H(\tilde{\sigma}) = H(\tilde{\eta}) + 2J_1 - h \leq H(\eta) + 2(J_1 - J_2) - h = H(\sigma) \tag{7.73}$$

and

$$H(\tilde{\sigma}) = H(\tilde{\sigma}) \vee H(\tilde{\eta}), \qquad H(\sigma) = H(\sigma) \vee H(\eta). \tag{7.74}$$

So the transition $\tilde{\eta} \to \tilde{\sigma}$, again falling into case (b), is energetically at least equivalent to $\eta \to \sigma$. On the other hand, either in the passage from η to $\tilde{\eta}$ we gain more than one broken bond, obtaining the strict inequality $H(\tilde{\eta}) < H(\eta) - 2J_2$, so that case (b) is strictly more convenient; or, as is immediately verified, $\tilde{\eta}$ cannot be a rectangle with horizontal side larger than one and then the minimal value of $H(\tilde{\sigma})$ is strictly larger than the value corresponding to the optimal case (b).

In conclusion, we can always restrict ourselves to considering only case (b).

This case was discussed explicitly after (7.68). The proposition is proved. □

Proposition 7.17 *The communication height between $-\underline{1}$ and $+\underline{1}$ is given by $H(\mathcal{P})$.*

Proof Consider the reference path ω^R introduced in (7.46) to study the standard Ising model. It is easy to see that even in the present, anisotropic case, it can be used as reference path: the maximal value of H in ω^R is achieved precisely in $\mathcal{P}$. Thus, by comparison, we get $\Phi(-\underline{1}, +\underline{1}) \geq H(\mathcal{P})$. On the other hand, since every path joining $-\underline{1}, +\underline{1}$ has to perform the transition $\mathcal{X}_{l_2^*-1} \to \mathcal{X}_{l_2^*}$, by Proposition 7.16, we have $\Phi(-\underline{1}, +\underline{1}) \leq H(\mathcal{P})$ which concludes the proof of the proposition. □

The following proposition shows that any configuration in $\mathcal{X}$ different from $-\underline{1}$ or $+\underline{1}$ is Γ_0-reducible (see definition after (6.234) of Chapter 6) for some $\Gamma_0 < \Gamma$.

Proposition 7.18 *There exists $\Gamma_0 < \Gamma$ such that $\mathcal{X}_{\Gamma_0} \subseteq \{-\underline{1}, +\underline{1}\}$.*

Proof We show that there exists $\Gamma_0 < \Gamma$ such that

$$\forall\, \eta \notin \{-\underline{1}, +\underline{1}\}\ \exists \eta' \in \mathcal{X}: H(\eta') < H(\eta),\ \Phi(\eta, \eta') \leq H(\eta) + \Gamma_0. \tag{7.75}$$

We take $\Gamma_0 = 2J_1 - h$ which, in our hypothesis, is certainly smaller than Γ; to get (7.75) it is sufficient to construct, for any $\eta \in \mathcal{X} \setminus \{-\underline{1}, +\underline{1}\}$, a suitable path ω leading to some η' with $H(\eta') < H(\eta)$ and $\Phi_\omega(\eta, \eta') \leq H(\eta) + \Gamma_0$ (see Definition 6.9).

We can suppose that η is a local minimum otherwise there exists a downhill path going from η to a local minimum η' with strictly smaller energy and the proposition would be proven in this case. So we know that η can only contain stable non-interacting rectangles and strips encircling the torus Λ. If $\eta \neq -\underline{1}$ and it contains a subcritical rectangle $R(l_1, l_2)$ with $l = l_1 \wedge l_2 < l_2^*$, consider the configuration η' obtained from η by cutting one row or column of minimal length l from $R(l_1, l_2)$; consider the path $\omega : \eta \to \eta'$ corresponding to the sequence of corner erosions in the row or column to be eliminated. Certainly, by subcriticality $\Phi_\omega(\eta, \eta') - H(\eta) = h(l-1) < \Gamma_0$ and $H(\eta') < H(\eta)$. Thus we are left with the case when $\eta \neq +\underline{1}$ is supercritical, namely it contains either a rectangle $R(l_1, l_2)$ with $l = l_1 \wedge l_2 \geq l_2^*$ or a strip encircling the torus Λ. In both cases there exists a path ω leading to the configuration η' obtained from η by increasing by one unit one side of $R(l_1, l_2)$ or the width of the strip. We have $\Phi_\omega(\eta, \eta') - H(\eta) \leq \Gamma_0$ and, by supercriticality, $H(\eta') < H(\eta)$. □

Alternative proof of Theorem 7.14 Point (i) of the theorem follows immediately by Theorem 6.38 of Chapter 6. To prove point (ii) write

$$P_{-\underline{1}}\big(\theta_{-\underline{1},\mathcal{P},+\underline{1}} > \tau_{+\underline{1}}\big) \leq P_{-\underline{1}}\big(\tau_{\mathcal{X}_{l_2^*-1}\setminus\mathcal{P}} < \tau_{+\underline{1}}\big); \tag{7.76}$$

by part (i) of the theorem, we get, for all $\delta > 0$,

$$P_{-\underline{1}}\big(\tau_{\mathcal{X}_{l_2^*-1}\setminus\mathcal{P}} < \tau_{+\underline{1}}\big) \leq P_{-\underline{1}}\big(\tau_{\mathcal{X}_{l_2^*-1}\setminus\mathcal{P}} < e^{\beta(\Gamma+\delta)}\big) + o(1). \tag{7.77}$$

Choosing δ sufficiently small and using Lemma 6.21 we conclude the proof. □

7.9 The ferromagnetic Ising model with nearest and next nearest neighbour interactions in two dimensions

Another model that turns out to be interesting for the question of relevance of the Wulff shape for low temperature nucleation is the ferromagnetic two-dimensional Ising model with nearest and next nearest neighbour interaction. We can think of using this model to describe the growth of a crystal. Again at first sight one could suppose that growth takes place through crystals that minimize the surface energy under fixed instantaneous volume. For the model under consideration it is easy to see that this Wulff shape is given by an octagon with coordinate and oblique sides proportional to the nearest neighbour and next nearest neighbour interactions, respectively. The main result of the paper [185] says that the typical growth of subcritical and supercritical crystals is through a sequence of particular shapes that differ significantly from the equilibrium Wulff octagons (see also [183]).

Let us now define the model. Again we consider an Ising system enclosed in an $L \times L$ torus Λ so that the configuration space is given by $\mathcal{X} := \{-1, +1\}^{\Lambda}$. The *energy* of a configuration σ is

$$H(\sigma) = -\frac{\tilde{J}}{2} \sum_{<x,y> \subseteq \Lambda} \sigma(x)\sigma(y) - \frac{K}{2} \sum_{\ll x,y \gg \subseteq \Lambda} \sigma(x)\sigma(y) - \frac{h}{2} \sum_{x \in \Lambda} \sigma(x), \tag{7.78}$$

where we suppose that $0 < h < K/7 < \tilde{J}/70$. Here $<x, y>$ denotes an unordered pair of nearest neighbours in $\Lambda : |x - y| = 1$ and $\ll x, y \gg$ denotes a pair of next nearest neighbours in $\Lambda : |x - y| = \sqrt{2}$. We refer to the model described by (7.78) as the JK model.

Recalling the definition of contour given after (7.12), it is easy to see that if a configuration σ has the boundary $\partial C(\sigma)$ consisting of a single contour γ, the energy of σ is

$$H(\sigma) - H(-\underline{1}) = J|\gamma| - K|A(\gamma)| - h|I(\gamma)|. \tag{7.79}$$

Here

$$J = \tilde{J} + 2K, \tag{7.80}$$

$|\gamma|$ is the length of γ, $|I(\gamma)|$ is the cardinality (area) of the interior $I(\gamma) \equiv C(\sigma)$; finally, $|A(\gamma)|$ is the number of corners (right angles) of γ. Notice that when four unit segments converge on the same point of γ we count four corners.

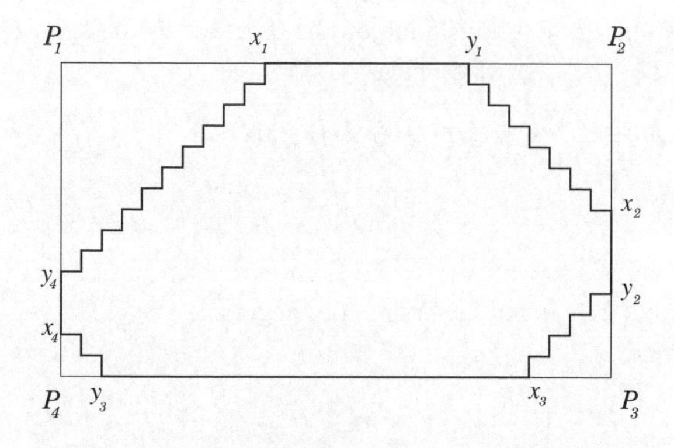

Figure 7.8

A contour γ is said to be *isolated* if it lies at a Euclidean distance at least $\sqrt{2}$ from other contours.

A relevant role will be played by a particular class of contours that we call octagons. An *octagon* is a closed convex contour inscribed in a rectangle R with edges parallel to the lattice axes. Call P_1, P_2, P_3, P_4 the vertices of R. The octagon contains four straight edges with extremes x_i, y_i, $i = 1, \dots, 4 : (x_1, y_1) \subseteq P_1P_2$, $(x_2, y_2) \subseteq P_2P_3$, $(x_3, y_3) \subseteq P_3P_4$, and $(x_4, y_4) \subseteq P_4P_1$, called *coordinate edges*, and four *oblique edges*, at $45°$ with respect to the coordinate axes, that have a local staircase structure with extremes y_1x_2, y_2x_3, y_3x_4, y_4x_1; see Figure 7.8.

An octagon is said to be *stable* if $|x_i - y_i| \geq 2$, $|y_i - x_{i+1}| \geq \sqrt{2}$, $i = 1, \dots, 4$. We use

$$L_i = |x_i - y_i|, \quad i = 1, \dots, 4, \tag{7.81}$$

to denote the lengths of its coordinate edges and

$$l_i = 1 + \frac{1}{\sqrt{2}}|y_i - x_{i+1}|, \quad i = 1, \dots, 4\,(x_5 \equiv x_1), \tag{7.82}$$

to denote the 'lengths' of its oblique edges. We set

$$\begin{aligned} D_1 &= \overline{P_1P_2} = L_1 + l_4 - 1 + l_1 - 1 = \overline{P_4P_3} = L_3 + l_3 - 1 + l_2 - 1, \\ D_2 &= \overline{P_2P_3} = L_2 + l_1 - 1 + l_2 - 1 = \overline{P_4P_1} = L_4 + l_3 - 1 + l_4 - 1, \end{aligned} \tag{7.83}$$

for the lengths of the sides of its *rectangular envelope* $R(Q) = P_1P_2P_3P_4$. We use

$$Q(D_1, D_2, l_1, l_2, l_3, l_4) \tag{7.84}$$

to denote the corresponding octagon.

By an abuse of notation we use the same symbol Q also to denote the set of all spin configurations σ giving rise to a unique closed contour (full of plus spins) consisting, up to a translation on the torus, of the octagon Q. Clearly, for $Q = Q(D_1, D_2, (l_i)_{i=1,\dots,4})$ we have

$$H(Q) - H(-\underline{1}) = 2J(D_1 + D_2) - hD_1D_2 - \sum_{j=1}^{4}\left[K(2l_j - 1) - \frac{1}{2}hl_j(l_j - 1)\right]. \tag{7.85}$$

We suppose that $2K/h$ and $2J/h$ are not integers.

Let us define

$$\begin{aligned} l^* &= \left[\frac{2K}{h}\right] + 1, \quad \text{i.e. } h(l^* - 1) < 2K < hl^* \\ D^* &= \left[\frac{2J}{h}\right] + 1, \quad \text{i.e. } h(D^* - 1) < 2J < hD^* \\ L^* &= D^* - 2(l^* - 1) = \left[\frac{2J}{h}\right] + 1 - 2\left[\frac{2K}{h}\right]. \end{aligned} \tag{7.86}$$

It follows from our assumptions that $L^* > l^*$.

We define a *standard octagon* as an octagon with

$$\min_{i=1,\dots,4} L_i \geq l^* \text{ and } l_j = l^*, \quad j = 1, \dots, 4.$$

In this case we simply write

$$Q(D_1, D_2) \equiv Q(D_1, D_2, l^*, l^*, l^*, l^*). \tag{7.87}$$

We call an octagon *regular* when

$$L_1 = L_2 = L_3 = L_4 = l_1 = l_2 = l_3 = l_4 = l$$

and denote it by $Q(l)$. We are only interested in regular octagons with $l \leq l^*$.

It is easy to see that among non-winding configurations the minima of the energy are given by sets of isolated octagons.

It will happen, similarly to the case of the nearest neighbour Ising model, that small octagons have a tendency to shrink whereas large ones tend to grow. Again, the dynamical mechanism responsible for this behaviour relies on the competition between the creation of a suitable stable protuberance and the erosion of an edge. The main difference here, with respect to the nearest neighbour Ising model, comes from the presence of the oblique edges in stable isolated clusters – the octagons. The new phenomenon, namely the growth or shrinking in the oblique direction, is still governed by competition between the creation of a certain protuberance and the erosion of an edge. However, three new features should be mentioned:

(i) the corresponding time scales are different with respect to those of the growth–shrinking mechanism in the coordinate directions;

(ii) for sufficiently large octagons, a sort of 'equilibrium' in the oblique direction is established;

(iii) for 'relevant' octagons, the global tendency to shrink or grow, to disappear or to invade the whole volume, is governed by the size of the 'circumscribed rectangle'.

It happens that, for sufficiently large sizes, it is interesting to consider not only the jumps between neighbouring generic octagons (with related basins of attraction and saddle configurations) but also the transitions between standard octagons $Q(D_1, D_2)$ with corresponding 'domains of attraction' that group together all the basins of attraction of octagons inscribed in the same rectangle R as $Q(D_1, D_2)$. These transitions at 'intermediate level' (corresponding to the exit from domains of attraction, i.e. suitable cycles, instead of basins of attraction) involve an intermediate time scale between that referring to the transition between neighbouring generic octagons and the global time scale referring to the transition between $-\underline{1}$ and $+\underline{1}$.

In the present case the set of global saddle configurations $\mathcal{S}(-\underline{1}, +\underline{1})$ coincides with the set $\mathcal{P}^{JK}$ of all configurations obtained from a standard octagon $Q(D^*, D^* - 1)$ or $Q(D^* - 1, D^*)$ (D^* was defined in (7.86)) by attaching, to one of its longer coordinate edges, a unit square protuberance.

For any $\bar{\sigma} \in \mathcal{P}^{JK}$ one has

$$\begin{aligned} H(\bar{\sigma}) - H(-\underline{1}) =: \Gamma^{JK} &= H(Q(D^*, D^* - 1)) + 2\tilde{J} - h \\ &= H(Q(D^*, D^*)) + h(D^* - 1) - 4K \end{aligned} \tag{7.88}$$

for the 'height' of the global saddle point.

Again it is possible to prove that the first excursion from $-\underline{1}$ to $+\underline{1}$ typically passes through a configuration from $\mathcal{P}^{JK}$ and the time needed for this to happen is of the order $\exp(\beta\Gamma^{JK})$.

With the same definitions and notation as for the standard Ising model we have the following.

Theorem 7.19 [185] *Let* $0 < h < K/7 < \tilde{J}/70$. *For every* $\delta > 0$,

$$\lim_{\beta\to\infty} P_{-\underline{1}}(\exp[\beta(\Gamma^{JK} - \delta)] < \tau_{+\underline{1}} < \exp[\beta(\Gamma^{JK} + \delta)]) = 1, \tag{7.89}$$

$$\lim_{\beta\to\infty} P_{-\underline{1}}(\theta_{-\underline{1},\mathcal{P}^{JK},+\underline{1}} < \tau_{+\underline{1}}) = 1. \tag{7.90}$$

In addition, we have much more detailed information about a typical path followed by our process σ_t during its first excursion from $-\underline{1}$ to $+\underline{1}$.

Roughly speaking, the typical trajectories during the first excursion begin by following a sequence of regular and almost regular octagons up to an edge l^*; after that the oblique edge stays almost constant at the value l^* while the coordinate

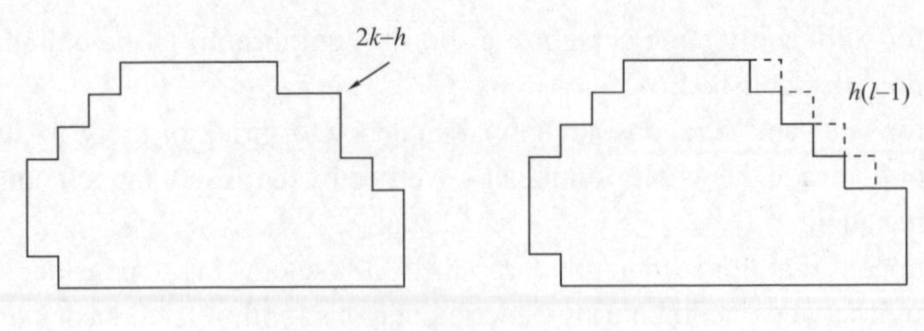

Figure 7.9

edges grow, keeping the rectangular envelope almost square, from the value l^* up to the value L^* corresponding to the critical nucleus.

Finally, the oblique edges stay almost constant at the value l^* whereas the coordinate edges continue to grow, until the whole volume is invaded by plus spins. This 'supercritical growth' is, on average, a descent in energy.

Let us explain, in words, what are the mechanisms responsible for the above described behaviour during the first excursion between $-\underline{1}$ and $+\underline{1}$. Consider a sufficiently large octagon $Q(D_1, D_2, l_1, l_2, l_3, l_4)$. We first notice that we have two mechanisms, still active as in the case of standard nearest neighbour Ising model, of growth and contraction, respectively: the formation of a unit square protuberance on a coordinate side, with cost $2\tilde{J} - h$, and the corner erosion of a coordinate side of length L, with cost $h(L-1)$. Then we observe that we have two new but similar mechanisms: the formation of a unit square protuberance on an oblique side, of cost $2K - h$, and the corner erosion of an oblique side of length l with cost $h(l-1)$ (see Figure 7.9).

From this competing process the critical length $l^* = [2K/h] + 1$ emerges: for $l \geq l^*$ the erosion of an oblique unit strip requires an energy larger than that needed to create a unit square protuberance leading to growth; the converse is true for $l < l^*$.

Notice now that the erosion of a unit oblique strip induces an increase of one unit of the length of the corresponding oblique side, whereas the growth of a new unit oblique strip induces its decrease by one unit. This compensation produces an equilibrium situation for $l = l^*$. Indeed when an oblique side is equal to l^* an erosion leads to $l^* + 1$ which is supercritical, so that the growth of a protuberance takes place before further erosion, thus reestablishing the length l^*. On the other hand a growth would induce a decrease of the length to the value $l^* - 1$ which is subcritical, so that an erosion reestablishing the length l^* will typically take place before another growth.

It is clear that this equilibrium in the length of oblique facets continues to hold even during supercritical growth; then, after a sufficiently long time, the coordinate sides L_i become much larger than l^*, so that the shape of the supercritical nucleus becomes almost rectangular. We refer to [185] for more details on the tube of typical trajectories as well as for complete proofs.

We want now to make some comments on the above results on $J_1 J_2$ and JK models. First we recall that, due to reversibility, the part of the optimal path in $(-\underline{1} \to +\underline{1})_{opt}$ from $-\underline{1}$ up to the global saddle $\mathcal{S}(-\underline{1}, +\underline{1})$ is the time reversal of a typical contraction path starting from $\mathcal{S}(-\underline{1}, +\underline{1})$ so that instead of analysing the typical growth during nucleation we can analyse the typical contraction; on the other hand, as we have already remarked in the case of the $J_1 J_2$ model, the relevant quantity, to determine the typical sequence of contracting droplets, is the speed of contraction and certainly not the surface energy. During the contraction the droplet just loses the facets corresponding to the maximal speed of contraction. For instance, as has already been noted, in the $J_1 J_2$ model the speed of contraction of a facet of length l has an isotropic expression in terms of l given by $\exp[-\beta h(l-1)]$ so that if we start from a rectangle $l \times l'$ or $l' \times l$ with $l < l'$, during the evolution the system is attracted by an $l \times l$ square which, subsequently, disappears following a sequence of quasi-squares.

For the JK model, suppose we start from a generic subcritical octagon $Q(D_1, D_2, l_1, l_2, l_3, l_4)$ with a circumscribing rectangle of minimal side $D := D_1 \wedge D_2 > 3l^*$ (so that it can be the rectangle circumscribing a standard octagon); we are first attracted (in a relatively short time of order $\exp[\beta(2K-h)]$) by the standard octagon with the same circumscribing rectangle; then, in a much larger time, of order $\exp[\beta h(D-1)]$, we are attracted by another standard octagon, with a square rectangular envelope $Q(D, D, l^*, l^*, l^*, l^*)$. Then we continue the contraction following a sequence of standard octagons with square or almost square circumscribing rectangles, up to the value $D = 3l^*$ that corresponds to a regular octagon with side length l^*. From this moment the contraction preserves the regular (or quasi-regular) shape.

The behaviour of the system during a subcritical contraction does not have anything to do with minimizing the energy at fixed volume, that would lead to the Wulff shape.

In the above two models, the Wulff shape appears instantaneously during the nucleation process; for the $J_1 J_2$ model it appears during supercritical expansion. For the JK model, the fact that the critical droplet is Wulff can be considered as an irrelevant accident.

It is interesting to notice that the above discussion is able to explain why in nature crystals do not in general have Wulff shape. We can summarize by saying that growth is, typically, a non-equilibrium phenomenon since growth is too fast to allow the equilibrium shape to be established.

7.10 The ferromagnetic Ising model with alternating field

We want now to describe the dynamical behaviour at low temperature of an Ising system with alternating field.

More precisely we consider, for very large inverse temperature β, discrete-time Metropolis dynamics for a two-dimensional ferromagnetic Ising system with nearest neighbour interaction (coupling constant $J/2 > 0$) enclosed in a sufficiently large but finite torus Λ and subject to a positive (negative) external magnetic field $h_1 > 0$ ($-h_2 < 0$) on the even (odd) horizontal lines.

We call this model the h_1, h_2 model. We take $\Lambda = \Lambda_L \equiv \{1, \ldots, L\} \times \{1, \ldots, L\}$ (with periodic boundary conditions); L is supposed to be even. The configuration space is $\mathcal{X} = \{-1, +1\}^\Lambda$. Let $\mathcal{L}_1$ ($\mathcal{L}_2$) be the set of horizontal even (odd) rows in Λ:

$$\mathcal{L}_1 = \{x = (x_1, x_2) \in \Lambda | x_2 \text{ is even}\},$$
$$\mathcal{L}_2 = \Lambda \backslash \mathcal{L}_1.$$

To any configuration $\sigma \in \mathcal{X}$ we associate the energy:

$$H(\sigma) = -\frac{J}{2} \sum_{<x,y> \subseteq \Lambda} \sigma(x)\sigma(y) - \frac{h_1}{2} \sum_{x \in \mathcal{L}_1} \sigma(x) + \frac{h_2}{2} \sum_{x \in \mathcal{L}_2} \sigma(x) \tag{7.91}$$

where $<x, y>$ denotes a generic unordered pair of nearest neighbour sites in Λ.

For different values of the magnetic fields, we have very different dynamical behaviour related strictly to the peculiar features of the equilibrium phase diagram of our model at low temperature.

The phase diagram in the (h_1, h_2)-plane at $\mathrm{T} = 0$ that corresponds to the characterization of the different sets of ground states associated with the different values of (h_1, h_2), is easily found. Three main different configurations appear as ground states in different regions: all plus $+\underline{1}$, all minus $-\underline{1}$, and an alternate configuration $\pm\underline{1}$ with all plus (all minus) on even (odd) horizontal lines. They are separated by the coexistence lines $h_1 = h_2,\ h_1 = 2J,\ h_2 = 2J$ that converge in the triple point $\tilde{P} \equiv (2J, 2J)$ (see Figure 7.10).

Using the specular symmetry with respect to the diagonal $h_1 = h_2$, it is sufficient to consider the region $h_1 \geq h_2$. On the line $h_1 = h_2 < 2J$ only two ground states, namely $+\underline{1}$, $-\underline{1}$, coexist whereas on the line $h_2 = 2J, h_1 > 2J$ we have many coexisting ground states (their number diverges as $\Lambda \to \mathbb{Z}^2$); they are given by $+\underline{1}$, the staggered configuration $\pm\underline{1}$ and all intermediate configurations that are obtained from $\pm\underline{1}$ by changing into $+1$ an arbitrary number of horizontal -1 lines in $\mathcal{L}_2$. In $\tilde{P}$ we have coexistence of these ground states together with those obtained from $\pm\underline{1}$ by changing into -1 an arbitrary number of horizontal $+1$ lines in $\mathcal{L}_1$; thus, in particular, in $\tilde{P}$ we have as ground states $-\underline{1}$, $+\underline{1}$, $\pm\underline{1}$.

We analyse the metastable behaviour by choosing the parameters h_1, h_2 in the region

$$\mathrm{I} := h_2 < h_1 < 2J$$

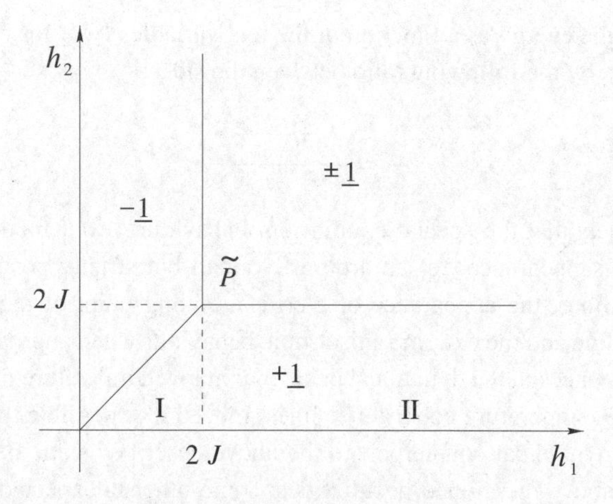

Figure 7.10

close to the line $0 < h_1 = h_2 < 2J$ or in the region

$$\mathrm{II} := 2J < h_1, \quad h_2 < 2J$$

close to the line $h_2 = 2J,\ h_1 > 2J$.

In [224A] the equilibrium (infinite volume Gibbs) states were studied for low but positive temperature T. It has been proven that, in this case, the coexistence lines $h_1 = 2J, h_2 > 2J; h_2 = 2J, h_1 > 2J$ disappear since in that region the system behaves like a collection of almost independent one-dimensional standard zero field Ising systems that are known to exhibit uniqueness of Gibbs state for any positive temperature. On the other hand for $h_1 = h_2 < 2J$ at low enough temperature T, we have coexistence of two phases and, given such a T, on the $h_1 = h_2$ line the coexistence ends at a critical point $P^* \equiv (2J - \delta, 2J - \delta)$ with $\delta \sim \exp(-\mathrm{const}\frac{1}{\mathrm{T}})$. It is not hard to realize that on the segment $0 < h_1 = h_2 < 2J - \delta$ our model shares many features with a zero field ferromagnetic anisotropic Ising model with coupling constants, along horizontal and vertical directions given, respectively, by $J_1 = J,\ J_2 = J - (h_1 + h_2)/4$ so that it becomes more and more anisotropic as we approach the triple point $\tilde{P}$.

Indeed let us call *stable* a rectangle $R(l_1, l_2)$ with horizontal (vertical) side l_1 (l_2), such that $l_1 \geq 2,\ \ l_2 \geq 3$ and $R(l_1, l_2)$ starts and ends in $\mathcal{L}_1$. The energy formation of a stable rectangle $R(l_1, l_2)$ is:

$$\begin{aligned} H(R(l_1, l_2)) - H(-\underline{1}) &= 2J(l_1 + l_2) - h_1 l_1 \left(\frac{l_2 + 1}{2}\right) + h_2 l_1 \left(\frac{l_2 - 1}{2}\right) \\ &= 2J(l_1 + l_2) - \frac{(h_1 - h_2)}{2}(l_2 + 1)l_1 - l_1 h_2, \end{aligned} \tag{7.92}$$

namely the expression valid for the $J_1 J_2$ model with $J_1 = J,\ J_2 = J - (h_1 + h_2)/4$. It is interesting to compute the $\mathrm{T} = 0$ Wulff shape, namely the shape minimizing the energy as a function of the real variables l_1, l_2 for fixed area; it is characterized by the following ratio between the sides:

$$\frac{l_2}{l_1} = \frac{J - \left(\frac{h_1+h_2}{4}\right)}{J}. \tag{7.93}$$

One expects that the special equilibrium behaviour of our model and in particular the disappearance, for an arbitrarily small but strictly positive T, of two coexistence lines, the appearance of a critical ending point of the $0 < h_1 = h_2$ coexistence line and the extreme anisotropic behaviour when approaching $\tilde{P}$ must have consequences in the dynamical behaviour at low temperature in the vicinities of these zero temperature coexistence lines. Indeed it is possible to prove results providing a dynamical counterpart to the above described static behaviour, confirming the above mentioned peculiarities; some other unexpected features also show up.

Let us now just summarize the main results on the metastable behaviour of the h_1, h_2 model. We refer to [224] for precise definitions and detailed results.

In the above defined regions I, II, where the unique ground state is $+\underline{1}$, we start from $-\underline{1}$ (or $\pm\underline{1}$) and consider the first hitting time $\tau_{+\underline{1}}$ to the configuration $+\underline{1}$; we are interested in computing its asymptotics for large β and describing the typical mechanism of transition (first excursion) from $-\underline{1}$(or $\pm\underline{1}$) to $+\underline{1}$.

(1) Near the $T = 0$ coexistence line given by $h_2 = 2J, h_1 > 2J$ (disappearing at $T > 0$) we have absence of metastability in the sense that the protocritical droplet is microscopic and the 'lifetime' does not diverge as we approach the coexistence line.

(2) Near the $T = 0$ coexistence line given by $0 < h_1 = h_2 < 2J$ we have 'usual' metastable behaviour. The transition is driven by the formation of a critical nucleus whose size diverges as we approach the coexistence line. The unexpected behaviour concerns the shape of the critical nucleus as well as the nucleation pattern.

Looking at the expression (7.92) one would be led to the conclusion that the critical droplet is the unique saddle point corresponding to

$$l_1 = \frac{4J}{(h_1 - h_2)}, \qquad l_2 = \frac{4J - h_1 - h_2}{(h_1 - h_2)} \tag{7.94}$$

defining a particular Wulff rectangle: this conclusion turns out to be erroneous. We find that the shape characterizing the critical nucleus and all the droplets appearing before its formation is rectangular but non-Wulff. It is characterized by the relation

$$l_2 = 2l_1 - 1$$

Notice that the non-Wulff nature of the nucleation pattern is even enhanced with respect to what has been found in the anisotropic Ising model. The relevant rectangles during the nucleation actually have a vertical side *larger* (almost twice as much) than the horizontal side whereas for the Wulff shape we have the opposite situation and the horizontal side becomes even arbitrarily larger than the vertical side if we are sufficiently close to $\tilde{P}$. This fact cannot be easily predicted on static grounds but is rather related to a genuine dynamic effect.

(3) In the vicinity of $\tilde{P}$ the mechanism of nucleation depends on the angle from which we approach it. In particular it turns out that, given a halfline emerging from $\tilde{P}$, lying inside the region I, and forming an angle α with the line $h_1 = 2J$, no matter how close we go to $\tilde{P}$, staying on this line, the size of the critical droplet remains constant: it depends on α and tends to infinity when α tends to the value $\pi/4$ corresponding to the line $h_1 = h_2$.

The critical droplet is a rectangle with edges:

$$l_1 = l_b^* = \left[\frac{2J - h_2}{h_1 - h_2}\right] + 1 \tag{7.95}$$

and

$$l_2 = l_h^* = 2l_b^* - 1. \tag{7.96}$$

The protocritical droplet representing the global saddle point between $-\underline{1}$ and $+\underline{1}$ in region I is a cluster 'close' to the rectangle $R(l_b^*, l_h^*)$ slightly different in different subregions of I; we refer to [224] for more details.

Let us say a few words to explain the apparently paradoxical result concerning the critical droplet. Again the explanation of the nucleation pattern comes from the elementary mechanisms of growth and contraction of stable droplets, in this case our stable rectangles. It is possible to show that the mechanism of growth of a stable rectangle consists, after the passage through a suitable contiguous saddle configuration, in the formation of a square 2×2 stable protuberance on a horizontal side or the formation of a 1×3 stable rectangular protuberance on a vertical side.

The first mechanism, after the filling of the two rows containing the 2×2 protuberance, leads from the stable rectangle $R(l_1, l_2)$ to another stable rectangle, $R(l_1, l_2 + 2)$, whereas the second mechanism leads to the stable rectangle $R(l_1 + 1, l_2)$. The typical contraction mechanisms are the reverse ones. A direct computation shows that the horizontal and vertical speeds of contraction are equal for a standard shape given by $l_1 = 2l_2 - 1$; thus we have that any subcritical rectangle $R(l_1, l_2)$ is first attracted by a maximal standard rectangle contained inside $R(l_1, l_2)$ and it shrinks successively, maintaining the standard shape.

7.11 The dynamic Blume–Capel model. Competing metastable states and different mechanisms of transition

We want now to discuss the so-called 'competing metastable states'. We refer to a situation where three or more stable thermodynamic phases coexist at equilibrium. Changing the thermodynamic parameters slightly, we can arrive at a situation characterized by a unique stable state with two or more metastable states. In particular we want now to analyse metastability and nucleation in the framework of a dynamic Blume–Capel model. This is a ferromagnetic lattice system where the single spin variable can take three possible values: $-1, 0, +1$. It was originally introduced to study the He^3–He^4 phase transition (see [22], [43]).

One can think of it as a system of particles with spin. The value $\sigma(x) = 0$ of the spin at the lattice site x will correspond to the absence of particles (a *vacancy*), whereas the values $\sigma(x) = +1, -1$ will correspond to the presence, at x, of a particle with spin $+1, -1$, respectively.

We suppose the system enclosed in a two-dimensinal torus Λ. The configuration space is $\mathcal{X}_{BC} = \{-1, 0, +1\}^\Lambda$. The energy associated with a configuration $\sigma \in \mathcal{X}_{BC}$ is

$$H(\sigma) = J \sum_{<x,y>} (\sigma(x) - \sigma(y))^2 - \lambda \sum_x \sigma(x)^2 - h \sum_x \sigma(x), \tag{7.97}$$

where λ and h are two real parameters, having the meaning of chemical potential and external magnetic field, respectively; J is a real positive constant (ferromagnetic interaction) and $<x, y>$ denotes a generic unordered pair of nearest neighbour sites in Λ.

We say that two configurations σ, η are *nearest neighbours* if they differ only in the value of the spin at a single site x and $|\eta(x) - \sigma(x)| = 1$. As usual, the dynamics is given by a reversible Markov chain with transition probabilities given, for $\sigma \neq \eta$, by

$$P(\sigma, \eta) = \begin{cases} \frac{1}{2|\Lambda|} e^{-\beta(H(\eta) - H(\sigma))^+} & \sigma, \eta \text{ nearest neighbours} \\ 0 & \text{otherwise.} \end{cases} \tag{7.98}$$

In [63] a slightly different dynamics was introduced, using a different notion of nearest neighbour configurations, that turns out to be, for some respects, more complicated to treat.

Let $-\underline{1}$, $\underline{0}$ and $+\underline{1}$ denote the configurations with all spins in Λ equal to $-1, 0, +1$, respectively. It is easily seen that the set of ground states (absolute minima for H) corresponding to different values of λ and h is characterized as follows:

- for $\lambda = h = 0$ the ground state is three times degenerate, the configurations minimizing the energy are $-\underline{1}$, $\underline{0}$ and $+\underline{1}$,
- for $h > 0$ and $h > -\lambda$ the ground state is $+\underline{1}$,
- for $h < 0$ and $h < \lambda$ the ground state is $-\underline{1}$,
- for $\lambda < 0$ and $\lambda < h < -\lambda$ the ground state is $\underline{0}$.

For $h = 0$, $\lambda > 0$, $+\underline{1}$, $-\underline{1}$ coexist. For $h = \lambda < 0$, $-\underline{1}$, $\underline{0}$ coexist. For $h = -\lambda > 0$, $+\underline{1}$, $\underline{0}$ coexist. See Figure 7.11.

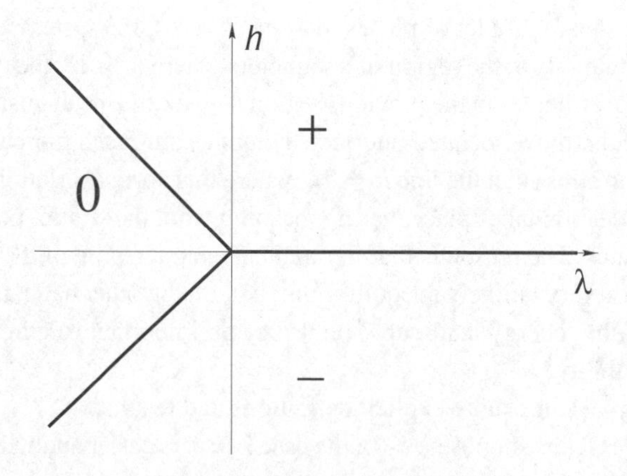

Figure 7.11

It has been shown, using Pirogov–Sinai theory, that this phase transition persists at positive temperature $T = 1/\beta$ in the thermodynamic limit (see [37]).

We are interested in the regime where λ and h are very small but fixed, the volume is large and fixed and T is very small.

The interesting region of parameters is

$$0 < |\lambda| < h \tag{7.99}$$

with sufficiently small h.

In the whole region (7.99) we have (in any finite box):

$$H(-\underline{1}) > H(\underline{0}) > H(+\underline{1}).$$

In other words $-\underline{1}$ and $\underline{0}$ are metastable whereas $+\underline{1}$ is stable. The main question that can be posed is, when starting from the highest metastable configuration $-\underline{1}$, whether or not the system reaches the intermediate metastable configuration $\underline{0}$ before relaxing to the stable configuration $+\underline{1}$.

In [63] an answer was found in the asymptotic regime corresponding to Λ, h, λ fixed, in the limit of large β. It was shown, in [63], that there is a change in the mechanism of transition from $-\underline{1}$ to $+\underline{1}$ when crossing the line $h = 2\lambda$. On the right side of this line ($h < 2\lambda$) the transition is *direct*, i.e. the system is driven to $+\underline{1}$ via the formation of a special critical droplet: a square 'picture frame' with suitably large size, made by a square of plus spins encircled by a unit layer of zeros in a sea of minuses. It is easy to see that a direct interface between minus and plus is unstable in our region of parameters. This explains the persistence of a thin layer of zeros between the pluses and the minuses.

In the other region ($h > 2\lambda$) the transition is *indirect* in the sense that the system first reaches the $\underline{0}$ configuration via the formation of a supercritical square droplet of zeros in a sea of minuses. Successively, via the formation of a supercritical square droplet of pluses in a sea of zeros, the system is driven to the final stable state $+\underline{1}$. In the region of parameters where $h > 2\lambda$, the two transitions are 'Ising-like' whereas in the previous region $h < 2\lambda$ the mechanism of transition and in particular the associated interface dynamics are much more complicated.

We want to stress that the line $h = 2\lambda$, where this abrupt variation of the mechanism of nucleation takes place, has no meaning from the 'static' point of view of the Gibbs states. The reason is that we are analysing a region of the configuration space which is very unlikely at equilibrium; but, on the other hand, this region and the shape of the 'energy landscape' on it play an important role in the relaxation from metastability.

Let us now give some more explicit definitions and results.

To fix ideas let us suppose $\lambda > 0$; the case $\lambda < 0$ is easier and will be discussed briefly later on.

We make use of some notions already introduced in the context of the standard Ising model: the notion of interacting rectangles (see Definition 7.8) and the notion of non-winding configuration of zeros and pluses in a unique 'sea' of minuses.

We often collect together classes of configurations modulo natural symmetries. By an abuse of notation we often denote by the same symbol a single configuration or the corresponding equivalence class.

A *plurirectangle* is a set of non-winding configurations where the zeros form a unique cluster whose external boundary is a stable (i.e. with minimal side length ≥ 2) rectangle R and the pluses lie inside non-interacting rectangular clusters contained in the interior of R. Notice that, given a plurirectangle, the internal pluses are separated from the sea of minuses by at least a unit layer of zeros (see Figure 7.12).

It is possible to show (see [63]) that the set of non-winding minima coincides with the set of families of non-interacting plurirectangles.

We call a *birectangle* $R(L^0, l^0, L^+, l^+)$ the particular plurirectangle given by the set of configurations containing a single rectangle $L^0 \times l^0$ (where $L^0 \geq l^0$) of

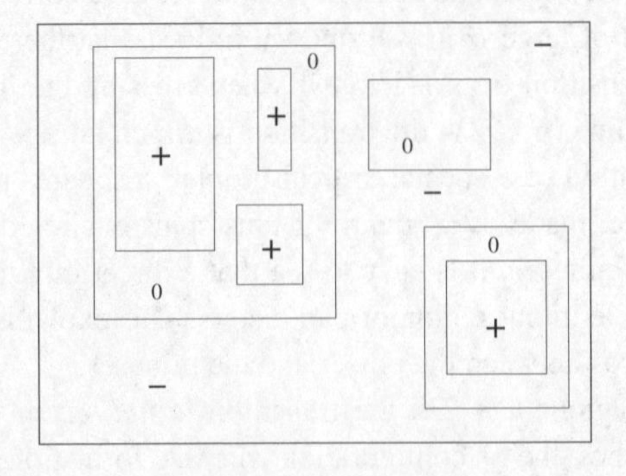

Figure 7.12

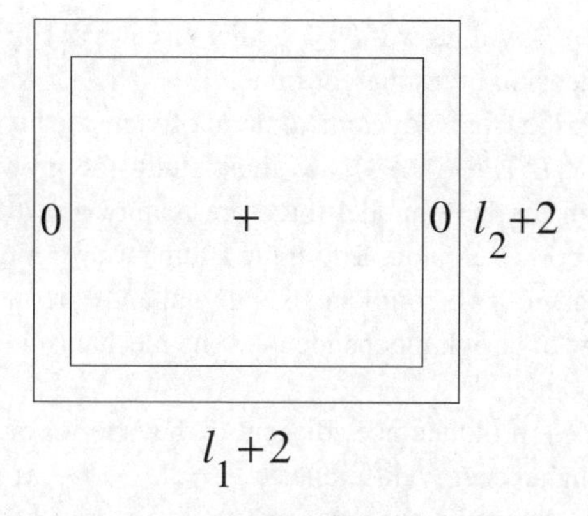

Figure 7.13

zeros with a single rectangle $L^+ \times l^+$ (where $L^+ \geq l^+$) of pluses in the interior (clearly $L^0 \geq L^+ + 2, l^0 \geq l^+ + 2$).

We call a *picture frame* $F(L^+, l^+) := R(L^+ + 2, l^+ + 2, L^+, l^+)$ the set of all configurations consisting of a single rectangle $L^+ \times l^+$ of pluses surrounded by a unitary layer of zeros (see Figure 7.13).

The energy of a birectangle $R(L^0, l^0, L^+, l^+)$ is

$$H(R(L^0, l^0, L^+, l^+)) - H(\underline{-1}) = (2L^0 + 2l^0)J + (2L^+ + 2l^+)J \\ - L^0 l^0 (h - \lambda) - L^+ l^+ (h + \lambda). \tag{7.100}$$

Now we consider a squared birectangle $Q(L, M) := R(L, L; M, M)$, whose energy $E(M, L) := H(Q(L, M)) - H(\underline{-1})$ is given by

$$E(M, L) = 4MJ + 4LJ - M^2(h - \lambda) - L^2(h + \lambda). \tag{7.101}$$

The graph of this function $E : \mathbb{R}^2 \to \mathbb{R}$ is a paraboloid with elliptical section and downhill concavity, the coordinates of the vertex are

$$M = M^* := \frac{2J}{h - \lambda}, \qquad L = L^* := \frac{2J}{h + \lambda}. \tag{7.102}$$

Let us first analyse the tendency to grow or shrink of a birectangle from a naïve, static, point of view.

Consider the energy of a squared frame $E(l) := H(F(l, l)) - H(\underline{-1})$. Using (7.101) we have

$$E(l) = -2hl^2 + l[8J - 4(h - \lambda)] + [8J - 4(h - \lambda)]; \tag{7.103}$$

the graph of this function is a concave parabola, with vertex in

$$l^* = \frac{2J - (h - \lambda)}{h}. \tag{7.104}$$

We expect that $F(l,l)$ will grow if $l \geq l^*$ otherwise it will shrink; hence l^* should be the critical dimension of a squared frame.

In order to describe from a dynamical point of view the behaviour of a general birectangle $R = R(L^0, l^0; L^+, l^+)$, we must study the growth and contraction mechanisms. As in the Ising model these are mainly growth of a (unit square) protuberance and corner erosion. But in the Blume–Capel model the relevant local minima are made of two components, internal and external, and both of them can generally grow or shrink independently. The mechanisms of growth and contraction correspond to:

(1) creation of a $+$ protuberance adjacent to the exterior of the internal rectangle, involving an energy difference $\gamma_+ := 2J - h - \lambda$;
(2) creation of a 0 protuberance adjacent to the exterior of the external rectangle, involving an energy difference $\gamma_0 := 2J - h + \lambda$;
(3) erosion ($+ \to 0$) of all but one $+$ spin in a row or column of length l of the internal rectangle, involving an energy difference $\phi_+(l) := (l-1)(h+\lambda)$;
(4) erosion ($0 \to -$) of all but one 0 spin in a row or column of length l of the external rectangle, involving an energy difference $\phi_0(l) := (l-1)(h-\lambda)$.

This analysis leads to introduction of the critical lengths

$$L^* := \frac{2J}{h+\lambda}, \qquad \tilde{L} := \frac{2J+2\lambda}{h+\lambda}, \qquad M^* := \frac{2J}{h-\lambda}, \qquad \tilde{M} := \frac{2J-2\lambda}{h-\lambda}. \tag{7.105}$$

Indeed the meaning of the critical lengths in (7.105) is the following:

$$\begin{aligned}
\gamma_+ > \phi_+(l^+) &\Longleftrightarrow l^+ < L^*,\\
\gamma_+ > \phi_0(l^0) &\Longleftrightarrow l^0 < \tilde{M},\\
\gamma_0 > \phi_0(l^0) &\Longleftrightarrow l^0 < M^*,\\
\gamma_0 > \phi_+(l^+) &\Longleftrightarrow l^+ < \tilde{L}.
\end{aligned}$$

We say that a cluster of non-minus is subcritical if it has a tendency to shrink, in the sense that its evolved will reach, with high probability for large β, $-\underline{1}$ before $\underline{0}$ or $+\underline{1}$; otherwise, if $\tau_{-\underline{1}}$ is typically larger than $\tau_{\underline{0}}$, $\tau_{+\underline{1}}$ it is called supercritical.

We have the following

- $l^+ < L^* \Leftrightarrow (h+\lambda)(l^+-1) < 2J - (h+\lambda)$, internal contraction is faster than growth, that is the internal component is (relatively) *subcritical*;
- $l^+ < \tilde{L} \Leftrightarrow (h+\lambda)(l^+-1) < 2J - (h-\lambda)$, internal contraction is faster than external growth;
- $l^0 < \tilde{M} \Leftrightarrow (h-\lambda)(l^0-1) < 2J - (h+\lambda)$, external contraction is faster than internal growth;
- $l^0 < M^* \Leftrightarrow (h-\lambda)(l^0-1) < 2J - (h-\lambda)$, external contraction is faster than growth, that is the external component is (relatively) *subcritical*.

The behaviour of our birectangle $R = R(L^0, l^0, L^+, l^+)$ depends on its dimensions, the possible cases are described below:

- $l^+ < L^*$ and $l^0 < M^*$, both internal and external components are subcritical, R is subcritical;
- $l^+ < L^*$ and $l^0 > M^*$, the internal component is subcritical whereas the external one is supercritical; R is supercritical and the system starting from R will typically reach $\underline{0}$ before $+\underline{1}$;
- $l^+ > L^*$ and $l^0 > M^*$, both internal and external components are supercritical; R is supercritical and the system starting from R will reach $+\underline{1}$ by passing through $F(L^0, l^0 - 2)$ (internal growth is faster than external growth since $\lambda > 0$);
- $l^+ > L^*$ and $l^0 < M^*$, the internal component is supercritical while the external one is subcritical, the future evolution of the system starting from R is controlled by a complex growth-and-contraction mechanism that involves both the zero and the plus and, in particular, depends on the relation $l^+ \gtrless \tilde{M}$.

It is reasonable to expect that when $l^+ > L^*$, $l^0 < M^*$ the system is attracted by a suitable picture frame. Thus it is certainly interesting to establish the tendency to grow or shrink of a picture frame. One can analyse the possible mechanisms of growth and contraction of a picture frame in the direction of smaller or larger picture frames; in particular, comparison of the two barriers that we have to overcome starting from a squared picture frame $F(l, l)$, in order to contract or grow, leads to the same critical value l^* that we found on static grounds. We refer to [63, 206] for more details.

It is clear that a nucleation pattern involving a sequence of growing picture frames leads directly from $-\underline{1}$ to $+\underline{1}$, whereas nucleation along a sequence of growing squares of zeros will first lead to $\underline{0}$ and then, after the subsequent formation of a critical squared droplet of pluses there will be the final decay to the ground state $+\underline{1}$.

It is then relevant to compare the energy formation, starting from $-\underline{1}$, of a rectangle $R(0, M^*)$ of zeros with side M^* with that of a picture frame $F(l^*, l^*)$.

A comparison which is interesting for small λ, h reads:

$$\Phi(-\underline{1}, R(0, M^*)) > \Phi(-\underline{1}, F(l^*, l^*)) \quad \Leftrightarrow \quad \frac{4J^2}{h - \lambda} > \frac{8J^2}{h} + \text{const.} \tag{7.106}$$

The above equation leads to

$$\Phi(-\underline{1}, R(0, M^*)) > \Phi(-\underline{1}, F(l^*, l^*)) \quad \Leftrightarrow \quad h < 2\lambda \tag{7.107}$$

so that for $h < 2\lambda$ the direct mechanism of transition via the formation of $F(l^*, l^*)$ is energetically convenient.

A simple computation shows that for $h > 2\lambda > 0$, when, during the transition between $-\underline{1}$ to $+\underline{1}$, the system typically visits $\underline{0}$, one has $\Phi(-\underline{1}, \underline{0}) > \Phi(\underline{0}, +\underline{1})$.

Thus, the tunnelling time between $-\underline{1}$ and $+\underline{1}$ is of the same order as that between $-\underline{1}$ and $\underline{0}$.

The above analysis was restricted to the region

$$\text{I} \quad 0 < \lambda < h.$$

In the region

$$\text{II} \quad 0 < -\lambda < h,$$

we have the same local minima for the energy as in region I namely they are sets of non-interacting plurirectangles; but now the comparison between the energy barriers corresponding to the various growth and contraction mechanisms, namely the quantities $\gamma_+, \gamma_0, \phi_+(l^+), \phi_0(l^0)$, changes totally. The main difference with respect to region I is that now, in II, we have

$$M^* < L^*,$$

and so we cannot even consider a possible mechanism of nucleation along a sequence of picture frames. Indeed one has that a birectangle is supercritical if and only if the minimal external side is not smaller than M^*. Then, as in the region $0 < 2\lambda < h$ but in a much easier way, we can prove that the escape from $-\underline{1}$ starts with the nucleation of an Ising-like protocritical droplet leading to $\underline{0}$. But now, contrary to the region $0 < 2\lambda < h$, the typical time $T^{-\underline{1}\to\underline{0}}$ for going from $-\underline{1}$ to $\underline{0}$ is much shorter than the typical time $T^{\underline{0}\to+\underline{1}}$ for going from $\underline{0}$ to $+\underline{1}$ so that the asymptotics of the time $T^{-\underline{1}\to+\underline{1}}$ of the transition from $-\underline{1}$ to $+\underline{1}$ is dominated by $T^{\underline{0}\to+\underline{1}}$.

We shall see that this feature has an implication for the behaviour of the system at infinite volume.

7.12 Metastability in the two-dimensional Ising model with free boundary conditions

Up to now we have studied the escape from a metastable phase using periodic boundary conditions; in this way we neglected the effects of the boundary.

Such effects can be modelled by considering a finite lattice with free boundary conditions.

The case of the standard (ferromagnetic, isotropic, n.n. interaction) two-dimensional Ising model with free boundary conditions (at low temperatures and small magnetic fields) was considered in [61A].

It was shown in [61A] that, although the main features of nucleation of the stable phase are not changed, some interesting new aspects arise: in particular one can say a priori where the nucleation of the stable phase will start, that is where the critical droplet will show up.

One can show rigorously in the limit $\beta \to \infty$ that the system exhibits metastability with a lifetime $\tau \sim \exp(\beta J^2/h)$ (we are using standard notation introduced in (7.1), namely we set the n.n. coupling constant equal to $J/2$). This exit time is much smaller than in the case of periodic boundary conditions. This is a consequence of the fact that the tendency of a droplet to grow is favoured when such a droplet has one of its sides on the boundary of the box or at a distance one from it.

Indeed, we have to introduce two different critical lengths, λ_1 and λ_2, that refer respectively to droplets close to and far from the boundary:

$$\lambda_1 = \left[\frac{J}{h}\right] + 1 \quad \text{and} \quad \lambda_2 = \left[\frac{2J}{h}\right] + 1, \tag{7.108}$$

The meaning of λ_1 and λ_2 is the following: a rectangular $l_1 \times l_2$ droplet with minimal side length $l = l_1 \wedge l_2$ located inside the bulk of Λ has a tendency to grow precisely if $l \geq \lambda_2$, whereas a rectangular droplet with minimal side length l located at a distance zero or one from the boundary has a tendency to grow precisely if $l \geq \lambda_1$.

The formation of a λ_1 critical nucleus adjacent to the boundary or of a λ_2 critical nucleus in the bulk correspond to different escape mechanisms; among these different mechanisms the one typically chosen at very low temperature by the system will correspond to the minimal energy cost. It is immediate that the minimal energy barrier corresponds to the formation of a λ_1 critical nucleus at a corner.

With an analysis similar to that developed for the standard Ising model with periodic boundary conditions, we see that the global essential saddle configurations $\mathcal{P}$ correspond to a droplet placed at a corner of Λ given by a $\lambda_1 \times \lambda_1 - 1$ rectangle with a unit square protuberance attached to the longest free side. Its energy formation is given by

$$\Gamma = H(\mathcal{P}) - H(-\underline{1}) = 2J\lambda_1 - h(\lambda_1^2 - \lambda_1 + 1). \tag{7.109}$$

With the usual notation the main results are summarized in the following.

Theorem 7.20 *For any* $\epsilon > 0$,

(i) $P_{-\underline{1}}(e^{\beta\Gamma - \beta\epsilon} < \tau_{+\underline{1}} < e^{\beta\Gamma + \beta\epsilon}) \overset{\beta\to\infty}{\longrightarrow} 1$,

(ii) $P_{-\underline{1}}(\theta_{-\underline{1},\mathcal{P},+\underline{1}} < \tau_{+\underline{1}}) \overset{\beta\to\infty}{\longrightarrow} 1$.

It is possible to define the tube $\mathcal{T}_\epsilon$ of all ϵ-standard paths in complete analogy with the case of periodic boundary conditions; now the typical nucleation pattern will involve a sequence of growing droplets whose circumscribed rectangle is a quasi-square placed around a corner of the box Λ. A theorem with the same statement as Theorem 7.13 holds true. We refer to [61A] for more details.

7.13 Standard Ising model under Kawasaki dynamics

As we said at the beginning of Chapter 4, the two typical examples of metastable states are (i) a supersaturated vapour and (ii) a magnetic system with magnetization opposite to the external magnetic field. When describing these systems by lattice Ising-like models, it is natural to use occupation number variables $\eta(x) \in \{0, 1\}$ for the first case and spin variables $\sigma(x) \in \{-1, +1\}$ for the second case. Of course, given any particular model we can interchange the representation via the transformation $\eta(x) = (1 + \sigma(x))/2$ (see Chapter 3). This trivial change of variables allows two different interpretations for the same model. For example, standard stochastic Ising models that we introduced in magnetic language can also be used to model the time evolution of a fluid. The corresponding dynamics is *non-conservative* in the sense that it does not preserve the number of particles. To describe a fluid it is much more natural to use a *conservative* dynamics.

We want now to discuss metastable behaviour for a class of lattice gas models subject to stochastic dynamics that conserve the number of particles (in the spin representation the conserved quantity would be the magnetization).

We want, in particular, to analyse the Kawasaki dynamics, i.e. a family of Markov chains (in continuous or discrete time) that are reversible with respect to the Gibbs measure for the Ising Hamiltonian (7.1), whose elementary process, instead of being a single spin-flip, like in Glauber dynamics, consists of a double spin-flip, interchanging the values of spin of two neighbouring sites. In lattice gas language this corresponds to a jump of a particle to an empty neighbouring site.

Let us now give some preliminary definitions. Let Λ_β be a large finite box centred at the origin with periodic boundary conditions. We consider simultaneously the two-dimensional and three-dimensional cases where $\Lambda_\beta \subseteq \mathbb{Z}^2$, $\Lambda_\beta \subseteq \mathbb{Z}^3$, respectively. The subscript β in Λ_β is to recall that the side length of Λ_β will depend on β. Indeed it will grow at a suitable exponential rate with β.

With each $x \in \Lambda_\beta$ we associate an occupation variable $\eta(x)$, taking the values 0 or 1, indicating absence or presence of a particle. A lattice gas configuration is denoted by $\eta \in \mathcal{X}_\beta = \{0, 1\}^{\Lambda_\beta}$. The equilibrium properties in a conservative context are described by a canonical Gibbs measure (see (3.43) of Chapter 3) that we are going to define for the Ising lattice gas. Let Λ_β^* be the set of bonds (equivalent to unordered pairs of n.n. sites) in Λ_β; we consider the interaction defined by the following canonical Hamiltonian:

$$H^{can}(\eta) = -U \sum_{(x,y) \in \Lambda_\beta^*} \eta(x)\eta(y), \tag{7.110}$$

i.e. there is a binding energy $-U < 0$ between neighbouring occupied sites and no interaction otherwise. For $A \subseteq \Lambda_\beta$, we denote by $N_A(\eta)$ the number of particles in A of the configuration η:

$$N_A(\eta) = \sum_{x \in A} \eta(x). \tag{7.111}$$

We fix the number of particles in Λ_β at the value

$$N = e^{-\Delta\beta}|\Lambda_\beta|, \tag{7.112}$$

where $\Delta > 0$ is an activity parameter. This corresponds to a density equal to

$$\rho \equiv \frac{1}{|\Lambda_\beta|}\sum_{x\in\Lambda_\beta}\eta(x) = e^{-\Delta\beta}. \tag{7.113}$$

As in (7.45) we denote by $\mathcal{V}_N$ the set of configurations with N particles:

$$\mathcal{V}_N = \{\eta \in \mathcal{X}_\beta\colon\ N_{\Lambda_\beta}(\eta) = N\}. \tag{7.114}$$

On $\mathcal{V}_N$ we define the *canonical Gibbs measure*

$$\nu_N(\eta) = \frac{e^{-\beta H^{can}(\eta)}\mathbf{1}_{\mathcal{V}_N}(\eta)}{Z_N^{can}}, \qquad Z_N^{can} = \sum_{\eta\in\mathcal{V}_N} e^{-\beta H^{can}(\eta)}. \tag{7.115}$$

We see from (7.113) and ((7.112)) that in order to have particles at all we must pick $|\Lambda_\beta|$ at least exponentially large in β. This means that the regime where the volume is fixed, considered up to now in this chapter and in particular in the case of non-conservative Glauber dynamics, has no relevance here.

We define a stochastic dynamics, called Kawasaki dynamics, in terms of a continuous time Markov chain $(\eta_t)_{t\geq 0}$ with state space $\mathcal{V}_N$, and with generator L given by:

$$(Lf)(\eta) = \sum_{(x,y)\in\Lambda_\beta^*} c((x,y),\eta)[f(\eta^{(x,y)}) - f(\eta)], \tag{7.116}$$

where

$$\eta^{(x,y)}(z) = \begin{cases} \eta(z) & \text{if } z \neq x, y \\ \eta(x) & \text{if } z = y \\ \eta(y) & \text{if } z = x \end{cases} \tag{7.117}$$

and

$$c((x,y),\eta) = e^{-\beta(H^{can}(\eta^{(x,y)}) - H^{can}(\eta))^+}.$$

It is easily verified that the reversibility condition holds:

$$\nu_N(\eta)c((x,y),\eta) = \nu_N(\eta^{(x,y)})c((x,y),\eta^{(x,y)}).$$

Discrete time versions are also possible.

Let us start with a heuristic discussion based on static grounds.

Consider the lattice gas at low temperature at its condensation point. Let

$$\rho_l(\beta) = \frac{1 + m^*(\beta)}{2}, \quad \rho_g(\beta) = \frac{1 - m^*(\beta)}{2}, \tag{7.118}$$

denote the densities of the liquid and gas phases. Here $m^*(\beta)$ is the spontaneous magnetization in the spin language given by $m^*(\beta) = \lim_{h\to o^+} m(\beta, h)$ (see (3.120) in Chapter 3). Taking the Hamiltonian (7.1) where the ferromagnetic coupling constant is $J/2$ and the magnetic field is $h/2$, we have, in d dimensions,

$$m^*(\beta) = 1 - 2e^{-2dJ\beta}[1 + o(1)] \text{ so that } \rho_g(\beta) = e^{-dU\beta}[1 + o(1)], \quad \beta \to \infty, \tag{7.119}$$

as follows easily from a perturbative argument, based on low temperature expansions, since the energy cost to invert the spin at the origin, with zero magnetic field, is $2dJ$ and $J = U/2$ (see Chapter 3); thus we see that $e^{-dU\beta}$ can be identified as the leading term of the density of the saturated gas at the condensation point (in the sense that $\lim_{\beta\to\infty} -(1/\beta)\log \rho_g(\beta) = dU$).

When describing a pure gaseous phase outside the coexistence region ($\rho < \rho_g(\beta)$) we know that the canonical Gibbs measure is equivalent to the grand canonical Gibbs measure given by

$$\mu(\eta) = \frac{e^{-\beta H(\eta)}}{Z}, \qquad Z = \sum_{\eta\in\mathcal{X}_\beta} e^{-\beta H(\eta)} \tag{7.120}$$

where

$$H(\eta) = H^{can}(\eta) - \lambda N_{\Lambda_\beta}(\eta), \tag{7.121}$$

$\lambda \in \mathbb{R}$ is the chemical potential taking a suitable value, in order to fix the average grand canonical density at the value $\rho = e^{-\Delta\beta}$; at very low temperature we have to choose $\lambda = -\Delta + o(1)$ as follows again by simple perturbative arguments since for large β the gas is very rarefied and for a free gas the density ρ equals the fugacity $z = \exp(\beta\lambda)$. The magnetic field appearing in spin representation (see (7.1)) is $h = 2U + \lambda$ (see Chapter 3); thus we find again that the chemical potential fixing the density ρ at the value $\rho_g(\beta)$ corresponds to $h = 0$.

Suppose that, starting from the condensation point, we increase the density slightly, but avoiding the appearance of droplets of the liquid phase. Then we get a supersaturated gas that can be described in terms of a *restricted ensemble* (see Chapter 4, [44, 240]) namely the grand canonical Gibbs measure restricted to a suitable subset of configuration space, for instance, where all sufficiently large clusters are suppressed. At low temperature this supersaturated gas will stay rarefied, so that its metastable state can be described as a pure gas phase with strong mixing properties.

We can suppose that, for the description of metastability, the canonical and grand canonical Gibbs measures are equivalent, provided they are suitably restricted as described above. Under these conditions let us make, in the limit of large β, a rough calculation of the probability, at the metastable equilibrium, of finding a cubic droplet of occupied sites centred at the origin. In two dimensions let us denote by $\mu^*(l \times l)$, the probability, under the restricted ensemble, of seeing a square droplet of side l centred at the origin; and in three dimensions let us denote by $\mu^*(m \times m \times m)$ the probability, under the restricted ensemble, of seeing a cubic droplet of side m centred at the origin. We have:

$$\mu^*(l \times l) \approx \rho^{l^2} e^{2l(l-1)U\beta}, \tag{7.122}$$

since ρ is the probability of finding a particle at a given site and $-U$ is the binding energy between particles at neighbouring sites. Substituting $\rho = e^{-\Delta\beta}$ we obtain

$$\mu^*(l \times l) \approx e^{-\beta E_2(l)}, \tag{7.123}$$

where

$$E_2(l) = 2Ul - (2U - \Delta)l^2. \tag{7.124}$$

Let $2U > \Delta$. The maximum of $E_2(l)$ is at $l = U/(2U - \Delta)$. This means that if this ratio is non-integer, droplets with side length $l < l_c$ have a probability decreasing in l while droplets with side length $l \geq l_c$ have a probability increasing in l, where

$$l_c = \left[\frac{U}{2U - \Delta}\right] + 1. \tag{7.125}$$

Then, in two dimensions, the choice $\Delta \in (U, 2U)$ corresponds to $l_c \in (1, \infty)$, i.e. to a non-trivial critical droplet size. Use $J = U/2,\ h = 2U + \lambda,\ \lambda = -\Delta$ to see that (7.125) is in agreement with (7.7).

The metastable behaviour for the non-conservative case in spin language occurs when $h \in (0, 2J)$. This corresponds precisely to $\Delta \in (U, 2U)$.

In physical terms, $\Delta \in (0, U)$ corresponds to the unstable gas, $\Delta = U$ to the instability threshold commonly called the *spinodal point*, $\Delta \in (U, 2U)$ to the metastable gas, $\Delta = 2U$ to the condensation point, and $\Delta \in (2U, +\infty)$ to the stable gas.

Similarly, in three dimensions we get:

$$\mu^*(m \times m \times m) \approx \rho^{m^3} e^{\beta[3Um^3 - 3Um^2]}. \tag{7.126}$$

Substituting $\rho = e^{-\Delta\beta}$, we obtain

$$\mu^*(m \times m \times m) \approx e^{-\beta E_3(m)}, \tag{7.127}$$

with

$$E_3(m) = -(3U - \Delta)m^3 + 3Um^2. \tag{7.128}$$

Let $\Delta < 3U$. The maximum of $E_3(m)$ occurs at $m = 2U/(3U - \Delta)$. If this ratio is non-integer, then cubic droplets with side length $m < m_c$ have a probability decreasing in m, while cubic droplets with side length $m \geq m_c$ have a probability increasing in m, where

$$m_c = \left[\frac{2U}{3U - \Delta}\right] + 1. \tag{7.129}$$

The metastable regime in three dimensions is $\Delta \in (U, 3U)$ corresponding to $m_c \in (1, +\infty)$; the analogue in the non-conservative Glauber case is $h \in (0, 4J)$.

7.13.1 Dynamics in two dimensions

Let us now consider the metastable behaviour from a dynamic point of view and see what happens locally. Let us first consider the two-dimensional case. We want to compare the probabilities of growing or shrinking for a square cluster of particles. The argument will be very rough. Suppose we pick a large finite box Λ, centred at the origin, and start with an $l \times l$ droplet inside Λ. Suppose that the effect on Λ of the gas in $\Lambda_\beta \setminus \Lambda$ may be described in terms of the *creation* of new particles with rate $\rho = e^{-\Delta\beta}$ at sites on the interior boundary of Λ and the *annihilation* of particles with rate 1 at sites on the exterior boundary of Λ. So the cost of creating a new particle is Δ whereas the cost of annihilating a particle is zero.

In other words, suppose that inside Λ the Kawasaki dynamics may be described by a Metropolis algorithm with energy given by the *local grand canonical Hamiltonian*:

$$H(\eta) = H^{can}(\eta) + \Delta N_\Lambda(\eta). \tag{7.130}$$

Then the energy barriers for adding respectively removing a row or column of length l are given in terms of the local saddles of H:

$$\text{energy barrier for adding} = 2\Delta - U,$$

which corresponds to a configuration η obtained by creating two particles (cost of 2Δ) one of which is attached to the droplet (gain of $-U$); indeed it is immediate to construct two declining paths (see the beginning of Chapter 6) emerging from η and going respectively to the original rectangular droplet and to the rectangular droplet obtained by adding a row or column.

$$\text{Energy barrier for removing} = (2U - \Delta)(l - 2) + 2U, \tag{7.131}$$

which corresponds to a configuration η' obtained by the consecutive removal and annihilation of $l - 2$ corners from the original droplet at a cost $2U - \Delta$ per step, followed by a final removal without annihilation at a cost $2U$. η' is similar to η: indeed it can be obtained from the contracted rectangle $(l - 1) \times l$ by creating two particles one of which is attached to the rectangle; see Figure 7.14. The balance of the two barriers indeed gives the critical size l_c in (7.125).

7.13.2 Dynamics in three dimensions

Let us next discuss the metastable behaviour from a dynamic point of view in three dimensions. We want to compare the probabilities of growing respectively

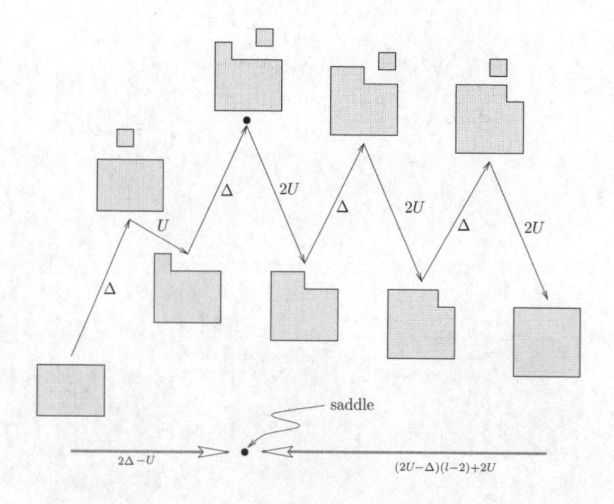

Figure 7.14

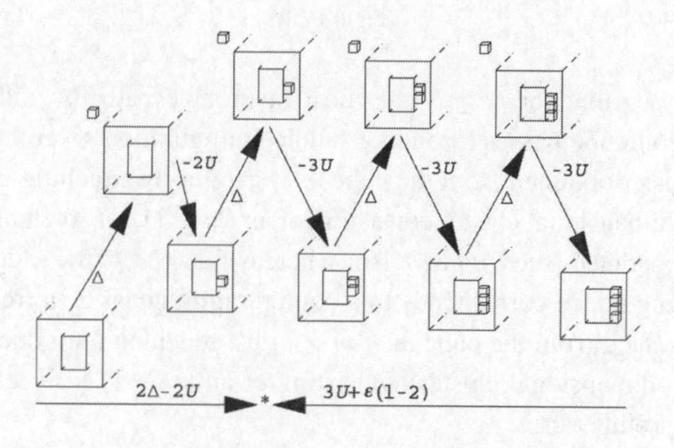

Figure 7.15

shrinking for a cubic droplet of particles with a quadratic droplet attached to one of its faces. Again, the argument will be very rough.

Suppose $\Delta < 2U$. Let

$$\epsilon := 3U - \Delta. \tag{7.132}$$

The energy barriers for adding respectively removing a bar (i.e. row or column) of length l from a two-dimensional droplet on the face of a three-dimensional droplet are given in terms of the local saddles of H (see Figure 7.15):

$$\begin{aligned} \text{adding bar: } & 2\Delta - 2U = 4U - 2\epsilon, \\ \text{removing bar: } & 3U + (3U - \Delta)(l-2) = 3U + \epsilon(l-2). \end{aligned} \tag{7.133}$$

The two barriers in (7.133) balance at $l = U/\epsilon$, which represents the two-dimensional critical droplet size on a face.

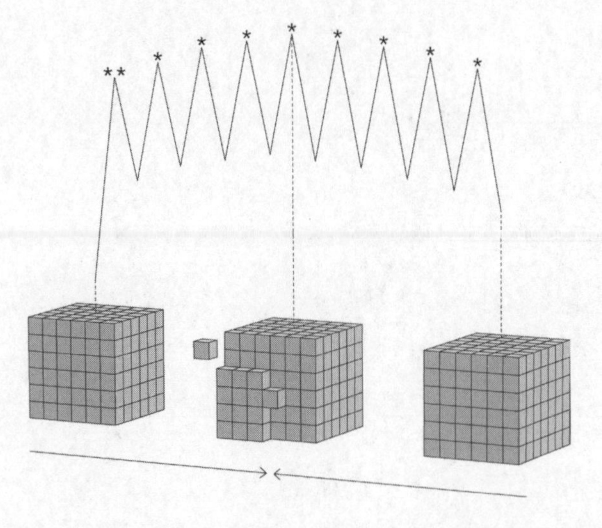

Figure 7.16

Then we see that for $m \geq l_c$ the most favourable path for adding a face to an $m \times m \times m$ cube passes through a saddle configuration given by a free particle and a cluster obtained from the cube $m \times m \times m$ by attaching onto one of its faces a two-dimensional cluster consisting of an $(l_c - 1) \times l_c$ rectangle plus a unit square protuberance (see Figure 7.16); whereas for $m < l_c$ the saddle configuration for adding a face corresponds to a configuration given by a free particle and a cluster obtained from the cube $m \times m \times m$ by attaching onto one of its faces a droplet (two-dimensional cluster) consisting of an $(m - 1) \times m$ rectangle plus a unit square protuberance.

Then, the energy barriers for adding respectively removing a face of side length m are:

$m < l_c$

$$\begin{aligned} \text{adding face: } & U(2m+3) - \epsilon(m^2 - m + 2), \\ \text{removing face: } & 3U + \epsilon(m-2); \end{aligned} \tag{7.134}$$

$m \geq l_c$

$$\begin{aligned} \text{adding face} &= U(2l_c+3) - \epsilon(l_c^2 - l_c + 2), \\ \text{removing face} &= -U(2m - 2l_c - 3) + \epsilon(m^2 - l_c^2 + l_c - 2). \end{aligned} \tag{7.135}$$

The two barriers in (7.135) balance at $m = 2U/\epsilon$, so (7.129) again appears as the three-dimensional critical droplet size.

Let us discuss briefly the main difficulties arising in the attempt to develop the above ideas rigorously and underline the main differences with the non-conservative case.

First of all we can say that conservative dynamics are much more difficult with respect to non-conservative dynamics such as Glauber dynamics since the conservation law induces strong correlations. Moreover, as we already remarked, in the conservative case the Markov chain $(\eta_t)_{t\geq 0}$ is not in the Freidlin–Wentzell regime since the volume Λ_β needs to grow exponentially fast in β. Already for this reason, we cannot apply the theory developed in Chapter 6, so we need new ideas. The real difficulty is to find the correct way to treat the gas in $\Lambda_\beta \setminus \Lambda$. The heuristic discussion given above was based on the assumption that the dynamics inside Λ is effectively described by the local grand canonical Hamiltonian H in (7.130). However, unlike the non-conservative dynamics, the conservative dynamics is not really local: particles must arrive from or return to the gas, which acts as a reservoir. It is therefore not possible to decouple the dynamics of the particles inside Λ from the dynamics of the gas in $\Lambda_\beta \setminus \Lambda$. This means that the gas must be controlled in some detail in order to prove that the above assumption is indeed a good enough approximation.

In [158] the authors considered, for the two-dimensional case, a simplified but still conservative model where the dynamics is Kawasaki in a finite box Λ (independent of β) whereas outside Λ the dynamics is given by a set of independent random walks; we refer to [158] for further details. It turns out that the main conceptual difficulties, concerning the behaviour of the gas, are still present in the simplified model, even though in the latter there are some technical simplifications. From the analysis developed in [158] one can see that the above local description provides a good approximation for the local aspects of the global dynamics in Λ_β. In other words there is a sort of dynamical equivalence between canonical and grand canonical descriptions. Locally the gas behaves like a reservoir producing particles independently in time at rate $e^{-\Delta\beta}$ and absorbing particles at rate 1.

Let us now give, for both $d = 2$ and $d = 3$, a precise definition of the *local Kawasaki dynamics* on Λ, with a boundary condition that mimicks the effect of an infinite gas reservoir outside Λ with density

$$\rho = e^{-\Delta\beta}.$$

Here we only analyse this local version of the dynamics, without providing a justification based on the fact that it is a good approximation of the 'true' dynamics in Λ_β. For this we refer to [158, 159].

Let $\Lambda \subseteq \mathbb{Z}^2$ or $\Lambda \subseteq \mathbb{Z}^3$ be a cubic box, let

$$\partial^-\Lambda = \{x \in \Lambda\colon\ \exists y \notin \Lambda\colon\ |y - x| = 1\},$$

$$\partial^+\Lambda = \{x \notin \Lambda\colon\ \exists y \in \Lambda\colon\ |y - x| = 1\},$$

be the internal respectively the external boundary of Λ, and let $\Lambda_- = \Lambda \setminus \partial^-\Lambda$ be the interior of Λ.

We set $\mathcal{X} := \{0, 1\}^{\Lambda}$. Each configuration $\eta \in \mathcal{X}$ has an energy given by the following Hamiltonian:

$$H(\eta) = -U \sum_{(x,y)\in\Lambda_-^*} \eta(x)\eta(y) + \Delta N_{\Lambda}(\eta), \tag{7.136}$$

where

$$\Lambda_-^* = \{(x, y):\ x, y \in \Lambda_- : |x - y| = 1\} \tag{7.137}$$

is the set of unoriented bonds in Λ_-. The interaction, which is acting only inside Λ_-, is a binding energy $-U < 0$ for each pair of neighbouring particles. In addition, there is an activity energy $\Delta > 0$ for each particle in Λ. Notice that $H - \Delta N_{\partial^-\Lambda}$ can be viewed as the Hamiltonian, in lattice gas variables, for an Ising system enclosed in Λ_-, with 0 boundary conditions. Thus the grand canonical Gibbs measure corresponding to the Hamiltonian (7.136) is given as the product of a grand canonical Gibbs measure describing an Ising system enclosed in Λ_- with 0 boundary conditions, times the equilibrium measure corresponding to a free system with chemical potential $-\Delta$, in $\partial^-\Lambda$.

Let $b = (x \to y)$ denote an oriented bond, i.e. an ordered pair of nearest neighbour sites. Define

$$\begin{aligned} \partial^*\Lambda^{out} &= \{b = (x \to y):\ x \in \partial^-\Lambda, y \in \partial^+\Lambda\}, \\ \partial^*\Lambda^{in} &= \{b = (x \to y):\ x \in \partial^+\Lambda, y \in \partial^-\Lambda\}, \\ \Lambda^{*,\,orie} &= \{b = (x \to y):\ x, y \in \Lambda\}, \end{aligned}$$

and put $\bar{\Lambda}^{*,\,orie} = \partial^*\Lambda^{out} \cup \partial^*\Lambda^{in} \cup \Lambda^{*,\,orie}$. Two configurations $\eta, \eta' \in \mathcal{X}$ with $\eta \neq \eta'$ are now called *communicating states*, written $\eta \leftrightarrow^K \eta'$, if there exists a bond $b \in \bar{\Lambda}^{*,\,orie}$ such that $\eta' = T_b\eta$, where $T_b\eta$ is the configuration obtained from η as follows:

$b = (x \to y) \in \Lambda^{*,\,orie}$

$$T_b\eta(z) = \begin{cases} \eta(z) & \text{if } z \neq x, y \\ \eta(x) & \text{if } z = y \\ \eta(y) & \text{if } z = x, \end{cases}$$

$b = (x \to y) \in \partial^*\Lambda^{out}$

$$T_b\eta(z) = \begin{cases} \eta(z) & \text{if } z \neq x \\ 0 & \text{if } z = x, \end{cases}$$

$b = (x \to y) \in \partial^*\Lambda^{in}$

$$T_b\eta(z) = \begin{cases} \eta(z) & \text{if } z \neq y \\ 1 & \text{if } z = y. \end{cases}$$

Notice that for $b = (x \to y) \in \Lambda^{*,\, orie}$, $T_b\eta$ is invariant under change of orientation of b while for $b \in \partial^*\Lambda^{out}$ and $b \in \partial^*\Lambda^{in}$ it is not.

The local Kawasaki dynamics is defined to be the discrete time Markov chain $(\eta_t)_{t\in\mathbb{N}}$ on $\mathcal{X}$ given by the transition probabilities

$$P^K(\eta, \eta') = \begin{cases} \frac{1}{|\Lambda^{*,\, orie}|} e^{-\beta(H(\eta')-H(\eta))^+} & \text{if } \eta \neq \eta', \eta \leftrightarrow^K \eta' \\ 0 & \text{if } \eta \neq \eta'\eta \not\leftrightarrow^K \eta', \end{cases}$$

and $P^K(\eta, \eta) = 1 - \sum_{\eta'\neq\eta} P^K(\eta, \eta')$. This is a standard Metropolis dynamics with an open boundary: along each bond touching $\partial^-\Lambda$ from the outside, particles are created with rate ρ and annihilated with rate 1, while inside Λ_- particles are conserved. Note that any change of particles inside $\partial^-\Lambda$ does not involve change in binding energy because the interaction acts only inside Λ_-. We want to observe that it is also possible to introduce a Glauber dynamics defined as in (7.2) with (7.136) in place of (7.1), namely with zero interaction inside $\partial^-\Lambda$ and between $\partial^-\Lambda$ and $\Lambda_- \cup \partial^+\Lambda$. This is useful in order to make a comparison with the local Kawasaki dynamics that we have introduced. Let us call it *special Glauber dynamics* as opposed to the *standard Glauber dynamics* with periodic boundary conditions and translationally invariant couplings. On Λ_-, local Kawasaki dynamics exchanges particles between neighbouring sites, while special Glauber dynamics creates or annihilates particles at single sites. Thus, on Λ_-, local Kawasaki dynamics is *conservative*, while special Glauber dynamics (as well as standard Glauber dynamics) is *non-conservative*. It is easy to verify that both are reversible with respect to the grand canonical Gibbs measure with Hamiltonian (7.136).

7.13.3 Results for local Kawasaki dynamics in two dimensions

Remember, from (7.136), that a pair of n.n. particles feel the interaction only if they both belong to Λ_-; thus we analyse in terms of clusters only the part of the configuration η that lies in Λ_-.

Let us adapt the geometric description of configurations in terms of clusters to the present case.

A *free particle* in $\eta \in \mathcal{X}$ is a site $x \in \eta \cap \partial^-\Lambda$ or a site $x \in \eta \cap \Lambda_-$ such that $\sum_{y\in nn(x)\cap\Lambda_-} \eta(y) = 0$, i.e. particle not in interaction with any other particle, where $nn(x)$ is the set of nearest neighbours of x.

Given a configuration $\eta \in \mathcal{X}$, consider the set $C(\eta) \subseteq \mathbb{R}^2$ defined as the union of the closed unit cubes centred at the sites inside Λ_- where η has a particle. Decompose $C(\eta)$ into maximal connected components $C_1, \ldots, C_m$ called clusters. A configuration $\eta \subseteq \Lambda$ is characterized by a set $C(\eta)$, depending only on $\eta \cap \Lambda_-$, plus possibly a set of free particles in $\partial^-\Lambda$, namely, $\eta \cap \partial^-\Lambda$.

$\mathcal{R}(l_1, l_2)$ is the set of configurations whose single cluster in Λ_- is an $l_1 \times l_2$ rectangle (see Definition 7.3 and eq. (7.16)).

Given integers $l_1, l_2 \geq 2$, we define $\mathcal{R}^{pr}(l_1, l_2)$ the set of configurations where the occupied sites form a unique cluster in Λ_- given by an $l_1 \times l_2$ rectangle with a unit square protuberance. Writing, as in (7.45) and in (7.114), for $n \in \mathbb{N}$,

$$\mathcal{V}_n := \left\{ \eta \in \mathcal{X} : \sum_{x \in \Lambda} \eta(x) = n \right\} \tag{7.138}$$

and recalling (see Definition 6.9 in Chapter 6) that for $\mathcal{Q} \subseteq \mathcal{X}, \mathcal{Q} \ni \eta, \eta'$, $\Phi_{\mathcal{Q}}(\eta, \eta')$ denotes the communication height between η and η' in $\mathcal{Q}$, we define

$$\begin{aligned} \mathcal{D}^{pr}(l_1, l_2) = \{\eta' \in \mathcal{X}:\ \exists \eta \in \mathcal{R}^{pr}(l_1, l_2):\ \Phi_{\mathcal{V}_p}(\eta, \eta') \\ \leq H(\eta) + U,\ H(\eta) = H(\eta')\}, \end{aligned} \tag{7.139}$$

with $p = l_1 l_2 + 1$.

In words, $\mathcal{D}^{pr}(l_1, l_2)$ is the set of configurations η' that can be connected to some $\eta \in \mathcal{R}^{pr}(l_1, l_2)$ by a path $\omega = \omega_1, \ldots, \omega_k \subseteq \mathcal{V}_p$ $(k \in \mathbb{N})$ such that

$$\omega_1 = \eta', \quad \omega_n = \eta, \quad \max_{1 \leq i < k} H(\omega_i) \leq H(\eta) + U, \quad H(\eta) = H(\eta'). \tag{7.140}$$

It is not hard to see that $\mathcal{D}^{pr}_{(l_1, l_2)}$ contains only configurations giving rise to a single monotone contour.

$\mathcal{D}^{pr,fp}(l_1, l_2)$ is the set of configurations obtained from a configuration in $\mathcal{D}^{pr}(l_1, l_2)$ by adding a free particle.

A particularly important set of configurations, which play the role of 'critical configurations', is given by

$$\mathcal{C}_2^* = \mathcal{D}^{pr,fp}(l_c - 1, l_c) \tag{7.141}$$

with

$$l_c = \left[\frac{U}{2U - \Delta} \right] + 1. \tag{7.142}$$

Recall that in the simple static and dynamic analysis developed above, this value came out as the critical droplet size. We denote by $\Gamma_2^K = \Gamma_2^K(U, \Delta)$ the energy of the critical configuration:

$$\Gamma_2^K = H(\mathcal{C}_2^*) = -U(2l_c^2 - 4l_c + 2) + \Delta(l_c^2 - l_c + 2). \tag{7.143}$$

We denote by $\underline{0}$ the configuration where Λ is empty and by $\underline{1}$ the set of configurations where Λ_- is full. Notice that $H(\underline{0}) = 0$.

The following theorem contains the main results on the asymptotic of tunnelling time, on the 'gate to stability' and on subcriticality or supercriticality of quasi-squares.

Theorem 7.21 [158] *Let $\Delta \in (\frac{3}{2}U, 2U)$ and suppose that $U/(2U - \Delta)$ is not integer, and let Λ be sufficiently large.*

(*a*) *Let* Γ_2^K *be the quantity defined in* (7.143). *Then*

$$\lim_{\beta\to\infty} P_{\underline{0}}\Big(e^{(\Gamma_2^K-\delta)\beta} < \tau_{\underline{1}} < e^{(\Gamma_2^K+\delta)\beta}\Big) = 1 \qquad \forall\delta > 0. \tag{7.144}$$

(*b*) *Let* $\mathcal{C}_2^*$ *be the set of critical configurations defined in* (7.141) *and*

$$\theta_{\underline{0},\underline{1}} = \max\{0 \le t < \tau_{\underline{1}}\colon\ \eta_t = \underline{0}\}, \quad \theta_{\underline{0},\mathcal{C}_2^*,\underline{1}} = \min\{t > \theta_{\underline{0},\underline{1}}\colon\ \eta_t \in \mathcal{C}_2^*\}. \tag{7.145}$$

Then

$$\lim_{\beta\to\infty} P_{\underline{0}}(\theta_{\underline{0},\mathcal{C}_2^*,\underline{1}} < \tau_{\underline{1}}) = 1. \tag{7.146}$$

(*c*) *Let* $\mathcal{Q}$ *be the set of configurations whose single contour is a quasi-square. Then, for all* $\eta \in \mathcal{Q}$ *if* l_1, l_2 *are the side lengths of the quasi-square in* η *and* $l = l(\eta) = l_1 \wedge l_2$,

$$\begin{aligned} &\text{if} \quad l < l_c \quad \text{then} \quad \lim_{\beta\to\infty} P_\eta(\tau_{\underline{0}} < \tau_{\underline{1}}) = 1,\\ &\text{if} \quad l \ge l_c \quad \text{then} \quad \lim_{\beta\to\infty} P_\eta(\tau_{\underline{1}} < \tau_{\underline{0}}) = 1. \end{aligned} \tag{7.147}$$

In words, Theorem 7.21 says the following.

(a) The nucleation time from $\underline{0}$ to $\underline{1}$ is $e^{[\Gamma_2^K+o(1)]\beta}$.

(b) The set $\mathcal{C}_2^*$ is a gate for nucleation: during the first excursion from the metastable configuration $\underline{0}$ to $\underline{1}$, all paths pass through this set with a probability tending to 1 as $\beta \to \infty$.

(c) Subcritical quasi-squares ($l < l_c$) shrink to $\underline{0}$, supercritical quasi-squares ($l \ge l_c$) grow to $\underline{1}$.

7.13.4 Results for local Kawasaki dynamics in three dimensions

Again we start with some geometrical definitions

Given integers $m_1, m_2, m_3 \ge 2$ and $l_1, l_2 \ge 1$, with $l_1 \le l_2$, $m_3 \ge l_2$, $m_2 \ge l_1$ we denote by $\mathcal{R}_{l_1,l_2}(m_1, m_2, m_3)$ the set of configurations without free particles whose single cluster in Λ_- is an $m_1 \times m_2 \times m_3$ parallelepiped plus an $l_1 \times l_2$ rectangle attached to one face (a rectangle is a parallelepiped with side lengths $1, l_1, l_2$). A *quasi-cube* is a parallelepiped with side lengths $m, m+\delta, m+\theta$ with $\delta, \theta \in \{0, 1\}$, $\delta \le \theta$. A cube is a quasi-cube with $\delta = \theta = 0$. A *quasi-square* is a parallelepiped with side lengths $1, l, l+\alpha$ with $\alpha \in \{0, 1\}$. A square is a quasi-square with $\alpha = 0$.

Given integers $m_1, m_2, m_3 \ge 3$ and $l_1, l_2 \ge 2$, $\mathcal{R}_{l_1,l_2}^{fp}(m_1, m_2, m_3)$ is the set of configurations obtained from a configuration in $\mathcal{R}_{l_1,l_2}(m_1, m_2, m_3)$ by adding a free particle; $\mathcal{R}_{l_1,l_2}^{pr}(m_1, m_2, m_3)$ is the set of configurations obtained from a configuration in $\mathcal{R}_{l_1,l_2}(m_1, m_2, m_3)$ by attaching a particle to one of the sides of the rectangle, so that it becomes a protuberance. $\mathcal{D}^{pr}_{l_1,l_2}(m_1, m_2, m_3)$ is the set of configurations given by

$$\mathcal{D}^{pr}_{l_1,l_2}(m_1, m_2, m_3) = \Big\{\eta' \in \mathcal{V}_p\colon\ \exists \eta \in \mathcal{R}^{pr}_{l_1,l_2}(m_1, m_2, m_3)\colon \\ \Phi_{\mathcal{V}_p}(\eta, \eta') \le H(\eta) + 2U,\ H(\eta) = H(\eta')\Big\} \tag{7.148}$$

with $p = m_1m_2m_3 + l_1l_2 + 1$. In words, $\mathcal{D}^{pr}_{l_1,l_2}(m_1, m_2, m_3)$ is the set of configurations η' that can be reached from some $\eta \in \mathcal{R}^{pr}_{l_1,l_2}(m_1, m_2, m_3)$ by a path $\omega = \omega_1, \ldots, \omega_k$ ($k \in \mathbb{N}$) in $\mathcal{V}_p$ such that

$$\omega_1 = \eta, \quad \omega_k = \eta', \quad \max_{1\le i<k} H(\omega_i) \le H(\eta) + 2U, \quad H(\eta) = H(\eta'). \tag{7.149}$$

Let $\mathcal{D}^{pr,fp}_{l_1,l_2}(m_1, m_2, m_3)$ be the set of configurations obtained from a configuration in $\mathcal{D}^{pr}_{l_1,l_2}(m_1, m_2, m_3)$ by adding a free particle.

A particularly important set of configurations, which plays the role of 'critical configurations' in the regime $\Delta \in (2U, 3U)$, is given by

$$\mathcal{C}^*_3 = \mathcal{D}^{pr,fp}_{l_c-1,l_c}(m_c - 1, m_c - \delta_c, m_c) \tag{7.150}$$

with l_c defined in (7.125), m_c defined in (7.129), and

$$\delta_c = 1 \ \text{ if } \epsilon\left(2\left[\frac{2U}{\epsilon}\right]\right) > 2U + \sqrt{(2U)^2 + \epsilon^2}, \quad 0 \text{ otherwise} \tag{7.151}$$

(recall $\epsilon = 3U - \Delta$). The δ_c comes from a fine tuning: depending on the value of U and Δ, either the prolate quasi-cube $(m_c - 1, m_c - 1, m_c)$ or the oblate quasi-cube $(m_c - 1, m_c, m_c)$ has the lowest energy. By definition, $\mathcal{C}^*_3 \subseteq \mathcal{V}_{n^*}$ with

$$n^* = m_c(m_c - \delta_c)(m_c - 1) + l_c(l_c - 1) + 2. \tag{7.152}$$

The 'prototype' configurations in $\mathcal{C}^*_3$ are $\mathcal{R}^{pr,fp}_{l_c-1,l_c}(m_c - 1, m_c - \delta_c, m_c)$ represented in Figure 7.17. We denote by $\Gamma^K_3 = \Gamma^K_3(U, \Delta)$ the energy of the critical configurations:

$$\begin{aligned}\Gamma^K_3 = H(\mathcal{C}^*_3) = &- \epsilon[m_c(m_c - \delta_c)(m_c - 1) + l_c(l_c - 1) + 2] \\ &+ U[m_c(m_c - \delta_c) + m_c(m_c - 1) \\ &+ (m_c - \delta_c)(m_c - 1) + 2l_c + 3].\end{aligned} \tag{7.153}$$

Now we can state the three-dimensional analogue of points (a), (b) of Theorem 7.21.

Theorem 7.22 [160] *Fix $\Delta \in (2U, 3U)$ such that $2U/(3U - \Delta)$ is not integer, and let Λ be sufficiently large.*

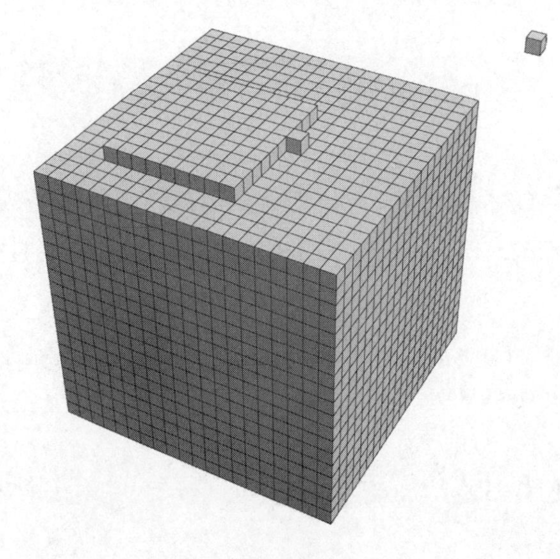

Figure 7.17

(*a*) *Let* Γ_3^K *be the quantity defined in* (7.153). *Then*

$$\lim_{\beta\to\infty} P_{\underline{0}}\Big(e^{(\Gamma_3^K-\delta)\beta} < \tau_{\underline{1}} < e^{(\Gamma_3^K+\delta)\beta}\Big) = 1, \qquad \forall\delta > 0. \tag{7.154}$$

(*b*) *Let* $\mathcal{C}_3^*$ *be the set of critical configurations defined in* (7.150) *and*

$$\theta_{\underline{0},\underline{1}} = \max\{0 \le t < \tau_{\underline{1}}:\ \eta_t = \underline{0}\}, \qquad \theta_{\underline{0},\mathcal{C}_3^*,\underline{1}} = \min\{t > \theta_{\underline{0},\underline{1}}:\ \eta_t \in \mathcal{C}_3^*\}. \tag{7.155}$$

Then

$$\lim_{\beta\to\infty} P_{\underline{0}}(\theta_{\underline{0},\mathcal{C}_3^*,\underline{1}} < \tau_{\underline{1}}) = 1. \tag{7.156}$$

We refer to [160] for a more complete statement including the analogue of point (c) of Theorem 7.21.

Results for Glauber dynamics in three dimensions

Let us now give the analogous result for the three-dimensional standard Ising model with Glauber dynamics. It is defined in perfect analogy with the two-dimensional standard Ising model (see (7.1), (7.2)) Now Λ is a three-dimensional torus of side length L. We recall that the correspondence rule between the lattice parameters and spin parameters is $U = 2J$, $\epsilon \equiv 3U - \Delta = h$. Let

$$l^* := \left[\frac{2J}{h}\right] + 1, \qquad m^* := \left[\frac{4J}{h}\right] + 1 \tag{7.157}$$

and ζ_c be defined like δ_c in (7.151) with $2J$ in place of U and h in place of ϵ.

Let

$$\mathcal{P}_3^G = \mathcal{R}^{pr}_{l^*,l^*-1}(m^* - 1, m^* - \zeta_c, m^*) \tag{7.158}$$

and

$$\begin{aligned}\Gamma_3^G + H(-\underline{1}) = H(\mathcal{P}_3^G) = &-h[m^*(m^* - \zeta_c)(m^* - 1) + l^*(l^* - 1) + 1] \\ &+ 2J[m^*(m^* - \zeta_c) + m^*(m^* - 1) \\ &+ (m^* - \zeta_c)(m^* - 1) + 2l^*].\end{aligned} \tag{7.159}$$

Theorem 7.23 [16] *Given J, for $1/h$ and L sufficiently large and such that $2J/h$, $4J/h$ are not integers,*

(*a*) *we have*

$$\lim_{\beta\to\infty} P_{-\underline{1}}\Big(e^{(\Gamma_3^G-\delta)\beta} < \tau_{+\underline{1}} < e^{(\Gamma_3^G+\delta)\beta}\Big) = 1, \qquad \forall\delta > 0; \tag{7.160}$$

(*b*) *let*

$$\begin{aligned}\theta_{-\underline{1},+\underline{1}} &= \max\{0 \le t < \tau_{+\underline{1}}:\ \eta_t \in -\underline{1}\}, \\ \theta_{-\underline{1},\mathcal{P}_3^G,+\underline{1}} &= \min\{t > \theta_{-\underline{1},+\underline{1}}:\ \eta_t \in \mathcal{P}_3^G\},\end{aligned} \tag{7.161}$$

then

$$\lim_{\beta\to\infty} P_{-\underline{1}}(\theta_{-\underline{1},\mathcal{P}_3^G,+\underline{1}} < \tau_{+\underline{1}}) = 1. \tag{7.162}$$

Proof of Theorem 7.22 (Sketch) We give the main lines of the proof, referring to [16, 160] for more details. Of course the proof of points (a), (b) of Theorem 7.21 is much easier and can be obtained along the same lines; point (c) of Theorem 7.21 comes as a byproduct of the proof of points (a), (b); we refer to [160] for more details. Moreover, as will appear clear, we can also use the argument of the proof of Theorem 7.22 to obtain the equivalent result (Theorem 7.23) for the three-dimensional Ising model under Glauber dynamics. Indeed the Glauber case, as far as the results like Theorem 7.22 are concerned, is much easier.

We remark that much more detailed results, including the tube of typical trajectories realizing the transition, were obtained in [16].

Let us observe that $\Gamma_3^K \sim 4U^3/\epsilon^2$ as $\epsilon \to 0$ (as follows from (7.153)) which identifies the nucleation time in the limit of weak supersaturation.

We note at this point that common methods used in several works on metastability like [16, 62, 63, 184, 185, 224] are not applicable in the present case of three-dimensional local Kawasaki dynamics. Indeed these methods needed the introduction of a set G like the one appearing in Claim 7.12 for the case of the two-dimensional standard Ising model with Glauber dynamics, together with good control of the typical tube of trajectories realizing the transition. Here we do not have enough control of the energy landscape to apply these methods, in particular we have a too approximate knowledge of the gate for the transition: we are able to prove that $\mathcal{C}_3^*$ is a gate but we are not able to extract from $\mathcal{C}_3^*$ a minimal gate. Nor are we able to characterize the typical descent paths from $\mathcal{C}_3^*$ to $\underline{1}$, as in the case of the standard two dimensional Ising–Glauber model. These are the reasons why we base the proof of part (a) of the theorem on the application of Theorem 6.38 of Chapter 6. As in the alternative approach to the study of the $J_1 J_2$ model that we presented above, it is sufficient (1) to find the communication height $\Phi(\underline{0}, \underline{1})$ and (2) to prove that there exists $\Gamma_0 \leq \Gamma_3^K$ such that all configurations with the exception of $\{\underline{0}, \underline{1}\}$ are Γ_0-reducible, i.e. $\mathcal{X}_{\Gamma_0} \subseteq \{\underline{0}, \underline{1}\}$. We want to stress that these two items are not necessary to get the result: they are a *sufficient* relatively weak condition implying relatively weak results.

Proof that $\Phi(\underline{0}, \underline{1}) = \Gamma_3^K$ The basic idea is to partition the configuration space into manifolds with constant number of particles:

$$\mathcal{X} = \bigcup_{n=0}^{|\Lambda|} \mathcal{V}_n \tag{7.163}$$

(see (7.138)) and to observe that, due to the form of the Ising Hamiltonian, the problem of determining the configurations of minimal energy in $\mathcal{V}_n$ coincides with the discrete 'isoperimetric problem' of minimizing the surface of a polyhedron with given volume. We recall some definitions and results from Alonso and Cerf [3].

(a) A *minimal polyomino* is a configuration whose single contour has minimal surface among all those with the same volume.
(b) A *principal polyomino* is a configuration whose single contour is a quasi-cube with a quasi-square attached to one face of the quasi-cube and with a bar attached to one side of the quasi-square (a bar is a parallelepiped with side lengths $1, 1, l_1$).
(c) A *standard polyomino* is a principal polyomino whose quasi-square is attached to one of the largest faces of the quasi-cube and whose bar is attached to one of the largest sides of the quasi-square.

Here the polyomino is the natural extension to the three-dimensional case of the analogous concept that we have already introduced in two dimensions.

In the following we report some results that are proven in [3].

For each $n \in \mathbb{N}$ there exists a unique 6-tuple $(m, l, k, \delta, \theta, \alpha)$ such that:

(i) $m, l, k \in \mathbb{Z}_+, \delta, \theta, \alpha \in \{0, 1\}$,
(ii) if $m = 0$ then $\delta = \theta = 0$, if $l = 0$ then $\alpha = k = 0$,
(iii) $\delta \leq \theta, k < l + \alpha, l(l + \alpha) + k < (m + \delta)(m + \theta)$, and

$$n = m(m + \delta)(m + \theta) + l(l + \alpha) + k. \tag{7.164}$$

Because of the previous result,it is natural to associate with each $n \in \mathbb{N}$ a principal polyomino whose quasi-cube has side lengths $m, m+\delta, m+\theta$, whose quasi-square has side lengths $l, l+\alpha$, and whose bar has length k.

The following discrete isoperimetric inequalities, proven in [3], are a key ingredient in our analysis.

(a) All principal polyominoes are minimal polyominoes.

(b) The set of minimal polyominoes of volume n coincides with the set of principal polyominoes of volume n if and only if n is of the form 'quasi-cube + quasi-square' or 'quasi-cube -1'.

We next construct a particular optimal nucleation path ω^K, i.e. an element of $(\underline{0} \longrightarrow \underline{1})_{opt}$. This path, which we call the *reference path for the Kawasaki dynamics*, goes from $\underline{0}$ to $\underline{1}$ through a particular sequence of growing standard polyominoes.

To define ω^K, we first define the analogous *reference path for the special Glauber dynamics*, which we denote by $\omega^G = \{\omega_n^G\}$ with $n = 0, \ldots, |\Lambda_-|$, namely,

$$\omega_0^G = \underline{0}, \omega_1^G = \{x_0\}, \ldots, \omega_{|\Lambda_-|}^G \in \underline{1}, \tag{7.165}$$

where x_0 is any site in Λ_- and $\{\omega_n^G\}$ is a growing sequence of standard polyominoes with $|\omega_n^G| = n$, such that $P^G(\omega_n^G, \omega_{n+1}^G) > 0$ and $\omega_n^G \subseteq \Lambda_-$ for any n. The parallelepiped circumscribing ω_n^G is always a quasi-cube, while the rectangle circumscribing the piece attached to a face of the quasi-cube contained in ω_n^G is always a quasi-square. The order of the directions of growth of the standard polyominoes may be chosen arbitrarily, depending on the starting point x_0, but is fixed.

Given a choice for ω^G, we can construct the path $\omega^K = \{\omega_{n,i}^K\}$ with $n = 0, \ldots, |\Lambda_-| - 1$ and $i = 0, 1, \ldots, i_n$ as follows:

$$\omega_{n,0}^K = \omega_n^G, \qquad \forall n, \tag{7.166}$$

and insert between each pair $(\omega_n^G, \omega_{n+1}^G)$ a sequence of configurations $\omega_{n,i}^K$, $i = 0, 1, \ldots, i_n$, belonging to $(\omega_n^G)^{fp}$ ($\equiv$ the set of configurations (ω_n^G) + a free particle), that creates a free particle at $\partial^- \Lambda$ and brings it to the droplet, i.e.

$$\omega_{n,i}^K = \omega_n^G \cup \{x_i^{(n)}\} \tag{7.167}$$

with $x_1^{(n)}, \ldots, x_{i_n}^{(n)}$ a sequence of neighbouring sites from $\partial^- \Lambda$ to $\omega_{n+1}^G \backslash \omega_n^G$. We obviously have $|\omega_{n,i}^K| = n + 1 - \delta_{i,0}$, where $\delta_{i,0}$ is the Kronecker symbol. For brevity we renumerate the reference path $\omega_{n,i}^K$ with a single index s, $s = s(n,i)$, i.e. $\omega_{n,i}^K = \omega_{s(n,i)}^K$.

Claim $\omega^K \in (\underline{0} \longrightarrow \underline{1})_{opt}$. *Moreover* ω^K *is a self-minimax path, that is for any* $0 \le s_1 < s_2$

$$\Phi(\omega^K_{s_1}, \omega^K_{s_2}) = \max_{s \in [s_1, s_2]} H(\omega^K_s) \tag{7.168}$$

and, for any $0 \le n_1 < n_2 \le |\Lambda_-|$

$$\Phi(\mathcal{V}_{n_1}, \mathcal{V}_{n_2}) = \max_{s \in [s(n_1,0), s(n_2,0)]} H(\omega^K_s). \tag{7.169}$$

Proof By definition, we have

$$\Phi(\underline{0}, \underline{1}) \le \max_s H(\omega^K_s) = H(\omega^K_{s_0}), \tag{7.170}$$

where $s_0 = \min\{s\colon H(\omega^K_s) \ge H(\omega^K_{s'})\ \forall s'\}$. Since $\omega^K_{s_0}$ is either a standard polyomino or a standard polyomino plus a free particle, we have $H(\omega^K_{s_0}) = H(\omega^K_{s_0 - 1}) + \Delta$.

Let $n_0 = |\omega^K_{s_0}|$. Since any path $\omega\colon \underline{0} \to \underline{1}$ has to cross the n_0-manifold $\mathcal{V}_{n_0}$ for a first time, we have

$$\Phi(\underline{0}, \underline{1}) \ge \min_{\eta \in \mathcal{V}_{n_0 - 1}} H(\eta) + \Delta = H(\omega^K_{s_0 - 1}) + \Delta = H(\omega^K_{s_0}). \tag{7.171}$$

Indeed, the first equality follows from the fact that $\omega^K_{s_0-1}$ is a standard polyomino, so that by the isoperimetic inequalities proven in [3], it is a configuration of minimal energy in $\mathcal{V}_{n_0-1}$. Combining (7.170) and (7.171), we get $\Phi(\underline{0}, \underline{1}) = H(\omega^K_{s_0})$, proving that $\omega^K \in (\underline{0} \to \underline{1})_{opt}$.

With the same argument we prove (7.168) and (7.169).

From the above claim by an easy calculation determining the maximal energy reached by ω^K, also via a comparison between the energy of the oblate quasi-cube $m_c - 1, m_c, m_c$ and the prolate quasi-cube $m_c - 1, m_c - 1, m_c$ we get that $\Phi(\underline{0}, \underline{1}) = \Gamma^K_3$. □

The following proposition implies, by virtue of Proposition 6.37 of Chapter 6, that from any configuration in $\mathcal{X} \setminus (\{\underline{0}\} \cup \underline{1})$ the Kawasaki dynamics hits $\underline{0}$ or $\underline{1}$ with an overwhelming probability in a time much shorter than the nucleation time.

Proposition 7.24 *There exists* $\Gamma_0 < \Gamma^K_3$ *such that* $\mathcal{X}_{\Gamma_0} \subseteq \{\underline{0}\} \cup \underline{1}$.

We show an equivalent statement: there exists $\Gamma_0 < \Gamma^K_3$ so that

$$\forall \eta \notin \{\underline{0}\} \cup \underline{1}\ \exists \eta' \in \mathcal{X}\colon \quad H(\eta') < H(\eta),\ \Phi(\eta, \eta') \le H(\eta) + \Gamma_0. \tag{7.172}$$

Suppose that $\eta \notin \{\underline{0}\} \cup \underline{1}$. Then

$$\exists\, x_0 \in \Lambda_-,\ y_0 \in \Lambda,\ |x_0 - y_0| = 1\colon \quad \eta(x_0) = 0,\ \eta(y_0) = 1. \tag{7.173}$$

If $y_0 \in \partial^- \Lambda$ and $\Lambda_- \cap \eta = \emptyset$, then we pick $\eta' = \underline{0}$ and a path $\omega\colon \eta \to \underline{0}$, decreasing in energy, consisting of a subsequent annihilation of the particles in $\eta \cap \partial^- \Lambda$. We may therefore suppose that there is at least one particle in Λ_-, i.e. $\eta \cap \Lambda_- \neq \emptyset$.

We first give the proof pretending that the dynamics is Glauber, i.e. particles can be created and annihilated everywhere in Λ. Afterwards we discuss briefly how the proof can be modified when the dynamics is Kawasaki.

Glauber (1) Let ω^G be the reference path for the special Glauber dynamics. We have $|\omega_i^G| = i$ and, by the already quoted results of [3], it follows that

$$\omega_i^G \in F(\mathcal{V}_{|\omega_i^G|}), \qquad \forall i. \tag{7.174}$$

Define

$$\omega_i = \omega_i^G \cup \eta \quad \forall i, \qquad p = \inf\{i \geq 1\colon\ H(\omega_i^G) \leq 0\}. \tag{7.175}$$

We have $\omega_0^G = \underline{0}$, and without loss of generality we may pick $\omega_1^G = \{x_0\}$ and start growing from there.

(2) For $1 \leq i \leq p$, write (recall (7.136))

$$\begin{aligned} H(\omega_i) - H(\eta) = &\left[H_{\Lambda_i}(\omega_i) + H_{\Lambda_i^c}(\omega_i) + W_{\Lambda_i,\Lambda_i^c}(\omega_i) \right] \\ &- \left[H_{\Lambda_i}(\eta) + H_{\Lambda_i^c}(\eta) + W_{\Lambda_i,\Lambda_i^c}(\eta) \right], \end{aligned} \tag{7.176}$$

where Λ_i is the support of ω_i^G, i.e. $\Lambda_i = \{x \in \Lambda\colon\ \omega_i^G(x) = 1\}$, $\Lambda_i^c = \Lambda \setminus \Lambda_i$, H_{Λ_i} is the Hamiltonian in (7.136) restricted to Λ_i, and

$$W_{\Lambda_i,\Lambda_i^c}(\eta) = -U \sum_{\substack{x \in \Lambda_i,\, y \in \Lambda_i^c \\ (x,y) \in \Lambda_-^*}} \eta(x)\eta(y). \tag{7.177}$$

Recall that $\Lambda_i \cap \partial^- \Lambda = \emptyset$.

(3) We have

$$\begin{aligned} H_{\Lambda_i}(\omega_i) &= H_{\Lambda_i}(\omega_i^G) = H(\omega_i^G), \\ H_{\Lambda_i^c}(\omega_i) &= H_{\Lambda_i^c}(\eta), \\ W_{\Lambda_i,\Lambda_i^c}(\omega_i) &\leq W_{\Lambda_i,\Lambda_i^c}(\eta), \end{aligned} \tag{7.178}$$

which hold, respectively, because ω_i and ω_i^G coincide on Λ_i, and because ω_i and η coincide on Λ_i^c, and because ω_i contains η. Substitution of (7.178) into (7.176) gives

$$H(\omega_i) - H(\eta) \leq H(\omega_i^G) - H_{\Lambda_i}(\eta). \tag{7.179}$$

Moreover,

$$H_{\Lambda_i}(\eta) > 0, \qquad \forall 1 \le i \le p. \tag{7.180}$$

Indeed, by (7.174) we have

$$H_{\Lambda_i}(\eta) \ge \min_{\xi \in \mathcal{V}_j} H_{\Lambda_i}(\xi) = H(\omega_j^G), \qquad \text{for } j = j(i) = |\eta \cap \Lambda_i|. \tag{7.181}$$

But $0 \le j < p$ for $1 \le i \le p$, since in $\eta \cap \Lambda_i$ there is at least the empty site x_0. Since p is the smallest integer k with $H(\omega_k^G) \le 0$, it follows that $H(\omega_j^G) > 0$. Thus, via (7.179) and (7.181),

$$H(\omega_i) - H(\eta) \le H(\omega_i^G) - H(\omega_j^G) < H(\omega_i^G), \qquad \forall 1 \le i \le p. \tag{7.182}$$

Denoting by Φ^G the communication height for Glauber dynamics, we see that (7.182) in turn yields

$$\Phi^G(\eta, \omega_p) - H(\eta) \le \max_{1 \le i \le p} H(\omega_i) - H(\eta) < \max_{1 \le i \le p} H(\omega_i^G) = \Gamma_3^G, \tag{7.183}$$

where we recall Γ_3^G is the analogue of Γ_3^K for Glauber. Since $\omega_0 = \eta$ and $H(\omega_p^G) \le 0$, (7.182) gives

$$H(\omega_p) < H(\eta). \tag{7.184}$$

This proves (7.75) with $\eta' = \omega_p$ and $\Gamma_0 < \Gamma_3^G$.

Notice that an almost identical argument would be valid also for the standard three-dimensional Glauber dynamics with the same value Γ_3^G for $\Phi(-\underline{1}, +\underline{1})$

Kawasaki We have to see how to modify the above argument when the dynamics is Kawasaki. The additional obstacle under Kawasaki is that, when we are growing the configuration by considering the union of η with the standard polyominoes $\Lambda_i = \text{supp}(\omega_i^G)$, as in (7.175), particles cannot be created arbitrarily but have to arrive from $\partial^-\Lambda$. We have to make sure that at any time the configuration is such that a particle coming from the boundary can be moved to where it is needed.

To prove (7.75) we may restrict ourselves to $\eta \in \mathcal{X}_{3U} = \{\eta \in \mathcal{X} : \gamma(\eta) > 3U\}$, with $\gamma(\eta) = \Phi(\eta, \{\eta' : H(\eta') < H(\eta)\})$ (see (6.227), (6.228) and (6.234) of Chapter 6).

Otherwise, since $3U < \Gamma_3^K$, we can take $\Gamma_0 = 3U$ to get (7.172). Thus, we need to prove that any $\eta \in \mathcal{X}_{3U}$ has the property that when the union is taken with any monotone sequence of principal polyominoes, this union never contains closed off regions. This property is proved in [160] by exploiting the regularity of configurations in $\mathcal{X}_{3U}$. We refer to [160] for more details: by (7.172) and the equality $\Phi(\underline{0}, \underline{1}) = \Gamma_3^K$, using Theorem 6.38, we conclude the (sketch of the) proof of Theorem 7.22. □

7.14 Metastability for reversible probabilistic cellular automata

We want to analyse now the metastable behaviour of a class of probabilistic cellular automata (PCA). These are families of Markov chains σ_t, $t \in \mathbb{N}$, on the space $\mathcal{X} = \{-1, +1\}^\Lambda$ (Λ is a torus of side L in $\mathbb{Z}^2$), with transition probabilities of the form

$$P(\sigma, \eta) := \prod_{x \in \Lambda} p_x(\eta(x)|\sigma), \qquad s, \eta \in \mathcal{X} \tag{7.185}$$

where p_x are non-negative and normalized in the sense that $p_x(\eta(x)|\sigma) + p_x(-\eta(x)|\sigma) = 1$. In words, at each integer time, simultaneously, each spin in Λ evolves independently according to single site transition probabilities p_x. These PCA generalize the dynamical systems called deterministic cellular automata where, likewise, the dynamics is given by a parallel updating rule but via a deterministic local map.

Particularly interesting is the case of *reversible* PCA satisfying the detailed balance condition with respect to a Gibbs measure. We make the choice

$$p_x(a|\sigma) := \frac{1}{2}[1 + a \tanh \beta S_x(\sigma)] \tag{7.186}$$

with $S_x(\sigma) := \sum_{y \in \Lambda \setminus \{x\}} J_{x,y}\sigma(y) + h_x$ and $J_{x,y}, h_x$ real numbers. It is easy to prove (see [64, 86]) that this PCA is reversible with respect to the Gibbs measure associated with the Hamiltonian

$$H_\Lambda(\sigma) = -\beta \sum_{x \in \Lambda} h_x \sigma(x) - \sum_{x \in \Lambda} \log\cosh[\beta S_x(\sigma)]. \tag{7.187}$$

Let us now consider the case $J_{x,y} = 1$ if and only if $|x - y| = 1$ (zero otherwise) and $h_x = h$ for all $x \in \Lambda$, see [62], which is particulary interesting because the single site flipping probabilities $p_x(-\sigma(x)|\sigma)$ are exactly those of the two-dimensional stochastic Ising model with nearest neighbour interactions and heat bath dynamics. This is a serial dynamics, also called Gibbs sampling, defined as follows. At each time t we choose a site x at random and then we set, at time $t+1$, the spin $\eta(x)$ to a value distributed according to equilibrium, conditional to the values (at time t) of the neighbouring spins. The above PCA can thus be considered as the parallel version of the heat bath dynamics.

We call ground states the minimizers of

$$E_\Lambda(\sigma) := \lim_{\beta \to \infty} \frac{H_\Lambda(\sigma)}{\beta} = -h \sum_{x \in \Lambda} \sigma(x) - \sum_{x \in \Lambda} |S_x(\sigma) + h|. \tag{7.188}$$

For $h = 0$ they are $+\underline{1}$, $-\underline{1}$ and the two chessboard configurations C^e and C^o

where, for $x = (x_1, x_2) \in \Lambda$,

$$C^e(x) = +1 \text{ if and only if } x_1 + x_2 \text{ is even}, \qquad C^o(x) = -C^e(x). \tag{7.189}$$

For $h > 0$ ($h < 0$), $+\underline{1}$ ($-\underline{1}$) is the unique ground state. Fix a sufficiently small positive h and take as initial condition $\sigma_{t=0} = -\underline{1}$; we want to analyse the transition to $+\underline{1}$. We want to study the asymptotic behaviour of the first hitting time $\tau_{+\underline{1}}$ and the tube of typical trajectories realizing the transition. In particular we are interested in the role of C^e and C^o.

Let $\ell^* := [2/h] + 1$ and $\mathcal{P}_{-\underline{1}}$ the set of configurations with a rectangular droplet of chessboard of sides ℓ^* and $\ell^* - 1$ in the sea of minuses with a unit plus protuberance, with four neighbouring minuses, placed on the longest side of the rectangle; in [62] the authors prove the following result.

Theorem 7.25 [62] *Let* $C := \{C^e, C^o\}$ *and* $\Gamma := -2h\ell^{*2} + 2\ell^*(4+h) - 2h$. *Then*

(*i*) *the system visits* $\mathcal{P}_{-\underline{1}}$ *before visiting* C, *namely*

$$\lim_{\beta\to\infty} P_{-\underline{1}}(\tau_{\mathcal{P}_{-\underline{1}}} < \tau_C) = 1$$

(*ii*) *the system visits* C *before visiting* $+\underline{1}$, *namely*

$$\lim_{\beta\to\infty} P_{-\underline{1}}(\tau_C < \tau_{+\underline{1}}) = 1,$$

(*iii*) *for any* $\varepsilon > 0$

$$\lim_{\beta\to\infty} P_{-\underline{1}}(e^{\beta\Gamma-\beta\varepsilon} < \tau_C < e^{\beta\Gamma+\beta\varepsilon}) = 1,$$

(*iv*) *for any* $\varepsilon > 0$

$$\lim_{\beta\to\infty} P_{-\underline{1}}(e^{\beta\Gamma-\beta\varepsilon} < \tau_{+\underline{1}} < e^{\beta\Gamma+\beta\varepsilon}) = 1.$$

Thus the asymptotics of τ_C and $\tau_{+\underline{1}}$ coincide in the sense of logarithmic equivalence.

The above theorem states that during the exit from the metastable $-\underline{1}$ state, the system visits the 'competing metastable state' C and, finally, reaches the stable state $+\underline{1}$.

To have a heuristic justification of this behaviour we notice that the direct transition from $-\underline{1}$ to $+\underline{1}$ is very low in probability; indeed the formation energy of a critical droplet of plus in a sea of minus is much larger than that of a critical droplet of chessboard.

It is also possible to describe in detail the typical paths followed by the system during the exit from the metastable phase. We refer to Theorem 3.13 in [62].

Let us discuss now the main features of the model and give an idea of the proof of Theorem 7.25. We refer to [62] for more details.

First of all we observe that in the PCA case, any pair of configurations is linked by a non-zero transition probability so that every sequence of configurations is an allowed path.

The PCA that we are considering is a reversible Freidlin–Wentzell Markov chain but it is not Metropolis: it satisfies condition R but not condition M at the beginning of Chapter 6. Indeed it is easy to exhibit examples of pairs of configurations σ, η such that both $P(\sigma, \eta)$, $P(\eta, \sigma)$ are exponentially small for large β. Thus is due to the non-Metropolis nature of the dynamics, the fact that the quantity Γ appearing in Theorem 7.25 is not equal to the energy difference between protocritical configurations in $\mathcal{P}_{-\underline{1}}$ and the starting configuration $-\underline{1}$. We recall that in the usual Metropolis dynamics $P(\sigma, \eta)$ can be exponentially small in β only if $H(\eta) > H(\sigma)$. Here the relevant quantity is the 'communication height' between a pair of configurations σ, η given by

$$\Phi_\Lambda(\sigma, \eta) = H_\Lambda(\sigma) - \log P(\sigma, \eta). \tag{7.190}$$

Reversibility implies $\Phi_\Lambda(\sigma, \eta) = \Phi_\Lambda(\eta, \sigma)$; moreover if either $P(\sigma, \eta)$ or $P(\eta, \sigma)$ is of order one for large β then $\Phi_\Lambda(\sigma, \eta) = H_\Lambda(\eta) \vee H_\Lambda(\sigma)$. Notice that, given a configuration σ there is at most one configuration η such that $P(\sigma, \eta)$ is of order one for large β. Indeed each p_x goes eiher to one or to zero exponentially fast for large β.

We say that a configuration σ is stable if

$$\Phi_\Lambda(\sigma, \eta) - H_\Lambda(\sigma) > 0, \qquad \forall \eta \in \mathcal{X} \setminus \{\sigma\} \tag{7.191}$$

Given $\sigma \in \mathcal{X}$ we denote by $T\sigma \in \mathcal{X}$ the unique configuration that can be reached from σ with a probability of order one: $\eta = T\sigma$ is the unique configuration such that $p_x(\eta(x)|\sigma) \sim 1, \quad \forall x \in \Lambda$. If σ is stable then $T\sigma = \sigma$.

We call a 'stable loop' a pair σ, η such that $T\sigma = \eta, \; T\eta = \sigma$.

It is easy to see that stable loops with more than two configurations cannot exist and that if σ, η form a stable loop then $E_\Lambda(\sigma) = E_\Lambda(\eta)$. The set of the two chessboard configurations C^o and C^e is a simple example of a stable loop namely $P(C^o, C^e) = P(C^e, C^o) \sim 1$. Let us call 'traps' the stable configurations and the configurations belonging to a stable loop.

The first step in the proof of Theorem 7.25 consists in the characterization of the stable traps as collections of suitable 'droplets'. The second step concerns the analysis of the 'destiny' of the traps i.e. the tendency to grow or shrink of the corresponding droplets. The third and main step consists in finding the communication height between $-\underline{1}$ and $+\underline{1}$.

Let $C \in \mathcal{C} := \{C^o, C^e\}$. Let $\mathcal{X}_{-\underline{1}}$, $\mathcal{X}_{\mathcal{C}}$ be the set of configurations with a well defined 'sea' of minuses, chessboard, respectively, We set $\mathcal{X}_{\mathcal{C}} := \mathcal{X}_{C^o} \cup \mathcal{X}_{C^e}$.

It is shown in [62] that the stable traps in $\mathcal{X}_{-\underline{1}}$ are characterized in the following way. There is a set of non-interacting (see Definition 7.8) rectangles $\bar{R}_1, \ldots, \bar{R}_p$, $R_1, \ldots, R_q$ which are stable in the sense that their minimal side length is $l \geq 2$.

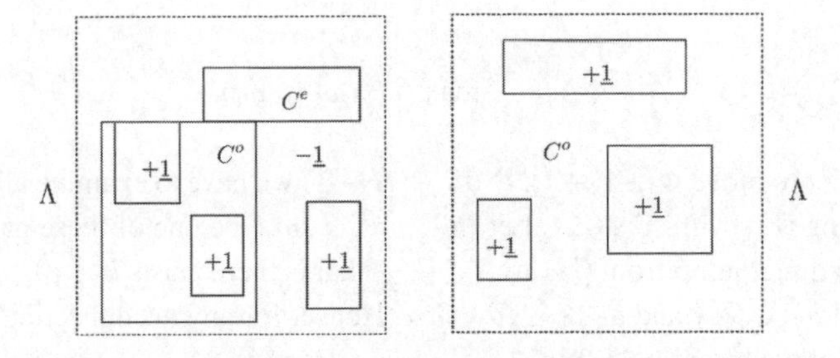

Figure 7.18

$\bar{R}_j$ are either full of chessboard or in their interior there is a collection of non-interacting, stable rectangles $\tilde{R}_1, \ldots, \tilde{R}_k$ surrounded, in $\bar{R}_j$, by chessboard. $\tilde{R}_l$, R_i are full of plus spins and outside $\bar{R}_j$, R_i all spins are -1.

In the case when there are no rectangles $\bar{R}_j$ (R_i), we set $p = 0$ ($q = 0$). In $\mathcal{X}_{-\underline{1}}$ the stable configurations can be obtained only with $p = 0$, corresponding to the case when in the sea of minuses there are only non-interacting stable rectangles full of pluses. Suppose $p > 0$, i.e. there is some rectangle containing chessboard in its interior: these configurations are not stable but elements of a stable loop. Intuitively, we can think that before any move at exponentially small cost these configurations are 'flip-flopping', in their chessboard part between C^o and C^e.

In $\mathcal{X}_{\mathcal{C}}$ the traps are given by collections of non-interacting stable rectangles full of pluses; because of the presence of the sea of chessboard there are only stable loops and no stable configurations; see Figure 7.18.

It is also shown in [62] that in $\mathcal{X}_{-\underline{1}}$ the rectangles of chessboard and the rectangles of $+1$ are subcritical, in the sense that they will typically shrink towards $-\underline{1}$ before touching $\mathcal{X}_{\mathcal{C}} \cup \{+\underline{1}\}$, if and only if their minimal side length is strictly less than l^*. Similarly in $\mathcal{X}_{\mathcal{C}}$ the $+1$ small rectangles are subcritical in the sense that they typically shrink visiting $\mathcal{C}$ before visiting $+\underline{1}$.

We notice that the sum of the quantity Γ appearing in Theorem 7.25 and $H(-\underline{1})$ represents the communication height between the largest subcritical droplet and the smallest supercritical droplet in $\mathcal{X}_{-\underline{1}}$.

Let us outline, now, the way to compute the communication height between $-\underline{1}$ and $+\underline{1}$.

The upper bound $\Phi(-\underline{1}, +\underline{1}) \leq \beta\Gamma + H(-\underline{1})$ is, as usual, relatively easy to find, using a suitable reference path. The lower bound will be based on the construction of a suitable set $\mathcal{G}_{-\underline{1}}$ playing the role of the generalized basin of attraction of $-\underline{1}$.

Given $\eta \in \mathcal{X}$ we define the *downhill* path starting from η as the unique path $\omega = \{\omega_0, \ldots, \omega_n\}$ such that $\omega_0 = \eta$, $T\omega_{i-1} = \omega_i$ for any $i = 1, \ldots, n$, and ω_n belongs to (or coincides with) a trap; we also set $\hat{\eta} := \omega_n$. We set

$$\mathcal{G}_{-\underline{1}} := \{\sigma \in \mathcal{X}_{-\underline{1}} : \hat{\sigma} \text{ subcritical}\} \qquad (7.192)$$

and

$$\Phi_\omega := \max_{i=1,\ldots,|\omega|} H(\omega_{i-1}, \omega_i). \tag{7.193}$$

In order to prove $\Phi(-\underline{1}, +\underline{1}) \geq \beta\Gamma + H(-\underline{1})$ we have to examine all the paths connecting $\mathcal{G}_{-\underline{1}}$ with $\mathcal{X} \setminus \mathcal{G}_{-\underline{1}}$. Let $\{\omega_0, \omega_1, \ldots, \omega_n\}$ be one of these paths; it has at least a direct jump from $\mathcal{G}_{-\underline{1}}$ to $\mathcal{X} \setminus \mathcal{G}_{-\underline{1}}$, that is there exists $k \in \{0, \ldots, n-1\}$ such that $\omega_k \in \mathcal{G}_{-\underline{1}}$ and $\omega_{k+1} \in \mathcal{X} \setminus \mathcal{G}_{-\underline{1}}$. Hence, for any ω connecting $\mathcal{G}_{-\underline{1}}$ with its exterior we have $\Phi_\omega \geq \min_{\sigma \in \mathcal{G}_{-\underline{1}}, \eta \in \mathcal{X} \setminus \mathcal{G}_{-\underline{1}}} H(\sigma, \eta)$.

We have to prove that

$$\min_{\sigma \in \mathcal{G}_{-\underline{1}}, \eta \in \mathcal{X} \setminus \mathcal{G}_{-\underline{1}}} H(\sigma, \eta) \geq \beta\Gamma + H(-\underline{1}). \tag{7.194}$$

This is indeed a very delicate point.

Given $\sigma \in \mathcal{G}_{-\underline{1}}$, consider a configuration $\eta = \eta(\sigma) \in \mathcal{X} \setminus \mathcal{G}_{-\underline{1}}$ that can be reached with the smallest energetic cost, namely $\eta \in \mathcal{X} \setminus \mathcal{G}_{-\underline{1}}$ is such that

$$H(\sigma, \eta) = \min_{\zeta \in \mathcal{X} \setminus \mathcal{G}_{-\underline{1}}} H(\sigma, \zeta) \tag{7.195}$$

The estimate (7.194) is achieved via a detailed analysis of the properties of $\eta(\sigma)$. We refer to [62] for more details.

7.15 Discussion of the results of Bovier and Manzo

In [31] (see Chapter 4) the authors introduce an approach to metastability valid in a quite general setting, that does not adopt the point of view of large deviations; it is not 'pathwise' and does not even take into account the description of typical trajectories.

The paper by Bovier and Manzo [29] contains the application of the powerful method developed in [31] to stochastic Ising models.

The results obtained in [29, 31], even though restricted to the asymptotic behaviour of the tunnelling times are, in some respects, stronger than those presented above.

Contrary to the approach discussed in our Chapter 6, where L^1 estimates are deduced by estimates in probability via good control of the tails, in [31] the results in probability are obtained via a very sharp control of the Laplace transforms of the tunnelling times.

Let us specify to the context of Markov chains satisfying condition M of Chapter 6. The mean value of the tunnelling time is described in terms of the quantities

$$P(\tau_\xi^\eta \leq \tau_\eta^\eta) \tag{7.196}$$

where we use the notation $\tau_\xi^\eta = \inf\{t > 0 : X_t^\eta = \xi, X_s^\eta \neq \eta \text{ for some } s \leq t\}$, $\eta, \xi \in S$, $S \equiv$ state space. X_t^η denotes the process starting at η.

In [31] metastability is characterized by the existence of a set $\mathcal{M}_N$ with the property

$$\sup_{\substack{\eta,\eta'\in\mathcal{M}_N \\ \zeta\in\mathcal{M}_N^c}} \frac{P(\tau^\eta_{\eta'} < \tau^\eta_\eta)}{P(\tau^\zeta_{\mathcal{M}_N} < \tau^\zeta_\zeta)} \longrightarrow 0 \quad \text{as } N\to\infty, \tag{7.197}$$

where N is a large parameter of the system (in the case of the application to stochastic Ising models we have $N = \beta$).

Let

$$\mathcal{I}_\zeta := \{\eta : H(\eta) < H(\zeta)\} \tag{7.198}$$

Under the non-degeneracy hypothesis that for any pair of distinct $\eta, \eta' \in \mathcal{M}_N$ the ratio

$$\frac{P(\tau^\eta_{\mathcal{I}_\eta} < \tau^\eta_\eta)}{P(\tau^{\eta'}_{\mathcal{I}_{\eta'}} < \tau^{\eta'}_{\eta'})} \tag{7.199}$$

either tends to zero or tends to infinity as N goes to infinity, the following theorem is proven in [31] (among other results characterizing the spectrum of the Markov generator).

Theorem 7.26 [31] *Let $\mathcal{M}_N$ be a set with properties* (7.197), (7.199). *For any $\eta \in \mathcal{M}_N$, let $\mathcal{M}_N(\eta) := \mathcal{M}_N \cap \mathcal{I}_\eta$. Then there exists $c > 0$:*

$$E(\tau^\eta_{\mathcal{M}_N(\eta)}) = c\left(P(\tau^\eta_{\mathcal{M}_N(\eta)} < \tau^\eta_\eta)\right)^{-1}(1 + o(1)) \tag{7.200}$$

and, for any $t > 0$

$$P\left(\tau^\eta_{\mathcal{M}_N(\eta)} > tE(\tau^\eta_{\mathcal{M}_N(\eta)})\right) = e^{-t(1+o(1))}(1 + o(1)). \tag{7.201}$$

The constant c is given explicitly and can be read as the degeneracy of the bottom of the cycle $\mathcal{C}_{\mathcal{I}_\eta}(\eta)$ where, for $\mathcal{A} \subseteq S$, $\eta \in S$, $\mathcal{C}_\mathcal{A}(\eta) = \{\zeta \in S : \Phi(\eta, \zeta) < \Phi(\eta, \mathcal{A})\}$.

All model dependent aspects of the problem are somehow hidden in $P(\tau^\eta_{\mathcal{M}_N(\eta)} < \tau^\eta_\eta)$ and the strength of this method relies on the precision in the computation of $P(\tau^\eta_{\mathcal{M}_N(\eta)} < \tau^\eta_\eta)$. In the case of Metropolis dynamics in the Freidlin–Wentzell regime, quantities like this can be computed directly using the well known Dirichlet representation (see e.g. [199], Theorem 6.1)

$$P(\tau^\eta_\xi < \tau^\eta_\eta) = \frac{1}{2}e^{\beta H(\eta)} \inf_{h\in\mathcal{H}^\eta_\xi} \sum_{\zeta,\zeta'\in S} q(\zeta, \zeta')e^{-\beta\max\{H(\zeta),H(\zeta')\}}(h(\zeta) - h(\zeta'))^2, \tag{7.202}$$

where $\mathcal{H}^\eta_\xi := \{h : S \to [0, 1];\ h(\eta) = 0,\ h(\xi) = 1\}$ and $q(\zeta, \zeta) = 0$.

By taking $h(\zeta)$ as the characteristic function of $\mathcal{C}_\xi(\zeta)$, we immediately get

$$P\left(\tau_\xi^\eta < \tau_\eta^\eta\right) \le C_1 e^{-\beta(\Phi(\eta,\xi)-H(\eta))}, \tag{7.203}$$

where $C_1 \le |\partial\mathcal{C}_\xi(\zeta)| \max_{\zeta,\zeta'\in S} q(\zeta,\zeta')$. On the other hand, using comparison with a suitable one-dimensional process, it is proved in [29], equation (4.5), that

$$P\left(\tau_\xi^\eta < \tau_\eta^\eta\right) \ge C_2 e^{-\beta(\Phi(\eta,\xi)-H(\eta))}, \tag{7.204}$$

with $C_2 \ge |S|^{-1} \min_{\zeta,\zeta'\in S}\{q(\zeta,\zeta') \,|\, q(\zeta,\zeta') > 0\}$.

We remark that from (7.203), (7.204) it is not difficult to show that (7.197) is the dynamical counterpart of the 'absence of deep wells', i.e. point (ii) of Theorem 6.38 in Chapter 6 which is the content of Proposition 7.24 in the Kawasaki case. The other hypothesis of Theorem 7.26 corresponds to $\eta_0 = S^m$, the set of metastable states, i.e. the set of most stable minima for the energy (see (6.235) in Chapter 6); it is needed to ensure the asymptotic exponentiality of the law of $\tau_{F(S)}^{\eta_0}/E(\tau_{F(S)}^{\eta_0})$ (recall that $F(S)$ is the set of ground states).

The estimate on the mean tunnelling time given by (7.203), (7.204) via Theorem 7.26 is considerably stronger than the usual Freidlin–Wentzell estimates and can be pushed to $C_1 \simeq C_2$ with a finer analysis.

As noticed in [29], in principle this computation can always be done and in particular it was carried out in the case of a unique minimal gate with the further simplifying hypothesis that the states in the gate are not connected to each other in one step. The result is the following.

Theorem 7.27 [29] *If* $\{\eta_0\} = S^m$,

$$P\left(\tau_{F(S)}^{\eta_0} < \tau_{\eta_0}^{\eta_0}\right) = C_{\eta_0} e^{\beta(\Phi(\eta_0,F(S))-H(\eta_0))}(1+o(1)). \tag{7.205}$$

With

$$C_{\eta_0}^{-1} = \sum_{\zeta\in S(\eta_0,\mathcal{X}^s)\cap\partial\mathcal{C}_{\eta_0}(F(S))} c_\zeta, \tag{7.206}$$

where the c_ζ are explicit constants depending on the one-step transition probabilities between ζ and its neighbours.

It is immediate to see that $c_\zeta = 0$ if ζ is unessential, showing that the sum in (7.206) involves only the states in the minimal gate $\mathcal{W}$, supposed unique (see Definition 6.10, equation (6.16)).

It is a very remarkable fact that the set $\mathcal{W}$ emerges in this context without any connection with the gate property.

We wish to notice that the analysis of the properties of the minimal gates becomes particularly interesting when settled in connection with the result for precise estimates of the mean tunnelling time given in [29].

A natural conjecture arises: is it the case that unessential saddles never give any contribution to the prefactor C_{η_0} in (7.206). As is proven in [207], the answer is negative. In other words, quite surprisingly, unessential saddles, not characterizing the typical behaviour of the process, may contribute to the prefactors and a weaker notion of essentiality is needed to control the L^1 estimates.

We want now to quote some results connected with the main argument of this chapter, but referring to different asymptotic regimes.

7.16 Metastability at infinite volume and very low temperature for the stochastic Ising model

The asymptotic regimes that we have considered up to now in this chapter, for various stochastic Ising (or similar) models, i.e. finite volume (for instance with periodic boundary conditions) and vanishing temperature, are physically motivated by the study of local aspects of nucleation. In the above asymptotic regime, mathematically characterized by a Freidlin–Wentzell reversible Markov chain, the transition between metastability and stability is driven by the formation of a unique critical nucleus which grows rapidly to invade the whole volume. For very large or infinite volumes many nuclei will form, almost independently in different locations, then they will grow and coalesce, driving the system to the stable state. The correct question to ask in this asymptotic regime (∞ volume, $\beta \to \infty$) is: what is the typical time to see the arrival of the stable phase in a given location, say the origin? We expect a very different mechanism of transition with respect to that of finite (independent of β) volume. In an infinite volume we expect to see the formation of many supercritical droplets inside a large volume surrounding the origin with a subsequent long growth of these droplets. This mechanism, involving spatial entropy, is more convenient with respect to a nucleation close to the origin without significant growth. In order for this more efficient mechanism to take place, the system must have enough room at its disposal and this is certainly the case in an infinite volume.

The case of the two-dimensional standard, continuous time, stochastic Ising model (see Chapter 4) in infinite volume, small but fixed (say positive) magnetic field in the limit of vanishing temperature was studied in [76, 77]. (See also [268, 269].) Suppose we start from all minus. The following heuristics is at the basis of the determination of the relaxation time t_{rel}, namely the typical time needed by the spin at the origin to become stably plus (see [76, 77]). At very low temperature, droplets of plus with square shape play an outstanding role. The growth of a square droplet is ruled by two distinct mechanisms:

(1) adding a unit square protuberance (with a plus spin) from the exterior to a face at the rate $e^{-\beta\gamma}$, β being the inverse temperature and $\gamma = 2J - h$ being the energy formation of this protuberance, namely the energy needed to flip a minus spin with one nearest neighbour plus;

(2) filling up, at rate one, of the exterior layer to which the protuberance belongs, starting from the sites nearest neighbour to the protuberance; the rate of this process is one since the energy needed to flip a minus spin with at least two positive nearest neighbours is non-positive.

Suppose that Γ is the energy formation of a critical droplet so that $e^{-\beta\Gamma}$ is the probability per unit volume and per unit time of the formation of this critical droplet. Let v be the speed of growth of the supercritical droplets. As we shall see it is possible to show that $v = e^{-\beta\gamma/2}$.

The relaxation time t_{rel} can be deduced by the following argument: consider the space-time cone, contained in the Cartesian product of $\mathbb{Z}^2$ with the time axis, with height t_{rel} and basis given by the square centred at the origin of side vt_{rel} and time coordinate 0. The order of magnitude of t_{rel} can be deduced by:

$$e^{-\beta\Gamma}(vt_{rel})^2 t_{rel} \cong 1. \tag{7.207}$$

This argument can be extended to an arbitrary dimension $d \geq 1$. We get

$$e^{-\beta\Gamma}(vt_{rel})^d t_{rel} \cong 1. \tag{7.208}$$

t_{rel} can also be seen as the typical time needed for a complete invasion of $\mathbb{Z}^d$ by the stable phase; indeed within that time any fixed location will be invaded. In two dimensions, using $v = e^{-\beta\gamma/2}$ we get

$$t_{rel} \cong e^{\beta\frac{(\Gamma+\gamma)}{3}}. \tag{7.209}$$

Notice that a similar argument applied to the displacement of a (one-dimensional) side yields the relation $v \cong e^{-\beta\gamma/2}$ that we had assumed preliminarily without justification. Indeed now the dimension is $d = 1$ and the speed of growth of a protuberance (inside the new layer) is of order one and, finally, Γ must be replaced by the energy formation of a unit square protuberance, namely γ. Thus, by (7.208) with γ in place of Γ, $v = 1$, $d = 1$ we see that in a time $e^{\beta\gamma/2}$ the new layer of thickness one has been invaded so that the speed of growth of the squared droplet is $e^{-\beta\gamma/2}$.

More generally, in dimension d if γ is the activation energy for a unit protuberance, by a similar argument we get $v \cong e^{-\beta\gamma/d}$ which leads to

$$t_{rel} = e^{\beta\frac{(\Gamma+\gamma)}{d+1}}. \tag{7.210}$$

The factor $1/(d+1)$ in the exponent on the right-hand side of (7.209) drastically reduces the lifetime with respect to the finite volume situation; see [76] for more details.

There is another slightly different way to understand the above heuristic.

The typical time needed to see a nucleation inside the cube $\Lambda(L)$, centred at the origin with side length L is given by $t_L^{nucl} \sim e^{\beta\Gamma}/L^d$ whereas the time needed to grow up to the invasion of the whole volume $\Lambda(L)$ is $t_L^{growth} \sim L/v$, v being the linear growth velocity for a supercritical nucleus. We have to optimize with respect to L, namely to find the L for which the sum $e^{\beta\Gamma}/L^d + L/v$ is minimal. This occurs when the two terms are of the same order of magnitude. By equating them we find $L = L^* = (e^{\beta\Gamma} v)^{1/(d+1)}$ from which $t_{rel} = L^*/v$. Using $v = e^{-\beta\gamma/d}$ we get the previous result.

Let $\kappa_c = (\Gamma + 2J - h)/3$: in [77] the main result is the following theorem.

Theorem 7.28 *Let f be a cylinder function. If* $t = e^{\beta\kappa}$

(1) $\lim_{\beta\to\infty} E(f(\sigma_t)) = f(-\underline{1})$ *if* $\kappa < \kappa_c$,

(2) $\lim_{\beta\to\infty} E(f(\sigma_t)) = f(+\underline{1})$ *if* $\kappa > \kappa_c$.

Thus $e^{\beta\kappa_c}$ is the relaxation time in agreement with the above heuristic.

Let us now give an outline of the proof; we shall do it in the framework of a simplified nucleation-and-growth model that was introduced in [76]. The state space is $\mathcal{X} = \{0, 1\}^{\mathbb{Z}^d}$; for $\eta \in \mathcal{X}$, if $\eta(x) = 0$ (1) we say that the site x is empty (occupied). Starting with all sites unoccupied, a site becomes occupied at rate $e^{-\beta\Gamma}$ if it has no occupied neighbour, $e^{-\beta\gamma}$ if it has one occupied neighbour and at rate 1 if it has two or more occupied neighbours. Occupied sites remain occupied forever. We suppose Γ, γ fixed with Γ sufficiently larger than γ.

The above described dynamics is manifestly irreversible; this is a relevant difference with respect to the stochastic Ising model. However, as shown by Dehghanpour and Schonmann, the most important features of the stochastic Ising model in the same asymptotic regime are still present in the simplified model.

Statement (2) of the theorem can be obtained by constructing a suitable event of nucleation and subsequent growth inside a suitable cube centred at the origin within a suitable time. The definition of such an event is based on the heuristic that we have discussed before. More delicate is the proof of point (1) of the theorem.

It is easy to convince oneself that different nuclei will interact. Recall that $v = e^{-\beta\gamma/d}$ is the speed of growth and we are looking at times of order $t = e^{\kappa_c\beta}$; suppose that two nuclei are close enough, i.e. at a distance smaller than $ve^{\beta\kappa_c}$ to each other; then, once they meet the remaining sites in the rectangular envelope of their union become occupied at an effective rate 1 as a result of a bootstrap percolation mechanism (see Section 7.4 and [2]). Since the only a priori upper bound for the speed of growth is 1, we have to look at a volume $\Lambda(e^{\kappa_c\beta})$. Inside this box, during a time interval $e^{\kappa_c\beta}$ a number of nuclei will be formed of order $(e^{\kappa_c\beta})^d e^{\kappa_c\beta} e^{-\beta\Gamma} = e^{\beta\gamma}$. Then the minimal distance between a pair of formed nuclei, in an optimal situation, is of order $e^{\beta\kappa_c}/(e^{\beta\gamma})^{1/d} = ve^{\beta\kappa_c}$ or less so that coalescence will take place with overwhelming probability. However, a result by Aizenman and Lebowitz will allow control of this interaction for times $t = e^{\kappa\beta}$, $\kappa < \kappa_c$.

Fix $t = e^{\kappa\beta}$, $\kappa < \kappa_c$.

The first step in the proof of point (1) of the theorem consists in substituting the evolution in $\mathbb{Z}^d$ with an analogous evolution restricted to the volume

$\Lambda := \Lambda(e^{\kappa_c \beta})$. Indeed it is easy to see that the probability that within a time $t = e^{\kappa\beta}$ an influence on the origin can come from outside is extremely small. We next consider a smaller cube $\Lambda' := \Lambda(L')$, $L' := Ce^{-\beta\gamma/d}e^{\kappa_c\beta}$, where C is a suitable positive constant. We set

$$\Lambda'_j = \Lambda' + L'j, \qquad j \in \mathbb{Z}^d. \tag{7.211}$$

The disjoint union of the Λ'_j constitutes a partition of $\mathbb{Z}^d$. The rescaled lattice is

$$\Lambda_{res} = \{j : \Lambda'_j \cap \Lambda \neq \emptyset\}. \tag{7.212}$$

Now we define a dichotomous stochastic field η on the rescaled lattice. We look at the boxes Λ'_j where a nucleation takes place within a time $e^{\kappa\beta}$; in this case we assign to $\eta(j)$ the value 1 and say that the corresponding site of Λ_{res} is occupied; let p be the probability that $\eta(0) = 1$. Given a configuration η we occupy iteratively sites of the rescaled lattice that have two or more neighbours. This deterministic procedure stops after a finite number of iterations giving rise to the bootstrap configuration $\bar{\eta}$. In [2] the authors prove that if p is sufficiently small in terms of the size of Λ_{res} then

$$P(\bar{\eta}(0) = 1) \to 0 \text{ as } \beta \to \infty. \tag{7.213}$$

It is shown in [76, 77] that this condition on p is verified in the present case.

Thus we know that with high probability the origin is not included inside a box Λ' where there has been a nucleation within a time $e^{\kappa\beta}$ or there has been the effects of a coalescence nearby, followed by an anomalous growth at speed one: this coalescence followed by fast growth can occur but only in very sparse subsets of Λ_{res}. So, to conclude the proof of point (1) of the theorem, we are left to prove that for a process 'without nucleation' the probability that there is a growth from outside Λ' that is able to reach the origin within a time $e^{\kappa\beta}$ is very small for large β.

7.17 Metastability at infinite volume and very low temperature for the dynamic Blume–Capel model

We want to discuss, now, metastability and nucleation at infinite volume and very low temperature in the framework of the dynamic Blume–Capel model in two dimensions. In other words we want to analyse a system with competing metastable states in the same asymptotic regime considered by Dehghanpour and Schonmann for the standard Ising model. This problem was studied in [205] for a simplified nucleation-and-growth model, analogous to the one considered in [76] and for the full model in [206].

The dynamic Blume–Capel model has already been introduced and analysed in a finite volume context (see Secton 7.11).

In the infinite volume situation the configuration space is given by $\{-1, 0, +1\}^{\mathbb{Z}^2}$.

The formal Hamiltonian is given (as in (7.97)) by

$$H(\sigma) = \sum_{<x,y>} (\sigma(x) - \sigma(y))^2 - \lambda \sum_x \sigma(x)^2 - h \sum_x \sigma(x), \tag{7.214}$$

where $\sigma(x) \in \{-1, 0, +1\}$ represents the spin at the site x; λ and h are the chemical potential and the magnetic field and $<x, y>$ denotes a generic pair of nearest neighbour sites in $\mathbb{Z}^2$. We keep λ and h fixed and take the limit of large inverse temperature β. In [206] a continuous time dynamics was introduced, directly at infinite volume. The corresponding transition rates are $c(\sigma, \sigma') := e^{-\beta[\Delta H(\sigma,x,\sigma'(x))]^+}$ if σ and σ' are nearest neighbour configurations (see before (7.98)) with $\sigma(x) \neq \sigma'(x)$, $c(\sigma, \sigma') = 0$ otherwise. $\Delta H(\sigma, x, \sigma'(x))$ is given formally by $\Delta H(\sigma, x, \sigma'(x)) = H(\sigma) - H(\sigma')$.

We have already described in Section 7.11 metastability and nucleation for a two-dimensional Blume–Capel system enclosed in a finite torus Λ, with given λ, h, in the limit of zero temperature. The interesting region of parameters was

$$0 < |\lambda| < h, \qquad \text{with } h \text{ sufficiently small.} \tag{7.215}$$

We recall that $-\underline{1}$, $\underline{0}$ and $+\underline{1}$ denote the configurations, of local minimum for the energy, with all spins equal to $-1, 0, +1$, respectively. In the whole region (7.215) we have (in any finite volume):

$$H(-\underline{1}) > H(\underline{0}) > H(+\underline{1}).$$

Even in the infinite volume situation, the main question that arises is whether or not, during the transition from $-\underline{1}$ to $+\underline{1}$, the system typically visits $\underline{0}$.

As we reported in Section 7.11, in [63] the answer was found in the finite volume scenario. In that case there is a change in the mechanism of transition from $-\underline{1}$ to $+\underline{1}$ when crossing the line $h = 2\lambda > 0$ in the h, λ plane. On the right side of this line ($h < 2\lambda$) the transition is 'direct', along a sequence of growing droplets almost full of pluses, whereas in the other region ($h > 2\lambda$) the transition is 'indirect' in the sense that the system first reaches the $\underline{0}$ configuration via the formation of a supercritical squared droplet of zeros in a sea of minuses. Subsequently, via the formation of a supercritical squared droplet of pluses in a sea of zeros, the system is driven to the final stable state $+\underline{1}$ (see Section 7.11 for more details).

It is natural to pose the following question: does the sort of 'dynamical phase transition' that has been detected in a finite volume persist in an infinite volume? If yes, in which form? One easily realizes that it is reasonable to expect a change in the mechanism of transition over the line $\lambda = 0$. Indeed, when passing from $\lambda < 0$ to $\lambda > 0$ two simultaneous effects take place.

(1) The local energy barrier, strictly related to the nucleation rate, between $-\underline{1}$ and $\underline{0}$ becomes higher than the local energy barrier between $\underline{0}$ and $+\underline{1}$.
(2) The speed of growth of a supercritical droplet of pluses in a sea of zeros becomes larger than the speed of growth of a supercritical droplet of zeros in a sea of minuses.

In other words one expects that, starting from $-\underline{1}$ and looking at an observable localized close to the origin, for $\lambda < 0$ one first sees a large droplet of zeros coming from a large distance and after a much larger time one observes the arrival of a large droplet of pluses; on the contrary, for $\lambda > 0$ the time of the first arrival of the zero phase near the origin is much longer than the time interval needed for the subsequent arrival of the plus phase.

In [205], for the irreversible simplified model and in [206] for the full model the authors, in addition to other results, proved a change in the asymptotic behaviour of the ratio between the time of first appearance, say, of a stable non-minus situation at the origin, denoted by $\tau_{\oplus}$, and the time interval, denoted by $\tau_{\oplus +}$, between $\tau_{\oplus}$ and the first appearance of the $+$ phase at the origin. For $0 < -\lambda < h$ one has that, typically, $\tau_{\oplus +}/\tau_{\oplus}$ goes to infinity whereas for $0 < \lambda < h$, $\tau_{\oplus +}/\tau_{\oplus}$ goes to zero.

When $\tau_{\oplus +} \ll \tau_{\oplus}$, information on the shape of large droplets is also given: it is shown that large droplets of zeros tend to be invaded by pluses in their interior so that, asymptotically, they become completely full of pluses with only a relatively thin layer of non-minus (typically zero) spins between the internal pluses and the sea of minuses.

7.18 Metastability for the infinite volume stochastic Ising model at $T < T_c$ in the limit $h \to 0$

We want to quote now another asymptotic regime that has been studied in [267, 271] infinite (or sufficiently large) volume, temperature T sufficiently small but fixed, vanishing magnetic field. This is certainly the most interesting situation from a physical point of view and the hardest one from the mathematical point of view. Indeed in this asymptotic regime, droplets of macroscopic size will enter into the game and the entropic effects will be relevant. In [271], where more complete results appear, the authors consider a wide class of single spin-flip dynamics that are reversible with respect to the Gibbs measures corresponding to the two-dimensional standard stochastic Ising model at a temperature $T < T_c$ and at a magnetic field $h/2 \neq 0$ (say $h > 0$) when the initial distribution ν is taken stochastically lower than the equilibrium minus state μ_- (see Theorem 3.35 of Chapter 3). We write $\nu \prec \mu_-$. For example we can take the continuous time Metropolis dynamics (see (4.50) of Chapter 4) with the flip rates $c(x, \sigma)$ as in condition M of Chapter 6 and, as initial state ν, a Dirac mass on the configuration with all minus. Call $\sigma^{\nu}_{h,t}$ the corresponding process (t is the time). The main result of [271] is the following theorem.

Theorem 7.29 *Suppose* T < T$_c$. *For every probability distribution* $\nu \prec \mu_-$ *the following happens. If we let* $h \searrow 0$ *and* $t \to \infty$ *together, then for all local observable* f

(*i*) $E(f(\sigma^{\nu}_{h,t})) \to \mu_-(f)$ *if* $\limsup h \log t < \lambda_c(\mathrm{T})$,
(*ii*) $E(f(\sigma^{\nu}_{h,t})) \to \mu_+(f)$ *if* $\limsup h \log t > \lambda_c(\mathrm{T})$ *where*

$$\lambda_c(T) = \frac{w(T)^2}{12Tm^*(\mathrm{T})}, \tag{7.216}$$

$m^*(\mathrm{T})$ *is the spontaneous magnetization and* $w(\mathrm{T})$ *is the integrated surface tension of the Wulff body of unit volume.*

The Wulff shape represents the macroscopic equilibrium shape of a droplet of plus (minus) phase plunged inside the minus (plus) phase; it corresponds to minimal surface free energy for a given volume. The precise definition can be found in [271].

Theorem 7.29 says that for times $t = e^{\lambda/h}$ with $\lambda < \lambda_c$ the state of the process is close to the $(-)$-phase whereas for $\lambda > \lambda_c$ it is close to the $(+)$-phase. In [271] the authors also prove that the difference between the state at time $t = e^{\lambda/h}$ with $\lambda < \lambda_c$ and the (–)-phase can be described in terms of an asymptotic expansion in powers of h.

Let us say a few words about the heuristics behind Theorem 7.29. We refer to [271] for more details. The starting point is the computation of the free energy barrier that the system has to overcome to create the stable phase locally. This barrier is associated with the creation, in a fixed location, of a suitable droplet of the stable phase (roughly the $(+)$-phase) in a background of the metastable phase (roughly the $(-)$-phase). Let S be the shape of the droplet and l^2 its volume. The free energy associated with this droplet can be written as

$$\Phi_S(l) = -(m^*)^2 hl^2 + w_S l. \tag{7.217}$$

Indeed the relevant difference in the bulk free energy per site between the minus and the plus phases is $2m^*h/2$. The contribution coming from the interface is proportional to the length of the boundary of the droplet, therefore to l. w_S represents the constant of proportionality; it is minimal for the Wulff shape. In this case we write w for w_S and Φ for Φ_S. The two terms in (7.217) are of the same order of magnitude when l is of the order $1/h$. We write $l = b/h$. We have $\Phi = \phi(b)/h$. We notice that ϕ is maximal for $b = w/2m^*$ where it takes the value $w^2/4m^*$.

Thus the Wulff shape is naturally associated with the minimal free energy barrier for exiting locally from metastability. Indeed this barrier corresponds to the formation of an optimal critical droplet of the correct scale proportional to $1/h$.

However we want to stress that there is absolutely no reason, even in this $h \to 0$ regime, for the system to follow a Wulff nucleation pattern. Indeed, during growth of the nucleus of the stable phase, the system does not have any reason to minimize the surface free energy for the given value of size of the nucleus.

To conclude the heuristics we observe that, since we are at infinite volume, the same mechanism that we have described in the infinite volume, $\beta \to \infty$ regime, introduced by Dehghanpour and Schonmann, is still active. Using a suitable estimate of the speed of growth of supercritical droplets, one can again see that the effect of the infinite volume is the appearance of the factor $d + 1$ in the denominator of the exponent of the relaxation time. As in the case of the infinite volume $\beta \to \infty$ regime, this drastic decrease of the lifetime is due to the effect of formation of a critical droplet far away from the support $\Delta(f)$ of the local function f (exploiting the 'spatial entropy') followed by growth up to the invasion of $\Delta(f)$. In this way, in our $d = 2$ case we get (7.216).

7.19 Applications

A possible title of this Section could be obtained by transposing a sentence that Dobrushin formulated in the case of Gibbs fields: metastability, from physics through mathematics to all sciences. We want to quote, in the following, some examples of applications of the general ideas of the pathwise approach to metastability to various situations, going beyond statistical mechanics.

The mathematical formulation allows us to move from the domain of dynamical phase transitions to several different scientific contexts, the characteristic features of metastable behavior.

We give a brief list of topics.

- *Interacting particle systems.* A first example that can be considered as an application to the theory of interacting particle systems is the Harris contact process which is discussed in Chapter 4.
- *Biology.* Interacting species and metastability, see [112].
- *Dynamical systems.* Occurrence of rare events for mixing dynamical systems, see [130].
- *Geomagnetism.* Randomly perturbed dynamical systems: flip-flop of the stochastic disc dynamo, see [170].
- *Linguistics.* Language acquisition and change in generalized GW model, see [45].
- *Sociology.* The dynamics of social dilemmas, see [139].

The following example is emblematic and will be briefly reported from the original paper.

- *Palaeonteology.* Neo-darwinian evolution implies punctuated equilibria, see [229].

> The two central elements of neo-darwinian evolution are small random variations and natural selection. In Wright's view, this led to random peaked adaptive landscape, with long periods spent near fitness peaks... the transitions between peaks are rapid and unidirectional even though (indeed because) random variations are small and transitions initially require movement against selection. Thus punctuated equilibrium, the palaeontological pattern of rapid transition between morphological equilibria, is a natural manifestation of the standard wrightian evolutionary theory... The simplest neo-darwinian model for evolution of the population mean x of a genetically determined character expresses the change dx in time interval dt as the sum of a natural selection term and a random variation term:
>
> $$dx(t) = F'(x(t))dt + \alpha dW(t). \tag{7.218}$$
>
> $F(x)$, a one-dimensional genotypic or phenotypic adaptive landscape, describes the mean population fitness. If, for example, the slope F' is positive (a rising landscape) at $x(t)$, natural selection pushes x towards larger values. The parameter α gives the magnitude of random variation relative to that of natural selection. The random process $W(t)$, a standard brownian movement with zero mean drift, represents, for example, genetic drift and short-term environmental fluctuations showing no obvious trend....

7.20 Related fields

We want to quote now some fields related to low temperature metastability. We quote first some non-homogeneous (in time) Markov chains similar to Freidlin–Wentzell chains, with applications to optimization problems. The typical example is *simulated annealing*. Consider the problem of finding the minima of an energy function H defined in a finite state space S and taking values in $\mathbb{R}$.

Consider an inhomogeneous Markov chain $(X_t)_{t\in\mathbb{N}}$ with transition probabilities at time t given by Q_{β_t} where

$$Q_\beta(x, y) := q(x, y)e^{-\beta[H(y)-H(x)]^+} \tag{7.219}$$

with $x, y \in S$, q being an irreducible symmetric Markov kernel. The sequence of diverging inverse temperatures $(\beta_t)_{t\in\mathbb{N}}$ is given; it is called a *cooling schedule*. Recall that $F(S)$ is the set of absolute minima of S. If the schedule varies sufficiently slowly, it is possible to show that

$$\sup_{x\in S} P_x(X_t \notin F(S)) \to 0, \quad \text{as } t \to \infty. \tag{7.220}$$

The study of this kind of Markov chain can be carried out using the conceptual cathegories of stationary Freidlin–Wentzell Markov chains. In particular, the decomposition of S into maximal cycles is a very powerful tool. We refer the interested reader to [53, 54, 289, 290].

Another related class of inhomogeneous Markov chains, also useful for optimization problems and whose theory uses Freidlin and Wentzell's ideas, are the so-called *genetic algorithms*. A basic reference for that is [56].

Finally, as a related subject, we quote some numerical results. The literature concerning numerical studies of metastability and nucleation is very large; we want just to quote the paper [288] by Tomita and Miyashita where the relaxation from metastability to stability is analysed via computer experiments using the standard two-dimensional stochastic Ising model. It is a remarkable fact that in [288] the authors find at low, fixed temperature behaviour that can be rigorously predicted, in the limit of low temperature, using the results of our Chapters 6 and 7. In particular, the size dependence of the relaxation time that is found numerically in [288] reflects the analysis performed by Dehghanpour and Schonmann. According to the size of the container, two different types of relaxation are found. For example, the ratio of the mean to the variance of the relaxation time changes considerably with the size. This can be explained by looking at the threshold after which the infinite volume Dehghanpour–Schonmann regime starts to hold. For volumes smaller than this threshold the distribution of the normalized relaxation time is almost exponential (variance $\simeq$ expectation) whereas for larger volumes it is much more deterministic.

REFERENCES

[1] A. de Acosta. 1990. Large deviations for empirical measures of Markov chains. *J. Theor. Probab.* **3** (3), 395–431.

[2] M. Aizenman, J. L. Lebowitz. 1988. Metastability effects in bootstrap percolation. *J. Phys. A* **21**, 3801–3813.

[3] L. Alonso, R. Cerf. 1996. The three-dimensional polyominoes of minimal area. *Electron. J. Combin.* **3**.

[4] E. D. Andjel. 1988. The contact process in high dimensions. *Ann. Probab.* **16**, 1174–1183.

[5] V. I. Arnold. 1978. *Mathematical Methods of Classical Mechanics*. Graduate Texts in Mathematics, 60. New York: Springer.

[6] V. I. Arnold, A. Avez. 1968. *Ergodic Problems of Classical Mechanics*. New York: Benjamin.

[7] R. Azencott. 1980. Grandes déviations et applications. In *École d'Été de Probabilités de Saint-Flour, 1978*, ed. P. L. Hennequin. Lecture Notes in Mathematics 774. Berlin: Springer, 1–176.

[8] T. A. Azlarov, N. A. Volodin. 1986. *Characterization Problems Associated with the Exponential Distribution*. New York: Springer.

[9] R. R. Bahadur. 1971. *Some Limit Theorems in Statistics*. CBMS-NSF Regional Conference Series in Applied Mathematics, vol. 4. Philadelphia, PA: SIAM.

[10] R. R. Bahadur, R. R. Rao. 1960. On deviations of the sample mean. *Ann. Math. Stat.* **31**, 1015–1027.

[11] R. R. Bahadur, S. L. Zabell. 1979. Large deviations of the sample mean in general vector spaces. *Ann. Probab.* **7**, 587–621.

[12] P. Baldi. 1988. Large deviations and stochastic homogenization. *Ann. Mat. Pura Appl.* **151**, 161–177.

[13] S. Barnett. 1990. *Matrices: Methods and Applications*. Oxford: Oxford University Press.

[14] D. J. Barsky, G. Grimmett, C. M. Newman. 1991. Percolation in half-spaces: equality of critical probabilities and continuity of the percolation probability. *Probab. Theory Relat. Fields* **90** (1), 111–148.

[15] R. Becker, W. Döring. 1935. Kinetische Behandlung der Keimbildung in übersättigten Dämpfern. *Ann. Phys. (Leipzig)* **24**, 719–752.

[16] G. Ben Arous, R. Cerf. 1996. Metastability of the three-dimensional Ising model on a torus at very low temperatures. *Electron. J. Probab.* **1** (10).

[17] O. Benois, T. Bodineau, E. Presutti. 1998. Large deviations in the van der Waals limit. *Stoch. Proc. Appl.* **75**, 89–104.

[18] C. Bezuidenhout, G. Grimmett. 1990. The critical contact process dies out. *Ann. Probab.* **18** (4), 1462–1482.

[19] P. Billingsley. 1965. *Ergodic Theory and Information*. New York: Wiley.

[20] P. Billingsley. 1968. *Convergence of Probability Measures*. New York: Wiley.

[21] D. Blackwell, J. L. Hodges. 1959. The probability in the extreme tail of a convolution. *Ann. Math. Stat.* **30**, 1113–1120.

[22] M. Blume. 1966. Theory of the first-order phase change in UO_2. *Phys. Rev.* **141**, 517–524.

[23] T. Bodineau. 1999. The Wulff construction in three and more dimensions. *Commun. Math. Phys.* **207** (1), 197–229.

[24] T. Bodineau, D. Ioffe, Y. Velenik. 2000. Rigorous probabilistic analysis of equilibrium crystal shapes. Probabilistic techniques in equilibrium and nonequilibrium statistical physics. *J. Math. Phys.* **41** (3), 1033–1098.

[25] E. Bolthausen, U. Schmock. 1989. On the maximun entropy principle for uniformly ergodic Markov chains. *Stoch. Proc. Appl.* **33**, 1–27.

[26] A. A. Borovkov. 1967. Boundary-value problems for random walks and large deviations in function spaces. *Theory Probab. Its Appl.* **12**, 575–595.

[27] A. A. Borovkov, B. A. Rogozin. 1965. On the multi-dimensional central limit theorem. *Theory Probab. Its Appl.* **10** (1), 55–62 (*Teor. Veroyatn. Ee Primen.* **10**, 61–69).

[28] M. Boué, P. Dupuis, R. S. Ellis. 2000. Large deviations for small noise diffusions with discontinuous statistics. *Probab. Theory Relat. Fields* **116**, 125–149.

[29] A. Bovier, F. Manzo. 2002. Metastability in Glauber dynamics in the low-temperature limit: beyond exponential asymptotics. *J. Stat. Phys.* **107**, 757–779.

[30] A. Bovier, M. Zahradnik. 1997. The low temperature phases of Kac–Ising models. *J. Stat. Phys.* **87**, 311–332.

[31] A. Bovier, M. Eckhoff, V. Gayrard, M. Klein. 2001. Metastability in stochastic dynamics of disordered mean field models. *Probab. Theory Relat. Fields* **119** (1), 99–161.

[32] S. Brassesco. 1991. Some results on small random perturbations of an infinite dimensional dynamical system. *Stoch. Proc. Appl.* **38**, 33–53.

[33] S. Brassesco. 1996. Unpredictability of an exit time. *Stoch. Proc. App.* **63**, 55–65.

[34] S. Brassesco, E. Olivieri, M. E. Vares. 1998. Couplings and asymptotic exponentiality of exit times. *J. Stat. Phys.* **93** (1/2), 393–404.

[35] S. Brassesco, E. Presutti, V. Sidoravicius, M. E. Vares. 2000. Ergodicity and exponential convergence of a Glauber+Kawasaki process. In *On Dobrushin's way. From Probability Theory to Statistical Physics.* American Mathematical Society Translation Series 2, vol. 198. Providence, RI: American Mathematical Society, 37–49.

[36] S. Brassesco, E. Presutti, V. Sidoravicius, M. E. Vares. 2000. Ergodicity of a Glauber+Kawasaki process with metastable states. *Markov Process. Relat. Fields* **6** (2), 181–203.

[37] J. Bricmont, J. Slawny. 1989. Phase transitions in systems with a finite number of dominant ground states. *J. Stat. Phys.* **54**, 89–161.

[38] A. Brønsted. 1964. Conjugate convex functions in topological vector spaces. *Mat. Fyz. Medd. Dan. Vid. Selk.* **34** (2), 3–27.

[39] W. Bryc. 1990. Large deviations by the asymptotic value method. In *Diffusion Processes and Related Problems in Analysis*, ed. M. Pinsky, vol. 1. Boston, MA: Birkhauser, 447–472.

[40] L. Bunimovich, C. Liverani, A. Pellegrinotti, Yu. Suhov. 1992. Ergodic systems of n balls in a billiard table. *Commun. Math. Phys.* **146** (2), 357–396.

[41] P. Buttà, A. De Masi, E. Rosatelli. 2003. Slow motion and metastability for a non local evolution equation. *J. Stat. Phys.* **112** (3/4), 709–764.

[42] P. Calderoni, S. Pellegrinotti, E. Presutti, M. E. Vares. 1989. Transient bimodality in interacting particle systems. *J. Stat. Phys.* **55**, 527–577.

[43] H. W. Capel. 1966, 1967. Physica **32**, 96 (1966); **33**, 295 (1967); **37**, 423 (1967).

[44] D. Capocaccia, M. Cassandro, E. Olivieri. 1974. A study of metastability in the Ising model. *Commun. Math. Phys.* **39**, 185–205.

[45] M. Cassandro, A. Galves. 1995. Language acquisition and change in generalized GW model, Preprint MOL4.

[46] M. Cassandro, E. Olivieri. 1977. A rigorous study of metastability in a continuous model. *J. Stat. Phys.* **17**, 229–244.

[47] M. Cassandro, E. Presutti. 1996. Phase transitions in Ising systems with long but finite range of interactions. *Markov Process. Relat. Fields* **2**, 241–262.

[48] M. Cassandro, A. Galves, E. Olivieri, M. E. Vares. 1984. Metastable behaviour of stochastic dynamics: a pathwise approach. *J. Stat. Phys.* **35**, 603–634.

[49] M. Cassandro, E. Olivieri, P. Picco. 1986. Small random perturbations of infinite-dimensional dynamical systems and nucleation theory. *Ann. Inst. H. Poincaré Phys. Theor.* **44** (4), 343–396.

[50] M. Cassandro, E. Orlandi, E. Presutti. 1993. Interfaces and typical Gibbs configurations for one-dimensional Kac potentials. *Probab. Theory. Relat. Fields* **96** (1), 57–96.

[51] M. Cassandro, E. Orlandi, P. Picco. 1999. Typical configurations for one-dimensional random field Kac model. *Ann. Probab.* **27** (3), 1414–1467.

[52] M. Cassandro, E. Orlandi, P. Picco, M. E. Vares. 2003. One dimensional random field Kac model: localization of phases. Ar Xiv: math. PR/0402174.

[53] O. Catoni. 1991. Sharp large deviation estimates for simulated annealing algorithms. *Ann. Inst. H. Poincaré* **27** (3), 291–383.

[54] O. Catoni. 1992. Rough large deviation estimates for simulated annealing. Application to exponential schedules. *Ann. Probab.* **20**, 1109–1146.

[55] O. Catoni, R. Cerf. 1995. The exit path of a Markov chain with rare transitions. *ESAIM Probab. Stat.* **1**, 95–144 (electronic).

[56] R. Cerf. 1998. Asymptotic convergence of genetic algorithms. *Adv. Appl. Probab.* **30** (2), 521–550.

[57] R. Cerf, A. Pisztora. 2000. On the Wulff crystal in the Ising model. *Ann. Probab.* **28**, 947–1017.

[58] R. Cerf, A. Pisztora. 2001. Phase coexistence in Ising, Potts and percolation models. *Ann. Inst. H. Poincaré* **37** (6), 643–724.

[59] N. Chafee, E. F. Infante. 1974. Bifurcation and stability for a nonlinear parabolic partial differential equation. *Bull. Am. Math. Soc.* **80**, 49–52.

[60] J. W. Chen. 1994. The contact process on a finite set in higher dimensions. *Chin. J. Contemp. Math.* **15**, 13–20.

[61] H. Chernoff. 1952. A measure of asymptotic efficiency for tests of a hypothesis based on sums of observations. *Ann. Math. Stat.* **23**, 493–507.

[61A] E. N. M. Cirillo, J. L. Lebowitz. 1998. Metastability in the two-dimensional Ising model with free boundary conditions. *J. Stat. Phys.* **90**, 211–226.

[62] E. N. M. Cirillo, F. R. Nardi. 2003. Metastability for a stochastic dynamics with a parallel heat bath updating rule. *J. Stat. Phys.* **110** (1/2), 183–217.

[63] E. Cirillo, E. Olivieri. 1996. Metastability and nucleation for the Blume–Capel model: different mechanisms of transition. *J. Stat. Phys.* **83**, 473–554.

[64] E. N. M. Cirillo, F. R. Nardi, A. D. Polosa. 2001. Magnetic order in the Ising model with parallel dynamics. *Phys. Rev. E* **64**, 057103.

[65] F. Comets. 1986. Grandes déviations pour des champs de Gibbs sur $\mathbb{Z}^d$. *C. R. Acad. Sci. Ser. I* **303**, 511–513.

[66] F. Comets. 1987. Nucleation for a long range magnetic model. *Ann. Inst. H. Poincaré* **23**, 135–178.

[67] J. T. Cox. 1984. An alternate proof of a correlation inequality of Harris. *Ann. Probab.* **12**, 272–273.

[68] J. T. Cox, A. Greven. 1990. On the long term behavior of some finite particle systems. *Probab. Theory. Relat. Fields*. **85**, 195–237.

[69] H. Cramér. 1937. Sur un nouveau théorème limite de la théorie des probabilités. *Colloquium on Theory of Probability*. Paris: Hermann.

[70] I. Csiszár. 1984. Sanov property, generalized I-projection, and a conditional limit theorem. *Ann. Probab.* **12**, 768–793.

[71] D. Dacunha-Castelle (ed.). 1979. *Grandes Déviations et Applications Statistiques*. In *Astérisque* **68**.

[72] D. A. Dawson, J. Gärtner. 1987. Large deviations from the McKean–Vlasov limit for weakly interacting diffusions. *Stochastics* **20**, 247–308.

[73] M. Day. 1983. On the exponential exit law in the small parameter exit problem. *Stochastics* **8**, 297–323.

[74] M. Day. 1983. Exponential levelling of stochastically perturbed dynamical systems, *SIAM J. Math. Anal.* **13**, 532–540.

[75] M. Day. 1990. Large deviations results for the exit problem with characteristic boundary. *J. Math. Anal. Appl.* **147** (1), 134–153.

[76] P. Dehghanpour, R. Schonmann. 1997. A Nucleation-and-growth model. *Probab. Theory Relat. Fields* **107**, 123–135.

[77] P. Dehghanpour, R. Schonmann. 1997. Metropolis dynamics relaxation via nucleation and growth. *Commun. Math. Phys.* **188**, 89–119.

[78] A. De Masi, E. Presutti. 1991. *Mathematical Methods for Hydrodynamic Limits*, Lecture Notes in Mathematics 1501. New York: Springer.

[79] A. De Masi, N. Ianiro, A. Pellegrinotti, E. Presutti. 1984. *A Survey of the Hydrodynamical Behavior of Many-Particle Systems. Non Equilibrium Phenomena II. From Stochastics to Hydrodynamics*. Amsterdam: North-Holland.

[80] A. De Masi, P. Ferrari, J. L. Lebowitz. 1986. Reaction–diffusion equations for interacting particle systems. *J. Stat. Phys.* **44** (3/4), 589–644.

[81] A. De Masi, S. Pellegrinotti, E. Presutti, M. E. Vares. 1994. Spatial patterns when phases separate in an interacting particle system. *Ann. Probab.* **22** (1), 334–371.

[82] A. De Masi, E. Orlandi, E. Presutti, L. Triolo. 1994, 1996. Glauber evolution with Kac potentials. I. *Nonlinearity* **7**, 633–696 (1994); II. **9**, 27–51 (1996); III. **9**, 53–114 (1996).

[83] A. De Masi, E. Olivieri, E. Presutti. 1998. Spectral properties of integral operators in problems of interface dynamics and metastability. *Markov Process. Relat. Fields* **4**, 27–112.

[84] A. De Masi, E. Olivieri, E. Presutti. 2000. Critical droplet for a non local mean field equation. *Markov Process. Relat. Fields* **6**, 439–471.

[85] A. Dembo, O. Zeitouni. 1993. *Large Deviations Techniques and Applications*. Boston, MA: Jones and Bartlett. (1998. 2nd edition. New York: Springer.)

[86] B. Derrida. 1990. Dynamical phase transition in spin models and automata. In *Fundamental Problems in Statistical Mechanics VII*, ed. H. van Beijeren. Amsterdam: Elsevier 273–309.

[87] J. D. Deuschel, D. W. Stroock. 1989. *Large Deviations*. Boston, MA: Academic Press.

[88] J. D. Deuschel, D. W. Stroock, H. Zessin. 1991. Microcanonical distributions for lattice gases. *Commun. Math. Phys.* **139**, 83–101.

[89] I. H. Dinwoodie. 1991. A note on the upper bound for i.i.d. large deviations. *Ann. Probab.* **19**, 1732–1736.

[90] R. L. Dobrushin. 1970. Prescribing a system of random variables by conditional distributions. *Theor. Prob. Appl.* **15**(3), 458–486.

[91] R. L. Dobrushin, M. R. Martirosyan. 1988. Possibility of high-temperature phase transitions due to the many-particle nature of the potential. *Theor. Math. Phys.* **75**, 443–448.

[92] R. L. Dobrushin, R. Kotecky, S. Shlosman. 1992. *Wulff Construction: a Global Shape From Local Interaction*. American Mathematical Society Translation Series. Providence, RI: American Mathematical Society.

[93] M. D. Donsker, S. R. S. Varadhan. 1975. Asymptotic evaluation of certain Markov process expectations for large time I. *Commun. Pure Appl. Math.* **28**, 1–47.

[94] M. D. Donsker, S. R. S. Varadhan. 1975. Asymptotic evaluation of certain Markov process expectations for large time II. *Commun. Pure Appl. Math.* **28**, 279–301.

[95] M. D. Donsker, S. R. S. Varadhan. 1976. Asymptotic evaluation of certain Markov process expectations for large time III. *Commun. Pure Appl. Math.* **29**, 389–461.

[96] M. D. Donsker, S. R. S. Varadhan. 1983. Asymptotic evaluation of certain Markov process expectations for large time IV. *Commun. Pure Appl. Math.* **36**, 183–212.

[97] P. Dupuis, R. S. Ellis. 1997. *A Weak Convergence Approach to the Theory of Large Deviations*. New York: Wiley.

[98] R. Durrett. 1980. On the growth of one dimensional contact process. *Ann. Probab.* **8** (5), 890–907.

[99] R. Durrett. 1984. Oriented percolation in two dimensions. *Ann. Probab.* **12** (4), 999–1040.

[100] R. Durrett. 1991. The contact process, 1974–1989. In *Proceedings of the 1989 AMS Seminar on Random Media*, ed. W. E. Kohler, B. S. White. American Mathematical Society Lectures in Applied Mathematics **27**. Providence, RI: American Mathematical Society, 1–18.

[101] R. Durrett, D. Griffeath. 1983. Supercritical contact process on $\mathbb{Z}$. *Ann. Probab.* **11** (1), 1–15.

[102] R. Durrett, X.-F. Liu. 1988. The contact process on a finite set. *Ann. Probab.* **16** (3), 1158–1173.

[103] R. Durrett, C. Neuhauser. 1994. Particle systems and reaction–diffusion equations. *Ann. Probab.* **22**, 289–333.

[104] R. Durrett, R. H. Schonmann. 1988. The contact process on a finite set II. *Ann. Probab.* **16** (4), 1570–1583.

[105] R. Durrett, R. H. Schonmann. 1988. Large deviations for the contact process and two dimensional percolation. *Probab. Theory Relat. Fields* **77**, 583–603.

[106] R. Durrett, R. H. Schonmann, N. I. Tanaka. 1989. The contact process on a finite set III. The critical case. *Ann. Probab.* **17**, 1303–1321.

[107] T. Eisele, R. S. Ellis. 1983. Symmetry breaking and random waves for magnetic systems on a circle. *Z. Wahrsch. Verw. Geb.* **63** (3), 297–348.

[108] R. S. Ellis. 1985. *Entropy, Large Deviations and Statistical Mechanics*. New York: Springer.

[109] R. S. Ellis. 1988. Large deviation for the empirical measure of a Markov chain with an application to the multivariate empirical measure. *Ann. Probab.* **16**, 1496–1508.

[110] R. S. Ellis, A. D. Wyner. 1989. Uniform large deviation property of the empirical process of a Markov chain. *Ann. Probab.* **17**, 1147–1151.

[111] A. C. D. van Enter, R. Fernández, A. D. Sokal. 1993. Regularity properties and pathologies of position-space renormalization-group transformation: scope and limitations of Gibbsian theory. *J. Stat. Phys.* **72**, 879–1167.

[112] V. R. Eston, A. Galves, C. M. Jacobi, R. Langevin. 1988. Dominance switch between two interacting species and metastability. *Atas do II Simpósio dos Ecossistemas da costa sul Brasileira*. São Paulo: CACIESP.

[113] W. Faris, G. Jona-Lasinio. 1982. Large Fluctuations for a nonlinear heat equation with noise. *J. Phys. A* **15**, 3025–3055.

[114] W. Feller. 1943. Generalization of a probability limit theorem of Cramér. *Trans. Am. Math. Soc.* **54**, 361–372.

[115] W. Feller. 1968. *An Introduction to Probability Theory and Its Applications*, Vol. I. 3rd edn. New York: Wiley.

[116] P. A. Ferrari, A. Galves. 1997. *Acoplamento e Processos Estocásticos. 21 Colóquio Brasileiro de Matemática*. IMPA. Rio de Janeiro: IMPA.

[117] M. E. Fisher. 1964. The free energy of a macroscopic system. *Arch. Ration. Mech. Anal.* **17**, 377–410.

[118] H. Föllmer, S. Orey. 1988. Large deviations for the empirical field of a Gibbs measure. *Ann. Probab.* **16** (3), 961–977.

[119] L. R. Fontes, P. Mathieu, P. Picco. 2000. On the average dynamics of the random field Curie–Weiss model. *Ann. Appl. Probab.* **10** (4), 1212–1245.

[120] C. Fortuin, P. Kasteleyn, J. Ginibre. 1971. Correlation inequalities on some partially ordered sets. *Commun. Math. Phys.* **22**, 89–103.

[121] D. Freedman. 1971. *Brownian Motion and Diffusion*. San Francisco, CA: Holden-Day.

[122] M. I. Freidlin, A. D. Wentzell. 1984. *Random Perturbations of Dynamical Systems*. New York: Springer. (1998. 2nd edition. New York: Springer.)

[123] S. Friedli. 2003. *On the Non-Analytic Behaviour of Thermodynamic Potentials at First Order Phase Transitions*. Thèse 2784. École Polytechnique Fédérale de Lausanne.

[124] S. Friedli, C.-E. Pfister. 2004. On the singularity of the free energy at first order phase transition. *Commun. Math. Phys.* **245**, 69–103.

[125] S. Friedli, C.-E. Pfister. 2004. Non-analyticity and the van der Waals limit. *J. Stat. Phys.* **114** (3/4), 665–734.

[126] D. Gabrielli, G. Jona-Lasinio, C. Landim, M. E. Vares. 1997. Microscopic reversibility and thermodynamic fluctuations. *Proceedings of Boltzmann's Legacy 150 years after his birth. Atti Convegni Lincei* **131**, 79–87.

[127] G. Gallavotti. 1972. Instabilities and phase transitions in the Ising model. A review. *Riv. Nuovo Cim.* **2**, 133–169.

[128] G. Gallavotti. 1983. *The Elements of Mechanics*. New York: Springer.

[129] G. Gallavotti. 1999. *Statistical Mechanics. A Short Treatise*. New York: Springer.
[130] A. Galves, B. Schmitt. 1990. Occurrence of rare events for mixing dynamical systems. *Ann. Inst. H. Poincaré Phys. Theor.* **52**, 267–281.
[131] A. Galves, E. Olivieri, M. E. Vares. 1987. Metastability for a class of dynamical systems subject to small random perturbations. *Ann. Probab.* **15** (4), 1288–1305.
[132] J. Gärtner. 1977. On large deviations from the invariant measure. *Theory Probab. Its Appl.* **22**, 24–39.
[133] B. Gaveau, L. S. Schulman. 1989. Metastable decay rates and analytic continuation. *Lett. Math. Phys.* **18**, 201–208.
[134] H. O. Georgii. 1988. *Gibbs Measures and Phase transitions*. Berlin: Walter de Gruyter.
[135] H. O. Georgii. 1993. Large deviations and maximun entropy principle for interacting random fields on $\mathbb{Z}^d$. *Ann. Probab.* **21**, 1845–1875.
[136] H. O. Georgii. 1994. Large deviations and the equivalence of ensembles for Gibbsian particle systems with superstable interaction. *Probab. Theory Relat. Fields* **99**, 171–195.
[137] G. Giacomin. 1995. Onset and structure of interfaces in a Kawasaki+Glauber interacting particle model. *Probab. Theory Relat. Fields* **103**, 1–24.
[138] I. I. Gihman, A. V. Skorohod. 1972. *Stochastic Differential Equations*. Berlin: Springer.
[139] S. Glance, B. Huberman. 1994. The dynamics of social dilemmas. *Sci. Am.* **76**, 1994.
[140] L. Gray. 1985. Duality for general attractive spin systems, with applications in one dimension. *Ann. Probab.* **14**, 371–396.
[141] D. Griffeath. 1978. Limit theorems for nonergodic set-valued Markov Processes. *Ann. Probab.* **6**, 379–387.
[142] D. Griffeath. 1979. Pointwise ergodicity of the basic contact process. *Ann. Probab.* **7**, 139–142.
[143] D. Griffeath. 1981. The basic contact process. *Stoch. Proc. Appl.* **11**, 151–185.
[144] R. Griffiths. 1964. Peierls proof of spontaneous magnetization in the two-dimensional Ising model. *Phys. Rev. A* **136**, 437–439.
[145] R. Griffiths. 1967. Correlations in Ising ferromagnets. I. *J. Math. Phys.* **8**, 478–483; II. **8**, 484–489.
[146] R. Griffiths, C. Y. Weng, J. S. Langer. 1966. Relaxation times for metastable states in the mean-field model of a ferrromagnet. *Phys. Rev.* **149**, 301–305.
[147] P. Groeneboom, J. Oosterhoff, F. Ruymgaart. 1979. Large deviations theorems for empirical probability measures. *Ann. Probab.* **7**, 553–586.
[148] T. Harris. 1960. A Lower bound for the critical probability in a certain percolation process. *Proc. Cambridge Philos. Soc.* **56**, 13–20.
[149] T. Harris. 1974. Contact interactions on a lattice. *Ann. Probab.* **2**, 969–988.
[150] T. Harris. 1977. A correlation inequality for Markov processes in partially ordered state spaces. *Ann. Probab.* **5**, 451–454.
[151] T. Harris. 1978. Additive set-valued Markov processes and graphical methods. *Ann. Probab.* **6**, 355–378.
[152] D. Henry. 1981. *Geometric Theory of Semilinear Parabolic Equations*. Lecture Notes in Mathematics 840. Berlin: Springer.
[153] T. Hida. 1980. *Brownian Motion*. Applications of Mathematics 11. New York: Springer.
[154] A. Hinojosa. Exit time for a reaction diffusion model. *Markov Process. Relat. Fields* (to appear).

[155] A. B. Hoadley. 1967. On the probability of large deviations of functions of several empirical cdf's. *Ann. Math. Stat.* **38**, 360–381.

[156] W. Hoeffding. 1965. On probabilities of large deviations. In *Proceedings of the Fifth Berkeley Symposium on Mathematical Statistics and Probability*. Berkeley, CA: University of California Press, 203–219.

[157] F. den Hollander. 2000. *Large Deviations*. Fields Institute Monographs 14. Providence, RI: American Mathematical Society.

[158] F. den Hollander, E. Olivieri, E. Scoppola. 2000. Metastability and nucleation for conservative dynamics. *J. Math. Phys.* **41**, 1424–1498.

[159] F. den Hollander, E. Olivieri, E. Scoppola. 2000. Nucleation in fluids. Some rigorous results. *Physica A* **279**, 110–122.

[160] F. den Hollander, F. Nardi, E. Olivieri, E. Scoppola. 2003. Droplet growth for three-dimensional Kawasaki dynamics. *Probab. Theory Relat. Fields*, **125** (2), 153–194.

[161] R. Holley. 1972. Markovian interaction processes with finite range interactions. *Ann. Math. Stat.* **43**, 1961–1967.

[162] R. Holley. 1974. Remarks on F. K. G. inequalities. *Commun. Math. Phys.* **36**, 227–231.

[163] K. Huang. 1963. *Statistical Mechanics*. New York: Wiley.

[164] N. Ikeda, S. Watanabe. 1989. *Stochastic Differential Equations and Diffusion Processes*, 2nd edn. Amsterdam: North-Holland.

[165] D. Ioffe. 1994. Large deviations for the 2D Ising model: a lower bound without cluster expansions. *J. Stat. Phys.* **74**, 411–432.

[166] D. Ioffe. 1995. Exact large deviation bounds up T_c for the Ising model in two dimensions. *Probab. Theory Relat. Fields* **102**, 313–330.

[167] D. Ioffe, R. H. Schonmann. 1998. Dobrushin–Kotecky–Shlosman theorem up to the critical temperature. *Commun. Math. Phys.* **199** (1), 117–167.

[168] S. N. Isakov. 1984. Nonanalytic features of the first order phase transition in the Ising model. *Commun. Math. Phys.* **95**, 427–443.

[169] I. Iscoe, P. Ney, E. Nummelin. 1985. Large deviations of uniformly recurrent Markov additive processes. *Adv. Appl. Math.* **6**, 373–412.

[170] H. M. Ito, T. Mikami. 1996. Poissonian asymptotics of a randomly perturbed dynamical system: flip-flop of the stochastic disk dynamo. *J. Stat. Phys.* **85** (1/2), 41–53.

[171] G. Jona-Lasinio, C. Landim, M. E. Vares. 1993. Large deviations for a reaction diffusion model *Probab. Theory Relat. Fields* **97**, 339–361.

[172] M. Kac, G. Uhlenbeck, P. Hemmer. 1963. On the van der Waals theory of the vapor–liquid equilibrium. I. Discussion of a one-dimensional model. *J. Math. Phys.* 4, 216–228.

[173] I. Karatzas, S. E. Shreve. 1988. *Brownian Motion and Stochastic Calculus*. New York: Springer.

[174] D. G. Kelly, S. Sherman. 1968. General Griffiths' inequalities on correlations in Ising ferromagnets. *J. Math. Phys.* **9** (3), 466–484.

[175] H. Kesten. 1981. Analyticity properties and power law estimates in percolation theory. *J. Stat. Phys.* **25**, 717–756.

[176] R. Z. Khas'minskii. 1980. *Stochastic Stability of Differential Equations* (English translation) Alphen aan den Rijn: Sijthoff & Noordhoff.

[177] C. Kipnis. 1981. Processus de champ moyen: existence, unicité, mesures invariantes et limites thermodynamiques. *Stochastics* **5** (1/2), 93–106.

[178] C. Kipnis, C. Landim. 1999. *Scaling Limits of Interacting Particle Systems*. Grundlehren der Mathematischen Wissenschaften 320. Berlin: Springer.

[179] C. Kipnis, S. Olla, S. R. S. Varadhan. 1989. Hydrodynamics and large deviations for simple exclusion processes. *Commun. Pure Appl. Math.* **42**, 115–137.

[180] H. Kneser. 1952. Sur un théorème fondamental de la théorie de jeux. *C. R. Acad. Sci.* **234**, 2418–2420.

[181] F. B. Knight. 1981. *Essentials of Brownian Motion and Diffusion*. Mathematical Surveys 18. Providence, RI: American Mathematical Society.

[182] R. Kotecky, I. Medved. 2001. Finite-size scaling for the 2D Ising model with minus boundary conditions. *J. Stat. Phys.* **104** (5/6), 905–943.

[183] R. Kotecky, E. Olivieri. 1992. Stochastic models for nucleation and crystal growth. In *Probabilistic Methods in Mathematical Physics* (*Siena 1991*). Riverside, NY: World Scientific, 264–275.

[184] R. Kotecky, E. Olivieri. 1993. Droplet dynamics for asymmetric Ising model. *J. Stat. Phys.* 70, 1121–1148.

[185] R. Kotecky, E. Olivieri. 1994. Shapes of growing droplets – a model of escape from a metastable phase. *J. Stat. Phys.* **75**, 409–506.

[186] H. Kunita, S. Watanabe. 1967. On square integrable martingales. *Nagoya Math. J.* **30**, 209–245.

[187] P. Lancaster. 1969. *Theory of Matrices*. New York: Academic Press.

[188] C. Landim, M. E. Vares. 1996. Exponential estimate for reaction diffusion models. *Probab. Theory Relat. Fields* **106**, 151–186.

[189] O. E. Lanford III. 1971. Entropy and equilibrium states in classical statistical mechanics. *Lect. Notes Phys.* **20**, 1–113.

[190] O. E. Lanford III, D. Ruelle. 1969. Observables at infinity and states with short range correlations in statistical mechanics. *Commun. Math. Phys.* **13**, 194–215.

[191] J. S. Langer. 1967. Theory of the condensation point. *Ann. Phys.* **41**, 108–157.

[192] J. L. Lebowitz, A. Martin-Löf. 1972. On the uniqueness of the equilibrium state for Ising spin systems. *Commun. Math. Phys.* **25**, 276–282.

[193] J. L. Lebowitz, O. Penrose. 1966. Rigorous treatment of the van der Waals–Maxwell theory of the liquid–vapor transition. *J. Math. Phys.* **7**, 98–113.

[194] J. L. Lebowitz, R. H. Schonmann. 1988. Pseudo-free energies and large deviations for non-Gibbsian FKG measures. *Probab. Theory. Relat. Fields* **77**, 49–64.

[195] L. Le Cam. 1986. *Asymptotic Methods in Statistical Decision Theory*. New York: Springer.

[196] T. Lee, C. Yang. 1952. Statistical theory of equation of state and phase transitions. II. Lattice gas and Ising model. *Phys. Rev.* **87**, 410–419.

[197] P. Lévy. 1948. *Processus Stochastiques et Mouvement Brownien*, 2nd edn. Paris: Gauthiers-Villars.

[198] J. T. Lewis, C.-E. Pfister, W. G. Sullivan. 1995. Entropy, concentration of probability and conditional limit theorems. *Markov Process. Relat. Fields* **1** (3), 319–386.

[199] T. M. Liggett. 1985. *Interacting Particle Systems*. New York: Springer.

[200] T. M. Liggett. 1999. *Stochastic Interacting Systems: Contact, Voter and Exclusion Processes*. Berlin: Springer.

[201] T. Lindvall. 1992. *Lectures on the Coupling Method*. New York: Wiley.

[202] T. Lindvall, L. C. G. Rogers. 1986. Couplings of multidimensional diffusions by reflection. *Ann. Probab.* **14**, 860–872.

[203] R. S. Liptser, A. N. Shiryayev. 1977. *Statistics of Random Processes I. General Theory*. New York: Springer.

[204] J. Lynch, J. Sethuraman. 1987. Large deviations for processes with independent increments. *Ann. Probab.* **15**, 610–627.

[205] F. Manzo, E. Olivieri. 1998. Relaxation patterns for competing metastable states: a nucleation and growth model. *Markov Process. Relat. Fields* **4**, 549–570.

[206] F. Manzo, E. Olivieri. 2001. Dynamical Blume–Capel model: competing metastable states at infinite volume. *J. Stat. Phys.* **104** (5/6), 1029–1090.

[207] F. Manzo, F. R. Nardi, E. Olivieri, E. Scoppola. 2004. On the essential features of metastability: tunnelling time and critical configurations. *J. Stat. Phys.* **115**, 591–641.

[208] C. Marchioro, A. Pellegrinotti, E. Presutti, M. Pulvirenti. 1976. On the dynamics of particles in a bounded region: a measure theoretical approach. *J. Math. Phys.* **17**, 647–652.

[209] F. Martinelli, E. Scoppola. 1988. Small random perturbations of dynamical systems: exponential loss of memory of the initial condition, *Commun. Math. Phys.* **120**, 25–69.

[210] F. Martinelli, E. Olivieri, E. Scoppola. 1989. Small random perturbations of finite and infinite dimensional dynamical systems: unpredictability of exit times. *J. Stat. Phys.* **55**, 477–504.

[211] F. Martinelli, E. Olivieri, E. Scoppola. 1990. Metastability and exponential approach to equilibrium for low temperature stochastic Ising models. *J. Stat. Phys.* **61**, 1105–1119.

[212] F. Martinelli, E. Olivieri, E. Scoppola. 1991. On the Swendsen–Wang dynamics I. Exponential convergence to equilibrium. *J. Stat. Phys.* **62**, 117–133.

[213] F. Martinelli, E. Olivieri, E. Scoppola. 1991. On the Swendsen and Wang dynamics II: Critical droplets and homogeneous nucleation at low temperature for the two dimensional Ising model. *J. Stat. Phys.* **62**, 135–159.

[214] A. Martin-Löf. 1973. Mixing properties, differentiability of the free energy and the central limit theorem for a pure phase in the Ising model at low temperature. *Commun. Math. Phys.* **32**, 75–92.

[215] A. Martin-Löf. 1979. *Lectures on Statistical Mechanics and Fundations of Thermodynamics*. Lecture Notes in Physics 101. Berlin: Springer.

[216] P. Mathieu, P. Picco. 1998. Metastability and convergence to equilibrium for the random field Curie–Weiss model. *J. Stat. Phys.* **91**, 679–732.

[217] T. Mikami. 1991. Large deviation theorems for empirical measures in Freidlin–Wentzell exit problems. *Ann. Probab.* **19** (1), 58–82.

[218] H. D. Miller. 1961. A convexity property in the theory of random variables defined on a finite Markov chain. *Ann. Math. Stat.* **32**, 1260–1270.

[219] R. A. Minlos, Ya. G. Sinai. 1967. The phenomenon of 'separation of phases' at low temperatures in certain lattice models of a gas I. *Mat. Sb.* **73** (115), 375–448.

[220] T. S. Mountford. 1993. A metastable result for the finite multidimensional contact process. *Can. Math. Bull.* **36** (2), 216–226.

[221] T. S. Mountford. 1999. Existence of a constant for finite system extinction. *J. Stat. Phys.* **96** (5/6), 1331–1341.

[222] C. Mueller. 1993. Coupling and invariant measures for the heat equation with noise. *Ann. Probab.* **21**, 2189–2199.

[223] S. V. Nagaev. 1979. Large deviations of sums of independent random variables. *Ann. Probab.* **7** (5), 745–789.

[224] F. R. Nardi, E. Olivieri. 1996. Low temperature stochastic dynamics for an Ising model with alternating field. *Markov Process. Relat. Fields* **2**, 117–166.

[224A] F. R. Nardi, E. Olivieri, M. Zahradnik. 1999. On the Ising model with strongly anisotropic field. *J. Stat. Phys.* **97**, 87–144.

[225] E. J. Neves, R. H. Schonmann. 1991. Critical droplets and metastability for a Glauber dynamics at very low temperatures. *Commun. Math. Phys.* **137**, 209–230.

[226] E. J. Neves, R. H. Schonmann. 1992. Behaviour of droplets for a class of Glauber dynamics at very low temperatures. *Probab. Theory Relat. Fields* **91**, 331–354.

[227] C. M. Newman, L. S. Schulman. 1977. Metastability and analytic continuation of eigenvalues. *J. Math. Phys.* **18**, 23–30.

[228] C. M. Newman, L. S. Schulman. 1980. Complex free energies and metastable lifetimes. *J. Stat. Phys.* **2**, 131–148.

[229] C. Newman, J. E. Cohen, C. Kipnis. 1985. Neo-darwinian evolution implies punctuated equilibria. *Nature* **315**, 400.

[230] E. Olivieri. 2003. Metastability and entropy. In *Entropy*, ed. A. Greven, G. Keller, G. Warnecke. Princeton, NJ: Princeton University Press, 233–250.

[231] E. Olivieri, E. Scoppola. 1995. Markov chains with exponentially small transition probabilities: First exit problem from a general domain I. The reversible case. *J. Stat. Phys.* **79**, 613–647.

[232] E. Olivieri, E. Scoppola. 1996. Markov chains with exponentially small transition probabilities: First exit problem from a general domain II. The general case. *J. Stat. Phys.* **84** (5/6), 987–1041.

[233] E. Olivieri, E. Scoppola. 1998. Metastability and typical exit paths in stochastic dynamics. *European Congress of Mathematics* (*Budapest, 1996*), vol. II. Progress in Mathematics 169. Basel: Birkhauser, 124–150.

[234] S. Olla. 1988. Large deviations for Gibbs random fields. *Probab. Theory Relat. Fields* **77**, 343–357.

[235] K. R. Parthasarathy. 1967. *Probability Measures on Metric Spaces*. New York: Academic Press.

[236] R. Peierls. 1936. On Ising's model of ferromagnetism. *Proc. Cambridge Philos. Soc.* **36**, 477–481.

[237] C. Peixoto. 1995. Metastable behavior of low temperature Glauber dynamics with stirring. *J. Stat. Phys.* **80** (5/6), 1165–1184.

[238] O. Penrose. 1989. Metastable states for the Becker–Döring cluster equations. *Commun. Math. Phys.* **124**, 515–541.

[239] O. Penrose. 1995. Metastable decay rates, asymptotic expansions and analytic continuation of thermodynamic functions. *J. Stat. Phys.* **78**, 267–283.

[240] O. Penrose, J. L. Lebowitz. 1971. Rigorous treatment of metastable states in the van der Waals-Maxwell theory. *J. Stat. Phys.* **3**, 211–236.

[241] O. Penrose, J. L. Lebowitz. 1987. Towards a rigorous molecular theory of metastability. In *Fluctuation Phenomena*, 2nd edn., ed. E. W. Montroll, J. L. Lebowitz. Amsterdam: North-Holland.

[242] C.-E. Pfister. 1991. Large deviations and phase separation in the two-dimensional Ising model. *Helv. Phys. Acta* **64**, 953–1054.

[243] C.-E. Pfister. 2002. Thermodynamical aspects of classical lattice systems. In *In and Out of Equilibrium. Probability with a Physics Flavor*, ed. V. Sidoravicius. Progress in Probability 51. Basel: Birkhauser, 393–472.

[244] S. A. Pirogov, Ya. G. Sinai. 1976. Phase diagrams of classical lattice systems. *Teor. Mat. Fiz.* **26** (1), 61–76.

[245] A. Pisztora. 1996. Surface order large deviations for Ising, Potts and percolation models. *Probab. Theory Relat. Fields* **104**, 427–466.

[246] D. Plachky, J. Steineback. 1975. A theorem about probabilities of large deviations with an application to queuing theory. *Period. Math. Hung.* **6**, 343–345.

[247] C. Preston. 1976. *Random Fields*. Lecture Notes in Mathematics 534. Berlin: Springer.

[248] E. Presutti. 1975. A mechanical definition of the thermodynamic pressure. *J. Stat. Phys.* **13** (4), 301–314.

[249] E. Presutti. 1999. *From Statistical Mechanics to Continnum Mechanics*. Leipzig: Max Planck Insitute.

[250] E. Presutti. 2000. *Lezioni di Meccanica Statistica*. Preprint.

[251] A. A. Pukhalskii. 1991. On functional principle of large deviations. In *New Trends in Probability and Statistics*, ed. V. Sazonov, T. Shervashidze. Moskow: VSP, 198–218.

[252] J. Quastel, S. R. S. Varadhan. 1997. Diffusion semigroups and diffusion processes corresponding to degenerate divergence form operators. *Commun. Pure Appl. Math.* **50**, 667–706.

[253] M. Reed, B. Simon. 1972. *Methods of Modern Mathematical Physics*, vol. 1, *Functional Analysis*. New York: Academic Press.

[254] F. Riesz, B. Sz. Nagy. 1955. *Functional Analysis* (English translation). New York: Ungar.

[255] R. T. Rockafellar. 1973. *Convex Analysis*. Princeton, NJ: Princeton University Press.

[256] W. Rudin. 1973. *Functional Analysis*. New York: McGraw-Hill.

[257] D. Ruelle. 1967. A Variational formulation of equilibrium statistical mechanics and the Gibbs phase rule. *Commun. Math. Phys.* **5**, 324–329.

[258] D. Ruelle. 1969. *Statistical Mechanics. Rigorous results*. New York: Benjamin.

[259] D. Ruelle. 1971. Existence of a phase transition in a continuous classical system. *Phys. Rev. Let.* **27** (16), 1040–1041.

[260] D. Ruelle. 1972. On the use of 'small external fields' in the problem of symmetry breakdown in statistical mechanics. *Ann. Phys.* **69**, 364–374.

[261] L. Russo. 1978. A note on percolation. *Z. Wahrsch. Verw. Geb.* **43**, 39–48.

[262] S. R. Salinas, W. F. Wrezinski. 1985. On the mean field Ising model in a random external field. *J. Stat. Phys.* **41**, 299–313.

[263] I. Sanov. 1961. On the probability of large deviations of random variables. *Select. Trans. Stat. Probab.* I, 213–244.

[264] M. Schilder. 1966. Some asymptotic formulae for Wiener integrals. *Trans. Am. Math. Soc.* **125**, 63–85.

[265] R. H. Schonmann. 1985. Metastability for the contact process. *J. Stat. Phys.* **41** (3/4), 445–484.

[266] R. H. Schonmann. 1987. Second order large deviation estimates for ferromagnetic systems in the phase coexistence region. *Commun. Math. Phys.* **112**, 409–422.

[267] R. H. Schonmann. 1992. The pattern of escape from metastability of a stochastic Ising model. *Commun. Math. Phys.* **147**, 231–240.

[268] R. H. Schonmann. 1994. Slow droplet driven relaxation of stochastic Ising models in the vicinity of the phase coexistence region. *Commun. Math. Phys.* **161**, 1–49.

[269] R. H. Schonmann. 1994. Theorems and conjectures on the droplet driven relaxation of stochastic Ising model. In *Probability and Phase Transition*, ed. G. Grimmett. NATO ASI Series. Dordrecht: Kluwer, 265–301.

[270] R. H. Schonmann, S. Shlosman. 1996. Constrained variational problem with applications to the Ising model. *J. Stat. Phys.* **83** (5/6), 867–905.

[271] R. H. Schonmann, S. Shlosman. 1998. Wulff droplets and the metastable relaxation of kinetic Ising models. *Commun. Math. Phys.* **194** (2), 389–462.

[272] E. Scoppola. 1993. Renormalization group for Markov chains and application to metastability, *J. Stat. Phys.* **73**, 83–121.

[273] E. Scoppola. 1994. Metastability for Markov chains: a general procedure based on renormalization group ideas. In *Probability and Phase Transition*, ed. G. Grimmett. NATO ASI Series. Dordrecht: Kluwer, 303–322.

[274] E. Scoppola. 1995. Renormalization and graph methods for Markov chains. In *Advances in dynamical systems and quantum physics* (*Capri, 1993*). River Edge, NJ: World Scientific, 260–281.

[275] E. Seneta. 1981. *Non Negative Matrices and Markov Chains*, 2nd edn. New York: Springer.

[276] G. L. Sewell. 1980. Stability, equilibrium and metastability in statistical mechanics. *Phys. Rep.* **57**, 307–342.

[277] P. D. Seymour, D. J. A. Welsh. 1978. Percolation probabilities on the square lattice. In *Advances in Graph Theory*, ed. B. Bollobás. *Ann. Discrete Math.* **3**, 227–245.

[278] G. L. Sievers. 1969. On the probability of large deviations and exact slopes. *Ann. Math. Stat.* **40** (6), 1908–1921.

[279] A. Simonis. 1996. Metastability of the *d*-dimensional contact process. *J. Stat. Phys.* **83** (**5/6**), 1225–1239.

[280] Ya. G. Sinai. 1982. *Theory of Phase Transitions. Rigorous Results*. Oxford: Pergamon.

[281] M. Sion. 1958. On general minimax theorem. *Pacific J. Math.* **8**, 171–176.

[282] M. Slaby. 1988. On the upper bound for large deviations of sums of i.i.d. random vectors. *Ann. Probab.* **16** (3), 978–990.

[283] H. Spohn. 1991. *Large Scale Dynamics of Interacting Particle Systems*. Berlin: Springer.

[284] C. Stone. 1974. Large deviations of empirical probability measures. *Ann. Stat.* **2**, 362–366.

[285] D. Stroock, R. Varadhan. 1979. *Multidimensional Diffusion Processes*. New York: Springer.

[286] C. J. Thompson. 1972. *Mathematical Statistical Mechanics*. New York: MacMillan.

[287] H. Thorisson. 2000. *Coupling, Stationarity, and Regeneration. Probability and its Applications*. New York: Springer.

[288] H. Tomita, S. Miyashita. 1992. Statistical properties of the relaxation process of metastable states in the kinetic Ising model. *Phys. Rev. B* **46**, 8886–8893.

[289] A. Trouvé. 1996. Cycle decompositions and simulated annealing. *SIAM J. Control Optim.* **34** (3), 966–986.

[290] A. Trouvé. 1996. Rough large deviation estimates for the optimal convergence speed exponent of generalized simulated annealing algorithms. *Ann. Inst. H. Poincaré Probab. Stat.* **32** (3), 299–348.

[291] S. R. S. Varadhan. 1966. Asymptotic probabilities and differential equations. *Commun. Pure Appl. Math.* **19**, 261–286.

[292] S. R. S. Varadhan. 1984. *Large Deviations and Applications*. Philadelphia, PA: SIAM.

[293] M. E. Vares. 1996. *Large Deviations and Metastability*. Collection Travaux en Cours. Paris: Hermann.

[294] A. D. Ventcel, M. I. Freidlin. 1969. Small random perturbations of a dynamical system with a stable equilibrium position. *Sov. Math. Dokl.* **10** (4), 886–890 (Dokl. Akad. Nauk. SSSR **187** (3), 506–509).

[295] A. D. Ventcel, M. I. Freidlin. 1970. On small random perturbations of dynamical systems. *Russ. Math. Surveys* **25**, 1–55 (*Usp. Mat. Nauk.* **25** (1), 3–55).

[296] A. D. Wentzell. 1990. *Limit Theorems on Large Deviations for Markov Stochastic Processes*. Dordrecht: Kluwer.

[297] E. Wong, M. Zakai. 1969. Riemann–Stieltjes approximations of stochastic integrals. *Z. Wahrsch. Verw. Geb.* **12**, 87–97.

[298] H. T. Yau. 1994. Metastability of Ginzburg–Landau model with a conservation law. *J. Stat. Phys.* **74**, 705–742.

Index